CONCEPTS'

DICTIONARY OF AGRICULTURAL SCIENCES

CONCEPTS'

DICTIONARY OF AGRICULTURAL SCIENCES

Dr. I.C. GUPTA
PRINCIPAL SCIENTIST
CENTRAL ARID ZONE RESEARCH INSTITUTE, JAISALMER

Dr. S.K. GUPTA
SR. DRAINAGE ENGINEER
CENTRAL SOIL SALINITY RESEARCH INSTITUTE, KARNAL

CONCEPT PUBLISHING COMPANY, NEW DELHI 110059

ISBN 81-7022-301-6

First Published 1992
Reprinted 2002

Published and Printed by
Ashok Kumar Mittal
Concept Publishing Company
A/15-16, Commercial Block, Mohan Garden,
NEW DELHI-110059

Phones : 5351460, 5351794
Fax : 091-11-5357103
Email : publishing@conceptpub.com

PREFACE

Agriculture is an applied science and it utilizes scientific principles and concepts from various disciplines. Each discipline has contributed its own share of terms and thus there are wide varieties of terms in the literature related to agriculture. Within this field, several specialised disciplines exist, yet the boundaries between these fields are quite flexible. Therefore, specialists in one discipline, often come across terminology that is less understood to them and they would like to know the exact meaning of many terms which they come across in their daily routine. Therefore, a definitional dictionary of this nature will be useful reference to researchers, planners and common men related to the field of agriculture.

Although no claim is made to include all the terms yet the scope of the compilation has been widened by including commonly used terms from various disciplines related to agriculture. The major disciplines which have been covered are agronomy, agricultural engineering, botany, breeding, chemistry, climatology, entomology, genetics, geology, social sciences and soil science. Thus, specialists of various disciplines should find this compilation a handy reference. The words in this dictionary have been arranged alphabetically without consideration of discipline so that it is easier to locate the word.

In compiling this dictionary a large number of books, journals, similar dictionaries, glossaries and encyclopaedias have been used. It is difficult to name and acknowledge each and every reference used in this compilation. We sincerely thank the auhtors known and unknown for whatever help we have received from their works.

Finally, we acknowledge the help received from our family members who have been of consistent help in technical editing and arranging the words. We also thank their forbearance during the period when this dictionary was prepared.

No attempt of this nature could be complete in itself and there could be many ommissions and editorial mistakes. The readers are requested to kindly let us know the mistakes such that these could be taken care of in the next revision. All communications and suggestions will be thankfully acknowledged.

I.C. GUPTA
S.K. GUPTA

A

Aapamoor A moor with elevated areas or mounds supporting dwarf shrubs and sphagnum interspread with low areas containing sedges and sphagnum thus forming a mosaic.

Abaca A plant of the banana family native to Borneo and the Philippines, valuable for its hard fibre. Also known as manila hemp.

Abapical On the opposite side to, or directed away from the apex.

Abaxial Situated on the side away from the stem or main axis, as a sporangium.

ABC soil A soil with a distinctly developed profile, including A, B and C horizons.

ABC system A procedure in seismic surveying to determine the effect of irregular weathering thickness.

Aberrant Differing from the normal type of a species, genus, or higher group in one or more characters, but not readily assignable to another group.

Abelmoschus esculentus See lady's finger

— **manihot** See ranbhindi

— **ficulneas** See ranbhindi

— **moschatus** See musk mallow

Abi Refers to a crop season of rice sown in May-June and harvested in Nov-Dec.

Abies The fir, a genus of trees in the pine family characterised by erect cones, absence of resin canals in the wood and flattened needle like leaves.

Abiogenesis The obsolete concept that plant and animal life arise from non-living organic matter.

Abiotic Refers to the absence of living organism.

Abiotic environment All physical and non-living chemical factors, such as soil, water and atmosphere which influence living organism.

Abiotic substances Non-living components of the environment, such as water, sun, oxygen, organic compounds, and soil nutrients.

Ablation factor The rate at which a snow or ice surface wastes away.

Ablation form A feature on a snow or ice surface caused by melting or evaporation.

Ablation till A surface deposit of loose, permeable somewhat stratified, sandy, and stony till overlying denser till.

Abnormal (1) Deviating from the normal by a considerable amount. The term does not carry any quantitative meaning, since its employment depends on good judgement based on consideration of the past records of similar occurrences. (2) Above or below the normal value of an element, regardless of the degree of deviation.

Abnormal seedlings The seedlings which do not show the capacity for continued development into normal plants when grown in good quality soil, and under favourable conditions of water supply, temperature and light.

Abohm The unit of electrical resistance in the centimetre gram second system, abohm equals 10^{-9} ohm in metre kilogram second system.

Aboiteaux Dike and drainage structure protecting marshlands; an outlet structure with flap gate which permits outflow of fresh water, but prevents inflow of tidal salt water.

Abortive Imperfectly formed or developed.

Abrasion The process of reducing material by grinding instead of cutting with tools.

Abrasive Grinding material such as sandstone, emery, carborundum, etc. The natural abrasives include the diamond, emery, carborundum, sand crushed granite and quartz, tripoli, and pumice. The artificial abrasives are in general either silicon carbide or aluminium oxide, and are marketed under many trade names.

Abruptly acuminate The term as applied to leaf pertains to having a broad extremity from which a point arises.

Abruptly pinnate Having the main axis of apopodium not winged, but bearing a number of secondary axes which are winged.

Abrupt textural change is a change from an ochric epipedon or an albic horizon to an argillic horizon. There is, in the zone of contact, a very appreciable increase in clay content within a very short distance in depth. If the clay content of the ochric epipedon or the albic horizon is < 20 per cent, the clay content should double within a distance in depth of 7.5 cm or less. If the clay content exceeds 20 per cent, the increase in clay content should be at least 20 per cent of the fine earth fraction, for example from 22 per cent to 42 per cent, within a distance of 7.5 cm in depth and the clay content in some part of the argillic horizon should be at least double that of the horizon above.

Abscess Circumscribed collection of pus.

Abscission A physiological process promoted by abscisic acid whereby plants shed a part, such as leaf, flower, seed or fruit.

Abscisson layer A zone of cells formed at the base of a leaf or other plant part.

Absolute Refers to the highest or lowest recorded value of any parametre on meteorological element, whether at a single station or over an area, during a given period.

Absolute humidity (1) It is defined as the mass of water vapour per unit volume of space or the gaseous pressure exerted by water vapours. (2) It is the ratio of the mass of the water vapour in a sample of air to the volume of sample.

Absolute instability The state of column of air in the atmosphere when it has a superadiabatic lapse rate of temperature, that is greater than the dry adiabatic lapse rate.

Absolute isohypse A line that has the properties of both constant pressure and constant height above mean sea level.

Absolute monopoly The control of the entire output of a commodity or service, for which there is no substitute, by a single producer or supplier. This kind of situation rarely occurs in real life.

Absolute stability The state of a column of air in the atmosphere when its lapse rate of temperature is less than the saturation adiabatic lapse rate.

Absolute water requirement See consumptive use.

Absorbed water Water held mechanically in a soil mass and having physical properties similar to ordinary water at the same temperature and pressure.

Absorbent A substance which absorbs liquid or gases and distributes them throughout its body; for example cotton.

Absorbent cotton Treated cotton fibre to make it able to absorb more water or any other liquid.

Absorption Act or process of absorbing or being absorbed.

Absorption coefficient Represents the volume of gas, reduced to standard conditions (0°C and 760 mm pressure), dissolved in unit volume of liquid when the pressure of the gas (without the vapour pressure of the liquid) is 760 mm.

Absorption losses Losses of water from a canal or reservoir by capillary action and percolation, and, in the case of canal, during the process of priming.

Abstract intelligence Persons having the ability with ideas, as in language, mathematics, science, words, numbers. Professional people are generally high in abstract ability.

Abstraction Stream piracy resulting from the shifting of divides.

Abutilon asiaticum See Kanghi

— **avicennae** see Indian mallow

— **indicum** see country mallow

— **polyandrum** Family Malvaceae. Shrub whose stem fibre is used for ropes and cordage. Found in north-west India, the Western

peninsula and tropical Himalayas
— **theophrasti** see Indian mallow

Abyssal zone The cold, dark, water zone of the ocean depths, below the euphytic zone q.v.; it corresponds roughly to the profundal zone q.v. of the lake habitat. No photo-synthesis q.v. can occur at this level and animal life is sparse.

A.C. The common abbreviation for alternating current.

Acacia aneura F. Miiell. Family Mimosaceae. An introduced tree or shrub from Australia whose leaves are used as fodder.
— **arabica** See *babul.*
— **catechu** See black cutch.
— **concinna** See soap pod.
— **dealbata** See silver wattle.
— **decurrens** See green wattle.
— **decurrens var. deal bata** See silver wattle.
— **decurrens var. mollis** See black wattle.
— **ferruginea** See son-khair.
— **jacque-montii** See kinbar.
— **leucophloea** See *safed kikar.*
— **lebbek** See *kokko*
— **melanoxylon** See Australian black- wood
— **modesta** Wood used for persian wheels, cane crushers, agricultural implements, and for fuel; gum similar to gum arabic. Used for canal plantation, drought resistance.
— **mollissima** See black wattle
— **nilotica** See *babul*
— **pycnantha** See golden wattle
— **senegal** See gumarabic acacia
— **tortilis** See Israeli *babul*
— **verek** See gumarabic acacia.

Acanthaceous Bearing thorns or prickles.

Acanthiophilus helianthi See safflower fly.

Acanthocarpus Having fruits covered with spines or prickles.

Acanthospermum hispidum See star burr.

Acariasis Condition of plants or animals infested with ticks or mites.

Acaricide A pesticide used to destroy mites on domestic animals, crops and man. Also known as mitcide.

Acarids Ticks and mites.

Acarocecidium A gall caused on a plant by gall-mites.

Acarology The study of ticks and mites.

Acarpous Devoid of fruits.

Accelerated erosion The erosion taking place at a much faster rate than the normal because of human and animal activities such as denudation of forest and grazing places.

Accelerating pump A plunger type pump attached to the carburetor for the purpose of increasing the richness of the mixture and to give a quick 'pick-up' under load.

Accelerator A mechanical device for regulating the amount of gas mixture which is fed to the engine, usually operated by the foot. In plastics, a chemical that hastens reaction particularly one that speeds resin hardening. Also called promoter.

Acceptability, Animal Readiness with which animals select and eat a forage; sometimes used interchangeably to mean either palatability or voluntary intake. See also Palatability.

Accessibility The degree of convenient arrangement of parts to permit easy adjustment, repair, assembling disassembling etc.

Accessory (1) A mechanical part or attachment not necessarily a part of the machine with which it is to be used but does make possible a wider range of work or better performance. (2) Additional or adventitious.

Accessory bud An embryonic shoot occurring above or to the side of an auxiliary bud; a bud formed on a leaf. Also known as supernumerary bud.

Accessory cell A morphologically distinct epidermal cell adjacent to and apparently functionally associated with guard cells on the leaves of many plants.

Accessory cloud A cloud form that is dependent, for its formation and continuation upon the existence of one of the major cloud genera; may be an appendage of the parent cloud or an immediately adjacent cloudy mass.

Accessory food factors Vitamins.

Accidental error, or Random error An error due to chance causes for which it is equally probable that the sign of the error is plus or minus. In a series of measurements accidental errors are compensating in their total effect.

Acclimated micro-organism Any micro-organism that is able to adopt to environment changes such as a change in temperature or a change in the quantity of oxygen or other gases.

Acclimation A reversible change in the morphology or physiology of an organism in response to environmental change when such alteration occurs in the laboratory under controlled conditions involving only a single variable.

Acclimatization In the course of several generation, adjustment or increase in tolerance shown by a species to the changed environment, to a new habitat or to a medium or set of conditions.

Accounting The recording of the transactions of a business to reveal the financial position of the business, permit the computation of tax liabilities, and provide management with information essential to the efficient control of the business.

Accounting costs Costs considered as recorded costs only; economic costs may be larger or smaller when alternative costs or opportunity costs and external effects are taken into account.

Accretion The gradual addition of new land to old by the deposition of sediment carried by the water of a stream.

Accumbent Describing an organ that leans against another, specifically referring to cotyledons having their edges folded against the hypocotyl.

Accumulative area A term used in U.A.E. All the area served downstream of a given section of the canal, the supply of which must pass through this section.

Accumulative escaping power Progressive totals at the end of each reach of the escaping power from the first reach down stream. The design section must take provision for this discharge in addition to calculated theoretical discharge.

Accumulator A cylinder into which water is forced in order to furnish the motive power in hydraulic machines of various kinds. A storage battery.

Accumulator plant A plant or tree that grows in a metal bearing soil and accumulates an abnormal content of the metal.

Accuracy The absolute nearness to the true value as compared to 'precision' which denotes relative or apparent nearness to the true value. In a more specialised sense, the word also occurs as meaning (i) in relation to an estimator, unbiasedness; (ii) in relation to the reciprocal of standard error.

Acephalous Having the style originate at the base instead of at the apex of the ovary.

Acervulus (*pl.* Acervuli) A "little heap", an erumpent, cushion-like mass of hyphae bearing conidiophores and conidia, sometimes with setae; characteristic of the Melanconiales.

Acetarious plants Salad plants.

Acetic acid A colourless organic liquid with pungent odour and used for destroying warts. Germicide, chief active component of vinegar. Important constituent of flavour in many milk products. Also called pyroligneous acid.

Acetic acid bacteria Acetic acid producing bacteria from alcohol.

Acetous Sour in taste and having qualities similar to vinegar.

Achatina fulica African giant snail. Often causing much damage to paddy and vegetable crops.

Achene (1) A small dry, hard, one chambered, one seeded indehiscent fruit, as in back wheat, lettuce and spinach, commonly referred to as a seed or (2) A dry one-seeded, two-carpeled fruit in which the seed coat is free of the pericarp.

Achoea janata See castor semi-looper, citrus fruit sucking moth.

Achoria grisella See lesser wax moth.

Achras Sapota See *chiku*

— **Zapota** See chiku.

Achyranthes muricata L. family Amaranthaceae. A herb whose tender twigs and inflorescence are used as vegetables.

Achyrontias styx See hawk moth.

Acicular Needle-like.

Acid Organic or inorganic substances with sour taste forming salts when allowed to mix with bases. A term used in chemistry for hydrogen compounds soluble in water.

Acid-fast Able to take and hold an acid stain.

Acid-fast bacteria Bacteria that do not lose their stain when treated with an acid after being stained with an aniline dye (example, Mycobacterium tuberculosis).

Acid flavour A sharp, sour taste.

Acid forming fertiliser Fertiliser which are capable of increasing the residual acidity of soils.

Acid fruits Lime and lemon used for making citric acid and seasoning of food. As a food article, for making pickles etc.

Acidimetre A device which helps in measuring the amount of acid present in a solution, also a chemical used for the purpose.

Acid number See acid value

Acidophilic Pertaining to bacteria tolerant of an acid solution. Also called aciduric.

Acidopill Any substance tissue or organism having an affinity for acid stains.

Acid phosphate Most widely used phosphoric fertilizer. Phosphoric acid content wholly water soluble. Brownish grey mixture of monocalcium phosphate and calcium sulphate.

Acid rocks Rocks containing free silica in abundance, e.g. Granite.

Acid Soils Soil with a preponderance of hydrogen ions, and probably of aluminium in proportion to hydroxyl ions. Specifically, soil with a pH value less than 7.0. However, for most practical purposes, a soil with a pH value less than 6.6. The pH values obtained vary greatly with the method used, consequently, there is no unanimous agreement on what constitutes an acid soil. The term is usually applied to the surface layer or to the root zone unless specified otherwise.

Acid soot or acid smut An agglomeration of carbon particles held together by moisture which has become acidic through combination with sulphur trioxide; soot particles range in size from about 1 mm to about 3 mm in diametre.

Acme thread A screw thread, the section of which is between the square and V threads. Used extensively for feed screws. The included angle of space is 29° as compared to 60° of the National Coarse or U.S. Thread.

Acid tolerance Tolerance of certain crops to acidity of the soil. Acid sensitive crops suffer injury when grown in acidic soils.

Acidulous water Mineral water either with dissolved carbonic acid or dissolved sulphur compounds such as sulphates.

Acid value The acidity of a solution expressed in terms of normality. Also known as acid number.

Acid wash A solution of phosphoric acid applied to steel parts that removes and neutralises the alkaline solutions used for grease removal after machining. Also leaves a metallic phosphate coating which accepts paints well and of itself, provides a degree of protection against rust.

Aciniform 1. Having the shape of a bunch of grapes. 2. Containing small grape-like kernels.

Ackerman steering gear Differential gear or linkage that turns the two steered road wheels of self-propelled vehicle so that all wheels roll on circles with a common centre.

Acolpate Refers to pollen grains lacking furrows or grooves.

Aconite Any plant of the genus Aconitum.

Acorn The nut of the oak tree usually surrounded at the base by a woody involucre.

Acorn disease A virus disease of citrus plants characterised by malformation of the fruit which is some what acorn shaped.

Acorus calamus See sweet flag.

— **gramineus** See sweet flag.

Acotyledon A plant with no distinct seed lobes.

Acoustic reflex The mechanism by which the ear protects itself from extra loud sounds by in fact reducing them; just as the eye protects itself from extra bright light by contracting the pupil.

Acquired character The character which has been acquired by the plant in the course of its life. It is not hereditory and also cannot be passed on to other generations.

Acquired immunity Acquired ability to resist a disease to which an individual would be normally susceptible. Active immunity is through exposure to or inoculation with pathogens whereas passive immunity is when these antibodies are received from another individual who has become immune.

Acre A unit of land measurement, 43,560 square feet, 10 square chains, or a square of 208.7 feet (63.612 m) side. It is equivalent to 0.4047 ha.

Acre-foot The volume of water that will cover 1 acre (0.4086 ha) to a depth of a foot (0.3048 m); equivalent to 325, 850 gallons of water.

Acre-foot per day The unit of volume rate of water flow.

Acre-inch The volume of water that will cover 1 acre (0.40406 ha) to a depth of 1 inch (2.54 cm).

Acre-yield The average amount of water taken from one acre of a reservoir.

Acric Ferrasols Real old Ferrasols having a very low CEC of the clay fraction ($\leq$ 1.5 me per 100 g).

Acrisols Real old tropical outbased clay soils with a low base saturation and a distinct argillic B horizon. They occur mainly in older land types with a monsoon climate.

Acrocarpous In some mosses of the sub-class Eubrya having the sporophyte at the end of a stem and therefore exhibiting the erect habit.

Acrocarpous fraxinifolius See red cedar.

Acrocercops syngramma See cashewnut leaf - miner.

Acrodomatia Specialised structures on certain plants adapted to shelter mites; relationship is presumably symbiotic.

Acrogynae Liverworts of the Jungermanniales in which the apical cell ultimately becomes an archegonium.

Acrohumox Humox that have a cation - retention capacity (from NH_4 Cl) of 1.5 me or less per 100 g clay in some subhorizon of the oxic horizon (or have 1.5 me or less extractable bases plus extractable aluminium per 100 g clay); do not have sombric horizon and do not have sheets that contain 30 per cent or more gibbsite and have 20 per cent by volume gravel size aggregates containing 30 per cent or more gibbsite in all subhorizons within 1 m of the surface.

Acronychia A small tree found in Uttar Pradesh, West Bengal, the Western Ghats and Khasi Hills. The leaves yield an essential oil and the tender ones used as condiment. Also called as Acronychia Pedunculata and belong to family Rutaceae.

Acronychia laurifolia See acronychia.

Acropetal succession Order of development in which the youngest structures are at the apex and the oldest at the base.

Acroscopic Facing or on the side towards the apex.

Across the slope system, or Keythrope system A system of tile drains consisting of running the lines of the tiles across rather than down the slope. Also called as keynote system.

Acrotthox Are the Orthox that have in some subhorizon of the oxic horizon a cation retention capacity of 1.5 me or less (from NH_4C1) per 100 g clay (or 1.5 me or less of extractable bases plus extractable aluminium per 100 g clay); do not have discernible structure in the major part of the oxic horizon or have only weak blocky or prismatic structure; and do not have cemented sheets that have 30 per cent or more gibbsite or a sub horizon that has 20 per cent or more by volume aggregates of gravel size that contains 30 per cent or more gibbsite within a depth of 1.25 m below the soil surface.

Acrustox Are the Ustox that have a cation-retention capacity (from NH_4Cl) or 1.5 me or less per 100 g clay in some subhorizon of the oxic horizon (or have 1.5 me or less of extractable bases plus extractable aluminium per 100 g clay).

AC soil A soil having a profile containing A and C horizons with no clearly developed B horizon.

Actinocarpous Having flowers and fruits radiating from one point.

Actinomorphic Manifesting radial symmetry, as flowers that have the floral parts in each series of equal size and position.

Actinomycetes Unicellular filamentous micro-organisms that branch monopodially or more rarely dichotomously and from radiating colonies; mainly found in the soil and cause of its characteristic odour.

Actinostele A protostele in which the xylem arms are star shaped in cross-section.

Activated charcoal Specially treated to make the charcoal more absorptive for gases, odours and colouring matter.

Activated sludge Sludge settling down in a septic or settling tank when allowed to stand or allowed to undergo fermentation by blowing air through it. Contain readily available forms 3-6% nitrogen, 2% P_2O_5 and 1% K_2O when such sludge is applied to soils.

Active acidity The activity of hydrogen ion in the aqueous phase of a soil. It is measured and expressed as a pH value.

Active agent The component part of pesticide formula that has toxic properties.

Active balance A favourable balance of payments when receipts from the export of goods and services exceed expenditure on the import of goods and services.

Active immunity See acquired immunity.

Active market A market for a particular class of stock or shares in which there are frequent and regular dealings.

Active principle That part of a substance which has the effect, not the carrier or other substance added to it. Also see active agent.

Activity rate The proportion of civilian employees in a given population age group. Low activity rates may be regarded, to some extent, as an indication of 'concealed' unemployment.

Actual water income Water received from precipitation as well as that entering the area as underground flow available for transpiration within the area.

Actuary A member of the insurance profession, specially skilled in the application of mathematical and statistical techniques to the problems of insurance of all types.

Actuate To move, cause to move, or bring into action. Automobile engine valves are actuated by motion of the camshaft.

Acuminate Pointed; tapering or drawn out into a long point.

Acute (1) Not chronic (2) Ending in sharp point.

Adansonia digitata See monkey bread tree.

Adapical Near or toward the apex or tip.

Adaptation (1) Change in an organism resulting from the action of natural selection on variation so that the organism is fitted more perfectly for existence in its environment. (2) The occurrence of genetic changes in population or species as the result of natural selection so that it adjusts to new or altered environment conditions.

Adapter Coupler that has been adapted for attachment of hose or threaded pipe to the irrigation pipe for sprinkle irrigation.

Adaptive enzyme An enzyme produced by an organism in response to the presence of its substrate or a related substance. It is also called an induced enzyme.

Adaxial (1) Situated on the side towards the stem or main axis, as a sporangium. (2) On the same side as or facing toward the axis of an organ or organism.

Additive Any substance added to a reaction mixture for increasing its efficiency or to give it some desirable characteristics.

Adenophyllous Having leaves with glands.

Adhatoda vasica See Malabar nut.

Adhesion Attraction of unlike materials where the substances are more or less firmly attached together by their adjacent surfaces; the sticking together of separate tissues or the organs of the body to each other as a result of an inflammation.

Adhesive agent Material that acts as a cementing agent between two solids.

Adiabatic atmosphere A model atmosphere characterised by a dry adiabatic lapse rate throughout the vertical extent.

Adiabatic equilibrium A vertical distribution of temperature and pressure in an atmosphere in hydrostatic equilibrium such that an air parcel displaced adiabatically will continue to possess the same temperature and pressure as its surroundings so that no restoring force acts on a parcel displaced vertically.

Adiabatic process A thermodynamic change of state of a system in which there is no transfer of heat or mass across the boundaries of the system.

Adiabatic rate The rate of temperatures drop that gases experience with rising elevations due to the decrease in air pressure, about 1°C per 100 m vertical rise of air; if water vapour is condensing, rate is about 0.6°C per 100 m.

Adisura atkinsoni See pod borer.

Adjust To bring about a proper arrangement of parts as regards relation, position, fit, etc.

Adjustable boring tool A tool in which the cutter can be set for different jobs, avoiding the necessity of changing both cutter and holder.

Adjustable parallels Wedge shaped bars of iron placed with the thin end of one, on the thick end of the other. The top face of the upper and the bottom face of the lower remain parallel, but the distance between the two faces can be increased or decreased, and the bars locked in position by means of a screw to prevent shifting.

Adjustable speed motor A motor whose speed can be varied over a wide range and when once adjusted remains constant regardless of load.

Adjusted value A value of a quantity derived from observed data by some orderly process which eliminates discrepancies arising from error in those data.

Adjusting screw A set screw, by the use of which the position of machine parts is adjusted or regulated more accurately than would be possible by the mere setting to dimensions.

Adjustment The placing and setting of engine or machine parts in related position.

Adjuvant A additive which intensifies or modifies the action of a chemical.

Ad libitium feeding Where animals are permitted to eat daily as much as they desire. See also Orts; Voluntary intake.

Administered pricing Pricing by 'price makers', not 'price takers' as under perfect competition, the pricing policies of monopolistic and oligopolistic sellers.

Adnate A condition of being closely attached to side of petiole or stalk, as stipules or leaves; attachment of any organ by its whole length.

Adobe Term used to describe a structural condition of soils, usually of high clay content, which crack into roughly cubical blocks when dried.

Adolescent river A river with a graded bed and a well cut channel that reaches base level at its mouth, its waterfalls and lakes of the youthful stage having been destroyed.

Adoption process It is a mental process through which an individual passes from first hearing about a new idea to its final adoption.

Adoretus duvauceli See apple leaf and fruit eating beetles.

— **lasiophygus** See grapevine chafer beetle.

— **versutus** See apple leaf and fruit eating beetles.

Adsorb Removal of a substance in the solution to a solid surface or a separate phase; to accumulate on surface.

Adsorbed water Water held in a soil mass by physico-chemical forces and having physical properties substantially different from absorbed water or chemically combined water at the same temperature and pressure.

Adsorbent Substances which are able to adsorb other substances on its surface.

Adsorption The adhesion of thin layers of gases, liquids or dissolved substances to the surface of solids.

Adsorption complex The group of substances in the soil capable of adsorbing other materials. Organic and inorganic colloidal substances form the greater part of the adsorption complex. The non-colloidal materials, such as silt and sand, exhibit adsorption to a much lesser extent than the colloidal materials.

Adulteration Making impure by the addition of other (usually inferior) ingredients or foreign substances.

Ad valorem 'According to value'.

Ad valorem tax A duty imposed on commodities in proportion to their value.

Advance To adjust the timing of an event for early occurrence such as the spark to occur earlier with regard to time of top dead centre of piston position. A copper, nickel alloy used for electrical instrument work.

Advance refunding A technique of the treasury to hold traditional bondholders; the treasury offers holders of expiring bonds especially attractive terms to buy its new bond issues. In this way it does not compete with private borrowers for the inflow of new savings.

Advection The movement of air, water, and other fluids in a horizontal direction.

Advection fog A type of fog caused by the horizontal movement of moist air over a cold surface and the consequent cooling of that air to below its dew point.

Advective hypothesis The assumption that local temperature changes are the result only of horizontal or isobaric advection.

Advective thunder storm A thunderstorm resulting from static instability produced by advection of relatively colder air at high levels or relatively warmer air at low levels or by a combination of both conditions.

Adventitious Out of place; not at the expected place. Also see aerial roots.

Adverse effect An adverse reaction of an ecosystem to a disturbance.

Advice A note sent by one merchant to another advising him of the despatch of goods ordered.

Aeciospore The spore produced in the aecial or cup stage of a rust.

Aecium A spore stage in rusts, e.g. in wheat rust the aecial stage develops on the barberry leaf.

Aeginetia indica Sugarcane bunga. A plant parasite on sugarcane roots.

Aeolian soil Soils which have been deposited by wind and are not of much value in agriculture.

Aerate To impregnate with a gas, usually air.

Aerated water Water purified by passing air through it.

Aeration Change of air at intervals or periodically; consideration of aeration in building design.

Aeration irrigation A method of applying water by sprinklers, based on very low application rates, approximately one half of the normal intake rate of the soil, combined with good nozzle pressure.

Aeration porosity Proportion of the bulk volume of soil that is filled with air at a given moisture tension.

Aeration, soil The process by which air in the soil is replaced by air from the atmosphere. In a well-aerated soil, the soil air is similar in composition to the atmosphere above the soil. Poorly aerated soils usually contain a much higher percentage of carbon dioxide and a correspondingly lower percentage of oxygen than the atmosphere. The rate of aeration depends largely on the volume and continuity of pores in the soil.

Aerator A device for introducing air into a liquid, either submerged (perforated pipe) or surface (mechanical paddles).

Aerial dusting Spreading insecticides, fungicides or herbicides from an aeroplane in the form of powder.

Aerial photogrammetry Photogrammetry with the aid of aerial photograph.

Aerial photograph A photograph of the earth's surface taken from a position above the ground, sometimes called aerial photo or air photograph.

Aerial propagation The production of roots on a cutting without a soil cover but using such material as moss. Also see air layering.

Aerial roots Those rising from the stem above the ground.

Aerial spraying Spreading insecticides, fungicides or herbicides from an aeroplane in the form of liquid sprays.

Aerial stem A stem which grows above

the soil, as opposed to a rhizome.

Aerial survey A survey utilising aerial photographs; the taking of aerial photographs for surveying purposes; and the photographs taken of an area for surveying purposes.

Aerobic (1) Having molecular oxygen as a part of the environment (2) Growing only in the presence of molecular oxygen, such as aerobic organisms (3) Occurring only in the presence of molecular oxygen, as applied to certain chemical or biochemical processes such as aerobic decomposition.

Aerobic bacteria Those needing free oxygen in order to live and function.

Aerobic composting A process of composting organic waste. Organic waste may include bedding, sweeping from animal sheds, urine earth and çattle dung etc. Compost is ready in about 4 months time.

Aerobiosis Existing in the presence of oxygen.

Aerophyte A plant growing attached to an aerial portion of another plant. Also epiphyte.

Aerosols Finely divided antiseptic solutions sprayed into the air.

Aesculus indica See Indian horse chestnut.

Aetiology The science of the investigation of the cause or origin of disease.

Afforest Establishment of a forest by artificial means; also includes re-establishment of such forests.

A - frame ridger The frame of such ridger is of the shape of letter A. Implement for making ridges and bunds which are usually larger than the bunds made from bund-former.

African oil palm Elaeis guineensis Jacq. Family Palmaceae. An oil bearing palm native of West Africa. Oil extracted from its nuts used in making soap and fuel for diesel engines. Grown in Kerala. Also known as red oil palm.

Agar Gelatinous substance prepared from Japanese seaweed used as a base for solid culture media.

Agaric Any fungus in the family Agaricaceae.

Agati grandiflora See agati sesbania.

Agati sesbania *Sesbania grandiflora* L. Pers. (*Robinia grandiflora* L.; *Agati grandiflora* Desv.) Family Papilionaceae. A tree native of tropical Asia. Flowers, fruit and leaves used as vegetable. Plant used as support for betelvine. Also called agastoya.

Agave *Agave americana* L. Family Agavaceae. A hardy, vigorous, xerophytic plant native of south America; cultivated in waste lands along fences and embankments. Propagation by suckers. Yield clean, white silky fibres. Also called American aloe, century plant, Kantala.

— **cantala** See cantala.

— **sisalana** See sisal agave.

— **vera cruz** See blue elephant aloe.

Age rotation The age at which the stand is considered ready for harvesting under the adopted plan of management.

Agglomerate The term as applied to a mass of protozoa; dustered as a head of flower.

Agglutination Visible clumping of cells suspended in a fluid.

Agglutinins Antibodies that cause agglutination.

Aggradation Filling in or levelling by deposition.

Aggregate (1) A group of soil particles cohering in such a way that they behave mechanically as a unit; (2) Many fine particles of soil held together in a single mass of cluster, such as a crumb or clod.

Aggregate demand The total effective demand or expenditure of all purchasers of capital and consumer goods within a given market.

Aggregate fruit A fruit developed in a cluster from the separate overies of a single flower.

Aggregate supply The total physical volume of goods and services currently coming on to a given market.

Aggregation The bringing together or summing of primary data. For example, the national income is an aggregate, in contrast with the income of an individual.

Aggregative index An index number which is constructed by aggregating a number of items.

Agistment The feeding and taking care

of cattle for reward.

Agitation dredging A method of dredging that involves overboard discharge of spoil with a high induced rate of dispersal.

Aglaia roxburghiana Miq. Family Meliaceae. A large tree whose fruits are edible and the fragrant flowers used in perfumary.

Agmark National insignia for quality and purity of agricultural and animal husbandry products. Given to products which confirm to certain standards and grade specifications as laid down in Agricultural Produce (Grading and Marketing) Act, 1937.

Agmark grade Grades as laid down in Agricultural Produce (Grading and Marketing) Act, 1937 based on the result of analysis of quality factors.

Agrarian Pertaining to agriculture or actions for the benefit of the rural community.

Agreement area An area of land which has been brought under an irrigation agreement extending over a period of years.

Agrestal Dwelling, or growing wild, in cultivated fields.

Agric horizon Is an illuvial horizon formed under cultivation that contains significant amounts of illuvial silt, clay, and humus. When a soil is brought under cultivation, the vegetation and the soil fauna as a rule are changed drastically. The plow layer is mixed periodically and, in effect a new cycle of soil formation is started.

Agricultural bolt A bolt, the body of which has a series of helical bands and grooves which are formed by a rolling process; used in farm machines and appliances.

Agricultural chemistry A branch of the science of chemistry dealing with the composition and transformations pertaining to soil, plants and animals; includes all the life processes through which food and fibre are obtained and control of these processes to increase yields, improve quality and reduce costs.

Agricultural climatology The study of climate as to its effect on crops; it includes, the relation of growth rate and crop yields to various climatic factors and hence the optimum and limiting climates for any given crop.

Agricultural commodity A common term used for any product of agriculture.

Agricultural economics Branch of economics as applied to farm management and production.

Agricultural Engineering (1) Application of knowledge, techniques and disciplines of various fields of engineering to the solution of problems arising in the field of agriculture and rural living with the object of reducing labour, improving agricultural productivity per worker, raising the standard of living by increasing the earning per worker. (2) A course of study dealing in various disciplines such as design and fabrication of agricultural implements, soil and water management, buildings design, use of electricity on farms and processing of agricultural products to achieve the above objectives.

Agricultural geology A branch of geology that deals with the nature and disciplines of soils, the occurrence of the mineral fertilisers and the behaviour of underground water.

Agricultural investigations Investigations and studies including soil surveys, land use planning, data on consumptive use by crops, men and animals employed on the farms, rotation of crops and crop patterns, treatment of deteriorated soils and ground water control.

Agricultural land All lands devoted to raising crops and livestock; includes farmstead, roads, irrigation and drainage channels, ponds and water supply etc.

Agricultural marketing Buying and selling of agricultural products. A complex system in which several intermediaries from government agencies to private businessmen are involved.

Agricultural meteorology The study and application of relationships between meteorology and agriculture, involving problems such as timing the planting of crops. Also known as agro-meteorology.

Agricultural productivity The growing conditions of crops as determined by six main regimes, namely moisture, nutrient, biological, temperature, air and mineral. These regimes are closely inter-related and inter-dependent, and have to be studied carefully in assessing soils for agricultural production.

Agricultural regions Well defined regions of a country having similar rainfall, temperature, altitude, natural vegetation, soil, crops and stocks. Parts which have some different conditions of the region are designated as sub-regions. Helps in coordinating agricultural research in the country.

Agricultural science A discipline dealing with the selection, breeding and management of crops and domestic animals for more economic production.

Agricultural subsidies Payments to the farmers by the government for the purpose of encouraging home production of foodstuffs while maintaining reasonable retail prices to consumers.

Agricultural wages board A statutory body responsible for negotiating the wages of agricultural workers.

Agricultural wastes Those liquid or solid wastes that result from agricultural practices such as cattle manure, crop residue, pesticides and fertilisers.

Agriculture The production and processing of plants and animals for mains use and enjoyment, involving soil cultivation and the breeding and management of crops and livestock.

Agriculture labourer One who neither owns any land or gets any land on hire but works for wages in agriculture.

Agriculturist A person identified with the profession of agriculture.

Agrobiology The science of plant nutrition and growth, crop production and soil management.

Agroecology The study dealing with the relation of agricultural crops to that of the environment.

Agrology Study of the applied aspects of soil science and management, also called pedology, edaphology.

Agromyza obtusa See gram pod fly, pigeonpea pod fly.

— **phaseoli** See cowpea stemfly.

Agronomy The branch of agriculture that deals with the theory and practice of field crop production and the scientific management of soil.

Agrophilous Having a natural habitat in grain fields.

Agrostis procera See *Eriochloa procera*.

— **flammatra** See gram cutworm.

— **ypsilon** See gram cutworm.

Agrostology A part of botany which studies grasses, their classification, management and utilisation.

Aftermath (1) Recovery growth of forage plants after harvesting by either animal or machine. (2) A second mowing; the crop of grass cut from the same soil after the first crop of the season.

A Horizon A mineral surface horizon in which humified organic matter has accumulated. Therefore this horizon is somewhat darker than the underlying B horizon. The decomposed organic material is mixed with mineral material by biological activity.

Aid-tying An obligation which may be imposed by a donor government upon a recipient to use aid funds for purchasing goods and services in the contributing country.

Ailanthus excelsa See tree of Heaven.

— **malabarica** See *guggula dhup*.

— **triphysa** See *guggula dhup*.

Air The atmosphere. The mixture of gases that forms the apparent blue envelope of the earth, although technically the term 'air' is usually applied to a limited portion of the atmosphere.

Air blast sprayer See mist sprayer.

Air bleed A small manually adjusted air valve by which the amount of air in fuel mixture can be varied.

Air-borne plant disease Plant disease that is due to the spores of fungi which are air borne, and alight on the plants leaves and stems.

Air bound Condition in a pipeline wherein air entrapped in summit prevents the free flow of water.

Air chamber A vessel installed on piston pumps to minimise the pulsating discharge of the liquid pumped. The chamber contains air under pressure and is fitted with an opening on its underside into which some of the liquid from the pump is forced upon the delivery stroke of the piston. The air acts as cushion to lessen the fluctuation of the liquid flow between the suction and delivery strokes of the piston.

Air-cleaner A device designed to separate dirt and other foreign substances from the air before it enters the combustion space.

Air-conditioning Relates to three different operations, namely air washing, humidifying, and dehumidifying.

Air control The means employed to operate the control surfaces of the aircraft.

Air-cooled transformer A transformer in which the heating of the coils sets up a ventilating draught.

Air curing Method of curing tobacco wherein whole tobacco is cured for a short time in the open air and then in a shed. Leaves are stripped after the curing.

Air drainage Renewal of soil air by diffusion and meteorological factors as soil temperature changes, barometric variations and action of wind, in the soil.

Air dried Lumber seasoned by drying in the air as opposed to kiln-dried.

Air dry (1) The state of dryness of a soil at equilibrium with the moisture content of the surrounding atmosphere. The moisture content depends on the relative humidity and the temperature of the surrounding atmosphere. (2) To allow to reach equilibrium in moisture content with the surrounding atmosphere.

Air dry soil The soils which have been dried in the air; contain only hygroscopic moisture and chemically combined water.

Air duct A tube, usually of fabric, supplying air for filling or for maintaining pressure in air filled parts of an aerostat.

Air-fuel ratio The quantity of gasoline and air, expressed as a ratio of their respective weights that go to make up the fuel mixture of an internal combustion engine.

Air layering The method of layering in which a ring of bark of about an inch in length below the bud from the portions of selected shoots is removed. Covering it with earth, moss or such other material and than covering and tying with a gunny cloth. Treated shoots kept moist. Method used for propagating fruits.

Air lift Means of raising water by the buoyancy of injected air.

Airlift pump A device composed of two pipes one inside the other used to extract water from a well, the lower end of the pipes is submerged and air is delivered through the inner pipe to form a mixture of air and water which rises in the outer pipe above the water in the well.

Air lock A stoppage in a water or gas line caused by a bubble of air being pocketed somewhere along the line.

Air mass An extensive body of the atmosphere which approximates horizontal homogeneity in its weather characteristics, particularly with reference to temperature and moisture distribution.

Air mass analysis In general, the theory and practice of syneptic surface-chart analysis by the so called Norwegian methods, which involve the concepts of the polar front and of the broad scale air masses which it separates.

Air mass climatology The representation of the climate of a region by the frequency and characteristics of the air masses under which it lies; basically, a type of synoptic climatology.

Air mass precipitation Any precipitation that can be attributed only to moisture and temperature distribution within an air mass when that air mass is not, at that location, being influenced by a front or by orographic lifting.

Air mass source region An extensive area of the earth's surface over which bodies of air frequently remain for a sufficient time to acquire characteristic temperature and moisture properties imparted by that surface.

Air pollution The presence in the outdoor atmosphere of one or more contaminants such as dust, fumes, gas, mist, odour, smoke, or vapour in quantities and of characteristics and duration such as to be injurious to human, plant or animal life or to property, or to interfere unreasonably with the comfortable enjoyment of life and property.

Air porosity The portion of the bulk volume of soil that is filled with air at any given time or under a given condition such as a specified soil water potential. Usually, this portion is made up of large pores, that is those drained by a tension of less than about 100 cm of water. See also moisture tension.

Air regime The air regime is an important aspect of the soil, because roots need oxygen, and produce carbon dioxide, which has to escape to the air. Penetration of oxygen and diffusion of carbon dioxide decrease with soil depth. The total volume of all pores and the type of pores in the root zone influence the growth of roots and consequently the crop yield.

Air relief valves Valves installed at major high points of longer main lines to avoid air collection.

Air Sacs (1) Dilatations of tracheae in many insects; sac representing tracheal system and having hydrostatic function in certain insect larvae. (2) Air cavities in the stems of plants.

Air screen cleaner It is the basic seed processing machine. Basic cleaning of all seeds is done by an air screen cleaner.

Air slaked lime Burnt lime which has absorbed carbon dioxide and water on exposure to air.

Air, soil The soil atmosphere, the gaseous phase of the soil, is the volume not occupied by solid or liquid.

Air sounding The act of measuring atmospheric phenomena or determining atmospheric conditions at altitude, especially by means of apparatus carried by balloons or rockets.

Air spora Air borne fungus spores, pollen grains, and micro-organisms.

Air temperature (1) The temperature of the atmosphere which represents the average kinetic energy of the molecular motion in a small region and is defined in terms of a standard or calibrated thermometre in thermal equilibrium with the air. (2) The temperature that the air outside of the aircraft is assumed to have as indicated on a cockpit instrument.

Air tight So constructed or sealed as to prevent any inlet or outlet of air.

Air turbulence Highly irregular atmospheric motion characterised by rapid changes in wind speed and direction and by the presence, usually, of up and down currents.

Air valve Device that releases air from a pipeline automatically without permitting loss of water. It may also admit air to the pipe automatically if the internal pressure becomes less than atmospheric.

Aitken nuclei Small particles a few hundredths of a micron in diametre, which normally exist in the atmosphere q.v. in concentrations varying from a few thousand to a few hundred thousand per millimetre. They are produced in large number by most combustion processes.

Aizoaceae A family of flowering plants in the order caryophyllales; members are unarmed leaf succulents chiefly of Africa.

Ajmud (1) Trachyspermum roxburghianum (DC) Craib. (Carum roxburghianum benth. ex Kurz.). A herb of family Umbelliferae. Seeds are used for flavouring and medicinal purposes. (2) See celeriac and celery.

Akee Blighia sapida Koenig (Cupania sapida Voigt). A tree native of west Africa and produces edible fruit.

Ak grasshopper Poecilocerus pictus Fab. Pest of papaya; also damages fig, brinjal, castor, tomato and ak (calotropis); nymph and adult feed on leaves and thus skelotonizing them.

Akinete A non-motile, thick-walled resting spore in which the vegetative wall of the cell becomes the wall of the spore.

Akola hoe An interculture implement developed at Akola for black cotton soils of Maharashtra State, has sharp triangular steel shares fixed on tines.

Akund Calotropis procera (Ait.) R.Br. (Asclepias gigantes L.) Family asclepiadaceae. Plant is very useful and is found in Punjab and Andhra Pradesh. Fibre used for cordage; floss from seed as stuffing material; stem and leaves as green manure. Also called as swallow-wart, safedak.

Al-aged Applied to a stand in which, theoretically, trees of all ages up to and including those of the seedling age are found. See also even-aged and uneven-aged.

Alandi Pomegranate type grown in west India; bears medium size fruit with deep red or deep pink flesh, hard seed, sweet but acidic juice.

Albaqualfs Are the aqualfs that have an abrupt textural change between an ochric epipedon or, more commonly, an albic horizon and the argillic horizon and have slow or very slow hydraulic conductivity in the argillic horizon; have no tonguing of albic materials in the argillic horizon; have no fragipan, duripan, or natric horizon; have mean summer and mean winter temperatures at 50 cm that differ by 5°C or more; and have no plinthite that forms a continuous phase or constitutes more than half of the matrix within some subhorizon between 30 cm and 1.25 m below the surface of the soil.

Albaquults The aquults that have an abrupt textural change between the ochric epipedon or the albic horizon and the argillic horizon, and they have low hydraulic conductivity in the argillic horizon. They do not have a fragipan, and they do not have plinthite that forms a continuous phase or constitutes half or more of the matrix of any subhorizon within 1.25 m of the soil surface.

Albedo The ratio of the amount of visible light reflected by a body to the amount incident upon it.

Albic Arenosols Arenosols consisting of albic material, at least 50 cm thick.

Albic E horizon A very light coloured (whitish or light gray) sandy horizon, mostly consisting of uncoated sand and silt grains. It overlies a spodic B, B, argillic B or natric B horizon or an almost impervious clay layer. It is a highly degraded soil horizon from which clay particles and iron oxides have been removed. In podzols, it overlies a spodic B horizon, which, in tropical regions, may be found at a depth of some metres and all the coarse quartz sand to a few metres depth consists of albic material. Such very infertile soils are often called 'white sands'. Albic E horizons are diagnostic for planosols.

Albic Luvisols Luvisols with an albic E horizon.

Albic material The material of an albic E horizon, mainly light-coloured, uncoated quartz sand material.

Albino A plant lacking chlorophyll.

Albizzia amara See lallei.
— **chinensis** See siran.
— **lebbeck** See kokko.
— **lucida** See gal wang.
— **odoratissima** See black siris.
— **procera** See safed siris.
— **stipulata** See siran.
— **wightii** See lallei.

Albumen Substance found between skin and embryo of many seeds, usually the eatable part.

Albuminoides (Sing. albuminoid) Class of heterogeneous proteins found in skeletal structures and connective tissues of animals, insoluble in neutral solvents. Includes gelatin and collagen found in cartilage and bone, elastin in ligaments and keratins of epidermal tissues such as hair wool. Chief part of organs and tissues of plants.

Albuminoid ratio See nutritive ratio.

Albumins Water soluble simple proteins which coagulate on heating; found in milk, blood, egg white, milk and vegetables.

Alcides mali See apple shoot borer.

Aldehydes A class of chemical compounds intermediate between alcohols and acids; most are colourless, volatile fluids with suffocating odours.

Aldrin 1,2,3,4, 10, 10-hexacholoro - 1,4,4a, 5,8,8a-hexahydro - 1, 4-endoexo - 5, 8-diamethanonaphthalene; an insecticide more poisonous than DDT, could be applied as dust or spray and kills the insect by contact or entering the stomach, can be mixed with fertiliser. It is effective against locusts, grasshoppers, termites, white grubs on potato, maize, wheat, sugarcane and against cut worms.

Aletophyte A weedy plant growing on the roadside or in fields where natural vegetation has been disrupted by man.

Aleurites fordii Hemsl. Family enphorbiaceae. A tree, native of China. Tung oil obtained from seed. Grown in Assam, Bengal, Bihar and Mysore. Called Tung oil tree or China wood oil tree.

— **moluccana** See candlenut.

— **triloba** see candlenut.

Aleurocanthus spiniferus See citrus white fly.

— **Woglumi** See citrus white fly.

Aleurolobus barodensis See sugar-cane white fly.

Aleuron Protein grain found in general protoplasm and used as reserve food material; layer containing protein of endosperm in monocotyledons. Also called aleurone.

Aleurone layer The outer layer of cells of the endosperm of seed, such as corn and wheat. This layer is filled with aleurone grains.

Alexander Early variety of peach. Fairly prolific, large sized fruit with skin greenish white nearly covered with deep red. Flesh greenish white, juicy and sweet.

Alexandrian laurel Mammea longifolia Planch. and Triana (Ochrocarpus longifalius (Wight) Benth and Hook f.). A tree of the family Guttiferae producing edible oils. Dry flower buds are used as a source of red dye used for colouring silk.

Alfalfa (1) Special valves attached to the top of risers pipes to control the flow. Also called as orchard valves. (2) See lucerre.

Alfisols It is one of 10 orders in Soil Taxonomy. Soils in this order have marks of processes that translocate silicate clays without excessive depletion of bases and without dominance of the processes that lead to the formation of a mollic epipedon. The unique properties of Alfisols are a combination of an ochric or an umbric epipedon, an argillic horizon, a medium to high supply of bases in the soil, and water available to mesophytic plants more than half the year or more than 3 consecutive months during a warm season. Because these soils have water and bases, they are as a whole, intensively used, but problems with tilth are very common.

Algae (Alga Sing). Microscopic or larger plants with chlorophyll; mostly living submerged in fresh or salt water; grouped into green, blue, yellow-green and diatoms. In soils found in constantly moist soils such as rice fields. Improve soils aeration and add organic matter to soils. Some of them capable of fixing atmospheric nitrogen.

Algal bloom A preponderance of algae in water which is turned completely green.

Algal scum Fresh or decaying floating surface layer of algae.

Algin A mucilaginous substance, alginic acid, obtained from certain algae.

Algology The study of algae. Also called phycology.

Alhagi camelorum See camel thorn

— **maurorum** See camel thorn

— **pseudalhagi** See camel thorn

Aliar Vilayati Mehndi. See Jamaica switch sorrel.

Alidade An instrument used in fire towers to locate forest fires. The alidade is equipped with sights for determining direction of fire.

Aliform Wing-shaped; applied to muscles as in insects.

Alimentation The process of nourishing or of being nourished.

Alismataceae A family of flowering plants belonging to the order Alismatales characterised by schizogenous secretory cells, a horseshoe shaped embryo and one or two ovules.

Alismatales A small order of flowering plants in the subclass Alismatidae, including aquatic and semiaquatic herbs.

Alismatidae A relatively primitive subclass of aquatic or semiaquatic herbaceous flowering plants in the class liliopsida, generally having apocarpous flowers, and trinucleate pollen and lacking endosperm.

Alkali In chemistry, any substance having marked basic properties in contradiction with acid, that is being capable of furnishing to its solution or other substances the hydroxyl ion (OH negative). The important alkali metals are sodium and potassium. In a less scientific sense the term is applied to the soluble salts, especially the carbonates and bicarbonates of sodium and potassium, which are present in some soils of arid and semi-arid regions in sufficient quantities to be detrimental to ordinary agriculture.

Alkali claypan A claypan containing ≥ 15 per cent exchangeable sodium or having sodium adsorption ratio ≥ 15.

Alkaline soil Generally, a soil that is alkaline throughout most or all of the parts of it occupied by plant roots, although the term is commonly applied to only the surface layer or horizon of a soil. Precisely, any soil horizon having a pH value greater than 7.0; practically, a soil having a pH above 7.3.

Alkalinity The quality or state of being alkaline. The concentration of OH negative ions.

Alkalinity, soil The degree or intensity of alkalinity of a soil expressed by a value greater than 7.0 on the pH scale.

Alkali seepweed Suaeda fruticosa L. Forsk (Chenopodium fruticosum L.) Family Chenopodiaceae. An undershrub used as fodder. Commonly found in Punjab and Rajasthan. Also called as lonia, or lunak.

Alkali tolerance Tolerance of certain crops to alkalinity of the soil. Alkali sensitive crops suffer injury when grown in alkaline soils.

Alkalization The process whereby the exchangeable sodium content of a soil is increased.

Alkaloid (1) Basic substance found in plants that is usually the part of the plant that has medicinal properties. (2) An alkaline nitrogenous substance usually poisonous to animals.

Allantoid Sausage shaped.

Allele, allel, allelomorph adjective forms: allelic, allelmorphic. One of a pair, or of a series of factors, that occur at similar loci of homologous chromosomes and for this reason are inherited in alternative pairs. One alternative form of a gene.

Allium ampeloprasum See leek.
- **ascalonicum** See shallot.
- **cepa** See onion.
- **porrum** see leek.
- **sativum** See garlic.
- **schoenoprasum** see chive.
- **tuberosum** see chinese chive.
- **uliginosum** see chinese chive.

Alligator pear See avocado.

Allocarpy The production of fruit following cross-fertilization.

Allogamy Cross-fertilization.

Allogenic Caused by external factors as in reference to the change in habitat of a natural community resulting from draught.

Allometry Study of relative growth; change of proportions with increase of size; growth rate of a part differing from a standard growth rate or from the growth rate of the whole.

Allophane An amorphous hydrated aluminosillicate of variable composition.

Alloploid or allopolyploid A polyploid having chromosome sets from different sources, such as different species. A polyploid containing genetically different chromosome sets; for example, from two or more species.

Allostratus cloud A principal cloud type in the form of a grey or bluish (never white) sheet or layer of striated, fibrous, or uniform appearance; very often totally covers the sky and may cover an area of several thousand square miles; vertical extent may be from several hundred to thousands of metres.

Allotropus flower A flower shaped in a manner such that its nectar is easily accessable to insects.

Allowable cut The volume of timber which may be cut from a forest under optimum sustained yield management.

Alloy A metallic substance consisting of two or more elements, one of which at least is a metal. It is generally obtained by melting together the constituent elements.

Alloying element An element added deliberately to a metal to modify its mechanical or physical properties.

All-spice The dried, unripe berries of a small, tropical evergreen tree, Pimenta officinalis, of the Myrtle family; yields a pungent, aromatic spice.

Alluvial Pertaining to material that is transported and deposited by running water.

Alluvial cone Cone or fan shaped deposit of alluvium.

Alluvial fan A fan shaped deposit of alluvium laid down by a stream where it emerges from an upland into less steeply sloping terrain.

Alluvial land Areas of unconsolidated alluvium, generally stratified and varying widely in texture, recently deposited by stream, and subject to frequent flooding. A miscellaneous land type.

Alluvial soil Most extensive and agriculturally important among the soil groups of India. Includes Indo-gagetic and deltaic alluvium. Soil derived mainly from mountain and hill washes brought down by the great rivers and deposited in the plains below. Sometimes subdivided into old and recent alluviums. Old alluvium red in colour. Recent alluviums get periodic deposits of fresh silt and are more fertile.

Alluvium A general term for all material deposited or in transit by streams, including gravel, sand, silt, clay and all variations and mixtures of these. Unless otherwise noted, alluvium is unconsolidated.

Almond Prunus amygdalus Batsch. (P. communis L. Arcang.; Amygdalus communis L.). Family Rosaceae. Deciduous nut tree. Native of the E. Mediterranean region. Highly nourishing and have medicinal value. *Badam roghun* (Oil of almond) used in perfumery. Propagated by seed though buding and grafting also possible.

Almond moth Ephestia cautella walker. Pest of stored wheat, rice, almonds, walnut, niger, coconut, pistachio, raisins, cashewnuts, and tamarind seed. Caterpiller weaves silken tubes in flour and over food materials, reducing their keeping quality.

Alocasia cucullata See giant taro.

— **indica** See large leaved caladium.

— **macrorhiza** See giant taro.

Alopecurus pratensis See meadow foxtail.

— **typhoides** See bulrush millet.

Alpage A summer grazing area composed of natural plant pasturage in upland or mountainous regions.

Alpha humus Mixture of indefinite composition of dark-coloured organic substances precipitated upon making strongly acid a dilute alkali extract of soil.

Alphonso The mango variety of par excellence of west India, shipped in cold storage.

Alpine soil Mountain soil occurring above the timberline.

Alpine strawberry Perpetual strawberry. Fragaria vesca L. Family Rosaccae. A herb. Fruits edible.

Alpine tundra Large, flat or gently sloping, treeless tracts of land above the timberline.

Alpinia galanga See greater galangal.

Alternanthera amoena See *kanchari*.

Alternate (1) Of the arrangement of leaves on opposite sides of the stem at different levels. (2) Of the arrangement of the parts of one whorl between members of another whorl.

Alternate bearing Habit of bearing heavy and sparse crops in alternate years, inherent in mango and some other species of fruits. Also called biennial bearing; on and off yearly rhythm. Giving good cultural treatment annually can keep biennial trees reasonably healthy, fit and fruitful in the off years.

Alternating current motor An electric motor of either single or polyphase type operated by an alternating current.

Alternate grazing Grazing two pastures in alternate succession. Also called as rotational grazing and intermittent grazing.

Alternate host (1) One or other of the two unlike hosts of a heteroecious rust. (2) A plant or animal upon which a disease organism exists for only a part of its life cycle.

Alternate leaves Leaves each of which arises singly at a node.

Alternation of generations A pattern in which two distinct plant bodies or organs produced in the same life cycle alternate with each other, as the gametophytic generation produces gametes and the sporophytic generation arising from the zygote produces the spores which produce the gametophytic plant.

Alterne A community exhibiting alternating dominance with other communities in the same area.

Altherbosa communities of tall herbs, usually succeeding where forests have been destroyed.

Altitude The vertical distance between a point and a datum surface or plane, such as mean sea level.

Altitudinal vegetation zone A geographical band of physiognomically similar vegetation correlated with vertical and horizontal gradients of environmental conditions.

Altocumulus cloud A principal cloud type, white or grey or both white and grey in colour; occurs as a layer or patch with a waved aspect, the elements of which appear as laninal, randed masses or ralls; frequently appears at different levels in a given sky. Abbreviated AC.

Alumina Aluminium trioxide, an impurity in phosphate rock which causes superphosphate to be wet and sticky.

Alluminium A very light silvery-white metal used independently or in alloys with copper.

Aluminium phosphide Fumigant for grains; available as tablets and exposure of 4-5 days allowed.

Alluminium toxicity Below pH 5 in many soils, aluminium compounds mainly in exchangeable term, occur in quantities sufficient to harmfully affect sensitive plants.

Alvar Dwarfed vegetation characteristic of certain Scandinavian steppelike communities with a limestone base.

Alysicarpus longifolius See *jangaligailia*

— **rugosus** DC. Family Papilionaceae. A herb used as fodder. Called shevra in Hindi.

Amalonado Cocoa variety; a hardy prolific yielder.

Amanthophilous Of plants having a habitat in sandy plains or hills.

Amantmul Melothria heterophylla Cogn. A prostrate or climbing herb of family cucurbitaceae grown throughout India. Leaves and fruits edible.

Amaranth (1) Any of the herbs of the genus Amaranthus and family amranthaceae whose leaves are edible; normally cultivated but also growing wild. (2) Amaranthus blitum L. var. oleracea Hook. f.

Amaranthaceae The characteristic family of flowering plants in the order caryophyllales; they have a syncarpous gynoecium, a monochlamydeous perianth that is more or less scarious, and mostly perfect flowers.

Amaranthus blitum var. oleracea See amaranth

— **caudatus** See *chaulai.*

— **cruentus** See *chaulai.*

— **gangeticus** See edible amaranth.

— **paniculatus** See *chaulai.*

— **spinosus** See thorny amaranth.

— **tricolour** See amaranth.

— **viridis** See *jangli chaulai.*

Amaryllidaceae The former designation for a family of plants now included in the liliaceae.

Amatangulu Carissa arduina Lamk. (Carissa bispinosa Desf., Arduina bispinosa L.). A spiny shrub of family Apocynaceae. Produces edible fruits in clusters. Hardy species of Carissa. Also called as martigula.

Ambia Kashmiri Apple variety indigenous to the Kashmir valley; medium to large, oblong, elliptical and tapering fruit. Good cropper, regular bearer.

Ambient quality standards Maximum permissible limits, maximum allowable concentrations, maximum acceptable levels of pollutants in specified media other than the receptor q.v. designed to ensure that under specified circumstances a primary or secondary standard q.v. is not exceeded. Also called environmental quality standards.

Ambient temperature Air temperature at a given time; not radiant temperature.

Ambivert Ambiverts are partly introverts and partly extroverts. They have a liking for people as well as liking for thought. It is very difficult to draw a line between introvert and extrovert. The same individual have certain qualities of the one and also of other.

Ameba, amoeba (*pl.,* amebas, amoebae) Protozoon that moves by extruding fingerlike processes (pseudopods).

Amebiasis Infection with pathogenic amebas; acute amebiasis is known as amebic dysentery.

Ameboid Resembling an ameba.

Amendment, soil (1) Chemical substances such as lime, gypsum; biological or physical substances used to make the soil more suitable for the growth of plants. (2) Fertilisers constitute a special group of soil amendments.

American beach Evodia fraxinifolia Hook. f. A tree of family Rutaceae found in the sub-tropical Himalayas and Khasia hills. Fruits are edible and also grown for fodder.

American Egyptian cotton A type of cotton developed by hybridization of Egyptian and American plants.

American lotus See East Indian lotus.

Ametoecious Of a parasite that remains with the same host.

Amide An organic compound containing nitrogen.

Amino acid Organic acid containing one or more amino groups ($-NH_2$) and at least one carboxyl groups ($-COOH$). Some amino acids such as cystine and methionine contain sulphur. Many amino acids linked together in a definite pattern form a molecule of protein.

Amitosis Direct cell division.

Ammonia Gas containing about 80 per cent nitrogen; used as fertiliser as anhydrous ammonia or as aquous ammonia. Soluble in water. Also a product of decomposition of urine.

Ammoniated superphosphate A compound fertiliser with 2-4 per cent nitrogen and 14-20 per cent phosphoric acid.

Ammonification The biochemical process whereby ammoniacal nitrogen is released from nitrogen-containing organic compounds.

Ammonifiers Fungi, or actinomycetous bacteria, that participates in the ammonification part of the nitrogen cycle and releases ammonia (NH_3) by decomposition of organic matter.

Ammonium chloride White crystaline compound containing 26 per cent nitrogen. Used as fertiliser but not recommended for those crops which are sensitive to chlorine. Used for industrial purposes.

Ammonium fixation The adsorption or absorption of ammonium ions by the mineral or organic fractions of the soil in such a way that the ions are relatively insoluble in water and relatively unexchangeable by the usual methods of cation exchange.

Ammonium nitrate White crystalline salt containing 33-35 per cent nitrogen, nitrate and ammonium nitrogen in almost equal proportion, highly hygroscopic and has acidic effect on soils. Used as fertiliser.

Ammonium phosphate A product obtained when phosphoric acid is treated with ammonia, consists mainly of monoammonium phosphate, diammonium phosphate or a mixture of the two; used as fertiliser.

Ammonium phosphate sulphate A fertiliser mainly consisting of ammonium phosphate and ammonium sulphate; is obtained when a mixture of phosphoric and sulphuric acids reacts with ammonia.

Ammonium sulphate A white crystalline fertiliser most commonly used in India. Contains 20-21 per cent nitrogen. As it is in the ammonium form, it is not easily leached with drainage as it is adsorbed on the exchange complex; can be used successfully in wetland rice. Increases acidity of the soils and hence recommended with bulk manures.

Ammonium sulphate nitrate Mixture of ammonium sulphate and nitrate, available in crystalline or granule form. Contains 26 per cent nitrogen; 75 per cent in ammonical and 25 per cent as nitrate form; quick acting and readily dissolved in water. It is recommended fertiliser with all the crops.

Ammophos Fertiliser with 16 per cent nitrogen and 20 per cent phosphoric acid.

Amnicolous Describing plants having a habitat on sandy river banks.

Amnion (1) Inner embryonic membrane in insects. (2) Viscous envelope in certain ovules.

Amniotic sac The embryo-sac, female gametophyte or megaspore.

Amoeba (pl. Amoebae) Microscopic, one celled protozoan, parasite.

Amoeboid Not having a cell wall and changing in form like an amoeba.

Amomum aromaticum see Bengal cardamom.

— **subulatum** see greater cardamom.

Amoora rohituka See Aphanamixis polystachya.

Amorphophallus campanulatus See elephant foot yam.

Amorphophyte Plants which bear irregular flowers.

Amorphous Not crystalline and of no definite shape.

Amorophous compound This term is commonly used in soil science for compounds of aluminium, silicon and iron, amorphous to X-rays, highly reactive and believed to be largely responsible for the fixation of compounds of, for example phosphorus molybdenum and boron, and for the high buffer capacity of soils in which they occur.

Amorphous frost Hoar frost which possesses no apparent simple crystalline structure; opposite of crystalline frost.

Amorphous matrix The intercellular substance usually lacking definite form and shape, in which certain unicellular algae or bacteria are imbedded.

Amorphous mineral (1) A mineral that has no definite crystalline structure. (2) A mineral that has a definite crystalline structure, but appears amorphous because of the small crystalline size.

Amorphous sky A sky characterised by an abundance of fractus clouds, usually accompanied by precipitation falling from a higher, overcast cloud layer.

Ampelocissus latifolia See panibelt.

Ampere The unit of measurement of electric current strength. A volt will force one ampere through a resistance of one ohm.

Ampere hour Used in rating storage batteries, representing a current of one ampere flowing for one hour.

Ampere metre Same as ammetre. A metre to register the density of flow of current.

Amphibious Capable of living at least at some time during the life span in both water and on land.

Amphicarpogenous Fruits which bury themselves below ground before maturity e.g. groundnut.

Amphicribral Having phloem surrounding the xylem.

Amphigastrium One of the reduced ventral leaves of a leafy liverwort.

Amphiphloic Having phloem both internal and external to the xylem.

Amphiphyte A plant growing on the boundry zone of wet land.

Amphiploid A plant possessing the sum of the somatic chromosome numbers of two species. Also called as amphidiploid.

Amphithecium The outer layer of cells that in some plants develops into the young sporophyte.

Amplifier A device for increasing the strength of weak currents or sounds.

Amplitude (1) An angle upon which the value of some mathematical function depends. (2) The distance covered by a particle in making a complete vibration.

Amplitude of the tide The semi range of the tides.

Amplixicaul Leaf which embraces the stem at its base.

Amputations Cutting of the limbs of a tree.

Amsacta albistriga See red hairy caterpiller.

— **moorei** See red hairy caterpiller.

Amygdalus communis See almond.

— **persica** See peach.

Amylase An enzyme which converts starch into dextrin and dextrin into maltose.

Amylopsin The pancreatic enzyme which breaks down starch into sugars; the chief carbohydrate of pancreatic juice.

Amylose The substance forming starch.

Amylum ($C_6H_{10}O_5$); the vegetable starch.

Anabatic wind An upslope wind; usually applied only when the wind is blowing up a hill or mountain as the result of a local surface heating, and apart from the effects of the larger scale circulation. Also called as mountain wind.

Anaberoga See arecanut foot rot.

Anabiosis The ability of a plant to remain in quiescent state when faced with unfavourable growing conditions e.g., Jowar (sorghum).

Anabolism Metabolic process by which simple substances are synthesised into complex materials of living tissue.

Anacardiaceae A family of flowering plants, the sucmacs, in the order sapindales; many species are allergenic to man.

Anacardium occidentale See cashew.

— **orientale** See marking nut tree.

Anacrogynae Liveworts in which the archegonia develop from behind the apical cell and never from the apical cell itself.

Anacystis A genus of blue green algae in the class cyanophycea.

Anadyomenaceae A family of green marine algae in the order siphonoeladales characterised by the expanded blades of the thallus.

Anaerobes Anerobic organism which are capable of getting oxygen for their growth by, reducing oxygen containing compounds in the soil and are not dependent upon supply of the free oxygen.

Anaerobic (1) Having no molecular oxygen in the environment (2) Growing in the absence of molecular oxygen, such as anaerobic bacteria (3) Occurring in the absence of molecular oxygen, as in a biochemical process.

Anaerobic composting A method of composting farm wastes in pits of convenient sizes (4.5 x 4.5m x 1 m). Raw material is filled in these pits to heights 40-50 cm above the edge and plastered with mud and cowdung mixture. The compost is ready within 4-5 months and contains about 0.8 to 1.0 per cent nitrogen.

Anaerophyte A plant that does not need free oxygen for respiration.

Anafront A front at which the warm air is ascending the frontal surface up to high altitudes.

Analogous Of similar function and often similar in appearance, but not of similar origin.

Analogy Resemblance in function though not in structure or development.

Analysis Resolution into simple elements; examining minutely the constitution of.

Analysis of variance The total variation displayed by a set of observations, as measured by the sum of squares of deviations from the mean, may in certain circumstances be separated into components associated with defined sources of variation used as criteria of classification for the observations. Such an analysis is called the analysis of variance.

Ananas comosus See pineapple.

— **sativus** See pineapple.

Anandrous Without stamens.

Anaphase A stage in mitosis in which the divided chromosomes move towards the opposite poles of the cell.

Anar butterfly Virachola isocrates Fabr. Insect pest of pomegranate. Short, hairy, stout, dark brown caterpiller with white patches; tunnels inside fruits and eats pulp; such fruits rot and falls; also attacks tamarind, orange etc.

Anarchic theory (intelligence) According to this theory, the mind is a host of highly particularised and independent faculties. The theory maintains that from a man's ability to do one kind of work we can infer absolutely nothing as to his ability to do another kind of work. Average of all energies is intelligence.

Anastatic water That part of the subterranean water in the capillary fringe between the zone of aeration and the zone of saturation in the soil.

Anatomy The division of morphology concerned with the internal structure of organisms including plants and animals, usually refers to parts seen without a microscope.

Anatvopous Having the main body inverted and straight, with the micropyle bent downward parallel with the funiculus.

Anaxial Lacking an axis, therefore being irregular in form.

Ancipital Having two edges, specially referring to flattened stems, as of certain grasses.

Andaman dwarf An early bearing dwarf coconut from Andaman; planted as ornamental plant.

Andaman giant A tall-growing late-maturing coconut palm from Andaman.

Andepts Are the inceptisols that go to a depth of 35 cm or more, or to a lithic or paralithic contact if one is shallower than 35 cm, one or both of the following: (a) A bulk density (at 1/3 - bar water retention) in the fine-earth fraction of the soil that is 0.85 g per cubic centimetre and the exchange complex is dominated by amorphous material; or (b) sixty per cent or more of the soil (by weight) is vitric volcanic ash, cinders, or other vitric pyroclastic material; and do not have an aquic moisture regime or do not have the characteristics associated with wetness that are defined for aquepts. They are approximately equivalent to Andosols.

Andosols Occur in volcanic regions, often on mountain slopes. They are formed in volcanic ash material, and have a dark mollic or umbric. A horizon often over a cambic B horizon. The ash material is very ligent, its bulk density being less than 0.85 g per cubic cm. Much material is amorphous, not well crystallised, giving the soil a smeary consistence (soils are thixotropic). The clay is characterised by dominance of allophane (amorphic hydrated aluminium silicates of various composition). They are approximately equivalent to Andepts.

Andrews cotton A sea island cotton variety commonly grown in Kerala and Mysore. Yield varies from 200-350 kg lint/ha.

Androecium (pl. Androecia) (1) Male reproductive organs of a plant. (2) The aggregate of stamens in a flower.

Androgen A male hormone.

Androgynous A plant having both stamens and pistils; hermaphrodite.

Andorpogon caesius See ginger grass.

— **citratus** See west Indian lemon grass.

— **contortus** See spear grass.

— **flexuosus** See east Indian lemon grass.

— **gryllus** See kufri.

— **martinii** See ginger grass.

— **muricatus** See vetiver.

— **nervosus** See pavna.

— **pertusus** See jarewa.

— **purpureo - sericeus** sea fodder sorghum.

— **vavennae** See plume grass.

— **sorghum** See sorghum.

Androspore A zoospore that produces a male filament as in Oedogonium.

Anemology Scientific investigation of winds.

Anemometre An instrument for measuring the speed of an air current, a wind gauge.

Anemometry The study of measuring and recording the direction and speed (or force) of the wind, including its vertical component.

Anemophilous Pollinated by wind-carried pollen.

Anemoscope A device used to indicate wind direction. Includes those which only records or indicate and record simultaneously.

Anethum foeniculum See fennel.

— **graveolens** See dill.

— **sowa** See Indian dill.

Ancuploid or heteroploid An organism or cell having a chromosome number other than an exact multiple of the monoploid or basic number. Hyperploid - higher. Hypoploid - lower.

Angaru Diospyros ferrea (willd.) Bakh. Maba buxifolia (pers.). Tree of family Ebanaceae found in western peninsula. Orissa and along the Coromondel Coast and produces edible fruits.

Angelica Angelica archangelica L. (Archangelica officinalis Hoffm.). Family Umbelliferae. An aromatic herb, native of Syria. The roots and fruits of which are used for flavouring purposes.

Angiosperm The common name for members of the plant division Magnoliophyta.

Angiospermic parasite Flowering parasitic plants; either partial parasite or produces normal green leaves and manufacture part of their food requirement themselves and obtain nutritional materials from hosts; total parasites are devoid of green leaves and obtain their complete requirement of food from hosts.

Angiospermous Having the seed in a closed case; the overy.

Angle iron A strip of structural iron, the section of which is in the form of a right angle.

Angle of repose or angle of friction The angle of a plane surface, inclined relatively to the horizon, upon which a body will, under specific conditions, just begin to slide. It varies with the nature of the particular materials placed in contact.

Angola grass See para grass.

Angoumois grain moth Sitotroga cerealella oliv. Serious pest of stored unmilled grain like wheat, barley, rice, maize and millets. Infestation passed on from field through grain and straw. larva bores into grains, eats kernels inside. Damage is restricted to surface layer.

Anguina tritici See wheat earcockle.

Angular Having or pertaining to an angle; refers to leaf originating at forking of stem as in many ferns.

Angular cobbly A descriptive term applied to coarse fragments. It is used for irregular and angular rock or mineral particles 7.5 to 25 cm (3 to 10 inches) in diametre. See also coarse fragments.

Angular velocity The ratio which the arc described in one second, by a body or point rotating about a centre, bears to the radius.

Anhydrous Dry or without water.

Animaloule Minute animals such as protozoa which cannot be seen with naked eye.

Animal day One day's tenure upon range or pasture by one animal.

Animal unit A unit represented by one mature cow (454 kg) or horse, or by as many animals as consume an equivalent amount of feed, such as 2 heifers, 2 colts, 7 sheep, 100 laying hens, or 5 broodsows, or the equivalent based upon average daily forage consumption of 12 kg dry matter per day.

Animal unit conversion factor A numerical figure expressing the forage requirement of a particular kind or class of animal relative to the requirement for an animal unit. A conversion factor is satisfactory with respect to the amount or forage required to maintain an animal, but may not be applicable in determining stocking rates for range and pasture use for particular kinds of classes of animals because of different grazing preferences.

Animal unit month (AUM) The amount of feed or forage required by an animal unit for one month.

Anion An ion that carries a negative charge of electricity.

Anion exchange capacity The total amount of exchangeable anions that a soil can adsorb. It is expressed as milliequivalents per 100 g of soil or other adsorbing material such as clay.

Anise Pimpinella anisum L. Family umbelliferae. A herb of the parsley family which produces aromatic seeds used in flavouring and in medicines. Native of mediterranean region but now cultivated in Uttar Pradesh, Punjab, Assam and Orissa. Oil also used in perfumery and soap making.

Aniseed Seed of anise.

Anisoearpous Referring to a flower whose number of carpels is different from the number of stamens, petals, and sepals.

Anisogamous Characterised by gametes which are morphologically similar, but differ in size.

Anisogamy Union of gametes which are morphologically similar, but differ in size.

Anisomeles Indica See gopali.

— **ovata** See gopali.

Anisomorphic Differing in shape and size or structure.

Anisotropic mass A mass whose properties have different values when they are measured in different directions at any given point.

Anisotropy The property of a plant that assumes a certain position in response to an external stimulus. The characteristic of a substance for which a physical property, such as index of refraction, varies in value with the direction in or along which the measurement is made. Also known as aeolotropy; eolotropy. The property of an egg that has a definite axis or axes.

Anjan See bunch grass; Anjan grass.

Anjan grass Cinchrus setigerus Vahl. Family Gramineae. Grass cultivated for fodder and for soil conservation.

Anjiri Ficus pailmata Forsk. (Ficus virgata Roxb.) Tree of the family Moraceae found in north-west India and Rajasthan. Yielding edible fruits. Young shoots used as vegetable; latex used in the preparation of curd.

Anjun Hardwickia binatta Roxb. Family Caesalpiniaceae. Large deciduous tree, yielding a dark red and extremely hard wood for bridge and house construction purposes. Cart wheels, farm implements and mine props are also made. Leaves used as fodder and manure. Oleoresin tapped from heartwood used for varnishes and bark for ropes etc. Planted for afforestation of arid and semi-arid regions. Grows on sandy loam and also on black cotton soils.

Annatto Dye or colour prepared from the seeds of Bixa orellana and used to impart colour to edible materials.

Annatto tree Bixa orellana L. Family Bixaceae. Huge shrub or small tree, native to tropical America, long clusters of pods containing seeds at the end of branches. Yields yellow dye called annatto dye used for colouring food materials and wool, paint and soap etc. Cultivated in south India.

Annona cherimola See cherimoyer.

— **diversifolia** See llama.

— **muricata** See soursop.

— **reticulata** See bullocks heart.

— **squamosa** See custard apple.

Annual A plant which completes its life cycle from seed in one year. Also a general term implying to structures or features that are marked after completion annually.

Annual bearing Producing fruit each year.

Annual closure The closure of an irrigation system for carrying out annual repairs.

Annual cut Forest products cut from a forest in one year.

Annual flood The highest peak discharge in a water year.

Annual layer The layer of wood produced in one year by a tree or a perennial. Also annual ring.

Annual precipitation The water deposited on earth in one year over a unit area. Includes rain, snow, dew, hail fog, mist and other forms of precipitation. The units of expression are centimetres of water depth.

Annual rainfall See annual precipitation; includes only rain water.

Annual ring A ring of wood put on each year by a growing tree; that is, the line indicating the growth for the period of one year. From the annual rings the age of the tree may be determined.

Annual storage The capacity of a reservoir that can handle a watersheds annual runoff but cannot carry over any portion of the water for longer then the year.

Annual variation The general pattern of a particular element throughout the year, obtained by plotting the normal values of the element for each month and connecting the points by a smooth curve.

Annual, winter Plant species that initiates growth in the fall, lives over winter, and dies after producing seed the following season.

Annuation The annual variation in the presence, absence, or abundance of a member of a plant community.

Annular auger A ring shaped boring tool which cuts an annular channel, leaving the core intact.

Annular budding Budding by replacement of a ring of bark on a stock with a ring bearing a bud from a selected species or variety.

Annular drainage pattern (1) Ringlike drainage pattern, subsequent in origin, associated with maturely dissected dome of basin structures. (2) Drainage pattern in which ringlike tributaries flow into the radial streams.

Annulus A ring; ringlike partial veil around stipe in the mushrooms.

Anode Opposite of cathode. The point where, or path by which a voltaic current enters an electrolyte; the positive voltaic pole or the plate constituting it.

Anogeissus latifolia See gum ghatti.

Anomala dimidiata See white grubs.

— **lineatopennis** See apple fruit and leaf eating beetles.

— **polita** See white grubs

— **rugosa** See white grubs; apple leaf and fruit eating beetles.

Anomaly A deviation from a norm for which an explanation is not apparent on the basis of available data.

Anomis sabulifera See jute semi-looper.

Anoxia A deficiency of oxygen, e.g. in the tissues or in the blood, or in a body of water. In the Baltic, anoxic bottom-water conditions tend to persist in the deeper parts.

Antagonism (1) The mutual killing, injury, or inhibition of growth of dissimilar organisms occupying the same ecological niche. (2) The power of one ion to inhibit the absorption power of another.

Antagonistic medicine Medicine which counteracts the action of another medicine.

Antecedent precipitation Rainfall that occurs a given number of days prior to the particular rain storm under consideration.

Antecedent precipitation index A weighted summation of daily precipitation amounts, used as an index of soil moisture. The weight given to each day's precipitation is usually assumed to be an exponential or reciprocal function of time with the most recent precipitation receiving the greatest weight. Such indices are used in estimating runoff and as measure of drought severity.

Antecedent soil water (Antecedent moisture) Degree of wetness of the soil prior to irrigation or at the beginning of a runoff period; expressed as an index or as total centimetres of soil water.

Antecedent stream Drainage established prior to deformation of beds by folding and faulting.

Antenna (pl. antennae) A organ of sensation on the head of an insect; jointed and movable.

Anterior Towards the front or head end.

Anther The sac-like portion of the stamen that contains the pollen.

Antherea assama Muga silkworm. Silkworm which produces muga silk.

Antheridiophore The structure on which an antheridium is borns.

Antheridium A sperm-producing structure (male sex organ) in the thallophytes, bryophytes, and lower tracheophytes.

Anthesis The period or act of flowering, the duration of life of a flower from the opening of the bud to the setting of the fruit.

Anthistiria imberbis See themeda grass.

Anthocyanin Any of a group of pigments found in blue, red, and violet flowers, leaves and stems; they are glycosides yielding anthocyndin and a sugar on hydrolysis; also called anthocyn.

Anthophilous Attracted by flower; feeding on flowers.

Anthophorous Bearing flowers.

Anthophyte A flowering plant; phaenogram; phanerogram; spermatophyte.

Anthorea mylitta See tussar silkworm.

Anthotaxy Arrangement of flowers.

Anthoxanthin Pigment in the sap of yellow colour and crystalline in nature, responsible for pale yellow to ivory-white colouration in plants. Also anthocyanin.

Anthrecnose A disease with limited necrotic lesions, caused by a fungus producing non-sexual spores in acervuli.

Anthropic soils Soils which are under cultivation for a long period.

Anti-bacterial serum A serum containing antibodies capable of destroying or rendering inert the microorganisms causing a disease.

Antibiotic A mold or fungus that is produced by a species of microorganism or synthesized chemically and, in dilute solution, has the capacity to kill, retard or completely repress the growth of other organisms, usually without harming higher orders of life.

Antibody Agent in the body that destroys or inactivates certain foreign substances that gain access to body, particularly microbes and their products for example Aglutinin, antitoxin, precipitin etc.

Anticinal Formed at right angles or perpendicular to the outside surface, as a cell wall.

Anticoagulant Agent that prevents coagulation.

Anticycloysis Any weakening of anticyclonic circulation in the atmosphere.

Anti cyclone Area of relatively high pressure surrounded by closed isobars, the pressure gradient being directed from the centre so that the wind blows spirally outward in a clockwise direction in the northern hemisphere, counterclockwise in the southern hemisphere.

Anticylonic shear Horizontal wind shear of such a nature that it tends to produce anticyclonic rotation of the individual air particles along the time of flow.

Anticyclonic wind The winds associated with a high pressure area and constituting part of an anticyclone.

Antidesma acuminatum See kalo

bilauni.
— **bunius** See chinese laurel.
— **ghesembilla** See black currant.
— **paniculata** See black currant.

Antidotes Medicines which counteract the effect of poison.

Antienzyme A substance which stops or retards the activity of an enzyme.

Anti-gungal chemicals Chemicals used for their fungus killing properties such as for washing fruits etc.

Antigastra catalaunalis See sesame leaf and pod caterpiller.

Antigen Substance that, when introduced, causes the body to produce antibodies.

Antimetabolite Prevents the formation or utilization of organic compounds in particular metabolic processes in living tissue.

Antipodal At the opposite end, as the nuclei or cells in the embryo sac which are at the end opposite the egg.

Antiseptic Substance that prevents the growth of bacteria.

Antiserum A serum containing antibodies which neutralise a particular toxin.

Antitoxin Immune serum that neutralises the action of a toxin.

Antitrades A deep layer of westerly winds in the roposphere above the surface trade winds of the tropics.

Antitiptic wind A wind for which the pressure force exactly balances the viscous force in which the vertical transfers of momentum predominate.

Antivenin Antitoxic serum for snake venom.

Antu Alpha naphthyl thio urea. A powerful rat killer. Also kills hogs, dogs and other meat eating animals. Apparently not poisonous to domestic fowls, plant-eating animals and to some extent to man.

Anuraphis helichrysi See peach leaf curling aphid.

Anvil A steel or iron block upon which forging is done.

Anvil block A massive block of cast iron which is placed beneath the anvils of steam and other heavy hammers, for the absorption of the vibration due to the blow. It is often embedded in masonry or concrete.

Apatite (CaF) Ca_4 $(PO_4)_3$. Crystalline calcium phosphate which includes some chlorine and flourine. Resistant to weathering and present in many common rocks though in small quantities. Primary source of soil phosphorus.

Apetalous Without petals, as a flower. An apetalous flower may have sepals.

Aphanamixis polystachya A tree of family Meliaceae. Seeds are used to extract oil. Also called sohaga.

Aphid A plant louse, a very small insect (family Aphidae) that lives on plants and sucks plant juices, some secret honeydew which ants are fond of; some of them are common agents for carrying virus diseases to plants. Also aphis.

Aphis See aphid.

Aphis craccivora See cowpea aphid.
— **gosspyii** See cotton aphid.

Aphosphorosis Deficiency of phosphorus.

Aphyllous Without foliage leaves.

Apiarist bee-keeper.

Apiary Colonies of bees in hives and other bee-keeping equipment in a location and used for the purpose of producing honey, worker bees or queen bees.

Apical dominance Inhibiting effect of a terminal bud upon the development of lateral buds.

Apical growth Growth occurring at the tip of an organ, as the meristem region of a root or stem tip.

Apical meristem (1) Undifferentiated tissue at the tip of root or stem composed of actively dividing cells. (2) A region of embryonic tissue occurring at the tips of roots and stems.

Apiculate Forming abruptly to a small tip.

Apiculture Science and art of raising honey bees, bee-keeping.

Apidae Family and sub-families of bees.

A.P.I. gravity It is the scale in degrees for gravity devised by American Petroleum Institute.

$$\text{A.P.I.} = \frac{141.5}{\text{Specific gravity at } 60^\circ \text{ F.}} - 131.5$$

Apiology Scientific study of bees; especially of honey bees.

Apion corchori See jute apion.

Apis dorsata See rock bee.

— **florea** Little bee. The frequently migrating plains species of honeybee in India.

— **indica** See Indian honey bee.

Apium graveolens var. dulce See garden celery.

— **graveolens var. rapaceum** See celeriac.

— **rapaceum** See celeriac.

Aplanospore A non-mobile spore in which the wall of the spore is not derived from the wall of the vegetative cell in which it is formed.

Apluda aristata See *bhanjura* grass.

— **mutica** See *bhanjura* grass.

— **varia** See *bhanjura* grass.

Apob An observation of pressure, temperature, and relative humidity taken aloft by means of an aerometeorograph; a type of aircraft sounding.

Apocarpous Having separate or partially united carpels.

Apocarpy The condition of having separate carpels.

Apogamic seedlings Seedlings produced vegetatively by the seed of certain species of plants. Three or four seedlings emerge from each seed, one is of sexual origin and other produced from cells of nucellus. All seedlings are identical to the parent in growth and production,

Apogamy The development of a sporophyte from gametophytic tissues without sexual fusion.

Apogean range The range of tide at the time of apogean tides.

Apogean tides The decreased range tides occurring at the time the moon is in apogee.

Apogee The position of the moon farthest from the earth in the elliptical orbit around the earth.

Apogeny Loss of the function of reproduction.

Apogestropism Negative geotropism; growth up or away from the soil.

Apomixis The development of an individual from an unfertilised egg without sexual fusion, whether the egg be normally haploid or abnormally diploid through failure of reduction division. Broadly may include also a sexual propagation by various methods.

Apophysis An enlargement at the base of the capsule of some mosses.

Apospory The formation of a gametophyte directly from sporophyte tissues, rather then from a spore.

Apothecium Apothecia pl. The cup or saucer-like ascus-bearing fruiting body; in the Discomycete section of the Ascomycetes.

Apparent density See bulk density, soil.

Apparent specific gravity See bulk specific gravity.

Appendage An organ or part attached to a trunk as a limb or branch.

Apple (1) Malus sylvestris L. Mill. (Malus pumila Mill.; Pyrus malus L.). Tree or fruit of species of the genus Malus or Pyrus of family Rosaceae. Native of South-west Asia, under cultivation in temperate region of the world. Progated by budding or grafting. (2) A name applied to other plants, the fruits of which resembles the common apple, such as custard apple.

Apple brown rot Disease of apple resulting from Selerotinia fructigena Aderh. and Ruhl. Symptoms include appearance of brown spots and tiny yellow or buff pustules in concentric zones on them later. Falling and mummifications of fruits occurs.

Apple fruit and leaf eating beetles Melolontha nepalensis Bl.; Adcretus duvauceli Bl.; A versutus Har.; Anomala polita Bl.; A lineatopennis Bl. Beetal pests of apple, feeding on leaves and fruit at night.

Apple fruit borer Dyscerus fletcheri Mshll.; D. Malignus Mshll. Weevil pests of apple whose grubs make tunnel in fruits.

Apple pink disease Disease caused by Pellicularia salmonicolour (B & Br.) Dastur. Characterised by formation of longitudinal strips. Finally the branch dies. May also occur in citrus or mango.

Apple powdery mildew Disease of apple caused by podosphaera leucotricha (Ell. and Everh.) Salm. Characterised by formation of white powdery growth on leaves, twigs, blossoms and fruits.

Apple root borer Dorysthenes hugelii Redt. Insect pest of apple; grub attacks root, bores and feeds inside and tree withers. Also attacks roots of pear, peach, cherry and apricot.

Apple scale A disease of apple caused by Venturia inaequalis (cooke) wint. Characterised by appearance of brownish or grey spots on leaves, brown or black and corky spots on fruits, shedding of leaves.

Apple shoot borer Alcides mali marshal. Weevil pest of apple; lays eggs in tender shoots; grubs bores into shoots, causing gall-like swelling and makes tunnel down the stem.

Apple sooty-blotch and fly-speck Disease of apple caused by Leptothyrium pomi (Mont. Fr.) Sacc. Characterised by appearance of circular rooty blotches on ripe fruits and of fly-specks in the form of number of tiny cushion like structures in groups.

Apple stem black Serious disease of apple caused by Coniothecium chomatosporum Corda. Symptoms include drying up of branches, appearance of longitudinal cracks in the bark with black powdery substance inside, and the turning black and drying of branches.

Apple stem borer (1) Linda nigroscutata Fairm. Insect pest of apple; beetle whose grub bores into twigs in which it makes holes to thrust out pallets of frass. (2) Zeuzera sp. Insect pest of apple; caterpillar bores into twigs and young branches, especially on top of tree, making them wither and dry off; tree often dies; also attacks pear, plum, walnut, cherry.

Apple stem brown Disease of apple caused by Botryosphaeria ribis G. and D. Symptoms include lossening of bark, which turns papery and rolls outward, upper limbs being more prone to attack.

Application rate Rate of the water that is applied to a given area.

Applied Economics Branch of economic science devoted to the study of practical problems, utilising the principles and tools of economic analysis provided by theoretical economics.

Applied geophysics Geologic exploration of prospecting using the instruments and applying the methods of physics and engineering, exploration by observation of seismic or electrical phenomena or of the earth gravitational or magnetic fields or thermal distribution.

Applied insect control The use of chemical sprays and dust and other methods in insect control.

Appreciation An increase in the value of an asset. For example, an increase in the value of property in relation to other assets during a period of growing demand.

Apprentice A learner in a trade. Usually one, who, by contract or indenture, is taught the rudiments of the trade during the term of the agreement.

Apprenticeship The system under which young persons training in a particular occupation have to 'serve their time', usually for five years, before being recognised as skilled.

Appressorium A swelling on a fungus germ tube for attachment to host in early stage of infection; found especially in anthracnose fungi and rusts.

Approach grafting See inarching.

Approximate Approaching, or near to, exactness, but not completely accurate.

Apricot Prunus armeniaca L. (Armeniaca vulgaris Lamk.) Family Rosaceae. Temperate fruit indigenous to India. Oil used for culinary purposes, fruit eaten fresh or after drying, propagated by shield budding. Fruit velvety when young, peach-like in colour and shape; flesh yellowish orange. Cultivated in Simla hill, Kashmir, Kulu and Kumaon. Also *Khubani*.

Apricot pink disease Physiological disease of apricot, for cause and symptoms. See apple pink disease.

Apricot sun scard Physiological disease of apricot, symptoms same as in peach sun scald.

Apron Floor or lining of material to protect a surface from erosion; for example lining below chutes, spillway or at the toes of dams.

Apsylla cistellata See mango shoot gall psyllid.

Aqua Liquid, solution.

Aqua-culture See hydroponics.

Aquatic Pertaining to water, living in or frequenting water. A plant growing in water.

Aqueduct A water conduit or channels, steel pipes, syphons, or tunnels, etc., through which a community is supplied water.

Aquents Entisols that either have sulfidic materials within 50 cm of the mineral surface or are permanently saturated with water and have a dominant hul in all horizons below a depth of 25 cm that is neutral or bluer than 10 Y and that changes on exposure to air; or are saturated with water at some time of the year or are artificially drained and have at some depth within 50 cm of the surface specified dominant colours (moist) in the matrix.

Aqueous tissue Plant tissue composed of thin-walled cells, having neither air spaces nor chloroplast.

Aquepts Inceptisols that either have an SAR that is 13 (or sodium saturation that is 15 per cent or more) in half or more of the upper 50 cm that decreases with depth below 50 cm and the soils are saturated with water at some time of the year within a depth of 1 m; or have artificial drainage or an aquic moisture regime and have one of the following:

(a) A histic epipedon composed of mineral soil materials;

(b) A mollic or umbric epipedon that is underlain immediately or at a depth 50 cm below the mineral soil surface by a horizon that has dominant colour; moist, on ped faces, or in the matrix if peds are absent as follows:

  i If there is mottling, chroma of 2 or less;

  ii If there is no mottling, chroma of 1 or less;

(c) a histic epipedon composed of organic soil materials or an ochric epipedon that is underlain at a depth 50 cm below the mineral soil surface by a cambic horizon or a subhorizon above a fragipan that has dominant colour; moist, on ped faces or in the matrix if peds are absent, that meets the requirements of b(i) or b(ii);

(d) A sulfidic horizon that has its upper boundary within 50 cm of the soil surface.

Aquiclude A formation which, although porous and capable of absorbing water slowly, will not transmit it fast enough to furnish an appreciable supply of water.

Aquifer A geologic formation or structure that transmits water in sufficient quantity to supply the needs of a water development. The term water-bearing is sometimes used synonymously with aquifer when a stratum furnishes water for a specified use. Aquifers are usually saturated sands, gravel, fractures, cavernous and vesicular rock.

Aquicherbose Herbaceous plant communities in wet areas, such as swamps and ponds.

Aquiprata Communities of plants which are found in areas such as wet meadows where groundwater is a factor.

Aquods Spodosols that have an aquic

moisture regime or are artificially drained (if a placic horizon, fragipan, or duripan is present, the soil need not be saturated below the placic horizon, fragipan or duripan) and that have characteristics associated with wetness, namely, one or more of the following: (1) A histic epipedon; (2) Mottling in an albic horizon or in the upper part of the spodic horizon; (3) A duripan in the albic horizon or a placic horizon that rests on a spodic or on a fragipan or an albic horizon that rests on a fragipan; (4) If free iron and manganese are not present, of if colour value, moist, is less than 4 in the upper part of the spodic horizon, either (a) Do not have coatings of iron oxide on the individual grains of silt and sand in the materials in or immediately below the spodic horizon when ever the value, moist, is 4 or more; or (b) Have fine or medium mottles of iron or manganese immediately below the spodic horizon.

Aquolls Mollisols that do not have an albic horizon, that have an aquic moisture regime or are artificially drained, and that have one or more of the following characteristics associated with wetness: (1) A histic epipedon; (2) An SAR 13 (or sodium saturation of 15 per cent) in the upper part of the mollic epipedon and decreasing SAR (or sodium saturation) at increasing depth below 50 cm; (3) One of the specified combinations of colour, moist.

Aquults Ultisols that are saturated with water at some time of the year or are artificially drained; have immediately below the plow layer or below any A 1 horizon that has a colour value, moist, less than 3.5 when rubbed, one or more of the following characteristics: (a) Chroma, moist, of 2 or less; (b) mottles due to segregation of iron; or (c) Iron-manganese concretions 2 mm in diametre; and (d) Have one or more of the specified characteristics of chorma and hue.

Arabian coffee Coffea arabica L. Family Rubiaceae. A shrub or small tree; native of Ethiopia. Commonly grown in Nilgiris, Mysore and Kerala.

Arabian cotton See Asiatic cotton.

Arable land that is fit for cultivation, tillable.

Arable soil Soil suitable for ploughing and cultivation.

Arachis hypogea See groundnut.

Arachnids Organisms belonging to the arthropods; differ from insects in having 4 pairs of limbs. Simple eyes (if any) and head and throat fused together; ticks and mites.

Arbitrage Dealing in commodities, currencies or securities between two or more markets to take profitable advantage of any differences in the prices quoted for such commodities, currencies or securities.

Arboreal Relating to trees; hence an arborist is a person who studies trees, and arboriculture is concerned with the forestry and the culture of trees.

Arborescent Treelike in size and form.

Arboriculture The science and art of growing trees, especially as ornamental or shade trees.

Arbour A bower; small structure of lattice work to support vines and provide a shady retreat.

Arbour system of training Method of training grapevines in which vines are spaced nearly 1.8 m apart on both sides of arbour with a flat roof with a metre wide and 2.5 m high. Rooted cuttings planted in February and allowed to branch into three main branches at about 90 cm from ground. Main branches fastened fanlike against 2 horizontal wires, stretched across sides of arbour and spread over roof as they grow. Branches by these arms pruned annually into short spurs containing 2 or 3 buds. Weak branches removed and strong ones retained. This system is continued from year to year. Also called pergola system.

Arc brazing An electric brazing process

wherein the heat is obtained from an electric are formed between the base metal and an electrode, or between two electrodes.

Arc cutting A cutting process wherein the severing of metals is effected by melting with the heat of an arc between the electrode and the base metal.

Archaeozoic The earliest era in geological history during which life is believed to have originated.

Archangelica officinalis See angelica.

Archegoniophore Structure on which archegonia are borne.

Archegonium The multicellular egg producing structure of the lower members of the Embryophyta.

Architectural tree model See model.

Architecture The visible, morphological expression of the genetic blueprints of organic growth and development.

Arc of contact In tooth gearing the space included between those two points where the contact of a single pair of wheel teeth begins and ends.

Arctic alpine Of or pertaining to areas above the timberline in mountainous regions.

Arctic front The semipermanent, semicontinuous front between the deep, cold arctic air and the shallower, basically less cold polar air of northern latitudes.

Arctic haze A condition of reduced horizontal visibility (but unimpeded vertical visibility) encountered by aircraft in flight (up to more than 30,000 feet or 9140 metres) over arctic regions.

Arctic high A weak high that appears in mean charts of sea level pressure over the Arctic Basin during late spring, summer, and early autumn. Also known as Arctic anticyclone; polar anticyclone; polar high.

Arctic mist A mist of ice crystals; a very light ice fog.

Arctic tree line The northern limit of tree growth; the simuous boundary between tundra and boreal forest.

Arcus A dense and horizontal rall-shaped accessory cloud, with more or less tailered edges, situated on the lower front part of the main cloud.

Arc welding The piece to be welded is usually made the positive terminal. Direct current is used. The welding rod is the negative. The work is touched with the rod and withdrawn slightly causing an arc.

Ardunia bispinosa See natal plum.

— **grandiflora** See natal plum.

Area An open space or court; an uncovered space. The measured content of a surface.

Area irrigated The area to which water has been applied.

Area leased Area for which cultivators have executed an agreement for taking canal water for certain periods.

Area matured The area irrigated upon which crops have matured.

Area non-perennial The area served by a non-perennial canal.

Area of dominance The region occupied by an organism so as to restrict or exclude organisms of other types.

Area sampling Area sampling is an application of cluster sampling with other design features interwoven, the idea being that by drawing probability samples of specific areas and sampling appropriately within them, it is possible to obtain a probability sample of the population.

Area of environmental concern Areas which, because of their environmental significance, require special management considerations.

Areca catechu See arecanut.

Arecaceae The palms, the single family of the order Arecales.

Arecanut Produce of Areca catechu L. Family Palmaceae. A slender stemmed palm. Nut largely used as a masticatory, chewed with betal leaves (piper betal), strengthen the gums and helps digestion. Galls caused by inscets (mochras) used as tanning material. Also called betal nut, supari.

Arecanut anthracnose Fungal disease of arecanut caused by Colletotrichum catechu.

Arecanut foot rot Disease of arecanut caused by Ganoderma Lucidum (Leyss ex. Fr.). Characterised by lower leaves turning yellow and buckling up, appearance of water-soaked soft patches on sides of stem near base, brownish gummy exudation from infected portion of stem.

Arecanut fruit rot Serious fungal disease of arecanut caused by Phytophthora arecae (coleman) Pethi: Symptoms include appearance of small water-soaked spots on young fruits and tender buds, shrivelling, rotting and complete diintegration of fruits, collapsing of growing parts, dropping off of diseased fruits and dying of trees. Also called Arecanut Peoleroga, mahali.

Arecanut leaf blight Disease of arecanut caused by Gonisthyrium arecase, characterised by appearance of spots on leaves which cause their shedding.

Arecanut mites Raoiella indica Hirst. Oligonychus sp. major pest of arecanut which suck sap resulting in yellowish patches on leaves. Later these patches gets enlarged and give bronze appearance to them. Checks growth of seedlings.

Arecanut stem bleeding Disease of arecanut caused by Ceratostomella paradoxa (de saynes) Dade. Characterised by appearance of discoloured depressions on tree base, followed by deep fissure gradually widening and deepening, with exudation of dark fluid in advanced cases and dwindling of crown to cause cessation of fruit bearing.

Arecanut stem-breaking Physiological disease due to sun injury; seen as crack on upper half of trees, which widen and elongate and central portion of stem turning hollow and breaking of stem. Control by protecting stem part by covering with leaf sheaths or coating it with zinc white, and tying split pieces of areca stem to extend below and above damaged area.

Arecolina An alkaloid used as anthelmintic and eathartic, derived from arecanut.

Areg A sand desert.

Arenga pinnata See gomuti palm.

— **Saccharifera** See gomuti palm.

— **Wightii** See Wight's saga palm.

Arenosols Very light coloured (light grey, very pale or white) coarse-texture sandy soils with a high proportion of almost pure quartz; have a lower water retention, are more permeable, have a low natural fertility and a lower CEC than other soils. Deficiencies of various minor elements are common. They may have an ochric A horizon. Arenosols are poor for agriculture. Natural vegetation is often forest. As soon as trees are cut, soils may erode and land becomes rather poor grazing land.

Areola (pl. areolae) Very small area as that between veins of a leaf; circular spot; cell-nucleus of plant

Areole Space occupied by a group of hairs or spines, as in cactus.

Argemone Mexicana See Mexican poppy.

Argemonin See Mexican poppy.

Argids Aridisols that have an argillic or a natric horizon.

Argillation Translocation of fine clay from the A into the B horizon, a complicated process influenced by various factors such as mineralogical composition, type of clay and organic matter, chemical, physical, biological, and environmental conditions. In the tropics and subtropics, argillic B horizons often occur in older soils that are alternately moist and dry.

Argillic B horizon An important subsurface horizon, because it is diagnostic for various major soil units. It is a horizon in which clay particles have accumulated from the upper soil layer (A or B), and that have filled pores and other openings. The main properties are: the increase in clay content occurring within a vertical distance of 30 cm; in sandy soils it contains at least 3 per cent more clay than the eluvial (upper) horizon; in clay soils it contains at least 8 per cent more clay than the eluvial horizon; in sandy soils it is 15 cm and in clay soils 7.5 cm thick; and there are clay skins on ped surfaces, and in pores. In most tropical clay soils (Kaolinitic clays) clay skins occur at least in the lower part of the B horizon.

Argillic horizon Is an illuvial horizon in which layer-lattice silicate clays have accumulated by illuviation to a significant extent. The process of illuviation of course, does not preclude concurrent formation of clay in the illuvial horizon. If the horizon is illuvial, it must have formed below the surface of the mineral soil though later it may have been exposed at the surface by erosion.

Argyroploce illepida See litchi fruitstone borer.

Aahar See pigeonpea.

Arid An area or climate that lacks sufficient moisture for agriculture without irrigation. According to Thornthwaite areas having moisture index below —40 Thornthwaite; annual rainfall is less than 250 mm.

Aridity A bioclimatic property of the earth which may be expressed by the relative amount of water gained from rainfall and lost by evaporation and transpiration i.e. P/ETP. It rises as precipitation decreases and as evapotranspiration increases. The water is not available for mesophytic plants for long periods. Most of the time, water is held at 15- bar tension, so that it is not available for plant growth.

Arid region The regions in which potential evapotranspiration far exceeds the total water available from precipitation.

Aril An additional integument formed on some seeds after fertilisation.

Arista (pl. Aristae) Awn, beard, of grain and grasses.

Arm (1) The main branches or extensions of the trunk of a tree or shrub; (2) A branch of a grapevine which is two years old or older.

Armeniaca vulgaris See apricot.

Armoracia lapathifolia See horse radish.

Arm palisade The fingerlike projections of the inner walls of the epidermis of certain leaves.

Army worm See paddy cutworm.

Arnebia hispidissima Lehm. Dc. A small herb which yield an dye. Family Barginaceae.

Aromatic Substances having pleasant odour; often used as a flavour agent.

Arrhizal Without true roots, parasites as example.

Arrowroot See Queensland arrowroot; East Indian arrowroot; Indian arrowroot; West Indian arrowroot.

Arsenic A chemical element commonly used in pesticides. Occur in soils as trace element.

Artemisia absinthium See wormwood.

— **dracunculus** L. A perennial herb or family compositae producing a flavouring extract.

Arterial drainage Drainage pertaining to main channels, such as rivers, streams and large watercourses.

Artesian aquifer An aquifer that is bounded above and below by impermeable beds and that contains artesian water. Also known as confined aquifer.

Artesian basin A geologic structural feature or combination of such features in which water is confined under artesian pressure.

Artesian leakage The slow percolation of water from artesian formations into the confining materials of a less permeable, but not strictly impermeable character.

Artesian pressure Pressure within a groundwater aquifer developed as a result of hydrostatic head.

Artesian spring A spring whose water issues under artesian pressure, generally through some fissure or other opening in the confining bed that overlies the aquifer.

Artesian water Groundwater that is under sufficient pressure to rise above the level at which it is encountered by a well, but which does not necessarily rise to or above the surface of ground.

Artesian well The well which supplies water continuously without pumping. The well penetrates a confined aquifer and due to the pressure of the water, the water rises above ground level.

Arthoniaceae A family of lichens in the order Hysteriales.

Arthrodonteae A family of mosses in the sub-class Eubrya characterised by them, membranous peristame teeth composed of cell walls.

Arthropod A jointed legged invertebrate, such as an insect or a crustacean.

Artichoke Cynar scolymus. A herbaceious perennial plant belonging to the order Asterales; the flower head is edible. Also see globe artichoke and Jerusalem artichoke.

Articles of association A document which must be submitted to the Registrar of Companies when seeking registration of a company.

Articular Relating to a joint.

Articulated growth Rhythmic growth which results in shoot units separated by morphological discontinuities, e.g., bud-scale scars.

Articulated shoot Shoot produced by articulated growth, usually with distinct budscale scars.

Articulation A joint between segments of a stem.

Artificial drainage Drainage instituted by man.

Artificial fertiliser Same as fertiliser.

Artificial manure See compost. In European usage artificial manure may denote commercial fertilisers.

Artificial reef A method of solid waste disposal, which comprises the tipping into deep water of junk and debris, rusting car bodies, used car tyres, and other material. The reef provides cover and breeding sites for various marine fish.

Artificial regeneration Renewal of forest crop by sowing, planting or artificial mean. Also the crop so obtained.

Artocarpus altilis See bread fruit.

— **communis** See bread fruit.

— **heterophyllus** See jack fruit.

— **hirsuta** Family Moraceae. An uncultivated species of the common jack fruit growing wild and useful as a timber tree. Family Moraceae.

— **incisa** See breadfruit.

— **integra** See jack fruit.

— **integrifolia** See jack fruit.

Arum See colocasia.

Arum companulatum See elephant foot yarn.

Arundinaria falcata See Himalayan bamboo.

— **spathiflora** Trin. Family Gramineae. A small bamboo growing in the undergrowth of western Himalayan forests; used for baskets.

— **wightiana** See Indocalamus wightianus

Arusa See malabar nut

Arvi See colocasia

Arviculture The cultivation of field crops.

Arytaina punctipennis See indigo psyllid.

Asafetida See asafoetida.

Asafoetida Dried latex obtained mainly from living rootstocks or tap roots of Ferula assafoetida L. A perennial herb found in Punjab and Kashmir. Latex acrid bitter taste; used to flavour food and in medicines. Also Narthex asafetide, asafetida, *hing*.

Ascidium A Pitcher-shaped plant organ or part.

Asclepiadaceae A family of tropical and subtropical flowering plant in the order Gentianales characterised by a well developed latex system; milweed (Asclepias) is a well known member.

Asclepias gigantea See akund.

Ascocarp A fruiting in which asci are borne.

Ascochyta phaseolorum See bean rot; bean blight.

— **pinodella** See pea foot and blight.

— **pinodes** See pea foot and blight.

— **pisi** See gram blight; pea foot and blight.

— **sorghi** See Johnson grass ascochyta leaf spot; sudan grass diseases.

Ascogenous hyphate Hyphae that bear asci.

Ascogonium The female sex organ of the Ascomycetes.

Ascomycetes One of the three main groups of the fungi, bearing sexual spores in asci.

Ascocarp or **Ascoma** Any structure producing asci, as in apothecium, perithecium.

Ascorbic acid $C_6H_8O_6$. Vitamin C, hexauronic acid. Water soluble antiscorbutic vitamin.

Ascospore Produced in ascus by free cell formation.

Ascribed status It comes by birth. In primitive and static societies social mobility is restricted. The majority of persons neither rise nor fall from the status ascribed to them by accident of birth.

Ascus (pl. Asci) Saclike, usually clayate cell containing ascospores, typically.

Aseptate Without cross-walls.

Aseptic Free from living microorganisms.

Asexual Vegetative, having no sex organ or sex spores; the imperfect stage of a fungus. In Botany, flowers that have no stamens, pollen, pistils ovules.

Asexual reproduction Reproduction without requiring the usual union of male and female sex cells.

Ash gourd Benincasa hispida (Thunb) Cogn. (B. cerifera Savi) Family Cucurbitaceae. Trailing or climbing herb grown for its fleshy fruit. Immature fruit used as vegetable. Contains vitamin A; also B and C. Prefers a warm climate and does well in plains and hills of low elevation. Also white gourd, wax gourd, chinese preserving gourd, *petha*.

Asian cotton See Asiatic cotton.

Asiatic cotton Gossypium herbaceum L. Family Malvaceae. Cotton. Also called levant cotton, Arabian cotton, Asian cotton, *Kapas* or *Rui*.

Asiatic tree cotton Gossypium arboreum L. (G. nanking Meyen, G. indicum Tod, G. neglectum Tod). Old species of cotton belonging to family Malvaceae. Also called simply tree cotton.

Asparagine A compound important in nitrogenous metabolism of plants; obtained from amino acids in leguminous and other seeds.

Asparagus Asparagus officinalis L. A dioecious, perennial monocot belonging to the family Liliaceae. Small herb, native of Europe and west Asia. The tender new shoots are used as vegetable. Also called Garden asparagus, halyum, *seet muli, merchuba*.

Asparagus officinalis See asparagus.

— **racemosus** Willd. A vegetable and herb of family Liliaceae.

Aspect The direction towards which a slope faces; exposure.

Aspection Seasonal changes in appearance as constitution of a plant community.

Aspen Any of several species of poplars (populus) characterised by weak, flattened leaf stalks which cause the leaves to flutter in the lightest breeze.

Asperifoliate Rough-leaved.

Asphalt Mineral pitch; a bituminous composition used for paving, flooring, roofing etc.

— **oil** Oil containing asphalt or having an asphalt base, as distinguished from oil having a paraffin base.

Asphyxiation (1) suffocation; death due to respiratory difficulty resulting from inadequate exchange of O_2 and CO_2. (2) A systematic condition of plants brought about by O_2 deficiency.

Aspidomorpha miliaris See sweet potato insect-pests.

Aspirate Removal of fluid or air from body cavities or from an abscess with the use of suction.

Aspirator A seed processing machine in which seed separations are made by use of air, on the basis of difference in terminal velocity of seed. It has a fan, at the discharge point, which creates a vacuum or negative pressure within the machine. Air rushing to fill in the vacuum creates a stream of air which is used to separate seeds.

Aspondylia sesami See sesamum insect-pests.

Aspongopus brunneus See cucurbit bugs.

— **janus** See cucurbit bugs.

Assam rubber See India rubber.

Assam tea An important tea cultivated in India with tall plants and soft leaves. Grows at elevations 1500 m above sea level.

Assembling The putting together in correct relation to the parts which comprise a piece of mechanism.

Assignable causes The variations due to other then chance or random causes, where the basic conditions and, thus, the population have changed, reflect the presence of assignable causes.

Assimilation Conversion of substances taken in from the outside into living tissues of plants and animals.

Assimilating Capable of the absorption of nutrients into its body.

Associate, soil A non-taxonomic but cartographic grouping of soils or land segments, which combines related soils into units having similarity in geomorphic position, landstorm, edaphic and mechanical properties of soils (climate, drainage, particle size etc.) and to some degree similarity in the geological nature of the soil materials and taxonomic classes.

Association (1) The degree of dependence or independence, which exists between two or more variates whether they be measured quantitatively or qualitatively. (2) An association is not a community but an organization within a community.

Association, plants A natural grouping of plants having common growth requirements or association identified according to the dominant species.

— **Soil** A natural grouping of soil associates based on similarities in climatic or physiographic factors and soil parent materials. It may include a number of soil associates provided that they are all parent materials. It may include a number of soil associates provided that they are all present in significant proportions.

Assurgent upward curving of a stem of the plant.

Astelic Lacking a slele or having a discontinuous arrangement of vascular bundles.

Asteridae A large sub-class of dicotyledenous plants in the class Magnoliopsida; plants are sympotalous, with wiitegmic, tenuinuclellate ovules and with the stamens usually as many as, or fewer than, the cordla lobes and alternate with them.

Astomatal Lacking stomata. Also known as astomous.

Astrobiology The study of living organisms on celestial bodies other then the earth.

Atactostele A structure in which the vascular bundles appear scattered throughout the parenchymatous ground tissue.

Atavism The reappearance of a

character after a lapse of one or more generations. Sometimes called throwing back or reversion. The individual is called throw back.

Ataxia Loss of ability to control bodily movements.

At best In relation to the stock market, an abbreviation for 'at the lowest possible price' in respect of a buying order, 'at the highest possible price' in respect of a selling order.

At discretion A term qualifying an instruction given by a client to his stockbroker to buy or sell stock or shares, it gives the broker discretion as to price.

Atemoya Fruit of Annonaceae family; resembles cherimoyer; cross between pumpkin and cherimoyer.

Athalia proxima See mustard saw-fly.

Athel tamarisk Tamarix aphylla L. Karst. (Tamarix articulata Vahl) Family Tamaricaceae. Tree of arid regions used for ploughs, persian wheels, ornamental articles and as fuel. Bark and gall used for tanning. Useful for afforestation and soil conservation of arid catchments, capable of growing in saline soils if surface irrigation is available.

— **dioica** See jhau.

— **traupoi** See French tamarisk.

At limit A term qualifying an instruction given by a client to his stockbroker to buy or sell stock or share, it places a specific limit on either the highest price that may be paid, or the lowest price at which a sale may be made.

Atmometre The general name of an instrument that measures the evaporation rate of water into the atmosphere.

Atmosphere (1) The mantle of grass that envelopes the earth all around. (2) Unit for expressing the pressure and is equal to 15 lb per sq. inch.

Atomise To divide a stream or jet of any liquid into a fine spray resembling smoke or dust in appearance.

Atomizer A device for reducing liquids to a fine spray.

At par Stocks and Shares the market price of which is the same as the nominal or face value.

Atriplex hortensis See orach.

Atrophy Reduction in size of an entire plant or part or suppression of development of certain organs; results of disease or disuse.

Attachment Any device used in correction with a machine in order to give a wider range or better quality of work.

Attacus ricinus See eri silkworm.

Attenuate Thin, reduce in density, strength or pathogenic activity; to destroy the power of disease-producing organism by chemical action.

At the base A term which means at the ground level while measuring the diametre of a standing tree as contrasted to the term at the stump or at breast height.

At the market A term qualifying an instruction given by a client to his stockbroker to buy or sell stock or shares, it indicates that the broker may buy or sell at about the market price prevailing at the time the instruction is given.

Atterberg limits See liquid limit and plastic limit.

Attractants Materials that attract insect pest so as to enable them to eat or come in contact with chemicals used for killing them.

Attribute A qualitative characteristics of an individual or item, usually employed in distinction to a variable or quantitative characteristics. Hence, 'attribute data' and 'enumeration data' applied to qualitative observations.

Atypical Not typical; disagreeing with the form, state or situation usually found under similar circumstances.

Auction A method of determining prices employed in the sale of objects which cannot be standardised.

Auctioneer An agent employed by a seller to sell his goods at an auction.

Audio visual aid It is an instructional device that can be heard as well as seen.

Audit The examination of the account books of an organisation, by an accountant or other competent persons, to ascertain their correctness.

Auger A boring tool of large size with attached handle at right angles to the tool line. Several types are made for different purposes.

— **soil** A tool for boring into the soil and withdrawing a small sample for observation in the field or laboratory. The different kinds of augers include those having worm-type bits; unenclosed; those having worm type bits enclosed in a hollow cylinder; and those having a hollow half cylinder with cutting edge on the side that rotates around a stabilizing vane.

Augite Minerals made up of silicates of calcium, magnesium, iron and sodium in varying proportions.

Aulacophora foveicollis See red pumpkin beetle.

— **intermedia** See red pumpkin beatle.

Aurantium decumana See pummelo.

— **maximum** See pummelo.

— **sinensis** See sweet orange.

Aureomycin An antibiotic drug.

Auricle (1) Earlike lobe at the base of leaf blades of certain grass species. (2) the external ear. (3) Anterior chamber.

Auriculate Refers to a leaf with expanded base surrounding the stem; leaf with lobe separate from the rest of the blade.

Auriform Ear shaped.

Aus Autumn rice.

Austausch coefficient See exchange coefficient.

Australian black wood Acacia melanoxylon R. Br: Family Mimosaceae. Tree from Australia and grown for its leafy phyllodes. Also black wood acacia.

Australian pea See Indian bean.

Austral region A North America biogeographic region including the region between transitional and tropical zones.

Austroriparian life zone The zone in which occurs the climate and biotic communities of the south eastern coniferous forest of North America.

Autecology The ecological relations between a plant species and its environment.

Autestic thinking It deals with fantasy, dreams and wishful thinking. This kind of thinking is an end in itself.

Authoritarian leadership Authoritarianism is a social process where the making of decisions is placed in the hands of one man who is presumed to know better than the others as to what the group should believe or do. The autocratic leaders dictate the way that are reached. They refuse to share in leadership because of fear of losing power.

Autoblast An Independence cell or micro-organism.

Autocarp A fruit produced as a result of self-fertilization.

Autochthonous flora Native microflora present in the soil in relatively constant number, presumed to subsist on the more resistance soil organic matter and hence unaffected by the additional organic matter added to the soil.

Autoclave Apparatus for sterilizing by steam under pressure; pressure steam steriliser.

Autoconsequent stream A stream in the process of building a fan or an alluvial plain, the course of which is guided by the slopes of the alluvium the stream itself has deposited.

Autoconvection The phenomenon of the spontaneous initiation of convection in an atmosphere layer in which the lapse rate is equal to or greater than the autoconvective lapse rate.

Autoconvective lapse rate The largely hypothetical environmental lapse rate of temperature in an atmospherc in which the density is constant with height (homogenous atmosphere);

3°C per 100 metres in dry air. Also known as autoconvection gradient.

Autocorrelation The internal correlation between members of series of observations ordered in time or space.

Autoecious Completing life cycle on one host; term used in rusts.

Autogamous Self-fertilizing.

Autogamy A process of self fertilization that results in homozygosis; occurs in some flowering plants, fungi, and photozoans.

Autogenetic drainage A self-established drainage system developed solely by headwater erosion.

Autogenous vaccine Vaccine made from a culture of bacteria obtained from the patient himself.

Autoimmune disease Disease such as rheumatic fever or glomerulonephritis wherein it is thought that autoimmunization against certain body proteins is the underlying mechanism for disease production.

Autoimmunity State resulting from the production (autoimmunization) by the body of antibodies against its own proteins; normally the anti-body forming system recognises its own body proteins as 'self' and does not react to them, but under certain conditions this normal inhibition of antibody production may fail.

Autoinfection Infection of one part of the body by bacteria derived from some other parts of the body.

Autolysis (1) Self-destructive process through hydrolysis after the death of a cell. (2) Natural softening process of vegetables and fruits after these have been picked.

Automatic centre punch A centre punch so constructed that, when pressure is applied, a spring controlled hammer contained within the handle is released with sufficient force to cause the point to leave its mark on metal.

Automatic circuit breaker An automatic device for breaking the circuit when the strength of current exceeds a pre-determined limit.

Automatic clutch A clutch operated by vacuum from the intake manifold and controlled by the accelerator pedal.

Automatic planter A automatic seed-cum-fertiliser planter. Keeps row to row and plant to plant distance constant. Could be either tractor or bullock drawn.

Automatic tide gauge An instrument consisting of a float connected to a recording drum which automatically records tidal heights with respect to time.

Automatic weather station A weather station of which the services of an observer are not required; usually equipped with telemetric apparatus.

Auto mechanics The practical science of repair and maintenance of automobiles.

Automobile A vehicle propelled by steam, gasoline or electrical power. The term, as generally used, includes pleasure and business cars, delivery cars, and trucks.

— **battery** A storage battery used to operate the starting and lighting systems of an automobile or to supply the current for an electrically operated vehicle.

Autonacrosis State of being poisoned due to self-produced carbon dioxide, rendered dormant and arrested in growth.

Autoparasite A parasite growing on a parasite.

Autophyte A plant not parasite or saprophyte; manufactures its food by photosynthesis.

Autopolyploid or sutoploid A polyploid arising through the multiplication of the complete genome complement of a species, e.g., an autotetraploid has four identical sets of chromosomes.

Autopsy Examination of internal organs of the dead body.

Auto-regression The generation of a series of observations, whereby the value of each observation is partly dependent upon the values of those which have immediately preceeded it, i.e. each observation stands in regression relationship with one or more of the immediately preceding terms.

Autosomic chromosomes Chromosomes other than the sex chromosomes.

Autotetraploid Organism with more than two seats of chromosomes in its body cells; both sets are derived from the same species.

Autotoxin Poisonous substances produced by changes within the organism.

Autotrophic Capable of utilizing inorganic carbon as the main source of carbon and of obtaining energy for life processes from the oxidation of inorganic elements (chemotrophic) or from radiant energy (phototrophic).

Autotrophic bacteria Organisms capable of forming their proteins and carbohydrates out of inorganic salts and carbon dioxide, self-sustaining. Nitrifying bacteria as example.

Autumn The season of the year in temperate zones that is transitional between summer and winter (September-November in the Northern Hemisphere; March-May in the Southern Hemisphere).

Auxanometre Instrument for measuring growth in plants.

Auxiliary Giving or furnishing aid or support, generally in a subordinate or secondary manner; supplementary; as an auxiliary view in drawing.

Auxiliary air valve carburettor A carburettor with an air valve designed to control the richness of the mixture.

Auxiliary cell Any of a group of cells produced in certain red algae in early formation of the cystocarp.

Auxiliary pump Any pump designed to perform a service to the main pump or power unit; e.g. a cooling water pump.

Auxiliary spillway Spillway built to carry runoff in excess to that carried by principal spillway.

Auximone An accessory growth - promoting factor in plant food.

Auxin A plant hormone or growth regulating substance, $C_8 H_{32} O_5$.

Auxospore A reproductive cell in diatoms by the union of the gametes or by asexual development.

Available energy Metabolizable energy.

Available moisture See available water.

Available nutrient The portion of any element or compound in the soil that can be readily absorbed and assimilated by growing plants (Available should not be confused with "exchangeable").

Available phosphoric acid Water-soluble plus citrate-soluble phosphoric acid in the soil.

Available water The portion of water in a soil between the field capacity and the permanent wilting point that can be readily absorbed by plant roots. Most workers consider it to be the water held in the soil against a pressure of up to approximately 15 bars. See also field capacity and moisture tension, soil.

Available water supply In water resources development projects, the total quantity of water available from the source within a certain time period, usually a year, and the rate of availability at any time within a year.

Avalanche wind The rush of air produced in front of an avalanche of dry snow or in front of a landslide.

Avena A genus of grasses (family Gramineae) including oats, characterised by an inflorescence that is loosely paniculate, two toothed lemmas, and deeply furrowed grains.

Average deviation or mean deviation A measure of dispersion derived from the average deviation of observations from central value, such deviations being taken *absolutely* i.e. without reference to algebraic sign.

Average infiltration rate The rate obtained by dividing the mass infiltration, that occurs during the time rainfall intensity exceeds the infiltration capacity, by the time during infiltration occurs at capacity rate.

Avena fatua See wild oats.

— **ludoviciana** See fodder oats.

— **sativa** See oats.

— **sterilis** See Indian oats.

— **sterilis var. culta** See oats.

— **strigosa** See fodder oats.

Average peasant One who employs most of the home labour in his farming and owns some of his cultivated land. He cultivates less than the capitalist farmer.

Average supply, mean supply, or period average The sum of the daily discharges at the canal head divided by the number of days in the base period.

Averrhoa acida See otaheite gooseberry.

— **bilimbi** See bilimbi.

— **carambola** See carambola.

Avian Pertaining to birds.

Avicennia officinalis See white mangrove.

— **resinifera** See white mangrove.

— **tomentosa** See white mangrove.

Avitaminosis A condition or disease as a result of vitamin deficiency.

Avocado Evergreen sub-tropical fruit grown in regions with mild winters. Commonly propagated by seed though budding and grafting also possible. A salad fruit with high fat content, good source of vitamin A, B, C and G and 1-4 per cent proteins. Originally from tropical America, grown in some parts of South India. Also butter fruit, alligator pear, ahuocate, aguocate, avocat, avacato, persea americana Mill; persea gratissima; P. drymifolia Cham and Schlect.

Avocat See avocado.

Avacato See avocado.

Avoidable losses The conveyance losses, delivery losses and farm losses.

Awn A bread or bristle like appendage extending from the tip of the lemma, point of leaf in certain plants, extension to a glume, abene, anther etc.

Axenic culture The growth of organisms of a single species in the absence of cells or living organisms of any other species.

Axial force A force which acts uniformly over the section of prismatic body so that its resultant coincides with the axis of the body.

Axial pitch More commonly referred to as 'pitch' may be described as the amount of advance per revolution of a single screw thread.

Axil The upper angle where a leaf or twig comes off the main axis of the stem.

Axile placentation Position of the placentae on the axis of ovary with more then one carpel.

Axillary bud The bud borne in the axil of a leaf.

Axis (1) A central line considered in relation to certain geometrical or mechanical relations. A point or line about which or on which a body revolves. (2) Supporting part of an organ or complex e.g. leaf axis, stem. (3) The second cervical vertebra.

Axis of symmetry An imaginary central line around which a symmetrically developed body is formed, and in which the centre of gravity is found.

Axle A shaft or device which carries the driving travelling, or truck wheels of a vehicle, such as a locomotive, wagon, trolley, automobile, etc.

Axogomy Having sex organs on a leafy stem.

Axoneme The central filament of a flagellum.

Axonopus compressis See carpet grass.

Ayirulent strain A type of normally pathogenic organism that has no disease producing power on a particular host species or variety.

Azadirachta indica See neem tree.

Azores high The semi-permanent subtropical high over the North Atlantic Ocean, especially when it is located over the eastern part of the ocean; when in the western part of the Atlantic, it becomes the Bermuda high.

Azonal soil Soil without distinct genetic horizons. Such soils may have non-Chernozemic Ah horizons, or thin chernozemic Ah horizons, but lack B horizons. In Canada they are included with Regosolic soils.

Azotobacter Azotobacter chroococum. A genus of bacteria including those species capable of fixing free nitrogen, strongly aerobic; susceptible to phosphorous deficiency.

Azygous Not existent in pairs.

B

Babul Acacia arabica wild. [Acacia nilotica (L.) Del]. Family Minosaceae. A medium sized tree with golden yellow flowers, long white thorns and dark brown rough bark. Wood reddish brown, hard, strong, and useful for making agricultural implements, cartwheels and as fuel. Bark yield tannin and gum excellent substitute for gum arabic of commerce. Pods eaten by animals and suited to dry and arid conditions. Also suitable for alkaline lands.

— **gum** Gum obtained from *babul*, is of great commercial use.

Baccate (1) Bearing berries. (2) Having pulps like a berry.

Bacciferous Producing berris.

Baciliform Rod-shaped.

Bacillariophyceae The diatoms, a class of algae in the division chrysophyta.

Bacillus (pl. Bacilli) A rod shaped bacterium, single celled family Bacillaceae; Bacteria which are cylindrical and do not form spores.

Back bulb A pseudobulb on certain orchid plants that remains on the plant after removal of the terminal growth, and that is used for propagation.

Backcross The cross of a hybrid to one of the parental types. The offspring of such a cross is referred to as the backward generation.

— **method of breeding** A system of breeding carried out by several generations of backcrossing and subsequent selection. The characters of the recurrent parent are retained for the most part, and characters from the non-recurrent parent are added.

— **parent** The parent of a hybrid with which it is again or repeatedly crossed; recurrent parent.

Back dune A stable dune behind the shifting frontal beach dune, often characterised by heavier vegetation.

Backfill (1) Earth or other material placed behind linings or structures. (2) Replacement of excavated material.

— **filling** Fill of broken stone or other coarse material outside a foundation or basement well to provide drainage.

Backfire (1) An incomplete explosion taking place in the intake of a gas engines, produced by faulty valves, late timing or lean fuel mixture. The vacuum produced causes fire to travel back in the manifold. (2) A fire intentionally set along the inner edge of a control line located ahead of an advancing fire. The backfire is set against the fire to be fought, so that when the two fires meet, both out.

Back flow Flow of water or sewage opposite to the normal direction of flow.

Back furrow A raised ridge left at centre of the strip of the land when ploughing is started from centre to side is called 'Back Furrow'.

Back gear An arrangement of gear wheels by which the power of the driving belt is proportionally increased, as on the head of a lathe.

Backgrounding Growing of replacement cattle, usually on high-forage systems. Backgrounding may take place anytime during the post-weaning period until the animal goes into the feedlot or the breeding home.

Backing brick The inner part of a brick wall, often of cheaper and less perfect brick then is used for the face of the wall.

Backlash The backward movement of a pair or train of gears when the driving pressure is variable. It equals the amount of clearance between the flanks of the teeth.

Back pressure (1) An opposing atmospheric force built up in the cylinders of an engine due to faulty timing, carbon deposits, dirty or foaled muffler, which prevents an engine from delivering its full power. (2) Air pressure in pipes greater than atmospheric pressure.

Back radiation The downward flux of atmospheric long wave radiation.

Backteremia Condition in which bacteria are in the bloodstream but do not multiply there.

Bacteria (Bacterium, sing.) Simplest form of plant life, unicellular, devoid of chlorophyll and nuclei. Usually about 1/1000 mm. Secrets enzymes and toxins to kill host tissues. Elongate and divide into two separate cells in 15-20 min if conditions are favourable. Forms spores. Also referred as micro-organisms or germs. Spherical (cocci), spiral (spirilla), wavy or bent or rod shaped, rod-shaped with hair like projections (flagellates); cylindrical which do not form spores (bacilli), those which form spores (bacteria).

— **count** The number and kind of bacteria in a substance per unit of volume.

Bacteria bean blight Fungal disease of kidney beans, mung, etc. caused by Xanthomonas phaseoli indicus. Symptoms are formation of large, irregular, dry, sunken areas, often brick-red or brownish in colour, and usually covered with yellowish-white crusts of exudate appearing on pods, and long reddish streaks on stem. Plants look wilted.

Bacteria denitrificans Denitrifying bacteria. Bacteria that live side by side with the azatobacter and clostridium and reduces nitrogen compounds in the soil and liberate nitrogen in the free state. Thrive under anaerobic conditions.

Bacterial ooze A discharge, exudate of sap or gun etc. from a plant as a result of bacterial disease.

Bactericide A germicide or an agent that kills bacteria.

Bacterine (Bacterin) A product which contains dead disease-producing organisms for introduction into the body of a animal to induce production of antibodies and thus increases resistance to infection.

Bacteriology The study of bacteria.

Bacteriophage A virus that infects bacteria and causes lysis of bacterial cell.

Bacteriostatis Preventing or hindering the growth and multiplication of bacteria.

Bacteroid An irregular form of certain bacteria.

Bactrocera persicae See custard apple fruit fly.

Bad land A land type generally devoid of vegetation and broken by an intricate maze of narrow ravines, sharp crests, and pinnacles resulting from serious erosion of soft geologic materials. This type is most common in arid or semiarid regions. It is grouped under miscellaneous land type.

Badrang Fagara budranga Roxb. (Zanthoxylum rhetsa (Roxb.) D.C.) A large tree of family Rutaceae. Fruit used as spice. Seed to get an aromatic oil. Wood useful for several purposes.

Bael Aegle marmelos (L.) Corr. (Crataeva marmelos L.). Family Rutaceae. A tall tree with edible ripe fruits. Unripe fruit used as astringent, stomachic and also for curing diarrhoea. Mucilaginous substance secreted round seeds used as cement and as varnish. Also called bel apple, elephant apple or Bengal quince and bael fruit.

Baffle A cross wall, set of vanes or guides, a grid, grating or similar device placed in a conduit to check eddy currents below them, and to effect a more uniform distribution of velocities and to dissipate energy. Also called as baffle wall.

Bag closer The machine used for sewing filled seed bags.

Bagasse The crushed stalks of sorgo and sugarcane after extracting the juice by heavy grooved rollers.

Bagger weigher These are small machines which, when properly mounted beneath a bin, will fill and weigh a bag in a single operation.

Baghu gosha Pear variety with round conical medium sized fruit with moderate sweet flavour. Also called Citron Des curmes.

Bagrada cruciferarum See painted bug.

Bahia grass Paspalum notatum Fluegge. Family Gramineae. A grass introduced from America and used as fodder.

Bajoon Corypha elata Roxb. Family Palmaceae. A tall palm growing in Bengal. Used for making starch from the trunk; vegetable from the buds. Seeds are also eaten.

Bajra See bullrush millet.

Bajra ergot Disease of *bajra* caused by Claviceps microcephala (Wallr.). The disease is recognised by the presence of light cream colour sugary exudation from inflorescence glistening against sunlight; during severe infection fall on leaves to form white excrustations. Insects attracted to exudation spread the disease. The ears infected are poisonous to man and animal.

Bajra green ear disease Bajra disease caused by Sclerospora graminicola (Sacc). Schroter. Characterised by transformation of ear either wholly or in part into loose head or small, twisted green leaf like structure. Affected leaves become white and later on brown. On lower surface of leaves whitish downy growth of sporangia appears.

Bajra leaf spot Disease in *bajra* caused by Curvularia penniseti. Characterised by appearances of dirty brown spots on edges of leaves. Also attacks leaf-sheath and ears.

Bajra rust Disease caused by Puccinia penniseti. Characterised by formation of small brownish yellow rust pustules in groups on bajra leaves and stem which turn blackish later.

Bajra smut Fungal disease of bajra caused by Tolyposporium penicillariae and T. senegalense. Symptoms include scattering of affected grains in ears, projecting prominently like pear shaped bodies beyond glumes with chocolate brown or bright green colour which turns dirty black later, enclosing deep brown or black spore mass.

Bakar Premna latifelia Roxb. A small tree of family Verbenaceae. Leaves of the tree are eaten in curries. Found in Bengal and South India.

Baked soil Soil having hard glazed surface due to wetting and intense drying in the sun.

Bakhar A cultivator used in south India. Also a blade harrow with plane or serrated blades. See harrow.

Baking soda Sodium bicarbonate.

Balance When a co-sharer turn at the end of a main or branch watercourse finishes, the main or branch watercourse is closed off and the water that has been left in the channel between the point of close off and the field of the co-sharer, is known 'balance'. Known in some states in India and in West Pakistan as 'nikal'.

Balanced fertiliser A fertiliser containing suitable proportions of necessary mineral elements required for proper crop growth or plant.

Balance equation An equation expressing a balance of quantities in the sense that the local or individual rates of change are zero. A diagnostic equation expressing a balance between the pressure field and the horizontal field of motion of the atmosphere.

Balance of payments The relation between the payments of all kinds made from the country to the rest of the world and its receipts from all other countries.

Balance of trade The relationship between the values of a country's imports and its exports.

Balance sheet A list of assets and liabilities of an individual or business.

Balancing depth Depth of a canal cross-section that the quantity of excavation is just equal to the earthwork required for banks.

Balancing item An item entered in the balance of payments accounts to make the account balance; it is the net total of the errors and omissions in other items.

Balanopaceae A small family of disecious dicotyledomous plants in the order Fagales characterised by extipulate leaves, seeds with endosperm, and the pistillate flower solitary in a multibracteae involucre.

Balanophoraceal A family of decotyledonous terrestrial plants in the order Santalales characterised by dry nut-like fruit one to five aules, unisexual flowers, attachment to the stem of the host, and the lack of chlorophyll.

Balanophora indica See coffee plant parasite.

Balanopsidales An order in some systems of classification which includes only the Balanopaceae of the Fagales.

Balansia oryzae See rice udbatta disease.

Bald An elevated treeless, often grassy, region in a forest vegetation zone.

Baldwin An American apple variety grown in Kulu valley of India. Fruit large, deep red. Fruit of good quality with flesh crisp, juicy and mildly subacid.

Bale A large tightly pressed and bound bundle of produce. Example bale of cotton, hay or straw.

Ball Compact earth attached to the roots of a plant in transplanting.

Ballast Crushed stone or broken brickbats; hence 'brick ballast' and 'stone ballast'.

Ball bearing Any one of the various bearings consisting of hardened steel balls in some sort of retainer, so constructed that the load is carried by the balls.

Ball check valve The active part of such a valve consists of a ball resting on a seat. Under pressure, the ball rises from the seat and permits flow in one direction; when pressure is removed or reversed, the ball returns to its closed position.

Ball copra Copra prepared from whole coconut. Matured coconut stored for 8-12 months during which all the water is absorbed due to slow desiccation. Kernel shrinks and detached from the shell. Copra is taken out by breaking the shell.

Balling Measurement of sugar per centage at fixed temperature in simple syrup. Also called brix.

Ballistic The fruits with explosive dehiscence and discharge of seeds.

Ball peen hammer The type of hammer commonly used by machinists; one end of the head is rounder or ball shaped for riveting or peening; the surface of the other end is flat and is used for striking a chisel, or in other such work.

Ball planting Transplanting with a compact ball of earth around the roots. Helps in nourishment and growth initially at the new location.

Balm Melissa officinalis L. A perennial herb of family Labiatae native of south Europe. Used for flavouring food products and oil used for perfumery and for flavouring beverages.

Balsa A common, second growth tree in the West Indies and in Central America. Its structure somewhat resembles basswood and poplar. Its extreme lightness combined with good strength makes it a desirable wood for use in airplane construction.

Balsam (1) A resinous or oleoresinous aromatic exudate from plants. (2) Any plant or tree yielding balsam.

Balsam apple Momordica balsamina L. A herbaceous vine of family Cucurbitaceae producing edible fruits.

Balsaminaceae A family of flowering plants in the order Geraniales, including touch-me-not (Impatient); flowers are irregular with five stamens and five carpels, leaves are simple with pinnate venation, and the fruit has dehiscent capsule.

Balsamodendron caudatum See hill mango.

— **Mukul** See Indian bdellium tree.

— **roxburghii** See *gugal.*

Balsam of tolu See tolu balsam tree.

Balsam Pear See bitter gourd.

Baluksag Gisekia pharnaceoides L. A herb of family phytolacaceae, the leaves of which are edible.

Bamboo Woody stemmed perennials of the genus Bambuseae family Gramineae. Used in building construction, paper manufacturing, baskets and mats making etc. Young shoots are eater as well.

Bambusa arundinaceae See thorny bamboo.

— **baccifera** See plains bamboo.

— **falcata** See Himalayan bamboo.

— **polymorpha** See betua.

— **spinosa** See thorny bamboo.

— **tulda** Roxb. Bamboo family Gramineae.

— **Vulgaris** See feathery bamboo.

Banana Musa paradisiaca L. (M. sapientum L.) Family Musaceae. Occupies large fraction of area under fruit trees in India. Tropical plant and can grow upto an altitude of 1700 m. A very useful plant.

Banana anthracnose Fungal disease of banana caused by Gloeosporium musarum, characterised by spots on fruits.

Banana beetle Nodostoma subcostatum Jac. Insect pest of banana; dark blue adult feed inside central leaf funnel and also damages skin of young fruit.

Banana black finger Fungal disease of banana caused by Macrophoma musae. Characterised by appearance of raised lesions on fingers.

Banana bunchy top and mosasic Disease of banana caused by virus. Symptoms include bunching together of leaves at the head of plant to give rosetted appearance, appearance of chloratic spots on leaves. Affected plants continue to live, but not producing fruit. Bunchy top spread by black banana aphis and mosaic by cotton aphis.

Banana cigar-end disease Caused by fungus stachylidium theobromae and results in rotting of distal end of fingers; on drying giving the appearance of a lighted cigar from which name is derived.

Banana fibre Fibre extracted from the leaf sheaths of banana.

Banana fig Preparation of banana of the Nendran variety. Fruits thoroughly ripened till skin is blackened, then pealed, spread on bamboo mats and exposed to sun on platforms. Dried product resembles dates or figs.

Banana flour Meal made from fully mature unripe bananas. First dipped in scalding water to help remove the skin. Skin removed banana split into two halves and sundried. Dried fruit grounded and sieved through 120 mesh sieves. Also called banana meal.

Banana freckle A fungus disease of the banana caused by Macrophoma musae, producing brown or black spots on the fruit and leaves.

Banana leaf spot Fungal disease of Banana caused by Mycospherella musicola and Cercospora musae resulting in leaf spots.

Banana main stalk and finger stalk rot Fungal disease of banana caused by Botryodiplodia sp. and Glocosporum sp. Characterised by rotting of main stalks of the bunch and stalks of fingers.

Banana meal See banana flour.

Banana Panama wilt Fungal disease of banana caused by Fusarium oxysporum var. Cubense (E.F.S.) Wr. and Reink. Characterised by leaves turning yellow and breaking of leaf-stalks close to trunk, dying of plants, production of bottle necked fruits with pithy and yellow flesh.

Banana stem weevil Odioporus longicollis Ol. Weevil pest of banana. Reddish to dark brown grub with long curved snout bores inside stem and thus killing the plant.

Banana sucker Young plant growing round a mother plant of banana during its life times separated along with its own underground bulbs for fresh transplanting. Two types are (1) sword sucker with pointed leaves; (2) Water sucker. The former is better for planting.

Banana weevil borer Cosmopolites sordidus Germ. Insect pest of banana. Grub bores into rhizomes, tunnelling the root stock and travelling up the stem.

Banco A meander or obow lake separated from a river by a change in its course.

Banding application (of fertilisers) The placing of fertilisers in the soil in continuous narrow bands or ribbons, generally at specific distances from the plants or seeds.

Bandicoot Bandicota indica. Biggest rat found in field and slums damages crops in fields and weakens foundation of buildings. Omnivorous in habit, feeding on grains, vegetable and fruits. Feeds and destroys potato and peas. Also called as bandicoot indica.

Banding The placing of a collar or band around the plant stem or tree trunk to poison, repel or trap insects or pests.

Band seeding Placing forage crop seed in rows directly above but not in contact with a band of fertiliser.

Bangalora Mango variety known for its heavy cropping, excellent keeping quality and plenty of fibreless hard bulb. Also known as Totapuri, Neelum and Rumani.

Bangiophyceae A class of red algae in the plant division Rhodophyta.

Bankalmi Ipomea maxima (L.f.) G. Don ex Sweet (Ipomea sepieria koenig ex

Roxb.). A twining herb of family convolvulaceae. Used as vegetable.

Bankanas disease See rice foot rot.

Bankapas Hibiscus vitifolium L. A herb of the family Malvaceae found in north India and Madhya Pradesh. The bark is a source of strong fibre.

Bank erosion Erosion of banks of river, canal lakes or ditches resulting from currents, wave cutting, caving, under-mining or gullying of slopes.

Bank-full stage The flow stage of a river in which the stream completely fills its channel and the elevation of the water surface coincides with the bank margins.

Banking The business of holding deposits and landing money.

Bank interest Interest paid and charged by the commercial banks for money deposited with or borrowed from them.

Bank loan A loan of a fixed amount for a specific purpose, carrying a definite date for repayment. Such a loan is usually made on approved security, the interest charge tending to be lower for larger concerns.

Bank overdraft A loan from a bank made on current account. Often a means of financing trading requirements on a short-term basis, an overdraft is of no definite duration and is repayable on demand.

Bankruptcy The inability to meet the demands of creditors. Bankruptcy may be followed by adjudication in which a bankrupts property is divided among his creditors.

Banks (1) A general term for certain financial institutions. (2) The activity of the sides of a channel, river, canal, drains, beginning at the margin of the water bordering a lake. (3) The upper surface of a furrow slice. (4) To pile up earth around plants.

Bank storage Water entering the banks of stream channels during high stages of stream flow, most of which returns to stream flow during falling stages.

Banner The fifth or posterior petal of a butterfly shaped (papilionaceous) flower.

— **Cloud** A cloud plume often observed to extend downwind from isolated mountain peaks, even on otherwise, clouded days. Also known as cloud banner.

Banti Echinochola stagnina (Retz.) Beaw. Family Gramineae. A herb found throughout India. Used as fodder. Stems sweet and palatable; grains also edible.

Banyan Ficus bengalensis L. Moraceae. A large evergreen tree usually starting its life as an epiphyte. Drought hardy. Produces aerial roots which when reaches ground, takes roots and grow into new trunks. Good for afforesting sand dunes. Also bargad, bar or nyagrodha.

Bao Medium and deep water winter rice.

Bar A unit of pressure equal to one million dynes per square centimetre.

Bara gokhru Pedalium murex L. Family pedaliaceae. Herb whose leaves are used as vegetable. Found in Gujarat, Konkan and Andhra Pradesh.

Baramasi A variety whose flowering and cropping is spread throughout the year, for example *Baramasi* mango, lemon and *chiku*.

Barani Rainfed crop.

Barb (1) A means of representing wind speed in the plotting of a synoptics chart being a short, straight line drawn obliquely towards lower pressure from the end of a wind direction shaft. Also known as feather. (2) A side branch on the shaft of a birds feather.

Barbados cherry Malpighia glabra L. A shrub or small tree of malpighiaceae. family producing edible fruits.

Barbarea praecox See Belle Isle cress.

— **Verna** See Belle Isle cress.

Barbate Bearded; having hair tufts.

Barbed (1) Having sharp points or hooks as on awns of wheat and barley. (2) One of the delicate thread-like structure extending obliquely from a feather rachis and forming the vane.

Barber A severe storm at sea during which spray and precipitation freeze on to the decks and rigging of ships.

Barber chair In loggers slang, a stump on which is left standing a slab that splintered off the tree as it fell. Generally it indicates careless felling.

Barcelona nut See hazel nut.

Barn chart The graphical representation of frequencies or magnitudes by bars drawn with lengths proportional to the frequencies or magnitude concerned. The frequency scale has no other interpretation than number and locations; thickness and areas of the bars have no significance.

Bar harrow A peg tooth harrow which has triangular shape.

Barium carbonate A white, water insoluble powder used as rodenticide.

Bark In woody stems, the tissue outside the functional phloem, derived chiefly from cork combium and outer non-functional phloem layers.

Bark-eating caterpillar Indarbela quadrinotata wlk. Insect pest of guava. Caterpillar feeds on bark and destroys tissue. Badly damaged plants stop growing and dieoff. Also attacks citrus, loquat, pomegranate, mango and litchi.

Bark grafting Method of grafting for rejuvenating old and diseased plants. The method consists in cutting stock at right angles and preparing the scion either in the form of wedge or sliced straight. Inserting the scion in-between the bark and the wood of stock. Usually two or more scions are used at the tying and sealing are done.

Barley Hordeum vulgare L. (Hordeum sativum Pers.) Family Gramineae. A herb cultivated and does best where winter is cool. Needs less water than wheat or oats. Generally grown on light soils. Optimum sowing time mid-October to mid-November Food for poor sections in India, used as *sattu*. Also used in beer and whiskey manufacture. Leaves and stem used as fodder. Dehusked varieties also available. Also known as *sattu*, *jav*, *jau*, *jaun* etc.

Barley black rust See barley stem rust.

Barley covered smut Fungae disease of barley caused by Ustilago hordei (perh.) Lagerh. Characterised by smutted heads, with grains replaced by black agglutinized spore mass covered by white papery membrane.

Barley foot rot and root rot Fungal disease of barley caused by Helminthosporium sativum P. Ket B. Characterised by root and collar browning and appearance of elongated brown spots on leaves glumes and spiklets. Poor filling of grains and forming of shrivelled grains.

Barley leaf rust Fungal disease caused by Puccinia hordei. Symptoms include small, round, light, yellowish brown pustules on leaves which later turn black and get covered by epidermis.

Barley loose smut Fungal disease caused by Ustilago nuda. Characterised by smutted heads where grains are replaced by black powdery mass of spores, leaving the rachis later naked.

Barley net blotch Fungal disease caused by Helminthosporium teres Sace. Characterised by formulation of large number of chocolate brown oval or elongated spots on leaves, killing seedlings.

Barley powdery mildew Fungal disease of barley caused by Erysiphe graminis. Light grey cushion-like cottony colonies above ground parts appear.

Barley sclerotial disease Fungal disease of barley caused by Pellicularia rolfsil. Characterised by white weft of mycelium forming near the base of the plant on which white to reddish-tan spherical sclerotia appear.

Barley stem rust Caused by Puccinia graminis tritici (Pers.) Eriks and Henn. Reddish brown to dark brown oblong pustules on clums and leaf sheath appear. Pustules turn black and grains get shrivelled later on. Also called black rust disease of barley.

Barley stripe Fungal disease caused by Helminthosporium gramineum Rebenh. Characterised by long, parallel, brownish steaks appearing on leaf-sheaths and leaves. These are continuous from base to tip. Affected leaves gradually wither or tear into shreds and gets dried up.

Barley water Squash of fruit in which part of added water is extract to dehusked and polished pearl barley.

Barley yellow rust See barley stripe.

Barn (1) A farm building used far storage of agricultural products and equipments or for housing favour animals. (2) A unit of area equal 10^{-24} square centimetre. Used in specifying nuclear cross section. Symbolised b.

— **Cured** As against sun or air cured, a substance cured in barn.

Barnyard Land adjacent to a barn usually fenced to enclose livestock.

— **grass** See barnyard millet.

Barnyard millet Echinochloa cruss-galli (L) Beauv. (Panicum cruss-galli L.). Family Gramineae. Herb chiefly grown for forage and food crop. Also called barnyard grass, water grass, sanwak.

Baroclinic field A distribution of atmospheric pressure and mass such that the specific volume, or density, of air is a function not solely of pressure.

Baroclinic instability A hydrodynamic instability arising from the existence of a meridional temperature gradient (and hence of a thermal wind) in an atmosphere in quasi-geostrophic equilibrium and possessing static stability.

Barometre An instrument for indicating atmospheric pressure; used for forecasting the weather and for measuring elevations.

Barometric wave Any wave in the atmospheric pressure field; the term is usually reserved for short period variations not associated with cyclonic-scale motions or with atmospheric tides.

Barolaxis Orientation movement of an organism in response to pressure change.

Barotrepic field A distribution of atmospheric pressure and mass such that the specific volume, or density, of air is a function solely of pressure.

Bar point share In this share the point of the share is provided by an additional bar which also serves the purpose of landslide of the share.

Barrel A cylinderical wooden or metal container.

Barren: A land which does not support any vegetation.

Barrier islands Elongated seafront islands of sand formed by the action of the sea.

Barrier theory of cyclones A theory of cyclone development, proposed by F.M. Exner, which states that a slow-moving mass of cold air in the path of rapidly eastward moving warmer air will bring about the formation of low pressure on the side of the cold air; analogous to the formation of a dynamic trough on the side of an orographic barrier. Also known as drop theory.

Barter The exchange of goods for goods, without the use of money.

Basal Pertaining to or located at the base. Being the minimal level for, or essential for maintenance of, vital activities of an organism, such as basal metabolism.

Basal cell The lowermost cell or the lower of two cells formed following initial division in the development of a gametangium or sporangium.

Basal disc The expanded basal portion of the stalk of certain sessile organisms, used for attachment to the substrate.

Basal growth Development from axis of a plant as opposed to growth at its extremities.

Basal leaf One of the leaf produced near the base of the stem; a radical leaf.

Basal metabolic rate Rate of metabolism of a organism at rest and expressed as percentage of normal heat production per hour per square metre surface area.

Basal metabolism The set of chemical and physical changes within the body which are necessary and just sufficient to maintain life.

Basal nerved A leaf which has veins emanating from the base.

Basalt See igneous rock.

Base circle In gear drawing, the circle on which the involute tooth is constructed.

Base exchange A physico chemical process in the soil in which cations absorbed can be replaced by other cations in chemically equivalent quantities.

— **capacity** The cations that a soil can absorbs, express in milliequivalents per 100g of soil. Also known as cation exchange capacity

Base level The level below which a land surface cannot be reduced by running water.

— **line** A survey line established with more than usual care, to which surveys are referred for coordinations and correlation.

Basella See Indian spinach.

Basella alba See Indian spinach.
— **rubra** See Indian spinach.
Base plate or bed plate The foundation plate or support of a piece of machinery.
Base saturation percentage The extent to which the absorption complex of a soil is saturated with exchangeable cations other than hydrogen and alluminium. It is expressed as a percentage of the total cation exchange capacity.
Base sedimentation survey Sedimentation survey normally made prior to the completion of a new dam before the storage of water has begun. In case of a reservoir in operation on which no base sedimentation survey was originally made, the first survey made is termed, 'Post construction first sedimentation survey'.
Basic cleaning Refers to the separations of material larger and smaller than good seed; general size grading and cleaning, such an air screen machine.
Basic crops Crops which because of acreage, value, climate etc. are considered important in agricultural economy of a region or country.
Basic data Records of observations and measurements of physical facts, occurrences, and conditions as to when they have occurred, excluding therefore any material or information developed by means of computation or estimate.
Basic slag By product of steel factories, may contain 6-20 per cent phosphoric acid. In India it is not used as fertiliser because of poor phosphoric acid content in iron ore. Must be pulverised before use. Alkaline in reaction.
Basic wind law See Buys Ballot's law.
Basidiomycetes Class 3 in the Fungi, characterised by septate mycelium, sometime with clamp-connections, and sexual spores on basidia.
Basidiospore Spore produced on a basidium in the Basidiomycetes.
Basidium (pl. Basidia) Club-shaped structure, which after fusion of two nuclei, produces four basidiospores.
Basifixed Attached at or near the base.
Basil The common name for any of the aromatic plants in the genus Ocimum of the mint family; leaves of the plant are used for food flavouring. Sheep hide tanned with bark.
Basiler Of pertaining to, or situated at the base.
Basin (1) The area drained by a river. (2) A level plot or field, surrounded by dikes, which may be flood irrigated, also a basin formed around a tree by forming a ridge all round to irrigate.
Basin irrigation A method of irrigation in which a level or nearly level area surrounded by an earth ridge or dike is flooded with water. See check irrigation.
Basin lister Tillage tool that dams lister furrows at regular intervals to create small basins, useful for moisture conservation in dry farming areas.
Basipetal Order of development in which the youngest structures are at the base of the developing axis and the oldest are at the apex.
Basophilous (1) Staining readily with basic dyes. (2) Of plants, growing best in alkaline soils.
Basswood A common name for trees of the genus Tilia in the linden family of the order Malvales. Also known as linden.
Best Phloem; especially, the fibrous cells in the phloem.
Bastard cedar Guazuma tomentosa Kunth. A tree of family sterculiaceae. A native of America whose fruits are edible. Decoction of fibre used for clarifying sugar. Also known as *rudrasksha.*
Bastard myrobalan See belleric myrobalan.
Bastard oleaster Elaeagnus latifolia L. Family Elaeagnaceae. A small evergreen tree or shrub found growing throughout hilly parts of India. Fruits edible when ripe, pleasantly acid and refreshing. Used for making jellies.
Bastard tree See coral tree.
Bast fibre Any fibre stripped from the inner bark of plants, such as flax, hemp, jute and sanice; used in textile and paper manufacturing, for example sunhemp and jute.
Batales A small order of dicotyledomous plants in the sub-class Caryophillidae of the class Magnoliopside containing a single family with

only one genus, Bates.

Batates edulis See sweet potato.

Batch (1) A quantity of anything considered as whole, as a batch of castings. A quantity of a mixture, as a batch of concrete. (2) A quantity of harvested crop put into bin or container on a repetitive basis, specifically for treatment such as drying.

Batch production The production of articles in batches; to be distinguished from job production and mass production.

Bat guano Partially decomposed bat manure (excreta).

Bath-tub farming See hydroponics.

Batocera rufomaculata See mango stem borer.

— **rugra** See mango stem borer.

Batten A strip of wood placed across a surface of one or more boards to prevent warping, to strengthen, etc., A cleat is also termed a batten. A strip used for attaching laths where studs are not accessible.

Batter board One of a series of horizontal boards set across or to one side of a trench line to indicate a desired level or reference grade from which ditch or trench elevations are determined.

Battery A number of primary or storage cells, condensers, or dynamos, grouped together as a single source of electricity. The term is incorrectly used when applied to a single cell.

Battery capacity The number of ampere hours that can be obtained from a storage battery.

Battery charger An electrical device for charging a battery. It usually converts alternating current to direct current for charging purposes.

Battery knapsack sprayer Sprayer similar to compressed air sprayer except that air and water discharged into them through hand-operated or power-driven control charge pump. Also known as pressure-retaining sprayer.

Baule unit Named after the scientist Baule, a unit of growth factor that will produce 50 per cent of the maximum possible yield or yield-producing capacity.

Bay A large estuary with a relatively high degree of flushing.

Bay-berry : See Box myrtle.

Bayonet-joint Articulation, usually in orthotropic shoots, where substitution growth has occurred; may be the result of sequential branching (in the model) or reiteration (after damage).

Bayou (1) A distributary channel for flood water, common to streams that flow through alluvial valley land. (2) Channels through swamps and marsh lands, including those that afford passage for tidal water. (3) Abandoned channels of the parent streams.

BC Soil A soil with a profile which has no A horizon.

Beach deposits Sediments that are modified in their degree of sorting, or surface relief, or both, by the action of waves in forming beaches.

Beach drifting Lateral movement of beach sediments parallel to the beach.

Beading The forming of a bead on a piece of work for ornamenting or stiffening.

Beak The awn on the outer chaff of wheat.

Beali Autumn rice.

Beam (1) The central shaft of the plough by which it is drawn. (2) A structural element in building etc.

Bean A vegetable grown for its highly nutritive value and eaten green, dried or ground into flour. The most important beans of India include: blackgram, chickling vetch, cowpea, field pea, red gram, Bengal grain, green gram, horse gram, kidney bean, cluster bean, lentil and pigeon pea.

Bean blight Fungal disease of kidney bean, mung and urad, caused by Ascochyta phaseolorum. Characterised by appearance of large brown spots on leaves, and black, minute dots on neucrotic lesions. Also known as bean rot.

Bean dry root rot Disease of kidney bean, mung and urad, caused by Macrophomina phaseoli Symptoms as in cotton dry root rot.

Bean leaf spot Disease of mung and *urad* and kidney bean caused by Cercospora cruenta and C. dolichi. characterised by appearance of angular, brown or red spots on leaves, stalks and pods.

Bean powdery mildew Fungal disease

of kidney bean, mung and *urad* characterised by white powdery patches on the leaves and other green parts which later get studded with black dots. The disease is called by Erysiphe polygoni.

Bean rot Fungal disease having symptoms of appearance of large brown spots on leaves and black minute dots on necrotic lesions. Caused by Ascochyta phaseolorum.

Bean rust A fungal disease caused by Uromyces appendiculatus. Characterised by appearance of small yellowish to brown and dark-brown spots on leaves and other parts of plant. Affects kidney bean, *mung* and urad.

Bean yellow mosaic A disease caused by virus and appearing in mung, urad and kidney bean. Appearance of yellow and green areas on leaves occur and badly affected leaves turn completely yellow.

Bear A person who, believing prices are about to fall, sells shares in the hope of buying them again at a lower price thus making a profit.

Beard Barbed or bristly hair-like outgrowths on grain; awn.

Bearing (1) That portion of a beam, truss etc. which rests on the supports. The support or carrier for a rotating shaft. (2) The direction or course of a line in relation to the cardinal points of the compass.

Bearing cap The top half of a journal bearing.

Bearing capacity The average load per unit area that is required to rupture a supporting soil mass.

Bearing metal Antifriction metals, brass, gun metal, and the various alloys used for making or lining the bearings of journals.

Bearing plate A plate of the thickness and area required to distribute a given load, such as a plate under a beam flange resting on a wall. If the plate is 5 cm. or more in thickness, it is called a slab.

Bearing surface The surface over which a load is distributed.

Bear market A market in which prices are falling.

Beaten rice Popular and important rice product. Prepared by first steeping paddy in water for two or three days and then boiling for some minutes. After cooling water is drained and paddy heated over a strong fire to burst open the husk. Wooden pestle is used to dehusk and flatten the kernel and husk is removed by winnowing.

Beating up Replacement of casualties in an artificially regenerated forest area by fresh sowing and planting. Also called infilling.

Bed depth The distance between the bottom of the dead furrow and the highest point of the bed area.

Bedding A method of surface drainage of narrow-width plough land in which the dead furrows (depth varying from 0.5 to 1.5 ft or 15.24 to 45.72 cm) run parallel to the prevailing land slope. Suitable on flat slopes of less than 1.5 per cent. Also known in England as 'ridges and furrows'.

Bed material sampler An instrument to sample the bed material, whether in motion or at rest.

Bed rock The solid rock that underlies soil and the regolith or that is exposed at the surface.

Bed width (1) In rectangular or trapezoidal section, the distance between the extreme points at the bottom of a cross-section. Also called as bottom width. (2) The distance between two adjacent dead furrows.

Bee Any insect of the families Prospidae, Megachillidae, Xylocopidae Andrenidae, Anthophoridae, Bombidae, Apidae. Generally beneficial, being insect parasite, predators, pollenizes or producers of honey and beeswax.

Bee bread Pollen gathered by bees, which is mixed with a little nectar or honey and deposited in the cells of the comb.

Bee candy Artificial food for bees made from sugar and substances like honey.

Bee culture See apiculture. Also called bee-keeping.

Bee forage See bee pasturage.

Beefwood See casuarina.

Bee glue Resinous substance from buds or leaf axils of certain trees utilized by bees to form combs.

Bee gum Beeswax.

Bee-keeping The management and

maintenance of colonies of honey-bees.

Bee pasturage Plants yielding nectar and pollen and within flying distance of a hive. Also called bee forage.

Beeswax A mixture of organic compounds secreted by special glands on the last four segments on the ventral side of the worker bees abdomen, used by bees for building comb. Used for cosmetics, candles and a number of industrial products.

Beet Beta vulgaris thread or garden beet, a cool season biennial of the order caryophyillales grown for its edible, enlarged fleshy root. Also see beet root and sugarbeet.

Beet heart rot and internal black spot Beet disease caused by born deficiency characterised by young unfolding leaves falling to develop normally, turning black and drying. Dying of buds in axils of older leaves occur. Application of borax with fertiliser is recommended for cure.

Beet root (1) Beta vulgaris L. Family chenopodiaceae. Biennial herb producing thickened root which is the edible portion. Top portion also used. (2) Beta zulgaris L. var. bengalensis Roxb. Beet root grown for its leaves used as vegetable. (3) See sugar-beet.

Beet sugar Sugar obtained from sugar beet.

Beggerweed Desmodium tortuosum DC. A herb of family Papilionaceae introduced into south India for use as fodder.

Bego-molasses Mixture of equal parts of molasses and begasse screening. Its nutritive value is nearly twice that of wheat bhusa.

Begonia A genus, family Begoniaceae of succulent herbs or undershrubs grown mainly as ornamental; leaves sometimes eaten as vegetable.

Begoniaceae A family of decolyledonous plants in the order violates characterised by an inferior ovary, unisexual flowers, stipulate leaves and two to five carpels.

Belat A strong land wind from the north or northwest which sometimes blows across the southeastern coast of Arabic and is accompanied by a hazy atmosphere due to sand blown from the interior desert.

Belladonna Atropa belladonna. A perennial poisonous herb that belongs to the family Solanaceae. Atropine is produced from the roots and leaves; used as an antispasmodic, as a cardiac and respiratory stimulant, and to check secretions. Also known as deadly nightshade.

Belle Isle cress Barbarea verna Aschers (B. praecox R. Br.). Family Curciferae. A herb native of Europe and used as a vegetable.

Belleric myrobalan Terminalia bellerica (Gaertn.) Roxb. (Myrobalanus bellerica Gaertn.). Family Combretaceae. A Commonly grown avenue tree the fruit of which are used for dyeing and tanning. Myrobalan fruit pulp used as adhesive.

Bellows type thermostat value These valves are provided with bellows containing a liquid like alcohol or ether. When these liquids expands, the valve opens because it raises off its seat.

Bell pepper Capsicum frutescens var. grossum. An annual herb native to America and of family Solanaceae. It is grown for its edible, variously coloured bell shaped mild flavoured fruits which are used in salad.

Belt conveyor A wide belt running over pulleys, transporting materials laid upon it.

Belt dressings Any of the various compounds used on machinery belting to prevent slipping.

Bemisia gossipiparida See tomato leaf curl.

— **tabaci** See cotton white fly.

Bench terrace A terrace designed to facilitate soil conservation, erosion control, and cropping on steep slopes by the development of level or flat gradient strips normal to the slope.

Benders Cotton staple of intermediate length.

Bending moment Bending moment for a section of loaded beam as the algebraic sum of all the loads, including weight of the beam, and reactions to the left or right of the section with respect to any point in the section.

Bending pin A tool used for straightening or expanding lead pipe. Also called as bending iron.

Beneficial water use Beneficial water use shall be held to mean the use of such quantities of water when reasonable intelligence and reasonable diligence are exercised in its application for a lawful purpose, as is economically necessary for that purpose.

Bengal cardamom Amonium aromaticum Roxb. Family Zingiberaceae. A perennial herb, used as condiment. Also known as *Badi elaichi* or Nepal cardamom.

Bengal sage Meriandra benghalensis benth. A herb of family Labiatae grown all over India. Used for culinary purposes. A native of Abyssinia.

Benign Harmless; when refers to diseases or disease processes, mild or non-malignant.

Benign disease A disease that does not cause severe injury to host.

Benincase cerifera See ash gourd.

— **hispida** See ash gourd.

Benthos The community of bottom-dwelling life.

Bent spike Pennisetum Pennisetum nervosum Trin. Family Gramineae. A drought resistance grass introduced from Uruguay and grown for fodder.

Ber fruit fly Carpomyia vesuviana Costa. Insect pest of jujube. Maggot feeds on fruit pulp due to which fruit falls off.

Bergera koenigii See curry leaf.

Bergeron Findeisen theory The theoretical explanation that precipitation particles form within a mixed cloud (composed of both ice crystals and liquid water drops) because the equilibrium vapour pressure of water vapour with respect to ice is less than that with respect to liquid water at the same temperature. Also known as ice crystal theory; Wedner Bergeson process.

Berm A formation of sand deposited as a ridge on the shore just above the normal high watermark by wave action.

Bermuda butter-cup Oxalis pes-caprae L. (Oxalis cernua Thunb). A herb of family oxalidaceae native of South Africa whose leaves are eaten as vegetable. Also known as *khatmitthi*.

Bermuda grass Cynodon dactylon (L.) Pers. Family Gramineae. A perennial grass with much brached aerial stems. Spreads by rooting at nodes of aerial stems and by underground rhizomes. Good for lawns, pastures and for checking erosion. Grass good for fodder.

Bermuda high The semi-permanent substropical high of the North Atlantic Ocean, especially when it is located in the Western part of that ocean area.

Berry (1) A simple fruit, the entire pericarp of which is fleshy. (2) Small pulpy fruit with the seed immersed. (3) The dry seed of certain plants such as the coffee berry.

Berseem Egyptian clover. Taifolium alexandrinum Juslen. Family Papilionaceae. A rabi fodder sown in October in North India. Rich in phosphate and calcium. Produces excellent hay.

Best profit equilibrium The point at which the marginal costs of a production and its marginal revenue from sales balance.

Beta vulgaris See beet root.

— **vulgaris var. bengalensis** See beet root.

— **vulgaris var. rapa** See sugar beet.

Betel nut a dried seed of the palm tree Areca catechus in the family Palmae; contains a narcotic. Also see arecanut.

Betel pepper See betelvine

Betelvine Piper betle L. Family Piperaceae. Perennial creeper cultivated for its leaf used as *Pan*. Plant prefers tropical forest conditions and regular supply of moisture.

Betelvine bug Disphinctus politus W. Insect pest of betelvine. Reddish brown bug and its nymph suck juice from tender leaves in between veins and results in black spot development.

Betelvine leaf rot See betelvine foot rot.

Betelvine wilt See betelvine foot rot.

Betterment An increase in the value of land and buildings arising from the action of other developers, or from restriction imposed on the use of land elsewhere by the public authorities.

Betua Bambusa polymorpha Munro. Family Gramineae. Bamboo used for making paper.

Betula The birches, a genus of deciduous trees composing the family Betulaceae.

Betilaceae A small family of dicotyledonous plants in the order Fagales characterised by stipulate leaves, seeds without endosperm, and by being monoecious with female flowers mostly in catkins.

Beushan Interculture of broadcast rice in standing water. Also called *biasi*.

B.F. The abbreviation for 'board feet' (also B.M., board measure).

Bhalia bush Moghania macrophylla Erect shrub with silky branches occurring in Himalayan and sub-Himalayan forest. Lac host and is raised from seeds.

Bhang See hemp.

Bhanjura grass Apluda mautica L. (Apluda varia Hack., A. aristata L.) Family Gramineae. A perennial grass with a glaucous appearance. Used to give bulk to more palatable grasses. Also called *tachula*.

Bharangi Clerodendrum serratum spreng. Family verbenaceae. Shrub found in Assam, Bengal, Madhya Pradesh. Its flowers and leaves used as vegetable.

Bhasa badha fishery Low-lying mashy salt water areas under reclamation in Sunderbans (Bengal) affected by tides. embanked and provided with sluices devised to trap fish seed to grow.

BHC Benzene hexachloride, an insecticide, available as dusting powder as well as water dispersible powder and as fumigant. Insects killed by contact or by stomach action or fumigant action. Can be used with other insecticides and fungicides.

Bhemal Grewia oppositifolia Buch. Ham. ex. Roxb. A small tree of family Tiliaceae whose fruits are edible. Bark yields fibre and leaves and young tree used as fodder.

Bhilmora Rumex hastatus D. Don. a herb of Family Polygonaceae, the leaves of which are used as a condiment.

B Horizon See horizon, soil.

Bhusa An Indian name for chaffed straw of cereals in particular of wheat.

Bias Generally, an effect which deprives a statistical result of representativeness by systematically distorting it, as distinct from a random error which may distort on occasions but balances out on the average.

Biasi An Indian term for interculture of broadcast rice. Another term is beushan.

Bibulous Capable of absorbing liquids as a blotter absorbs ink or water.

Bicostate Of a leaf, having two principal longitudinal ribs.

Bichloride of mercury $HgCl_2$ Mercuric Chloride, corrosive sublimate. A poisonous white powdery chemical used as disinfectant; water soluble.

Bicollateral Having the phloem both internal and external to the xylem, as a vascular bundle.

Bid The price a person is willing to pay.

Bid filling After a contract has been allotted, an arrangement whereby firms who composed for the contract notify a central agency of the prices submitted on tender.

Biennial The kind of plant that produces vegetable growth during the first year of growing season. After a period of storage or over-wintering out of doors, flowers, fruits and seeds are produced during the second year and the plant dies.

Biennial bearing See alternate bearing.

Biflagellate Having two flagella, as a gamete.

Bignay See Chinese laurel.

Bihar hairy caterpillar Diacrisia obliqua W. An orange-coloured hairy caterpiller that feeds on the leaves of sweet potato and a number of other crops.

Bilaikand Ipomea digitata L. A common climbing herb of family convolvulaceae, used as fodder.

Bilateral monopoly A market situation in which a single seller confronts a single buyer.

Bilateral oligopoly A market situation in which a few sellers confront a few buyers.

Bilateral symmetry The condition of having distinct and similar right and left sides, as in the flowers of sweet pea.

Bilateral trade Trade between two countries; strictly interpreted, the value of the goods and services exported by country A to country B must be exactly matched by the value of the exports from country B to country A.

Bilimbi Averrhoa bilimbi L. Family Oxalidaceae. A tree native of South-east Asia cultivated for its edible fruit.

Bill hook A sharp edged curved blade tool used for cutting and harvesting crop-plants.

Bill of exchange An order for the making of payment, mainly used in international trade.

Bill of loading A document of title to goods received for shipment. It usually gives a brief description of the goods which have been received for shipment and states the terms on which they are to be carried.

Bimetallic type thermostat valve It is a bimetallic strip which helps in opening the valve because of unequal expansion of two metallic strips.

Bimetallism A monetary system in which the units of currency are expressed in terms of two metals, usually gold and silver. The metals exchange in a specific ratio and are accepted in unlimited quantities for coinage.

Bin An enclosed structure used for storage of seed.

Binder (1) A machine which cuts and binds a crop into bundles. (2) A fibre connecting the staples of wood so as to form a fluce. (3) Materials which help in improving the cohesiveness of soils etc.

Binder tobacco Tobacco wrapped around the filter leaves and segments of a cigar.

Binomial (1) The nomenclature system of naming that makes use of two names the generic and specific names, for each type of organism. (2) A series expansion in mathematics known as Binomial expansion.

Bin sampler The large sized trier, used for drawing samples from bins. These are constructed on the same principles as bag triers but are much larger.

Binucleate Having two nuclei.

Bioassay The finding out of the potency of various substances such as drugs, hormones and vitamins by testing upon animals and micro-organisms. Also standardization of such preparations.

Biochemical activity Chemical process in living organism, tissues and cells.

Biochemical oxygen demand (BOD) A measure of the amount of dissolved oxygen required in biochemical processes to oxidize wastes in water.

Biocide A chemical preparation used for the control or eradication of a pest or a plant, including insecticides, herbicides, fungicides, and rodenticides.

Bioclimatics The study of the effect of climate on living organisms. Also bioclimatology.

Biolog A type of steam fog caused by contact between extremely cold air and the warm, moist air surrounding human or animal bodies or generated by human activity.

Bio-gas plant See gobar gas plant.

Biogeography The science concerned with the geographical distribution of animal and plant life.

Biological balance Dynamic equilibrium that exists among members of a stable natural community.

Biological control The destruction or suppression of undesirable living things such as insects, plants or animals through the introduction, propagation and dissemination of predators, parasites and diseases.

Biological interchange The interchange of elements between organic and inorganic states in a soil or other substrata through biological activity. See mineralization and immobilization.

Biological magnification A process through which toxins become concentrated as they are passed up the food chain.

Biological mineralization See mineralization.

Biologicals Products derived from a living process or living matter such as serums, vaccines, bacterines, antitoxin etc. Also called as biologics.

Biological shield A thick wall or shield, usually consisting of a three to four metre thickness of concrete, surrounding the core of a nuclear reactor to absorb neutrons and gamma radiation q.v. for the protection of operating personnel.

Biologic erosion Erosion of soil by water or wind as a result of the soil being exposed by the burrowing of rodents, destruction of vegetation by insects etc.

Biologics See biologicals.

Biology The scientific study of life and living organisms.

Biolysis The decomposition of organic matter resulting from activity of living organisms, disintegration of life.

Biomass The part of a given habitat consisting of living matter, expressed either as dry or wet weight of organisms per unit area or as volume of organisms per unit volume of habitat.

Biome A large, natural assemblage of associated plants and animals, extending over a large region of the earth surface.

Biometry The application of statistical methods to the study of biological problems.

Bionomics The study of the relations among organism, and the relationship between them and their environment. Also called bionomy and ecology.

Bionomy See bionomics.

Biophysical Pertaining to living organisms and water conditions of the environment.

Bioplasm Protoplasm.

Biopsy Diagnostic examination of a piece of tissue removed from the living organism.

Biosequence A sequence of related soils that differ from one another primarily in the kinds and numbers of soil organisms that play a part in soil formation.

Biosphere The life zone of the earth, including the lower part of the atmosphere, the hydrosphere, soil, and the lithosphere to a depth of about 2 kilometres.

Biosynthesis Synthesis of complex compounds in a living body using simple substances.

Biota Animal and plant life of a particular region considered as a total ecological entity.

Biotic Related with life; biological.
- — **influence** Biological influences on vegetable life as contrasted to climatic influences.
- — **potential** Inherent ability of an organism to reproduce and survive.

Biotin One of the members of the vitamin B complex (vitamin H); considered to cure baldness.

Biotite mica See mica.

Biotope (1) The living space occupied by an organism. (2) An area of uniform environmental conditions and biota.

Biotype A population of individuals with identical genetic constitution. A biotype may be homozygous or heterozygous. See genotype.

Bipinnate Having leaflets growing in pairs on paired stems.

Bird Any member of the class Aves of the phylum vertebrata, the body of which is more or less covered with feathers; domesticated birds.

Bird scarer A mechanical device perched on tree-tops to scare away birds by producing loud sound at regular interval. Calcium carbide and water are used to make such sounds. Other types also available.

Birds eye of grape See grape anthracnose.

Birds foot trefoil Lotus corhiculatus L. A perennial herb of family Papilionaceae used as fodder.

Bischofia javanica See bishop wood.

Bisect (1) A profile of plants and soil showing the vertical and lateral distribution of roots and tops in their natural position. (2) To cut into two equal halves.

Bisequa Two sequa in one soil; that is two sequences of an eluvial horizon and its related illuvial horizon.

Bisexual A individual having both male and female reproductive organs. Also hermaphrodite.

Bishop's weed See omum.

Bishop wood Bischofia javanica Blume. Family Euphorbiaceae. A tree, from the bark of which a red dye is extracted. Bark also used as a tan.

Bisporangiate Having two kinds of sporangia in the same structure, as a strobilus of Selaginella.

Bispore In certain red algae, an asexual spore that is produced in pairs.

Bissetia steniella See sugarcane stem borer.

Biston supp essaria See tea looper.

Bit (1) The blade of an axe. (2) The part which is used for drilling in a drill (drilling bit).

Bitmara Gardenia campanulata Roxb. A large shrub or small tree of the family Rubiaceae found in Assam and Bihar. Leaves and fruits are edible.

Bitter gourd Momordica charantia L. Family Cucurbitaceae. Climbing or trailing herb. Fruits ridged, warty bitter and used as vegetable. Warm seasons crop and rich in iron and vitamin A, B and C. Sandy loam and slit loam soils suitable for crop growth.

Bitter orange Sour orange.

Bivalent bacterin A bacterin prepared from two different types of disease producing bacteria; gives protection against both the diseases.

Bivalent chromosome A pair of synapsed or associated homologous chromosomes.

Bivoltine Silkworm which gives two crops a year.

Bixa orellana See annatto tree.

Black alkali Alkali soil, where the excessive sodium carbonate dissolves the organic matter present in the soil. When evaporation of moisture takes place from the surface a brown or black incrustation is left. See alkaline soil, sodic soil.

Black-arm See cotton bacterial blight.

Blackberry Any of the upright or trailing shrubs of the genus Rubus in the order Rosales; an edible berry is produced by the plant.

Black body A hypothetical "body" that absorbs all of the electromagnetic radiation striking it; that is, one that neither reflects nor transmits any of the incident radiation.

Black caraway See earth nut.

Black cutch Acacia catechu (L.f.) Willd. A tree of family Mimosaceae good for afforestation. Yields gum and is a host for lac insects.

Black chaff A bacterial disease of wheat caused by Xanthomnas transplucens undulosa and characterised by dark, longitudinal strips on the chaff.

Black currant Antidesma ghesaembilla Gaertn. (Antidesma paniculata Roxb. Tree found in tropical Himalayas, Assam and Bengal. Fruits are edible and bark used for making rope. Also called umtao.

Black end (1) A disease of the pear marked by blackening of the epidermis and flesh in the region of the calyx. Believed to be a result of a disturbed water relation. (2) A fungus disease of the banana caused by several species, especially Gloesporium musarum, characterised by discolouration of the stem of the fruit.

Black fire A bacterial disease of tobacco caused by Psendomonas angulata and characterised by angular leaf spots which gradually darken and may fall out, leaving ragged holes.

Black gram Phaseolus mungo var. radiatus L. Family Papilionaceae. Highly important pulse. Contain phosphoric acid. Prefers water retentive stiff, loamy or heavy soils. Grown as rainfed crop. Two types one large black seeds and the other small green seeds. Known as urd.

Black head disease (1) A parasitic disease of the banana caused by eelworms of the family Tylenchidel. (2) A rot disease of the banana root stock caused by the fungus Thielaviopsis paradcra.

Black kernel A fungus disease of rice caused by Carvanlaria lunata and characterised by dark discolouration of the kernels.

Black knot A fungus disease of certain fruit and nut trees that is characterised by black exercences on the branches; destructive to plum, cherry gooseberry, filbert, and hazel.

Black line A disease of walnuts, especially of English varieties grafted to black walnuts, characterised by a black line of dead tissue at the graft union, eventually leading to death of the tree.

Black market The buying or selling of goods at prices above prices fixed by a Government during emergency conditions.

Black mold Any dark fungus belonging to the order mucorates. A fungus disease of rose grafts and union bulbs marked by black appearance due to the mold.

Black mulberry Morus nigra L. Family Moraceae. Introduced from Iran.

Produces fruits bigger and more juicy than other varieties.

Black musale Curculigo orchioides Gaertin. Family Amaryllidaceae. A perennial herb whose root is black in colour. Root is grounded and eaten as flour. Also known as Kalimusli or tallura.

Black mustard Brassica nigra Koch. A oil seed crop of family Cruciferae seeds used as condiment and for preparations of plasters and poultices; oil content varies from 25-30 per cent. Oil used for lighting, lubrication, and culinary purposes. Also *kali rai.*

Black nose A physiological disease of the date with the distal end of the fruit becoming dark, cracking, and shrivelling.

Black pateh A fungus disease of red clover characterised by simultaneous backening of plants in groups.

Black pepper Dry, unripe fruit of the plant Piper nigrum L. of family Piperaceae. Native of Indo-Malayan region. Used as a spice and condiment. Cultivated in the Western Ghats, Mysore and Maharashtra.

Black plum See Jambolana.

Black pod A pod rot of cacao caused by the fungus Phytophthora faveri.

Black point A disease of wheat and other cereals characterised by blackening of the embryo ends of the grains.

Black sing (1) A virus disease of cabbage and other members of the family Cruciferae characterised by dark necrotic and often sunken rings on the surface of the leaf. (2) A virus disease of the tomato characterised in the early stage by small black rings on young leaves.

Black root Any plant disease characterised by black discolourations of the roots.

Black root rot (1) Any of several plant diseases characterised by dark lesions of the rot. (2) A fungus disease of the apple caused by Xylaria mali. (3) A fungus disease of tobacco and other plants caused by Thielaviopsis basicola.

Black rot Any fungal or bacterial disease of plants characterised by dark brown discolouration and decay of a plant part.

Black shank A black rot disease of tobacco caused by Phytophthora parasitica var. nuotianae.

Black soil A great group of soils in the chernozemic order. The soils occur in the cold to cold subhumid grassland and parkland regions. They have a very dark surface (Ah or Ap) horizon and ordinarily a brownish B (Bm, Btj, or Bt) horizon, which may be absent, over a highly base-saturated, usually calcareous C horizon. In India it covers a large portion next only to alluvial soils.

Black spot Any bacterial or fungal disease of plants characterised by black spots on a plant part.

Black tea See tea.

Black thread A fungus disease of the para rubber tree at the point where the tree is tapped, caused by Phytophthora meadii and characterised by black stripes. Also known as black stripe; stripe canker.

Black water Water rich in humic acids and poor in nutrients, especially common of tropical rivers draining white sandy soil.

Black-wood acacia See Australian black-wood.

Black zedoary Curcuma caesia Roxb. Herb of family Zingiberaceae. Its tubers are used in cosmetics. Also called *kali haldi.*

Bladder A membranous sac filled with air or fluid. Also cyst; vesica.

Bladderdock Rumex vesicarius L. A herb of family polygonaceae whose leaves are edible. Grown throughout India. Also known as *chuka.*

Blade (1) The flat, active working part of a tool instrument, or device, as the blade of a knife, the blade of an oar, the blade of a propeller, etc. (2) The flat portion of a grass leaf above the sheath that extends from the stem.

Blade harrow Harrow or intercultivating implement fitted with a curving blade for cutting soil horizontally.

Blank A piece of timber cut to a specialised size and shape from which the finished article is made.

Blanket A cover of one type (pervious or impervious) material by a layer of another type of material.

Blanketing Covering the entire field

with overhead sprinkling irrigation at night to protect plants and fruit trees from the damaging effects of frost.

Blastodermic vessel See embryo sac.

Blattaria Order of insects to which cockroaches belong.

Blaze A mark made on the trunk of a standing tree by painting or chipping off a spot of bark with an ax. It is used to indicate a trial, boundary, location for a road, trees to be cut, and so on.

Bleaching Any process which lightens the colour in a plant by the destruction of the chlorophyll or also from preventing the green colour to develop.

Bleaching powder Chlorinated lime.

Bleeding (plants) Exudation of watery sap from vessels at a cut surface due to root pressure.

Bleeding canker A fungus disease of hardwoods caused by phytophthora cattorum and characterised by cankers which exude a reddish ooze on the trunk and branches.

Bleeding disease A fungus disease of the coconut palm caused by Ceratostomel la paradeca, characterised by a rust-coloured exudation from cracks on the stem.

Blemish Anything that mars the appearance of wood but is not serious enough to be classed as defect.

Blend (1) To mix two or more ingredients to obtain a standardised product. (2) In plants or animals to get an off-spring having characteristics of each of the plant or animal in which cross-breeding is done.

Blended inheritance Transmission of characteristics intermediate to those of the parents.

Blighia sapida See akee.

Blight A disease with sudden, severe leaf damage and often with general killing of flowers and stems. See necrosis.

Blight canker A phase of fire blight characterised by cankers.

Blind Placement of loose soil around a tile or conduit to prevent damage or misalignment when the trench is backfilled. Allows water to flow more freely to the tile.

Blind drain Type of drain consisting of an excavated trench, refilled with pervious materials such as coarse sand, gravel or crushed stones, through whose voids water percolates and flows towards an outlet. Often referred to as a French drain because of its initial development and widespread use in France.

Blinding material Material placed on top of and around a closed drain to improve the flow of water to the drain and to prevent displacement during backfilling of the trench.

Blind inlet Surface water inlet to a drain in which entrance of water is by percolation rather than open flow channels.

Blind seed A fungus disease of forage grasses caused by Phealea temullenta, resulting and abortion of the seed.

Blind weed plant See swamp cabbage.

Blister (Plants) See malformation.

Blister beetles Beetles belonging to the family Meloidae, which feed on plants. Have acid, cantharides, in their bodies.

Blister (1) A fungus disease of the tea plant caused by Exobasidium vexans and characterised by blisterlike lesions on the leaves. (2) A rust disease of Scotch pine caused by Cronartium asclepiadeum and characterised by blister-like lesions on the twigs.

Blister canker A fungus disease of the apple tree caused by Nammlaria discreta and charaterised by rough, black cankers on the trunk and large branches. Also known as apple pex.

Blister spot A bacterial disease of the apple caused by Pseudomonas papulants and characterised by dark-brown blisters on the fruit and cankers on the branches.

Blizzard A severe weather condition characterised by low temperatures and by strong winds bearing a great amount of snow (mostly fine, dry snow, picked up from the ground).

Blob A radar, oscilloscope evidence of a fairly small scale temperature and moisture inhomogeneity produced by turbulence within the atmosphere.

Block (1) A territorial division in states (2) A territorial division of forest to which a local popular name is given and is bounded by natural features.

Blocking high Any high (or anticyclone) that remains nearly stationary or moves slowly compared to the west to east motion upstream from its location, so that it effectively blocks the movement of migratory cyclones across its latitudes. Also known as blocking anticyclone.

Block tariff A method of charging for gas or electricity in which the price per therm or unit is highest for a specified initial number, or block, of therms or units consumed within a prescribed period, and lower for additional quantities of therms or units consumed within the same period.

Blocky See structure types, soil.

Blood (1) The fluid that carries nourishment and oxygen to all parts of the animal body and carries back waste products. (2) In plants and animal breeding, descended from parents of proven ability; relationship by blood.

Blood disease A bacterial disease affecting the vascular tissues of the banana in the celebes; believed to be caused by Xanthomonas celebensis and characterised by blighting of the leaves and reddish brown rot of the fruit.

Blood line Pedigree; relationship.

Blood meal A fertiliser prepared from animal blood by drying and grounded. Comprises of 80 per cent proteins which are not easily digestible.

Blood rain Rain of a reddish colour caused by dust particles containing iron oxide that were picked up by the rain drops during falling.

Blood red A variety of apple also called Cascoyne's Scarlet.

Blood serum Fluid or plasma left after removal of corpuscles and fibrin.

Bloom (1) Blossom; flower. (2) Seasonal dense phytoplankton. (3) Freshness and vigour approaching perfection.

— **of fruit** A layer of wax particles on external surface of certain fruits like grapes, peaches.

Blossom end rot (1) Any rot disease of fruit that originates at the blossom end. (2) A physiological disease of tomato believed to be caused by extreme fluctuations in available moisture, characterised by shallow leathery depressions with a water-soaked appearance around the tip end of the fruit.

Blotch A blot or spot, usually superficial, used to describe the diseased area of a plant.

Blow ball A fluffy seed ball, as of the dandetion.

Blower A device for increasing draft; used for ventilation or the production of blast.

Blowing sand Sandy material that is or has been subjected to wind action.

Blowing snow Snow lifted from the surface of the earth by the wind to a height of 6 feet (1.8 metres) or more (higher than drifting snow) and blown about in such quantities that horizontal visibility is restricted.

Blowing spray Spray lifted from the sea surface by the wind and blown about in such quantities that horizontal visibility is restricted.

Blow off A valve or drain connection on a steam or hot water boiler so arranged to draw off water and steam with any accumulated oil, grease, and dirt.

Blow out An area from which sand has been removed by wind; appearing as barren, shallow depressions, or gaps in dunes.

Blue butterfly Cosmolyce (Polyommatus) baeticus L. Insect pest of pulses. Short pale green caterpillar feeding on leaves, flowers and pods.

Blue elephant aloe Agave vera-cruz Mill. Family Agavaceae. Herb from the leaves of which fibre is obtained and used for making ropes, mats and cordage. Also called *Kuwarbuti.*

Blue grass The common name for several species of perennial pasture and lawn grasses in the genus poa of the order Cyperales.

Blue-green algae The common name for members of the Cyanophyceae.

Blue lotus of India Nymphaea stellata Willd. Family Nymphacaceae. Native of South East Asia. Rhizomes and seeds edible. Also called *Nilkamal.*

Blue panicum Panicum antidotale Retz. Family Gramineae. Tall perennial grass with creeping root stock. Used as fodder and for fixation and reclamation of sand dunes. Also called blue panic grass.

Blue rock pigeon Columba live. Field and house pest which feeds on seeds of maize, pulses and groundnut sown in fields.

Blue rot A fungus disease of conifers caused by members of the genus cesatostomella, characterised by blue discolouration of the wood. Also known as bluing.

Blue stone See copper sulphate.

— **vitriol** See copper sulphate.

B.M. Abbreviation for 'Board measure' meaning board foot or feet.

Board Thin timber as distinguished from planks or strips. Long pieces less than 2.54 cm thick and more than 5.08 cm wide are called boards. Stiff paper composite, used to reinforce various cover materials, and in box making.

Body (1) A collective name of all the organs of an animal. (2) Part of a mouldboard plough; base to which handle, beam, share and mouldboard are fixed.

Boehmeria nivea See ramie.

Bog Permanently wet land having low bearing strength.

Bog iron Impure iron deposits that develop in bogs or swamps by the chemical or biochemical oxidation of iron carried in solution.

Bog mass Moss of the genus sphagnum occurring as the characteristic vegetation of bogs.

Boiling point Temperature at which change from liquid to vapour state occurs.

Boil smut A fungus disease of corn caused by Ustilago maydis, characterised by galls containing black spores.

Bole (1) The Stem or trunk of a tree, usually the lower, usable or merchantable portion of the tree trunk. (2) The lowermost region of the coconut stem where roots are localised also boll.

Bole area The surface area of the main stem of a tree which is calculated by multiplying circumference by height.

Boll The leathery capsule which contains the seed of cotton plant; the seed pod of flax. Also bole.

Bolling A tree whose branches have been removed.

Boll period That phase in the growth when cotton bolls have been formed but not opened.

Boll rot A fungus rot of cotton bolls caused by Glomerella gossypii and Xanthomonas malvacearum.

Bolson A basin or depression having no outlet.

Bolt (1) Formation of an elongated stem or seed stalk suddenly under certain conditions. In the case of biennial plants, this generally occurs during the second season of growth. (2) Small logs or sections of larger logs that have been split. A bolt is usually less than 1.4 metre long. (3) A iron piece having threads on one or both sides for nuts.

Bombacaceae A family of decolyledonous tropical trees in the order Malvales with dry or fleshy fruit usually having woolly seeds.

Bombax ceiba See red silk cotton.

— **malabaricum** See red silk cotton.

— **mori** See mulberry silkworm.

Bombay locust Patanga succincta L. Greenish pale locust found in Indo-Pakistan and Malayan region. Highly damaging to crops. Also see locust.

Bonavist bean See lablab bean.

Bonding The Process of uniting two of more pieces of weed or veneers by means of an adhesive. This process is characterised by continuity of the union over the whole of the area of contact.

Bone products Products obtained from bone and includes Gelatine, glue, phosphatic manures, ossein, bone-meal, buttons, combs, paper-cutters, cigarette holders, toys, cattle and poultry feeds and di-calcium phosphate etc.

Bonus-penalty contract A type of contract under which the contractor is guaranteed a bonus, usually a fixed sum of money, for each day the project he has undertaken to construct is completed ahead of a specified date, while he agrees to pay a penalty for each day of delay in completing the works after the specified date.

Boodleaceae A family of green marine algae in the order Siphonocladales.

Boom A period of expansion of business activity. Opposite of slump or recession. A boom reaches a peak when the economy is working at full capacity.

Boom sprinkler Sprinkler having extended arms well over the field with rotating sprinklers spaced along these arms, thus reducing the length of path of travel by the water stream through the air or in other words reducing wind effect, and extending the range covered by one machine.

Booster A generator connected in series with a circuit for the purpose of increasing the voltage of that circuit. Generally used in connection with a system where a storage battery carries part of the load. The booster increases the voltage to a point where it becomes necessary to charge the battery.

Booster brake An auxiliary air chamber operated from the intake manifold vacuum, connecting with regular brake pedal and increasing breaking effects with less pedal pressure.

Boot The sheath and partially opened blade of the upper leaf of grass; the stage at which the inflorescence expands the boot.

Boot stage (1) Growth stage at which the grain plant is at inflorescene. (2) Growth stage of grasses at the time the head is enclosed by the sheath of the uppermost leaf.

Boracic acid White powder used in medicine as a very weak antiseptic. Also called boric acid.

Borassus flabellifer See palmyra palm.

— **flabelliformis** See palmyra palm.

Borax Sodium tetraborate; used in medicine as a weak antiseptic. Applied to soils directly or with commercial fertilizer to supply boron. Toxic to plants in high concentrations, boron higher than 3 ppm.

Bordeaux mixture Popular copper fungicide prepared by mixing copper sulphate solution with I mm water in different proportions. Usual proportions and 4:4:50 (copper sulfate, quicklime and water) or 5:5:50.

Bordeaux paste Popular fungicide of copper sulphate and quicklime of the consistency of paint and applied on bark or cut ends of plant using brush.

Border gate A farm structure used to assist in distributing the water from one border, check, or basin to another lower down the slope. Under some conditions a border gate may be used to distribute the water from the field lateral to the land.

Border parenchyma Thin walled cells, partially enclosing the veins in leaves.

Border rows The recommended number of rows of the male parental line grown on all the sides of hybrid seed field growing two different parents.

Border strips Long, narrow, transversally level, parallel irrigation units, laid out in the direction of the ground slope and separated by borders.

Border strip irrigation See irrigation methods.

Borecole Brassica oleracea L. var. acephala DC. Family Cruciferae. A herb, the young shoots and leaves of which are used as vegetable.

Boric acid See boracic acid.

Boro Spring rice.

Boron Plant food element; absorbed by plants as borate (BO_3) ion and required in very small amounts. Harmful if quantity of boron exceeds. Primarily concerned with the uptake of calcium by the roots and its efficient utilization. Also concerned with protein synthesis and potassium metabolism. Also see boron deficiency.

Boron deficiency Diseased condition in plants specially fruit trees; brought about by boron deficiency and may result in the death of growing plants. Considerable distortion of leaves and stems with margins of leaves turning brown, thickening, corking and splitting of veins occur. Pith is affected. This results in hollow stems and severely deformed fruits. Control by applying borax.

Boron toxicity Diseased condition of plants brought about by excess of boron in the soil. It may result in death of the plants.

Boron water Water with or without the conditions of salinity and alkalinity but containing limiting or excess of boron as well, usually more than 3 ppm.

Borrow pit A source from where material is borrowed to complete a section or make fills.

Boswelia serrata See Indian frankincense.

Botanical garden An institution for the

culture of plants collected chiefly for scientific and educational purposes.

Botany The science of plant life. It is branch of biology and includes anatomy, cytopology morphology, mycology, paleobotany, pathology, physiology, phytogeography, taxonomy etc.

Bothriochloa insculpta (Hochst.) Family Gramineae. A perennial grass introduced from Brazil and useful in soil conservation.

Botryodiplodia theobromae See cinchoma stem canker; tea root rot.

Botryose Term for plant producing flowers in clusters.

Botryosphaeria ribis See apple stem brown.

Botrytis disease Any of various fungus diseases of plants caused by fungi of the genus Batrytis; characterised by soft sotting.

Bottle gourd Lagenaria siceraria (Mol.) Standl (Lagenaria leucantha; Lagenaria vulgaris Seringe). Important climbing or trailing herb: vegetable fruit crop used for vegetable, sweetmeats and pickles. Contains vitamin B. Warm season crop growing best in warm and humid climate. Also called *lauki, ghia, kaddu,* white flowered gourd.

Bottle gourd plume moth Sphenarches caffer Z. A small slender spiny caterpillar that feeds on cucurbitaceous plants.

Bottle graft A plant graft in which the scion is a detached branch and is protected from wilting by keeping the base of the branch in a bottle of water until union with the stock.

Bottle silt sampler A kind of silt sampler used for collection of suspended silt load in a stream. It consists of a frame suspended from a pipe, holding a one-litre bottle fitted with a rubber stopper which is operated by a lever at the top of the suspension pipe.

Bottom The section of a plough consisting of the mouldboard, share, frame and land side.

Bottom bar The lower wooden bar of a frame for a comb in a beehive.

— **board** Floor of a beehive on which the hive body rests.

Bottom break A branch that arises from the base of a plant stem. The break or crack that separates a block of stone from a quarry floor.

Bottom clearance Vertical distance from the inlet of the pump suction pipe to the bottom of the sump.

Bottom-dwelling fauna Animals living near or at the bottom mud of ponds constitute food of fish which frequent pond bottoms.

Bottomland See floodplain.

Bottom rot (1) A fungus disease of lettuce, caused by Pellicularia filamentosa that spreads from the base upward. (2) A fungus disease of the trunks caused by pore fungi.

Bottom tier The tier below the middle tier. It is 40 cm (16 inches) thick or extends to a lithic or hydric contact. The sub-group classification of organic soils is based partly on this tier.

Botulism A food poisoning caused by a bacterium in preserved foods or feeds of animals. Can be fatal.

Boulders Rock fragments over 60 cm (2 ft) in diametre. See also coarse fragment. In engineering practice boulders are greater than 20 cm (8 inches) in diametre.

Boundary bank The bank constructed out of the excavated material of the boundary trench and the outer toe of which marks the limit of right-to-way easement.

Boundary layer theory According to this theory reduction in friction due to application of lubricants is explained on the basis of sliding effect i.e. the lubricant is supposed to be soaked in rubbing surfaces forming a oily surface over it.

Boundary stone Stones marking boundary of right-of-way easement for the canal.

Boundary trench A trench excavated near the outer extremity of right-of-way easement to provide a safeguard against cultivators ploughing into the canal limits.

Bound moisture Moisture intimately associated with the finer wood elements of the cell wall by absorption and held with sufficient force to reduce the vapour pressure significantly.

Bound water Water that is hydrogen-bonded to cell colloidal solutes such as protein and that is not free to act as a solvent for an added crystallite. Also called unfree water.

Bounty A payment (i.e. a subsidy) given by the state to merchants or manufacturers to encourage particular branches of business activity.

Boussingaultia baselloides See Madeiravine.

Bowen's ratio At a water surface, the ratio of the energy flux upward as sensible heat to the energy flux used in evaporation.

Box drain A drain with upright sides, flat top and bottom.

Box gauge Tide gauge operated by float in long vertical box to which tide is admitted through opening in bottom.

Box myrtle Bay-berry. Myrica nagi Thunb. (Myrica rubra Sieb & Zucc.). A tree of family Myricaceae. Fruits edible. Bark used on tan and yields a yellow dye.

Boycott An economic sanction which may be used for political or economic ends. It takes the form of one party refusing to supply goods or service to, or to purchase goods and services from, another party.

Braccia A rock composed of coarse angular fragments cemented in a fine-grained matrix.

Brace Any device to strengthen or support. In a plough, a rod connected to the beam to reinforce the landside.

Brace root Aerial roots extending downward from the lower above-ground nodes. On coming in contact with the soil, true roots develop. They function as brace and feeder roots. Also called aerial roots or prop roots.

Brachiate Having opposite paired branches on alternate sides; branched.

Brachus andis See pulse beetle.

— **phaseoli** See pulse beetle.

Brachysm Plant dwarfing in which there is shortening of the internodes only.

Brachytrypes achatinus See common field cricket.

Bract A modified leaf subtending a flower or the branch of a flower.

Bracteate Provided with bracts.

Bracteole A secondary bract at the base of a flower; also called as bractlet.

Bractlet See bracteole.

Brake horse power The horse power of a machine taken off a brake attached to it and registered by a dynamometre.

Brake lever A lever to which brake rods are attached in a mechanical braking system.

Brake lining Woven material of cotton, asbestos and fine copper wire, or the like, applied to brake bands.

Brake pedal The foot pedal which controls the application of the brakes.

Brake ratchet Usually a notched quadrant which engaged a pawl attached to the brake lever thus permitting a setting of the brake lever in a fixed position.

Brake rod A rod terminating with an eye or clevis forming a part of the linkage in a mechanical baking system.

Brake shoe A metal casting fitted to the curve of a wheel and pressed against the wheel to secure braking action.

Braking surface The area of contact between the stationary and moving parts of a brake. The area of the face of contact of surfaces through which braking friction is obtained.

Bramble (1) Any prickly, clinging shrub. (2) Any plant of the genus Rubus, family Rosaceae, such as blackberry and raspberry.

Bran The seed-coat of cereal grains which is separated from the flour and used as stock feed.

Bran bait A mixture of bran, poison and a sweetening agent used as a bait attracting the insects to kill them.

Branch A lateral stem.

Branch gap An interruption in the vascular tissue of a stem at the point where a branch trace arises.

Branch order The numerical sequence of an axis; hence absolute branch order, the morphological order of a branch as determined by primary development. Apparent, relative or visible branch order, the actual branch order determined by secondary changes and available for simple analysis.

Branching sequential The sequence of appearance of branches as determined by the architectural model; hence reiterative branching, or branching by reiteration, the appearance of branches not determined by the architectural model.

Branch trace Vascular tissue which runs from a stem into a branch.

Branding (1) An attempt to create or enhance differences between similar products, or between groups of products, by the use of a distinctive name or symbol. (2) To make identification mark into the skin of cattles.

Bran mash Feed prepared by moistening bran with warm water.

Brashness A condition of wood characterised by low resistance to shock and by an abrupt failure across the grain without splintering.

Brashy soil Soil composed almost entirely of coarse, angular rock fragment with fine soil matter filling the interstices.

Brass An alloy of copper and zinc, or copper, zinc, and lead. The latter is not common.

Brazil cherry See surinam cherry.

Brazilian lucerne See stylo.

Breadfruit Artocarpus altilis (Park.) Fosb. (A. communis Forst.; A incisa. L. f.). Family Moraceae. Fruit rich in calcium. Tropical plant. Leaves good fodder, wood used for building purposes. Allso called bread tree.

Breadfruit fruit rot Disease of breadfruit caused by Phytophthora palmivora. Symptoms as in jack fruit rot.

Breadfruit pink disease Disease caused by Corticium salmonicolor. Characteristics same as in jack fruit pink disease.

Breadfruit soft rot Disease of breadfruit caused by Rhizopus artocarpi. Characterised by rotting and dropping of fruits at all stages of development, fruit developing a mouldy appearance due to fungus growth on surface.

Bread grain A cereal, such as wheat or rye, which yields flour for loaf bread.

Bread tree See breadfruit.

Bread wheat Triticum vulgare. Species of wheat most common in north India.

Bread yeast See yeast.

Break To plough a fallow or virgin land.

Breaker A mechanical device consisting of a movable arm and stationary post, both faced with metal contacts which interrupt the current passing through, at properly timed intervals.

Breaker arm A metal bar with a contact point of tungsten on one end and a round opening on the other, bushed with fibre, bakelite, or similar non-conductor, on which the arm can swing from a metal pin.

Breaker strip A strip of canvas in the side wall of a tire casing to increase wearing qualities.

Break even chart A chart which shows the profitability or otherwise of an undertaking at various levels of activity, indicating a point at which neither profit or loss is made.

Break grade To change grade; as in a tile line, ditch or field.

Breaking bulk The function of a wholesaler in dividing up a commodity into quantities or sizes to meet the particular requirements of individual retailers.

Breaking of a virus Loss of flower colour in a variegated pattern, especially in tulips.

Breast (1) The front of the body between the neck and the abdomen. (2) The forepart of a mould board on a plough.

Breast wall When applied to irrigation practice, a wall, generally of masonry or concrete, provided immediately above the face of a submerged orifice, opening or sluice. Also referred to as 'curtain wall', face wall or panel wall.

Breed (1) A group of individuals having a common origin and possessing certain distinguishing characteristics not common to other members of the same species. (2) To improve characteristics in plants and animals through control.

Breed character Details of information which distinguishes a breed of animal or plant.

Breeder (1) A specialist in breeding. (2) An animal or plant used for breeding.

Breeder seed The initial increase of seed of the strain which is developed by the breeder and from which foundation seed is derived.

Breeders stock seed The seed harvested from a nucleus or a breeders stock field.

Breed in and In See in and in breeding.

Breeding (1) The application of genetic principles to the improvement of farm animals and cultivated plants by controlled mating and selection, or hybridization of plants by and animals in order to improve the species. (2) The production of nuclear fuel, by absorption of neutrons in a nuclear reactor, at a rate exceeding that at which fuel is being consumed.

Breed out To eliminate undesirable characteristics. See in and in breeding; out breeding.

Breed true The ability to transmit a characteristic uniformly, to the offspring.

Breeze (1) A light gentle, moderate, fresh wind. (2) In the beawfort scale, a wind speed ranging from 4 to 31 miles (6.4 to 49.6 kilometres) per hour.

Bressia Rock composed of coarse, angular fragments cemented together.

Brewers yeast See yeast.

Brick Unit of baked or burnt (i) argillceous or clayey earth; (ii) other admixtures, such as lime and sand, cement and sand.

Brickfielder A hot dry, dusty north wind blowing from the interior across the southern coast of Australia.

Bridelia retusa See ekdania.

Bridge graft A plant graft in which each of several scions is grafted in two positions and the stock, one above and the other below an injury.

Bridge grafting Method of grafting practised to save trees with diseased or damaged trunks. Consists in inserting both ends of the scion shoots into the bark to bridge the damaged parts, and forming a sort of bridge over it. After union take place, scion will grow and replace diseased portion of the trunk.

Bridled pressure plate An instrument for measuring air velocity in which the pressure on a plate exposed to the wind is balanced by the force of a spring and the deflection of the plate is measured by an inductance type transducer.

Bright band The enhanced echo of snow as it melts to rain, as displayed on a range-height indicator scope.

Briischa A northwest wind in the Bergell valley, Switzerland.

Brilliant Gardenia See dikemali.

Brilliantly clear juice Fruit juice prepared by complete clarification as apple and grape juice.

Brimstone Sulphur.

Brinjal Solanum melongena L. var. esculentum. Family Solanaceae. A vegetable fruit having high calcium contain and contains vitamin C. Can be grown throughout the year. Severe frost is harmful. Also called egg plant, *baingan*.

Brinjal bacterial wilt Bacterial disease of brinjal caused by Pseudomonas solanacearum due to which dropping in lower leaves occur.

Brinjal blight Fungal disease caused by Pellicularia (Sclerotium) rolfsii and characterised by sudden wilting of foliage followed by yellowing of leaves and browning of branches.

Brinjal bud worm Gnorimoschema (phthorimaea) glapsigona M. Insect pest of brinjal; caterpillar, boring into and feeding on flower buds.

Brinjal damping off Disease of brinjal caused by Pythium spp., phytophthora spp. and Pellicularia filamentosa characterised by rotting of seedlings at ground level which fall over.

Brinjal dry root rot Disease of brinjal caused by Macrophomina phaseoli characterised by death of roots.

Brinjal early blight Disease of brinjal caused by Alternaria solani; characters same as in potato early blight.

Brinjal leaf caterpillars Eublema olivacea M., Psara bipunctalis F., Phyctia clientella Z. Insect pest of brinjal, rolling leaves and feeding on them.

Brinjal leaf eating beetle Epilachna spp., insect best of brinjal. Young one feeds on leaves and other tender parts.

Brinjal leaf spot Fungal disease of brinjal caused by Cercospora melongenae characterised by chlorotic spots, turning grey to brown and shedding of leaf.

Brinjal leaky fruit rot Disease of brinjal caused by Pythium aphanidermatum. Symptoms include infection of fruit at blossom end in touch with soil, tissues turning watery and getting covered by cottony mycelium.

Brinjal little leaf virus Virus disease of brinjal, characterised by plant producing tiny yellow leaves and no fruit.

Brinjal phomopsis blight and fruit rot

Disease of brinjal caused by Phomopsis vexans. Characterised by dark brown lesions on stem and watery lesions on fruits, fruits later getting black and mummified.

Brinjal powdery mildew Disease due to fungus Erysiphe cichoracearum DC. Characterised by ash-coloured powdery growth in patches on the leaves.

Brinjal shoot and fruit borer Leucinodes orbonalis. Serious insect pest of brinjal attaching top shoots and fruits; pale white caterpillar with violet marking bores into stem, eats away tissues, causing attacked parts to drop and wither away. Also cuts fruits at calyx.

Brinjal stem borer Euzophera perticella R. Insect pest of brinjal; pale white caterpillar that attacks stem.

Brinjal sucking insects Eurentins echinus D. and Phenacoccus insolitus G. Insect pest; bugs; the first feeding on leaves and turning them yellow and the second attacking isolated old plants.

Brisa (1) A northeast wind which blows on the coast of South America or an east wind which blows on Puerto Rico during the trade wind season. (2) The northeast monsoon in the Philippines. Also spelled briza.

Brisote The northeast trade wind when it is blowing stronger than usual on Cuba.

Bristles Stiff, sharp, hair or hair-like growth on plants. Also stiff, wiry hairs of pigs, hogs and wild boars.

Bristly foxtail Setaria verticillata (L.) Beauv. Family Gramineae. A grass grown for its grain.

Brix The percentage of total solids in sugarcane juice; read off from brixometre; ranges from 18-22; 15 in juice of young and immature cane.

Brixometre A special hydrometre which reads directly total solids in the sugarcane juice.

Broad bean Vicia faba L. (Faba vulgaris Moench.). A herb commonly grown in Northwest India. Erect growing type of bean producing very large and usually fat orbicular or angular seeds used as vegetable. Also called windsor bean, bakla.

Broad bean leaf spot Disease of broad bean caused by Alternaria brassicae phaseoli.

Broad bean powdery mildew Fungal disease caused by Erypsiphe polygani. Symptoms as in bean powdery mildew.

Broad bean rust Fungal disease of broad bean caused by Uromyces fabae. Characterised by yellow cup-shaped pustules in clusters on green parts of plants and rust coloured small round pustules turning jet black later.

Broadcast To sow or scatter seed on the surface of the land by hand or by machinery. The seed may or may not be covered by working the ground afterward.

Broadcaster A machine that utilises a rotating fan-like distributor for sowing grain, grass and clover seed as for spreading fertiliser.

Broad crested weir A overflow structure for measuring water, often rectangular in cross-section, in which the water adheres to the surface of the crest rather than springing clear. Contrast with sharp crested weir.

Broad irrigation The irrigation of crops with sewage. It differs from sewage farming in that sewage disposal is the primary object of broad irrigation, with the raising of crops being incidental, while the reverse is true of sewage farming.

Broadleaf A tree with two cotyledons, or seed leaves; it usually is deciduous, that is, it sheds all its leaves annually. The broadleaved trees, such as maple and oak, have relatively broad, flat leaves as contrasted with the conifers, such as pines, which have narrow leaves, or needles.

Broadleaf tree Any deciduous or evergreen tree having broad, flat leaves.

Broccoli Brassica oberacea var. Italica. A biennial crucifer of the order Capparales which is grown for its edible stalks and buds.

Broeboe A strong dry east wind in the south western part of the island of Celebes.

Broken (1) Descriptive of a sky cover of from 0.6 to 0.9 (expressed to the nearest tenth). (2) Rice kernels broken during milling.

Broken orange pekoe See tea processing.
— **pekoe** See tea processing.
— **tea** See tea processing.

Broker An intermediary between two or more persons engaged in a business transaction. Brokers include import brokers, issue brokers, insurance brokers and stock brokers.

Brokerage Commission charged by a broker for selling or buying shares, commodities or insurance on behalf of a client.

Bromelia comosa See pineapple.

Bromus catharticus See rescuegrass.

Brood capping A porous substance with which honey bees seal the brood cells in a beehive

Brood cell A cell in which the larvae of bees are reared.
— **chamber** That part of the beehive in which the brood is reared.

Brood comb A sheet of brood cells constructed in a wooden frame in a beehive.
— **hest** See brood comb.

Broussonetia papyrifera See paper mulberry.

Brown A great group of soils in the Chernozemic order. The soils occur in the cool, subarid to semiarid grassland regions, and consist of a brown [dry surface (Ah or Ap)] horizon and ordinarily a lighter-coloured brownish B (Bm, Btdj, or Bt) horizon, which may be absent over a highly base-saturated usually calcareous, C horizon.

Brown algae The common name for numbers of the Phaeophyta.

Brown bast A physiological disease of the part discolouration of the inner bark near the trapping cut.

Brown blight A virus disease of lettuce characterised by spots and streaks on the leaves, reduction in leaf size, and gradual browning of the foliage, beginning at the base.

Brown blotch A bacterial disease of mushrooms caused by Pseudomonas tolasi and characterised by brown blotches on the fruit.

Brown canker A fungus disease of roses caused by Cryptosparella embrina and characterised by lesions that are initially purple and gradually become buff.

Brown felt blight A fungus disease of conifers caused by several Ascomycetes, especially Herpotrichia nigra and Neopeckia coulteri, a dense felty growth of brown or black mycetia forms on the branches.

Brown hay Hay resulting from stacking in too moist a condition when it gets overheated and undergoes considerable fermentation. Less nutritive than green hay.

Brown hemp See mesta.

Brownian motion Peculiar dancing motion possessed by finely divided particles in colloidal suspension, usually seen through an ultra microscope.

Browning Any plant disorder or disease marked by brown discolouration of a part. Also known as stem break.

Brown leaf rust A fungus disease of rye caused by Puccinia dispersa.

Brown patch A fungus disease of grasses in golf greens and lawns caused by various soil inhabiting species, typically producing brown circular areas surrounded by a band of greyish black mycelia.

Brown rot A disease which destroys the fruit of peaches, plums, cherries and apricots and also of apples. The disease is usually recognised by the rotting of fruits, blasting of flowers and the killing of young stems. A brown spot appears on the fruit and may enlarge until the whole fruit is decayed.

Brown root A fungus disease of numerous tropical plants, such as coconut and rubber, caused by Hymenochaete noxia and characterised by defoliation and by incrustation of the roots with earth and stems held together by brown mycelia.

Brown root rot (1) A fungus disease of plants of the pea, cucumber, and potato families caused by Theilavia basicola and characterised by blackish discolouration and decay of the roots and stem base. (2) A disease of tobacco and other plants comparable to the fungus disease but believed to be caused by nematodes.

Brown sarson Brassica compestris L. var. Dichotoma Watt. Family Cruciferae. An oilseed crop. Oil content 42-45 per cent. Cake fed to cattles.

Brown seaweed A common name for the larger algae of the division Phaeophyta.

Brown snow Snow intermixed with dust particles.

Brown spot Any fungus disease of plants especially Indian Corn, characterised by brown leaf spots.

Brown stem rot A fungus disease of soyabeans caused by Cephalosporium gregatum in which there is brownish interval stem rot followed by discolouration and withering of leaves.

Brown sugar A powdery *gur* (Jaggery).

Browse That part of leaf and twig growth of shrubs, woody vines, and trees available for animal consumption; to consume browse.

Brush (1) A device for drawing off from the commutator the electric current generated in the armature windings of a generator. Brushes are usually carbon; sometimes copper. (2) Shrub vegetation and trees that do not produce timber. (3) The tuft of hair on the tip of a wheat kernal.

Brush breaker A large powerful mold-board plough used for ploughing waste land overgrown with brush. Also called as scrub breaker.

Brush drag See brush harrow.

Brush drain Covered field drain consisting of a trench, the lower part of which is filled with brush to form the drainage channel.

Brush harrow A make shift harrow of short, tough branches attached to a log. Also called brush drag.

Brushing The process of collecting hatched silkworms.

Brush matting A covering of branches placed on badly eroded lands to conserve moisture and reduce erosion while other vegetative cover is growing.

Brush wood A growth of low shrubs and bushes; name for fallen or cut branches and twigs of trees. Also see brush.

Brussels sprouts Brassica oleraceae L. var. gemmifera DC. Family Cruciferae. Vegetable grown for the edible buds or small cabbage-like heads. Young shoots, buds and leaves edible. Also called Bud-bearing cabbage.

Bryophyta A small phylum of the plant kingdom, including mosses liverworts, and hornworts, characterised by the lack of true roots, stems, and leaves.

Bryopsida The mosses, a class of small green plants in the phylum Bryophyta. Also known as Musci.

Buchanan's mango See cudappah almond.

Bucharian rhubarb Rheum rhabarbarum L. (Rheum undulatum L.). Family Polygonaceae. A herb whose leaves are used as a vegetable.

Buck As used in forestry: To saw felled trees into logs or bolts.

Bucket elevator This equipment is used to lift the seeds to the top of bins/machines for cleaning, sorting or stacking.

Bucket pump A lift pump clamped to a bucket or pail and used in spraying. See bucket sprayer.

Bucket sprayer Sprayer designed to be used with a bucket containing spray liquid; provided with single or double acting pump. Laborious and used only when areas to be sprayed are small.

Buckeya The common name for deciduous trees composing the genus Aesculus in the order sapindales; leaves are opposite and palmately compound, and the seed is large with a firm outer coat.

Buckle plate A plate with dished or corrugated deformations for the purpose of giving more strength and stiffness than could be had from a flat plate.

Buck scraper An implement used for proper grading and levelling of uneven land and for the removal of saline, alkali incrustation. Has scraper pan with plain or serrated cutting edge. Also called leveller or scoop.

Buckwheat Fagopyrum spp. Not a true cereal, but usually considered as one. Used for making flour and livestock feed. Also for bee-feeding when in bloom.

Bud (1) The shoot apical meristem and its immediate derivatives (leave primordia, but scales). (2) To graft by inserting a but in the slit of a stem of a different plant (budding). (3) An underdeveloped shoot or stem.

Bud bearing cabbage See brussels sprouts.

Bud chip A bud used in grafting.

Bud complex A group of closely juxtaposed buds; may be either primary, e.g., a series of buds all of the same branch order developed within a single leaf axil, or secondary, the products of repeated branching producing an aggregate of buds belonging to several branch orders.

Budding A form of grafting in which the scion consists of a bud which is inserted under the bark of the stock. The shot that develops from the bud which was the scion of a bud graft.

Budding cloth A strong grade of muslin cloth saturated with wax used to prevent grafted buds from drying. Also called wax cloth.

Budget An estimate of national revenue and expenditure for the ensuing fiscal year, announced by the Chancellor of the Exchequer in the April of each year.

Budgetary control An important aspect of management accountancy which consists of the comparison of achieved trading results with the results assumed in the budget in order to ascertain or ensure that the subjects of the plan are being achieved, or to provide a starting point for the investigation of discrepancies which could lead to revisions of the budget.

Budget Surplus An excess of Government income over Government expenditure. This may be achieved deliberately in order to drain off purchasing power for consumer goods and reduce inflationary pressure. The surplus may be retained by the Government or invested in productive activities.

Budget year The 1-year period beginning with the start of the accumulation season at the firm line of a glacier or ice cap and extending through the following summers ablation season.

Bud grafting see budding

Bud mutation See bud sport.

Bud rot Any plant disease or symptom involving bud decay.

Bud scale A modified leaf or other organ enveloping and protecting a resting bud.

Bud selection Choosing buds from suitable plants or tree for grafting.

Bud sport A branch, flower, or fruit that differs genetically from the remainder of the plant.

Budstick A shoot whose buds are removed for propagation.

Bud variation Abnormal variations in stems, leaves of fruit which can be transmitted by asexual production. Also see bud sport.

Buffer area A limited use area separating a developed area from a protected area.

Buffer compound soil The clay, organic matter, and materials such as carbonates and phosphates that enable the soil to resist appreciable change in pH.

Buffer stocks Stocks of raw materials and commodities held for the purpose of reducing fluctuations in the process of primary products. Buffer stocks may be used to match the supplies coming on to a market with the current demand in order to narrow the movement of prices.

Buffer strips Established strips of perennial grass or other erosion-resisting vegetation; usually on the contour in cultivated fields. These reduces runoff and erosion.

Building materials All materials which enter into building construction.

Built-up lining A type of lining built up in situ of alternate layers of bitumen or asphalt and a supporting fabric, such as jute, cotton, fibre glass.

Bulb (1) An enlarged flashy thick, underground part of a stem surrounded by a mass of leafy scale. Scales of a bulb are actually thickened and shortened leaves. Roots develop from the base of a bulb, e.g., onion. (2) A part of the plant resembling a bulb.

Bulbel See bulbil.

Bulbil (1) Multicellular body essentially meant for the reproduction of the plant; may be modification of the vegetative or floral bud. (2) A secondary bulb usually produced on the aerial part of a plant.

Bulb vegetables Group of vegetables where bulbs are edible. Onion and garlic important.

Bulk density, soil The mass dry soil per unit bulk volume. The bulk volume is determined before the soil is dried to constant weight at 105°C. It has been called apparent density.

Bulk fertiliser Commercial fertiliser delivered without packing in solid or liquid form.

Bulk method of breeding The growing of segregating generations of a hybrid of self-pollinated crops in a bulk plot, with or without mass selection, followed by individual plant selection in F_6 or later generations.

Bulk specific gravity The ratio of the bulk density of a soil to the mass of a unit volume of water. (It is also called apparent specific gravity).

Bulk volume The volume, including the solids and the pores, of an arbitrary soil mass.

Bull A person, who believing prices will rise, buys shares with a view to selling them at higher price. A 'tried' of stale bull is a person who has made a purchase but has found no early opportunity to sell at a profit.

Bulldozer (1) A heavy forming machine for bending iron and steel. (2) A crowler tractor with a blade in front which helps in moving the earth.

Bullet Round nosed cylindrical plough point of a mole drain plough which forms a rounded cavity as the plough is drawn through the soil. Also referred to as a torpedo. In older type ploughs, a ball or plug is attached to the rear of the bullet to eliminate crumbs and smooth out the walls of the mole cavity.

Bulletin board The material for display may be news sheets, announcements, booklets, bulletins, circular letters, newspaper cuttings, cartoons, pictures, charts, posters, maps, graphs, subject outlines, etc.

Bull's eye rot A fungus disease of apples caused by either Neafabrach malicorticies or Gloeosporium pernnans and characterised by spores resembling eyes on the fruit.

Bull's eye squall A squall forming in fair weather, characteristic of the ocean off the coast of South Africa; it is named for the peculiar appearance of the small, isolated cloud marking the top of the invisible vartex of the storm.

Bull gear The large gear which gives motion to the table on a metal planing machine. The large driven gear on a crank shaft.

Bull market A market in which prices are rising.

Bull stretcher A bull brick laid with the long edge exposed.

Bullocks heart Annona reticulata. A deciduous tree of family Annonaceae.

Bulrush millet Pennisetum typhoides (Burm. f.), (Pennisetum typhoideum L.C. Rich.; Alopecurus typhoides Burm. f.). A most important millet crop of India known as pearl millet or bajra in Hindi.

Bumper (1) A safety device of spring steel or wood placed on front and rear ends of an automobile to provide protection at those points. (2) Over flowing or filled up to the brim. (3) Well grown crop likely to give high yield.

Bumper blocks Rubber blocks used between axles and car frame to absorb effect of severe shocks.

Bunch A cluster of plants or fruits.

Bunch bean Kidney bean.

Bunch grass Pennisetum ciliare L. (Cenchrus ciliaris L.) Family Gramineae. A perennial, tufted grass, excellent fodder. It is highly nutritious, is cut before flowering. Seeds also edible and could be used in soil conservation programmes. Also called *Anjan raiba* in Hindi.

Bunch groundnut See groundnut.

Bunching In a systematically oscillating pattern, where signs (or symbols) occur in concentration, they are said to be bunching or clustering.

Bunchy top Any viral disease of plants in which there is a shortening of the internodes with crowding of leaves and shots at the stem apex e.g. in banana and manila hemp this disease results in stunting in the plant and bunching of the leaves, as a result the plants fail to bear fruit.

Bund A low ridge of soil along the field contour.

Bund former An implement used to form bunds. Essentially consists of two concave blades joined across in a converging manner.

Bundhis An Indian term for small bunds.

Bundle scars Scars left in leaf scars at the time of leaf fall by the breaking of vascular bundles passing from a stem into a petiole.

Bundle sheath The tissue, usually of thick-walled cells, surrounding the veins in a leaf or stem.

Bunium persicum See earth nut.

Buran A violent northeast storm of South Russia and Central Siberia, similar to the blizzard.

Bur clover Medicago hispida Gaertn. A herb of family Papilionaceae. Native of Europe and Asia. Used as green manure and fodder crop.

Burgandy mixture Copper fungicide prepared by mixing solutions of copper sulphate with washing soda. Usually 1 or 2 per cent mixture is used. Mixture tested for both acidity and alkalinity before use.

Buried soil Soil covered by an alluvial, fossial, or other deposit, usually to a depth greater than the thickness of the solum.

Burl A hard, woody growth on a tree trunk or on roots, more or less rounded in form. It is usually the result of entwined growth of a cluster of buds. In lumber, a burl produces a distorted and unusual (but often attractive) grain.

Burlan A coarse cloth of jute or flax used for wrapping balls and bundles.

Burlock clover See lappa clover.

Burn Scorched appearance due to excessive manuring of plants; also withered condition due to lack of moisture.

Burned bearing The black rough surface of a bearing usually caused by poor lubrication.

Burning prescribed Burning carried out under the direct supervision of crews especially trained in the methods of when, where, and how fire can be used beneficially to improve timber management.

Burr (1) The rough, sticky or prickly envelope of any fruit. (2) A mature hull on the cotton boll. (3) A weed with burrs. (4) A thin, ragged fin left on the edge of a piece of metal by a cutting or punching tool.

Burring reamer A tapered reamer used for counterinking and for removing burrs caused in cutting pipe.

Burr mill A mill for grinding crops, in which two ribbed plates or disks suck or crush the material between them.

Burrow Excavations into earth made by certain animals such as rats, rabbits etc. in which they live. Also Gallery.

Burr weed Xanthium strumarium L. Family Compositae. An annual herb whose leaves appear in rainy season and die out during hot weather. Produces fruits which are spiny and bear seeds. Also called *goghru* in Hindi.

Burseraceae A family of decotyledonous plants in the order sapindales characterised by an avary of two to five cells, prominent serin ducts in the bark and wood, and an intra-staminal disk.

Burseraa delpechianum See Mexican linaloe.

— **Penicillata** See mexican linaloe.

— **serrata** See mirtenga.

Bush (1) To fit a hollow sleeve into some part in order that it may serve as a bearing. When wear and tear occurs, a new bushing may be applied, avoiding the necessity of replacing an entire part. (2) In general the insertion of any piece to bring about a reduction of diametre.

Business economics The study of the ways in which commodities are produced and distributed.

Business saving Saving out of profits, or ploughing back and provision for depreciation, the undistributed profits of companies and corporations.

Butea frondosa See flame of the forest.

— **monosperma** See flame of the forest.

Butomaceae A family of monocotyledonous plants in the order Alismatales characterised by secretory canals, linear leaves, and a straight embryo.

Butt Base of tree or lower end of log.

Butte An isolated hill rising abruptly above the surrounding area and having steep sides and a flat top. Most of the top has been removed by erosion, and it has a smaller summit area than a mesa.

Butter bean Small seeded lima bean.

Butterfly pea Centrosema, Centrosema pubescens Benth. Family Papilionaceae. Relished herbacious fodder legume found wild in pastures. Also cultivated. Heavy seeder and used for soil conservation also.

Butter fly valve A winged damper supported on a fulcrum or shaft, movement of which controls passage of gas or air through an opening.

Butter-head lettuce Lettuce-type; forms more or less spongy or bunched heads with thin, inner leaves blanching to a very light yellow; oily and buttering to the touch.

Buttering The spreading of mortar on the edges of a brick before placing it in position.

Butt hinge A hinge secured to the edge of a door and the face of the Jamb which butts against the edge of the door when it is shut, as distinguished from strap hinge.

Butt joint Where the ends of two pieces of timber come together without over lapping.

Button (1) An early stage in the development of a basidiocarp. (2) Less than three months old coconut flowers. (3) An irregularly shaped berry.

Buttress roots Branch roots given off above ground, arching away from stem before entering the soil, forming additional props. Common in tropical rain forest species.

Butt rot A fungus decay of the base of a tree trunk caused by polypores.

Buxbaamiales An order of very small, atypical mosses (Byopsida) composed of three genera and found on soil, rock, and rotten wood.

Buyers Market A market in which producers, suppliers and dealers experience difficulty in selling the goods which are available.

By product A useful product which results from the production of some other commodity but is not the main purpose of the activity.

Buys Ballot's law A law describing the relationship of the horizontal wind direction in the atmosphere to the pressure distribution; if one stands with one's back to the wind, the pressure to the left is lower than to the right in the Northern Hemisphere; in the Southern Hemisphere the relation is reversed. Also known as basic wind law.

C

C_3 Plant A plant that fixes CO_2 directly into carbon 3-phosphoglycerate; the net rate of photosynthesis is much less than maximal, limited by low concentration of CO_2 in the air.

C_4 Plant A plant that fixes CO_2 first into four carbon oxalocetate and then into 3-phosphoglycerate, yielding a high concentration of CO_2, and synthesizing hexose much faster per unit leaf areas, growing much faster, and functioning better at higher light intensities than a C_3-plant.

Caatinga A sparse, stunted forest in areas of little rainfall in northeastern Brazil; trees are leafless in the dry session.

Cabbage (1) Brassica oleracea L. var. capitata L. Family Cruciferae. A widely cultivated plant with a mass of elongated leaves packed together around a short stem. Forming a large bud like structure called cabbage head. A native of Western Europe and the northern shores of Mediterranean. Thrives best in cool and moist climate. Sowing from August-September. in plains and March-July in Hills. Contains vitamin A, B and rich in C. Heavy soil preferred. (2) The soft and brittle bud in the centre of the crown of a coconut palm. Rich in phosphate and potash and has a nutty flavour.

Cabbage bacterial leaf spot Fungal disease of cabbage caused by Xanthomonas campestris var. armoriaceae. Characterised by appearance of water-soaked leaf spots and drying up of leaves.

Cabbage borer Hellula undalis F. A whitish brown caterpillar mining into cabbage leaves and feeding on shoots. Burrows later into stem of cabbage.

Cabbage butterfly Pieris brassicae L. Insect pest; green caterpillar with black head that feeds on leaves of cabbage.

Cabbage palm Livistona australis Mart. (Corypha australis R. Br.) A tall palm of family Palmaceae, the young unfolded leaves are consumed as vegetable.

Cable An insulated electrical conductor or group of conductors, protected by a waterproof coat.

Cable box A box which houses the splices or connections between cables.

Cacao Theobroma Cacao. A small tropical tree of the order Theales that bears capsular fruits which are a source of cocoa powder and chocolate. Also known as chocolate tree.

Cacao curing Process of curing the beans of cacao in which beans are taken out of the harvested pods and allowed to undergo regulated fermentation in special boxes. Bitter taste is removed and a chemical compound theobromine develops which is a stimulant. Sun drying is then allowed and shells are removed. Kernels are pressed to extract oils and fats which are used to manufacture cocoa. Alkaloid theobromine is obtained from the cakes remaining after oil extraction.

Cacao tree Theobroma Cacao L. family streculiaceae. Small tree which gives commercial cacao. Native in America and cultivated in Asia and Africa. Shells of the beans fed to cattles or used as manure. Warm, moist, tropical climate is desirable with rains varying from 150 to 200 cm and elevation 300-600 m. Well-drained virgin forest ideal for growth. Propagated by cuttings, starts yielding in about 3 years and peak reaching after 7-8 years.

Cactaceae The cactus family of the order Caryophyllales; represented by the American stem succulents, which are mostly spiny with reduced leaves.

Cactus The common name for any member of the family Cactaceae, a group characterised by a fleshy habit spines and bristles, and large brightly coloured, solitary flowers.

Cadang-cadang An infectious virus disease of the coconut palm characterised by yellow-bronzing of the leaves.

Cadastral surveys Surveys relating to the establishment of land boundaries and sub-divisions made to create or define limitations of titles. They also include surveys to retrace or restore property lines and corners. The term 'land survey' and 'property survey' are synonymous with cadastral survey.

Caducous A term relating to an early dropping off of parts of a plant such as leaves.

Caesalpinoidea A sub-family of dicotyledonous plants in the legume family, leguminosae.

Cage wheel It is a wheel or an attachment to a wheel with spaced cross bars for improving the traction of the tractor in a wet field.

Cajanus cajan See pigeon pea.

— **indicus** See pigeon pea.

Cake A residue from oilseeds after the oil has been removed. Used as feed material to cattles or as manure.

Caked Solidified into a dry hard mass.

Calamintha hortensis See summer savoury.

Calamus (1) A hollow reed-like stem without nodes. (2) The quill of a feather.

Calcareous algae (1) Marine algae that forms hard coverings of calcium compounds and frequently develop reefs. (2) Algae that grows on limestone or in soil impregnated with lime.

Calcareous soil Soil containing sufficient calcium carbonate, often with magnesium carbonate. It effervesces visible when treated with cold 0.1 N hydrochloric acid.

Calciferol $C_{28}H_{43}OH$, Vitamin D_2.

Calciferous Producing or containing carbonate of lime.

Calcification (1) Deposition of insoluble lime salts, especially calcium carbonate or phosphate, in tissues. (2) Process of accumulation of a layer of lime in the soil profiles in a dry climate.

Calcifuge See calciphobes.

Calcify To deposit or secret lime salt which hardens, as when cartilage becomes bone.

Calcimorphic See calomorphic (soils).

Calciphiles, or Calcicoles Plant which takes in large quantities of calcium.

Calciphobes Plants which demand less calcium and are intolerant of soils containing large quantities of lime. Also known as calcifuge.

Calciphytes Plants that require or tolerate rather large amounts of calcium or are associated with soils rich in calcium.

Calcite ($CaCO_3$). Calcium carbonate crystalline in form, present in sedimentary rocks as rombohydrous crystals. It is a source of calcium in soils.

Calcium (Ca). Chemical essential for plant and animal growth. Principle element in bone. Absorbed by plants as calcium (Ca^{++}) ion. In combination with pectine in the plant it forms calcium pectate which is an essential constituent of cell walls and hence causes stiffness in straw. Promotes activity of soil bacteria. Necessary for development of good root system. Excess retards some plant diseases but in potato promotes scab disease. Application of calcium neutralises acidity and ameliorates the conditions of heavy clay soils, making them more friable and easily cultivable. Essential for the reclamation of alkali lands wherein calcium is adsorbed on the exchange complex replacing sodium. Improves physical properties of such soils to a great extent.

Calcium arsenate [$Ca(AsO_4)_2$]. Insecticide applied as dust or spray for killing leaf eating insects. Action is through stomach.

Calcium carbonate ($CaCO_3$). Used as antacid and calcium source in mineral feeds. Also called limestone, chalk, precipitated chalk, natural lime. Used to reclaim non-calcarous alkali soils with H_2SO_4 or Sulphur.

Calcium chloride ($CaCl_2$) A chemical agent used for bleaching. Rarely as a source of calcium in alkali lands.

Calcium cyanide [$Ca(CN)_2$]. An effective fumigant to control insects and rodents.

Calcium deficiency Physiological condition in plants and animals due to calcium deficiency in the body. Plant roots die at the tips or remain stubby adversely affecting plant growth particularly of legumes. Mottled discolouration of leaves followed by death of tissues. Commonly associated with acidity and toxic accumulation of iron, aluminium and manganese salts.

Calcium hydroxide [$(CaOH)_2$]. Hydrated lime is commonly used term. Slightly soluble in water and raises the pH of the soil. Also termed as water-slaked lime.

Calcium hydroxide Solution Commonly known as lime water. Used in the preparation of insecticides spray and paste; wide ranging uses.

Calcium meta-phosphate [$Ca(PO_3)_2$]. A vitereous product resulting from the treatment of phosphorus rock with gaseous phosphorus pentaoxide (P_2O_5) at high temperatures. A carrier of phosphorus for use in commercial fertiliser.

Calcium metabolism The physical and chemical processes by which calcium is utilised by plant and animal cells.

Calcium oxide (CaO). Used as dust dilutent, calcium carrier and for raising pH of the soil. Also termed as burnt lime, quick lime or lime.

Calcium phosphorus metabolism Interchange of the two important bone forming elements of calcium and phosphorus within the living organism.

Calcium phosphorous ratio The ratio in which calcium and phosphorus are required by animals for maintaining optimum health. Optimum ration is 1:1 and 2:1.

Calcium sulphate [$CaSO_4$]. Chemical existing in various forms. Hydrated calcium sulphate is known as gypsum and dried plaster of Paris. Former used in reclamation and management of alkali soils. Later used for making toys and several other purposes.

Calendar of work It is a plan of work arranged chronologically according to the time when each step of work is to be done.

Calliche (1) A layer near the surface, more or less cemented by secondary carbonates of calcium or magnesium precipitated from the soil solution. It may be a soft thin soil horizons, a hard thick bed just beneath the solum, or a surface layer exposed by erosion. It is not a geologic deposit. (2) Alluvium cemented with sodium nitrate, sodium chloride or other soluble salts in the nitrate deposits of Chile and Peru.

California current The Ocean current flowing southward along the Western Coast of the United States to Northern Baja California.

California fog Fog peculiar to the coast of California and its valleys of the Coast, winds displace warm surface water, causing colder water to rise from beneath, resulting in the formation of fog in the coastal valleys, fog is formed when moist air blown inland during the afternoon is cooled by radiation during the night.

Calina A haze prevalent in Spain during the summer, when the air becomes filled with dust swept up from the dry ground by strong winds.

Caliper (or calipers) tree An instrument used to measure diametres of trees or logs. It consists of a graduated rule with two arms, one fixed at right angles to one end of the rule, the other sliding parallel to the fixed arm.

Call loan A loan which may be terminated or 'called' at any time by the lender or borrower.

Callus (1) Protective covering which forms over a cut or wounded surface in plants. (2) Small, hard, outgrowth at base of spikelet or of floret, in some grasses.

Calocarphum sapota See marmalade plum.

Calocoris angustatus See sorghum earhead bug.

Calomel Mercurous chloride Hg_2Cl_2. A purgative of tasteless, white powder.

Colomorphic (Soils) Soils which owe their chief characteristic to the high content of calcium available to plants in the parent material. Also calcimorphic.

Calonyction muricatum See michai.

Calophyllum apetalum See poonspar.

Calophyllum wightianum See poonspar.

Calopogonium mucunoides Desv. Family Papilionaceae. A herb used as fodder and protection against soil erosion.

Calorie (gram calorie) Unit for measuring chemical energy, defined as the heat necessary to raise the temperature of one gram of water from 14.5° to 15.5°C at standard pressure, one kilo calorie (k cal) raises the temperature of one kg of water by 1° C. Thus 1 k cal = 1000 cal and 1000 k cal = 1 mega calorie (M cal).

Calotropis gingantea See madar.

— **procera** See akund.

Calpe emarginata See citrus fruit - sucking moths.

Calycifloral Describing flowers which have their leaves and stamens inserted in the Calyx.

Calyciform Describing flowers shaped like a calyx; cup-shaped.

Calycle A group of co-axial leaves around the base of a calyx in some flowers; has the appearance of an outer calyx.

Calyptra In the sporophyte of a moss, a sheath which covers the capsule, the tissue develops from a portion of the archegonium.

Calyptrogen The histogen in a growing tip of the root from which the root cap is developed.

Calyptrogen dermatogen The outer histogen in certain young roots.

Calyx Collective term for the sepals of a flower.

Cam A device mounted on a revolving shaft used for transposing rotary motion into an alternating reciprocating or back and forth motion.

Cambium A laterally disposed sheath of generative tissues usually found between the xylem and phloem. It gives rise to secondary xylem (wood) and phloem, a part of the inner bark.

Camellia kissi See Wild tea.

Camel's foot tree Bauhinia purpurea L. Family Caesalpiniaceae. A tree whose bark is used for tanning, dyeing and for fibre; flower buds and fruits used as vegetable and leaves as fodder. Also called *Lal Kachnar* in Hindi.

Camel thorn Alhagi pseudalhagi (Bieb) Desv. (A. maurorum Baker; Alhagi camelorum Fisch). A thorny perennial shrub of the desert belonging to the family Papilionaceae. Used as camel fodder. It is a weed also in cultivated crops.

Cam lifters or cam followers Those fixed or adjustable metal parts fitted between the camshaft and the valve stems. The throw of the cams controls the opening and closing of the valves.

Campanulate Bell-shaped.

Camphor Gum camphor produced from cinnamomum camphora or synthetically, forms colourless or white crystals; soluble in ether, alcohol, vegetable oils but insoluble in water. Useful as a circulatory and respiratory stimulant, anti-spasmodic, nerve sedative and antipuritic. Also see camphor tree.

Camphor basil Ocimum kilimandscharicum Guerke. Family Labiatae. A camphor and camphor oil producing perennial shrub. Leaves produce camphor oil. Propagation by seed, root cuttings. It can also be cultivated in waste lands. The Indian term for this is *Karpur tulsi.*

Camphor tree Cinnamomum camphora (L) T. Nees and Eberm. Family Lauraceae. Native of China, Japan and Formosa. Fairly large tree with aromatic leaves. Usually grown as hedge plant.

Campylotropous Having the ovule turned so that the base and apex are close together.

Canadian thistle Cnicus arvensis. Family compositae. A perennial and obnoxious weed with characteristic spiny leaves and small lilac-purple or white flowers borne in clusters; has aerial shoots. Mostly found in wheat fields. Also called Canada thistle.

Canal (1) An artificial water channel used for navigation purposes for conveying water for irrigation, domestic or industrial purposes, etc. (2) A small channel in the tissues of plants.

Canal evaporation losses Losses due to evaporation from the water surfaces of canals. Generally accounted for in total losses.

Canalization Measures adopted for confining a stream to a clearly marked channel for preventing damage while in flood as by covering agricultural land with sand or rendering it unsuitable by spreading out of a torrent over the more level stretches in the lower reaches of its course. Includes the production of plant screens or low bunds parallel to the stream and covering the reclaimed land with vegetation that does well in sand.

Canarium commune See Java almond.

Canavalia ensiformis See sword bean.

— **gladiata** See sword bean.

Candlenut Aleurites moluccana (L) Wild. (A. triloba Forst.) Family Euphorbiaceae. A tree native to Malaysia, growing wild in Assam and South India. Oil obtained used in paints and varnishes. Bark used as tans. Also called varnish tree or *jangali akhrot*.

Candy (1) Fruit or vegetable product; essentially a preserve impregnated with sugar by dipping in a syrup highly concentrated with sugar and after properly drying can be kept without spoilage. (2) A paste-like mixture made by kneading sugar into honey fed to queen bees in transit.

Cane (1) Ripened shoot of the past season in certain fruit plants. (2) Any hollow or pithy, usually slender jointed stem, such as in sugarcane or the bamboo grasses. (3) A stem growing directly from the base of the plant as in most Rosaceae such as blackberry and roses.

Cane blight A fungus disease affecting the canes of several bush fruits, such as currents and raspberries; caused by several species of fungi.

Cane fruit Any small fruit such as blackberry, raspberry etc., which bears its fruit on canes.

Cane molasses The heavy, residual syrup after the crystallization of cane syrup.

Canescent Having a greyish epidermal covering of short hairs.

Cane Sugar See sucrose.

Cane system of training A system of training grapevines. Trunk developed as in head system, and arms selected in the plane of trellis, 2 to 3 arms developed in each direction. Old canes cut at pruning; fruiting spurs with 8-12 buds each selected from canes of medium thickness as close to head as possible. Renewal spurs of one or two buds each kept near base of the fruiting canes to provide shoots for next years crop system suitable for varieties having unfruitful buds near base of canes.

Canker An area of necrosis on a woody stem resulting in shrinkage and cracking followed by the formation of callus tissue around the area, ultimately killing the stem. A localised chronic inflammation of the ear in cats, dogs, foxes, ferrets, and others caused by the mite Otodectes cynotis. Also see malformation.

Canker stain A fungus disease of trees caused by Endoconidiophora fimbriata platani and characterised by bluish-black or reddish-brown discolourations beneath blackened cankers on the trunk and sometimes the branches.

Cannabaceae A family of dicotyledonous herbs in the order urticales, including Indian hemp (Cannabis sativa) and characterised by erect anthers, two styles or style branches and the lack of milky juice.

Cannabis A genus of tall annual herbs in the family Cannabaceae having erect stems, leaves with three to seven elongate leaflets, and pistillate flowers in spikes along the stem.

Cannabis sativa Indian hemp, the source of a commercial fibre, hashish, and marijuana.

Cannaceae A family of monocotyledonous plants in the order zingiberales characterised by one functional stamen, a single functional pollen sac in the staman, mucilage canals in the stem, and numerous aules in each of the one to three locules.

Canna edulis See queensland arrowroot.

— **Indica var. orientalis** See Indian shot.

— **orientalis** See Indian shot.

Canning Sealing in airtight containers freshly cooked food; a method of preservation.

Cannon theory (emotion) This theory holds that emotion is essentially a preparatory reaction of the organism of biological survival value in times of danger. According to it, all the physiological responses are associated with biological preservation of the organism, and with the defence of the organism against attack.

Canopy The cover of branches and foliage formed by the crowns of trees in a wood or by other woody plants.

Canopy inlet Entrance to a closed conduit which has been shaped to induce priming at minimum submergence. Having the upper end of an inclined conduit centred over the bottom forming a canopy, and placing a plate above a vertical riser are effective methods.

Canopy interception Plant cover that protects the soil surface by dissipating raindrop energy. Also the amount of rainfall intercepted by the canopy.

Cant Thick piece of timber with or without squared edges, sown from a flitch or a log and intended for further conversion into small sizes.

Cantala Agave cantala Roxb. Family Agavaceae. A short stemmed woody herb, native to Mexico which yields fibre from its leaves. The fibre is used for rope, cordage and twines manufacture. Also known as Manila maguey.

Cantaloupe Cucumis melo var. cantalupensis. Family Cucurbitaceae. A rounded, comparatively small variety of muskmelon, usually with netted rind and with a delicate flavour.

Canthium parviflorum See kirni.

Cantilever spring A spring which has one end fixed at the axle and the other end attached to the frame as in quarter elliptic springs.

Canvas Closely woven cloth of fine texture and made from jute fibre of high quality.

Canyon A relatively narrow, steep sided valley often formed by a river cutting through soft rock; typical of arid regions.

Cap To seal a cell in a honeycomb with wax, as by a bee.

Capability class A rating that indicates the capability of land for some use such as agriculture, forestry, recreation or wildlife. In the Canadian system, it is a grouping or lands that have the same relative degree of limitation or hazard. The degree of limitation or hazard is nil in Class-1 and becomes progressively greater to Class-7.

Capability subclass A grouping of lands that have similar kinds of limitations and hazards. It provides information on the kind of conservation problem or limitation. The class and subclass together provide information about the degree and kind of limitation, for broad landuse planning and for the assessment of conservation needs.

Capacitance Generally, the ability to store up electricity. The capacitance of a condenser is measured by the charge it will hold when the difference of potential across its terminals is one volt.

Capacitive reactances The measure of resistance to the flow of an a.c. current through a condenser.

Capacitor An electric condenser.

Capacity (1) The estimated maximum level of production from a plant on a sustained basis, allowing for all necessary shutdowns, holidays, etc. (2) The quantity of electricity which a condenser is able to store or condense is called its electrostatic capacity.

Capacity statement In designing irrigation canals a statement prepared for determining the capacity of each outlet and minor taking off from a distributing canal.

Cap cell One or more cells formed at the apex of a specialised reproductive structure; liberation of the gametes or spores usually occurs when the cap cells disintegrate or separate.

Cap cloud An approximately stationary cloud or standing cloud on or hovering above an isolated mountain peak; formed by the cooling and condensation of humid air forced up over the peak. Also known as cloud cap.

Cap doctor The strong southeast wind which blows on the South African Coast.

Cape gooseberry Physalis peruviana L.(P. edulis Sims.) Family Solanaceae. Native fruit tree of Peru; grows in Nilgiri, low-growing perennial, producing berries resembling cherries in size and shape. Delicious jam prepared. Propagation by seed; cutting and layering also possible.

Also called Peruvian Cherry, *rasbari*.

Caperbush Capparis spinosa L. Family Capparidaceae. Shrub native of the Mediterranean region. Now grown in Punjab, Gujarat and Western Ghats. Flower buds, known as capers, used for flavouring pickles, relishes, sauce. In Hindi it is called *kalvari*.

Capers Flower buds of caperbush.

Capillary (1) Hair-like; applied to moisture held between and around particles of soil. (2) One of the minute thin-walled vessels which form networks in various parts of the body such as blood, nymph or biliary capillaries.

Capillary actions The phenomenon by which water or other liquids are carried by minute pores through the soil or upwards into plants.

Capillary attraction Apparent attraction between soil and water considered to cause moisture to move to drier parts of soil.

Capillary capacity See field capacity.

Capillary conductivity See hydraulic conductivity.

Capillary fringe A zone essentially saturated soil just above the water-table. The size distribution of the pores determines the extent and degree of the capillary fringe.

Capillary porosity The small pores or the bulk volume of small pores that hold water in soils against a tension usually greater than 60 cm of water. See also moisture tension, soil.

Capillary potential See matric potential.

Capillary pressure head Height water will rise by surface tension above in free water surface in the soil. Also called capillary rise.

Capillary soil moisture Moisture in soil held by surface tension force against the force of gravity.

Capillary water The water held in the small pores of a soil usually with a tension greater than 60 cm of water left behind in soil after gravitational water drains off. When in maximum quantity it is said to have reached field capacity.

Capillitium A delicate, branched network in the sporangia of slime fungi.

Capital One of the factors of production and may be defined as wealth used for the production of further wealth.

Capital account An element in the balance of payments account which deals with financial transactions, the buying and selling of securities.

Capital accumulation An increase in man-made equipment such as machinery, tools, buildings and other structures and stocks of goods, used for or capable of being used for the production of goods and services.

Capital allowances Deductions from gross business profits for the purpose of covering the depreciation of assets permitted by the tax authorities when calculating taxable income.

Capital charges Charges which include interest on the amount of capital employed, and provision for depreciation or repayment of principal.

Capital cost of The cost of obtaining the total capital employed by a business, expressed as a rate of interest.

Capital employed The capital in use in a business; it consists of the total assets minus the current liabilities. Current liabilities are those liabilities providing resources for less than a year; their exclusion helps avoid fluctuations in total assets which might otherwise impair comparisons based on several years trading.

Capital expenditure Expenditure of a non-recurrent nature resulting in the acquisition of assets.

Capital formation and depreciation Capital formation in a given time period is the expenditure on fixed capital assets (buildings, plant and machinery, vehicles, etc.) either for replacing or adding to the stock of fixed assets.

Capital gains Profits made from the sale of investments or property, due allowance being made for any decline in the value of money during the period which elapsed between buying and selling.

Capital goods Goods made for the purpose of producing consumer goods, and other capital goods, e.g. machinery of all kinds. Synonymous with 'producer goods'.

Capital intensive Description of forms of production in which there is a considerable use of capital equipment per person employed.

Capital investment Value of inventory, often at the beginning of the fiscal year. In agriculture it includes all farm properties.

Capitalised value The capital equivalent of an annual payment. Capital values are determined by the rate of interest and the rent produced.

Capitalism A social system in which the means of production, distribution and exchange are wholly or substantially owned by private persons.

Capitalist farmer One who employes most of the hired labour in his farming. A farmer is also defined as capitalist farmer who owns most of his cultivated land. Generally the size of the land-holding is larger than the average peasant. He has also more resources than the average peasant.

Capital market A market comprising institutions which deal in the purchase and sale of securities.

Capital movement The movement of money capital from one country to another.

Capital recovery The process by which the original investment in a physical asset is recovered over its economic life. In the absence of inflation capital recovery is achieved through depreciation allowances.

Capital transactions The lending or borrowing of sums of money or the transfer of assets accumulated in the past.

Capitalised value The capital equivalent of an annual payment. Capital values are determined by the rate of interest and the rent produced.

Capitalism A social system in which the means of production, distribution and exchange are wholly or substantially owned by private persons.

Capitalist farmer one who employs most of the hired labour in his farming. A farmer who keeps a permanent "Chakar" is also defined as capitalist farmer and owns most of his cultivated land. Generally the size of the land holding is larger than the average peasant. He has also more resources than the average peasant.

Capitata Swollen at the tip. Forming a head, as certain flowers of the Compositae.

Capitular The growing of fruits or flowers in small cluster or heads.

Capnodium ramosum See mango sooty mould.

Capparaceae A family of dicotyledonous herbs, shrubs, and trees in the order capparales characterised by arieal placentatian; hypogynous, mostly regular flowers; four to many stamens and simple to trifoliate or palmately compound leaves.

Capparales An order of dicotyledonous plants in the sub-class Dilleniidae.

Capparis aphylla See karer.

Capparis decidua See karer.

Capparis horrida Capparis zeylanica L. (C. horrida L.f.) Family Capparidaceae. A thorny shrub used as hedge plant.

Capparis spinosa See caperbush.

Capparis zeylanica See capparis horrida.

Caprifoliaceae A family of dicotyledonous, mostly woody plants in the order Dipsacales, including elderberry and honeysuckle; characterised by distinct filaments and anthers, typically five stamens and five corolla lobes, more than one ovule per locule, and well-developed endosperms.

Capsicum The fruit of a plant of the genus capsicum, especially C. frutescens, cultivated in southern India and the tropics; a strong irritant to mucous membranes and eyes.

Capsicum annuum See chilli.

Capsicum frutescens See chilli.

Capsulated Surrounded by or enclosed in a sac of living tissue.

Capsularis jute Corchorus capsularis L. Family Tiliaceae. Commercial variety of jute. Hardy shrub and highly adaptable, growing well both in lowlands and uplands, has bitter leaves, round capsules, oval, deep brown seeds and white but relatively week fibres, common varieties are *Fanduk, Dhaleswri, Deshihar, Kakya, Bombai.* Also known as white jute, narcha.

Capsule A simple, dehiscent, dry fruit composed of two or more carpels; the spore case of a moss or liverwort; a slimy outer wall layer of bacteria. A sac like membrane enclosing organ.

Captive market A market in which the supplies of a good or service is in a monopolistic position, the consumer being unable to obtain suitable substitutes or to do without.

Carabine In France and Spain, a sudden and violent wind.

Carallia brachiata See kierpa.

Caralluma fimbriata See makedshingi.

Carambola Averrhoa carambola L. Family Oxalidaceae. Native to Southeast Asia. A small tropical tree grown for its light golden yellow fruit with thin smooth, translucent skin, ovoid in shape, pulp watery acid or sweet. Sweet variety eaten fresh, sour one for drinks. Propagation by seed; layering and inarching possible. Also called *Karmal* in Hindi.

Caraway Carum carvi L. Family Umbelliferae. A white-flowered perennial herb, native of Europe and West Asia. Fruits used as condiments flavouring agent, and for medical purposes. Also known as *shiajira*.

Carbeen Eucalyptus tesselaris. Family Myrtaceae. Eucalyptus species from Australia. Beautiful shaft-like tree of moderate to fairly large size with rough bark. It yields oil.

Carbohydrates Starches, sugar, cellulose, fibres, and woody materials. They are composed of carbon, hydrogen, and oxygen in the ratio of 2 molecules of hydrogen to 1 molecule of oxygen.

Carbohydrate-nitrogen ratio The relative proportion of accumulated carbohydrate and nitrogen in stems and leaves of plants. The ratio changes with age, vigour and fruit setting etc. of the plant denoted as C : N ratio.

Carbolic acid See phenol.

Carbon An essential chemical element and component of plants and animals. Present in soils, humus, plant residue etc.

Carbon 14 A radioactive isotope of carbon with atomic weight 4, produced by collisions between neutrons and atmospheric nitrogen.

Carbonaceous Pertaining to carbon, carbonous.

Carbonaceous feeds Feeds containing more starch such as cereals.

Carbonate A salt of carbonic acid.

Carbonation Chemical process during weathering that converts basic oxides into carbonates.

Carbon bisulphide (CS_2). A clear, highly inflammable liquid, used as fumigant for stored grains. Destructive to mice and rodents. Also used for fumigating insect galleries or holes in trees. Highly absorptive and penetrative.

Carbon cycle A cycle in which carbon from CO_2 in the atmosphere is fixed into organisms during photosynthesis and chemosynthesis, liberated by respiration, death, and decomposition, and ultimately returned to its original state or deposited in sediments such as coal or carbonate.

Carbon dioxide (CO_2). Carbonic acid gas. A heavy, colourless gas present in the atmosphere, assimilated by plants and produced by dry decomposition of organic substance.

Carbon disulphide See carbon bisulphide.

Carboniferous period A period of the paleozoic era characterised by the formation of great coal buds.

Carbonize To reduce to carbon. Also to coal with carbon.

Carbon tetrachloride (CCl_4). A chemical used as fumigant and toxic if inhaled.

Carbon-nitrogen ratio (1) The ratio of the weight of organic carbon to the weight of total nitrogen in a soil or in an organic material. It is obtained by dividing the percentage of organic carbon (C) by the percentage of total nitrogen (N). (2) A ratio of carbon to nitrogen, an effective measure of the percentage of proteins in food.

Carbon tool steel Steel containing little or no alloying metal. Used in tools for cutting operations at low speeds, where a hard cutting edge and high quality finish is desired.

Carbuncle Acute, suppurative inflammation of the sub-cutaneous tissues and the skin.

Carburetion It is the process of mixing air with liquid fuel to form a combustible mixture. (see also carburettor).

Carburettor A device for converting liquid fuel into vapour and mixing it with air in such proportions as to form the most efficient combustibel mixture.

Carburettor bowl Reservoir part of the carburettor which holds liquid gasoline or fuel.

Carburettor float Usually an airtight

metal container which floats on the surface of the fuel in the bowl of the carburettor and controls the flow of gasoline from the main fuel line.

Card A wire-toothed instrument used for cleaning and straightening wood, flax and cotton fibres.

Cardamom Elettaria cardamomum (L). Mato var. Minor Watt. Family Zingiberaceae. Crop cultivated for its capsule and is native to moist evergreen forests of India. Important spice used for flavouring or for medicinal or masticatory purposes. Commercially grown. Thrives best at altitudes of 750-1500 m having a well distributed rainfall of 150 to 625 cm and a temperature range of 50° to 90°F. Seed contains 2-8 per cent of a strongly aromatic volatile oil with pleasant taste.

Cardamom clump root Disease caused by Rhizoctonia bataticola (Taub) Butl. and Pythium spp. Usually leaves become pale and yellow due to rotting of rhizomes and roots. Diseased clumps conspicuous by fallen leafy shoots.

Cardamom foorki disease Virus disease of cardamom, characterised by complete phyllody of flowering spike, plants producing dwarf-shoots bearing no inflorescence.

Cardamom katte disease See cardamom marble disease.

Cardamom marble disease Caused by virus and transmitted by black banana aphid, Pentalonia nigronervosa. Characterised by general yellowing of entire leaf with slender, broken stripes of deep green tissue over surface, following veins and running parallel to mid-rib, subsequently following mosaic pattern. Deterioration and dying off of clumps. Also called cardamom mosaic, cardamom katte disease.

Cardamom mosaic See cardamom marble disease.

Cardamom thrips Toeniothrips cardamomi R. Pest of cardamom; attack fruits, which shrink and show corky incrustation of skin.

Cardinal winds Winds from the four cardinal points of the compass, that is north, east, south and west winds.

Carding (1) Drawing fine teeth through cotton to make all the fibres lie in one direction and so form a silver. (2) Also first step in preparation of scoured wool.

Carica candamarcensis See mountain papaya.

— **Papaya** See papaya.

Carinal canal In the stem of the Equisetales, canals which develop beneath the stem ridges in the stele.

Carnots principle The amount of work done by a heat engine is dependent on its temperature alone, being independent of the nature of the intermediary agent that is employed.

Carissa arduina See natal plum.

— **bispinosa** See natal plum.

— **carandas** See karanda.

— **edulis** Vahl. species of karanda; straggling shrub, producing oval berries, red, turning black on maturity. Also see karanda.

— **grandiflora** See natal plum.

— **ovata** R. species of karanda; native of Australia; open shrub; produces small fruits used in jam making.

— **spinarum** L. Hardiest species of karanda. Thriving even in poor and rocky soils; bears small edible fruits. Useful tree for afforestation of rocky ravines and alkaline soils.

Carob Ceratonia siliqua L. Family Caesalpiniaceae. Tree native of the East Mediterranean region. Pods used as fodder; seeds source of high protein flour; endosperm contains gum for rubber industries, paper-making and in tanning.

Carotenase An enzyme that effects by hydrolysis of carotenoid compounds, used in bleaching of flour.

Carotene Yellow orange or red pigments, compound of carbon and hydrogen that occurs in plants; a precursor of vitamin A. Alpha, beta and gamma carotenes may be converted into vitamin A in the animal body. Also carotin.

Carpel (1) The basic specialised leaf of the female reproductive structure in angiosperms; a megasporophyll. (2) A floral leaf bearing ovules. Pistil considered simple when there is only one carpel and when it is more than one, it is compound.

Carpet grass Axonopus compressis (Swrtz). P. Beauv. Family Gramineae. Grass introduced from southern U.S.A., Mexico and Brazil. Moist soils of Assam are good for growth. Good pasture and lawn grass. Also called tropical carpet grass.

Carpogonium The basal, egg-bearing portion of the female reproductive organ in some thallophytes, especially red algae.

Carpology The study of the morphology or structural anatomy of fruits and seeds.

Carpomyia vesuviane See ber fruit fly.

Carpophore The portion of a flower receptacle that extends between and attaches to the carpels. The stalk of a fruiting body in fungi.

Carpophyte A thallophyte that forms a sporocarp following fertilization.

Carpoptosis Abnormal fruit drop.

Carposporangium In red algae, a sporangium that forms the cystocarp and contains carpospores.

Carpospore In red algae, one of the non-motile spores produced on short filaments arising from the carpogonium following fertilization.

Carraggen chandrus orispus A cartilaginous red algae harvested in the Northern Atlantic as a source of carraggenin.

Carriage bolt An oval or buttonhead black bold with square neck which prevents the bold from turning while the nut is being tightened.

Carrier (1) Infected plant showing no marked symptoms but source of infection for other plants. (2) Any substance added to a chemical compound to facilitate its storage, transport or use in the field by increasing its bulk.

Carrier drain Open or subsurface drain which receives the water from the relief drain and may in addition be also functioning as a relief drain for the area through which it passes.

Carrot Daucas carota, L. var. sativa DC. Family Umbelliferae. Root crop grown all over India. Good source of vitamin A and iron. Contains vitamin B and C; rich in sugar. Tolerant to high temperatures but thrives best in cool weather called *gajjar* in Hindi.

Carrot pronginess Misshapen and unsightly root produced by carrot when its tap root meets a hard, dry or very wet layer of soil, excess of manure and fertiliser or some obstruction in its early growth.

Carrying capacity The limit to the amount of life that can be supported by any given habital; specifically, the number of individuals of any particular species that can be supported by a habitat. In another sense, the reasonable limits of human occupancy or use of a resource.

Carthamin One of the two main colouring matters found (0.03 to 0.06 per cent) in safflower; scarlet red, insoluble in water. Imparts bright red colour to cotton and silk. Also see safflower yellow.

Carthamus Oxycanthus See wild safflower.

— **tinctorius** See safflower.

Cartography A science of drawing maps.

Cartridge fuse A fuse enclosed in an insulating tube in order to confine the arc when the fuse blows.

Carum bulbocastanum See black caraway.

— **carvi** See caraway.

— **copticum** See omum.

— **petroselinum** See parsley.

— **roxburghianum** See ajmud.

Caruncle Protuberance on or about ilium of a seed.

Carya illinoensis Family Juglandaceae. A tree whose nuts are eaten. Also called pecan nut

— **olivaeformis** See carya illinoensis.

— **pecan** See carya illinoensis.

Caryophyllaceae A family of dicotyledonous plants in the order caryophyllales differing from the other families in lacking betalains.

Caryophyllidae A relatively small subclass of plants in the class Magnoliopsida characterised by trinucleata pollen, ovules with two integuments, and a multilayered nucellus.

Caryopsis A small, dry, indehiscent fruit having a single seed with such a thin closely adherent pericarp that a single body, a grin, is formed such as in most grasses or cereals.

Caryota urens Family Palmaceae. Fibre obtained from leaf-sheaths.

Juice used for making sugar. Also called wine palm.

Casaba melar Cucumis melo. A winter musk melon with a yellow rind and sweet flesh belonging to the family cucurbitaceae of the order violales.

Casein A constituent of milk and represents 75-80 per cent of total nitrogen content of milk.

Caseinogen Phosphoproteins. Combination of proteins with a phosphorus caontaining substance other than nucleic acid.

Cash Money in the form of coin and bank-notes.

Cash crop A crop sold directly to the market as contrasted to the one that is used for food, feed, fodder etc.

Cashew Anacardium occidentale L. Family Anacardiaceae. A commercially important crop of India grown for its fruit and nuts. Most important products are nuts and shell oil, juice used for making pies, preserves and wine. Propagated by seed; budding, grafting, air layering possible. Mostly grown in coastal districts. Known as *kaju* in Hindi.

Cashew apple Swollen stalk of cashewnut flower; soft juicy and edible; rich in vitamin C and contains carotene; juice made in syrup; fermented and made into liquor.

Cashew borer Plocaederus ferrugineus L. Insect pest of cashew; dark brown bug and its white grub bore into dark and feeds on soft tissues.

Cashew gum Gum exuded from the trunk of the cashewnut tree. On exposure to air solidifies into a light yellow to reddish brown mass. It dissolves easily in warm water and used in book binding and tanning.

Cashewnut leaf-eating caterpillar Cricula trifenestrata H. Insect pest of cashewnut. It damages leaves of young plants by sucking sap turning them pale, brown and crinkled.

Cashewnut leaf-miner Acrocercops syngramma M. Insect pest of cashew. The caterpillar mines into newly formed tender leaves, developing whitish patches.

Cashewnut leaf thrip Selenothrips rubrocinctus G. Insect pest of cashewnut. It damages leaves of young plants by sucking sap turning them pale, brown and crinkled.

Cash flow The passage of money into and out of a business enterprise.

Cash generation The net cash flow arising from an investment by a business enterprise, assuming this to be positive.

Cash programme A programme aiming at the shortest possible completion time for a project or projects.

Casimiroa edulis See white sapote.

Casparian strip A thin band of suberin or lignin like deposition in the radial and transverse walls of certain plant cells during the primary development phase of the endodermis.

Cassava Manihot esculenta. A shrubby perennial plant grown for its starchy, edible tuberous roots. Also known as manihot; manioe.

Cassia angustifolia See Indian senna.

— **fistula** See Indian laburnum.

— **siamea** See kassod tree.

— **tora** Family Caesalpiniaceae. A herb whose tender young leaves are used as vegetable.

Cassumunar ginger Zingiber cassumunar Roxb. A perennial herb whose rhizomes are used as condiment. Also known as *jungali adrak* in Hindi.

Castanea sativa See European chestnut.

— **vesca** See European chestnut.

— **vulgaris** See European chestnut.

Castanopsis indica See Indian chestnut.

Castellanus A cloud species with at least a fraction of its upper part presenting same vertically developed cumuliform protuberances (same of which are more tall than wide) which give the cloud a crenellated or turreted appearances. Previously known as castellatus.

Casting (1) The process of making pottery in molds from fluid clay. Metal part made pouring the metal, when liquid, into a mold. Pouring liquid, catalysed resin. Usually with fillers and pigment, into an open mould, where it hardens. (2) Excrement of earthworms.

Cast iron Iron of ordinary use, cast in molds. Has high carbon content; cannot be rolled, formed, or tempered. Cast iron nickel alloys have many of the properties of steel castings.

Castor Ricinus communis L. Family Euphorbiaceae. A small tree used for various purposes. Stalk used as fuel, in paper industry or thatching of roofs, leaves fed to eri silkworm, seed used for extracting oil which is used for lubricating, lighting and industrial purposes. Takes six months to mature. Called *arand* in Hindi.

Castor bean The seed of the castor oil plant (Ricinus communis), a coarse, erect annual herb in the family Euphorbiaceae of the Geraniales.

Castor butterfly Ergolis merione C. Insect pest of castor. A green spiny caterpiller of a snuff coloured butterfly that feeds on leaves.

Castor hairy caterpillar Euproctis lunata Wlk. Insect pest of castor. Dark reddish brown hairy caterpillar of a snuff coloured butterfly feeding on leaves.

Castor leaf spot Fungal disease caused by Cercospora ricinella. Characterised by minute black or brown spots with pale green margins on leaves, spots, changing to pale brown and greyish white.

Castor oil Vegetable oil extracted from the castor bean. Specific gravity about 0.960 to 0.966 at 15.56°C; used as purgative, lubricant and for dressing of leather.

Castor oilcake Residue of castor bean after oil has been extracted. Inedible and hence used as manure in conjunction with quick acting chemical fertilisers. Contains 4, 1.9 and 1.4 per cent nitrogen, P_2O_5 and K_2O and contains 2-15 per cent oil.

Castor pod borer Dichocrocis punctiferalis Guen. Caterpillar bores into shoots and pods, deposits balck mass of excreta on pods; webs pods together.

Castor powdery mildew Disease of castor caused by Leveillula taurica. Symptoms as in cotton powdery mildew.

Castor rust Fungal disease caused by Melampsora ricini (Bivi) Pass. Symptoms are: appearance of tiny, pink coloured pustules of rust on underside of leaves, with yellow spots on upper side.

Castor seedling blight Fungal disease caused by Phytophthora parasitica. Characterised by appearance of dull green round patches on leaves, spots turning yellow or brown with concentric zones on lower leaf surface and also on stems.

Castor semi-looper Achoea janata L. Serious insect pest of castor. Caterpillar feeds on leaves and makes holes; badly damaged leaves dry up and fall off sometimes attacks stem; when serious, defoliate plant completely.

Castrate To remove the stamens from a flower. In animals it is to remove the testicles or to destroy their use.

Casuarina Casurina equisetifolia L. (C. muricata Roxb). Family Casuarinaceae. Large, quick growing conifer like evergreen tree. Propagation through seeds. A good plant for fuelwood. Makes good charcoal can be planted under adverse conditions and used as shelter belt and windbreakes.

Casuarinaceae The single, monogeneric family of the plant order Casuarinales characterised by reduced flowers and green twigs bearing wholrs of scale like, reduced leaves.

Casuarina equisetifolia See casuarina.

— **muricata** See casuarina.

— **suberosa** See swamp oak.

Casuarinales A monofamilial order of dicotyledonous plants in the subclass Hamamelidae.

Catabolism Breaking down of complex bodies into products of simpler composition Also katabolism.

Catalysis Acceleration or retardation of reaction due to presence of catalyst.

Catalyst Substances that speeds up a chemical reaction but is not decreased in amount by the reaction.

Catch basin Basin designed to catch sediment.

Catch crop A crop planted on land where other crops have failed, usually a quick growing one. Also a quick growing crop planted and harvested between two regular crops in consecutive seasons or between patches of regular crop in the same season.

Catchment area The total area draining into given waterway, lake or reservoir.

Catchment basin A watershed or drainage basin.

Catchup A thick sauce made from tomatoes, mixed with spices, vinegar and sugar. Also called ketchup, catsup, katsup.

Catechu (1) See black cutch. (2) A stringent substance containing tannin from the bark, wood or fruit of certain plant.

Catena A non-taxonomic grouping of a sequence of soils of about the same age, derived from similar parent materials and occurring under similar climatic conditions, but having unlike characteristics because of variations in relief and in drainage.

Catenulate In chains, or in an end-to-end series.

Caterpillar (1) Worm like larva of a butterfly or certain other insects. (2) Vehicles fitted with supplementary wheels or a device on the endless chain principle and crawls like a caterpillar.

Catface (1) A scar on the surface of a log, generally elliptical in shape, resulting from wounds which have not healed over; also a fire scar at the base of a tree. (2) Physiological disorder in tomato characterised by distortion of the blossom end of fruits due to which these are unsalable.

Cation An ion that carries a positive charge of electricity.

Cation exchange The interchange of a cation in solution and another cation on the surface of any surface-active material such as clay colloid or organic colloid.

Cation exchange capacity The total amount of exchangeable cations that a soil can absorb. It is sometimes called 'total exchange capacity', 'base exchange capacity', or 'cation absorption capacity'. It is expressed in milliequivalents per 100 g of soil or of other absorbing materials such as clay.

Catkin A pendulons uni-sexual spike of flowers.

Cat's paw A puff of wind; a light breeze affecting a small area as one that causes patches of ripples on the surface of water.

Catsup See catchup.

Cat tail millet Setaria glauca (L). Beauv. (Panicum glaucum). A herb of family of Gramineae grown as fodder and whose grains are also eaten. Also known as *Bandra*.

Cattle Domesticated bovine animals, including cows, steers, and bulls, raised and bred on a ranch or farm.

Cattle urine Liquid excretion of cattles which is an excellent manure in which nitrogen is present in the form of urea.

Cattley-guava Psidium cattleyanum Sabine; P. littorale Raddi. Family Myrtaceae. Fruit tree, fruit resembling guava. Suitable for growing at higher elevation in the Nilgiris. Also called strawberry or Chinese guava.

Caudex (1) The main axis of a plant, including stem and roots. (2) The upright stem of a fern, palm etc.

Caudicle A slender appendage attaching pollen masses to the stigma in orchids.

Caulerpaceae A family of green algae in the order siphonales.

Caulescent Having an above ground stem, or producing a leafy stem above ground.

Caulicle The initial stem in a young seedling of an embryo.

Cauliflorous Producing flowers on the older branches or main stem.

Cauliflory Production of flowers on the trunk.

Cauliflower Brassica oleracea L. var. botrytis. Family Cruciferae. Probably a native of Italy or of Southern Europe on the Mediterranean; most important Brassica crop grown in India for its flower parts used as vegetable. It contains vitamin B and proteins. Needs a cool and moist growing season. Medium and heavy soils suitable for early production. Harvested when head is of proper size and right maturity. Also called *phul gobhi* in Hindi.

Cauliflower buttons Under-sized heads of cauliflower formed due to premature heading, caused by planting in poor soil, keeping seedlings too much crowded in nursery, insufficient supply of nitrogen, insufficient moisture, check in growth in early stages, planting poor seedlings in dry and hot weather.

Cauliflower disease (1) A disease of

the strawberry plant caused by the eelworm and manifested as clustered, puckered and malformed leaves. (2) A bacterial disease of strawberry and some other plants caused by corynebacterium fascians.

Cauliflower fuzzy head An off-quality head or curds of cauliflower showing a velvety appearance; due to high temperatures at maturity.

Cauliflower leafy head An off-quality heads or curds of cauliflower where small, green leaves are seen growing between the segments of the head; caused by high temperatures during maturity period.

Cauliflower ricey head An off-quality cauliflower head or curds, showing granular texture, lacking compactness. Caused by elongation of peduncles which bear individual flowers. Formed due to high temperature at maturity.

Cauline Belonging to or arising from the stem particularly if on the upper portion.

Caulis The stem or stock of a herbaceous plant.

Caulame The stem structure or stem axis of a plant as a whole.

Causal organism An organism such as fungus, bacterium, nematode etc. which produces a particular disease.

Causative agent Any substance, pathogens etc. which produces a disease.

Caustic An agent capable of corroding or eating away tissues; corrosive.

Caustic lime Calcium oxide, burnt lime.

Caustic potash Potassium hydroxide.

Caustic soda Sodium hydroxide.

Cautery Destructive effect produced by cauterizing agent used for searing or burning the skin or other tissues by means of caustic chemicals or heat.

Caver A gentle breeze in the Hebrides west of Scotland. Also spelled kaver.

Cavern flow Movement of water through large openings in rocks such as caves, or through coarse sorted granular material.

Cavity well A well pumping water in which it does not penetrate. Therefore called non-penetrating well. The name cavity well has been derived from the cavity which is formed in the aquifer from which water is pumped. The impermeable or semi-permeable layer above the aquifer serves as roof of the cavity.

CC A cubic centimetre. 1000 cc constitute one litre.

Cecidiology A division of biology which studies galls on plants caused by insects, mites and fungi.

Cecidium Plant gall produced either by insects or by fungi as a consequence of infection.

Cedar The common name for a large number of evergreen trees in the order Pinales having fragrant, durable wood.

Cedrela tree Toona ciliata Roem. (Cedrela toona Roxb. ex Rottl. and Willd.) Family Meliaceae. Ornamental and economic tree. Wood red, soft, shining and suitable for furniture etc. Frequently grows on moist ravines, on banks of streams and in damp situation. Flowers yield yellowish-red dye used for dyeing cotton. Bark used as tan. Also called red cedar.

Ceiba pentandra See kapok.

Celastraceae A family of dicotyledonous plants in the order Celastrales characterised by erect and basal ovules, a flower disk that surrounds the ovary at the base, and opposite or sometimes alternate leaves.

Celeriac Apium graveolens L. var. rapaceum DC. (A. rapaceum Mill.). Family Umbelliferae. A biennial herb grown for its thickened, celery flavoured roots eaten as vegetable; fruits used for extracting a much valued oil and seeds as spice. Also called Salari.

Celery Apium graveolens L. var. dulce. Family Umbelliferae. Biennial herb forming fleshy root and clump of compound leaves with long leaf stalks; used as vegetable; blanched during cultivation. Native of Europe.

Cell The living cell incudes protoplasm, nucleus, and cytoplasm within its walls. It is a unit in structure, heredity, physiological processes and development. (3) A hexagonal unit compartment of a honey comb.

Cell conjugation Union or fusion of cells.

Cell count Count of the number of red

and white blood cells of blood. Corpuscles in a given quantity of blood.

Cell sap The more fluid material of a cell.

Cellular Referring to a cell or cells.

Cellular plant A non-flower-producing plant without vascular tissue.

Cellulase Enzymes that attack cellulose.

Cellulose Major skeletal material in the cell wall of plants; chemically, A inert carbohydrate; an anhydride of beta-D glucose units. A cellulose molecule may contain between 1600 and 2700 beta-D glucose units.

Cellulose microfibrils Elements in the fine fibrillar composition of cellulose in the walls of plant cells, revealed by the electron microscope.

Cell wall The membranous covering of a cell secreted by the cytoplasm in growing plants; consist largely of cellulose, but may contain chitin in some fungi and silica in some algae.

Celosia argentina See cockscomb.

Celtis australis See hackberry.

— **orientalis** See charcoal tree.

Cementation Weakly cemented, strongly cemented and indurated. In engineering practice, consistency has essentially the same meaning as consistence.

Cemented indurated Having a hard, brittle consistence because the particles are held together by cementing substances such as humus, calcium carbonate, or the oxides of silicon, iron, and aluminium. The hardness and brittleness persist even when the soil is wet.

Cenchrus biflorus Roxb. Family Gramineae. Grass, whose young shoots are fed to cattle, seeds eaten at times of scarcity.

— **ciliaris** See bunch grass.

— **prieurii (Kunth)** Maire. Family Gramineae. An excellent fodder grass found in semi-desert regions.

— **setigerus** See *anjan* grass.

Censoring A sample is said to be censored when certain values are unknown (or deliberately ignored) although their existence is known.

Census of population An enumeration of inhabitants.

Central leader System of training a fruit tree in which it is trained to form a trunk which extends from the surface of the soil to the top of the tree thus growing indefinitely, more rapidly and vigorously than the side branches resulting in a robust, close centred and tall tree which bears the fruits more vigorously near the top.

Central placentation The placenta is borne on the axis of an ovary made up of but one carpel.

Central seed committee The central seed committee was constituted under sub-section (1) of Section 3 of the Indian Seeds Act, 1966. Its function is to advise the Central Government and the State Government on matters arising out of the administration of the Seeds Act, 1966, and to carry out the other functions assigned to it by or under the Seeds Act, 1966.

Central tendency The tendency of quantitative data to cluster round some variate value. The position of some central value is usually determined by one of the measures of location, such as mean, median or mode. The closeness into which values cluster round the central value is measured by one of the measures of dispersion, such as the mean deviation or the standard deviation.

Centre of action A semipermanent high or low atmospheric pressure system at the surface of the earth; fluctuations in the intensity, position, orientation, shape or size of such a centre are associated with widespread weather changes.

Centre drill A short drill used for centring work in order that it may be carried on the lathe centres. Centre drills are usually made in combination with a countersink, which permits a double operation with one tool.

Centre jump The formation of a second low pressure centre within an already well-developed low-pressure centre; the latter diminishes in magnitude at the centre.

Centre of resistance It is the point at which resultant of all the horizontal and vertical forces act and this lies at a distance equal to 3/4 the size of the plough from the share wing.

Centre riser outlet Riser outlet placed at or near the balance point or centre of a sprinkler action.

Centrifugal divider (Gamet type) This equipment is also commonly used for dividing a sample in the seed testing laboratory. It makes use of centrifugal force to mix and scatter the seeds over the dividing surface.

Centrifugal force That force by which a body moving in a curve tends to fly off from the axis of rotation. Serves to cause solids or liquids to separate from liquids of lower specific gravity.

Centrifugal separation The separation of substances of different densities by means of centrifugal force, as the separation of fat from milk.

Centrifugal switch A switch used on the single phase, split phase motor to open the starting winding after the motor has almost reached synchronous speed.

Centrifuge Instrument in which a container full of liquid is spun around at high speed so that the heavier part of the liquid or solids may be carried to the bottom by centrifugal force.

Centrioles Two or more granules contained in the centrosome and prominent during cell division.

Centripetal Developing or maturing from the outside towards the centre as maturation of metaxylem.

Centrosema See butterfly pea.

Centrosema pubescens See butterfly pea.

Centrosome (cell centre) Condensed portion of the cytoplasm of certain cells containing the centrioles; it plays an important part in cell division.

Century plant See agave.

Cepaceous Smelling or tasting like onion or garlic.

Cephaelis ipecacuanha See ipecac.

Cephalandra indica See little gourd.

Cephaleuros parasiticus See tea red rust.

Cephalewios Parasitic plant belonging to the group Myxophyceae (a low order of plants); parasitic particularly on tea leave.

Cephalosporium sacchari See sugarcane wilt.

Cephalostachyum pergracile Munro. Family Gramineae. A tufted bamboo frequently found in association with teak. Culms used for mat-making and buildings.

Ceratochloa cathartica See rescue-grass.

Ceratonia siliqua See carob.

Ceratophyllaceae A family of rootless, free floating dicotyledons in the order Nymphaceales characterised by unisexual flowers and whorled, cleft leaves with slender segments.

Ceratostomella fimbriata See sweet potato black rot.

— **paradoxa** See pineapple stem-end rot; pineapple storage rot; arecanut stem bleeding; coconut stem bleeding.

Cercospora An imperfect fungus genus, family Dematiaceae, with a large number of species, their distinguishing feature being largely based upon the species of the plant infected and the type of lesions produced.

Cercospora arachidicola See groundnut *tikka* disease.

— **batatae** See sweet potato leaf spot.

— **bataticola** See sweet potato leaf spot.

— **canaraliae** See sword bean leaf spot; cowpea leaf spot.

— **capsia** See chilli leaf spot.

— **carthami** See safflower leaf spot.

— **coffeicola** See coffee brown eye-spot.

— **cruenta** See bean leaf spot; cowpea leaf spot; lima bean leaf spot.

— **devisii** See sweet clover leaf spot.

— **dolichi** See bean leaf spot; lablab leaf spot; horsegram leaf spot.

— **hibisci** See lady finger leaf spot.

— **kopkei** See sugarcane diseases.

— **melongenae** See brinjal leaf spot.

— **musae** See banana leaf spot.

— **nicotinae** See tobacco frog-eye leaf spot.

— **oryzae** See rice narrow brown leaf spot.

— **personata** See groundnut tikka disease.

— **ricinella** See castor leaf spot.

— **sesami** See sesamum blight.

— **sesbania** See sesbania leaf spot.

— **sojina** See soyabean leaf spot.

— **sorghi** See maize leaf spot.

— **stizolobii** See velvet bean leaf spot.

— **traversiana** See fenugreek leaf spot.

Cercospora leaf spot Any of several fungus diseases of plants caused by Cercospora species and characterised by areas of discolouration on the leaves.

Cercosporalla theae See tea bird's eye spot.

Cereals (1) Derived from the latin word 'ceres', meaning goddess of tillage and corn; common foodgrains of the world; belong to the natural order Gramineae or family of grasses. Rich in starch but also provides small and variable proteins, fat mineral and vitamins. (2) A prepared foodstuff from grain, like oatmeal.

Cereal forage Cereal crop harvested when immature for either hay, silage, or green chop or as pasturage.

Cerebroid With brainlike convolutions or folds.

Ceriman See monstera.

Ceropegia hirsuta See khantodi.

— **lawii** See kharpudi.

— **oculata** See khantodi.

Cerotelium desmium See cotton rust.

— **fici** See fig rust.

Certification (seed) A system of maintaining the quality of seeds. The crops offered for certification are raised as per requirements for seed certification established by a seed certification agency. Several inspections are made to ensure purity and quality of seeds.

Certification agency The certification agency established under section 8 or recognised under section 18 of the Seed Act, 1966. Its major function is to certify seeds of any notified kind or varieties.

Certification sample A sample of seed drawn by a certification agency or by a duly authorised representative of a certification agency.

Certification tag Means a tag or label (of a specified design) of the certification agency. It constitutes a certificate granted by the certification agency.

Certified seed Seed for use in commercial crop production, produced from foundation registered, or certified seed. Usually grown, processed, and labelled under supervision and regulation of a public agency.

Certified seed producer Means a person who grows or distributes certified seeds in accordance with the procedure and standards of the certification agency.

Cespitis A turf horizon, or surface soil layer held together by a tangle of roots, mostly grass.

Cespitose Turf-like; having dense growth as in a turf.

Cess A levy or tax.

Cesspit A pit in which refuse, night soil etc. is dumped.

Cesspool A receptacle in which household sewage or other liquid waste is discharged.

Cevitamic acid Vitamin C.

Ceylone olive Elaeocarpus serratus L. Family Elaeocarpaceae. A tree whose fruits are edible and found in Mysore and Kerala states of India.

Ceylone raspberry Rubus lasiocarpus Sm. Family Rosaceae. Herb whose fruits are edible.

Ceylone spinach Talinum triangulare Willd. A herb of Family Portulaçaceae introduced from Ceylon and whose leaves and stems are used as vegetable.

Cumecs per Cu. m. Use for specifying air flow in cu.m. per second for each cu.m. of crop, obtained by dividing the cumecs of air by total bin volume in cu.m.

Cumecs per sq.m. Used for specifying air flow in cu.m. per second on the basis of floor area of one sq.m obtained by dividing the cumecs of air by the area of the floor in square metre.

Chablis A gap in the forest produced by the fall of a tree; the fall of the tree itself and the resulting forest damage (from the French).

Chaetochloa italica See Italian millet.

Chaetophoraceae A family of algae in the order Ulotrichales characterised as branched filaments which taper toward the apices, sometimes bearing terminal setae.

Chafer Kind of beetle usually the cock chafer. Many pests of cultivated plants.

Chaff The floral parts of cereals, generally separated from the grain in the threshing or winnowing; finely cut hay or green stalks used as feed for cattle.

Chaff cutter An implement consisting of a stand, a body and main wheel. Body consisting of rollers, side plates to hold the rollers, shaft, gears, front plate, spring and main wheel with knives. Also called circular fodder cutter.

Chain A length measure in land surveying; also a chain 66 ft. long with 100 rings of 0.66 ft. each. 10 square chains is one acre.

Chain harrow A harrow having no rigid body, but consisting of interlocked, round, square or different shaped links which make up a strong, flexible mesh to adjust to the inequalities of the ground surface. Used for rolling up field and collecting weeds uprooted by the other implements.

Chain pumps Persian wheel.

Chain tongs Used for holding pipe from turning or to turn pipe. It is a heavy bar with sharp teeth at one end. Which are held against the pipe by a chain wrapped around the pipe and attached to the bar.

Chain-pit method Method of removal of stumps by burning. It consists of clearing the bark around the stump, starting the fire, and covering with soil so that combustion takes place under conditions of deficient draft with burning or charring through the stump.

Chalazae The portion of an ovule just below the place where the integuments are united to the nucellus.

Chalazogamy A process of fertilisation in which the pollen tube passes through the chalaza to react the embryo sac.

Chalcid fly Any small insect of the superfamily Chalcidoidea which in its early stage exists in or on other larvae or pupae, thus destroying many undesirable pests.

Chalk Soft, white, fine-grained limestone which is highly fossiliferous. As a parent rock, gives distinctive character to soils. Impure chalk is sometimes termed as marl.

Challiho Strong southerly winds which blow for about 40 days in spring in some parts in India; sometimes the winds are violent causing blinding dust storms.

Chamaephyte Any perennial plant whose winter buds are within 25 centimetres of the soil surface.

Chamaerops ritchiena See mazari palm.

Chamaesiphonales An order of blue-green algae of the class cyanophyceae; reproduced by cell division, colony fragmentation and endospores.

Champac Michelia champaca L. Family Magnoliaceae. A tall tree common in evergreen forests and also planted for its fragrant flowers. Oil important in certain diseases and also used in perfumery.

Chain type tractor See crawler tractor.

Chance causes In the quality of a product, process or service, where there is a stable pattern of variations that can be considered natural and inherent in the process, the variations are said to be due to chance or random causes.

Chance seedling A fruit or nut tree which grows from seed scattered by wind, birds etc. Sometime bears fruits.

Chancre Initial lesion of syphilis.

Chandraki Spinning tray on which silkworms ripe for building cocoons are mounted to serve as support. Consists of rectangular bamboo tray on which bamboo strips of 1.25-5 cm width are fixed in concentrated circles or as a spiral.

Change It may be defined as any alternation in the position or the condition of anything from a state previously existent.

Change agent One who advocates or attempts to introduce another method; the new system itself. Teachers or agricultural extension worker are examples.

Change chart A chart indicating the amount and direction of change of some meteorological element during a specified time interval; for example, a height-change chart or pressure change chart. Also known as tendency chart.

Channel A natural stream that conveys water; a ditch or channel excavated for the flow of water.

Channel improvement Improvement of the hydraulic flow characteristics of channels by increasing the cross-section, straightening, or cleaning in order to increase its carrying capacity.

Channel net Stream channel pattern within a drainage basin.

Channel stabilisation Erosion prevention and stabilisation of velocity distribution in a channel using jetties, drops revetments, vegetation, and other measures.

Channel storage Water temporarily stored in channels while enroute to an outlet.

Channel terrace Type of contour terrace for soil conservation. Wide and relatively shallow channel excavated at suitable intervals on a falling contour with a suitable longitudinal gradient; excavated soil is deposited as a wide, low ridge along the lower edge of the channel. Recommended for relatively impervious soil and in heavy rainfall tracts.

Channel type terrace Terrace which is constructed by moving earth for the embankment form the uphill side only.

Channery A descriptive term used for thin and flat limestone, sandstone, or schist fragments up to 15 cm in length. See also coarse fragments.

Chaparral A vegetation formation characterised by woody plants of low stature, impenetrable because of tough, rigid, interlacing branches, which have simple, waxy, evergreen, thick leaves.

Character (1) One of the many details of structure, form, substance, or function that make up an individual organism. The Mendelian characters of genetics represent the end products of development, during which the entire complex of genes interacts within itself and with the environment. (2) Also a judging term for quality, attractiveness, etc.

Charales Green algae composing the single order of the class Charophyceae.

Charcoal rot A fungus disease of potato, corn and other plants caused by Macrophomina phaseoli; tissues of the root and lower stem are destroyed and blackened.

Charcoal tree Trema orientalis Blume (Celtis orientalish) family Ulmaceae. A shrub of small tree whose bark is used for ropes, cordage, rough cloth, wood source of gunpowder charcoal and fruits edible, wood pulp used for making paper.

Chareae A tribe of green algae belonging to the family Characeae.

Charge The quantity of electricity residing on an electrostatically charged body.

Charging The process of changing electro-chemically the plates and electrolyte of a storage cell or battery.

Charophyceae A class of green algae, division chlorophyta.

Charophyta A group of aquatic plants, ranging in size from a few centimetre to several metres in height, that live entirely submerged in water.

Charring Burning partly, scorching.

Charsa See mhote.

Chasmophyte A plant that grows in rock crevices.

Chassis The unit which includes all parts of the automobile except the body.

Chaubattia clay mud See grafting clay.

Chaubattia paste Fungicide preparation obtained by mixing thoroughly red lead powder, copper carbonate (finely powdered) and lanoline (1:1:1). Has long lasting fungicidal action.

Chaulai Amaranthus caudatus L. (Amaranthus paniculatus L., Amaranthus Cruentus, L.). Family Amarathaceae. A herb grown as a vegetable. Also see amaranth. Also called grain amaranth, velvet flower, ramdana.

Chaulumugra oil Oil obtained from the Kernels of the seeds of Hydrocarpus wightiana, a tree found in moist forests of India. It is used in the treatment of leprosy and skin disease.

Chayote Sechium edule Sw. Family Cucurbitaceae. A south American trailing or climbing vine, a cururbit introduced into the hill areas of north India for its fruit vegetable.

Chebulic myrobalan See yellow myrobalan; myrobalan.

Check (1) A lengthwise separation of the wood, which usually extends across the rings of annual growth, commonly resulting from stresses set up in wood during seasoning. (2) Cheek gate; a structure placed in an irrigation ditch to form an adjustable dam to control the water level in it, to facilitate irrigation. (3) Level or nearly level area surrounded by ridges in which the water applied may be ponded until absorbed.

Check Basin Irrigation See irrigation methods.

Check Dam Small dam or any low obstructive structure to decrease the streamflow velocity, minimise channel scour, and promote deposition of sediment.

Check drain Conventional drain altered by use of checks so that it can be used as a sub-irrigation system.

Check gate See check.

Check irrigation, block irrigation, or check flooding A method of flooding in which the field is divided into a number of checks. If the check are 'contour checks', the method is called 'contour-check method'.

Check plot A small area treated or untreated, which is used as a standard for comparison in an agricultural experiment.

Check valve A valve that checks or controls the flow through it, either automatically or it can be adjusted by hand.

Chemical A substance obtained by or used in a chemical process.

Chemical analysis Breaking-up of a substance into the simple substances or elements of which it is made.

Chemical eluviation Eluviation in which decomposition occurs and certain products liberated are removed in true colloidal solution to be deposited in other horizons.

Chemical fertiliser See fertiliser.

Chemical gardening See hydroponics.

Chemical pathology The study of disease by using chemical methods.

Chemical weed control The control of weeds in crops by the application of herbicides as pre-emergence or post-emergence sprays.

Chemistry, soil The division of soil science dealing with the chemical constitution, properties and reaction of soils.

Chemi-culture See hydroponics.

Chemoautotrophic Organisms capable of utilizing the energy released in chemical reactions involving inorganic compounds in synthesizing their food and protoplasm.

Chemolysis Decomposition of organic matter brought about by chemical agents.

Chemosphere The vaguely defined region of the upper atmosphere in which photochemical reactions take place; generally considered to include the stratosphere (or the top thereof) and the mesosphere, and sometimes the lower part of the thermosphere.

Chemotactically A plant's response to chemical stimuli.

Chemotherapy Treatment of disease by the administration of drugs that destroy the causative organisms of the disease but do not injure the patient.

Chemurgy Chemical research designed to promote the use of agricultural products for industrial purposes, or to obtain new, valuable products from plants.

Chenopodiaceae A family of dicotyledonous plants in the order caryophyllales having reduced, mostly greenish flowers.

Chenopodium album See pigweed.

— **fruticosum** See alkali seepweed.

— **murale** See kharatua.

Cheque A bill of exchange drawn on a banker to pay without delay a stated sum of money to the person to whom the cheque is payable.

Cherells Disease of cacao fruits met with in South India; characterised by drying up of pods on trees.

Cherimola See cherimoyer.

Cherimoya See cherimoyer.

Cherimoyer Annona cherimola Mill. Family Annonaceae. a medium sized tropical tree grown for its fruits.

Chernozemic An order of soils that have developed under xerophytic or mesophytic grasses and forests, or under grassland-forest trasnsition vegetation, in cool to cold sub-arid to sub-humid climates. The soils have a dark coloured surface (Ah, Ahe, Or, Ap) horizon and a B or C horizon, or both of high base saturation. The order consists of the brown, dark brown, black and dark gray great groups.

Cheroot Roll of tobacco, very much like a cigar, except that it is usually cut at

both ends.
Cherri See cherry.
Cherrophyte A plant that grows in dry waste lands.
Cherry Temperate fruit; species of the genus Prunus of family Rosaceae grown for its edible fruit or for ornament. Prunus cerasus L, cooking cherry: Prunus avium L, Sweet cherry. Also called *cherri* and known as *gilas* in Hindi.
Cherry apple See Siberian crab apple.
Cherry coffee The produce resulting from dry processing of the coffee beans. Also called native.
Cherry crab See Siberian crab apple.
Cherry leaf spot A fungus disease of the cherry caused by Coccomyces hiemalis spotting and chlorosis of the leaves occurs. With consequent retardation of tree and fruit development.
Cherry pit The stone of the cherry, in which the seed is enclosed.
Cherry sun scald Physiological disease of cherry; symptoms same as in peach sun scald.
Cheshunt compound Fungicide prepared by mixing finely ground copper sulphate and ammonium carbonate (1:5.5) stored in tightly closed bottles for 24 hours before use. Soluble in water; used as aqueous solution (1 oz - 2 gal); used for disinfection/ disinfestation of soil to control damping off disease. Metal containers are not used for storage.
Chestnut The common name for several species of large, deciduous trees of the genes castanea in the order Fagales, which bear sweet, edible nuts. Also the nuts produced by such trees.
Chestunt blight A fungus disease of the chestnut caused by Endothia parasitica, which attacks the bark and combium, causing cankers that girdle the stem and kill the plant. Also known as Chestnut canker.
Chian See nicker bean.
Chick pea See Bengal gram.
Chickweed Stellaria aquatica Scop. Family Caryophyllaceae. Small aquatic winter weed with white flowers; like mist and shady situations. Common in temperate regions. Leaves and tender stalks used as leafy vegetable.
Chicory Cichorium intybus L. Family Compositae. Plant of temperate areas whose roots yield the chicory of commerce, also used as a vegetable. Propagated by seed.
Chieseling See subsoiling.
Chigna See safed siris.
Chikling vetch See lathyrus.
Chiku Achras zapota L. (A sapota L.; Sapota achras Mill.; Manilkara Zapota (L) Van Royen). Family Sapotaceae, fruit grown in India, native of Mexico. Tree shallow rooted, slow growing, evergreen, flesh of ripe fruit honey brown, soft, mellow, granular, very sweet and delicious. Propagated by vegetative methods of which air layering is most common.
Chill (1) Sensation of coldness attended by shivering. (2) To reduce the temperature of a product and increase its life and value.
Chili A warm, dry, descending wind in Tunisia, resembling the sirocco; in southern Algeria it is called chichili.
Chilli Capsicum annuum L., Capsicum frutescens L. Family Solanaceae. A condiment crop used green or ripe. Adds pungency to foods and pickles.
Chilli anthracnose and dieback Disease caused by Colletotrichum capsici (Syd.) Butt. and Bisby. Characterised by formation of dark sunken spots formed on fruits with pink or dark-coloured dots in centre, rotting and falling off of fruits. May cause die-back of twigs also.
Chilli bacterial leaf spot Disease of chilli caused by Xanthomonas vesicatoria. Characterised by small, dark, greasy spots on leaf, petiole and tender part of stem, water soaked spots on green fruit.
Chilli blight Fungal disease of chilli caused by Alternaria solami characterised by dark and leathery spots on leaves, premature death and falling off of older leaves.
Chilli damping off Disease caused by Pythium aphanidermatum, phytophthora spp. and Pellicularia filamentosa. Symptoms include shrinking of cortical tissue of hypocotyl and falling of seedling.
Chilli leaf curl Disease of chilli brought about by thrips, virus and powdery mildew (Oidium sp.). Characterised

by plants generally getting stunted, bushy, with curled or crinkled leaves, which may be dark or yellowish green, reduction in size of leaves, with or without leaf-shedding, restriction in flowering and fruit formation.

Chilli leaf spot Disease caused by Cercospora capsici. Characterised by appearance of small circular water-soaked spots on leaves, stem, petiole and peduncle of fruits, leaves turning yellow and dropping off, sun scald of fruits.

Chilli mosaic Virus disease characterised by the appearance of light and green patches on leaves.

Chilling injury The injury to fruits and vegetable due to low temperature.

Chilli thrip Scirtothrips dorsalis Hood. Pest of chilli; lacerates leaf tissues, feeds on sap of foliage, curling up leaves.

Chilli wilt Disease of chilli caused by Fusarium annuum. Characterised by cortical decay of collar and roots.

Chilotraea auricilia See sugarcane insect-pests.

— **infuscatellus** See sugarcane insects pests.

Chilo tumidicostalis See sugarcane stem borer.

— **zonellus** See maize borer.

Chimera A plant or part of it, composed of two or more genetically different types. It may result from mutation, irregular mitosis, somatic crossing over, or artificial fusion (grafting). There are three main types viz. periclinal, sectorial and mericlinal.

Chimney cloud A cumulus cloud in the tropics that has much greater vertical than horizontal extent.

China clay Kaolin.

China grass See ramie.

China jute See Indian mallow.

China tea One of the two important types of tea in cultivation in India. A dwarf bush with leathery leaves and grown up to 2500 m; many hybrids also cultivated.

China wood oil tree See tung oil tree.

Chinese cabbage Brassica pekinensis (Lour.) Rupr. Family Cruciferae. Leaves used as vegetable.

Chinese chaste tree Vitex negundo L. Family Verbenaceae. Deciduous shrub common on the banks of streams and on wastelands. Twigs used for basket work. Suitable for training stream banks and general afforestation work. Also called Nirgandi and Samhalu.

Chinese cherry Muntingia calabura L. A tree of family Elaeocarpaceae which produces edible fruits; also called capulin.

Chinese chive Allium uliginosm Don (A. tuberosum Roxb.). Family Liliaceae. Plant of the onion family grown in Bengal and used as a flavouring substance.

Chinese Date See jujube.

Chinese guava See cattely guava.

Chinese laurel Antidesma bunius (L). Spreng. A small tree of family Euphorbiaceae found in Assam, Khasia Hills and Western Ghats; fruits are edible and bark used for making ropes.

Chinese layering See air-layering.

Chinese open-leaf sarson Brassica chinensis. A Brassica species grown as leafy vegetable in the hill regions of north India.

Chinese preserving gourd See ash gourd.

Chinese soap berry Sapindus mukorossi Gaertn. (Sapindus detergens Roxb.). Tree of family Sapindaceae, the fruits of which are used for washing silks etc. Also called Soapnut tree and *ritha* in Hindi.

Chinese sumac Rhus Chinesnsi Mill. (Rhus semialata Murr). A tree of family Anacardiaceae; produces edible fruits. Also called *tatri*.

Chinese Water chestnut Eleocharis dulcis (Burm. f.) Trin. ex Henschel (E. Plantaginea R. Br.; E. tuberosa Schult.). Family Cyperaceae. A herb found in Uttar Pradesh, Bihar, Bengal, Assam and Orissa; produces edible tubers.

Chini See *sathgudi*.

— **orange** See *sathgudi*.

Chip budding Method of budding wherein a single bud with a large piece of wood attached to it is removed from the scion wood and a similar cut is made on the stock. The bud is inserted and tied over; used in grapes. Also termed as yemma budding.

Chironomid An insect of the order Diptera whose young ones; blood red and worm like, are invariably found at the bottom of ponds living in tubes.

Chisel An implement, with points about 30cm apart, used to till the soil some 30-45cm deep.

Chisel sub-soil A tillage implement having one or more cultivator type feet to which are attached strong knife like tools that are used to shatter or loosen hard, compact layers, usually in the subsoil, to depths below normal plow depth. See also subsoiling.

Chitin A nitrogen containing polysaccharide present in the covering layer of insects and in the cell walls of many fungi.

Chi-square (c^2) Test A statistical comparison of observed with theoretical ratios.

— **Goodness of fit** Comparison of an observed mendelian ratio with a theoretical ratio.

— **Independence** A test for association between two series of variables.

Chive Allium Schoenoprasum L. Family Liliaceae. Plant of the family of onion, the leaves and bulbs of which are used as condiment. Found wild and also cultivated. Also called *piazi.*

Chlamydospore Thick-walled, asexual resting spore formed by the rounding up of any mycelial cell; also used for smut spores.

Chlorangiaceae A primitive family of colonial green algae belonging to the Tetrasporales in which the cells are directly attached to each other.

Chloranthus Relating to a plant with green flowers.

Chloranthy A reverting of normally coloured floral leaves or bracts to green foliage leaves.

Chlordane A chlorinated hydrocarbon insecticide; it is a viscous, volatile, amber-coloured liquid. Acts as stomach and contact poison. (2, 3, 4, 5, 6, 7, 8, 8-octachloro-2, 3, 3a, 4, 7,7a hexahydro- 4-7-methanoindene).

Chlorenchyma Chlorophyll-containing tissue in parts of higher plants, as in leaves, parenchyma tissue containing chloroplast.

Chloride of lime White powder; makes good deodorant when mixed with water (1 lb in 2.5 gal.). Disinfectant. Also called bleaching power, calcium chloride, chlorinated lime, calcium hypochlorite.

Chlorinated lime See chloride of lime.

Chlorination Disinfecting a substance, retarding decomposition, or oxidising organic matter, with chlorine or bleaching powder.

Chlorine Plant food element, absorbed by plants as chlorine (Cl^-) ion; exact role in plant nutrition not known. Chlorine compounds used as disinfectants and deodorants.

Chorine deficiency Condition in plants brought about by deficiency of chlorine observed specially in tomatoes. Wilting of tips of leaflet blades seen, followed by chlorosis and bronzing at the base of leaflets.

Chlorine solution Non-irritant disinfectant for washing udders of cow.

Chloris gayana See Rhodes grass.

Chlorite group A group of non-expanding platy clay minerals, similar chemically to the vermiculite group, but having a single hydroxide layer between the sheets in place of the exchangeable cations and water.

Chlorococcales A large, highly diverse order of unicellular or colonial mostly fresh-water green algae in the class Chlorophyceae.

Chlorodendrineae A suborder of colonial green algae in the order Volvocales comprising some general with individuals capable of detachment and mobility.

Chloroform Colourless, heavy, non-inflammable, very volatile liquid miscible with alcohol, ether, oils and other organic solvents; an inhalent, anaesthetic, antispasmodic, rubefacient.

Chloromonadophyceae A group of algae considered by some to be a class of the division Chrysophyta.

Chloromonadophyta A division of algae in the plant kingdom considered by some to be a class Chloromonadophyceae.

Chlorophyll The green colouring matter in plants. It is essential to the process of photosynthesis. Composed of two green chemical compounds chlorophyll a, and chlorophyll b, typically associated with two yellow compounds, carotine and xanthophyll. Enables plants to synthesise plant food from carbon dioxide and water in the presence of energy from light.

Chlorophyllous Containing chlorophyll.

Chlorophyta The green algae, a highly diversified plant division characterised by chloroplasts, having chlorophyll a and b as the predominating pigments.

Chlorophytum tuberosum Dalz.: Family Liliaceae. An undershrub found in Madhya Pradesh, Andhra Pradesh, Madras and Kerala whose leaves are used as vegetable. Also called as *kulai*.

Chloropicrin Clear to yellow, volatile soil fumigant; a nematocide, rodenticide and herbicide. Poisonous to humans. Also known as tear gas.

Chloropidae The chloropid flies, a family of myodarian cyclorrhaphous dipteran insects in the subsection Acalypteratae.

Chloroplast The body in the cell cytoplasm that contains the green chlorophyll pigment.

Chloroplast plastid A sponge-like structures in which chlorophyll is enmeshed; performs specific functions such as manufacturing food, building up starch grains, producing colours.

Chlorosis Yellowing or blanching of leaves and other parts of chlorophyll-bearing plants; usually caused by a mineral deficiency. Sometimes it may be due to disease infection as well.

Chlorotic Lacking chlorophyll.

Choke A fixed or variable device used to reduce the amount of air and thus increase the richness of the gas mixture in the carburettor.

Choke coil Turns of wires wound inductively. Usually on an iron core, which will tend to retard the rapid change of a varying electric current.

Choke weed Any weed or plant that crowds out other plants.

Cholesterol Fat-like alcohol ($O_{27}H_{45}OH$) occurring as crystals with notched corners; found in blood, bile, egg yolk, seeds of plants, and animal fats.

Choline A component of vitamin B complex; an organic compound present in plants and animals; essential for growth.

Chomophyte Plant that grows in crevices and fissures and in rock debris.

Chops Mixture of dehydrated leaves and stems as they come form the dehydrator.

Chorion The tender fleshy substance of the original nucleus of a plant seed.

Chorizon See horizon, soil.

Christophine See Chayote.

Christ's thorn Carissa carandas L. Family Apocynaceae. Protective hedge-like plant yielding berry like fruits, edible and attractive. Evergreen hardy shrub grown from seeds.

Chroma The relative purity, strength, or saturation of colour. It is directly related to the dominance of the determining wavelength of light. It is one of three variables of colour.

Chromatids Half chromosomes, resulting from longitudinal division that later become daughter chromosomes.

Chromatin The granular, deep staining protoplasmic material in the cell nucleus.

Chromogenic bacteria Bacteria which produces colours in the substance in which they are found. Also called pigment-forming bacteria.

Chromosomal aberrations Changes resulting from breaks in the chromosomes and abnormal fusion of the broken ends, causing change in structure.

Chromosome Dark-staining, rod-like or thread-like body in the nucleus of the plant cell at the time of cell division that carries the genes controlling the development of the different plant characters.

Chronic Long lasting form of disease, as opposed to acute form.

Chronosequence A sequence of related soils that differ from one another in certain properties primarily as a result of time as a soil forming factor.

Chrococcales An order of blue-green algae (Cyanophyceae) that reproduces by cell division and colony fragmentation only.

Chrysalis The state of a butterfly, moth etc., in which it becomes inactive and is enclosed in a hard container while changing from larva into the adult stage.

Chrysanthemum cinerariiafolium See pyrethrum.

— **coccineum** See persian insect flower.

Chrysobalanus icaco See coco-plum.

Chrysophyceae A Golden-brown algae, making up a class of fresh and salt water unicellular forms in the division Chrysophyta.

Chylophyllous Having succulent or fleshy leaves.

Chrysophyllum cainito See star apple.

— **roxburghii** See star apple.

Chrysopogon fulvus See goria.

— **gryllus** See kufri.

— **montanus** See goria.

Chuck In itself a very broad term, meaning a device for holding a rotating tool or work during an operation. There are many different kinds of chucks for various purposes.

Chuka Croton oblongifolius Roxb. Family Euphorbiaceae. Herb yielding oil used as purgative and insecticide.

Chuni An Indian term for the by-product obtained while preparing *dal* from pulses; contains husk and some broken pieces of grain; ground pulses and is fed to the animals.

Churning The beating, shaking or stirring.

Chute (1) A high-velocity open channel for conveying water to a lower level without erosion. (2) Any inclined trough or tube to guide sliding objects from a higher to a lower level.

Chute blocks Blocks provided at the entrance of the stilling basin to stabilise the formation of hydraulic jump, to increase effective depth, to break up flow into a number of water jets, to create turbulence and to lift the jets off the floor to reduce basin length.

Cicca acida See star gooseberry.

— **disticha** See otheite gooseberry.

Cicer arietinum See Bengal gram.

Cichorium endivia See endive.

— **intybus** See chicory.

C.I.F. Cost insurance and freight. Term used of goods shipped where the price includes shipping and insurance charges. A C.I.F. quotation means that the seller must ship the goods meeting all charges up to "on board" and paying freight insurance, etc. He must supply the buyer with all the documents required to enable the goods to be imported. The buyer must pay all expenses, customs duties, etc. on arrival of the goods at the port of destination.

Cigarette beetle Lasioderma serricorne Fab. Insect pest of stored tobacco. Also attacks coriander, mustard, cumin, dried turmeric, dried ginger, Grub feeds on tobacco leaves, making large galleries. During severe attack tobacco is reduced to powder. Adult also cuts holes into tobacco leaves. Also called tobacco borer beetle.

Cigar filler Tobacco leaves placed in the core of a cigar; used as long fragments or short and shredded.

Cigar wrapper Tobacco leaf wrapped spirally around the filler.

Ciliate A protozoan that move by means of cillia.

Ciliated Provided with cilia.

Cilium, pl. Cilia (1) Hairlike swimming organ on bacteria or zoospores. (2) Tender, sensitive hair forming a fringe along the edges of leaves.

Cimex rotundatus See bed bug.

Cinchona Cinchona spp. A large genus of evergreen trees of family Rubiaceae, from which quinine sulphate and other alkaloids are extracted. Quinine extracted from root, stem and twig bark. Requires deep well drained loamy soils rich in humus and *p*H 5-6. Grows at an elevation of 1100-2500 m. Propagation by seed and patch budding, cutting and layering. Other species which are grown in various parts of India are Cinchona calisaya, Cinchona ledgeriana, Cinchona succirubra.

Cinchona black root rot Disease caused by Resellinia spp. Symptoms as in tea black root rot.

Cinchona brown root disease Disease of cinchona caused by Fomes noxius. Symptoms as in tea brown root disease.

Cinchona calisaya See cinchona.

— **ledgeriana** See cinchona.

— **succirubra** See cinchona.

Cinchona pink disease Disease of cinchona caused by Corticium salmonicolour. Symptoms as in tea pink disease.

Cinchona seedling blight and damping off Fungal disease caused by

Phytophthora spp. and Macrophomina phaseoli characterised by damping off of seedling in nursery.

Cinchona stem cankar Disease of cinchona caused by Botryodiplodia theobromae. Characterised by cankerous growth in open wounds.

Cinnamomum camphora See camphor.

— **cassia** See cinnamon.

— **iners** See cinnamon.

— **obtusigorm** See cinnamon.

— **tamala** See Indian cassia.

— **zeylanicum** See cinnamon.

Cinnamon Cinnamomum zeylanicum Bl. (true cinnamon); Tree producing aromatic bark. Other cinnamon producing trees are C. iners, C. cassia, C. obtusiform. Family Lauraceae. Tree growing on west coast; light red wood, moderately hard scented. Bark used as spice. Best bark produced by trees in siliceous soil with an admixture of vegetable mould; fairly heavy rainfall conducive to tree growth. Propagation by seed, old stumps, layers. Ready for harvest in 3rd year. Bark peeled off in strips, heaped and covered with sacks for a day. Outer skin or bark scraped off, bark partially dried in shade for a day or two when it forms quills, which when finally dried is marketed. Oil which is used for several purposes is obtained from chips, waste bark or leaves. Also called *dalchini*.

Cinnamon bark The coarser pieces of bark of cinnamon usually obtained from the thicker portions of the shoots; not fit for chewing; used for distilling the essential oil of cinnamon.

Cinnamon oil Oil obtained by steam distilling the inner bark of cinnamon. Also see cinnamon.

Cireadian Pertaining to a period of about 24 hours; applied particularly to the rhythmic repetition of certain phenomena in organisms at about the same time each day.

Circinate Manifesting an unrolling development, like fern leaves.

Circinate vernation Uncoiling of new leaves from the base toward the apex, as in ferns.

Circium arvense See Canada thistle.

Circuit The course followed by an electric current passing from its source through a succession of conductors and back to its starting point.

Circular fodder cutter See chaff cutter.

Circular Vortex An atmospheric flow in parallel planes in which streamlines and other isopleths are concentric circles about a common axis: an atmospheric model of easterly and westerly winds is a circular vortex about the earth's polar axis.

Circulatory system In process engineering, the system consisting of circulation of air in a drier or in a building.

Circulation (1) The regular movement of any fluid within definite channels as in body. (2) The streaming movement of protoplasm of plant cells.

Circulation flux Flux due to mean atmospheric motion as opposed to eddy flux; the dominant flux in low latitudes.

Circulation index A measure of the magnitude of one of several aspects to large scale atmospheric circulation patterns; indices most frequently measured represent the strength of the zonal (east-west) or meridional (north-south) components of the wind at the surface or at upper levels, usually averaged spatially and often averaged in time.

Circulation pattern The general geometric configuration of atmospheric circulation usually applied, in synoptic meteorology, to the large scale features of synoptic charts and mean charts.

Circula trifenestrata See cashewnut leaf eating caterpillar.

Circumneutral Describing soil with pH value slightly above or below seven. Neither strongly acidic nor alkaline.

Cirriform Descriptive of clouds. Composed of small particles, mostly ice crystals, which are fairly widely dispersed, usually resulting in relative transparency and whiteness and often producing halo phenomena not observed with other cloud forms. Having the form of a cirrus; generally applied to a prolonged, slender process.

Cirrocumulus cloud A principal cloud type appearing as a thin, while path of cloud without shadows, composed of very small elements in the form of grains, ripples and so on.

Cirrostratus cloud A principal cloud type, appearing as a whitish veil, usually fibrous but sometimes smooth, which may totally cover the sky and often produces halo phenomena either partial or complete.

Cirrus Tendril; tendril-like structure.

Cirrus cloud A principal cloud type composed of detached cirriform elements in the form of white, delicate filaments, of white (or mostly white) patches or narrow bands.

Cistern A natural or artificial reservoir or receptacle for holding rain water or other liquids.

Citrange A hybrid tree (Poncirus trifoliata X citrus sinensis) of family Rutaceae, which bears fruit, more pronounced odour than the orange. It is mainly used as a stock by citrus breeders.

Citrate soluble phosphoric acid See reverted phosphoric acid.

Citric acid An acidic white crystalline substance present in lemons, limes, gooseberries, raspberries etc.; usually obtained from citrate of lime by decomposition and filtration. Used in medicines, dyeing, artificial lemonade.

Citriculture The cultivation of citrus fruits.

Citron Citrus Medica L. (Citrus tuberosa Mill., C. cedrata Raf., C. aurantium var. medica Wight and Arn., C. medica var. medica proper Hook f.). Family Rutaceae. A large thorny shrub grown in Khasia hills of Assam and other places in India for its highly aromatic fruit. Fruit edible. Also called *galgal.*

Citronella grass (1) Cymbopogon winterianus Jowitt. Family Gramineae. Grass from which essential oil, citronella oil, of the Java type is extracted. Used in the synthesis of organic compounds like synthetic menthol, because of high geraniol citronella and citronollol content. (2) Cymbopogon nardus (L.). Rendle (Andropogon nardus L.) Family Gramineae. Grass from which essential oil known as Ceylon type citronella oil is extracted; used in perfumery and cosmetics.

Citrullus vulgaris See water melon.

— **vulgaris var. fistulosus** See squash melon.

Citrus Important genus of trees of family Rutaceae. Cultivated in south China, Malaya and sub-Himalayan parts of Assam from time immemorial. Third most important fruit of India. Most important citrus fruits grown are sweet oranges, the *Santra* or Mandarin orange, the lime and the lemon, grape fruit.

Citrus acida See sour lime.

— **auriantiifoia** See sour lime.

— **aurantium** See sour orange; sweet orange.

— **aurantium var. sinensis** See sweet orange.

— **aurantium subsp. sinensis** See pumello. var. *decumana*

— **aurantium var. medica** See citron.

— **cedrata** See citron.

— **decumenus** See pummelo.

— **decumana var. paradisi** See grape fruit.

— **decumand var. racemosa** See grape fruit.

— **hystrix subsp. acida** See sour lime.

— **japonica** See kumquat.

— **lima** See sour lime.

— **limettioides** See sweet lime.

— **limon** See lime.

— **limonum** See lime.

— **morgarita** See nagami kumquat.

— **maxima** See pumello.

— **medica** See citron.

— **medica var. acida** See sour lime.

— **medica var. limetta** See sweet lime.

— **medica var. limonum** See lemon.

— **medica var. limon** See lemon.

— **medica var. medica proper** See citron.

— **nobilis** See mandarin.

— **paradisi** See grape fruit.

— **racemosa** See grape fruit.

— **reticulata** See mandarin.

— **sinensis** See sweet orange.

— **trifoliata** See frifoliate orange.

— **tuberosa** See citron.

Citrus anthracnose A fungus disease of citrus plants caused by colletotrichum gloeosporioides and characterised by tip blight, stains, on the leaves and spots, stains, or rot on the fruit.

Citrus aphids Toxoptera citricidus Fab., T. aurantii Fab. Dark aphids living on tender leaves and shoots of citrus fruit trees and sucking sap out of them.

Citrus bark eating caterpillar Indarabela quadrinotata Walk. Insect pest of citrus trees. Damage as in mango bark-eating caterpillar.

Citrus blast A bacterial disease of citrus trees caused by Pseudomonas syringae and marked by drying and browning of foliage and twigs and black pitting of the fruit.

Citrus brown rot See citrus gummosis.

Citrus canker A bacterial disease of citrus plants caused by Xanthomonas citri and producing lesions on twigs, foliage, and fruit; appearance of raised corky eruptions on all green parts.

Citrus decline Physiological disease of citrus caused by insufficient soil moisture and nutrition. Characterised by yellowing and gradual reduction of leaf, die-back of twigs, decline and gradual death.

Citrus diplodia gummosis Disease caused by Diplodia natalensis. Symptoms include the oozing out of gum from forks and bigger limbs, dying of bark on which gum dries.

Citrus exanthema Disease of citrus due to copper deficiency. characterised by development of multiple buds in axils, forming short misshapen shoots, dying of young twigs and exudation of gum from base of twigs.

Citrus fluted scale Icerya purchasi Mask. Also known as cottony cushion scale. Yellow, red or brown insect coated with whitish or yellowish mealy substance; pest on citrus trees. Adult females and nymphs suck sap from trees.

Citrus foliocellosis Physiological disease mainly due to zinc or manganese deficiency.

Citrus foot rot See citrus gummosis.

Citrus fruits Any of the edible fruits having a pulpy endocarp and a firm exocarp that are produced by plants of the genus Citrus and related genera.

Citrus fruit sucking moths Othreis fullonia Cl., O. ancillia Cl., O. materna L., Achoea janata L., Calpe emarginata Fabr. Destructive insect pest of citrus fruit, moth pierces fruits and sucks juice, causes premature fall of fruits. Also attacks grape fruit, pumello, *malta* orange, A. janata attacks guava.

Citrus green bug Rhynchocoris humeralis Thunberg. Bug, pest of citrus trees; nymph and adult suck juice from fruits causing premature shedding.

Citrus gummosis Disease of citrus caused by Phytophthora palmivora and Phytophthora parasitica Dastur. Characterised by appearance of water-soaked area in basal part of plant, later exudation of brownish gummy substance, girdling of tree, extension of infection to crown and roots, appearance of cracks in bark and roots and slow dying of plants.

Citrus leaf and fruit rot Disease caused by Phythophthora palmivora Butler. Characterised by leaves becoming water soaked, rotting and falling down, fruits turning soft and falling off, formation of a white, cottony growth on all infected parts; black canker later appearing on smaller branches and crown rot and girdling of stem.

Citrus leaf miner Phyllocnistis citrella Stainton. Major insect pest of citrus; greenish yellow caterpillar mines into leaf tissues, and makes shining galleries on leaves; causes serious setback in growth of trees. Also attacks orange, lime, pummelo, and all varieties of grapefruit and malta.

Citrus mottle leaf See citrus foliocellosis.

Citrus pink disease Caused by Corticium salmonicolor. Symptoms as in tea pink diseases.

Citrus powdery mildew Disease of citrus caused by Oidium tingitanium Carter. Characterised by appearance of white powdery growth on young fruit.

Citrus psylla Diaphorina citri Kuw. Insect pest of citrus, nymph and adult suck sap from tender parts of plants such as shoots, buds, flower. Clusters round shoots secretes honeydew, causes immature shedding of leaves and non-setting of fruits.

Citrus scab A fungus disease of citrus plants caused by Sphaceloma rosarum, producing scab like lesions on all plant parts.

Citrus sooty mould Disease of citrus caused by Meliola butleri. symptoms as in sapodilla sooty mould.

Citrus tristeza Virus disease of citrus. Characterised by gradual decline, trees showing profuse flowering and heavy setting of fruits, suppression of new growth, bronzing and yellowing of leaves, defoliation and dying back of twigs.

Citrus white fly Aleurocanthus spiniferus Quaint, (A. woglumi Ashby, Dialeurodes citri Ashm.). Insect pest of citrus; nymph and adult suck sap from leaves, secrete honey dew leading to sooty mould. In severe attack, leaves turn yellow and fall prematurely.

Citrus wither tip Disease of citrus caused by Colletotrichum gloeosporioides. Characterised by shedding of leaves and die-back of mature twigs.

Civet fruit See durian.

Cladode Branch arising from axil of leaf or green flattened stem, resembling a foliage leaf.

Cladoniaceae A family of lichens in the order Lecanorales, including the reindeer mosses and cup lichens, in which the main thallus is hollow.

Cladophorales An order of coarse, wiry, filamentous, branched and unbranched algae in the class Chlorophyceae.

Cladophyll A branch arising from the axil of a true leaf and resembling a foliage leaf. Also known as cladode.

Cladoptosis The annual abscission of twigs or branches instead of leaves.

Cladosporium fulvum See tomato leaf mould.

— **herbarum** See wheat black mould.

Clammy cherry See Sebsten.

Clamp A tool for holding portions of work together, both in wood and metal. A frequently used form of clamp shaped like the letter "C"; pressure is obtained by means of a thumb screw.

Clamp connection The connection, tubular in form, between two cells in the Basidiomycetes, providing a pathway for nuclei to achieve the dikaryon phase.

Clania crameri See tea bagworm.

Class It is the extreme fixation of class differences. When a class becomes fixed in its relations so that the individual born into it can not enter into another class, it becomes a caste. (2) A group that includes variates of similar magnitude. Division of phylum, group of soils, waters, plants, animals etc.

Class, Age One of the intervals into which the range of ages of vegetation is divided for classification and use.

Class, Soil A group of soils having a definite range in a particular property such as acidity, degree of slope, texture, structure, land use capability, degree of erosion, or drainage. See also structure, soil and texture, soil.

Classification The systematic arrangement of units according to respective groups.

Classification soil The systematic arrangement of soils into categories on the basis of their characteristics. Broad groupings are made on the basis of general characteristics and sub-divisions on the basis of more detailed differences in specific properties.

Clausena excavata See wampee.

— **lansium** See wampee.

— **wampi** See wampee.

Clavate Club-shaped.

Clavus See ergot.

Claw (1) The slender extended, lower part of the petal in some flowers. (2) Claw hammer.

Clay (1) As a particle size, a size fraction less than 0.002 mm in equivalent diametre or some other limit (Geologists and Engineers). (2) As a rock term, a natural earthy, fine grained material that develops plasticity with a small amount of water. (3) As a soil term.A textural class soil material that contains 40 per cent or more clay, less than 45 per cent sand and less than 40 per cent silt. See also Texture, soil. (4) As a soil separate material usually consisting largely of clay minerals but commonly also of amorphous free oxides and primary minerals.

Clay complex Complex of the soil constituted by clay and humus, resembling each other in many properties; also called clay-humus complex.

Clay composition The make up of common clays. Mainly made up of silicates of various types and called silicate clays and sesquioxide, when formed of aluminium and iron hydroxide which is not common. Silicate clays are divided into three main groups, *kaolinitic*, the montmorillionitic and the hydrous mica or illitic.

Clayey Containing large amounts of clay or having properties similar to those of clay.

Clayey soil Includes all clay textural classes such as sandy clay, silty clay and clay etc.

Clay films (Skin) Coating of oriented clays on the surfaces of soil peds and mineral grains, and in soil pores.

Clay loam Soil material that contains 27 per cent to 40 per cent clay and 20 per cent to 45 per cent sand. See also texture, soil.

Clay mineral Finely crystalline hydrous aluminium silicates and hydrous magnesium siiicates with aphyllosillicate structure.

Claypan A term used in the United States for a dense, compact layer in the profile having a much higher clay content than the overlying material from which it is separated by a sharply defined boundary. In the Canadian classification system, this pan is recognised as a clay enriched alluvial (B) (Bt) horizon.

Clay structure Structure of clay particle when viewed under the electron microscope, seen to be definitely crystalline and lathe-shaped or made up of irregular hexagonal slates.

Clean (1) Free from flaws, blemishes or foreign matter. (2) To remove undesirable matter from a substance. (3) To cultivate field to remove weed.

Clean boled Trunk free of branches, as a tree.

Clean bred Thorough bred.

Clean cotton Seed cotton free from extraneous hulls, bolls or trash.

Clean cultivation See clean culture.

Clean culture Intensive cultivation of a field so as to remove all weeds. Also called clean cultivation or clean tillage.

Cleaning operation In forestry, a tending operation done in a sapling crop, involving the renewal or topping of inferior growth including individuals of the favoured species, climbers etc., when they are interfering with the better-grown individuals of the favoured species.

Clean tillage See clean culture.

Clear (1) Timber which is free from defects and blemishes. (2) In land reclamation, to remove trees, bushes, stones or other obstacles to tillage.

Clear-air turbulence A meteorological phenomenon occurring in the upper troposphere and lower stratosphere, in which high-speed air craft are subject to violent updrafts and downdrafts. Abbreviated as CAT.

Clear cutting In forestry, the removal of the entire timber stand from an area.

Clearing house A banking institution where interbank debts are cancelled out.

Cleavage The cell divisions which transform a fertilised ovum into an embryo.

Cleavage edge The edge of the share which forms joint between mouldboard and share on the frog is known as cleavage edge.

Cleft grafting A top-grafting method in which the scion is inserted into a cleft cut into the top of the stock. Also wedge grafting.

Cleistogamy Production of small closed flowers that are self-pollinating and contain numerous seeds; fertilisation taking place within unopened flowers.

Cleistothecium (pl. Cleistothecia) A perithecium without a special opening; in powdery mildews; a closed ascocarp.

Cleome viscosa See sticky cleome.

Clerodendrum serratum See *Bharangi*.

Cleviceps microciphala See *bajra* ergot.

— **purpurea** See ergot.

Clevis The horizontal or vertical brackets provided at the end of the beam of a mouldboard plough for adjusting the depth and width of the plough and the place which attaches the plough to the yoke. Also called bridle.

Client system The individuals, groups, organisations communities with which the change agent works for introducing a change is called the client system.

Climate The sum total of all atmospheric or meteorological influences, principally temperature, moisture, wind, pressure, and evaporation, which combine to characterise a region and give it individuality by influencing the nature of its land forms, soils, vegetation and land use. Contrast with weather.

Climate classification The division of the earth's climates into a system of contiguous regions each one of which is defined by relative homogeneity of the climate elements.

Climate continental The type of climate characteristic of land areas separated from the moderating influence of oceans by distance, direction, or mountain barriers, marked by relatively large daily and seasonal change in temperature.

Climate control Schemes for artificially altering or controlling the climate of a region.

Climate cycle Actual or supposed cyclical recurrences of such weather phenomena as wet and dry years, hot and cold years, at more or less regular intervals, in response to long range terrestrial and solar influences, such as volcanic dust and sunspots.

Climate, oceanic The type of climate characteristics of land areas near oceans which contribute to the humidity and at the same time have a moderating influence on temperature and range of temperature variation. Synonym marine climate.

Climateric peak The maximum point of respiration rate of mature fruit, especially apples, term usually applied to harvested fruit, but it may apply to fruit still on tree.

Climatic analogues Areas that are enough alike with respect to some of the major weather characteristics affecting the production of crops.

Climatic change The long-term fluctuation in rainfall, temperature and other aspects of the earth's climate.

Climatic climax A climax community viewed, by same authorities, as controlled by climate.

Climatic controls The relatively permanent factors which govern the general nature of the climate of a portion of the earth, including solar radiation, distribution of land and water masses, elevation and large-scale topography and ocean currents.

Climatic cycle A long period oscillation of climate which recurs with some regularity, but which is not strictly periodic. Also known as climatic oscillation.

Climatic divide A boundary between regions having different types of climate.

Climatic forecast A forecast of the future climate of a region; i.e., a forecast of general weather conditions to be expected over a period of years.

Climatic index A simple, single numerical value that expresses climatic relationships; for example, the numerical value obtained in Transeau's precipitation-evaporation ratio.

Climatic prediction The estimation of the chances that specified climatic conditions will occur in a certain time interval.

Climatic province A region of the earth's surface characterised by an essentially homogeneous climate.

Climatic stratification The adaptation of a species to a particular climate.

Climatic year A continuous twelve-month period during which occurs a complete annual climatic cycle and which is selected for the presentation of data relative to meteorological phenomena.

Climatic zone A belt of the earth's surface within which the climate is generally homogeneous in some respect; an elemental region of a simple climatic classification.

Climatography A quantitative description of climate, particularly with reference to the tables and charts which show the characteristic values of climatic elements at a station or over an area.

Climatological forecast A weather forecast based upon the climate of a region instead of upon the dynamic implications of current weather, with consideration given to such synoptic weather features as cyclones and anticyclones, fronts, and the jet stream.

Climatological station elevation The elevation above mean sea level chosen as the reference datum level for all climatological records of atmospheric pressure in a given locality.

Climatological station pressure The atmospheric pressure computed for the level of the climatological station elevation, used to give all climatic records a common reference; it may or may not be the same as station pressure.

Climatological sub-station A weather-observing station operated (by an unpaid volunteer) for the purpose of recording climatological observations.

Climatology That branch of meteorology concerned with the mean physical state of the atmosphere together with its statistical variations in both space and time as reflects in the weather behaviour over a period of many years.

Climax A plant community of the most advanced type capable of development under, and in dynamic equilibrium with prevailing environment.

Climax dominant Any species which is dominant more or less in stable biotic community.

Climax state The final equilibrium community reached in the process of botanical succession.

Climax vegetation A compact, coherent mass of soil ranging in size from 5 to 10 mm to as much as 200 to 250 mm; produced artificially, usually by the activity of man, by plowing, digging, etc. especially when these operations are performed on soils that are either too wet or too dry for normal tillage operations.

Climber Any plant having long and slender branches, which may attach itself to other plants or supports by means of tendrils or by entwining.

Climber cutting A silviculture operation in which climbers are cut back.

Climosequence A sequence of related soils that differ from one another in certain properties primarily as a result of the effect of climate as a soil forming factor.

Clincher rim A wheel rim with turned edge or groove on both sides to engage the flange of a tire. Inflating the tire, locks the tire on the rim.

Cling Tendency of cotton fibres in a sample to stick together.

Clingstone (1) Any fruit especially the varieties of the peach, whose flesh adheres strongly to the seed or stone, even at maturity. (2) The stone of such a fruit.

Clinosequence A group of related soils that differs from one another in certain properties primarily as a result of the effect of the degree of slope on which they were formed. See also toposequence.

Clip (1) Shears. (2) To trim a plant. (3) To cutoff the tops of a crop at an early stage of growth. (4) To sheer the hair of an animal close to the skin. (5) (a) Short wool, (b) The process of shearing wool, (c) The year's production of wool.

Clisere The succession of ecological communities, especially climax formations, as a consequence of intense climatic changes.

Clisiocampa indica See tent caterpillar.

Clod A compact coherent mass of soil produced by digging or plowing. Clods usually slake easily with repeated wetting and drying.

Clod crusher See clod crushing-roller; also see float.

Clod-crushing roller An implement with a number of iron rings or rollers star-shaped; solid or with bars; frame holds the rollers; meant for crushing clods.

Cloddy Referring to soils with clods.

Clone Individual plant propagated vegetatively by rooting portions from a single original plant e.g. cuttings from rose.

Close-centred tree In pruning, a globe shaped tree in which the trunk is contained as a leader in the centre of the tree, produced by heading back the side branches and the trunk.

Close circuit A complete path from the source through an external circuit back to the source. Usually accomplished through the closing of a control in the form of a switch.

Close conduit flow Flow distinguished by the absence of a free water surface.

Close cropping The grazing of the animal on grass or other forage dangerously close to the crown of the plant, which can ruin the grazing land. Also called close grazing.

Closed cycle A self-contained steam-condenser cooling system for electric power plants that requires water input only to replace evaporative losses and to dilute residues.

Closed drains Sub-surface drain, tile, or perforated pipe that receives surface water through surface inlets.

Closed-drainage area, blind drainage area or non-contributing area Area in which surface flow collects in sinks or lakes having no surface outlet either to other streams or to the sea.

Closed economy No imports or exports; an assumption that may be made for the purposes of economic analysis.

Closed herd A herd in which no outside blood is introduced.

Closed high A high that may be completely encircled by an isobar or contour line.

Closed low A low that may be completely encircled by an isobar or contour line, that is an isobar or contour line of any value, not necessarily restricted to those arbitrarily chosen for the analysis of the chart.

Close-fertilization Self-pollination.

Close grazing See close cropping.

Clostridium Species of spore-bearing bacilli; grow anaerobically; occur frequently in wounds, commonly in animals. Many species are pathogenic.

Clostridium pasteurianum Non-symbiotic nitrogen-fixing bacterium, less efficient than Azotobacter; Works in anaerobic conditions.

— **septicum** See malignant oedima.

— **tetani** See tetanus.

— **welchii** See enterotoxaemia.

Clothesline effect Horizontal heat transfer from a warm upwind area to a relatively cooler crop field.

Cloud Suspensions of minute water droplets or ice crystals produced by the condensation of water vapour. The nucleus that are in the nucleus of an atom but not in closed shells. Any suspension of particulate matter, such as dust or smoke, dense enough to be seen.

Cloud band A fairly well defined mass of cloud observed at a distance; covers an appreciable portion of the horizon sky, but does not extend overhead.

Cloud base For a given cloud or cloud layer, that lowest level in the atmosphere at which the air contains a perceptible quantity of cloud particles.

Cloud burst In popular terminology any sudden and heavy fall of rain, usually of the shower type, and with a fall rate equal to or greater than 100 mm per hour. Also known as rain gush; rain gust.

Cloud classification (1) A scheme of distinguishing and grouping clouds according to their appearance and, where possible, to their process of formation. (2) A scheme of classifying clouds according to altitudes high, middle, or low clouds. (3) A scheme of classifying clouds according to their particulate composition; water clouds, ice crystals clouds, or mixed clouds.

Cloud cover The portion of the sky cover which is attributed to cloud usually measured in tenth of sky covered. Also known as cloudage; cloudiness.

Cloud deck The upper surface of a cloud.

Cloud droplet A particle of liquid water from a few micrometres to tens of micrometres in diametre, formed by condensation of atmospheric water vapour and suspended in the atmosphere with other drops to form a cloud.

Cloud-drop sampler An instrument for collecting cloud particles, consisting of a sampling plate or cylinder and a shutter to the cloud for a predetermined length of time; the sampling surface is covered with a material which either captures the cloud particles or leaves an impression characteristic of the impinging elements.

Cloud forest A broad leaved evergreen forest occupying a montane tropical area where regular condensation occurs.

Cloud formation (1) The process by which various types of clouds are formed generally involving adiabatic cooling of ascending moist air. (2) A particular arrangement of clouds in the sky, or a striking development of a particular cloud.

Cloud height The absolute altitude of the base of a cloud.

Clouding An undesirable translucent condition of liquids.

Cloud level (1) A layer in the atmosphere in which are found certain cloud genera; three levels are usually defined namely high, middle, and low. (2) At a particular time, the layer in the atmosphere bounded by the limits of the bases and tops of an existing cloud form.

Cloud modification Any process by which the natural course of development of a cloud is altered by artificial means.

Cloud particle A particle of water, either a drop of liquid water or an ice crystal, comprising a cloud.

Cloud-phase chart A chart designed to indicate and distinguish super cooled water clouds from ice-crystal clouds.

Cloud physics The study of the physical and dynamical processes governing the structure and development of clouds and the release from them of snow, rain, and hail.

Cloud shield The principal cloud structure of a typical wave cyclone, that is, the cloud forms found in the cold-air side of the frontal system.

Cloud street A line of cumuliform clouds frequently one cumulus element wide, but ranging upward in width so that it is sometimes difficult to differentiate between streets and bands.

Cloud system An array of clouds and precipitation associated with a cyclonic-scale feature of atmospheric circulation, and displaying typical patterns and continuity.

Cloud top The highest level in the atmosphere at which the air contains a perceptible quantity of cloud particles for a given cloud or cloud layer.

Cloudy The character of a day weather when the average cloudiness, as determined from frequent observations, is more than 0.7 for the 24-hour period.

Cloudy juice Fruit juice prepared by pressing, followed by straining through fine cloth or sieve.

Clove (1) The unopened flower bud of a small, conical, symmetrical evergreen tree, Eugenia caryophyllata, of the myrtle family (Myrtaceae); the dried buds are used as a pungent, strongly aromatic spice. (2) A small bulb developed within a larger bulb, as in garlic.

Clove Oil Oil distilled from cloves, leaves and unripe fruits of Eugenia caryophyllata Thunb. Used in medicines, perfumes, soaps.

Clover (1) A general term for low herbs of the genus Trifolium, family Leguminosae, with compound, usually trifoliate leaves; excellent forage plants serve as bee-forage and for nitrogen-fixation in the soil. (2) Name applied with a qualifying word of plants of genera Medicago, Lespedeza and Melilotus.

Clover crown wart Disease of clover caused by Physoderna trifolii. Symptoms as in lucerne crown wart.

Clover downy mildew Disease of clover caused by Peronospora trifolirepentis. Symptoms as in lucerne downy mildew.

Clover root rot Disease caused by lelicularia filamentosa. Symptoms as in cowpea root rot.

Clover rust Disease of clover caused by Uromycess trifolii and V. Minor. Characterised by appearance of small, brownish rust pustules on all green parts of the plant.

Clover sooty blotch Disease of clover caused by Cymadothia trifolii. Characterised by appearance of dark brown or black angular blotches on leaves which later turn black and warty; leaves looking yellow or brown.

Club moss The common name for members of the class Lycopodiatae.

Club root A disease of brassica crops occurring on acid soils and caused by a 'saline fungus' which infects the soil. Irregular swellings appear on the roots and later decay sets in.

Club wheat Triticum compactum Host. Family Gramineae. Annual cereal plant with flat leaves and club-shaped spikes grown for fodder.

Clump (1) A cluster of plants. (2) A mass of roots stocks.

Cluster (1) The form in which bees

cling together in the hive after swarming or during winter. (2) A larger number of plants grouped, arranged or growing in close proximity to one another. (3) An inflorescence.

Cluster bean Cyamopsis psoraoides DC. [Cyamopsis tetragonoloba (L.) Taub.; psoralea tetragonoloba (L.)] family Papilionaceae. An important leguminous crop in India. It is drought resistant and adapted to light soils. Grown for forage and for green manuring. Green pods in unripe stage used as vegetable. Called as *Guar.*

Cluster bean bacterial leaf spot Disease of cluster bean caused by xanthomonas cyamopsidis. Characterised by small, circular, water soaked spots, turning jet black and forming irregular lesions surrounded by chlorotic areas.

Cluster bean dry root rot Disease caused by Macrophomina phaseoli. Symptoms similar to cotton dry root rot.

Cluster bean leaf spot Disease caused by Alternaria brassicae. Symptoms as in mustard leaf spot.

Cluster bean powdery mildew Disease caused by Leveillula taurica. Symptoms as in cotton powdery mildew.

Cluster bean ozonium wilt Disease of cluster bean caused by ozonium texanum var. parasiticum. Characterised by rooting of collae region, white weft of mycelium on soil near base of plant, wilting and drying of plants.

Cluster bean root rot Disease caused by Pellicularia filamentosa. Symptoms as in groundnut root rot.

Cluster bean sclerotial disease Disease of cluster bean caused by pellicularia rolfsil. Symptoms as in cotton sclerotial disease.

Cluster fig Ficus glomerata Roxb. (Ficus goolereea Roxb.) Tree of the family Moraceae; leaves used as fodder; bark for tan. Fruits also edible. Also called *Gular, udumbara* or *domoor.*

Cluster primordium The rudimentary growth which develops into a cluster of grapes.

Cluster sampling Where clusters of observations are formed on some basis that reduces cast e.g. geographical compactness, but care is taken to ensure heterogeneity and to avoid redundancy of information.

Cluster thinning The selective removal of grape clusters after bloom soon after the berries have set. It helps improving the quality size and colouring.

Clutch It is a device, used to connect and disconnect the tractor engine from the transmission gears and drive wheels. It transmits power by means of friction between driving and driven members.

Clutch pedal The left foot pedal which connects or disconnects the clutch.

Coagulase Enzyme that hastens coagulation.

Coagulate To curdle, to clot.

Coagulation Change of liquids into more or less solid state due to clotting or curdling action of certain agents like ferments, heat and chemicals.

Coagulin Any agent capable of coagulating albuminous substances.

Coalesce Growing together into one body or spot.

Coalescence process The growth of raindrops by the collision and coalescence of cloud drops or small precipitation particles.

Coarse (1) Designating any large, disproportionate feature. (2) Referring to the roughness in appearance of a grain or particle. (3) A common term for light textured soils referred as coarse textured.

Coarse fragments Rock or mineral particles greater than 2.0 mm in diametre.

Coarse grains Grains whose flour cannot be used for leavened bread. Also called feed grains.

Coarse sand Twenty five per cent or more very coarse and coarse sand; less than 50 per cent any other one grade of sand. Sand - 25 per cent or more very coarse, coarse and medium sand and less than 50 per cent fine or very fine sand. Fine sand - 50 per cent of more fine sand, less than 25 per cent very coarse, more fine sand, or less than 25 per cent very coarse, find sand. Very fine sand - 25 per cent or more very fine sand.

Coarse sandy loam Twenty-five per cent or more very coarse and coarse sand and less than 50 per cent any other one grade of sand. See texture, soil.

Coarse texture (1) The texture exhibited by sands, loamy sands, and sandy loams except very fine sandy loam. A soil containing large quantities of these textural classes. (2) A textural condition of a fruit, vegetable or other plant or plant produce.

Coastal Of or pertaining to the sea coast; specifically, the waters, margins, or shorelands of estuarine basins, and the near shore ocean.

Coastal upwelling A process by which water moves shoreward along the bottom of water basin into shore zone; a reaction usually caused by wind forcing surface waters offshore.

Coastal waters As a management definition, territorial or interior waters that contain a measurable quantity or per centage of seawater (e.g., more than 0.5 parts per thousand).

Coastal watershed A drainage basin that drains directly into coastal waters; excludes drainage basins that drain wholly into freshwater channels tributary to coastal waters.

Coated ginger Ginger in which very little of the peel is removed in curing.

Coating Material covering the soil particles. See also clay films (skins).

Cob (1) The ear of maize; the female inflorescence or shoot arising from nodes about midway from the stalk; composed of a shank from which husks arise and terminate in the ear; fresh silk bifurcates at the top and acts as receptive for pollen. (2) Male swan.

Cobalt (Co) A trace element that occurs in soils; constituent of vitamin B; more significant in animals than in plant nutrition.

Cobble See cobblestone.

Cobblestone Rounded or partially rounded rock or mineral fragment 7.5 to 2.5 cm in diametre. See also coarse fragments. In engineering practice cobbles are greater than 7.5 cm but less than 20 cm in diametre.

Cobblestone drain See blind drain.

Cobbly Containing appreciable quantities of cobblestons. The term is used to describe both soil and land. 'Angular Cobbly' is used when the fragments are less rounded. See also coarse fragments.

Cob filburt See cobnut.

Cobnut Roundish, sometimes flattened hazelnut. Leafy cup shorter than nut.

Coca Erythroxylon (Erythroxylum) coca, Lam. Family Erythroxylaceae. Plant, native to Peru and Bolivia, from the leaves of which cocaine is obtained by distillation. Thrives at elevation of 600-1500m and needs very humid atmosphere. Propagated by seed and cuttings.

Cocaine See coca.

Cocaine plant See coca.

Coccidia A group of microscopic, parasitic protozoa that invade the organs of the digestive tract of animals.

Coccinia cordifolia See little gourd.

— **indica** See little gourd.

Coccolith One of the small, interlocking calcite plates covering members of the Coccolithophorida.

Coccidiosis Disease caused in animals and fowls by one or more of the different kinds of protozoa belonging to the group coccidia.

Coccolithophorida A group of unicellular, biflagellate, golden-brown algae characterised by a covering of coccoliths.

Coccomyxaceae A family of algae belonging to the Tetrasporales composed of elongate cells which reproduce only by vegetative means.

Cocculiferous Referring to plants which bear berries.

Coccus (pl.cocci) A sphere shaped bacterium.

Coccus viridis See coffee green scale.

Cochin oil See East Indian lemongrass oil.

Cochlearia armoracia See horse radish.

Cochliobolus miyabeanus See rice brown eye spot.

Cochlospermum gossypium See white sick cotton.

— **religiosum** See white sick cotton.

Cock (1) A small pile into which hay may sometimes be built in order to complete its curing or to protect it from rain and sun. (2) A part of the instrument showing wind direction.

Cockle A small gall on wheat formed due to the attack from a nematode, anguina tritici. See wheat earcockle.

Cockscomb Celosia argentea. Family Amaranthaceae. (1) An annual weed, occurring in dry crops on medium to light soils; can be serious enough to smother the crop. Sets seeds profusely, contaminating grain of wheat and other cereals. Seeds remain dormant in soil till the next rains. Also called quail grass. (2) A fleshy protuberance on a cock head.

Cockspur grass See barnyard millet.

Cocoa See cacao.

Cocoa butter Excess fat taken out of cocoa beans in the process of manufacturing cocoa powder. Along with cocoa powder and sugar forms the chocolate. Also used in confectionery, margarine-making and perfumery.

Cocoa nib Product obtained by cleaning, roasting and shelling cocoa beans. Crushed and mixed with sugar to form drinking chocolate; sold as flakes or powder.

Cocomyces prunophorae See plum leaf spot.

Coco-plum Chrysobalanus icaco L. Family Rosaceae. A shrub or small tree introduced into Kerala which yields edible fruit. Also called icaco plum.

Coconut Cocos nucifera L. Family palmaceae. A perennial tall-growing palm, grown for its nuts and other uses. Provides food and oil from the fruit kernel coir or fibre from the outer husk, shell for making activated charcoal, cups, boxes etc. Leaves used for mat making and thatching and roots for medicinal purposes. Young inflorescence when cut excudes sweet juice from which gur or toddy (alcoholic drink) is made. Tender fruit provides refreshing drink and kernel used for cookery and sweetmeats. Trunk used in building construction. *Copra* obtained after drying used in confectionery, making oil and its cake used as feed and manure. The edible palm oil is used for several other purposes. Tropical plant; grows luxuriantly along sea-coast; needs warm humid climate, bright sunshine and rainfall 100 cm or more. Needs well-aerated soils; grows best in coastal sandy soils, and tidal lands of the back waters. Grows up to an elevation of 1000 m. Nuts planted in nursery and one year old seedlings are transplanted. Also called *narial*.

Coconut beetle Oryctes rhinocerus L. Serious pest; beetle attacks crown, bores into tender parts, especially in the axils of leaves; causes twisting and distorting of entire shoot, and kills young trees. Also called rhinoceros beetle.

Coconut black-headed caterpillar Nephantis serinopa Meyr. Insect pest of palms. Caterpillar feeds on leaves, making leaf dry up.

Coconut bud rot A fungus disease of the coconut palm caused by Phytophthora palmivora and characterised by destruction of the terminal bud and adjacent leaves. Also called coconut red rot.

Coconut fruit rot See Coconut red rot.

Coconut grey leafy spot Disease of coconut caused by Pestalotia palmarum. Characterised by formation of small yellow spots on leaf which turn grey with small black dots at centre.

Coconut leaf rot disease of coconut palm caused by Helminthosporium halodes Dreck, Colletotrichum paucisetum Petch. and Gliocladium roseum Bain. Characterised by blackening and shrivelling up of distal ends of leaflets of younger leaves; appearance of reddish brown spots on tender leaves, development of soft rot and drying and falling off central shoot.

Coconut nut fall See coconut red rot.

Coconut oil Oil extracted from the dried kernel of the coconut fruit (Copra) used in making soap, hydrogenated oil. Also called copra oil.

Coconut oilcake Residue left after extraction of oil from copra. 21.3 per cent protein is present on an average basis. Cattle feed.

Coconut red palm weevil Rhyncophorus ferrugineus Fb. Insect pest of coconut, affecting 4 to 15 years' old tree. Weevil breeds into inside palm; larvae feed on plant tissue near crown, causing drying up of central shoot.

Coconut red rot Disease caused by phytophthora palmivora Butler. Characterised by wilting and withering away of younger leaves in central part of the crown, turning light greyish brown, then darker brown and breaking at the base due to rotting of basal portion, rotting of growing point and destruction of leaves of crown and apical bud. Tender nuts also rot and fall off. Also called bud rot, fruit rot and nut fall.

Coconut root disease Disease of coconut tree characterised by drooping of outer whorl of leaves, leaflets turning brown and dry, new leaves becoming progressively smaller and shedding of immature nuts.

Coconut root rot Disease caused by Ganoderma lucidum. Characterised by withering and drooping off of leaves which remain suspended form tree. Emerging new leaves are smaller and yellowish, tree turning barren.

Coconut stem bleeding Disease of coconut caused by the oozing out of reddish brown liquid out of cracks on stem, fluid drying and turning black.

Coconut wilt Virus disease of coconut characterised by general wilting of leaves, ripping of leaflets, necrosis of leaf-tips, yellowing of old leaves, crown getting smaller, roots dying from tip backwards.

Cocoon The silky protective case of many larval forms before they become pupae; silky or other covering formed by many animals for their eggs.

Coco peat The light fluffy refuse consisting of a mixture of cork pith and fibre shorts left after extraction of useful fibre from the coconut husk. Also coir dust.

Coco-plum Chrysobalanus icaco L. family Rosaceae. A shrub or small tree introduced into Kerala which yields edible fruit. Also called icaco plum.

Cocos nucifera See coconut.

Condensation cloud A mist or fog of minute water droplets that temporarily surrounds the fireball following an atomic detonation in a comparatively humid atmosphere.

Condensation nucleus A particle, either liquid or solid, upon which condensation of water vapour begins in the atmosphere.

Codiaceae A family of green algae in the order Siphonales having macroscopic thalli composed of aggregates of tubes.

Codominant (1) A tree which grows close enough to other trees as to receive little sunlight from the sides. (2) Any plant or animal species which shares dominance with other species in the area.

Coefficient of contraction Ratio between the decreased length, area of section, or volume and the original length, area of section or volume.

Coefficient of determination The square of the coefficient of correlation, the relative measure of the closeness of the association, is termed the coefficient of determination.

Coefficient of discharge A coefficient by which the theoretical discharges of water through orifices, weirs or other hydraulic structures must be multiplied to obtain the actual discharge.

Coefficient of drag or coefficient of friction The ratio of the wall shearing stress to the dynamic head of the stream.

Coefficient of multiple correlation The product moment correlation between the actual values of the dependent variate in multiple regression and the values as given by the regression equation. It measures the closeness of the combined association between a number of independent variables and the dependent variate. The coefficient is usually denoted by R.

Coefficient of multiple determination R^2 The square of the coefficient of multiple correlation R. The measure of the closeness of the combined association of multiple variates, is termed the coefficient of multiple determination.

Coefficient of traction It is the ratio of the total force output of the traction device in the direction of travel to the dynamic weight on the traction device.

Coefficient of variation The standard deviation of a distribution divided by the mean. It may also be expressed as percentage.

Coefficient of purity It is the ratio of

sucrose to total solids in sugarcane juice expressed as percentage. It increases with ripening of juice. When 85-90 per cent the cane is ready for milling.

Coenocyte The plant body in certain algae and fungi in which the vegetative filament is multinucleate and divided into separate cells.

Coenocytic Multinucleate.

Coenothecalia An order of the class Alcyonareae that forms colonies; lacks spicules but has a skeleton composed of fibrocrystalline or gonite.

Coentype An organism having the characteristic structure of the group to which it belongs.

Coenurus cerebralis See gid.

Co-enzyme A substance which activates an enzyme or accelerates its action; co-ferment.

Coffee Coffee sp. Family Rubiaceae. Perennial shrub with opposite leaves, cream coloured flowers, and a berry-like fruit. Grown for the coffee seeds inside the berries. Native to Asia and Africa and grown in south India, in foot hills of Western Ghats and sub-ranges of Eastern Ghats. Requires a humid climate with a well distributed rainfall of 200 cm or above and temperatures of 55°-90° C. Does well in deep, well-drained forest loam having gentle slope with a northern or eastern aspect. Propagation by cutting or grafting; preference to seed. Commences bearing in third or fourth year and continues for over 50 years. Seeds inside berries freed of the pulp, dried in the sun and used green or roasted and ground to make coffee. 5-6 lb of fresh coffee berries give one pound of cured beans. Caffein constitutes up to 3 per cent of the seeds. C. arabica, C. robusta and C. liberica are the three species.

Coffee arabica See Arabian coffee.
— **liberica** See Liberian coffee.
— **robusta** See Congo coffee.
— **stenophylla** G. Don. Family Rubiaceae-coffee variety native of Sierra Leone tried in India but not commercially important.

Coffee black bean Physiological disease of coffee, characterised by blackening of beans in fruit, varying from pin-head spot to complete rotting.

Coffee black root rot Disease of coffee caused by Rosellinia spp. Symptoms as in tea black root rot.

Coffee black rot The disease is caused by Pellicularia koleroga (Cooke). Characterised by leaves, twigs and berries turning black, with coarse, web-like structures on under-surface of leaves, petiole, stem and berries.

Coffee brown eye spot Disease of coffee caused by Cercospora coffeicola. Characterised by appearance of small circular spots on leaves, causing leaf shedding.

Coffee brown root disease Disease caused by Formes noxius and is similar in character as tea brown root disease.

Coffee die back Disease caused by Colletotrichum coffeanum. Characterised by shedding of leaves, blackening and drying from tip downwards of twigs, and premature drying up of berries.

Coffee green scale Coccus viridis Gr. Small green scale insect pest of coffee. Clusters around stems and leaves, sucks sap; also exudes honey-dew inviting mouldy growth.

Coffee koleroga See coffee blackrot.

Coffee leaf disease See coffee rust.

Coffee nursery damping off and collar rot Disease caused by Pellicularia filamentosa (Rhizoctonia saloni). Characterised by stem getting constricted and rot extending down into roots.

Coffee pink disease Disease of coffee caused by Pellicularia salmonicolor. Symptoms as in tea pink disease.

Coffee plant parasite Flowering plant Balanophora indica whose tuberous outgrowths parasitise lateral roots of coffee.

Coffee pruning Process of cutting of the branches of the coffee plant that has borne fruit. The pruning is followed every year in the month of Jan. or Feb.

Coffee red borer Zeuzera coffeae Niet. Caterpillar bores into and damages coffee stems.

Coffee rust Caused by Hemileia vastatrix Birk and Br. Appearance of small yellowish spots on undersurface of the leaves occurs. Covering of leaf surface with orange powder takes place. In severe cases leaf turns brown, dries and falls off. Also called coffee leaf disease.

Coffee species Three species under cultivation. See coffee.

Coffee topping Practice of cutting of the upper portion of the main stem when plant is 75cm high. The cutting is done flush with the top of the primaries to encourage the latter to elongate and produce secondary branches. Process repeated when plant is twice high than before.

Coffee triage Lower grade of coffee beans consisting of broken bits and dark beans sorted out by hand.

Coffee white stem borer Xylotrechus quadripes Chev. Pest of C. arabica coffee. Larva attacks tips of branches of young and old trees and tunnel down into trunk. Borer-affected plant shows yellowing and withering of leaves and drying of branches.

Cogon grass Imperata cylindrica (L.) P. Beauv. (I. arundinacea Cyrilli), family Gramineae. A perennial grass of which stem is used for thatching and paper making. Fruits for stuffing pillows and cushions. Grass used in soil conservation in areas of high rainfall.

Cohesion Attraction of substance of like characteristics such as that of one water molecule to another.

Coir Coconut fibre separated out from the husk, by striking with wooden mallets or by retting; retted coir of superior quality; yarn produced used for ropes, matting, bags and nets.

Coix lacryma See job's tears.

— **lacryma-jobi** See job's tears.

Cola nut A native of tropical Africa. Tree of which the nuts or seeds contain 2 per cent caffeine (kolatine). Used in the preparation such as kola wine, kola chocolate and medicines.

Cold wave A rapid fall in temperature within 24 hours to a level requiring substantially increased protection to agriculture, industry, commerce, and social activities.

Colemania sphenareoides See Deccan wingless grasshopper.

Coleochaetaceae A family of green algae in the suborder ulotrichineae; all occur as attached, disklike, or parenchymatous thalli.

Coleoptera Order of insects to which beetles belong.

Coleoptile (Ingraminae) The first leaf above cotyledon which encloses the stem tip and other leaves (sheath).

Coleorhiza The sheath surrounding the radicle in monocotyledons.

Coleus parviflorus See Madagaskar potato.

— **rotundifolius** See Madagaskar potato.

Cole vegetables Vegetable group which include cabbage, cauliflower and kholkhol. Belong to genus Brassica and family Cruciferae. Essentially cold weather crops; can withstand considerable frost.

Coliform bacteria Group of bacteria consisting of Escherichia coli and related intestinal inhabitants; these ferment lactose to form acid and gas.

Coliform test Test for estimating coliform bacteria.

Collapse The corrugated appearances of timber caused due to excessive but uneven shrinkage during drying.

Collar (1) Portion of plant which marks the transition between stem and roots, sometimes by a gall. (2) A piece of pipe for joining two pipes together as in underground irrigation system.

Collard Brassica oleracea var. acephala. A biennial crucifer of the order capparales grown for its rosette of edible leaves.

Collar rot Any of the several plant diseases in which lesions girdle the stem or trunk near the ground.

Collateral (Ovule) Side by side.

Collateral bud An accessory bud produced beside an axillary bud.

Collateral bundles Vascular bundles in which the phloem is external to the xylem.

Collateral relationship The relationship of two plants or animals of common ancestry.

Collecting ditch, or quarter ditch A field ditch which receives water from the dead furrows or small field ditches and conducts it to an outlet ditch.

Collective farming society A cooperative society for farming purposes where the owners pool the land in the name of society.

Collective fruit Fruit formed from a

complete inflorescence as mulberry and pine apple.

Collective villages These are communal, collective settlements, where all property is collectively owned and worked in an organised way on a collective basis. The land, and all property belongs to the villages, the members give their labour to the common pool and get all supplies of food, housing, clothing, education, cultural and social services and whatever is required and is within the means of the village.

Collenchyma A strengthening tissue composed of cells with walls usually thickened at the angles of the walls.

Colleter Secretory, usually non-vasculated organ (gland or multicellular hair) associated with a bud.

Colletotrichum agaves See sisal hemp anthracnose.

— **camelliae** See tea brown blight.

— **capsici** See chilli anthracnose and dieback, cowpea die-back, sword bean diseases, turmeric leaf blotch, sesbania diseases, horse gram die-back.

— **catechu** See arecanut anthracnose.

— **coffeanum** See coffee die-back.

— **falcatum** See sugarcane diseases.

— **gloeosporioides** See citrus wither tip; mango anthracnose.

— **graminicolum** See Johnson grass red leaf spot, Sudan grass diseases, sorghum diseases.

— **indicum** See cotton anthracnose.

— **lindemuthianum** See lima bean anthracnose.

— **paucisetum** See coconut leaf rot.

— **piperi** See pepper pollu.

— **psidii** See guava anthracnose.

Collision efficiency The fraction of all water drops which, initially moving on a callision course with respect to other drops, actually collide (make surface contact) with the other drops.

Colloid (1) An insoluble substance composed of very minute, particles less than one micron in size. (2) A mineral particle less than 0.0002 mm in diametre. (3) A substance which does not form a true solution in water nor diffuse readily through animal or vegetable membranes. Does not affect the boiling point or vapour tension of the solution.

Colloidal Referring to a state of matter in which particles are larger than those which ordinarily form crystals.

Colloidal clay The smallest fraction-particles in a clay fraction, 1 micron or 0.0001 mm or less in diametre. These determine some of the physical and chemical properties of the soil. Serve as a binding material.

Colloidal instability A property attributed to clouds, by which the particles of the cloud tend to aggregate into masses large enough to precipitate.

Colloidal sulphur Unstable form of sulphur. It is most effective form of all elemental sulphur forms as fungicide because of very small size of the particles.

Collosobrucus chinesis See pulse beetle.

Collusive tendering A restrictive practice in which firms tendering for a particular job agree to submit a common price, or to allow one of their number to submit a lower tender, thus allocating orders between themselves.

Colluvial soil material Soil material that has moved down hill and has accumulated on lower slopes and/or at the bottom of the hill. Colluvial material is moved downhill by the force of gravity and to some extent by frost action and local wash. Syn. colluvium.

Colluvium Deposit of rock fragments and soil material accumulated at the base of steep slopes by gravitational action. Soil that is formed of colluvium is known as colluvial soil.

Colocasia Colocasia esculanta (L.) Schott. (C. antiquorum Schott.) family Araceae. A herb cultivated throughout India but native of south-east Asia. Stem tuber edible and easily digestible leaves also used as vegetable.

Colocasia antiquorum See colocasia.

— **esculanta** See colocasia.

Colocasia blight Disease caused by Phytophthora colocasiae Rach. in the colocasia. Characterised by development of round or irregular oval spots of various sizes with dark brown centres and brown and yellow margins on leaves. Whitish fungus growth on undersurface of affected tissues, leaves falling down when leaf stalk affected.

Colocasia leaf eaters Monolepta signata O. and Prodenia litura F. Insect pest of colocasia feeding on its leaves.

Colon The second position of the intestine of insects; part of the large intestine in vertebrates.

Colon bacillus Escherichia coli.

Colony (1) cluster of micro-organisms grown on a culture medium originating from a single bacterium or spore. (2) An aggregate of worker bee, drones and a queen bee, living together as unit.

Color See Munsell colour system.

Coller See coulter.

Colter slit (Mole Drainage) Knife like cut or slit through soil between ground surface and mole cavity made when a mole plow is drawn through the soil.

Columella A sterile central axis in the sporangia of certain fungi; also the sterile central tissue in the sporophytes of liverworts and mosses.

Columnar (1) See structure types, soil. (2) Describing a plant whose outline resembles a column.

Columnar structure Particles which are prismatic with rounded tips, term commonly used in soil structure.

Colza oil See Yellow sarson.

Comatose Unconscious; in a coma.

Comb A back-to back arrangement of two series of hexagonal cells in a bee-hive.

Comb foundation A thin sheet of embossed or stamped beeswax on which the bees will construct combs.

Combination pliers Those adjustable for size of opening by means of a slip joint, having the outer grip scored and the inner grip notched for grasping round objects.

Combination spray A spray consisting of two or more insecticides or herbicides.

Combine A machine used for harvesting crop. This machines simultaneously threshes and cleans the crop. In improved machines even bagging arrangements are made.

Combined, complex, system or ramspol method A type of layout in subirrigation method in which during the subirrigation season alternate border ditches are used for the delivery and removal of water.

Combined water Water remaining after hygrospcopic water evaporates. Only bright red heat can drive out this water.

Combining ability The ability of an inbred to transmit desirable performance to its hybrid progenies.

Combining weight Of an element or radical is the atomic weight divided by its valency.

Combustion The chemical union of a combustible substance with oxygen, resulting in the production of heat.

Combustion chamber That space within the cylinder and in the cylinder head directly above the piston.

Command The height at an outlet or turnout site of the water level in canal above the general level of the land in the area to be irrigated from that outlet.

Command statement In designing irrigation canals a statement prepared to determine the levels at which water should be run in a distributing system so that it may be possible to irrigate by gravity flow all the area that is included in the capacity statement.

Commelinales An order of monocotyledonous plants in the subclass Commelinidae marked by having differentiated sepals and petals but lacking nectaries and nectar.

Commelina oblique Buch-Ham. Family Commelinaceae. A herb, young shoots are eaten as vegetable.

Commelinidae A subclass of flowering plants in the class Liliopsidae.

Commensal A non-parasitic organism which lives attached to a host and shares its food.

Commensalism A relationship between two kinds of organism living in the same cultural environment without harm to either species and in which one or both member(s) of the pair may obtain food, protection, or other benefits.

Commiphora caudata See hill mango.

— **mukul** See Indian bdellium tree.

— **roxburghii** See *gugul*.

Commission A brokers charge on the

shares which he sells and buys for his clients; the amount of commission is established by a fixed scale.

Commodity A goods or service which satisfied a particular distinct want felt by consumers.

Commodity currency The adoption of a commodity as a currency, e.g. tobacco, furs, cereals, sugar, rum, cotton, etc.

Common bond Similar to a running bond, but with every fifth, sixth, or seventh course a header course, either full or flemish, the former being all headers, the latter with headers and stretchers alternately.

Common brick Brick such as is used for rough work or for filling in or backing.

Common custard apple See bullock's heart.

Common field cricket Brachytrypes achatinus Stall. A pest of maize, *arhar* and jute. Attacks young seedlings and cuts them from base.

Common high brass An alloy of copper and zinc usually in 66-34 composition.

Common jujube Ziziphus sativa Gaertn (Ziziphus vulgaris Lamk). A shrub of family Rhamnaceae whose fruits are edible. Found in Punjab and Bengal States in India.

Common millet Panicum miliaceum L. Family Gramineae. A quick maturing highly drought resistant grain crop. Grown mostly as a rainfed crop. Also called proso-millet or hog millet.

Common millet downy mildew Disease of common millet caused by Sclerospora graminicola (Sacc.) Schroter. Characterised by plants growing dwarf, tillering excessively after crown and producing further branches from auxillary buds, turning of floral bracts into leaf-like malformations, kernels failing to form, leaves showing necrosis and browning followed by splitting and shedding.

Common millet head smut Disease caused by Sphacelotheca destruens. Characterised by entire inflorescence turning into a smutted gall, covered by a greyish white false membrane, with dark-brown powdery spore mass coming out when ruptured.

Common mulberry Morus australis poir (Morus acidosa Griff., Morus indica auct. non L.). Tree commonly found in sub-Himalayan tracts of India whose fruits are edible. Leaves fed to silk worms. Also see white mulberry.

Common pear Pyrus communis L. Cultivation restricted and most varieties self-unfruitful. Propagated by shield budding. Also known as European pear and *nakh*.

Common pricing The fixing of identical prices for goods by agreement between otherwise competing firms selling through wholesale or retail channels of distribution.

Common salt (NaCl) Sodium chloride most common chemical used in household. Used for preservation of foods. Also a good and effective herbicide.

Common sow thistle Sonchus oleraceus L. Family Compositeae. An all the year round weed in both dry and irrigated areas with a slender long stem. Seeds moderately. Also known as milk thistle and *Doodhi*.

Common thyme Thymus vulgaris L. Family Labiatae. A sub-shrub, well branched but with prostrate habit. Oil for perfumery is obtained. Thymol derived from this oil is used in medicines. Leaves used as condiment.

Common vetch Vicia sativa L. Family Papilionaceae. Grown for fodder and called *ankra*.

Common wheat Triticum vulgare L. Family Gramineae. One of the important cultivated species in India. Another specie is T. durm which has the largest number of varieties in India.

Communication It is a process of human interaction in which one person communicates a new idea to another person.

Community (1) A sub-group having many of the characteristics of society, but on a smaller scale, and with less intensive and coordinated common interests. (2) Aggregation of organisms or plants characterised by a distinctive combination of two or more ecologically related species, an example is a deciduous forest. Also known as ecological community.

Community classification Arrangement of communities into class with respect to their complexity and extent, their stage of ecological succession, or their primary production.

Community development It is the process of encouraging and educating the community to set about solving local problems using its own resources and local organization, although outside help may be provided in the way of expert advice, materials and finances.

Community organization It may be defined as a continuing process for obtaining the best integrated social interaction of individuals, groups and institutions within a community so as to enable it to act collectively and advance the common welfare.

Commutator A contrivance for reversing the direction of electric current in any circuit.

Compact (1) To press soil particles together so as to firm the soil before or after planting. (2) Pertains to a densely packed, firmly massed soil.

Compact body A region where the cells in tissue are arranged close together with a minimum of intercellular space.

Compacted earth lining Compaction of natural subgrade soil below bed and sides of the canal with the object of utilizing the natural bond between soil particles to achieve a soil structure of greater strength.

Compactors machines or equipment (rollers of tampers) used for developing suitable compaction of earth-fill in embankments, rolled-fill dams, etc.

Compact soil See consistency.

Companion cell One of the elongated cells adjoining sieve tubes in phloem tissue of most plants.

Companion crop Crop sown with another crop, usually a small grain with which a forge crop is seeded, preferred to the term "nurse crop"

— **Cropping** See inter-cropping.

Company reserves: Profit set aside for 'ploughing back' into a business.

Compartment A territorial unit of a forest, for the purpose of administration, description and record.

Compatible Readily capable of fertilization.

Compatibility (1) A new idea or practice that is consistent with existing ideas and beliefs. (2) In fruit plants, a condition in which the male gametes will fertilise the female gametes.

Compensation level An effective light penetration level in freshwater ecosystems, it is the depth at which photosynthesis balances respiration.

Compensation point The point at which the respiration rate equals the photosynthetic rate.

Competence The diametre of the largest particle a stream can move.

Competitive plant A plant which is capable of forcing out plants of other species from a field.

Complementary genes Genes that interact to produce a new character.

Complementary goods Goods which are necessarily used in conjunction with each other, e.g. pens and ink, knives and forks, petrol and motorcars.

Complementary factor A factor which only in conjunction with one or more genes controls a particular character.

Complemental feed Fodder given to livestock either in place of or in addition to grazing.

Complete fertiliser A fertiliser which contains three major nutrients to be supplied to the soil, viz. nitrogen, phosphoric acid and potash. Also mixed fertiliser.

Complete flower A flower which bears four types of floral organs, sepals, petals, stamen, and carpels.

Complete record The information which relates to the origin, variety, kind, germination and purity of seeds of any notified kind, or variety, offered for sale, sold or otherwise supplied.

Completely saturated (1) Referring to clay containing all the basic ions that it is capable of holding. (2) A soil whose all the pores are filled with water.

Complexity (Idea) Complex in understanding, new ideas that are relatively simple to understand and use will generally be accepted more quickly than more complex ideas.

Complex, Soil A mapping unit used in detailed and reconnissance soil surveys where two or more defined soil units are so intimate-intermixed geographically that it is impractical, because of the scale used, to separate them.

Complex virus Two or more viruses or virus strains which together cause a disease in a host with symptoms that are not predictable from the knowledge of individual components.

Components Manufactured goods which are incorporated into a final product without further processing.

Component materials of soil horizons Solid liquid and gaseous materials forming soil horizons. Inorganic material composed of substances formed from chemical and physical weathering of parent rocks. Organic materials consisting of living plants and animals and their residues and remains forming the solid portion, the soil solution and the gaseous components occupying a portion of the pore space.

Composital The single family of the order Asterales; perhaps the largest family of flowering plants, it contains about 19,000 species.

Composite demand The total demand for a product which arises when it is required for a number of different purposes.

Composite disease A disease caused by a virus complex.

Composite sample A sample formed by combining and mixing all the primary samples taken from the lot.

Composite supply The total supply of a number of products which satisfy a particular demand.

Composite system A combination of for example, the herringbone and gridiron system.

Composition (1) The relative production of various plant species in a given area. (2) The components used in any mixture of feed fertiliser etc.

Compost (1) Organic residues, or a mixture of organic residues and soil, that have been piled, moistened, and allowed to decompose. Mineral fertilisers are sometimes added. If it is produced mainly from plant residue, it is often called "artificial manure" or "synthetic manure". (2) To cause organic waste to become decomposed.

Composting Process of reducing vegetable and animal refuse to a quickly utilizable condition for improving and maintaining soil fertility. A good manure similar to farmyard manure.

Compound Made up of several elements.

Compound bar chart Bar chart showing two or more components in each bar.

Compound fertiliser Fertiliser containing more than one nutrient elements

Compound leaf A leaf the blade of which is subdivided into several distinct parts.

Compound structure The size, shape and distinctiveness of the aggregates of individual particles belonging to more than one type of soil.

Compressed air sprayer A sprayer used for spraying field crops wherein pressure inside the tank is created by the air using a hand pump. Due to this pressure liquid comes out and through the nozzle it is sprayed.

Compressibility The susceptibility of a soil to decrease in volume when subjected to load.

Compression gauge A metre which registers in kg per square cm the pressure on gases in the combustion chamber.

Compression ignition engine A type of engine in which the fuel is sprayed into the cylinder and ignited by the heat of compression of the air charge.

Compression ratio Ratio between the atmospheric pressure when the piston is at the end of the down stroke and the pressure when the piston is in its topmost position of the compression stroke. It is measured in atmospheres.

Concentrate All feed, low in fibre and high in total digestible nutrients, that supplies primary nutrients (protein, carbohydrate and fat), e.g., grains, cottonseed meal, and wheat bran.

Concentrated flow The flowing of a rather large accumulated body of water over a relatively narrow course. It often causes serious erosion and gullying.

Concentrated organic manures Concentrated materials of organic origin like oilcakes, bone meal, urine and blood containing plant nutrients in concentrated form.

Concentration The amount of active

material in a diluent.

Concentration of produce A term in marketing, includes operations concerned with the assembling and transport of produce from the field to a common place of market.

Concentric One circle within another with a common centre.

Conceptacle In algae such as fucus, a cavity with an apical pore and containing the reproductive bodies.

Concrete frost Ice in the soil in such quantity as to form virtually a solid block.

Concrete lined channel A irrigation channel lined with concrete. Initial investment high but maintenance cost minimum and seepage losses reduced considerably. Used *insitu* or preset concrete blocks.

Concrete mix A processed mixture of the ingredients of the concrete, i.e. aggregates, cement and water.

Concretion A mass or concentration of a chemical compound such as calcium carbonate or iron oxide, in the form of a grain or nodule of varying size, shape, hardness, and colour, found in soil and in rock. The term is sometimes restricted to concentrations having concentric fabric. The composition of some concretions is unlike that of the surrounding material.

Condensation A term referring to the process of turning vapour into liquid phase.

Condiment Spice or other seasoning material.

Conditioner A chemical or other substance which when applied to soils alone or mixed with fertiliser, improves their structure and prevents caking or coalescence of fine soil particles.

Conditioning (1) Any process in which animals, plants or their products are treated prior to processing of final use. (2) The process of controlling the moisture of stored grains.

Conditioning and precleaning This involves removing awns or hulls, breaking up clusters, scalping off large trash, and other operations which improve the condition and flowability of the seed.

Conditioning of wheat The conditioning process wheat is a made to undergo before products are made of it. Process depends upon the quality of wheat and the item to be made and also on the manufactures.

Conductance The opposite of resistance, i.e. the ease with which a conductor carries an electric current.

Conducting tissue The principal water and food transporting tissue in a plant, xylem and phloem.

Conduction (1) The movement of liquid nutrients and water etc. from one part of the plant to another. (2) Conduction of heat from one point to another in solid materials.

Conduit (1) A conductor as a large pipe for fluids. (2) A metal enclosure used to protect electric wiring. The tubing may be rigid or flexible. (3) Any channel open or closed, for the conveyance of water.

Cone A specialised branch bearing an aggregation of sporophylls.

Cone clutch A form of friction clutch in which the power necessary for driving is effected by the friction of smooth, turned conical surfaces. These surfaces are often faced with leather.

Cone of depression The inverted bell-shaped cone developed around a pumping well which increases size with time of pumping.

Cone of influence See cone of depression.

Cone wheat See rivet wheat.

Confined aquifer An aquifer enclosed by impermeable layer both at the top and the bottom and in which water is held under pressure.

Confidence belt The area between the upper and lower confidence limits is the confidence belt.

Confined water Groundwater, constrained by an overlying confining bed, which is under sufficient pressure to rise above the bottom of the confining bed.

Confinement The degree of closure of a coastal water basin; "confined" indicates shoreline length three times width of opening.

Confounding Where other variables render non-comparable two groups, the disturbances caused to the variables is said to be "confounding" and the variables are reckoned to be "confounded".

Confining clays Impervious or slowly previous clay layers in the subsoil which restrict or confine the movement of percolating water.

Confluence (1) A stream formed from the flowing together of two or more streams. (2) The place where such streams joins.

Congenial Able to cross-fertilise readily, to unite easily, as with stocks and scions.

Congenital Present at or dating from birth, but not, necessarily, inherited in the genetical sense.

Congo coffee Coffea robusta Linden. Family Rubiaceae. Native of congo now grown in India.

Conical divider (Boerner type) This equipment is commonly used for dividing samples in the seed testing laboratory

Conidiophore Simple or branched hypha on which conidia are produced.

Conidiosporangium A sporangium forming zoospores or aplanospores.

Conidiospore A spore formed asexually by constriction at the tip of a hypha.

Conidium (pl. Conidia) Any asexual spore except sporangiospore or chlamydospore.

Conifer A tree belonging to the order coniferae, usually evergreen, with cones and needle-shaped leaves, and producing wood known commercially as softwood.

Coniferous Cone-bearing.

Coniothecium chomatosporum See apple stem black.

Coniothyrium arecae See arecanut leaf blight.

Conjugation The fusion of two similar gamets, as in the conjugales.

Conk A definite, individual, woody, spore-bearing fruiting body of a wood-destroying fungus, which projects beyond the bark.

Connecting rod A rod or arm, with bearing at both ends, which makes the connection between crankshaft and pistons.

Connection Electrical conductor joining parts of an electrical circuit.

Connector Any device for holding in electrical contact the ends of conducting wires in such a manner that they may be readily released when it is desired to disconnect them.

Conservation reservoir, or **Conservation storage reservoir** A reservoir, impounding water for useful purposes, such as municipal supply, power, irrigation and recreation.

Conservation, Soil (1) Protection of the soil against physical loss by erosion or against chemical deterioration, that is, excessive loss of fertility by either natural or artificial means. (2) A combination of all methods of management and landuse that safeguard the soil against depletion or deterioration by natural or man-induced factors. (3) The division of soil science dealing with soil conservation (1) and (2).

Conservation The protection, improvement and use of natural resources according to principles that will assure their highest economic or social, benefits.

Conservative grazing The practice of limiting the number of livestock on a grazing area in such proportions as would not exhaust the area in successive seasons.

Conservatory A glasshouse which could accommodate plants for display.

Conserve To save, improve and to use intelligently, logically used in context with soil conservation, energy conservation etc.

Consistence (1) The resistance of a material to deformation or rapture. (2) The degree of cohesion or adhesion of the soil mass. Terms used for describing consistence at various soil moisture contents are as follows:

- **Wet-soil** non-sticky, slightly sticky, sticky, and very sticky, non-plastic, slightly plastic, plastic and very plastic.
- **Moist-soil** loose, very friable, friable, firm and very firm; compact, very compact, and extremely compact.
- **Dry Soil** loose, soft, slightly hard, hard, very hard, and extremely hard.

Consolidation The gradual reduction in volume of a soil mass resulting from an increase in compressive stress.

Constant returns Description of a

situation in which all the productive services are increased in a given production, the amount of product being increased in the same proportion. Factor substitutions and qualitative changes are not introduced.

Constituents The components or elements of a substance.

Constitutive enzyme An enzyme whose formation does not depends on the presence of a specific substrate.

Constraint A condition or restriction on how something may be achieved. For example, an objective such as the maximization of the national product may be subject to a constraint that the absolute or relative incomes of a particular group shall not be reduced.

Construction surveys This term refers to all types of surveying activities which are performed in connection with any construction operation or which are necessary immediately prior to the start of construction or during construction. These include surveying operations in connection with all types of excavation, embankment placing, structure erection and machinery installation.

Consumers (1) Plant-eating animals (herbivores). (2) The members of the general public who constitutes the buyers of items manufactured or produced.

Consumer credit Loans to members of the general public to finance the purchase of goods. These loans take the form of credit from supplies, hire-purchase and personal loans from banks.

Consumer durable A good of reasonably long life, such as a refrigerator or piece of furniture, as distinct from, say foodstuffs.

Consumer goods Products in the actual form in which they will reach domestic consumers.

Consumption The use of goods and services for the satisfaction of wants. Consumption was described by Keynes as the sole object of all economic activity.

Consumption function A schedule that relates aggregate consumption expenditure to the level of disposable income; called also the schedule of 'propensity to consume' or sometimes, the schedule of 'intended consumption'.

Consumption goods, Collective Goods and services provided collectively by the community and paid for out of taxes.

Consumptive use The sum total of water used in transpiration, growth of the plants and the evaporation from the soil surface. It is a function of crop, soil, climate and water availability etc.

Consumptive-use efficiency The ratio of consumptive water use by the crop of an irrigated farm or project and the irrigation water stored in the root zone of the soil on the farm or the project area.

Consumptive use of water See absolute water requirement.

Consumptive water requirement See absolute water requirement.

Contact herbicide A herbicide which kills the plants by coming in contact with plant tissue rather than by absorption.

Contact insecticides Insecticides which kill the insects directly on contact by penetrating through body pores or membranes through the outer part of their body.

Contact spray A solution containing contact herbicide or insecticide and used as a spray.

Contagious (highly) Communicable in common parlance, a disease that is easily 'caught'.

Contaminants The factors which affect the genetic and physical quality of seed such as off-types, foreign pollen, other crop plants, weed plants, plants affected by designated diseases, weed seeds, etc., are referred to as contaminants.

Continental air A type of air whose characteristics are developed over a large land area and which therefore has relatively low moisture content.

Continental tropical air A type of tropical air produced over subtropical arid regions, it is hot and very dry.

Continental high A general area of high atmospheric pressure which on mean charts of sea-level pressure is seen to overlie a continent during the winter. Also known as continental anticyclone.

Continentality The degree to which a point on the earth's surface is in all respects subject to the influence of a land mass.

Continental terrace The continental shelf and slope together.

Continuous cropping Planting of the same crop on the same land year after year. Also called continuous cultivation.

Continuous cultivation See continuous cropping.

Continuous drying Drying carried out with a continuous flow of grain and air, in contrast to batch operation.

Continuous grazing Grazing limited to the same land throughout the season.

Continuous growth Growth without visible rhythm producing shoots without distinct articulations (cf. articulated growth).

Continuous variable Where between pair of values, however close together, it is possible to have another value.

Contour (1) An imaginary line on the surface of the earth connecting points of the same elevation. (2) A line drawn on a map connecting points of the same elevation.

Contour basin A basin made by levee constructed along contours. Occasionally cross levees may also be present. Also see contour checks.

Contour bench levelling A method of soil conservation for preparing steep lands fit for efficient irrigation. The method consists in dividing land into a series of strips on the approximate contour and levelling each strip independently, forming a series of steps in successive elevations around the slope.

Contour border irrigation A method of irrigating gently sloping fields, the whole area being divided into strips by ridges along the contours and cross ridges. The ridges confine the water to a particular strip till it is completely full, before it is let to flow to the next lower strip. Also called contour check irrigation.

Contour bund In soil conservation, an embankment constructed on a contour.

Contour bunding The construction of small bunds across slope of the land on a contour so that the long slope is cut into a series of small ones and each contour bund acts as a barrier to the flow of water.

Contour checks Checks or plots made for irrigation which are prepared by contracting ridges along contours at vertical intervals of 6-12 cm, and connecting them with cross ridges at intervals, method used on sloping land needing little land levelling.

Contour ditch Irrigation ditch laid out approximately on the contour.

Contour farming Conducting field operations, such as plowing, planting, cultivating and harvesting, on the contour. It is a soil conservation technique.

Contour flooding Method of irrigating by flooding from contour ditches.

Contour furrow A furrow made along a contour or at a uniform grade.

Contour-furrow irrigation A type of furrow method of irrigation in which water is applied by adjacent furrows across the slope on controlled grade. Also known as 'cross-contour-furrow irrigation'.

Contour guidelines Lines laid out approximately on the contour to serve as a guide for farming operations.

Contour interval The difference in elevation between adjacent contours on a map.

Contouring The practice of planting or preparing a field according to the contour of the land.

Contour lines Lines placed on a map to show the topography of the ground and to give a reading of elevations.

Contour planting Planting so that the rows run-around the hill or slope on the same level, rather than up and down.

Contour strip cropping (1) A method of soil conservation in which alternate strips of erosion resisting crops are grown with not too erosion resistant crops. The width of the strips vary with crops, soils and climate of the area. (2) The production of crops in long variable width strips which are placed approximately on the contour, and crosswise to the slope.

Contour system of planting A soil conservation measure wherein rows of trees are planted along a uniform slope and usually at right angles to the slope. Reduces soil loss on hills with steep slopes.

Contour terrace A low terrace formed on steep mountain slopes with inadequate natural cover to prevent erosion until natural cover becomes established. Checks or low dams are placed in the trough to prevent any lateral flow. It is also used on permeable agricultural land.

Contour trenching Excavating trenches along the contour; undertaken in uncultivated lands in the upper portions of the catchment. Bunds are formed along the trenches. Creates more favourable moisture conditions in shallow soils for growth of natural vegetation or of planted tree seedlings. Breaks velocity of runoff.

Contract An agreement, either oral or in writing, whereby one party undertakes to do something for the other party to the contract.

Contract Note A legal document relating to the sale or purchase of shares. It contains full details of the transaction.

Contracted weir Weir having sufficiently sharp upstream edge to cause the nape to contract.

Contractile vacuole One or more spherical bodies forming in the body of certain primitive algae which are considered regulatory in water control and possibly play a role in the elimination of waste materials.

Control (1) Prevention of, or reduction of loss from, plant disease. (2) A standard used for comparative purposes in experiments.

Control schedule A summary programme by which a project is broken into its principal features, and shows in what years the work on each feature will be done and gives an estimate of the cost of that work for each year.

Control section, Soil The vertical section upon which the taxonomic classification of soil is based. The control section usually extends to a depth of 100 cm (40 inches) in mineral materials and (tentative, system of soil classification for Canada) to 160 cm (64 inches) in organic materials.

Control structure Water regulating structure, usually for gravity flow systems.

Control survey A survey to provide permanent and accessible benchmarks and monuments to be used as reference points for other surveys throughout the given area.

Control system Consists of steering device, throttle and spark control, brake levels, gear shift lever, foot pedals, and the brake system.

Control drainage Regulation of the water table by means of pumps, control dams or check drains, or a combination of these, for maintaining the water level at a depth favourable to best crop growth.

Controlled mosaic A mosaic in which the photographs have been adjusted by reference to ground control so as to produce a continuous photographic representation of a part of the earths surface.

Convection In general, mass motions within a fluid resulting in transport and mixing of the properties of that fluid. Distinction is made between: (1) free convection (or gravitational convection), motion caused only by density differences within the fluid, and (2) forced convection, motion induced by mechanical forces such as deflection by a large-scale surface irregularities, turbulent flow caused by friction at the boundary of a fluid, or motion caused by any applied external force.

Convection theory of cyclones A theory of cyclone development proposing that the upward convection of air (particularly of moist-air) due to surface heating can be of sufficient magnitude and duration that the surface inflow of air will attain appreciable cyclonic rotation.

Convective cloud A cloud which owes its vertical development, and possibly its origin to convection.

Convective instability The state of an unsaturated layer or column of air in the atmosphere whose wet bulb potential temperature (or equivalent potential temperature), decreases with elevation. Also known as potential instability.

Convective precipitation Precipitation

from convective clouds, generally considered to be synonymous with showers.

Convective region An area particularly favourable for the formation of convection in the lower atmosphere, or one characterised by convective activity at a given time.

Convergent evolution Evolution of unrelated or distantly related groups of organisms along similar lines, resulting in the development of similar traits or features in the unrelated groups.

Convergent improvement System of crossing back-pollinating, and self pollinating with accompanying selection in an effort to improve inbred lines of cross-pollinated plants without interfering with their behaviour in hybrid combinations.

Convergent precipitation A synoptic type of precipitation caused by local updrafts of moist air.

Conversion The process of sawing timber.

Conversion factor That factor by which the numerical value made in one system of units must be multiplied to arrive at the numerical value in another system of units.

Convertibility The freedom to exchange any currency for another currency at the ruling rate of exchange.

Conveyance Legal description for the transfer of property from a seller to a buyer.

Conveyance losses, or **transmission losses** Losses of water in transit from the source of supply to the point of service whether in natural channels or in artificial ones, such as canals, distributaries, ditches, or watercourses. They comprise evaporation from the water surface, seepage, and incidental transpiration by vegetation growing in the water or along the banks of natural channels, canals or watercourses. Also referred to as total losses or canal losses.

Conveyance structure Structures built in a canal system to help provide general control and conveyance of the irrigation water to a location of final control usage, or disposal.

Conveyors The equipment used for moving the seeds.

Convolute Rolled together coiled. Applied to leaves under cotyledons.

Convolvulus arvensis See field bindweed.

— **batatas** See sweet potato.

— **hispidus** Ipomea hispida (Vahl.) R & S. (Ipomea eriocarpa R. Br.) Herb of family Convolvulaceae.

— **reptans** Swamp cabbage, Ipomea reptans Poir. Family Convolvulaceae. An aquatic herb used in soil and water conservation.

Cookia punctata See wampee.

Cool-season grass Grass species adapted to rapid growth during the cool moist periods of the year; usually dormant during hot weather or injured by it.

Co-op A short of co-operative.

Co-operation In its simplest form, method of organization in which persons associate voluntarily and form themselves into a body for satisfying specific economic needs.

Co-operative movement An organisation consisting of co-operative societies engaged in the manufacture and distribution of consumer goods, run democratically for the mutual benefit of their member customers.

Co-operative villages Individual farmers may cultivate their land separately or in joint groups and may receive their common services from a co-operative stores, co-operative implement shed and common municipality services.

Co-partnership in Industry Schemes introduced by individual firms to give employees a bigger stake in the prosperity of the firms they work for.

Copper acetoarsenite Paris green.

Copper blight A leaf spot disease of tea caused by the fungus Guignardia camelliae.

Copper deficiency Condition in plants brought about by deficiency of copper. Produces whithering and yellowish grey chlorosis in the margins of tips of younger leaves in cereals. Leaves fail to unroll, dwarfed and distorted heads, associated with exantheme or die-back of citrus, pears, plums.

Copper-fungicide Copper compounds used for killing fungi causing plant diseases, act by contact, killing only those spores or other parts of disease contact. Examples are Bordeaux mixture, copper oxychloride, cuprous oxide, perenox, perelan etc.

Copper spot A fungus disease of lawn grasses caused by Gloeocercospora sorghi and marked by coppery-red areas.

Copper sulphate (C_4SO_4) A fungicide.

Coppice A growth of small trees that are repeatedly cut down at short intervals; the new shoots are produced by the old stumps.

Coppice forest A forest of second-growth sprouts. A forest area, recently cleared, in which new shoots have sprouted from the stems of the woody plants.

Coppice-mound Small mound of soil material stabilised around desert shrubs.

Coppice shoot A shoot arising from an adventitious bud at the base of a woody plant that has been cut near the ground or otherwise burnt back.

Copra Dried coconut kernel.

Copra oil See coconut oil.

Coprogenous earth A material in some organic soils that contains at least 50% by volume of fecal pellets less than 0.5 mm in diametre.

Coprophilous Living on a dung substrate, as certain fungi.

Coral tree Erythrina variegata L. (E. indica Lamk.). A multi-purpose tree of family Papilionaceae.

Cordon system of training A system of training grape vine. The truck is allowed to grow straight up to 60 cm and then trained in a horizontal direction along the wire of the trellis.

Coriander Coriandrum sativum (L.). Family Umbelliferae. An aromatic herb with mild aroma. Also known as *dhania* (Hindi).

Coriander powdery mildew Disease of coriander characterised by faint slightly discoloured specks from which white powdery spots spread. Caused by Erysiphe polygoni.

Coriander sativum See coriander.

Corn See maize.

Corona An appendage on the inner side of the corolla of some flowers.

Coronal root A root arising just at or slightly above the soil surface from the nodes of the crown of a plant of the grass family. Such roots makes the main system of such plants.

Corrosion The process of corroding such as rocks by the water due to their solvent action when it comes in contact with the rocks.

Corrugation irrigation A method of irrigation for close growing crops. A method in between border irrigation and furrow irrigation.

Cotton Any plant of the genus Gossypium, family Malvaceae. An important fibre crops of the world. Main varieties are Gossypium arboreum L. G. herbaccum L., G. hirsutum L. and G. herbadenre. A sub-tropical crop. Cotton seed oil is used for edible purposes. Known as *Kapas*.

Cotton anthracnose Disease caused by Colletotrichum indicum Dastur. Appearance of small, broken, sunken spots on bolls, shedding of bolls, formation of yellow/pink and brittle fibre and dying of seedlings.

Cotton aphid Aphis gossypii Glov. Insect pest of cotton. Suck sap from leaves, stems flowers and bolls.

Cotton areolate mildew Disease caused by Ramularia areola Atk. characterised by appearance of angular, yellow brown spots and whitish growth on older leaves.

Cotton bacterial blight See cotton black arm.

Cotton black arm Disease of cotton caused by Xanthomonas malvacearum (Smith) Dowson. Characterised by scattered angular water soaked spots on leaves, turning black and drying up of young shoots. Yellow lint formation.

Cotton bollworm See cotton spotted bollworm, cotton pink bollworm.

Cotton bugs Dysdercus cingulatus Fb. and the dusky cotton bug. Oxycarenus lactus Kirby. Insect pest of cotton which sucks the juice of flower buds and bolls and taint the lint.

Cotton damping off See cotton wet rot.

Cotton dry root rot Disease of cotton caused by Macrophomina phaseoli. Large brown lesions on collar portion, discolouration of vascular portion of roots, rotting of rootlets and drying of plants.

Cotton grey mildew See cotton areolate mildew.

Cotton jassid Empoasca devastans Dist. E. flavescenes Dist. Nymphs and adults such juice from leaves and turn them yellow and reddish brown, leaves fall off.

Cotton leaf roller Sylepta derogata Fabr. Insect pest whose caterpillar feeds on lower surface of leaves and rolls them, eats up the whole leaf. Sometime whole plant defoliated.

Cotton leaf spot Disease of cotton caused by Mycosphaerella gossypina. Characterized by appearance of round, irregular, dark brown spots with ashy centres on leaves; drying up of whole leaf.

Cotton powdery mildew Disease caused by leveillula taurica and Oidium sp. Characterised by white shiny powdery patches on leaf under-surface.

Cotton red leaf A physiological disease in which leaves turn yellow or red; drying up and shedding of leaves.

Cotton root rot Disease caused by Rhizoctonia bataticola and R. Solani. Sudden wilting of plants, rotting of roots and shredding of bark.

Cotton rust Disease caused by cerotelium desmium. Formation of small, yellowish brown rust pustules on leaves occur.

Cotton sclerotial disease Disease caused by Pellicularia rolf sci. Characterised by white weft of mycelium forming near the base of the plant of which white to reddish-tan spherical sclerotia appear.

Cotton shore shin See cotton dry root rot.

Cotton stem borer Sphenoptera gossypi G. Beetle grub of the insect pest bores into the stem.

Cotton stem weevil Pempheres offinis F. Insect pest whose white grub bores into stem near the base and causes gall like swelling.

Cotton stenosis Virul disease characterised by crinkling and deformation of leaves.

Cotton tirak A physiological disease characterised by early yellowing, drooping and shedding of leaves. Bolls crack and do not open properly.

Cotton wet rot Caused by Phytophthora parasitica and Phythium spp. Leaf rot and collar rot occur. Death of seedlings.

Cotton white fly Bemisia tabaci Genn. Serious insect pest; nymphs and adults suck sap from leaves, produces honey-dew like substance which attract sooty mould.

Cotton wilt Serious disease caused by Fusarium oxysporum and F. vasinfectum Atk. characterized by leaves turning pale and yellow and withering prematurely. Plants die in patches.

Cowpea Lobia. Vigna sinensis (L.) Savi ex Hassk. (Dolichos sinensis L.; vigna catjang Walp.). Family Papilionaceae. An annual legume generally found as spreading bush; grown for grain and green manuring.

Cowpea anthrachose Disease caused by Glomerella lindemuthianum. Characterised by infected portions showing dark brown sunken spots, with raised reddish or yellowish margins.

Cowpea aphids Aphis caraccivora K. Insect pest infesting leaves, stems and pods.

Cowpea bacterial blight Disease of cowpea caused by Xanthomonas vignicola. Characterised by appearance of light, yellow, irregular to circular spots with necrotic brown centres on leaves. Falling off of affected leaves early.

Cowpea die-back Disease caused by colletotrichum capsici. Drying up of twigs and branches from tip downwards. Shrivelling of pods.

Cowpea dry root-rot Disease caused by Macrophomina phaseoli. Symptoms as in cotton dry root-rot.

Cowpea leaf spot Disease caused by Cercospora canescens and C. cruenta. Symptoms include appearance of angular, brown or red spots on leaves, stalks and pods.

Cowpea mosaic virus disease; yellow and green patches on leaves with mottling, plants remain stunted.

Cowpea powdery mildew Disease caused by Erysiphe polygoni. Characterised by white powdery patches on leaves and other green parts which later on turn black.

Cowpea root rot Fungus disease

caused by Pellicularia filamentosa. Red brown discolouration on stems occur which spreads and causes plant to collapse.

Cowpea rust Fungus disease caused by Uromyces appendiculatus. Characterised by appearance of small yellowish to brown and dark-brown spots on leaves and other parts.

Cowpea stem fly Agromyza phaseoli C. A small blue black fly that lays eggs on young stems and maggots boring into them.

Crab Gecarcinucus jacquemontic. Crustacean; sometimes a serious pest of paddy which cuts plants at the base.

Crabgrass Any of the several species of genus Digitaria, family Gramineae. Serious weed pest of lawns and fields.

Crab stock A seedling apple plant used as a grafting stock.

Cracked stem A boron deficiency disease of celery characterised by brown mottling of leaves and brittleness and cracking of leaf stalks.

Cradle Footing structure shaped to fit the conduit it supports.

Cradle knoll A small knoll formed by earth that was raised and left by an uprooted tree.

Craftsman A skilled worker in a particular occupation, trade or craft, who is able to apply a range of skills to basically non-repetitive work with a minimum of direction and supervision.

Crag and tail An elongate hill that has at one end a steep face of ice-smoothed rock and at the leeward end a tapering streamlined tail of till.

Cramp An involuntary spasm or contraction of a muscular tissue in the body.

Crank A link in a mechanical linkage or mechanism that can turn about a centre of rotation.

Crank angle (1) The angle between a crank and some reference direction. (2) Specifically the angle between the crank of a slides crank mechanism and a line from crankshaft to the piston.

Crank arm The arm of a crankshaft attached to a connecting rod and piston.

Crank axle (1) An axle containing a crank. (2) An axle bent at both ends so that it can accommodate a large body with large wheels.

Crankcase The housing for the crankshaft of an engine, where, in the case of an automobile, oil from hot engine parts is collected and cooled before returning to the engine by a pump.

Crankcase ventilator Draft from the fan creates a suction which is used to carry off vapours formed in the crankcase thus avoiding a dilution of the oil which would otherwise result from water and gasoline vapours.

Crankshaft The shaft about which a crank rotates.

Crank-type duster Hand operated duster in which fan is rotated at high speed to draw dust from the dust chamber and discharged through the delivery pipes and through one or two rows of nozzles. Also called rotary duster.

Crash cost The cost associated with the shortest possible completion time for a project. The crash cost is higher than the normal cost due to the expense of using more manpower or more plant and equipment.

Crataeva marmelos See bael.

Crawler tractor A tractor usually used for heavy work. It is without pneumatic tyres and is provided with a endless track on which it moves.

Crede's method Instillation of a 2% silver nitrate solution into each eye of the newborn infant to prevent ophthalmia neonatorum.

Credibility (communication) Credibility is the degree to which a communication source is perceived as trustworthy and competent by the receiver.

Credit A loan made either directly, or indirectly through deferred payment goods received; thus credit may be extended by a creditor (lender) to a debtor (borrower) either through the transfer of cash with the promise of future payment, or through the sale of goods to be paid for later in an agreed manner.

Credit insurance A type of insurance affording indemnity to merchants and manufacturers against the risk of loss due to the insolvency or protracted default of customers to whom goods are sold on credit.

Credit rating An evaluation of the soundness of an individual or business firm as a credit risk.

Credit sale A procedure for purchasing goods under which the purchaser pays a deposit on receipt of the goods followed by a number of instalments until the debt is cleared.

Credit squeeze The contraction of credit as a result of government measures.

Credit transfers A simple means of paying money to somebody else, other than by direct payment in cash.

Credit union A group of people engaging in organised self-help who save their money together and make low interest loans to each other.

Creeper (1) A plant which spreads over or under the soil surface by roots or branch stem. (2) Vine. (3) Any plant which is prostrate or trailing in its habit of growth.

Creep erosion See erosion.

Creeping sorrel See Indian sorrel.

Creosote Oily, colourless liquid obtained from the tar produced by the destructive distillation of wood. Used as an insect repellent, wood preservative and in medicines; mild insecticide and antiseptic.

Creosote emulsion An emulsion of creosote and water; a wood preservative.

Crepitating Having a crackling sound like air under the skin.

Cresol An oily, colourless liquid or solid obtained from coal-tar or wood; used for making the antiseptic lysol.

Cress A name commonly applied to certain members of the family Cruciferae which have pungent, edible leaves as watercress. Occasionally applied to other plants having the same characteristics.

Crest (1) The top of a dam dike, spillway, or weir; frequently restricted to the overflow portion. (2) The summit of a wave; peak of a flood. (3) A toothed or irregular ridge on the petals or flower clusters in certain plants.

Crest cloud A type of standing cloud which forms along a mountain ridge, either on the ridge, or slightly above and leeward of it, and remains in the same position relative to the ridge. Also known as cloud crest.

Cretaceous Geological period of the Mesozoic represented by deposits frequently made up of rocks containing chalk.

Crevassee fillings Ridges or hummocks formed from glacial sediments that were deposited by water in the cracks and crevasses of the ice.

Criador The rain bringing west wind of northern Spain.

Cricus arvensis See Canada thistle.

Crimson clover Trifolium incarnatum L. Family Papilionaceae. Herb introduced from U.S.A. and used as a fodder.

Crippled (1) An imperfect flower. (2) An animal unable to move due to disease, injury or physiological condition.

Crisphead lettuce Lettuce type; forms solid cabbage-like heads.

Critical concentration The limiting concentration of an impurity in an irrigation water after which it adversely affects crop growth, impairs the quality of the crop products, or damages the soil on which crops are grown.

Critical day length Critical light period.

Critical depth When water flows at a given volumetric rate along an open channel, its total energy will vary with the depth. The depth corresponding to the minimum total energy is termed the critical depth.

Critical flow That velocity of flow at which the energy of flow is a minimum.

Critical hydraulic gradient Hydraulic gradient at which the intergranular pressure in a mass of cohesionless soil is reduced to zero by the upward flow of water.

Critical light period The minimum amount of daily light required by certain plants for their growth and flowering. Also called critical light period, critical photoperiod.

Critical photo period See critical light period.

Critical slope (1) Angle of repose. (2) That slope which will, sustain in a given discharge at uniform, critical depth in a given channel.

Critical storm period The duration of that storm which causes the greatest peak of runoff in a drainage basin.

Critical suction head Maximum suction lift of a pump.

Critical temperature Temperature of air when the heat lost by the animal body is equal to heat produced in the body.

Critical void ratio The void ratio corresponding to the critical density.

Crocus sativus See safforn crocus.

Crop (1) A group of plants grown collectively for their produce; any product of the soil. (2) The honey sac of the honey bee. (3) To raise plants. (4) To cut or trim or clip.

Crop dusting Applying fungicides or insecticides in powder form to a crop; usually done from a low flying aircraft.

Crop intensity The percentage of the area irrigated during a crop to the culturable commanded area.

Crop irrigation requirement Amount of water required to meet the consumptive use of crops minus the amount available from natural sources such as dew, snow, rainfall or ground water.

Crop land Land used primarily for the production of adapted cultivated, close-growing, fruit, or nut crops for harvest, along or in association with sod crops, except forest crops.

Cropping index The number of crops grown per year on a given land area times 100. Abbreviated as CI.

Crop plant A cultivated plant.

Croppy Term applied to jute fibre in which the top ends are comparatively hard and stiff.

Crop ratio The crop ratio (e.g. Kharif-Rabi ratio) is defined as the ratio between the anticipated areas to be irrigated of the two crops during a year.

Crop requirements A term sometimes used in considering agricultural water uses and refers both to net irrigation requirements and net water requirements.

Crop residue The unused part of the crop that is not harvested in the grain system of farming and is usually returned to the land. Examples are straw, cornstalks, clover, and soybean hulls.

Crop rotation A method of protecting the soil and replenishing its nutrition by planting a succession of different crops on the same land.

Crop trees Trees which are designated to make up the final or rotation timber crop.

Crop water requirement See consumptive use.

Crop year (1) The time between planting and harvesting of a crop. (2) The time between one harvest to the next.

Crop yield The yield obtained per unit of land area usually denoted as quintals per ha or any other similar unit.

Cross (1) A hybrid; an individual whose parents differ genetically. (2) To hybridize.

Cross bedding An arrangement in which thin layers of stratified sediment are transverse or oblique to the main plan of stratification.

Cross breeding The mating of pure bred animal or pollinating of plants belonging to different breeds; also mating of animals or plants belonging to different species.

Cross cultivation Ploughing or harrowing the land second time but at right angles to that of the first ploughing or harrowing.

Cross cutting Sawing timber across the grain.

Cross fertilise To fertilise the ovule or ovules of one flower with the pollen from another flower; commonly refers to the fertilising of flowers of one plant by pollen from another plant.

Cross fertility The ability of a plant to produce viable seed by cross pollination. Also called interfertility.

Cross fertilization Fertilisation resulting from the use of pollen from another plant.

Cross fruitfulness The ability of a variety of a fruit plant to mature fruit after pollination by another specific variety. Also interfruitfulness.

Cross grained timber Timber in which the fibre alignment deviates from a direction parallel to the long axis of the piece.

Crosshaul A method of loading log-transportation vehicles. One end of a line is passed over the load, around the log to be loaded, and made fast to the load. Power applied to the other end of the line imparts a rolling motion to the log.

Cross incompatibility The inability of the gametes from two different individuals to unite although the individual themselves are otherwise fertile.

Cross inoculation Inoculation of one legume species by the symbiotic bacteria from another.

Cross pollination Transfer of pollen from the anthers of one plant to the stigma of another plant.

Cross-section analysis The analysis of a series of records of economic data for different groups of people, firms or countries, at the same moment or in the same period of time.

Cross-sterility See inter-sterility.

Cross timber A strip of oak-dominated forest stretching across grassland.

Cross unfruitfulness The inability of a variety of a plant to mature fruit if only the pollen of another variety is used. Also called inter-unfruitfulness.

Crotch (1) The common joint of large branches in fruit trees from which sometimes more than one lateral branch arises. (2) The area behind the legs of an animal.

Crotch injury Injury caused to the fruit trees in which the bark of the crotch area gets killed because of cold weather.

Croton Croton tiglium L. Family Euphorbiaceae. A shrub or small tree native of South-east Asia used for croton oil of commerce. Also known as *Jamalgota.*

Croton oil A poisonous yellow liquid; purgative.

Crotovina A former animal burrow in one soil horizon that has become filled with organic matter or material from another horizon. It is also spelled *'krotovina'.*

Crowd To press against one another, as plants in a forest or field competing against each other for space.

Crowder An implement to make small bund for soil conservation usually across the slope. Similar to a bund former.

Crowd out To compete so vigorously for space as to kill other vegetation.

Crowfoot Dactyloctenium aegyptium (Desf). Beauv — (Eleusine aegyptiaca Desf.). Family Gramineae. A perennial herb whose seeds are consumed as food and straw as fodder.

Crown (1) The upper part of a tree, including the branches with their foliage. (2) The upper surface of a furrow slice. (3) The place where the stem and root join in a seed plant; coller. (4) Base of stem where roots arise. (5) The top of an animal's head.

Crown area index (CAI) Total projected crown area above unit area of ground surface; a measure of canopy density.

Crown bud A growing point located at the transition zone between the root and stem of a plant.

Crown canopy The canopy formed by the crowns of all the trees in a forest, or an uneven-edged forest, by the crowns of all trees in a specified class. Also crown covers leaf cover.

Crown cover See crown canopy.

Crown fire A forest fire which extends to and sweeps along in the tops and branches of trees.

Crown gall A bacterial disease of many plants induced by Bacterium tumefaciens and marked by abnormal enlargement of the stem near the root crown.

Crown grafting A method of vegetative propagation whereby a scion 8.15 centimetres long is grafted at the root crown, just below ground level.

Crown grafting Bark grafting.

Crown rot Any plant disease or disorder marked by deterioration of the stem at or near ground level.

Crown rust A rust disease of oats and certain other grasses caused by varieties of Puccinia caronata and marked by light-orange masses of fungi on the leaves.

Crozier The apex of a young fern leaf; a formation in the Ascomycetes.

Crucifer Any plant of the family Cruciferae (Brassicaceae). It includes cabbage, cauliflower, mustard, cress, radish, turnip, *knolkhol*, Brussels sprouts, *rai, toria* etc.

Crucifera black rot Disease of crucifera and much serious on cabbage is caused by Xanthomonas campestris (Pammel) Dowson. Characterised by blackening of leaf-veins, shedding of lower leaves and stunting of plant growth.

Crucifera black vein See Crucifera black rot.

Crucifera boron deficiency Deficiency of boron in soils cause this disease in which brown necrosis in fleshy pith of stem and core of plants appears. Internal breakdown of brown necrotic tissues. Can be controlled by applying 20 kg per hectare borax to soils.

Crucifera club root Caused by Plasmodiophora brassicae in which roots form enlarged spindle shaped or spheroid clubs. Temporary flagging of leaves occurs which leads to wilt or retarded growth.

Crucifera damping off Disease of early crops of cabbage and cauliflower caused by Pythium aphanidermatum (Eds.) Fitz.; Corticium solani (Prille Delacr) Bourd and Galz., Sclerotium rolfsii. See damping off.

Crucifera flea beetle Phyllotreta cruciferae Goeze. Insect pest of crucifers. White larva mines into roots; bluish-greenish adults feed on leaves and make holes, making young leaves dry up and also dying of plants.

Crucifera leaf spot and blight Disease caused by Alternaria brassicicola and A. brassicae. Characterised by small dark spots spreading to form circular lesions. Brown discolouration in cauliflower and black mouldy appearance of cabbage after harvest.

Crucifera white rust See crucifera white

Crucifera white blister Disease of cruciferous plants caused by Cystopus candidus. It is characterised by white shiny fustules on leaves and stems. It also causes deformation of parts of stems and flowers.

Crude fat An ether extract containing fat and fat-like substances which are contained in feeds.

Crude fibre Coarse fibrous portions of plants such as cellulose, partially digestible and relatively low in nutritional value. In chemical analysis it is the residue obtained after boiling plant material with dilute acid and when with dilute alkali.

Crude oil Crude mineral oil processed from crude petroleum. Has good adhesive and protective qualities.

Crude sap Sap stream containing absorbed plant nutrients travelling upward in a plant.

Cruise A survey of forest lands to locate and estimate volumes and grades of standing timber; also, the estimate obtained in such a survey. (Scaling is the measurement of the volumes of individual logs after the trees have been felled).

Crumb structure A very porous granular structure in soils; easily crushed aggregates.

Crusher (1) A machine used to crush agricultural products or fruits to extract oil or juice out of them. (2) Also machines to crush minerals etc. to reduce them in size.

Crushing strength The force required to crush a mass of dry soil, or conversely, the resistance of a mass of dry soil to crushing. It is expressed in units of force per unit area (pressure).

Crust A surface layer of soil, from a few millimetres to 2.5 cm thick, that when dry is much more compact, hard and brittle than the material just under it.

Crustose Manifesting a crustlike habit of growth, as certain lichens.

Cryic layer A perennially frozen layer.

Cryolite (Na_3AlF_6) Sodium fluoalumite or sodium aluminium fluoride. An insecticide used as dust or spray.

Cryology The study of the properties of snow, ice and frozen ground.

Cryophilic bacteria Bacteria which require low temperature for their growth and activity.

Cryoturbation Frost action, including frost heaving.

Cryptocotylar germination See hypogeal germination.

Cryptogam (1) A plant without seeds (All plants except spermatophytes). (2) A lower plant such as the fungus, alga, lichen, fern, moss etc. which reproduces by spores, and does not produce a proper flower.

Crytogeal germination Germination in which the plumule is buried by late extension of the cotyledonary axes.

Cryptophyceae A class of algae of the Pyrrhophyta in some systems of classification equivalent to the division Cryptophyta.

Cryptophyta (1) A division of the algae in some classification schemes; equivalent to the cryptophyceae. (2) A plant, specifically a cryptogam, whose buds, shoot tips or spores survive the unfavourable season in the ground or at the water level, if growing in water.

Cryptophychus gravis See mango nut weevil.
— **mangiferae** See mango nut weevil.

Cryptoxanthin A yellow carotenoid pigment found in maize and other grains.

Crystal A homogeneous inorganic substance of definite chemical composition bounded by plant surfaces that form definite angles with each other to give the substance a regular geometrical form. See also mineral, soil.

Crystallised fruit or **vegetable** Candy coated with crystals and granular sugar.

Crystal lattice See lattice structure.

Crystallization Formation of solid in the form of crystals.

Crystalloid (1) A protein crystal found in certain plant cell. (2) A substance which in solution readily diffuses through an animal membrane.

Crystal sugar Sucrose, pure sugar commonly used in household and made from sugarcane in sugar factories. It can also be made from sugarbeet.

Cuban bast Hibiscus tiliaceus L. (Paritium tiliaceum St. Hil.) A herb of the family Malvaceae commonly grown for its edible leaves in Sunderbans and along the coast of the Peninsula. Bark fibre used for ropes and mats. Also known as *Mahoe*.

Cubeb Piper Cubeba L.F. (Cubeba officinalis Raf.). Family Piperaceae. Climber belonging to the pepper family; berries important in indigenous medicines. Used in spices, condiments and in perfumery. Also called tailed pepper or *kabab chini.*

Cubeb officinalis See cubeb.

Cubic foot per second Rate of fluid flow at which 1 cubic foot of fluid passes a measuring point in 1 second (35.3 cu. ft = 1 cu.m.).

Cubic metre volume equivalent to that of a cube of sides equal to one metre.

Cubic yards Volume equivalent to that of a cube of sides equal to one yard and is equal to 0.7646 cubic metre.

Cubing Process of forming hay into high density cubes to facilitate transportation, storage, and feeding.

Cucumber Cucumis sativis L. An annual cucurbit in the family Cucurbitaceae grown for its edible immature fleshy fruit.

Cucumber mildew (1) A downy mildew of cucumbers and melons caused by Pseudoperonospora cubensis. (2) A powdery mildew of cucumbers and melons caused by Erysiphe cichoracearum.

Cucumber mosaic A virus disease of cucumbers and related fruits, producing mottling of terminal leaves and fruits and dwarfing of vines.

Cucurbita maxima See winter squash.
— **moschata** See pumpkin
— **pepo** See summer squash.
— **siceraria** See bottle gourd.

Cucurbit bug Aspongopus janus F. and A. brunneus N. Red and pale brown bugs that suck sap from cucurbits.

Cucurbit downy mildew Disease due to Pseudoperonospora cubensis. Lower leaf surface contains purplish spots whereas yellow spots appear on the upper. Fruit does not mature or right flavour does not appear.

Cucurbit fusarium root rot Caused by Fusarium solani and F. Cucurbitae. Symptoms include midseason wilt, stem base turning dark brown with soft mushy cortical wilt, rotting of fruits and damping off of young seedlings.

Cucurbit mosaic Virus disease characterised by mosaic on young leaves near growing tips, V-shaped necrotic area appearing on older leaves from margin to mid-rib, dwarfing of internode, mild mottling to extensive or watery malformation in young fruits.

Cucurbit powdery mildew Disease caused by Erysiphe cichoracearum DC and similar to pea powdery mildew in character.

Cucurbit stem rot Disease of cucurbit caused by Pellicularia (Sclerotium) rolfsii and characterised by cortical decay of stem at soil level, with wilting of young leaves, watery rot of fruits.

Cucurbit wilt A bacterial disease of cucumbers and related plants caused by Erwinia tracheiphila, characterised by sudden wilting of the plant.

Cudappahs and Vindhyans Ancient rock formations in India, which have resulted in matured soils.

Cuesta A hill or ridge that has a steep

cliff on one side and a more gradual slope on the other side, controlled more or less by the attitude of the rock strata.

Culm Jointed stem of a grass.

Cull A tree or log of merchantable size rendered unmerchantable because of poor form, limbyness, rot or other defects.

Culling Selecting and grading of fruits and vegetables with a view to discard those which do not satisfy the desired specification or standard.

Cultivar Name given to all cultivated variants of plants.

Cultigen Any cultivated variety or species of organism for which there is no known wild ancestor. Also known as cultivar.

Cultivated variety Cultivar.

Cultivation (1) Tillage to prepare land for seedling transplanting, and later to control weeds and loosen the soil. (2) The planting, tending, harvesting and improving of plants.

Cultivator A farm implement pulled behind a powered machine that is used to break up soil, kill weeds and create a surface mulch to conserve moisture.

Culturable command area That portion of the culturable irrigable area which is commanded by flow irrigation.

Culturable irrigable area The gross irrigable area less the area not available for cultivation, e.g. village areas, roads and isolated patches of unculturable lands.

Culturable waste A class under land revenue classification. Infertile land, uneconomical in cultivation, submarginal land which cannot be cultivated.

Cultural (1) Pertaining to or produced by plant breeding. (2) Pertaining to village practices.

Cultural practice The individual or the sum total of practices in tillage, tending, harvesting and improvement of crop plants.

Cultural variety See cultivar.

Culture (1) Growth of micro-organisms on a nutrient medium; to grow micro-organisms on such a medium. (2) The micro-organism so grown (3) The working of the ground in order to raise crop. (4) Attention given to the growth or propagation of plants or animals with a view to improvement.

Culture media (sing., medium) Artificial food material upon which microbes are grown (cultured).

Culture solution Solution containing chemicals for growing plants in containers without soil. It provides all the nutrients for growth. Also called nutrient solution.

Culturing Artificial propagation of organisms on nutrient media or living plants.

Culvert A transverse drain or waterway structure under a road,. railroad, canal, or through an embankment by means of a pipe or enclosed channel.

Cumatophyte A plant that grows under surf conditions.

Cumec Water flow at the rate of one cubic metre per second.

Cuminum cyminum See cumin.

Cumin Cuminum cyminum L. Family Umbelliferae. Slender annual herb of the Mediterranean region used for flavouring curries, soups etc. Also used in medicine also called as *safed jeera*.

Cumin powdery mildew Disease caused by Erysiphe polygoni DC. Shows symptoms similar to pear powdery mildew.

Cumulative error An error which, in the course of the cumulation of a set of observations, does not tend to zero. The relative magnitude of error does not then decrease as the number of observations increases.

Cumulative infiltration Summation of the depth of water absorbed by a soil in a specified elapsed time in reference to the time of initial water application.

Cumulative runoff The total summation of runoff over a specified period of time. Successive summations are frequently plotted against time.

Cumuliform cloud A fundamental cloud type showing vertical development in the form or rising mounds, domes or towers.

Cumulo layer A layer of sandy silty, or clayey material in an organic soil.

Cumulonimbus cloud A principal cloudy type in the form of individual, detached elements which are generally dense and possess sharp non-fibrous outlines.

Cumulose soil Sedimentary soil of recent origin. Made up largely or partly decomposed remains of plants growing in and around lakes, ponds and marshes.

Cuneate Wedge shaped, as an apical cell with four cutting faces.

Cuneiform wedge-shaped.

Cup See corona

Cupania sapida See akee.

Cup copra Copra produced for milling and extraction of oil. Coconut is broken into two halves and sun dried. Further sun dried after detaching from the shell and stored in air tight godowns.

Cuprocide A copper compound used as an insecticide.

Cupule The asexual body on the thallus of certain liverworts which produce gemmae.

Curculigo orchioides See black musale.

Curdle To coagulate.

Curds The heads of cauliflower.

Cure (1) To preserve a product by drying, smoking, salting etc. (2) To heal.

Curf The earth from which bricks are made.

Curie The unit of radioactivity q.v. It is equal to that quantity of any radioactive nuclide in which the number of disintegrations per second is 3.7×10^{10}.

Curing, or **maturing** The process of keeping concrete or mortar damp for the first week or month of its life to prevent or replenish the loss of necessary moisture during the early, relatively rapid stage of hydration. Some times, also referred to as maturing.

Curled dock Rumex crispus L. A herb of family Polygonaceae the leaves of which are used as vegetable.

Curling Abnormal bending of the leaf blade downward along the main vein.

Curly top A virus diseases of sugar-beets and certain other plants that is transmitted by a leafhopper. Affected plants are dwarfed and have curled upturned leaves.

Currant (1) Any of the certain plants of the genus Ribes, family Saxifragaceae, grown as ornamentals or for fruits. (2) Dried seedless raisins, varieties of Vitis vinifera, used in confectionary.

Currant leaf spot An angular leaf spot of currants caused by the fungus Cercospora angulata.

Currency The money of a particular country.

Current account An element in the balance of payments account dealing with commercial transactions.

Current annual increment The growth of a stand or tree for a specific year.

Current cross An offspring of a plant or animal developed in the first season of cross breeding.

Current curve Graphic representation of flow of tidal current. In the reversing of current, curve is referred to rectangular co-ordinates, the flood being considered positive and the ebb velocity negative; in general, approximates to a cosine curve.

Current diagram Graphic representation showing velocities of flood and ebb currents, and times of slack and strength over a considerable stretch of the channel of a tidal waterway, the times relating to tide or currents phases at some reference station.

Current difference Difference between time of slack water or strength of current in a locality, and time of corresponding phase of the current at a reference station for which predictions are given in a current table.

Current direction indicator, or direction finder A device to indicate the direction of flow of water.

Current expenditure Expenditure recurrent in nature and not resulting in the acquisition of assets.

Current fallow A revenue classification. A land under cultivation but not cultivated during the year under consideration.

Current ratio Current assets divided by current liabilities.

Current table Table which gives daily prediction of times and velocities of tidal currents, usually supplemented by current differences and constants by which additional prediction can be obtained for other places.

Current transactions The receipts and expenditure of income which forms a continuous process, year after year.

Curry leaf tree Murraya koenigii (L)

Spreng. (Bergera koenigii L.) Family Rutaceae. A large shrub or small tree grown for its leaves used for flavouring curries. Also called *mitha-neem*.

Curve fitting (1) The fitting of mathematical curve to any statistical data, e.g. regression data or time series. (2) The fitting of a mathematically specified frequency curve to a frequency distribution.

Curve of demand A graph showing the amount of water needed for irrigation at various times during the crop season based on elements of time and quantity.

Curve of supply A graph showing the water available based on elements of time and quantity.

Cuscuta Evolvulus alsinoides. Family Convolvulaceae. Flowering parasite with slender yellowish stems twining round host stem into which it sends suckers or haustoria. Also called dodders, *visnu krantha*.

Cusec A unit of discharge. Denotes that one cubic foot of water flowing per second. If it flows at this rate for a day it constitute near 1.98 acre feet (24.42 ha cm) of water which is approximately taken as 2 acre feet (24.67 ha cm).

Cushion plant A low-branched chamaephyte that forms a dense resilient mass, often in windy areas.

Cusp A sharp point.

Cuspidate With an apex somewhat abruptly, sharply and concavely constructed into an elongated short pointed tip.

Custard apple Annona squamosa (L.). Family Annonaceae. A table fruit; roots sensitive to water stagnation. Propagated by seed, budding, grafting; Inarching also practised. Known as Sharifa in Hindi.

Custard apple fruit fly Bactrocera persicae B. Fly pest of custard apple, whose maggot bores into fruit.

Custard apple mealy bug Pseudococcus virgatus C., P. Lilacinus C., Ferrisiana virgata ckll. Small reddish white mealy bug infesting leaves, shoots, buds and fruits of custard apple, sucking sap from them; secretes honey-dew which attracts fungus attack.

Custard apple pink disease Disease of custard caused by Corticium salmonicolour. Symptoms same as in tea pink disease.

Custard winds Cold easterly winds on the northeastern coast of England.

Customs The uniformly accepted ways of acting about some social aspects of life.

Cut The yield, during a specified period, of products that are cut, as of grain, timber, or, in saw milling, lumber.

Cutan A modification of the texture, structure, or fabric at natural surfaces in soil materials due to concentration of particular soil constituents or *in situ* modification of the matrix. Cutans may be composed of any of the component substances of the soil material.

Cutaneous Anything that has to do with skin.

Cut and fill (1) A term used to describe a cross-section of a canal which is partly in cutting and partly to be in embankment. (2) Also used in land levelling to designate soil that is cut and filled.

Cut back (1) To prune the plant by removing the terminal ends of shoots or branches. (2) To reduce to the original size or number etc.

Cutch dye Vegetable dye obtained from the wood of Acacia catechu.

Cuticle (1) The outer or epidermal cells covering a plant. (2) The imbricated scale layer of wool fibre formed by the flattened and overlapping outer cells; encloses the cortex of elongated cells.

Cutin Outer covering of plants composed of waxes and waxy polymers.

Cutis Skin.

Cutoff system or **intercepting system** A drainage system for draining seepy hillsides; tiles are placed along the hillside to intercept the seep water and prevent it from reaching the bottom land.

Cutoff wall, or **cutoff** A wall, collar, or other structure, intended to reduce percolation of water along otherwise smooth surfaces, or through porous strata. It is provided in or on the foundations.

Cutout relay An automatic magnetic switch attached to the generator to cut out generator circuit to prevent overcharging of battery.

Cut over An area from which some or all the trees have been removed.

Cuttage The making and tending of cuttings.

Cuttails and tules Aquatic perennial grass like herbs of family Typhaecae; weed.

Cutter bar A part of the mower or combine which cuts the grass or crop; the cutting mechanism.

Cutting A piece of plant stem with one or more nodes which when placed under suitable conditions will produce roots and shoots resulting in a complete plant.

Cutting edge It is the front edge of the share which makes horizontal cuts in the soil.

Cutting improvement A cutting made in a stand past the sapling stage for the purpose of improving its composition and character by removing trees of less desirable species, form, and condition in the main crown canopy.

Cutworm A larval stage of moths of the family Noctuidae, which are soft-skinned caterpillars. Some eat up leaves, some chew into fruits. Some of these feed at or below ground level.

Cyamposis psoraloides See cluster bean.

— **tetragonoloba** See cluster bean.

Cyanophyceae The blue green algae, a class of plants in the division Schizophyta.

Cyanogenetic plants Plants which when ingested by animals give hydrocyanic (prussic) acid, a potent poison, like sorghum, sudan grass etc.

Cyathium An inflorescence in which the flowers arise from the base of a cuplike involucre.

Cycadales An ancient order of plants in the class cycadopsida, characterised by tuberous or columnar stems that bear a crown of large, usually pinnate leaves.

Cydanthales An order of monocotyledonous plants composed of herbs or seldom composed of more or less woody plants with leaves that usually have an expanded blade.

Cyclas, formicarius See sweet potato weevil.

Cycle A systematic recurring fluctuation.

Cycle, cutting The planned interval between major felling operation in the same stand.

Cycle of erosion The succession of events involved in the evolution of landforms.

Cyclone wave A disturbance in the lower troposphere of wavelength 1000-2500 kilometres, cyclone waves are recognised on synoptic charts, a migratory high and low pressure systems.

Cyclophysis The organisational status of a meristem in a tree determined by its age.

Cyclostropic wind The horizontal wind velocity for which the centripetal acceleration exactly balances the horizontal pressure force.

Cydonia oblonga See quince.

— **vulgaris** See quince.

Cylindrocapsaceae A family of green algae in the sub-order ulotrichineae comprising thick-walled, sheathed cells having massive chloroplasts.

Cymadothia trifolii See clover sooty blotch.

Cymbopogon caesius See ginger grass.

Cyme (1) Any determinate inflorescence. (2) A flower cluster in which the central apex ceases growth early but lateral branches keep forming towards the base and grow upwards and outwards, forming a flat topped cluster.

Cynara scolymus See globe artichoke.

Cynodon dactylon See Bermuda grass.

— **plectostachyus** See giant star grass.

Cyperales An order of monocotyledonous plants in the subclass Commelinidae with reduced mostly wind-pollinated or self-pollinated flowers that have a unilocular two or three carpellate ovary bearing a single ovule.

Cyperus hexastachyos See nut grass.

— **rotundus** See nut grass.

Cypheliaceae A family of typically crustose lichens with sessile apothecia in the order Caliciales.

Cyphomandra betacea See tomato.

Cypress Any tree of the genus,

Cupressus, family Pinaceae.

Cyrtosis A virus disease of cotton characterised by stunting distortion and abnormal branching and colouration.

Cyst Abnormal structure containing liquid, air, or solid material; a stage in the history of certain protozoa during which time the encysted organism is protected by a surrounding wall.

Cystase An enzyme which can dissolve cellulose.

Cysticercus Larval form of the tapeworm in the tissues of the intermediate host, which upon entry into the intestinal canal of the definitive host develops into an adult worm.

Cystic stage The stage in the development of a parasite during which it is enclosed in sac which separates it from the tissues of the host.

Cystine A white crystalline aminoacid from proteins, necessary in an animal's diet to develop horn, hair etc.

Cystidium A large sterile structure in the hymenium of Basidiomycete.

Cystocarp In the red algae, a spore-producing body formed by the carpogonium, carpospores and other associated cells.

Cystolith A concretion of calcium carbonate arising from the cell walls of modified epidermal cells in some flowering plants.

Cystopus candidus See mustard white rust.

— **corandidus** See crucifer white blister.

Cytodites nudus See air-sac mite.

Cytogamy conjugation, union or fusion of cells.

Cytogene Factor in cytoplasm affecting hereditory behaviour.

Cytogenesis The formation, reproduction or development of cells.

Cytogenetics The interpretation of the phenomena of heredity in terms of cell structure.

Cytology Science treating of the study of cells, their origin, structure, physiology, development and reproduction of plants and animal cells, often with particular reference to the chromosome in the cells.

Cytolymph Cell sap; the more fluid constituent of a cell.

Cytolysin Antibody that can dissolve cells.

Cytophagy The engulfing of cells by other cells.

Cytoplasm All protoplasm of the cell except the nucleus.

Cytoplasmic male sterility A type of male sterility conditioned by the cytoplasm rather than by nuclear genes and transmitted only through the female parent.

D

Dactyl A digit, a finger or a toe.
Dactylis gromerata See orchard grass.
Dactyloctenium aegyptium See crowfoot.
Dacus ciliatus See melon fly.
— **cucurbitae** See melon fly.
— **dorsalis** See mango fruit fly.
— **zonatus** See peach fruit fly.
Dado plane A narrow plane for cutting flat grooves in wood work.
Dadoor In India, a wind blowing down the Ganges valley from the Siwalik hills at Hardwar.
Daily mean The average value of a meteorological element over a period of 24 hours.
Dairy A farm concerned with the production of milk. An establishment where milk products are made.
Dairy cattle A cattle breed selected, raised and bred for ability to produce large quantities of milk.
Dairy machinery Equipment, such as pasteurisers, used in the production and processing of milk and milk products.
Dairy product Any food made from milk or milk products.
Dakin's solution Neutral solution of sodium hypochlorite once used in the disinfection of wounds.
Dalbergia latifolia Rosewood. Tree of family Papilionaceae producing hard and close grained timber.
— **sissoo** See sissoo.
Dallis grass Paspalum dilatatum Poir. Family Graminea. A perennial grass introduced from the U.S.A. and grown for fodder.
Dalmaitan pyrethrum See pyrethrum.
Dam (1) A barrier to confine or raise water for storage or diversion, to create a hydraulic head, to prevent gully erosion, or for retention of soil, rock, or other debris. (2) A quadruped female parent.
Dammer bee Melipona spp., Trigona spp. Honey bee of smallest size. Builds nests in hollows of trees, rocks or walls. Honey difficult to extract and yield small. Also called stingless bee.
Damp Moisture in air, on surface or diffused throughout solid.
Damping off Sudden death of a seedling which falls of along discoloured and shrunken spot near ground level. Characteristic of certain diseases in plants due to fungus attack.
Dampness Moderate to high relative humidity; moisture.
Damp proofing The treating of a wall with some impervious material to prevent moisture from oozing through.
Damp wood termite Termite group which makes colonies in living trees or moist dead wood.
Dandruff An accumulation of bran-like scales on the skin, due to exfoliation of the epidermis or to the over activity of the sebaceous glands.
Dark Gray A great group of soils in the Chernozemic order. The soils occur in the cool to cold, subhumid grassland-forest transitional regions, and have a dark gray partially eluviated surface (Ahe or Ap) horizon and a brownish B (Bm, Btj, or Bt) horizon, which may be absent, over a highly base saturated, usually calcareous C horizon.
Dark Gray Gleysolic See Humic Gleysol.
Darling shower A dust storm caused by cyclonic winds in the vicinity of the River Darling in Australia.
Darnel grass See rye grass.
Dasheen Colocasia esculenta. A plant in the order Arales, grown for its edible corn.
Dasyneura lini See linseed gall fly.
Data Records of observations and measurements of physical facts, occurrences, and conditions, reduced to written, graphical or tabular form. Also see datum.
Data processing Any operation or

combination of operations on data, including everything that happens to data from the time they are observed or collected to the time they are destroyed. Also known as information processing.

Data processing machine A computer or a component of a data processing system, such as a card reader or tape unit.

Data record A collection of data items related in some fashion and usually contiguous in location.

Data reduction The transformation of raw data into a more useful form.

Date Delicious and nutritious fruit of the date palm, Phoenix dactylifera. Contains carbohydrate with 75-80 per cent sugars in dry flesh. Combined with milk, forms a balanced human diet. See date palm.

Date curing The artificial or sun curing of date. For sun-curing fruits are picked up at dung stage, spread on mats in the sun for 4-7 days. For artificial curing, fruits picked in one lot when few fruits reach dung stage; dipped in 1 per cent boiling caustic soda solution for one minute, then arranged on wire gauge trays and placed in drier at 54.5°C.

Date palm Phoenix dactylifera L. Family Palmaceae. A tall native of India and Arabia. Need long summer and high day and night temperatures. Grown for its fruit date and tree sap which is a source of intoxicating drink. Propagated by seed and by offshoot. Also called *khajoor*.

Date plum Diospyros lotus L. Family Ebenaceae. Tree found in the Punjab and Kashmir which produces edible fruits. Also called *Amlok*.

Datum (1) Any numerical or geometrical quantity or set of such quantities which may serve as a reference or base for other quantities. (2) Thing known or granted, assumption or premise from which inference are drawn.

Daucus carota var. sativa See carrot.

Daughter (1) A female off-spring. (2) The primary division of first generation off-spring or any plant regardless of sex.

Day length Number of hours for light for each 24 hours cycle; an important factor in the growth of plants.

Day light Colloquial term used to indicate the extension of a grade line to or above the ground surface.

Day lighting To light an area with daylight.

Daylight point The point of intersection between the grade line and the profile line.

Day neutral Reaching maturity regardless of relative length of light and dark periods.

Day neutral plant Capable of flowering under either long or short daylengths.

Dazed Term applied to jute fibre, which is dull, lifeless, and of inferior strength and poor spinning quality, resulting from over-retting; also due to storing jute while moist.

DBH Diametre (of a tree) at breast height, or 1.37 m above the ground.

DD (Dichlororopropane and Dichloropropylene) Soil fumigant and derived from petroleum. Liquid form, less volatile and safe for use. Effective in control of cyst-forming nematodes.

DDT (Dichlorodiphenyl trichloroetheane) A nerve poison. Kills insects by contact, but many also act as stomach poison. Toxicity due to presence of P, P isomer which may be up to 80% in the technical material. Used as dust as well as spray; cucurbits sensitive to strong dose. At least 15 days should elapse before sprayed crop is harvested.

Dead arm a fungus disease caused by Crysptosporella viticola in which the main lateral shoots are destroyed; common in grapes.

Dead furrow A double furrow left in the middle of the field or between two lands in ploughing.

Dead Horse Work which has been paid for but has not yet been completed; a situation in which services and goods are paid for in advance of performance or production.

Dead knot, or loose knot A knot which is not held firmly in place.

Dead ripe Pertaining to grain or fruits which are absolutely ripe.

Dead sea A body of water that has undergone precipitation of its rock salt, gypsum, other evaporites.

Dead seeds Seeds which, at the end of the test periods, are neither hard nor fresh and have not produced seedlings, are classified as dead seeds.

Dead storage level Level in a reservoir corresponding to dead storage.

Dead water water which is not flowing with a significant velocity.

Dealer A person carrying on a business of selling goods, whether by wholesale or by retail, or engaging in the buying and selling of shares.

De-alkalization Removal of exchangeable sodium (or alkali) from the soil usually by chemical treatment and leaching.

Deal wood A term loosely used to denote light timbers used for packing cases, crates and similar work.

Death Complete and permanent cessation of vital functions in an organism.

Death duties Duties payable when a property passes at death. Death duties are progressive, increasing in production with the value of the estate.

Debearder Debearder is a preconditioning machine used to debeard barley, clip oats, break up lucerne seeds pods, break up grass seeds doubles, separate and polish vegetable seeds, decorticate sugarbeet seeds and remove extra glumes and hulls.

Debility Weakness or impairment of strength.

Debit Entry in account of sum owing; left side of account in which these entries are made.

Deblossom To remove blossoms from a plant to increase its vigour.

Debris (1) A term applied to the loose material arising from the distintegration of rocks and vegetative material, transportable by streams, ice, or floods. (2) Dead stalks, leaves and other rubbish left on a field after harvest.

Debris cona Fan-shaped deposits of debris including soils, sand gravel and boulders built up at a point where a mountain stream meets at valley, or where the velocity of a stream is reduced sufficiently to cause such deposits.

Debris dam A barrier built across a stream channel to retain rock, sand, gravel, silt, or other material.

Debris screens, or **Trash screens** Devices in a sprinkler system or any other system to keep the system free of trash that might plug the sprinkler nozzles.

Debt A sum of money, or quantity of goods or services, owed by one individual or body to another.

Decalcification (1) Removal of calcium carbonate or calcium ions from soil by leaching, a natural process in soil formation. (2) Removal of calcium from the bones of animals in negative calcium balance.

Decantation A method for mechanical dewatering of a wet solid by pouring off the liquid without disturbing under lying sediment or precipitate.

Decanter Tank or vessel in which solids or immiscible dispersion in a carrier liquid or coalisce, with clear upper liquid withdrawn (dacanted) as overflow from the top.

Decapitation Method of burning trees for rejuvenation in which most branches are severely pruned.

Decay (1) Any chemical or physical process causing disintegration or deterioration. (2) Disintegration of wood substance due to action of wood destroying fungi. Also known as rot. (3) Decomposition of organic matter (4) The breaking up of rocks by weathering agents, with stress or chemical effects.

Decayed knot, or **Unsound knot** A knot softer than the surrounding wood, and containing decay.

Deccan trap One of the rock formations in India; comprises volcanic lava rich in ferromagnesium and alumina compounds; black cotton typical soil derived from it.

Deccan wingless Grasshopper Colemania sphenareoides Bol.; pest of sorghum, maize, groundnut, pear millet, blackgram, red gram, *ragi* etc. Adults and nymph feed on crop leaves and in severe attack, defoliate plants completely.

Decidous tree (1) A term applied to trees which lose their leaves each year at the end of the growing season. (2) Pertaining to falling of leaves; fruits and petals of flowers.

Deciduous forest Forest composed chiefly of deciduous dominants.

Decile A positional measure comprising

one of the nine variate values which divide the total frequency into ten equal parts.

Decimal currency Metallic and paper currency in which each successive denomination is ten times the value of that next below it.

Decimal fraction Any number written in the form; an integer followed by a decimal point followed by a (possibly infinite) string of digits.

Decimal number A number signifying a decimal fraction by a decimal point to the left of the numerator with the number of figures to the right of the point equal to the power of 10 of the denominator.

Decimetre One-tenth of a metre, a linear unit of measurement and denoted as 0.1 metre.

Deci-Semens per metre A unit to demote electrical conductivity and is equal to millimhos per centimetre.

Deck, log A pile of logs or a rollway.

Declinate Curved toward one side or downward.

Decline (1) A gradual wasting away; a gradual loss of vigour. (2) A physiological condition of citrus trees in which there is a die-back of small branches and twigs, yellowing of foliage and poor fruit yield and a lack of feeder, roots. (3) Slope downwards.

Declinous flower See imperfect flower.

Decoction Something obtained by boiling a substance to remove its soluble constituents.

Decomposers Bacteria that reduce dead matter to basic minerals.

Decomposition The act, process, or result of decomposing by natural decay or by chemical action, breaking of complex chemical compounds or rocks.

Decompression lever This lever releases the combustion chamber of the engine and hence helps in starting the engine.

Decompound Divided or compounded several times, with each division being compound.

Deconic acid See capric acid.

Decore To remove the core of a fruit.

Decorticate (1) Lacking a cortical layers. (2) To remove bark or cortex, husk, peel etc.

Decortication Removal of the pith and bark from fibrous and other tissues.

Decreaser Forage (range) plant that is gradually replaced by other species in a stand.

Decrepitate Crackle.

Decrescent Wanning; decreasing.

Decumbent Resting on substratum with ends turned up; denoting a plant whose stems lie horizontally on the grounds with only the extremities rising vertically.

Decurrent Running down the stipe or stem. Laying down on the ground but with an ascending tip, specifically referring to a stem.

Decussate (leaf) arrangement In pairs, the successive pairs at right angles; cross shaped (×).

Dedifferentiation Disintegration of a specialised habit or adaptation. Return of a specialised cell or structure to a more general or primitive condition.

Dedifferentiation (of axes) Change from one level of differentiation to a previous, "younger", level — often as an ecological response expressed by the apical meristem (rejuvenation), e.g., from plagiotropy to orthotropy if the model sequence goes from orthotropy to plagiotropy.

Deed of Transfer A formal agreement under seal bearing a government revenue stamp and signed by both the buyer and the seller.

Deep feeder A plant whose roots go quite deep.

Deepening A decrease in the central pressure of a pressure system on a constant height chart, or an analogous decrease in height on a constant pressure chart.

Deep freeze To preserve food products by rapidly reducing the temperature to about 0°F and maintaining well below 32°F.

Deep percolation Water which percolates below the root zone and cannot be used by plants. In rice it is of considerable magnitude in relation to the water requirement of the crop.

Deep soil (1) A thick humus surface layer of soil. (2) A very considerable thickness of soil and soil parent material over bedrock.

Deep water rice Special floating variety

of rice grown in low lying areas in 4.5 to 6 m deep water. Sown broadcast with pre-monsoon rains. Ears harvested from boats. Also called floating rice.

Deep well A well that draws its water from beneath shallow impermeable strata, at depths exceeding 6-7 metres.

Deep-well turbine pump Centrifugal pump adapted for deep well use. In this case pump is attached with a shaft to the motor placed above ground.

Deer's foot See field bindweed.

Defect Any irregularity or imperfection in a tree, log, piece product, or lumber that reduces the volume of sound wood or lowers its durability, strength, or utility value; also used for any animal, fowl or agricultural product.

Deferred shoot Shoots arising from dormant buds.

Deficiency A state of incompleteness; the absence, 'deletion', or inactivation of a segment of a chromosome.

Deficiency disease Disease arising out of lack of necessary nutritive element or substance in the food supply. May be in animals, human or plants. Also called hidden hunger.

Deficit An excess of liabilities over assets.

Deficit Financing The financing of a budget deficit by the Government by means of borrowing against the issue of Government Securities, that is by increasing the National Debt.

Definite Fixed, constant. Applied to inflorescences with primary axis terminating early in flower; also to stamens limited to 20 in number.

Definitive host The host in which a parasite reproduces sexually.

Deflation (1) A situation in which prices and money incomes are falling, accompanied by an increase in the value of the monetary unit. (2) The removal of fine soil particles from the coil by wind.

Deflation basin Basin formed by the removal of fine material by wind.

Deflator The price index used in the process of dividing value aggregates by price aggregates to obtain quantity aggregates.

Deflorate To remove the flowers from a plant.

Defoliants A chemical, or method or treatment, that causes only the leaves of a plant to fall or abscise. The fruits remain attached.

Defoliation Loss of a tree's leaves by insects, disease, or other causes.

Defoliator Any insect which destroys the foliage of a plant.

Deforest To remove or destroy most or all the trees in a stand so that natural reproduction is not possible.

Deformation Distortion of the whole body or part of it due to inheritance or by accident. Also deformity.

Degeneration (1) Deterioration of a tissue or an organ of the body causing it to cease performing its duties. (2) Deterioration of a variety, usually under commercial cultivation.

Degenerative disease See degeneration, diseases resulting in progressive deterioration.

Degeneriaceal A family of dicotyledonous plants in the order Magnoliales characterised by laminar stamens; a sobitary, pluriovulate, unsealed carpel; and ruminate endosperm.

Degerm (1) To remove bacteria from skin by mechanical cleaning or application of antiseptics. (2) To separate germ cells from cereal grains.

Deglutinate To remove gluten from a cereal.

Degradation The changing of a soil to a more highly leached and weathered state, usually accompanied by morphological changes such as the development of an eluviated, light-coloured A (Ae) horizon.

Degree brix Measure for sugar content of sugarcane or sugar solution, measured by a special instrument, saccharimetre.

Degree day (1) A measure of the departure of the mean daily temperature from a given standard. (2) Also an index commonly used in calculating the harvesting date.

Degree of drainage A qualitative measure of the efficacy of a drainage system in an area.

Degrees of freedom The degrees of freedom of a set of observations (which are subjected to sampling variation) is the number of independent values in the sample which could be assigned at will within the specification of the system.

Degree of polymerization The number of structural units in the average polymer molecule in a particular sample. Abbreviated as D.P.

Dehiscence (Dehisce, verb) Opening of an organ on an existing line of weakness, such as the opening of anthers to shed pollen and the splitting of a seed pod along a definite line to release seed.

Dehull To separate hulls from grains, seeds etc.

Dehulled seed Seed from which pods, glumes, or other outer coverings have been removed as sometimes with lespedeza and timothy; also often ambiguously referred to as "hulled" seed.

Dehusk To separate husk from seeds grains etc.

Dehydration Removal of water from a substance, as in the drying of fruits or preservation.

Dehydrator Any device which artificially removes moisture from a product for preservation or for reducing its bulk and weight.

Dehydrofreezing A dual process for preserving food which involves partial dehydration followed by quick freezing.

Dehydrogenage An enzyme which removes hydrogen atoms from a substrate and transfers it to an acceptor other than oxygen.

Delamination The separation of cells into layers. Gastrulation in which the endodermal layer splits off from the inner surface of the blastoderm and the space between this layer and the yolk represents the archenteron. Separation of a laminate into its constituent layers.

Deletion The absence of a segment of a chromosome involving one or more genes.

Delinting Removal of lint fibres that cling to cotton seed after ginning.

Deliquescence (1) The condition of repeated divisions ending in fine divisions; seen especially in venation and stem branching. (2) The absorption of atmospheric water vapour by a crystalline solid until the crystal eventually dissolves into a saturated solution.

Delivery box Structure diverting water from a canal to a farm unit often including the measuring devices. Also called turnout.

Delivery channel Watercourse, farm lateral, or field channel.

Delivery loss See conveyance loss.

Delta region A region in the atmosphere characterised by difluence.

Demand The amount of a commodity or service which will be bought at any given price per unit of time.

Demand curve A graphical presentation of a demand function, $x = \phi\ (P)$, showing how much of a commodity will be bought per unit of time at any given price, providing that the other influences governing demand remain unchanged.

Demand management policies Government policies aimed at restraining inflationary trends by restrictions or disincentives which regulate the level of demand for goods and services.

Demand schedule A table giving the demand for a commodity or service at various prices, while all other factors determining demand remain unchanged.

Demersal Living at or near the bottom of the sea.

Demeton An organo-phosphorus compound, an insecticide used against aphids and mites. Also called systox.

Democratic or **co-operative leadership** These democratic leaders are interested in every member as an individual and also assist each group member to satisfy himself through "Teamwork". They delegates responsibility to other members also.

Demonetisation The act of depriving a coin of its status as legal tender.

Dendriform Tree-like, branched.

Dendritic drainage pattern Drainage pattern characterised by irregular branching of tributaries in many directions, usually joining the mainstream at acute angles.

Dendrochronology The science of dating events and variations in environment in former times by comparative study of growth rings in trees and aged wood.

Dendrohydrology The science of determining hydrologic occurrences by the comparison of tree ring thickness with stream flow or precipitation.

Dendroid An organism with a branching habit, resembling a tree.

Dendrology The division of forestry concerned with the classification, identification and distribution of trees and other woody plants.

Dendrometre A device used to measure a tree's height and diametre using principles based on the relation of the sides of similar triangles.

Dendrophthoe A genus of evergreen shrubby, partial parasite, till recently included under Loranthus.

Denitri, fying bacteria The bacteria that cause biological reduction of nitrate or nitrite to gaseous nitrogen.

Dense (1) The close packing; (2) Designating a soil mass in which pore spaces form a relatively small proportion to the total volume and where any large pores or cracks are absent.

Density channel A channel used to investigate a density current; for example in experiments relating to the behaviour of cold masses of air in the atmosphere and related frontal structures.

Density, Crown The compactness of the crown cover of the forest, dependent upon (a) the distance apart and (b) the compactness of the individual crowns. A loose term combining the meanings of crown closure and shade density.

Density current Turbidity current. A highly turbid and relatively dense current which moves along the bottom slope of a body of standing water.

Density function A density function of a measure is a function which gives rise to m when it is integrated with respect to some other specified measure.

Density ratio The ratio of the density of the air at a given altitude to the air density at the same altitude in a standard atmosphere.

Dent, Dental, or **Tooth** A tooth-like projection on an apron, or other surface to deflect or break the force of flowing water; a form of baffle.

Dentine A hard elastic substance, chemically resembling bone, composing the greater part of teeth.

Denudation (1) The sum of the natural processes which wear down the surface of the earth, including natural or normal erosion. (2) Removal of all natural cover from an area.

Denutrition The failure to convert food elements into nutritional substances.

Deodorant A substance which removes and neutralises offensives odours, generally by oxidizing the products of decomposition of organic matter and thereby absorbing or marking the odours.

Deodorise Deprive of odour; disinfect.

Deoxyribonucleic acid (DNA) One of the two nucleic acids that have been identified; essential for biologic inheritance.

Depasture Put cattle to graze on a pasture.

Depauperate Plant which does not grow to its natural size and has a starved or underdeveloped appearance.

Depegram On a diagram having entropy and temperature as coordinates, a curve representing the distribution of the dew point as a function of pressure for a given surrounding of the atmosphere.

Dependent plants A plant which does not manufacture its own food but depends upon other living or dead things for its nutrition.

Depeq Strong winds over Loet Tawar (Sumatra, East Indies) during the southwest monsoon.

Depilatory A chemical preparation which removes hair from the skin, e.g. Barium Sulphate.

Depleted soil Soil that has lost most of its available nutrients and hence economic utility for crop production is impaired.

Depletion (1) Using a resource, such as water or timber, faster than it is replenished. (2) Reduction of the charge-carrier density in a semiconductor below the normal value for a given temperature and doping level. (3) The percentage reduction in the quantity of fissionable atoms in the fuel assemblies or fuel mixture that occurs during operation of a nuclear reactor.

Depletion curve Recession curve for

soil moisture, stream flow, ground water, etc. usually shown as decay function.

Deporaus marginatus See leaf-cutting weevil, mango leaf-cutting weevil.

Deposit (1) The product of combustion which forms on finished surfaces. (2) Solid matter which accumulates on a metal surface as by any of the various plating processes. Also the accumulation of scale in a boiler, etc. (3) Material left in a new position by a natural transporting agent such as water, wind, ice or by the activity of man.

Deposit account A banking account for money not likely to be required immediately. Such money is subject to several days' notice of withdrawal.

Deposition (1) The addition of fresh soil as by floods. (2) In pesticides application, the dust or spray which remains on the plant surface.

Deposit money Money created by the commercial banks constituting the greater proportion of the supply of money.

Depression (1) A hollow or cavity in the land surface, flesh of the animals. (2) A severe drop in income or prices in an area or country.

Depression storage Water retained in puddles, ditches, and other depressions in the surface of the ground and therefore not available for producing surface run-off.

Depreciation The diminution in the original value of an asset due to use (wear and tear) and/or obsolescence.

Depth The vertical distance between the bed of a canal and the water surface line.

Depth-area curve or **Rainfall Intensity-Area curve** A curve showing the relationship between the average intensity of the rainfall of a storm and the area of the storm.

Depth-duration area value The average depth of precipitation that has occurred within a specified time interval over an area of given size. Abbreviated as DDA value.

Depth-duration curve or **Rainfall Intensity duration curve** A curve showing the relationship between the average intensity of the rainfall of a storm and the duration of the storm.

Depth factor A depth which would contain enough grain so that if all the theoretical heat available for drying could be used, it would all dry to equilibrium in a period of time equal to the time of one half response.

Depth integrating sediment sampler A device to take a sample of suspended sediment in a stream. It is lowered from the steam surface to the bed and raised to the surface again at a constant rate. The sample of sediment-laden water is drawn in continuously thus obtaining a composite sample which is integrated over the depth of the stream.

Depth of irrigation The volume of water delivered to a given area in one irrigation divided by the area; depth of soil to be brought to field capacity.

Depth of runoff The total runoff from a drainage basin divided by the area.

Depulp To remove the fleshy substance from around the seeds of berries or fruits. Also called macerate.

Dequeen To remove the bee queen from a bee colony.

Derating A government technique for reducing the burden of taxation on particular branches of industry and commerce.

Derivative (1) A chemical compound derived from another related compound by simple modification of its formula. (2) A term in mathematics during differentiation of a function with respect to another function.

Derived data Records of observations and measurement of physical facts, occurrences, and conditions which have been developed from 'basic data' by means of standard methods of computation and estimate.

Derived demand The demand for a commodity not required for its own sake, but for its contribution to the manufacture of another commodity.

Derma See dermis.

Dermatocarpaceae A family of lichens in the order Pyrenulates having an umbilicate or squamulose growth fram; most members grow on limestone or calcareous soils.

Dermatogen The outermost histogen of a growing root or stem tip which differentiates into the epidermis.

Dermatomycosis Skin infection due to

fungus.

Dermatophytes Fungi which infect only the skin.

Dermatosis Any skin disease.

Dermis True skin; skin layer between epidermis and subcutaneous connective tissue. Also derma.

Derris Derris elliptica (Wall.) Benth. Family Papilionaceae. A climbing shrub whose roots are used to manufacture insecticide, derris powder.

Derris elliptica See derris.

— **ferruginea** Benth. Herb of the family Papilionaceae whose roots are used to manufacture insecticide.

— **trifoliata** Lour. (Derris uliginosa Benth.) Herb of family Papilionaceae whose roots are a source of an insecticide.

— **uliginosa** See Derris trifoliata.

— **powder** A commercial insecticide made of powdered root of the plant of Derris spp. The insecticidal property is due to poison rotenone. Also used as a fish poison.

Desalinization A process of removal of soluble salts from saline soils. This is usually accomplished by leaching.

Desert A land area having an arid climate which is barren or sparsely vegetated with herbs and shrubs, and occasionally with trees and grasses also.

Desert locust Schistocerca gregaria Forsk. One of the three species of locusts found in India. Visits India periodically in big swarms and causes heavy damage to crops. Threatens 64 countries in an area of about 31.10 million square km. Can mass multiply, migrate to long distances and destroy any vegetation that comes its way. Control include destroying of swarms, hoppers and eggs. See locusts.

Desert pavement Gravel accumulated at the soil surface after removal of the finer material by wind action.

Desert varnish A glossy sheen or coating on stones and gravel in arid regions.

Desiccant Compound that promotes dehydration or removal of moisture from plant tissue.

Desiccate To dry thoroughly, to remove moisture from an object to definitely below the normal level.

Desiccated coconut White, sweet and nutty flavoured desiccated meat of coconut used for culinary purposes. Prepared by dehusking the coconut, detaching of the fresh kernel from the shell with machinery, parring the brown outer skin, washing and passing it through grater, shredder and special cutters, drying out material at 160°-180°F, sifting, grading and packing the produce in lead-lined cases.

Designated diseases Refers to the diseases specified for certification of seeds and in whose regard certification standards must be met.

Designated weeds Refers to weed species specified for certification of seeds of a given kind and in whose regard certification standards must be met.

Designed duty of water Duty of water assumed in an irrigation project for designing capacities of channels.

Design flood (1) The maximum flood that any structure can safely pass. (2) The flood adopted to control the design of a structure. (3) The flood against which a given area is to be protected.

Design flood hydrograph The hydrograph of flow adopted to represent limiting volumes and concentration of runoff for use in determining design capacities of spillways for dams, etc. or other hydraulic studies.

Design storm The design storm is that estimate of a rainfall, amount and distribution over a particular drainage area which is acceptable for determining the design flood.

Desilting area Area of grass, shrubs or other vegetation used solely for inducing deposition of slit and debris from flowing water, located above a stock tank, pond, field or other area needing protection from sediment accumulation.

Desilting basin Structures built on a sprinkler system to trap sand or suspended silt when water supply comes from streams or open ditches.

Desi plough A tillage equipment used in India. The equipment include a wooden bottom having a steel share and a handle; plough botton connected through a wooden beam to yoke of bullocks, body made of hardwood like acacia arabica, makes V-shaped furrows, going 5-8 cm deep. It does not invert the soil and is available in various sizes and weights.

Desired water income The minimum average precipitation and underground flow available for transpiration within the area to cover the possible water expenditure during a crop season or annually.

Desmodium Desmodium paleaceum Guill. Perr. A perennial herb of the family Papilionaceae introduced from Nigeria and grown for fodder.

Desmodium paleaceum See Desmodium.

— **tortuosum** See beggarweed.

— **triflorum** (L.) DC. herb of the family Papilionaceae, grown as a soil binder and for fodder.

Desmos longitlorus (Roxb) Saff. (Unona longiflora Roxb.) A tree of family Anonaceae whose fruits are edible.

Desmostachya bipinnata See *kusa* grass.

Dessert fruit Any fruit, berry etc. which may be served after a meal.

Destructive distillation The decomposition of organic compounds by heat in a closed container and the collection of the resultant volatile liquids and oils.

Destructive erosion Accelerated erosion.

Detach To disconnect, to remove. In shop practice, frequently, in order to make adjustments or change the set up for other operations.

Detailed soil survey Soil survey where the soil boundaries are plotted accurately from observations made throughout the surveyed area.

Determinate growth The type of development in which a plant ripens all of its seeds in approximately the same time.

Detoxicant Substance used for neutralising the action of poisonous substances.

Detrainment The transfer of air from an organised air current to the surrounding atmosphere.

Detritivorous Feeding on freshly dead or partially decomposed organic matter.

Detritus (1) Fragmental material, such as sand, silt, and mud, derived from older rocks by distintegration. (2) Particles of plant matter in varying stages of decomposition.

Detritus tank A tank in which heavy suspended matter is removed in sewage treatment.

Deuterogamy Secondary pairing of sexual cells or nuclei replacing direct copulation in many fungi, algae and higher plants.

Devaluation and revaluation The determination of a new fixed exchange rate for a currency, lower in the former case and higher in the latter.

Developing country A poor or underdeveloped country which is making economic progress in respect of capital investment, education and training, productivity and general standard of living.

Development The process of growth and changes taking place in an individual animal or plant (from embryonic stage till maturity).

Development farm A farm to determine the optimum crop pattern and water requirements of crops, put into operation in connection with the development of irrigation from an irrigation project.

Deviate The value of a variate measured from a standard point of location, usually the mean. It is often understood that the value is expressed in standard measure, i.e. as a proportion of the parent standard deviation.

Deviation (1) The difference between a value and a value from which it is regarded as deviating. The absolute value of the deviation, i.e. the value taken without regard to sign, is called the absolute deviation. (2) The algebraic difference between the mean of a series of data and an individual member of the series.

Deviation standard (Statistics) A measure of the average variation of a series of observations or items of a population about their mean. In normally distributed sets of moderate size the interval of the mean plus or minus the standard deviation includes about two-thirds of the items.

Devil grass See Bermuda grass.

Devonian The geological period following the Silurian and preceding the Carboniferous.

Dew Water condensed onto grass and other objects near the ground, the temperatures of which have fallen below the dew point of the surface air because of radiational cooling during the night but are still above freezing.

Dewatering (1) Removal of water from solid material by wet classification, centrifugation, filtration, or similar solid-liquid separation techniques. (2) Removing or draining water from an enclosure or a structure, such as river bed, caisson, or mine shaft, by pumping or evaporation.

Dew point The temperature at which water vapour begins to condense. The temperature at which air becomes saturated when cooled without addition of moisture or change of pressure, any further cooling causes condensation. Also known as dew-point temperature.

Dew-point formula A formula for the calculation of the approximate height of the lifting condensation level; employed to estimate the height of the base of convective clouds; under suitable atmospheric and topographic conditions.

Dew-point hygrometre An instrument for determining the dew point by measuring the temperature at which vapour being cooled in a silver vessel begins to condene. Also known as cold-spot hygrometre.

Dew point pressure The gas pressure at which a system is at its dew point, that is the condition of gas temperature and pressure at which the first dew or liquid phase occurs.

Dew-point recorder An instrument which gives a continuous recording of the dew point, it alternately cools and heats the target and uses a photo-cell to observe and record the temperature at which the condensate appears and disappears. Also known as mechanised dew-point metre.

Dew-point spread The difference in degrees between the air temperature and the dew point.

Dextrins Bodies of varying composition formed when starch is hydrolised. Also produced in plant bodies by the action of cellular enzymes, and during cooking of starchy foods.

Dextrose Glucose or grape sugar.

Dhaincha Sesbania cannabina (Retz.) Pers. (S. aculeata Pers. var. cannabina). Family Papilionaceae. A green manure crop. A legume adapted to wide range of soils and climatic conditions. Tolerant to salinity and alkalinity.

Dhamani Grewia tiliifolia Vahl. A tree of family Tiliaceae produced for its edible fruits.

Dhenkli A water lifting device suitable for lifts up to 5 m. A bucket of two to four gallons is generally used. Also called *Dhenki*.

Dhol phulio Eragrostis tremula Hochst. Family Gramineae. A grass grown for fodder.

Dholsamudra Leea macrophylla Roxb. ex. Hornem. A herb of family Vitaceae. The roots have medicinal properties and fruits edible.

D-horizon A sub-surface layer or rock stratum which is neither parent material nor the source of parent material for the soil.

Diacrisia obliqua See Bihar hairy caterpillar.

Diagnosis Identification of the nature of cause of diseases, soils limitations etc.

Diagonal system of planting System of planting fruit trees by which an additional tree is planted in the centre of each square of permanent trees used as a filler tree. Kept for a small period. Also known as quincunx system.

Dialeurodes citri See citrus white fly.

Dialysis The filtration of substances through semipermeable membranes.

Diametre of knot The maximum distance between the points farthest apart on the periphery of a round knot, on the face of which it becomes visible. In the case of a spike or a splay knot, the maximum width of the knot visible on the face on which it appears shall be taken as its diametre.

Diammonium phosphate A compound fertiliser. It contains 21 per cent of nitrogen and 5 per cent of phosphoric acid. A good fertiliser for use in alkali lands.

Diamond back moth Plutella maculipennis Curt. Insect pest of crucifers; small pale green caterpillar which feeds on leaves making holes in them.

Diapause A temporary stoppage of development of an insect embryo or larva.

Diaphorina citri See Citrus psylla.

Diaphragm pump A pump used in automobiles to move oil under pressure.

Diary A book for keeping records of day to day happenings in agriculture, transactions and operations on a farm are noted.

Diaspore A unit of dispersal.

Diastase An enzyme that catalyzes the hydrolysis of starch to maltose- Also known as vegetable diastase.

Diatomaceous earth The earthy deposit of fine, greyish siliceous material composed chiefly or wholly or the remains of diaforms. It may occur as a powder or as a porous, rigid material. See diatoms.

Diatoms Algae having siliceous cell walls that persist as a skeleton after death. These microscopic unicellular or colonial algae belong to the class Bacillariaceae. They are abundant in both fresh and salt water their remains are widely distributed in soils.

Diazinon o, o-diethyl-o (-2-isoporpyl-4-melthyle-b-pyrimidinyl) Phosphorothioate. Insecticide, slowly hydrolysed in water and in presence of acidic material, used in control of San jose scale, woolly aphids, mustard aphids, thrips, jassids and bugs.

Dibble (1) A pointed tool to make holes for planting tree seedlings. (2) To sow or plant seed with a dibble, see dibbling.

Dibbling Method of sowing; seeds are put in holes dibbled by a iron or wooden stake at regular interval.

Dicalcium phosphate $CaHPO_4$. A fertiliser containing 34.0 per cent P_2O_5 and 32 per cent CaO. It has an equivalent basicity of 25 per cent.

Dice To cut into small cubes.

Dicot See dicotyledon.

Dichanthium annulatum See marvel grass.

— **caricosum** (L.). A Camus. Family Gramineae. A grass used as fodder

Dichloroethyl ether A soil fumigant and nemacide.

Dichloropropylene A soil fumigant and nemacide.

Dichocrosis punctiferalis See castor pod borer.

Dichogamy The absence of simultaneous maturation of stamens and stigmas in a flower so that their ripening does not overlap and cross pollination is ensured.

Dichopsis elliptica See Indian gutta percha tree.

Dichotomy A type of branching in plants in which the main axis or stem forks repeatedly into two branches, as in certain sea weeds, liverworts, etc.

Dicotyledon Plant that produces two seed leaves in each seed.

Dicotyostele A stele consisting of separate vascular bundles, or of a network of bundles, produced by overlapping leaf gaps.

Dicumbent Trailing on ground.

Die (1) An internal screw used for cutting an outside thread. (2) A formed piece used in stamping out similar parts for quantity production.

Die-back Progressive death of branches or shoots beginning at tips, characteristic of certain plant disease.

Die-casting A very smooth and accurate casting made by pouring molten metal or composition, under pressure, into a metal mold or die.

Die casting metal Light coloured alloys of aluminium, lead zinc or copper base of widely varying composition depending on the use of which they are to be put. For automobile hardware, an alloy of 90 per cent aluminium, 4 per cent copper, 4 per cent nickel, and 2 per cent silicon is much used.

Dieldrin 1, 2 3, 4, 10, 10-hexachloroexo-6-7, epoxy-1,4 4a, 5, 6, 7, 8, 8a-octahydro-1, 4-endo, exo-5, 8 dimethanonaphtahline. Insecticide chemically related to aldrin. It is contact and stomach poison and available as dust, water dispersible powder, emulsifiable concentrate and solution. Poisonous to mammals.

Die off To die and fall to the ground, as a leaf.

Die out Cease to exist.

Diesel engine An internal combustion engine of high efficiency, in which a high temperature is obtained through compression; the oil is ignited by the heat thus developed.

Differential Relating to a construction in which a movement is obtained by a difference in two motions in the same direction, as in the rear axle of an automobile.

Differential analysis Synoptic analysis of change charts or of vertical differential charts (such as thickness charts) obtained by the graphical or numerical subtraction of the patterns of some meteorological variables at two times or two levels.

Differential block A chain tackle which depend primarily on a pulley of two diametres fitted with lugs which fit the links of a chain, the larger diametre winds the chain in as the smaller diametre takes it out.

Differential chart A chart showing the amount and direction of change of a meteorological quantity in time or space.

Differential gearing The system of gearing which permits one rear wheel of an automobile to move independently of the other; thus avoiding excessive tire wear and strain on mechanical parts.

Differential host A plant that produces distinctive reactions to a specific virus or viruses, thus enabling the differentiation of viruses.

Differential stain Stain distinguishing between different groups of organisms (examples: acid fast stain, gram stain).

Differential Unit It is a special arrangement of gears to permit one of the rear wheels to rotate slower or faster than the other while turning on a curved point.

Differential water capacity The absolute value of the rate of change of water content with soil pressure. The water capacity at a given water content will depend on the particular desorption or adsorption curve used. Distinction should be made between volumetric and specific water capacity. Also see water, soil.

Differential weir A weir installed on a parent channel to control upstream water level. It may be diagonal or of other shapes in plan.

Differentiation (of axes) Morphological specialisation either as sequential expression of meristematic specialisation or growth correlation (e.g. from orthotropic to plagiotropic shoots).

Diffuse Widely spread, not localised not sharply defined.

Diffused runoff Rainfall which runs off simultaneously over the whole area of the slope contrast to runoff concentrated to a single channel.

Diffuse radiation Radiation reaching the Earth's surface after having been scattered from the direct solar beam by molecules or suspensions in the atmosphere.

Diffuse spring Springs in which flow is not concentrated at one outlet but spread out.

Diffusion The transport of matter as a result of the movement of the constituent particles. The movement due to concentration difference.

Diffusion diagram A diagram for displaying the comparative properties of various diffusion processes, with coordinates of the mean free path or mixing length and mean molecular speed or diffusion velocity, for molecular or eddy diffusion, respectively; each point of the diagram determines diffusivity.

Digera arvensis Digeria muricata (L) Mart. (Digera arvensis. For K., Achyranthes muricata L.). Family Amaranthaceae. A herb whose tender twigs and inflorescence are used as vegetables.

Digest (1) To convert food within bodies into a form that can be assimilated. (2) To soften, dissolve or alter a substance by heat and chemical action.

Digestible Feed that can be converted into a form absorbable from within the body.

Digestibility The readiness with which a substance may be converted into an absorbable form within the body. Also see digestibility, true.

Digestibility, true Actual digestibility or availability of a feed, forage, or nutrient as represented by the balance between intake and faeces loss of the same ingested material.

Digestible crude protein The total nitrogenous compound protein fed to livestock less the nitrogenous compound eliminated in the faeces.

Digestible dry matter (DDM) Feed-faeces difference expressed in dry matter (DM).

Digestible energy (DE) Feed faeces difference expressed in calories.

Digestible nutrients Portion of nutrients consumed that are digested and taken into the animal body. This may be either apparent or true digestibility; generally applied to energy and protein.

Digestion coefficient The average percentage of each nutrient in the feeding stuff digested by the animal.

Digitalis Digitalis purpurea L. Family Scrophulariaceae. A biennial and sometimes a perennial herb, native of west Europe. Leaves used as medicine. Cold, humid conditions ideal. Contains glycosides, principally digitoxin, gitoxin and gitalin. Also called Foxglove.

Digitalis lanata See Grecian foxglove.

— **purpurea** See digitalis.

Digitaria Digitaria cruciata (Nees) A. Camus var. esculenta Bor. Family Gramineae. A grass cultivated for grain and fodder.

Digitate Applied to a compound leaf, of which all the leaflets are borne on the apex of the petiole.

Digitinervate Having straight veins extending from the petiole like fingers.

Digitipinnate Having digitate leaves with pinnate leaflets.

Diisenwind The mountain-gap wind of the Dardonelles; a strong eastern wind which blows out of the Dardanelles into the Aegean Sea, penetrating as far as the inland of Lemons, and caused by a ridge of high pressure over the Black Sea.

Dikamali Brilliant gardenia, Gardenia resinifera Roth. (Gardenia lucida Roxb.). a common shrub grown for fragrant flowers belonging to family Rubiaceae.

Dikaryon A pair of closely associated nuclei, each of which has been derived from a different parent cell in most cases.

Dill Dill seed. Anethum graveolens L. (Peucedanum graveolens Benth. Hook f.). Family Umbelliferae and a native of Eurasia. Fruits are used as spice and condiment and also used as carminative. Also see Indian dill.

Dilleniaceae A family of dicotyledonous trees, woody vines, and shrubs in the order Dilleniales having hypogynous flowers and mostly entire leaves.

Dillenia indica See elephant-apple.

— **pentagyna** (1) Roxb. Family Dillieniaceae. Tree whose flowers are edible and bark is used for cordage making and old leaves as sand paper.

Dilleniales An order of dicotyledonous plants in the Subclass Dilleniidae characterised by separate carpels and numerous stamens.

Diluent Substances mixed in concentrated solutions to reduce their strength.

Dimefox Organo-phosphorus compound; N, N, N′, N′- tetramethyle phosphorodiamic flouride; a volatile contact insecticide and acaricide. Applied on trunk against sap-feeding insects or on soils as capsules. Not considered safe to apply on plant surfaces.

Dimethoate O,O dimethyl S (N. methyl carbamoyl methyl) phosphorothiolothionate; systemic organo-phosphorus inecticide, used for control of aphids, spider mites and leaf-miners.

Dimentionless unit graph A unit graph useful for comparing unit hydrographs of different drainage areas or those resulting from different storm patterns. It is derived either from an observed storm-flow hydrograph of a flood event or from a unit hydrograph or can be constructed from a summation graph.

Dimension stock Timber converted to specified widths, thicknesses and lengths up to certain limits.

Dimictic lake A lake which circulates twice a year.

Dimidiate Having one half smaller than the other of a perithecium, having outer wall covering only top half.

Diminution Increasing simplification of inflorescences on successive branches.

Dimmer A device for reducing the glare of headlights. The usual practice is to raise and lower the headlight beam by means of a foot switch. A rheostat placed in series with a lamp to control its brilliancy.

Dimmerfoehn A rare form of foehn where during a very strong upper wind from the south, a pressure difference of 12 millibars or more exists between the south and north sides of the Alps; a stormy foehn wind then overlaps the upper valleys in the northern slopes, reaches the ground in the lower parts of the valleys, and enters the foreground as a very strong wind; the foehn wall and the precipitation extend beyond the crest across the almost calm surface area in the upper valleys.

Dimonoecious Having perfect (bisexual) flowers as well as pistillate and staminate flowers on one plant.

Dimorphism (1) The occurrence of two distinct forms within a species. (2) The production of two distinct types of flowers on one plant.

Dimocarpus litchi See litchi.

Dimorphism Having two forms or types, as in the life cycle of certain algae with different gametophytic and sporophytic generations.

Dinoflagellate A member of the class Dinophyceae.

Dioecious Having male and female flowers borne on separate plants.

Diorite See igneous rock.

Dioscorea alata See white yam.

— **aculeata** See lesser yam.

— **bulbifera** See potato yam.

— **deltoidea** Wall. Family Dioscoreaceae, shrub grown for its edible fruit. Also yields a steroid hormone, cortison, used as medicine.

— **esculanta** See lesser yam.

— **fasciculata** See lesser yam.

— **pentaphylla** L. Family Dioscoreaceae. Herb found throughout India and grown for its edible roots.

— **prazeri** Prain and Burk. A herb of family Dioscoreaceae whose tubers are source of a hormone, cortisone. Rhizomes also used as substitute for soap for washing hair.

— **sativa** See potato yam.

Dioscoreaceae A family of non-ocotyledonous, leafy-stemmed, mostly turning plants in the order, liliales having an inferior ovary and septal nectaries and lacking tendrils.

Diospyros assimilis See Macassar ebony.

— **chinensis** See Japanese persimmon.

— **ebenum** See Macassar ebony.

— **embryopteris** See Indian persimmon.

— **ferrea** (Wild). Bakh, Maba buxifolia (Pera). Tree of family Ebanaceae which produces edible fruits.

— **kaki** See Japanese persimmon.

— **lotus** See date plum.

— **melanoxylon** See Cloromendel ebony persimmon.

— **peregrina** See Indian persimmon.

— **sapota** See Macassar ebony.

— **tomentosa** See Nepal ebony persimmon.

— **tupru** See Coromandel ebony persimmon.

Dip To submerge a substance living body in a liquid, e.g. to dip a produce or fruits in a preservative for a small period.

Diphyllons Having two leaves.

Diplococcus Bacteria that occur in groups of two.

Diplodia corchori See jute black band disease.

Diplohaplant An organism characterised by alternating, morphologically similar haploid and diploid generations.

Diploid Having the double chromosome number, as is characteristic of cells of the sporophyte generation in plants.

Diploid apple Apple varieties that is self-fruitful or productive from its own pollen.

Diploknema butyraceae See Indian butter tree.

Dipsacales An order of dicotyledonous herbs and shrubs in the subclass Asteridale characterised by an inferior ovary and usually opposite leaves.

Dipsacus fullonum See Fuller's teasel.

— **sylvestris** See Fuller's teasel.

Diptera Order of insects to which mosquitoes and flies belong.

Dipteral With double peristyle.

Dipterocaryaceae A family of dicotyledonous plants in the order Theales having mostly stipulate, alternate leaves a prominently exerted connective and a calyx that is mostly winged in fruit.

Dipper dredge An excavating machine which uses a boom-type shovel.

Dipterous (1) Of, related to or characteristic of Diptera. (2) Having two wings or winglike structures.

Direct area Area served with irrigation water directly from a main or principal canal.

Direct attack This attack is in the form of either an anger or an intelligent, rational attack on the difficulty. In anger, person loses his control. Intelligent adjustment is most desirable.

Direct bearing A direct vertical support in a structure.

Direct cell A closed thermal circulation in a vertical plane in which the rising motion occurs at higher potential temperature than the sinking motion.

Direct count In soil microbiology, any one of several methods for estimating by direct microscopic examination the total number of micro-organisms in a given mass of soil.

Direct drive Power transmission by direct connection between the power unit and pump, without belts, chains or gears; the unit so connected are called monoblocks.

Direct microscopic count Counting of all bacteria living or dead by smearing on slide and/or staining with suitable dyes.

Directional pull See draft.

Direction of irrigation Direction in which the water flows on surface of the field from the distribution pipe or ditch.

Direct relationship A parent offspring relationship between individuals in breeding.

Direct runoff The sum of surface runoff, interflow and channel precipitation.

Disaccharides (C_{12} $H_{22}O_{11}$) Group of carbohydrates, most important of which are sucrose, lactose, and maltose.

Disafforest Reduce forest land to ordinary land.

Disassimilation Conversion of assimilated substances into less complex or waste ones.

Disbudding The removal of flower buds and/or shoot buds from a plant. This helps in preventing the unwanted shoots and better growth of remaining buds.

Disc (Related to plough). It is a circular, concave revolving steel plate used for cutting and inverting the soil.

Disc-arm type corrugated furrow opener The implement having sharp concave discs in front and cylindrical drums behind; used for making uniform furrows and corrugations in light textured soils. Also used for smoothening and compacting furrows after cultivation.

Discharge (Hydraulics) (1) Rate of flow, specifically fluid or silt flow; as volume of fluid passing a point per unit time, commonly expressed as cubic feet per second, million gallons per day, gallons per minute, or cubic metres per second. (2) An exudate or abnormal material coming from a wound or from any other opening in the body.

Discharge adapter A fitting installed near the head of pipe-line system to act as an increase from the pumping discharge size to the mainline size.

Discharge bay Bay for the discharge of a pumping plant.

Discharge velocity The rate of discharge of water through a porous medium per unit of total area perpendicular to the direction of flow.

Discharge coefficient (Hydraulics) The ratio of actual rate of flow to the theoretical rate of flow through orifices, weirs, or other hydraulic structures.

Discharge curve Rating curve showing the relationship between stage and rate of flow or similar relations.

Discharge formula (Hydraulics) A formula to calculate rate of flow of fluid in a conduit or through an opening. For steady flow discharge, $Q = AV$, wherein Q is rate of discharge; common units are cubic metre per second per square metre, and metre per second, respectively. To calculate the mean velocity V, for uniform flow in pipes or open channels.

Discharge head Vertical distance between the intake level of a water pump and the level to which it discharges water freely to the atmosphere.

Discharge mass curve A curve in which cumulated value of recorded discharge are plotted against time. The slope of the curve at any time represents the discharge at that instant.

Discharge pipe Pipe leading from the discharge side of a pump; taking the discharged water.

Disclimax A climax community that includes foreign species following a disturbance of the natural climax by man or domestic animals. Also known as disturbance climax.

Disc harrow A harrow with varying number of high carbon steel discs. A secondary tillage implement. The discs fixed in two gangs whose direction and placement can be varied.

Discolichenes The equivalent name for lecanorales.

Discomycetes The cup fungi, a subclass of Ascomycetes; with apothecia.

Discount A reduction in a previously determined price granted to a customer or class of customers.

Discounting The process by which future costs and revenues are converted into a present net worth. Future money is considered to be worth less than present money, a rate of interest being used for discounting.

Discovery It may be defined as a finding of thing or principles that has already existed.

Discrepancy A difference between the results of duplicate or comparable measures of a quantity. The difference in computed values of a quantity obtained by different processes using data from the same source.

Discrete variable Where the possible values of the variable move in steps or jumps.

Discretion The amount by which a broker, on his clients instructions; may vary the price at which his client wishes to deal.

Disc-ridger An implement to make ridges. The disc gangs are used and these could be adjusted for making different sizes of the ridges.

Disc separator It is an upgrading machine which makes the separations on the basis of length differences. The disc separator basically consists of a series of cast iron discs which revolve together on a horizontal shaft inside a cylindrical body.

Disease control Any method which inhibit the activity or effect of disease-producing organisms. Also which modifies the conditions favourable to disease.

Disease cycle The complete chain of events in the manifestation of a disease. From infection to complete recovery.

Disease garden A special nursery for the study of reaction to specific pathogens.

Disease resistant Refers to plants and animals which are not readily susceptible to a particular disease.

Dishoarding The desire on the part of individuals to hold fewer assets in the form of money, thus reducing 'idle balances'.

Disifloral Having flowers with enlarged disclike receptacles.

Disinfectant An agent which destroys disease producing organisms and their spores when brought in contact with it. The agent may be chemical or physical.

Disinfection Freeing a diseased plant, organ or tissue from infection.

Disinfest To rid, destroy or free from insects, parasites or rodents etc.

Disinfestant Any substance which destroys or inactivates disease organism in the soil, seed surface, animals, birds or buildings.

Disinfestation (1) Mechanical or physical means of hindering, the development of parasites or destroying them. Usually pertains to external parasites. (2) Killing or inactivating disease organisms before they can cause infection; or surface of seed or plant part or in soil.

Disinflation The removal of inflationary pressure from the economy in order to maintain the value of the monetary unit.

Disintegration The breakdown of rock and mineral particles into smaller particles by physical forces such as frost action. See also weathering.

Disjunctor Cell or projection connecting spores of chain.

Disk wheel Made as steel stampings. Some have detachable rims, others have the wheel and rim made as a unit. The usual practice is to change the wheel when a tire change is to be made.

Disorder An unnatural physical condition. It may be of plant or animals.

Disperse (1) To break up compound particles, such as aggregates, into the individual component particles (2) To distribute or suspend fine particles, such as clay, in or throughout a dispersion medium, such as water.

Dispersed soil Soil in which the clay readily forms a colloidal soil. It is usually with a low permeability which upon drying shrinks, cracks and becomes hard and upon wetting slakes readily and is plastic.

Dispersed villages In the case of dispersed or non-nucleated villages, the farmers live separately in their respective farms. Their habitats being thus dispersed, their social life assumes a different form.

Dispersing agent Substances or chemicals added to get colloids completely and uniformly distributed in the solutes. Examples are the substances added to insecticides to get uniform concentration of sprays; in mechanical analysis of soils to get the clay dispersed etc.

Dispersion coefficient See dispersion ratio.

Dispersion of produce Term used in marketing to the process involved in the operations of wholeselling and retailing at the terminal market.

Dispersion medium The portion of colloidal system in which the dispersion phase is distributed.

Dispersion ratio This is measured by shaking a soil sample in water for a given time and measuring the proportion of the total mass of particles less than 0.02 mm in diametre that remains in suspension. It is a measure of soil erodibility. Pretreatment of the sample and the measurement procedure must be taken into account while evaluating results.

Disphinctus politus See betelvine bug.

Displacement volume of an engine It is the total swept volume of all pistons during power stroke occurring in one minute.

Disposable income The amount of income that individuals and households actually receive and have available for spending; all sources of income minus income taxes and other compulsory deductions on these earnings.

Dissect To divide, cut and separate into different parts.

Displacement metre Metre that measures the quantity of flow by recording the number of times a container of known volume is filled and emptied primarily used for low flow measurements.

Dissection The work of erosion in destroying the continuity of a relatively even surface by cutting ravines or valleys into it.

Dissemination (1) Transport of inoculum from a diseased to healthy plant. (2) The natural scattering or distribution of seeds.

Disseminule An individual organism or part of an individual adapted for the dispersal of populations of the same organisms.

Dissepiment (1) A partition which divides a fruit or an ovary into chambers. (2) One of the vertically positioned thin plates situated between the septa in extinct corals of the order Rugosa.

Distal Standing far apart. Pertaining to end of any structure farthest from the middle line of organism or from point of attachment.

Distance A non-negative number associated with pairs of geometric objects. The spatial separation of two points, measured by the length of a hypothetical line joining them.

Distichous Occurring in two vertical rows.

Distichous (leaf) arrangement Alternate, but restricted to a single plane.

Distil To drive of gas or vapour from a substance by heat and condense the gas by cooling.

Distillate Substances obtained by distillation.

Distillary waste The residue obtained

from the fermentation and distillation of beer and distilled spirit. It is used as fertiliser since it contains potash and also some ammonical nitrogen.

Distillation test In this test, curves between temperature and per cent of the fuel distilled are drawn. It is a measure of the volatility of a fuel.

Distilled liquor Alcoholic beverages obtained by distilling and alcohol-containing liquid such as wine or fermented fruit juice and then further treating the distillate to obtain a beverage of specific character. Also known as hard liquor.

Distillery The building where distillation of alcoholic beverages occurs.

Distributaries, or **secondary canals** Canals which take off from the main canals or branches and supply water to minors and outlets of runouts.

Distributary An irregular branch flowing out from a main stream and not returning to it as in a delta. Also known as distributary channel.

Distribution Range of a substance or group in biographical division of a country or the globe.

Distribution channel or supply channel An irrigating channel feeding outlets.

Distribution, marginal theory of A theory seeking to demonstrate that the price of each factor of production depends upon its marginal productivity.

Distribution graph A statistically derived hydrograph for a storm of specified duration, graphically representing the per cent of total direct runoff passing a point on a stream, as a function of time, usually presented as a histogram or table of per cent runoff within each of successive short time intervals.

Distribution of a random variable For a discrete random variable, a function (or table) which assigns to each possible value of the random variable the probability that this value will occur; for a continuous random variable x, the non-stop non-decreasing function which assigns to each real t the probability that x is less than or equal to t. Also known as distribution function; probability distribution, statistical distribution.

Distribution reservoir A service reservoir connected with the conduits of a primary water supply; used to supply water to consumers according to fluctuations in demand over short time periods and serves for local storage in case of emergency.

Distribution system (1) System of ditches and their appurtenances which conveys irrigation water from the main canal to the farm unit. (2) Any system which distribute substances to users.

Distributor (1) The device, operating in fixed time with the engine, which distributes the high tension current to the spark plugs. (2) A dealer who has been granted by a company an exclusive or preferential right to buy and sell in specified markets.

Distributor gear A gear at the lower and of a distributor shaft in mesh with a gear cut or mounted on the camshaft, which drives the distributor shaft.

Distribution loss Measure of the uniformity of irrigation water distributed over a field. Also See conveyance loss.

Disturbance A disruption, or perturbation, of an ecosystem resulting from human activity.

Disturbed profile A soil profile where the horizons have been mixed up as a result of breaking up of natural vegetation and the soil being brought under cultivation.

Disutility Opposite of utility; it implies that the consumption of a good or service gives rise to dissatisfaction.

Ditch An artificial excavation; a drain or irrigation channel.

Ditch storage See channel storage.

Ditylenchus angustus See paddy eelworm.

Diurnal Daily, recurring once each day, e.g., pertaining to a plant whose blossoms open during the day and close at night.

Diurnal inequality When applied to tidal currents, difference in height of the two high waters or of the two low waters of each day; also the difference in velocity between the two flood currents or of the two ebb currents of each day.

Diurnal migration The daily rhythmic

movements of organisms in the sea from deeper water to the surface at the approach of darkness and their return to deeper water before dawn.

Diurnal range The difference between the mean higher high waters and the mean lower low waters.

Diurnal tides Tides requiring full day to complete their cycle, i.e., only one high and one low water in each lunar day.

Diversification The practice of raising a number of crops or animals or both on a farm. Also diversified farming.

Diversified farming See diversification.

Diversion Channel constructed across the slope for the purpose of intercepting surface runoff; changing the accustomed course of all or part of a stream.

Diversion box See division box.

Diversion canal An artificial channel for diverting water from one place to another.

Diversion chamber A chamber designed to direct a stream into a channel or channels. Also diversion box.

Diversion dam A structure of barrier that diverts part of or all the water of a stream to a different course.

Diversion ditch A ditch constructed to intercept the water which would otherwise flow into a district from the land outside, and to carry it around the district to the outlet stream below.

Diversion gate A gate which may be closed to direct water from the main conduit or canal to a lateral or some other channel.

Diversion ratio The ratio of the flow diverted to the stream flow.

Diversion requirement, gross irrigation requirement or delta at head of main canal Irrigation requirement at the source of irrigation supplies. It is equal to net irrigation requirement plus water losses and operational waters in transit, and is the same as 'gross duty of water' when the latter is expressed in similar units.

Diversion terrace A terrace constructed to divert water from a terrace system, or to divert water from its natural drainage channel for some purpose, viz. to reduce the speed of the runoff water.

Diversity The variety of species present in a biological community.

Diversity index A measure of diversity.

Divided bar graph Each bar has subdivisions to indicate the respective divisions. These sub-divisions may be represented with different colours or different types of lines or signs.

Divide The line of separation between drainage system; the summit of a ridge between streams.

Dividend The portion of a company's profits which the Directors each year decide to distribute to ordinary shareholders. This is usually expressed as a percentage of the nominal value of the shares.

Dividend cover The number of times a divided is covered by a company's earnings.

Dividend warrant Payment of a dividend to an investor by means of a cheque.

Divisibility (idea) A practice that can be tried on a limited basis.

Division (1) A part of a substance divided into many parts. (2) Propagation of fruit plants from vegetative parts that separate out naturally from the parent plant or are readily separated from it.

Division board See dummy.

Division box, or proportional distributor on the field A farm structure to divide supply between the two or more ditches. Also known as 'outlet box, diversion box'.

Doab A term used in India to define a area lying between and immediately above the junction of the two streams.

Doctor A cooling sea breeze in the tropics.

Docynia indica See Indian crab apple.

Dodonaea viscosa See Jamaica switch sorrel.

Doister In Scotland, a severe storm from the sea. Also known as deaister; dyster.

Doisy unit A unit for standardisation of vitamin K.

Dolabriform Shaped like an ox head.

Doldrums A nautical term for the equatorial trough, with special reference to the light and variable nature of the winds. Also known as equatorial calms.

Dolichos biflorus See horsegram.

— **dissectus** See moth bean.

— **gladiatus** See jack bean.
— **lablab** See lablab bean.
— **lablab var. lignosus** See field bean.
— **lablab var. typicus** See garden bean.
— **sinensis** See cowpea.

Dolisform Barrel-shaped.

Dolomite ($MgCO_3CaCO_3$) Calcium magnesium carbonate; sedimentary rocks.

Dolomitic limestone A dolomite whose major constituent is limestone.

Domestic (1) Native, in contrast to foreign. (2) Pertaining to home or animal species commonly used by man around his home.

Domesticate To bring a wild animal, bird or plant under control and improve it by careful breeding, management etc.

Dominance The predominance of one of any pair of Mendelian characters.

Dominant (1) A heritable character possessed by one parent of a hybrid which, when pure, will express itself in the hybrid to the apparent exclusion of the opposite or recessive character in the other parent. (2) Designating a plant species which characterises an area.

Dominant trees Those in the forest that are tallest, largest, and most valuable.

Dorbank A hard to extremely hard layer of soil in certain soils of arid regions. It may simply be massive or it maybe laminated; the latter parallel to the soil surface. It may or may not be calcareous, or salty. Its colour is related to the soil in which it occurs. It is related to the duripan of other classification systems.

Dormancy (1) Condition of relative inactivity as applied to seeds, tubers, and perennial plants during the winter. (2) A reduction in protoplasmic activity due to carbon dioxide concentration.

Dormant Sleeping, resting or inactive stage of development in plants, animals or insects. Also called resting, sleeping, inactive or quiscent.

Dormant bud A bud which is normally inactive, but may begin to develop under certain given conditions favourable to it.

Dormant grafting Grafting conducted on a plant when it is physiologically inactive.

Dormant pruning Pruning carried out at a stage when the plant is physiologically inactive.

Dormant spraying A spraying of the plant done when the plant is inactive in its growth i.e., in dormant season.

Dorsal The upper or outer surface of a plant organ. In a leaf, the outer surface in position of development is the dorsal surface even though at maturity, it is usually the under surface when expanded.

Dorsiventral Having an upper surface structurally different from the lower surface. Many leaves, fern prothali, and liverwort thalli exhibit this type of differentiation.

Dorysthenes hugelii See apple root borer.

Double crop The raising of two successive crops on the same field in one irrigation season. The crop that is first out is called 'first crop' and the crop harvested later in the season is called 'second crop'. The term 'second crop lands' is sometimes used for 'double crop lands'.

Double cropping A form of multiple cropping in which two crops are grown on a field at different times of the year.

Double cross A hybrid seed produced by combining two single crosses; a hybrid between two heterozygous single crosses of parent lines, and is not uniform as the single cross. A term used particularly in corn where four inbred lines are used as parents. The double cross is the F_1 cross between two single crosses.

Double ditch or drain See W-ditch.

Double end bolt A bolt having no solid head, being threaded at each end for the reception of nuts, often called a 'stud' bolt.

Double entry A technique used in book-keeping whereby two separate entries are made of every transaction.

Double eye A plant cutting for vegetative propagation with two buds.

Double layer In colloid chemistry, the electric charges on the surface of the disperse phase, usually negative, and the adjacent diffuse layer, usually positive, of ions in solution.

Double-main system A system of drains used where seeps appear on both sides of a broad flat, also where a stream separates an open flat into two parts.

Double mass curve A plot of accumulated annual or seasonal precipitation at an individual station against the concurrent accumulated mean precipitation for a graph of surrounding stations.

Double roller gin Type of a commercial roller gin with 2 rollers and a box in between them through which raw material is fed. Has a higher capacity better ginning of lints and less breakage of seed in comparison to single roller gin.

Double top cross (1) A seed processing machine which separates seed by their relative ability to roll or slide, which is in turn determined by shape and surface texture. (2) A cross made in the production of hybrid seeds consists in crossing a high yielding, agronomically suitable single cross with an adapted variety from wide and diverse genetic make-up.

Double transplantation System of transplanting long duration rice seedlings twice. In the second nursery, seedlings are maintained for a month and tillers coming out during this period are also used for planting after separation.

Double-work In plant propagation, to graft or bud a scion to an intermediate variety that is itself grafted on a stock of still another variety.

Doubling time In demography q.v., the period of time during which a population q.v. density doubles.

Douche Treatment of any part of the cavity by means of a stream of water.

Dough stage The stage in the development of a cereal grain in which the inside of the kernel has a consistency like that of a dough.

Douglas fir Pseudotsuga menziesii A large coniferous tree in the order Pinales; cones are characterised by bracts extending beyond the scales. Also known as red fir.

Down draft carburettor A carburettor in which the air enters at the top and travels downward past the spray nozzle.

Downrush A term sometimes applied to the strong downward, flowing air current that makes the dissipating stages of a thunderstorm.

Downy mildew Disease of plants caused by fungi of families Pythiaceae and Peronosporaceae. Fungi infect leaves, stems, flowers and fruits of host plants, giving the infected part a downy or felty appearance.

Dowsoniales An order of mosses comprising rigid plants with erect stems rising from a rhizome-like base.

Draft (1) It is the horizontal component of the (total force required to pull an implement) parallel to the line of motion. Also called directional pull. (2) Also used in vertical drainage indicating components with which water is taken out of the aquifer.

Draft animal Any animal used for pulling heavy loads or used for tillage in fields e.g., ox.

Draft breed Any breed of animal which is suitable for pulling heavy loads.

Drag, or Total drag The force exerted on a body in the direction of its relative motion in a real fluid in which it is submerged. When the real fluid is water, it is referred to either 'drag' or 'hydraulic drag'.

Drag harrow A harrow with long curved teeth, with round chisel or duck-foot shaped points and used for combing the weeds from the lower layers of the soil to the surface.

Dragon fly A predatory flying insect belonging to the order odonata. Aquatic in habit while young.

Drag Scrapers, Levelling drag scraper, or blade scrapers Scrapers fitted with the cutting blade which during operation drags or pushes the earth.

Drain (1) Channels, such as open ditches or drain tile, so that excess water can be removed by surface or by internal flow. (2) To lose water from the soil by percolation. (3) An artificial or natural channel which receives and carries off excess or unwanted water from an area of land. Also called drainage channel.

Drainable pore space Volume of water that will be released per unit volume of soil when water table is lowered. It is often taken as constant although it is known that it varies.

Drainage (1) The removal of excess surface water or ground water from land by means of surface or subsurface flow or by internal flow. (2) To lose water (from the soil) by percolation.

Drainage area The area from which drainage water is collected and delivered to an outlet.

Drainage benefits The betterment accruing to land from the establishment of a system of drainage often used as a basis for taxation.

Drainage channel A conduit or channel, either artificial or natural, for carrying off surplus ground or surface water.

Drainage coefficient The runoff or surplus surface and ground water from an area to be removed in 24 hr. The usual units of expression are cm of water per day. It may also be expressed as flow rate per unit of area, which varies with the total size of area. Also called drainage modulus.

Drainage culvert, or **supply culvert** A culvert passing drainage water under canal, railroad, road, etc.

Drainage curves Design curves giving prescribed rates of surface runoff for different levels of crop protection, and which vary according to size of the drainage area.

Drainage density The average length of streams, natural or artificial per unit area within the basin.

Drainage furrows Furrows for carrying runoff water.

Drainage head The difference between the level of water in a drain at any given point and the lowest of the land (on a line drawn at right angles through the given point).

Drainage modulus See drainage coefficient.

Drainage pattern (1) Arrangement of drains or lines of drains, one with another. Arrangements of tributaries within a watershed. (2) An arrangement or composition of natural drainage system that suggests or reveals a design.

Drainage soil As a natural condition of the soil, soil drainage refers to the frequency and duration of periods when the soil is free of saturation; for example, in well-drained soils the water is removed readily but not rapidly; in poorly drained soils the root zone is waterlogged for long period unless artificially drained, and the roots of ordinary crop plants cannot get enough oxygen; in excessively drained soils the water is removed so completely that most crop plants suffer from lack of water. Strictly speaking, excessively drained soils are a result of excessive runoff due to steep slopes or low available water holding capacity due to small amounts of silt and clay in the soil material.

Drainage system A system comprising a main drain, its branches and subsidiary drain. It includes structures and pumps used to collect and dispose of excess surface or sub-surface water.

Drain tile Pipe of burnt clay, concrete, etc., in short lengths used as a conduit in subsurface drainage. Also known as 'field drain pipe', 'subsoil pipe' and 'land pipe'.

Drainage water The water that the soil is unable to hold against gravitational force. Also called gravitational water, free water, excess water. The water from surface or ground flow due to a storm or otherwise flowing in a drain is also called drainage water or excess water.

Drainage well (1) A vertical shaft in a masonary dam to intercept seepage before it reaches the down stream side. (2) A tubewell specifically meant for drainage as in vertical drainage.

Draper separator seed processing machine which separates seed by their relative ability to roll or slide, which is in turn determined by shape and surface texture.

Draught See draft.

— **animal** See draft animal.

— **breed** See draft breed.

Draw Colloquial term for shallow, open natural drainage way.

Drawbar (1) A bar used to connect a tender to a steam locomotive. (2) A beam across the rear of a tractor for coupling machines or other loads. (3) A clay block submerged in a glass marking furnace to define the point at which sheet glass is drawn.

Drawbar hitch It is a crossbar with suitable holes attached to lower hitch links and is used in hitching the trailing implement with the tractor.

Drawbar horse-power The horse-power available at the drawbar in the rear of a locomotive or tractor to pull the vehicles behind it.

Drawbar pull The force with which a locomotive or tractor pulls vehicles on a drawbar behind it.

Drawdown The magnitude of the change in water surface level in a well, reservoir, or natural body of water resulting from the withdrawal of water. The difference between the static and the flowing bottom hole pressure.

Draw hoe Garden hoe.

Draw off Discharge drawn by off takes at different points of the irrigation canal.

Draw-off statement In designing irrigation canals a statement prepared for finding out the capacity of a canal reach by reach.

Draw well A open well.

Dredge (1) A cylindrical or rectangular device for collecting samples of bottom sediment and benethic fauna. (2) A floating excavator used for widening or deepening channels, building canals, constructing levees, raising material from stream or harbour bottoms to be used elsewhere as fill, or mining. (3) The act of excavating with a dredge; 'dredging'.

Dredging Removing solid matter from the bottom of a water area.

Dressed timber Timber which has been sawn, planed and worked to exact required condition.

Dressing (1) Application of manure or compost used as a fertiliser. (2) Application of various materials for protecting a wound and encouraging healing. (3) Material so applied. (4) Covering a seed with a fungicide powder. (5) The trimming of excess fat and bone from a carcass.

Dried blood The collected blood of slaughter animals, dried, ground and sold as fertiliser. Effective and quick acting on all crops and soils. Contains 10-12 per cent nitrogen and 1-2 per cent phosphoric acid.

Drier A device to remove water. (1) A substance that absorbs water. (2) A substance that is used to hasten solidification. (3) Material such as salts of lead, manganese, and cobalt, which facilitates the oxidation of oils, used in paints and varnishes to speed drying.

Drift (1) Unconsolidated material deposited by geological processes at one place after having been removed from another. (2) The movement of spray or dust particle due to wind action away from the area of application.

Drift Sprayer See mist sprayer.

Drill (1) To sow in rows. (2) A line impressed on a seed bed for sowing seeds. (3) An implement for planting seeds, which forms a small furrow, puts the seed and covers the seed with the soil and also compacts it. Also fertiliser could be applied and the drill is then called seed-cum-fertiliser drill.

Drill planter Also called grain drill. See drill.

Drip of a tree The place at which most water runs off the tree during rains, the periphery of the tree.

Drive Logs or timbers that are being floated on a stream from the forest to a mill or shipping point.

Drive chain A form of heavy roller chain used to deliver power to the driving wheels. Such chains are used on trucks where great strength is required.

Driving pinion The input gear in the differential of an automobile.

Drives A drive is a form of motivation in which the organism is impelled by factors, essentially beyond its control to act without foresight of ends. These consist of inner energies, tendencies, and urges that interact with outside forces and become dynamic patterns of response as a result of this interaction.

Drive screw or screw nail A type of screw intended to be set by blows of a hammer but which can be removed with a screw driver.

Drive shaft The shaft which connects the transmission to the rear axle by the use of universal joints, and which carries the power of the motor to the rear axle. (Mech) Also used for a similar purpose on other machinery.

Driving wheel A wheel that supplies driving power.

Drizzle drop A drop of water of diametre 0.2 to 0.5 millimetre falling through the atmosphere, however, all water drops of diametre greater than 0.2 millimetre are frequently termed rain drops as opposed to cloud drops. Heavier than mist but less intense than shower.

Drone The male bee, hatched from an unfertilised egg. Does not gather honey and has no sting.

Droop To hang downwards due to lack of water, nourishment etc.

Drop Structure built to discharge water in an irrigation ditch from one level to a lower one, with special provision to prevent the force of falling water from undermining the structure on the downward side; a soil conservation measure. A drop may be vertical or inclined. The latter is called a chute.

Drop-inlet spillway Overfall structure in which the water is dropped through a vertical riser connected to a discharge conduit.

Dropper The downward outgrowth of the vegetative shoot in certain bulbous plants, which may form a new bulb at the apex.

Drop roots Buttress roots.

Drop spillway A spillway usually lesser than 6 metres high having a vertical downstream face, and water drops over the face without touching the face.

Dropsy Excessive accumulation of water fluid in any of the tissues or cavities of the body.

Drosicha mangiferae See mango mealy bug.

Drought A period of abnormally dry weather sufficiently prolonged so that the lack of water causes a serious hydrologic imbalances (such as crop damage, water supply, shortage and so on) in the affected area, in general the term should be reserved for relatively extensive time periods and areas.

Drowned flow, or **submerged flow** A condition of flow through or over a structure where such a flow is affected by submergence or the existence of tail water.

Drowning ratio The ratio between the depth over the outlet crest of water level downstream to that of water level upstream (H_d/H_s).

Drug A substance often of plant origin, used in the treatment of disease.

Drum brake A brake in which two curved shoes fitted with heat-and-wear resistant timings are forced against the surface of a rotating drum.

Drum cam A device consisting of a drum with a contoured surface which communicates motion to a cam followers as the drum rotates around an axis.

Drumhead cabbage A type of cabbage with heads that are broad but elliptical in cross-section.

Drumlin An elongate or oval hill of glacial drift, commonly glacial till, deposited by glacier ice and having its long axis parallel to the direction of ice movement.

Drum mixer A simple mixer commonly used for dry treatment of seeds. It consists of a drum, through which a pipe is run at an angle, mounted on two horses/bullocks.

Drumstick tree Moringa oleifera Lamk. (M. pterygosperma Gaertn. Guilandina moringa L.) Family Moringaceae. A small tree, native of India, whose green fruits are used as vegetable and oil from seeds in medicine. Also called *horseradish tree*.

Drupe A one-celled fruit, such as a cherry, having thin or leathery exocarp, a fleshy mesocarp, and a one or two seeds with a stony endocarp. Also known as stone fruit.

Drupelet An individual drupe of an aggregate fruit. Also known as grain. Also called drupel or drupeole.

Drupel See drupelet.

Drupeole See drupelet.

Drive pinion Drive the large bevel gear which is connected to the differential.

Drive shaft A shaft which transmits power from a motor or engine to the rest of a machine.

Dry (1) To preserve a product by dehydration. (2) To remove moisture by air drying or artificially drying by heating.

Dryad A high forest tree occupying a precise ecotype, a shade plant.

Dry adiabet A line of constant potential temperature on a thermodynamic diagram.

Dry-adiabatic lapse rate A special process rate of temperature, defined as the rate of decrease of temperature with height of a parcel of dry air lifted adiabatically through an atmosphere in hydrostatic equilibrium. Also known as adiabatic lapse rate, adiabatic rate.

Dry-adiabatic process An adiabatic process in a system of dry air.

Dry aggregate A compound or secondary soil particles that is not broken down by dry sieving.

Dry air Air that contains no water vapour.

Dry basis Analysis of a substance for its constituents and each expressed on the basis of the dry material.

Dry bulb temperature The temperature of the air as measured by an ordinary thermometre.

Dry bund A bunded tank which dries up in summer.

Dry climate (1) In W. Koppen's climatic classifications, the major category which includes steppe climatic and desert climate, defined strictly by the amount of annual precipitation as a function of seasonal distribution and of annual temperature. (2) In C.W. Thornwaite's climatic classifications, any climate type in which the seasonal water surplus does not counteract seasonal water deficiency, and having a moisture index of less than zero; included are the dry subhumid, semi-arid, and arid climates.

Dry date Type of date which has low moisture and high sugar content.

Dry disk clutch The driving disks are covered on both sides with a friction material and are driven by keys in the clutch ring which is bolted to the engine flywheel. Called 'dry' because no oil or grease must contact the clutch facings.

Dryer (drier) A unit which provides the conditions for removing moisture, generally by forced ventilation with or without the addition of heat.

Dry farming Producing crops, requiring some tillage, in subhumid or semiarid regions without irrigation. Also dry land farming.

Dry feeds Feeds of dry nature fed to animals. Wheat straw after threshing is an example.

Dry fog A fog that does not moisten exposed surfaces.

Dry freeze The freezing of soil and terrestrial objects caused by reduction of temperature when the adjacent air does not contain sufficient moisture for the formation of hoarfrost on exposed surfaces.

Dry fruit A fruit with a dry pericarp.

Dry heat Hot air which has a relatively low moisture content.

Dry gangrene (1) Form of gangrene in which necrosed part of the body is dried up and shrivelled due to lack of blood supply. Also see gangrene. (2) Death without bacterial infection. Also see mummification.

Drying The removal of moisture to a point at which the rate of deterioration from chemical and biological activity is slowed down.

Drying front In a bin of drying grain, the advancing zone which includes all of the grain that had just reached some arbitarily initial moisture content, the zone between the initial and final moisture content.

Drying oil Oil having iodine value of over 120. Hardens to resinous solids on exposure to air. Used in paints etc. Linseed oil is an example.

Dry land (1) Land in arid and semi-arid areas. (2) Land cropped with the help of natural precipitation, not irrigated.

Dry land farming The practice of growing crops in areas of low rainfall without irrigation. Also called as dry farming.

Dry mash A mixture of ground cereals or cereal products with supplements fed to poultry without addition of liquids.

Dry matter (DM) Total amount of matter in a feed less the moisture it contains.

Dry matter per cent Substance less water found by even-drying a weighted sample, weighting, and determining per cent lost. See also moisture, dry basis, moisture, wet basis.

Dry processing of coffee Processing of coffee beans after picking. The method consists in removing the covering of the bean, by sun-drying of berries and pounding.

Dry rot A rapid decay of seasoned timber caused by certain fungi which cause the wood to be reduced to a dry friable texture. Any of various set diseases of plants characterised by drying of affected tissues.

Dry salt-curing Method of curing hides. Consists in air-drying wet salt-cured hides after removing of excess salts.

Dry season In certain type of climate, an annually recurring period of one or more months during which precipitation is at a minimum for the region.

Dry sieving Particle size distribution analysis of powdered solids the sample is placed on the top sieve screen of a nest (stack), with mesh openings, decreasing in size from the top to the bottom of the nest.

Dry spell A period of abnormally dry weather, generally reserved for a less extensive, and therefore less severe conditions than a drought; in the United States, describes a period lasting not less than two weeks during which no measurable precipitation was recorded.

Dry weight percentage The ratio of the weight of any constituent of a soil to the oven day weight of the soil. Usually used for moisture content in the soil.

Dry wood termite Termite group whose colony is partly on dry wood.

Duabanga grandiflora See lampati.

— **sonneratioides** See lampati.

Dual-purpose canal A canal designed, constructed, operated and maintained for the combined purpose of navigation and irrigation or irrigation and domestic and industrial water supply.

Dual purpose cattle breeds Breeds of cattle, whose females are good breeder and males good for draft purposes.

Dual purpose pond A irrigation tank used for drainage as well as for irrigation.

Duchesnea indica See Indian strawberry.

Duck-foot plough A wide based wooden plough.

Duck weeds Group of floating plants belonging to the genera Lemma, Azolla, Wolffia.

Duck wheat Tartary buckwheat.

Duct (1) Any tube which conveys fluid air, vapour or other substances. (2) A series of cells which have lost their walls at the point of contact.

Ductule A minute duct; fine thread like terminal portion of a duct.

Ductless gland Glands which have no discharging ducts for their secretions, and which pour them directly into the blood or lymph which circulates through the glands. Also called Endocrine glands. Examples are thyroid, adrenal or pituitary glands.

Dulse Any of several species of red algae of the genus Rhodymenia found below the intertidal zone in northern latitudes, an important food plant.

Dumetose Bushy or bush-like.

Dummy Wooden partition in a bee-hive serving as a movable wall helping reduce the size of brood chamber in bee colony so that bee can have air conditioned hive and are protected from enemies. Also called division board.

Dune Wind-built ridges and hills of sand formed in the same manner as snowdrifts. They are started by some obstructions, such as a bush, boulder, or fence, that causes as eddy or otherwise thwarts the sand-laden wind. Once begun, the dunes themselves offer further resistance and they grow to form various shapes.

Dung The faeces or excrement of animals and birds used as manure.

Dunnage Mats, straws, gratings, sand etc. stowed under pilex of bags in storage.

Duplex-soils Refers to soils with a relatively permeable top soil abruptly overlying a very slowly permeable diagnostic horizon which is not a hardpan.

Duplicate genes Two separately inherited factors either alone or together, giving similar effects.

Duplication The occurrence of segment more than once in the same chromosome of genome.

Durability (wood) Resistance offered by wood to agents of natural destruction like insects and fungi.

Duration curve (Hydraulics) A graphical representation of the number of times given flows are equalled or exceeded during a certain period of record.

Duramen The hard, darker, central region of a tree stem heart wood.

Durian Durio zibethinus Murr. Family Bombacaceae. Tall tree (25 m-35 m) indigenous to Malaya, very spiny fruit with large seeds and cream coloured edible pulp. Propagation by seed, inarching successful. Also called civet fruit.

Duric A Bc horizon that is strongly cemented and usually has an abrupt upper boundary and a diffuse lower boundary. Cementation is usually strongest near the upper boundary. Ordinarily the colour is similar to that of the parent material and the structure is amorphous or coarse platy. Air-dried clods do not slake when immersed in water. The duric horizon does not meet the requirements of a podzolic B horizon but may meet those of a B horizon.

Durio zibethinus See durian.

Durra Sorghum vulgare Pers. var. durra (Forsk) Hubb. and Rehd. (Sorghum durra Stapf.). Family Gramineae. Also called *Kaffir corn.*

Durum wheat Triticum durum. Family Gramineae. Produces high quality flour suitable for macaroni production. Also called macaroni wheat.

Dusky cotton bug Oxycarenus laetus Kirby. Insect pest of cotton, nymphs and adult feed on open bolls and spoil lint. Also see cotton bug.

Dust (1) Seeds treating products formulated in the forms of dust. (2) Fine, dry particles of earth or other matter which can be wafted by wind. (3) An insecticide or pesticide applied in dry state, application of such dust.

Dust-bowl A name given early in 1935, to the region in the south-central United States afflicted by drought and dust storms, including parts of Colorado, Kansas, New Mexico, Texas and Oklahoma, and resulting from a long period of deficient rainfall combined with loosening of the soil by destruction of the natural vegetation, dust bowl describes similar regions in other parts of the world.

Dust chamber A chamber through which gases pass to permit deposition of solid particles for collection. Also known as ash collector, dust collector.

Dust devil A small but vigorous whirlwind, usually of short duration, rendered visible by dust, sand and debris picked up from the ground, diametre range from about 10-100 feet (3-30 metres) and average height is about 600 feet (180 metres).

Duster Apparatus for applying insecticides or fungicides in dry form. A hand or power driven mechanical equipment consisting of a blower, dust reservoir and nozzle outlet.

Duster-cum-sprayer A double purpose machine which could be used as a duster or sprayer. Also called spray-duster.

Dust formulation Mixing of an insecticide when it is to be used as a dust with a special quality talcum powder or other ingredient in a known proportion to form a fine powder. Insecticide 5-10 per cent of the mixture.

Dust horizon The top of dust layer which is confined by a low-level temperature inversion and has the appearance of the horizon when viewed from above, against the sky, the true horizon is usually obscured by the dust layer.

Dust mulch A loose, finely granular, or powdery layer on the surface of the soil, usually produced by shallow cultivation and also by deposition.

Dust storm A strong turbulent wind carrying large clouds of dust.

Dust whirl a rapidly rotating column of air over a dry and dusty or sandy area, carrying dust, leaves and other light material picked up from the ground when well-developed, it is known as dust devil, sand augar, sand devil.

Duty (of water) The number of acres covered by one cusec (2.832×10^{-2} cumecs) of water flowing continuously throughout the growing season of the crop. Varies with crop, nature of soil, season and also to some extent on the method of irrigation.

Duty at distributary head Duty of water reckoned or measured at the head of a distributary or tertiary canal.

Duty at outlet, or lateral duty Duty of water reckoned or measured at the outlet or head of a water-course or lateral.

Duty attained Duty of water as actually prevailing on an irrigation project under operation.

Duty of a well The average annual area irrigated by a well.

Duty on capacity The full supply factor attained by a project system or a canal after it has been opened for irrigation. Also called 'full supply duty' in Punjab and elsewhere.

Dwarf (1) An animal or plant which is much smaller when mature than others of its species. (2) A condition of retarded growth of plant or animal caused by disease, lack of water, mineral deficiency etc.

Dwarf wheat Triticum sphaerococcum Perciv. Family Gramineae.

Dwigh In Newfoundland, a sudden shower or snow storm. Also known as dwey dwoy.

Dy Finely divided partly decomposed organic material accumulated in peat soils in the transition zone between the pea and the underlying mineral material. Dy-peat also refers to amorphous material formed from humus soils that have settled in lake waters. Dy is poorer in nutrients than gyttja and is characterised by a high C:N ratio.

Dyes Dyes are sometimes used along with seed treatment. It serves three purposes (i) as a visible means of evaluating the completeness of treatment coverage, (ii) a warning that the seed has been treated and (iii) serves as 'colourbrand' of a seed company.

Dynamic (1) Changing with time such as dynamic properties, dynamic nature etc. (2) Producing or manifesting activity.

Dynamic balance That types of balance which permits parts to rotate without vibrations at any speed for which they have been balanced.

Dynamic head Specific energy in a flow system.

Dynamometre An instrument for measuring draft of tillage implements, and for measuring resistance of soil to penetration by tillage implements.

Dyne Force which, when applied to a body having a mass of one gram, gives it an acceleration of one centimetre/sec^2.

Dyscerus fletcheri See apple fruit borers.

— **malignus** See apple fruit borers.

Dysdercus koenigi See red cotton bug.

Dystric brunisol A great group of soils in the Brunisolic order. The soils may have mull Ah horizons less than 5 cm thick. They have Bm horizons in which the base saturation (NaCl) is usually 65 per cent to 100 per cent and the pH is usually 5.5 or lower.

Dystrophia See dystrophy.

Dystrophic Refers to soil that has suffered marked leaching, such that the sum of the exchangeable Ca, Mg, K and Na, expressed in me/100 g clay, is less than 5. The figure is calculated from the s-value and the clay content. Such soil is said to have a low base status.

Dystrophy Faulty nutrition. Also called dystrophia.

E

Ear (1) The auditory organ. (2) The spike of grasses, usually of cereals. (3) To produce or form such spikes.

Earias fabia See spotted bollworm.

— **insulana** See spotted bollworm.

Ear leaf The seed leaf or cotyledon.

Ear leaved The plant having ear-like leaves.

Early (1) Usually used for varieties which bear fruits, leaves etc. earlier than the normal varieties. (2) Designating a period for planting or harvesting which comes before the normal or usual dates.

Early maturity The completion in a plant or animal of development of form and function sooner than other species.

Earmark To place an identifying mark on a piece to indicate that it is to be set aside or used for some particular purpose.

Ear stage The stage of development of a plant in which the ear is being formed.

Earth channels Unlined earth channels commonly used for conveyance and distribution of irrigation water. Cheap but requires continuous maintenance, results in seepage loss which may be quite high and thus is a precious loss to the nation; could be damaged by rodents etc. Lining is more desirable.

Earth dam Dam constructed of compacted soil materials.

Earthen section Section of an irrigation channels having boundaries comprising softer material of the lithosphere, such as silt, clay.

Earthing up To cultivate in a manner that the soil is thrown towards the plants to form a ridge with the plants at the crest.

Earth nut Bunium persicum (Boiss) Fedts. (Carum bulbocastanum Clarke non Koch). Family Umbelliferae. A perennial herb, the starchy tubes of which are used as vegetable and the seeds as spice. Also called as groundnut.

Earth pressure The pressure or force exerted by soil on any boundary.

Earth quake A movement or tremor of the earths crust that originates below the surface as a result of slippage along faults or because of volcanic action.

Earth shadow Any shadow projecting into a hazy atmosphere from mountain peaks at times of sun rise or sunset.

Earth worms Family Lumbricidae. Elongate worms, mostly terrestial, living in burrows in damp or wet earth. Nocturnal; swallow earth and ejecting the same in veriform or pelleted form as surface or sub-surface castings. Help in improving physical conditions and fertility of soil.

Ear to row selections An obsolete method of plant breeding. Selection of goods ears of the crop; reselecting the best ears from high-yielding progenies and repeating process in subsequently years by continual growing of ear-to-row selected progenies.

Easker A winding ridge of irregularly stratified sand, gravel, and a rapidly flowing glacial stream.

East African star grass See giant star grass.

East Indian arrowroot Curcuma angustufolia Roxb. Family Zingiberceae. Plant cultivated for its rhizomes which yields starch.

East Indian lemon grass Cymbopogon flexuosus (Nees ex Steud) Wats. (Andropogon flexuosus Nees). Family Gramineae. Cultivated and leaves used for east Indian lemon grass oil.

East Indian lemongrass oil Lemon grass oil obtained from the lemon grass. The oil of commercial value with citral content 75-86 per cent. Readily soluble in 70 per cent alcohol and hence considered superior to west Indian lemon grass oil. Also called Cochin oil.

East Indian lotus Nelumbo nucifera Gaertn. (Nymphaea nelumbo L. Nelumbium speciosum Willd.). Family Nymphaeceae. A common aquatic herb whose rhizomes, seeds and young leaves are eaten as vegetable.

Ebb tide Tide during period between high water and following low water.

Ebenaceae A family of dicotyledonous plants in the order Ebenales, in which a latex system is absent and flowers are mostly unisexual with the styles separate, at least distally.

Ebenales An order of woody, sympetalous dicotylednous plants in the subclass Dilleniidae, having axile placentation and usually twice as many stamens as corolla lobes.

Ebony Any of several Africa and Asian trees of the genus Diospyros, providing a hard, durable wood.

Ebracteate Without bracts, or much reduced leaves.

Ecesis The adjustment of a plant to new habitat involving germination, growth and reproduction.

Echhornia crassipes See water hyacinth.

Echinate With spines or bristles.

Echinulate Having small, pointed spines; used of spores.

Ecobiotic adaptation Adaptation to a particular mode of life within a habitat.

Ecoclimatic adaptation Adaptation to the broad physical and climatic features of any particular region.

Ecocline A genetic gradient of adaptability to an environmental gradient; formed by the merger of ecotypes.

Ecofallow A system for destroying weeds and conserving soil moisture in crop rotation with minimum disturbance of crop residue and soil.

Ecological inversion point See inversion point, ecological.

Ecological inversion surface See inversion; See inversion surface ecological.

Ecology The phase of biology that treats the mutual relations of organisms and their environment.

Econometrics a branch of economics in which hypothesis are mathematically formulated and statistically tested, and in which economic data are systematically analysed with the aid of a computer.

Economical haul distance An arbitrary distance within which any piece of equipment and haul material operate most efficiently or profitably.

Economic appraisal The study of the economic implications of a project, as a first step in assessing its acceptability.

Economic behaviour The way in which people persue their material well-being individually and collectively.

Economic efficiency The efficiency with which scares resources are used and organised to achieve stipulated economic ends.

Economic entomology The branch of insect study directed towards preventing man's looses and increasing his gains through manipulation of insect populations.

Economic history The study of the struggle of man to relate scares resources to competing ends through time.

Economic man One who is completely rational in satisfying his wants and pays no regard to the interests of others.

Economic problem A problem of ensuring that all available resources are used to the best possible effect, in the resolving of which decisions must be made as to which ends are to be satisfied first and perhaps which wants will have to be left unsatisfied.

Economics A social science concerned with how people, either individually or in groups, attempts to accommodate scarce resources to their wants through the processes of production, substitution and exchange.

Economic significance A quality possessed by a good or service when it commands a price and enters into the circle of exchange.

Economist A social scientist who has become an expert in dealing with the economic aspects of social phenomena.

Economic water duty Duty of water which results in (1) the maximum yield or maximum net profit per unit area when land is the limiting factor, (2) the maximum yield or maximum net profit per unit of irrigation water when water is the limiting factor.

Economiser An arrangement of pipes or reservoirs in which the feed water for steam boilers is heated up to or higher than the boiling point. Feed water heaters, therefore, are economisers.

Economy (1) The economic system of a country, or part of a country. (2) The optimum use of resources so that the maximum satisfaction is gained from any given input of resources.

Economy of high wages A view that if a man is well paid, so that he enjoys a high standard of living, his working efficiency will be raised and with it his productivity.

Ecosystem Open system including plants, animals, organic residues, atmospheric gases, water, and minerals that are involved in the flow of energy and circulation of matter.

Ecotone The transition area, or border area, between two ecological communities, as between a marsh system and a forest system.

Ecotope Combinations of niche and habitat occupied by a plant.

Ecotype Plant type or strain within a species, resulting from exposures to a particular environment.

Ectogony The influences of pollination and fertilization on structures outside the embryo and endosperm; effect may be on colour, chemical composition, ripening, or abscission.

Ectoparasite A parasite that lives on the exterior of its host.

Ectophloic Having outer phloem only.

Ectophyte A plant which lives externally on another organism.

Ectosymbiont A symbiont that lives on the surface of or is physically separated from its host.

Ectotrophic Obtaining nourishment from outside; applied to certain parasitic fungi that live on and surround the roots of the host plant.

Ectotrophic mycorrhiza A mycorhizal association in which the fungal hyphae form a compact mantle on the surface of the roots. Mycellial strands extend inward between cortical cells and outward from the mantle of the surrounding soil.

Edaphic (1) Of or pertaining to the soil. (2) Resulting from, or influenced, by factors inherent in the soil or other substrate rather than by climatic factors. Roots are edaphic structures since they grow in the soil.

Edaphic community A plant community that results from or is influenced by soil factors as salinity and drainage.

Edaphology The science that deals with the influence of soils on living things, particularly plants, including man's use of land for plant growth.

Edaphon The whole living community of the soil.

EDCT A fumigant; mixture of Ethylene dichloride and carbon tetrachloride in ratio of 75:25. Available as liquid. Volatile and vapourises at normal temperatures.

Eddy diffusivity The exchange coefficient for the diffusion of a conservative property by eddies in a turbulent flow.

Edge-grained timber, or vertical grained timber: Quarter sawn or radial sawn timber.

Edgeworthia buxifolia Reptonia buxifolia A. DC. (Monotheca muscatensis A. DC., Edgeworthia buxifolia Folc.) A small tree of family Sapotaceae which produces edible fruits.

Edible Amaranth Amaranthus tricolour L. (Amaranthus gangeticus L.) Family Amaranthaceae. A native of India grown in summer months. Tenders shoots consisting of profuse whorl of leaves cooked like spinach. Good source of vitamins B and C. Seeds very nutritious. Repeated pickings possible and harvested when 20-30 cm high.

Edible oil Oil fit for human consumption.

Eelworm See nematode.

Eelworm disease Root-knot disease.

Effective outgoing radiation The difference between the outgoing long wave radiation of the earth surface and the downward long wave radiation from the atmosphere.

Effective precipitation See effective rainfall.

Effective rainfall (1) Rain that produces runoff. (2) In irrigation practice that portion of total precipitation which is retained by the soil so that it is available for use for crop production.

Effective water-use Consumptive use, less evaporation from rainfall. It includes transpiration and evaporation from irrigation and transpiration from effective rainfall.

Efficiency, animal Animal productivity relative to feed use.

Efficiency of the hydraulic jump The ratio of the specific energy after the jump to that before the jump.

Efficiency of irrigation See irrigation efficiency.

Efficiency of statistical design The reliability of statistical techniques for a given cost, or the relative economy for a given reliability.

Efficiency, or **Relative efficiency of estimator** One estimator is more efficient than a second if its sampling distribution clusters more tightly, so that sampling variability is less. Thus, efficiency, or smallness of sampling variability, is a desirable characteristic of an estimator, and unbiasedness alone is not sufficient.

Effloresce To produce blossoms; turn to fine powder on exposure to air (Chemistry); comes to the surface and there crystallises (of salts).

Efflorescence (1) The salt affected soils, a crust of salts. (2) The period or state of full bloom.

Effluent seepage Diffuse discharge of ground water to the grounds surface.

Effluent stream Stream or reach of a stream which receives water from groundwater or seepage.

Effluent water The water which flows out of the ground; a spring. Commonly used for water pumped from the ground as drainage effluent.

Effoliation Shedding or removal of leaves.

Effuse Expanded; spread out in a definite form.

Egg apparatus A group of three cells, consisting of the egg and two synergid cells, in the micropylar end of the embryo sac in seed plants.

Egg cell The female germ cell. On being fertilised by the male sperm it develops into the embryo.

Egg grader A machine used for grading eggs. Ungraded eggs are carried along a moving belt, they pass over a weighting device which automatically sorts them into separate containers.

Egg membrane The layer of tough tissue lining an egg cell.

Eggplant Solanum melongena. A plant of the order Polemoniales grown for its edible egg shaped, fleshy fruit. Also called brinjal.

Egg tree Garcinia xanthochymus Hook. F. (Garcinia tinctoria Dunn.). Tree of the family Guttiferae whose fruits are edible. Exudations from under ripe fruits used for drying.

Egyptian clover See berseem.

Egyptian cotton Gossypium arboreum L. var. nadam (Watt) Prokh; (G. barbadense L.) Family Malvaceae. A woody plant grown as an annual for its long staple lint cotton.

Egyptian henna See henna.

Egyptian indigo Indigofera articulata Gouan (I. argentea L. var. caerulea Baker). Family papilionaceae. A shrub grown for fodder.

Egyptian onion Allium cepa L. var. viviparum. A variety of onion, family Liliaceae. Produces small bulbs. Also called top onion.

Egyptian privet See henna.

Ehretia acuminata See heliotrope tree.

Einstein A unit for measuring the energy output of light that is equal to one mole of quanta (6.02×10^{23} quanta).

Ejaculate To emit or emitted substance.

Ekdania Bridelia retusa Spreng. Family Euphorbiaceae. A tree whose bark is used for tanning; fruits are edible and leaves are used as fodder.

Elaeagnaceal A family of dicotylednous plants in the order Proteales, noted for pellate leaf scales which often give the leaves a silvery gray appearance.

Elaegnus angustifolia See oleaster.

— **hortensis** See oleaster.

— **latifolia** See bastard oleaster.

— **umbellata** See ghain.

Elaeis guineensis See African oil palm.

Elaeocarpus floribundus See jalpai.

— **ganitrus** See Utrasum bead tree.

— **serratus** See Ceylon olive.

— **sphaericus** See Utrasum bead tree.

Elastin See albuminoides.

Elater A spiral, filamentous structure that functions in the dispersion of spores in certain plants, such as liverworts and slime molds.

Elaterophare A tissue bearing elaters, found in some liverworts.

Elate sylvestris See wild date.

Electrical brooder A brooding machine that is heated by electrical energy.

Electrical resistance blocks Small blocks consisting of a pair of electrodes set in an absorbent material, such as plaster of Paris, used to estimate soil moisture content. Resistance between the electrodes is a function of the moisture content of the block and the surrounding soil.

Electrical storm A popular term for a thunder storm.

Electrokinetic potential (1) In a colloidal system, the difference in potential between the immovable layer attached to the surface of the dispersed phase and the dispersion medium. (2) The work done in bringing a unit charge from infinity to a reference point in the liquid layer attached to the solid phase in a colloidal system.

Electrophoresis Application of an electric current to separate substances that move faster in an electrical field from those that move more slowly; the serum proteins can be so separated.

Electrotaxis Movement of an organism in response to stimulation by electric charges.

Electrotropism Orientation response of a sessile organism to stimulation by electric charges.

Elemental sulphur Widely used as dusting powder; fungicide to control powdery mildew and seed treatment of jowar for controlling loose smut.

Eleocharis dulcis See Chinese water chestnut.

— **plantaginea** See Chinese water chestnut.

— **tuberosa** See Chinese water chestnut.

Elephant apple (1) Dillenia indica L. Family Dilleniaceae. A medium sized tree, native of the moist region and cultivated in many parts of India. Fruits and flowers edible. Wood used for packing cases and plywood etc. (2) Also see bael and wood apple.

Elephant ear See colocasia.

Elephant foot yam Amarphophallus campanulatus (Roxb.) Blume ex. Done. (Arum campanulatum Roxb.) Family Aracaceae. An undershrub yielding edible corm. Used as vegetable. Propagated from the buds of corm. Also called *Jimmy kand*.

Elephant grass See Napier grass.

Elettaria cardamomum See cardamom.

Eleusine aegyptiaca See crow foot.

— **coracana** See finger millet.

— **indica** See goose grass.

Elevation (1) Height above sea level or above a datum. (2) A land feature which is at greater height than the surrounding land.

Elevation head Energy possessed by a fluid due to its position above some datum.

Elevator (1) A device for raising materials from a lower to a higher level. Could be of moving belt type, bucket type or of any convenient shape. (2) A building designed for the handling and storing of grains. (3) The equipment used for moving seeds.

Elfin forest A stunted forest of higher elevations in a moist warm region typified by gnarled trees having abundant epiphytes and mosses.

Elkington system A system of tile drains similar to the cutoff system and used for the same purpose, namely to intercept seepage water, but has, in additions wells dug at regular intervals to let the water come up from a lower stratum and enter the drain. Also known as 'sink-hole drainage'.

Elliptical Of the form of an ellipse, oval-shaped.

Elongation factor Any of several proteins required for elongation of growing polypeptide chains during protein synthesis.

Elsasser's radiation chart A radiation chart developed by W.M. Elsasser for the graphical solution of the radiative transfer problems.

Eluvial horizon A soil horizon that has been formed by the process of eluviation. See also illuvial horizon.

Eluviation The transportation of soil material in suspension or in solution within the soil by the downward or lateral movement of water.

Eluvium The residual deposits of rock weathering; the soil itself or the parent material of the soil.

Emasculate To remove the anthers or male organs from a flower.

Embargo Legal prohibition against the import or export of a specified good or service.

Embayment A relatively small and shallow estuary with rather restricted flushing (differs from lagoon by having significant freshwater inflow).

Emblic Phyllanthus emblica (Emblica officinalis Gaertn.). Family Euphorbiaceae. Found in tropical forests of India. Medium sized tree sometimes going tall. Fruit used fresh or as preserve. Rich source of vitamin C. Bark leaves and fruits used for tanning. Grown from seeds; inarching and budding also done. Also called as *amla*.

Emblica myrobalan See myrobalan.

— **officinalis** See emblic.

Embryo (1) Young sporophyte of a seed plant. (2) An early stage of development in multicellular organisms. (3) The product of conception up to the third month of human pregnancy. (4) Redimentary plantlet within a seed; the germ.

Embryobionta The land plants, a subkingdom of the Plantae characterised by having specialised conducting tissue in the sporophyte (except bryophytes), having multi-cellular sex organs, and producing an embryo.

Embryology That part of biology which deals with the formation and development of the embryo.

Embryonated Term used in reference to a parasite egg containing a small, immature parasite.

Embryonic cells A class of undeveloped cells which make up the structure of the embryos; non-specific in function in the very early stage of development, but become more specific as embryo matures, making up tissues and organs.

Embryo sac The female gametophyte of a seed plant, containing the egg, synergids; polar and antipodal nuclei; fusion of the antipodals and a pollen generative nucleus forms the endosperm.

Emergence To appear; usually used in two ways (1) The appearance of the first leaves of the crop plant above ground. (2) The escaping of the insect from its cocoon or pupal case.

Emergency operation An operation of regulation of water supply in an irrigation system, whether planned or improvised, which is carried out as the result of an emergency condition arising.

Emergency service pumping plant Pumping plant designed to operate when the gravity outlet of a drainage system is obstructed by high water.

Emergency spillway A spillway used to carry runoff exceeding a given design flood.

Emergent An isolated tree standing above the average forest canopy, but which did not necessarily emerge.

Emetic An agent which cause vomiting; copper sulphate.

Emetine See ipecacaunha.

Eminence (1) A position of high prominence (2) Swellings on the parts of the body.

Emissary sky A sky of cirrus clouds which are either isolated or in small, separated groups; so called because this formation is often one of the first indications of the approach of the cyclonic storm.

Emissivity The ratio of the emittance of a given surface to the emittance of an ideal black body.

Emmalocera depressella See sugarcane root borer.

Emmer wheat Triticum dicoccum Schrank. (T. dicoccum Schiibl.). Family Gramineae. One of the 5 species of wheat grown in India. Has tough glumes and has to be pounded to separate grain. Suitable for preparing semolina and vermicilli.

Empirical Elementary; based on trial and error methods, e.g. empirical formulae.

Employment rate The percentage of a body of persons available for employment at any time actually in employment at that time.

Employment volume The number of persons actually in employment at any time in any industry or region.

Empoasca devastans See cotton jassid, ladies finger leaf hopper bug.

— **flavescenes** See cotton jassid.

— **karri var. motti** See potato jassid.

Emulsifiable concentrate A liquid formulation, which will form an emulsion when mixed with water or certain other liquids.

Emulsifier Also called emulsifying agents. Substances when added to emulsion of insecticides helps resist tendency of oil and other immiscible liquids to coaksce when mixed with water. This helps in stabilizing emulsions.

Emulsifying agents See emulsifiers.

Emulsion (1) A finely divided, or resinous substance held in suspension in an aqueous liquid (2) A mixture of liquids that do not dissolve in each other.

Emulsion spray Liquid concentrate containing a known percentage of insecticide, usual constituents being the insecticide dissolved, the mineral oil and the emulsifier, the last helping give an emulsion in which the insecticide is dispersed, in water giving a milky white liquid; used as dormant spray.

Emulsoid An emulsion whose suspended particles are colloids which absorb some of the liquid in which they are suspended; e.g. starch in boiling water.

Enarthrosis Ball and socket joint.

Enation An outgrowth from the surface of an organ of a plant, usually caused by parasitic invasion of fungi, bacteria or virus.

Encephalic Of the brain.

Encrustation A crust or hard coating on or in the body.

Encyst To become enclosed in a sac, bladder or cyst.

Encystment The process of forming or becoming enclosed in a cyst or capsule.

Endarch Formed outward from the centre, referring to xylem or its development.

End cap Female fitting used to plug the end of a pipe.

Endemic Species native to a particular environment or locality; not introduced.

Endive See winter lettuce.

End moraine, or **Terminal moraine** A ridgelike accumulation of drift built chiefly along the terminal margin of a valley glacier or the margin of an ice sheet. It is mainly the result of deposition by ice, or deformation by ice thrust, or both.

Endobiotic Referring to an organism living in the cells or tissues of a host.

Endocarp The inner layer of the wall of a fruit or pericarp.

Endochrome An colouring matter or pigment within a cell.

Endocommensal A commensal that lives within the body of its host.

Endoconidium, (pl. Endoconidia). Conidium formed within a hypha.

Endodermis The innermost tissue of the cartex of most plant roots and certain stems consisting of a single layer of at least partly suberized or cutinized cells; functions to control the movement of water and other substances into and out of the stele.

Endoenzyme, or **Intracellular enzyme** An enzyme formed with in the cell and not excreted into the medium.

Endogamy Sexual reproduction between organisms which are closely related. Pollination of a flower by another flower of the same plant.

Endolithic Living within rocks, as certain algae and coral.

Endomorph A mineral enclosed in another.

Endo-parasite A parasite which feeds on its host from inside. In its larval stage, may at first feed on the surface of the host before penetrating inside.

Endophyte A plant that lives within, but is not necessarily parasitic on, another plant.

Endoplasm The inner softer layer of protoplasm.

Endosmosis The passage of a fluid inwards through a porous septum.

Endosepsis A fungus disease of figs caused by Fusarium moniliforme Fici; fruits rot internally.

Endosperm The nutritive tissue formed within the embryo sac in seed plants. It commonly arises following the fertilization of the two primary endosperm nuclei of the embryo sac by one of the two male sperms. In a diploid organism the endosperm is triploid.

Endosperm nucleus The triploid nucleus formed within the embryo sac of most seed plants by fusion of the polar nuclei with one sperm nucleus.

Endospore The inner coat of a spore; a spore formed in a case.

Endothecium Inner tissue of sporophyte which gives rise to spores in some cases.

Endothelium The layer of cell lining blood vessels etc.

Endotrophic Nourished or receiving nourishment from within, for example, fungi or their hyphae receiving nourishment from plant root in mycorrhizal association.

Endotrophic mycorrhiza A mycorrhizal association in which the fungal hyphae are present on root surfaces as individual threads that may penetrate directly into root hairs, other epidermal cells, and occasionally into cortical cells. Individual threads extend from the root surface outward into the surrounding soils.

End plug Plug or cap at the end of the sprinkler lateral or other pipe lines to close the line and often fitted with facilities for flushing the line.

Endrin 1, 2, 3, 4, 10, 10 — hexachloro — exo- 6, 7 — epoxy — 1, 4, 4a, 5, 6, 7, 8, 8a — octahydro — 1, 4, 5, 8 — endo, endo dimethano — naphthalene; an insecticide acting as a contact or stomach poison. Available as dust or emulsifiable concentrate.

Energy flow Transfer of energy within the tree or between the tree and its environment.

Energy gradient Change in energy per unit length in the direction of flow or motion.

Energy head The elevation of the hydraulic grade line at any section plus the velocity head corresponding to the mean velocity of water in the section. The energy head may be referred to any datum, on to an inclined plane, such as the bed of a conduit.

Energy line A line joining the elevation of the energy heads of a stream. The energy is above the hydraulic grade line a distance equivalent to the velocity heads at all sections along the stream.

Energy loss in hydraulic jump The difference in specific energies before and after the jump.

Engineering surveys Surveys (field or ground surveys, aerial surveys, preparation of maps and charts) executed or to be executed for the purpose of obtaining information for planning an engineering project or development and estimating its cost. The information obtained or to be obtained may, in part, be recorded in the form of an engineering map or plot.

English walnut See walnut.

Engraft To make a graft.

Enology The art and science of making wine.

Enphytotic A plant disease causing constant damage from year to year. Disease peculiar to a location.

Enrich Add valuable materials; add manure and fertiliser to soils.

Enrichment culture A technique in which environmental, including nutritional, conditions are controlled to favour the development of a specific organism or group organisms.

Ensemble See set.

Ensilage See silage.

Entada phaseoloides See nicker bean.

— **scandens** See nicker bean.

Ensile To store forage as silage.

Ensiling The anaerobic fermentation process used to preserve immature green corn, legumes, grasses, and grain plants; the crop is chopped and packed while at about 70-80% moisture and put into silos or other containers to exclude air.

Enteric Pertaining to the intestines.

Enterolobium saman See rain tree.

Enteron The alimentary tract.

Entire Pertaining to leaf-margin; continuous margin.

Entomology The science that deals with insects; a part of zoology.

Entomologist A person who is specialist in the study of insects.

Entrainment Removal of suspended life from an aquatic ecosystem by withdrawal of water.

Entomophagous Insect-eating; insectivorous.

Entomophilous Pertaining to flowers that are fertilised by insects which carry the pollen from plant to plant.

Entozoa Internal animal parasite, such as stomach worms.

Entrance head, or entry head The head required to cause flow into a conduit or other structure: it includes both entrance loss and velocity head.

Entrance loss The head lost in internal friction and external friction at the inlet to a conduit.

Entrails Internal organs, especially those of the cavities of the body, as the digestive tract, heart, liver, kidneys, spleen and lungs. Also called Viscera.

Entyloma oryzae See rice leaf smut.

Envelope An outer covering of an egg. Any surrounding structure, as floral envelop.

Environment The external conditions that may act on soils or organisms to influence their development, which include sunlight, temperature, moisture, and other organisms.

Environmental impact An environmental change that affects human needs (ef. ecologic effect).

Envrionmental lapse rate The rate of decrease of temperature with elevation in the atmosphere. Also known as atmospheric lapse rate.

Enzymatic Refers to a process or reaction involving the catalytic action of an enzyme or group of enzymes.

Enzyme Specialised protein compound occurring in both plant and animal bodies that is capable of producing chemical transformations without itself being changed or destroyed; functions as a biochemical catalyst.

Enzyme repression The process by which the rate of synthesis of an enzyme is reduced in the presence of a metabolite, often the end product of a chain of reactions in which the enzyme in question operates near the beginning.

Eolation The destructive or constructive action of the wind by which land surfaces are modified.

Eolian Pertaining to the action or the effect of the wind, as in eolian mounds or eolian deposits (of dust). Also spelled aeolian.

Eolian deposit Sand, or silt, or both, deposited by the wind.

Eosinophil Granular leukocyte whose granules stain with eosin.

Epacridaceae A family of dicotyledomous plants in the order Ericales, distinguished by palmately veined leaves, and stamens equal in number with the carolla lobes.

Ephedra A genus of low, leafless, green-stemmed shrubs belonging to the order Ephedrales; source of the drug ephedrine.

Ephedra gerardiana See ephedrine

— **intermedia** See ephedrine

Ephedrine (1) Ephedra gerardiana wall. Family Ephedraceae. A small herb found in drier regions of the temperate and Alpine Himalayas from Kashmir to Sikkim. Stem source of drug. (2) Ephedra intermedia Schrenk and May. A shrub from which a drug to cure Asthma is extracted. (3) The drug produced from (1).

Ephemeral plant An annual plant that completes its life cycle in one short moist season; desert plants are examples.

Ephemeral stream A stream or portion of a stream which flows only in direct response to precipitation and whose channel is completely dry at certain times.

Ephestia cautella See almond moth.

Epibiotic Living, usually parasitically, on the surface of plants or animals; used especially in case of fungi.

Epicalyx Stipules, fused in pairs, producing an apparent outer or extra calyx; structure just before calyx produced by aggregation of bracts or bracteoles.

Epicarp The outer layer of the pericarp. Also known as exocarp.

Epicentre The part of the earth's surface directly above the focus of an earthquake.

Epicormic Originating from the trunk, as in an epicormic shoot developing from latent meristem.

Epicotyl The growing point of the embryo, which gives rise to the shoot, or the above ground part of the plant.

Epicotyledonary axis Stem immediately above the cotyledonary node.

Epiderm See epidermis.

Epidermis External layer of cells. Also called epiderm.

Epigeal Plants in which the cotyledons appear above the surface of the soil.

Epigeal germination In which the cotyledon(s) escape from the seed coat and usually appear above ground as photosynthetic organs.

Epigean Pertaining to a plant or plant part that grows above the ground surface. Living near or on the ground surface, applied especially to insects.

Epigenous Developing or growing on a surface, especially of a plant or plant part.

Epigyny A condition in which the ovary is imbedded in the receptacle, so that the other floral organs seem to originate at a point above the ovary.

Epilachna circularis See potato beetle.

— **dodecastigma** See potato beetle.

— **ocellata** See potato beetle.

— **vigintioctopunctata** See potato beetle.

Epilimnion The upper stratum of water in lake which usually has the highest oxygen concentration and is characterised by a temperature gradient of less than one degree C per metre of depth.

Epinasty The effect of one branch on another which determines their final orientation.

Epipetalous Having stamens located on the carolla.

Epiphyll A plant that grows on the surface of leaves.

Epiphyte A plant which gains physical support, but not sustenance, from another plant, by growing on it.

Epiphytotic (1) Any infectious plant disease that occurs sporadically in epidemic proportions. (2) Of or pertaining to an epidemic plant disease.

Episepalous Having stamens growing on or alternate to the sepals.

Episodic growth Rhythmic growth.

Episperm The outer coat of seed; testa.

Epistasis The suppression of a character dependent upon the action of a gene or genes by a gene or genes not allelic to those suppressed. Those characters suppressed are said to be hypostatic. Distinguished from dominance that refers to the members of one allelic pair.

Epithelioma Malignant tumour consisting of epithelial cells.

Epithelium Tissue that covers all body surface (skin and mucous membranes) and consists of one or more layers of cells which vary in shape and arrangement.

Epitrophy Generalised response leading to the formation of erect axes.

Eppley pyreheliometre An instrument for measuring direct and diffuse radiation.

Epsom salts ($MgSO_4 7H_2O$). Hydrated magnesium sulphate; used as laxative or purgative.

Equaliser brake An arrangement of levers and rods, so designed that pressure applied to the brake pedal or brake lever will be transmitted through one rod to two or more rods in such a manner as to give equal braking pressure on the wheels.

Equivalent diametre In sedimentation analysis, the diametre assigned to a non-spherical particles. It is numerically equal to the diametre of a spherical particle having the same density and velocity of fall.

Equivalent potential temperature The potential temperature corresponding to the adiabatic equivalent temperature.

Equalizing ditch Secondary ditch, usually parallel to a field ditch, used to furnish irrigation water to two or more furrows.

Equalization of produce Refers to in marketing the carrying forward of goods for future distribution.

Equal pay Equal material reward for men and women doing equal work.

Equator The latitude 0° on the earth's surface that is equidistant from the poles and separates the northern hemisphere.

Equatorial dry zone An arid region existing in the equatorial trough; the most famous dry zone is situated a little south of the equator in the central Pacific. Also known as arid zone.

Equatorial tide The tide occurring when the moon is on, or close to the plains of the earth's equator.

Equatorial vortex A closed cyclonic circulation with the equatorial trough.

Equilibrium moisture content The moisture content of the product when it is in equilibrium with the surrounding atmosphere is called the equilibrium moisture content or hygroscopic equilibrium.

Equinoctial plant A plant whose

flowers open and close at a particular hour of the day.

Equinoctial rains Rainy seasons which occur regularly at or shortly after the equinoxes in many places within a few degree of the equator.

Equinoctial storm In semipopular belief, a violent storm of wind and rain which is supposed, both in the United States and in Britain to occur at or near the time of the equinox. Also known as line gale; line storm.

Equinox The time when the sun passes directly above the equator; in the Northern Hemisphere the vernal equinox is about March 21 and the autumnal equinox is about September 22; these dates are reversed in the Southern Hemisphere.

Equipaarte In Mexico, heavy cold rains during October to January, which last for several days. Also known as equipatos.

Equipotential Line or surface having a constant potential energy.

Equitant Refers to leaves which overlap each other at the bases.

Equivalent area, or Equivalent valley area Comprises the entire area or valley (except for areas consuming no stream flow) reduced to an area consuming water at a rate equivalent to that by the cropped area.

Equivalent acidity Refers to the kilograms of pure calcium carbonate required to neutralise the acids produced in the soil from the quantity of fertilisers used, on the basis of which acid-forming nature of a fertiliser is decided.

Equivalent basicity The basic residue of any fertiliser measured in terms of calcium carbonate required to neutralise it. Term for expressing the basicity of a fertiliser.

Eradicant (1) Any agent which destroys the causative organism of a disease after the disease is established. (2) Any chemical or compound used for destroying plants.

Eradicant fungicide One which destroys a fungus at its source.

Eradication (1) Control of disease by eliminating the pathogen after it is already established. (2) Eliminating potential host plants. (3) To control, pests, weeds etc.

Eragrostis obessinica See Teff love-grass.

— **curvula** See weeping love grass.

— **lehmanniana** See lehman love-grass.

— **tef** See teff love grass.

— **tremula Hochst** Family Gramineae. A grass grown for fodder.

Erect Designating upright plants, in contrast to prostrate ones.

Erect stem A stem that stands, having a vertical or upright habit.

Eremacausis The natural burning out of organic matter, which results in the near absence or very low content of humus in soils or arid regions.

Eremium A desert plant community.

Ergolis merione See castor butter fly.

Eremascoideae A monogeneric subfamily of ascosporogenous yeasts characterised by mostly seprate mycelia, and spherical asci with eight oval to round ascospores.

Ergone A term used to denote any substance with vitamin like or hormonic properties which cannot with certainty, be classed either as a hormone or a vitamin.

Ergot (1) A fungus disease of cereals caused by Claviceps purpurea Tal. (2) The drug obtained from the fruiting body of the fungus.

Ergotism A disease in man and lower animals due to eating of ergot contaminated grain or its products.

Erianthus arundinaceus See pin reed grass.

— **munja** See *munj* grass.

— **ravennae** See plume grass.

Ericaceae A large family of dicotyledonous plants in the order Ericales distinguished by having twice as many stamens as carolla lobes.

Ericales An order of dicotyledonous plants in the subclass Dilleniidae, plants are generally sympetalous with unitegmic ovules and they have twice as many stamens as petals.

Eriobotrya japonica See loquat.

Eriochloa procera (Retz.) C.E. Hubb. [(Eriochloa polystachya Hook. F. non H.B. & K.; Agrostis procera (Retz.)]. Family Gramineae. A herb used as fodder.

— **polystachya** See Eriochloa procera.

Eriodendron anfractuosum See kapok.

Erioglossum Erioglossum rubiginosum (Roxb.) Blume (E. edule Blume). Family Sapindaceae. A tree which produces edible fruits.

Eriophyes See mango mites.

Eriosoma lanigerum See wooly aphids.

Ericophyte A plant that grows on a heath or moor.

Erineum An abnormal growth of hairs induced on the epidermis of a leaf by certain mites.

Eriocaulaceae The single family of the order Eriocaulales.

Eriocaulales An order of monocotyledonous plants in the subclass Commelinidae, having a perianth reduced or lacking and having unisexual flowers aggregated on a long peduncle.

Eriophyllous Having leaves covered by a cottony pubescence.

Eri silk Creamy white silk obtained from the cocoons of the eri silkworm.

Eri silkworm Attacus ricinis. Domesticated silk worm feeding on castor leaves.

Erode (1) To wear away or remove the land surface by wind, water, or other agents. (2) To wear away, to eat into cells or layer of cells.

Erodible Susceptible to erosion. It is expressed by terms such as highly erodible and slightly erodible.

Erosion (1) Erosion is the term applied to wearing away of the land surface by any of the forces such as raindrops, water movement, wind or any other geological agent. It also includes erosion resulting from such processes as gravitational creep. (2) All processes by which earthy material or rocks are loosened and moved from place to place. (3) Gradual destruction of body tissues due to pressure, inflammation or irritation.

Erosion classes A grouping erosion conditions based on the degree of erosion or on the characteristic patterns. The classes apply to accelerated erosion, but not to normal natural, or geological erosion, Four erosion classes are recognised for water erosion and three for wind erosion.

Erosion payment A layer of coarse fragments such as sand, gravel, cobbles, or stones that remains on the surface of the ground after the removal of the particles by erosion.

Erosion potential Amount of erosion that may be expected under given climatic, topographic, soil, crop and cultural conditions.

Erosion resistant crop A crop, because of its dense foliage, root system, etc. provides effective protection against soil loss by erosion.

Erosive Tending to cause erosion. Applies to the eroding agents, such as wind or water. See erodible.

Erosive velocity Velocity of the erosive agent necessary to cause erosion of the material in question.

Erratic A transported rock fragment different from the bedrock where it lies. The term is generally applied to fragments transported by glacier ice or by floating ice.

Eruca sativa See roquette.

Eructation Act of belching or giving off gas from the stomach.

Erumpent Breaking through surface of Substratum.

Eruum lens See lentil.

Erwinia carotovora See potato black leg and sift rot.

Erythrina indica See coral tree.

— **variegata var. orientalis** See coral tree.

Erythrocyte Red blood cell.

Erythroxylaceal A homogeneous family of dicotyledonous woody plants in the order linales characterised by petals that are internally appendiculate, three carpels, and flowers without a disk.

Escape (1) A channel through which surplus or excess water may be removed from a canal. It is at the tail, it is called 'tail escape' or 'outfall'. (2) A plant originally cultivated, now found wild.

Esculent (1) Edibility. (2) Plant suitable for human food.

Espalier Method of training a fruit tree in which the tree is planted against a wall and the main branches trained in a plane parallel to the wall in a geometric design.

Essence A substance that contains a

high degree of essential oils or other stored product of a plant or plant product.

Essential amino acid An amino acid that cannot be synthesised in the body from other aminoacids or substance or one that cannot be made in the quantities needed for body use.

Essential element Any element necessary for growth and development of a plant or animal. Also see trace elements.

Essential floral organs Organs primarily responsible for reproduction in plants; stamens and pistils.

Essential host A host for one stage of development of a parasite, without which it cannot develop to maturity.

Essential oil Any aromatic volatile oil composed of various constituents and contained in plant organs, with characteristic odour, and sometimes other properties. Also called ethereal oil or volatile oil.

Establishment growth Post-seedling growth of plants, usually without secondary thickening meristems (cambiums), resulting in gradual development of massive primary axes (e.g., monocotyledons).

Ester An organic salt formed by the interaction of an alcohol and an organic acid.

Estimate A statement showing the probable cost of a proposed work, and generally comprising a report, plans, specifications and a statement of measurements with an abstract showing the total estimated cost of each item. Also a judgement on the crop yield made with incomplete data.

Estimation Estimation is concerned with statistical inference about the numerical value of unknown population values from incomplete data, such as a sample.

Estivate To become and remain dormant in summer.

Estivation (1) Dormancy. (2) The arrangement of the parts of a flower in the bud.

Estrogenic Pertaining to any hormonal substance capable of producing estrus in a female animal.

Estuary A passage where the tide meets a river current especially an arm of the sea at the lower end of a river; a 'firth'.

Ethereal oil See essential oil.

Ether extract Fats, oils, waxes, and similar plant components that are extracted with dry ethyl ether in chemical analysis.

Ether extractive See fat.

Ethnobiology The study of plants and animals in relation to man.

Ethnocentricism Generally, the members of each group think of their culture as the best. This attitude is called ethnocentric.

Ethylene dibromide Soil fumigant. Heavy, colourless; less volatile, less toxic and hazardous for use. Injected into soil about 20 cm deep. Not used for bromine sensitive plants like onion. Used for root knot nematode control mixed with dichloropropane and dichloropropylene.

Ethylene oxide Chemical used as fumigant; odourless, heavier than air, with god penetrating properties. Used at low temperatures; toxicity enhanced by mixing with parts by volume of CO_2. Used for fumigating dried fruits and oil seeds or commodities containing vegetable oils.

Etiolated Refers to plants grown in reduced light.

Etiolation Failure of plants or parts of plants to produce chlorophyll due to darkness or disease.

Etiology Science that deals with the origins or causes of disease. Also called aetiology.

Etroplus suratensis See pearl spot.

Eubacteriales An order of the class Schizomycetes. True bacteria.

Eublema olivaceae See brinjal leaf caterpillar.

Eubrya A subclass of the masses (Bryopsida). The leafy gametophytes arise from buds on the protonemata, which are nearly always filamentous or branched green threads attached to the substratum by rhizoids.

Eubryales An order of mosses (Bryatae); plants have the sporophyte at the end of a stem, vary in size from small to robust, and generally grow in tufts.

Eucalypt See Eucalyptus.

Eucalyptus Any tree of the genus Eucalyptus family Myrtaceae. A native of Australia. Grown in India; can tolerate adverse conditions. Criticised widely for its vast potential of using ground water and its large scale plantations in sweet water belt has been attributed as one of the reasons of water table decline. No proof as yet is available to this criticism.

Eucalyptus canaldulensis See river red gum.
— **citriodora** See lemon-scented spotted gum.
— **coolabah** See flooded box tree.
— **globulus** See blue gum.
— **melonophloia** See silver-leaved ironbark.
— **obosa** See giant mallee.
— **ochrophloia** See napunyah.
— **rostrata** See red gum.
— **salmonophloia** See salmon gum.
— **stricklandi** See strickland's gum.
— **terminalis** See kulcha.
— **tesselaria** See carbeen.

Eucalyptus oil Volatile oil derived from Eucalyptus leaves. Expectorant and antiseptic colourless to pale yellow and has a characteristic odour and a cooling taste.

Euchlaena mexicana See teosinte.

Eugenia aquea See water rose-apple.
— **aromatic** See clove.
— **caryophyllata** See clove.
— **formosa** See jamrul.
— **fruticosa** See wild jaman.
— **jambolana** See jambolana.
— **jambos** See rose-apple.
— **malaccensis** See Malayan apple.
— **uniflora** See Surinam cherry.

Euglenophyceae The single class of the plant division Euglenophyta.

Euglenophyta A division of the plant kingdom including one-celled, chiefly aquatic flagellate organisms having a spindle-shaped or flattened body, naked or with a pellicle.

Euphorbiales An order of dicotyledonous plants in the subclass Rosidale, having single leaves and unisexual flowers that are aggregated and reduced.

Euphorbia prostrata See milk weed.

Euphoria longan See longan.
— **longana** See longan.

Euploid An organism or cell having a chromosome number which is an exact multiple of the monoploid or haploid number. Term used for a euploid series are haploid, triploid, terraploid, etc.

Euproctis lunata See castor hairy caterpillar.

Eurentins echinus See brinjal sucking insects.

European bee Apis mellifera. Honey bee found in Europe and similar to honey bee found in India in its habits.

European beech Fagus sylvatica L. A tall tree of family Fagaceae and produces edible seeds.

European black-berry Rubus fruticosus L. Family Rosaceae. A shrub producing edible fruits.

European black currant Ribes nigrum L. A shrub of family Saxifragaceae, native of Eurasia and produces edible fruits.

European canker (1) A fungus disease of apple, pear, and other fruit and shade trees caused by Nectria galligena and characterised by cankers with concentric rings of callus on the trunk and branches. (2) A fungus disease of poplars caused by Dothichiza populea.

European chestnut Castanea sativa Mill. (Castanea vesca Gaertn.; Castanea vulgaris Lamk.). A tall tree of family Fagaceae whose seeds are edible.

European elder Sambucus nigra L. Family Caprifoliaceae. A shrub native of Eurasia and north Africa whose fruits are edible.

European gooseberry Ribes grossularia L. A bush of family saxifragaceae raised for its fruits. It is native of Eurasia and now grown in India.

European hazel See hazel-nut.

European meal worm Tenebrio molitor L. A black beetle which with its grub feed on flour, meal, bran, broken grains etc. Usually found in dark, damp and dingy godowns and places.

European pear See common pear.

European penny royal Mentha pulegium L. Family Labiatae. A herb native of Europe and W. Asia. Leaves and flowering top yields an oil which is used in perfumery and cosmetics.

European white waterlily Nymphaea alba L. Family Nymphaeaceae. An aquatic herb, native of Europe and north Africa. Root stocks edible.

Euryale ferox See gorgan nut.

Eurybathic living at the bottom of a

body of water.

Eurytherm An organism that is tolerant of a wide range of temperatures.

Eusporangiate Characterised by the development of a sporangium from the inner segment of the initial cell.

Eustele Vascular stem structure with either collateral or bicollateral bundles arranged in a cylinder, with anastomosing vascular strands.

Eutely Having the body composed of a constant number of cells, as in certain rotifers.

Eutric brunisol A great group of soils in the Brunisolic order. The soils may have mull Ah horizons less than 5 cm (2 inches) thick, and they have Bm horizons in which the base saturation (NaCl) is 100 per cent.

Eutrophication Of bodies of water, the process of becoming better nourished either naturally by processes of maturation or artificially by fertilization.

Eutrophic water Water rich in nutrients with abundant plankton.

Eutropy The turning of a plant to the sun's rays.

Euxinic Of or pertaining to an environment of restricted circulation and stagnant or anaerobic conditions.

Euzophera perticella See brinjal stem borer.

Evaporation (1) The process by which water is changed from the liquid or the solid state into the gaseous state through the transfer of heat energy. (2) The product or result of evaporating.

Evapotranspiration The loss of water from a given area during a specified time by evaporation from the soil surface and by transpiration from the plants. Potential evapotranspiration is the maximum transpiration that can occur in a given weather situation with a low growing crop that is not short of water and does not completely shade the ground.

Even-aged A term applied to a stand in which relatively small age differences exist between individual trees. The maximum difference in age where the stand will not be harvested until it is 100 to 200 years old, upto 25 per cent of the rotation age may be allowed.

Evergreen Pertaining to a perennially green plant. Also known as aiophyllous.

Evergreen forest Forest composed chiefly of evergreen dominants.

Evodia fraxinifolia See American beach.

Evolute Turned back; unfolded.

Evolvulus alsinoides See cuscuta.

Exanthema (1) Eruption or discharge of gum or other substance from diseased tissues. Also refers to die-back disease in citrus. (2) Any disease accompanied by skin eruption.

Exarch Characterised by a centripetal development, that is, towards the centre of the stem.

Excavation (1) To dig the earth for channel, trench etc. (2) formation of a cavity.

Exceedance interval The average number of years between the occurrence of a given event and a still greater event.

Excess rainfall Direct runoff at the place where it originates.

Excess water Water above the crop need. commonly known as drainage water.

Exchange acidity The titratable hydrogen and aluminium that can be replaced from the adsorption complex by a neutral salt solution. It is usually expressed as milliequivalents per 100 g of soil (meq/100 g soil).

Exchange capacity The total ionic charge of the adsorption complex that is active in the adsorption of ions. See also anion exchange capacity and cation exchange capacity.

Exchangeable cation percentage The extent to which the adsorption complex of a soil is occupied by a particular cation. It is expressed as: exchangeable cation (me/100 g) × 100/cation exchange capacity (meq/100 g).

Exchange coefficient Coefficients of eddy flux in turbulent flow. Also called austausch.

Exchange control Control of rates of exchange of currencies by a Government.

Exchangeable phosphate The phosphate anion reversibly attached to the surface of the solid phase of the soil in such form that it may go into solution by anionic equilibrium reactions with isotopes of phosphorus or with other anions of the liquid phase without solution of the colloid phase to which it was attached.

Exchangeable potassium The potassium that is held by the adsorption complex of the soil and is easily exchanged with the cation of neutral non-potassium salt solution. Readily available to growing plant roots.

Exchange rate In relation to currency, the price of currency in terms of other currencies.

Exchange rate parity The fixed rate of exchange between one currency and another. If a currency has no parity, then it is described as having a floating exchange rate.

Exchequer The central account of the government, kept by the Treasury.

Excise duty A revenue duty imposed on goods and services produced.

Exclusion Control of disease by preventing its introduction into disease free areas.

Exclusive dealing agreement An agreement between two or more firms to deal exclusively with each other and refuse to deal with other parties in respect of a commodity or service, or class of commodities or services, or in respect of a specified technology or class of technology.

Excocarp Outer layer of the pericarp.

Excrescences See malformation.

Excreta Wastes material eliminated from body or any tissue thereof; deleterious substances formed in a body.

Excretion the act of eliminating waste material or the eliminated products.

Excument (1) Flowing out. (2) Having an undivided main stem or trunk. (3) Having the midrib extending beyond the apex.

Exelastis atmomosa See red gram plume moth.

Exfoliation Stripping off in layers; process by which concentric sheets peel off from bare rock surfaces; the shedding of leaves, or scales from a bud.

Exhaustion Loss of plant nutrients by crop removal or leaching to such a degree that crop yields are affected.

Exhaust manifold The hollow casting through which the gases from the various cylinders are conducted on their way through the exhaust pipe and through the muffler.

Exhaust valve The valve of an engine which provides an outlet for the gas or steam which has contributed its power to the operation of the engine.

Exine See exosporium.

Exobasidium vexans See tea blister blight.

Exobiology The branch of science concerned with the study of extraterestrial life; whether life exists beyond the earth is unknown.

Exoenzyme or Extracellular enzyme An enzyme excreted by a microorganism into the environment. An enzyme that acts outside the cell.

Exogonium purga See Jalap.

Exogynaus Having the style longer than and exerted beyond the corolla.

Exosmosis or **Exosmose** Osmotic diffusion towards the outside of a cell or vessel or the less concentrated solution.

Exosporium The outer of two layers forming the wall of spores such as pollen and bacterial spores. Also known as exine.

Exostome The opening through the outer integument of a bitegmic ovule.

Exotic Not native; foreign. Of trees and plants, those introduced from other climates or countries.

Exotic plant Introduced plant; not native.

Exotoxin Toxin secreted by a microorganism into the surrounding medium.

Expansive soils See swelling clays.

Expectation The expected value of a function of variate values is its mean value in repeated sampling.

Expeller Device where torsion force is used in perforated cages into which oilseed meal is fed by screw principle; oil flows out through perforations of cage, cake being forced out at opposite end.

Expiration Emission of carbon dioxide by plants and animals.

Explicit cost Costs in the form of a payment either in cash or kind.

Explosion The igniting and expanding of the gases in the combustion chambers of an engine.

Export bounty A premium or subsidy paid by the State to encourage exports.

Export duties Duties or taxes imposed on goods exported from a country.

Extended forecast In general, a forecast of weather conditions for a period extending beyond two days from the day of issue. Also known as extended-range forecast; long-range forecast.

Extension agents See change agents.

Extension education Extension education is an applied science consisting of contents derived from researches, accumulated field experiences and relevant principles drawn from the behavioural sciences, synthesised with useful technology in a body of philosophy principles, contents and methods focussed on the problems of out of school education for adults and youths. It helps them to solve their problems.

Extension evaluation The study of the degree of success we get in terms of our original objectives is called extension evaluation.

Extension programme An extension programme is a set of clearly defined, consciously conceived educational objectives derived from an adequate analysis of situation which are to be achieved through extension teaching.

External drainage The natural elimination of accumulation of precipitation water on the soil surface.

External economy A fall in the cost of any of the materials and services which a firm requires, which are obtained from outside sources.

External seed-borne disease Plant disease, where the disease organism in its living state or its spores are present superficially on the seed without any organic union with it, or as mycelium in superficial infection. Controlled by application of fungicides to the surface of the seed.

External seed-treatment Treatment of seed for external seed-borne fungal disease; consists of thoroughly mixing the seed, preferably in a rotating drum so as to cover the seed surface with a layer of the fungicide.

Extinction coefficient A measure of the space of diminution, or extinction, of any transmitted light.

Extract Preparation obtained by soaking a substance in a liquid solvent, then separating out the dissolved active ingredients from the inert matter, concentrating or evaporating it until almost or completely dried.

Extratropical cyclone Any cyclone scale storm that is not a tropical cyclone. Also known as extratropical low; extratropical storm.

Extremity The limb; the distal portion of the limb; the distal end of any limb like structure.

Exudate Liquid discharge from diseased tissues of animals or plants.

Exudation Any discharge through an incision or pore, such as gum, resin, moisture etc. or as in the case of animals, bloody discharge.

Exudation pressure See root pressure.

Eye (1) The organ of sight. (2) A pigment spot in various animals and in lower plants. (3) The bud of a tuber.

Eye-spots Ocilte; certain pigment spots in many lower plants and animals and also in some vertebrates which have a visual function.

F

F_1 The first filial generation. The first generation of a given mating.

F_2 The second filial generation, produced by crossing or by self-pollinating the F_1. Subsequent generations F_3, F_4 etc. Parents are denoted by P_1, P_2 etc.

Fabaceous Belonging to the legume or pea family; Leguminosae.

Faba vulgaris See broad bean.

Fabiform Having the shape of a bean.

Fabric soil The physical constitution of a soil material expressed by the spatial arrangement of the solid particles and associated voids.

Factor (1) The same as gene. (2) Any agent contributing to a result. It may be biotic, climatic or nutritional.

Factor cost What producers receive for the sale of their products and services. This is not synonymous with market prices.

Factorial experimentation, or **Factorial design** A method of experimentation where, in place of the classical method of varying only one experimental factor at a time, a comprehensive study is made of the number of experimental factors at the same time. This simultaneous appraisal is possible because of the balance built into the design which avoids confounding.

Factor incomes Incomes accruing to the factors of production wages, salaries, profits, interest and rent.

Factor mobility The ease or difficult with which factors of production can move between uses.

Factors of production The various agents which combine to produce goods and services. Land, labour and capital are the traditional categories; organisation or enterprise is sometimes added as a fourth category.

Factors, specific Factors of production of a specialised kind which cannot easily be adjusted to serve an alternative purpose.

Facultative organism A micro-organism that lives under either aerobic or anaerobic conditions.

Facultative parasite An organism that is usually saprophytic, but which under some conditions may become parasitic.

Faeces Excrement discharged from the bowels.

Fagus sylvatica See European beach.

Falcate Hooked or sickle shaped.

Falciform Sickle-shaped.

Fall (1) Dropping of a plant part. (2) Land slope. (3) Amount of rainfall.

Fallage The act of cutting trees.

Fallopian tubes Horizontally placed tubes to which the horns of the uterus are attached; open out like a funnel which are placed below the right and left overies.

Fallout The movement of solid matter through the air as dust; to natural fallout man adds, for example, Radioactive fallout.

Fallout winds Tropospheric winds that carry the radioactive fallout materials, observed by standard winds aloft observation techniques.

Fallow Cropland left idle to restore productivity, primarily through accumulating water or nutrients or both. The soil is tilled for at least one growing season to destroy weeds; to encourage moisture storage, and to promote decomposition of plant residues.

False blossom (1) A fungus disease of the cranberry caused by Exobasidium oxycocci; erect flower buds are formed which produces malformed flowers that set no fruit. Also known as false bloom. (2) A similar virus disease of the cranberry transmitted by the leafhopper, Scleroracusvaccinii. Also known as Wisconsin false blossom.

False fruit Fruits formed from the

receptacle or other parts of the flower, in addition to the ovary or from complete inflorescences; cashew-nut example.

False ring A layer of wood that is less than a full season's growth and often does not form a complete ring.

False saffron See safflower.

Family A term used in classification, signifying a group of related genera; families grouped into orders which are often descendants of a single individual.

Family of origin or orientation The family in which one is born.

Family of procreation The family which one sets up after one's marriage.

Family soil A category in the Canadian system of soil classification. Differences are primarily texture, drainage, thickness of horizons, permeability, minerology, consistence, and reaction.

Famine A widespread, acute scarcity of food and fodder. Usually a result of water scarcity or a widespread disease.

Fancier A breeder who shows particular interest in and the development of, a particular breed or type of animal or plant.

FAO Food and Agriculture Organization set up by U.N.O. Set up by allied nations to serve as a clearing house through which the member nations can compare their agricultural consumptions, production and trade programmes and suggest adjustments to meet the world food needs.

Farinaceous Starchy.

Farinose Yielding farina, a fine meal of vegetable matter. Covered with a white powdery substance.

Farm (1) A tract of land singally or in parcels (at least one hectare) used for cultivating crops or raising animals. (2) To cultivate the tract for agricultural purposes.

Farm animals Animals associated with agriculture.

Farm building Any service building at a farm.

Farm business The series of production and maintenance activities and financial transactions in connection with running a farm.

Farm economy The practical adjustment, organisation, administration of the affairs and business of a farm.

Farmer One who operates a farm or is engaged in the business of a farm; a agriculturist.

Farm forestry Practice of raising small woods on a farm in addition to normal cultivation to make the farm more self-sufficient.

Farm house Residential building on a farm.

Farming The skills and practices of agriculture.

Farming needs The gap between the farming practice adopted by the farmers at present and the farming practices recommended by the Research Workers at present.

Farm irrigation structures Devices, pieces of equipment and small hydraulic structures used by the individual irrigators to divert water from the supply ditch and to convey it to the several parts of his farm including control, regulation and measurement of water.

Farm loss (water) Water delivered to a farm which is not made available to the crop to be irrigated.

Farm losses Losses of water on the farm due to uneven distribution, poor handling, evaporation and percolation into the subsoil due to over-irrigation.

Farm manure See farm yard manure.

Farmstead Part of the farm occupied by the main set of buildings, adjacent yards, kitchen garden and family orchard.

Farmyard Small area adjacent to a cattle yard where animals can be kept in open.

Farmyard manure Most commonly used manure in India. Consisting of a mixture of cattle dung, bedding used in the stable and remnants of straw or stalks of plants fed to cattle. Average content 0.3 per cent N, 0.15 per cent of P_2O_5 and 0.3 per cent of K_2O. Partially rotted manure can be applied 3-4 weeks prior to crop sowing and enough moisture is given for its decomposition. Well-rotten manure can be applied just prior to sowing or planting of crops.

Farsetia jacquemontii Hook. F and Thoms. A herb of family Cruciferae whose tender twigs are used as condiment.

Fasciation Malformation of plant parts resulting from disorganised tissue growth.

Fascicle A small bundle or cluster, as of fibres or leaves.

Fastigiate (1) Having erect branches that are close to the stem. (2) Becoming narrower at the top; arranged in a conical bundle.

Fatigue (1) Tiredness. (2) Loss of productive capacity of a soil.

Fatty acid Organic, monobasic acid.

Fatuoids Mutants occuring in cultures of cultivated oats and possessing characters of Avena Fatua, wild oats.

Fauna A term refers to all indigenous animals of a place or a country or a period.

Feather rot A fungus rot of both dead and living tree trunks caused by Poria subacida and characterised by the white stingy or spongy natures of the rotted tissue.

Feathery bamboo Bambusa Vulgaris Schrader. Family Gramineae. Bamboo cultivated in the hotter parts of India. Used for mat and baskets. Also young buds eaten as vegetable.

Features The fixed physical objects, such as coral reefs, mud flats, or grass beds of coastal ecosystem.

Fecal index Indirect method of estimating indigestibility of dry matter by determining concentration of an indicator in faeces.

Fecund Fertile; prolific; fruitful.

Fecundation Pollination or fertilisation.

Fecundity (1) Power of a species to multiply rapidly. (2) Capacity of a species to form reproductive organs.

Feed (1) An agricultural product meant for use as a food for livestock. (2) To supply essential nutrients to animal or plant.

Feed supplement A feed or feed mixture containing one or more nutrients in a relatively higher order. Mixed with the normal feeds to supply these nutrients to improve the quality of the products, to meet the feeding standards or to make up the deficiency.

Feedback (1) A characteristic of interrelated markets in that a change in one market may affect many other markets, and the changes in these other markets may in turn affect the original market. (2) Communicating back to the other or "the return process".

Feeder root A root or rootlet with hairs which absorb food (water and nutrients) for the growth of a plant.

Feeding Ingestion of food by animals or plants. Also making feed available to them.

Feeding root See feeder root.

Feeding standard The standards which indicates the amounts of nutrients to be given to various species of farm animals and of various age groups so as to have the optimum growth and productive work.

Feeding value See Forage quality.

Feedstuff Material(s) consumed by animals that contributes nutrients to the diet.

Feijoa sellowiana See pineapple-guava.

Feldspar Important group of soil forming silicate minerals forming 48 per cent of the minerals of earth's crust; primary source of potassium, sodium and kaolinitic clay of soils. Has two important groups; potash feldspar forming the orthoclase group; albite or soda feldspar and anorthite or lime feldspar, together constituting the plagioclase group. Weathers easily and produces clay on hydrolysis.

Felspar See feldspar.

Female (animal) An egg or young producing animal.

— **(organ of plant)** Ovary, the organ which becomes fruit after fertilisation.

Female, thread A thread which is cut in a hole or on a hollow surface.

Fennel Foeniculum vulgare. A tall perennial herb of the family Umbelliferae; a spice is derived from the fruit. Also called *Saunf*.

Fenugreek Trigonella foenumgraecum L. Family Papilionaceae; a leafy vegetable and fodder crop cultivated throughout India. Seeds used as spice and condiments. Cool season crop. Also called *Methi.*

Fenugreek downy mildew Disease of fenugreek caused by Peronospora trigonellae; symptoms same as in lucerne downy mildew.

Fenugreek leaf spot Disease of

fenugreek caused by Cercospora traversiana. Characterised by brown, elliptic or round spots with dark margins on leaves.

Fenugreek powdery mildew Disease of fenugreek caused by Erysiphe polygoni. Symptoms as in bean powdery mildew.

Fenugreek rust Disease of fenugreek caused by Uromyces anthylidis. Symptoms as in incerne rust.

Feracious Fruitful.

Feral Descriptive of animals or plants that are no longer domesticated or cultivated and have reverted to a wild state.

Fermentation (1) Anaerobic oxidation of carbohydrates and carbohydrate such as compounds by enzyme action of microorganism; gaseous oxygen is not involved in this energy-yielding process. (2) Enzymatic curing of tobacco leaves. (3) Digestion in the rumen by micro-organisms.

Fern (1) Any flowerless plant of the order filicales; resembles seed plants by having roots, stems and leaves, but reproduces by spores. (2) Any fern-like plant.

Feronia elephantum See wood apple.

— **limonia** See wood apple.

Ferrisiana virgata See custard apple mealy bug.

Ferruginous soil The soils containing iron minerals specially limonite and haematite in abundance. Red colour or shades of yellow and brown due to their presence.

Fersiallitic A term used in tropical pedology to describe soils less weathered than ferralitic. The clay fraction SiO_2/Al_2O_3 ratio ranges from 1.7 to about 2.3. The consistence is firmer and the cation exchange capacity of the clay separate is higher than in ferralitic soils.

Fertility, soil The status of a soil in relation to the amount and availability to plants of elements necessary for plant growth.

Fertility, vitamin Vitamin E.

Fertilisation (1) The fusion of a male gamete (sperm) with a female gamete (egg) and of their nuclei, without which their later development is usually impossible. (2) The application to the land of nutrients (fertiliser) to promote the feeding of plants.

Fertiliser Any organic or inorganic material of natural or synthetic origin that is added to soil to supply certain elements essential to the growth of plants. Also called chemical fertilisers, artificial fertilisers.

Fertiliser analysis Quantitative analysis of a chemical fertiliser specially of elements of food value to soil.

Fertiliser application Application of the fertiliser on soil or plant. The methods includes, broadcasting, drilling, or liquid application to soil and spraying on plants.

Fertiliser applicator Equipment connected with the lateral line of a sprinkler system for injecting water soluble fertiliser into the water to be sprinkled.

Fertiliser attachment A fertiliser hopper with necessary arrangements of regulation and drilling of fertiliser to soil attached to a seed drill.

Fertiliser formula A formula according to which different fertiliser materials are mixed so as to prepare a mixed fertiliser of desired grade.

Fertiliser grade The minimum guarantee of the plant nutrient content of the fertiliser expressed as whole numbers in terms of total N, available P_2O_5 and water soluble potash (K_2O).

Fertiliser requirement The quantity of certain plant nutrient elements needed, in addition to the amount supplied by the soil, to increase plant growth to a designated optimum.

Ferula assafoetida See asafoetida.

— **narthex** See asafoetida; narthex asafoetida.

Fescue A group of grasses of the genus Festuca, used for both hay and pasture.

Festuca elatior See meadow fescue.

— **gigantea** See giant fescue.

— **ovina** See sheep fescue.

— **rubra** See red fescue.

Fetid Rotten; giving out an offensive colour. Also called foetid.

Fiat money Money which the State declares to be legal tender, although its content value may be little or nothing. Most of fiat money consists of notes, although token money may also be included in the term.

Fibratus A cloud species characterised by a fine hairlike or striated composition, the filaments of which are usually distinctly separated from each other; the extremities of these filaments are always thin and never terminated by tufts or hooks. Also known as filesus.

Fibre (1) A strand of nerve, muscle, connective tissue. (2) Elangoated elastic plant cell for mechanical strength; often tapers off at either end; serves as support to plant stem. (3) The insoluble residue left after sugars, acids, proteins and ash are accounted for in the analysis. Consists mainly of cellulose and lignified cell-walls.

Fibre board Board made of compressed fibre used as a building material.

Fibre crops Plants, such as flax, hemp, jute, and sisal cultivated for their content or yield of fibrous material.

Fibre flax The flax plant grown in fertile, well-drained, well-prepared soil and cool, humid climate, planted in the early spring and harvested when half the seed pods turn yellow; used in the manufacture of linen.

Fibric layer A layer of organic soil material containing large amounts of weakly decomposed fibre whose botanical origin is readily identifiable.

Fibril A small thread or fibre, as a root hair or one of the structural units of a striated muscle. One of the minute threadlike elements of a natural or synthetic fibre.

Fibrin An elastic, coagulated, insoluble, whitish protein found in blood, serous fluids and muscle tissues. See fibrinogen.

Fibrinogen A soluble protein of blood and body fluids of animals, which by the activity of thrombin, yields fibrin.

Fibrinous exudate A discharge containing fibrin.

Fibrisol A great group of soils in the organic order that are saturated for most of the year. The soils have a dominantly fibric middle tier, or middle and surface tiers if a terric, lithic, hydric, or cryic contact occurs in the middle tier.

Fibrosis Formation of fibrous tissues on the body organs.

Fibrous root A thread-like root, as in grasses.

Fibro-vascular bundles Strands of water and food conducting cells in the stem, leaves and other parts of the plant. Called veins in the leaves. Where they are easily separable in that case popularly known as fibres.

Ficus A genus of tropical trees in the family Moraceae including the rubber tree. A genus of gastropod molluses having pear-shaped, spirally ribbed scluptured shells.

Ficus auriculata See eve's apron.
— **benghalensis** See banyan.
— **carica** See fig.
— **elastica** See Indian rubber.
— **glomerata** See cluster fig.
— **goolereea** See cluster fig.
— **hispida** See *kathumbar.*
— **infectoria** See pilkhan.
— **lacor** See pilkhan.
— **macrophylla** See eve's apron.
— **Palmata** See *anjiri.*
— **roxburghii** See eve's apron.
— **rumphii** See eve's apron.
— **virgata** See *anjiri.*

Fiducial limits Limits between which a parametre is considered to lie on a given probability. The term also occurs as a synonym of 'confidence limits'. The two are often the same in commonly occurring cases, but their conceptual genesis is different.

Field bean (1) Hyacinth bean. Dolichos lablab L. var. lignosus (Lablab niger Medic). Family Papilionaceae. A bushy type of bean whose pods are fibrous, only seed used for cooking. Also called sem. (2) See Australian bean.

Field bindweed Convolvulus arvensis. Family Convolvulaceae. Perennial weed with creeping root-stock and trailing or twiney stem. White to light pink blossoms. Has deep root system and found commonly in wheat fields.

Field capacity The amount of water remaining in soil capillaries once the gravitational water has been drained away. Also called capillary capacity of the soil.

Field command The difference in water level, in a water course or lateral and the level of the highest point of parcel of land irrigated.

Field crop Any herbacious plant grown

in cultivated fields.

Field delivery point The point on the watercourse, or branch watercourse at which delivery of water to a co-sharer or group of co-sharers is given. The delivery structure can be a mere shovel cut or proper distribution or division structure.

Field ditches Shallow ditches with flat side slopes that can be crossed with farm machinery.

Field mint Mentha arvensis L. Family Labiatae. A perennial herb used for extracting essential oil used in the preparation of certain kinds of cigarettes and pharmaceuticals. Also used as carminative, refrigerant and stimulant.

Field pea See garden pea.

Field rat Nesokia indica, a common rat who stores grains in burrows. Also lives on roots of grasses.

Field strip cropping A soil conservation measure in which the field crops are planted in more or less parallel strips across fairly uniform slopes but not on exact contours.

Field Tee A fitting in the form of Tee installed in the pipeline system to guide flow of water in more than one direction in main line or sprinkler line.

Field water application efficiency The ratio of water stored in the root depth of the soil of a field to the water delivered to the field.

Field weirs Adjustable weirs built to regulate the water level and the flow of water in the ditches used in sub-irrigation.

Fifteen-atmosphere percentage See moisture tension, soil.

Fifteen-bar percentage See moisture tension, soil.

Fig Ficus carica L. Family Moraceae. Deciduous sub-tropical tree belonging to same family as mulberry and jack-fruit. Grown on variety of soils and propagated by vegetative methods. Nutritious fruit rich in vitamin A and Vitamin C.

Fig beetle Olenecambtus biolobus Fab. Beetle pest of fig, whose grub bores into the stem and kills the affected branch.

Fig mosaic A virus disease causing mottling of fig leaves.

Fig rust Disease caused by Cerotelium fici (Cast) Arth. Rusty brown raised pustules appear on lower leaf surface and discolouration occur on the upper one.

Fig sunburn Physiological diseases in which cracking and peeling off of fig bark occurs. Caused by sun exposure of young plants.

Fiji disease A virus disease of sugarcane; elongated swellings on the underside of leaves precede death of the plant.

Filament A colony of cells in linear series, as in certain algae; the stalk of a stamen; thread like structure; glowing portion in an electric bulb.

Filamentous algae An undesirable weed of ponds, Algae submerged in water often floating. Serves as a food of fish. Also called hair weed.

Filbert Either of two European plants belonging to the genus corylus and producing a thick-shelled, edible nut. Also known as hazelnut.

File (1) A hard steel instrument made in various shapes and sizes, for smoothing wood or metal. (2) To cut, trim, or finish with a file.

Filial generation See F_1.

Filiform An extreme condition of leaf in certain plants which develops into a sinuous ribbon or thread shape as a result of disease.

Fill Man-made deposits of geological, soil and diverse waste materials, or one or two of these.

Filler leaf The portion of the tobacco that makes up the central part of a cigar. Also the special type of tobacco leaf used for the purpose.

Filling Artificial elevation of land by deposit of soil or sediment.

Filling in The process of building in the centre of the wall between the face and back.

Filterable virus A virus which can pass through a filter through which ordinary bacteria cannot pass.

Filter press mud A good source of manure. These are suspended impurities in sugarcane juice retained in a plastic condition in the filter press in sugar factories. Can also be used as soil amendment.

Filtration The act or process of filtering, as the removal of foreign matter by passing water through sand beds.

Fimbriate Fringed, or toothed as petals, tube ducts.

Fimetarius Growing on heaps of dung or manure.

Fin A flat plate used as a coulter.

Final cut The quantity of material removed in harvesting the main crop of trees at the end of rotation.

Final felling The removal of the last seed or shelter trees after regeneration has been effected under a shelterwood system.

Final host The host in which a parasite develops to maturity.

Final intake of water The rate of water entry at the soil surface when the constant rate of intake has been established. Also called basic infiltration rate of the soil. Usually measured using double ring infiltrometres.

Finances The provision of money at the time it is wanted. It provides the means by which people are enabled to consume more than they produce in certain periods of time.

Finance bill A bill not involving goods, simply promising to pay a specified sum of money on a specified date.

Fine clay A clay fraction of specified size less than 2 mm, usually less than 0.2 or 0.08 mm.

Fine earth The fraction of mineral soil consisting of particles less than 2 mm in diametre.

Fine gravel The soil separates which include particles 2 mm to 1.25 mm in diametre.

Fineness modulus An empirical factor found by dividing the total of the percentages of materials retained on specified sieves by 100 and is an indicator of the fineness of an aggregate, paint, cement, etc.

Fine plucking The plucking of leaves of the tea bush restricted to the terminal bud and two leaves.

Fine pruning The removal of a large number of small branches and spurs from a tree.

Fine sand 50 per cent or more fine sand or less than 25 per cent very coarse, and medium sand and less than 50 per cent very fine sand.

Fine sandy loam Thirty per cent or more fine sand and less than 30 per cent very fine sand or between 15 and 30 per cent very coarse, coarse, and medium sand.

Fine texture Consisting of or containing large quantities of the fine fractions particularly of silt and clay. It includes all the textural classes of clay loams and clays; clay loam, sandy clay loam, silty clay loam, sandy clay, silty clay, and clay. Sometimes it is subdivided into clayey texture and moderately fine texture. See also texture, soil.

Fines A term used in soil mechanics for the portion of a soil finer than a No. 200 U.S. standard sieve.

Finest rate of discount Lowest rate of discount.

Finger millet Eleusine Coracana Gaerth. Family Gramineae. An important millet crop suited for dry cultivation in rainfall regions of 50-100 cm; can tolerate certain degree of alkalinity in soils. Grain is nutritious.

Fire bean Scarlet runner bean.

Fire surface A fire which runs over the forest floor and burns only the surface litter, the loose debris, and the smaller vegetation. Ground fire also is used.

Firm (1) A term describing the consistence of a moist soil that offers distinctly noticeable resistance to crushing, but can be crushed with moderate pressure between the thumb and forefinger. See also consistence. (2) To compact the soil, crushing and pulverizing the lumps to facilitate capillary movement.

Firminana colorata See scarlet sterculia.

First bottom The normal floodplain of a stream.

First generation hybrid Same as F_1 hybrid.

First-order climatological station A meteorological station at which aurographic records or hourly readings of atmospheric pressure, temperature, humidity, wind, sunshine, and precipitation are made, together with observations at fixed hours of the amount and form of clouds and notes on the weather.

First-order station After U.S. National Weather Service practice, any meteorological station that is staffed in whole or in part by National Weather Service (Civil Service) personnel, regardless of the type or extent of work required of that station.

Firth A narrow arm of the sea; also, the opening of a river into the sea. An estuary.

Fiscal policy the policy adopted by a Government for raising revenue to meet expenditure and for influencing the level of business activity.

Fish berry Anamirta cocculus L. Wight and Arn. (Menispermum cocculus L.) Family Menispermaceae. A climbing shrub, the fruits of which are used to stupefy fish and also used as an insecticide.

Fish manure A manure. Residue after fish oil is extracted. Contains 3-8 per cent nitrogen and 4-6 per cent phosphoric acid. Quick acting and applied as powder. Also called Fish meal, fish powder.

Fish meal See fish manure.

Fish pond fertiliser Inorganic or organic matter applied to a fish pond to promote the growth of a plankton and aquatic plants.

Fish powder See fish manure.

Fish screen A device intended to prevent the entrance of fish into a conduit.

Fissility A property of timber of splitting in the direction of the fibres when a tensile force is exerted at right angles to the grain.

Fission Method of reproduction common among protozoa and bacteria. It involves a gradual constriction of the cell on opposite sides and as it progresses, formation of two cell parts or sections into which the divided cell contents gather. This continues till the two sides meet, to separate to form two new cells.

Fission yeast The fungus which reproduce asexually by the division and separation of the cells into two equal parts.

Fistula Surgically established opening between a hollow organ and the skin for experimental purposes (such as an esophageal fistula).

Five-day forecast A forecast of the average weather conditions and large-scale synoptic features in a 5-day period; a type of extended forecast.

Fixation (1) The process or processes in a soil by which certain chemical elements essential for plant growth are converted from a soluble or exchangeable form to a much less soluble or nonexchangeable form for example, Phosphate fixation. See also nitrogen fixation. (2) The process of filling and coagulating a cell by means of chemical or physical agency.

Fixation of nitrogen The fixation of atmospheric nitrogen in the soil by symbiotic and non-symbiotic bacteria.

Fixation of phosphorus The conversion of soluble phosphorus in the soil into less soluble and available forms.

Fixed ammonium Ammonium held in a soil which is neither water-soluble nor easily exchangeable.

Fixed costs Production costs which tend to be unaffected by variations in the volume of output.

Fixed potassium Potassium held by a soil which is neither water-soluble nor readily exchangeable.

Flaccid Deficient in turgor. Soft, flabby, or relaxed.

Flacourtia cataphracta See Puneala plum.

— **indica** Flacourtia indica (Burm.f.) Merr. (F. ramontchi L' Herit.) A small tree of family Flacourtiaceae. Grown for its edible fruit.

— **jangomas** See Puneala plum.

— **ramontchi** See F. indica.

Flacourtiaceal A family of dicotyledonous plants in the order Violales having the characteristics of the more primitive members of the order.

Flag (1) A branch with dead leaves on an otherwise green tree. (2) To droop.

Flag leaf (1) The first leaf immediately below the inflorescence in cereals. (2) The large flag like leaf at the top of the maize plant just before the tassel emerges.

Flag smut A smut affecting the leaves and stems of cereals and other grasses, characterised by formation of sori within the tissues, which rupture releasing black spore masses and causing fraying of the infected area.

Flagella (Sing, flagellum) Long, hair-like processes that by their lashing activity cause the organism to move; one or more flagella may be attached to one or both ends of the organism or completely around it.

Flagellates Bacteria organisms, wavy or bent or rod-shaped; rod-shaped with hair like projections.

Flagellum A flexible, whiplike appendage on cells, used as an organ of locomotion.

Flagging Dying of the central leaf in grasses while other leaves stay green.

Flaggy See coarse fragments.

Flagstone A thin fragment of sandstone, limestone, state shale, or rarely of schist, 15 to 37 cm (6 to 15 inches) long see also coarse fragments.

Flail A fairly heavy, flat piece of wood fastened to a length of rope or leather for use in threshing seed.

Flame cultivator A flame thrower for destroying weeds between rows of crops.

Flame of the forest Butea monosperma (Lamk.) Taubert. (Butea frondosa Koenig ex Roxb.). A small deciduous tree of family Papilionaceae; produces orange scarlet inflorescence.

Flame thrower Compressed air type of sprayer filled with kerosene oil for producing flame. Used for burning swarms of locusts and scrub weeds.

Flap and patch budding A method of budding similar to T-budding. A square or rectangular patch from the stock is peeled on three sides and the bud shield pushed under the flap to secure perfect union. Flap is brought in position and wrapped around when the bud shield is fitted without the over-lying flap, it is patch budding.

Flashboard Plank, generally held horizontally in vertical slots on the crest of a dam or check structure to control the upstream water level.

Flash cards These are the series of cards which when presented before the audience in proper sequence, tell a complete story. Each card is of about 25 mm to 30 mm in size and contains a picture or diagram. Each individual card is "flashed" before the audience accompanied by the verbal commentary.

Flash point It is the temperature at which inflammable vapours are given off by an oil. (See also flash test).

Flash test In this test, oil sample is heated and the temperature at which distinct flash is obtained when flame is passed over the container is called flash point.

Flat (1) A description of bonds being sold in default, since no interest is being paid. (2) Used to describe relatively levelled lands or regions.

Flat grain beetle Laemophloeus minutes oliv. A reddish brown beetle with long feeders whose grub is yellowish white. It attacks grain flour and meals that is in bad condition. Uninjured grains are unaffected.

Flat grained timber Plain or tangential sawn timber.

Flat rate tariff A tariff for the supply of electricity consisting solely of a single-unit charge: used chiefly by very small users of electricity for lighting and power.

Flat type incubator An incubator with only single layer incubation with eggs laying flat. Generally of small capacity.

Flat zone The upper level portion of the border wherein the stream of water spreads evenly across the width of the border strips before the flow starts down the slope.

Flavescence Yellowing or blanching of green plant parts due to diminution of chlorophyll accompanying certain virus disease.

Flax (1) Linum usitatissimum. An erect annual plant with linear leaves and blue flowers; cultivated as a source of flaxseed and fibre. (2) The fibre extracted from linseed plant.

Flax rust A disease of flax caused by the rust fungus Melampsora lini.

Flax seed Linseed, the seed obtained from the seed flax plant; a source of linseed oil.

Flax wilt A fungus disease of flax caused by Fusarium oxysporum lini; diseased plants wilt, yellow, and die.

Flayer See horizon, soil.

Flea beetle Any small, jumping beetle, eating small holes in the leaves of plants. Family Chrysomelidae.

Fleece The fluffy mass of cotton remaining after seeds have been removed by ginning.

Flesh The pulpy or juicy portion of fruits

or plant storage organs such as potato.

Flexibility (1) The quality of bending without breaking. (2) Ability to adjust to changing conditions.

Flint A dense, amorphous or cryptocrystalline rock composed essentially of silica. Remains as residue in the weathering of chalks and limestones.

Flint maize Zeamays var. indurata. A group of maize of inferior quality.

Flinty It is hard, translucent endosperm of maize or wheat kernel, as compared to the more mealy portions.

Flip chart It is like an album of drawings, pictures, charts. It is an aid and assists an extension worker or a teacher to tell a story.

Flist In Scotland, a Keen blast or shower accompanied by a squall.

Flitch A piece of sawn or hewn timber intended for further conversion.

Float A wooden device consisting of one or more blades used to level a seedbed. It levels the minor irregularities in the land without changing the general topography of the area.

Floatation method A method in which wheat is soaked in ample water to remove floating seeds over the water. This is done to remove the diseased galls before sowing.

Float carburettor A carburettor in which a float of cork or hollow metal controls the height of gasoline.

Floating smoothing the surface of newly placed concrete or mortar with a trowel.

Floating power A method of mounting the engine in the channels frame so that its vibration is absorbed by rubber cushions.

Floating rice See deep water rice.

Flocculation The aggregation of suspended colloidal materials into aggregates.

Flocculation test Test dependent upon the coalescence of finally divided or colloid particles into larger particles that precipitate.

Flocculent Covered with a soft waxy substance giving appearance of wools.

Floculus A small, tufty or cloud like mass; a small cluster of fine particles in a soil which is flocculated by the addition of lime or other substances.

Flood (1) An overflow or inundation that comes from a river or other body of water and causes or threatens damage. (2) to irrigate; the method commonly referred as flood irrigation.

Flood control Methods or facilities for reducing flood flows.

Flood control project A project meant either solely or primarily for flood control purpose, although in the latter case it may incidentally serve other purposes. It may utilise one or several of the flood control methods. Also called 'flood mitigation project'.

Flood frequency Over a period of years, the average number of times a flood of a given magnitude is likely to occur.

Flood irrigation To run water to a depth of few centimetre without any control of water during spreading in the plots.

Flood peak The highest value of the stage or discharge attained by a flood, thus, peak stage or peak discharge.

Flood plain storage Volume of water which can be temporarily stored in a flood plain.

Flood routine The process of computing the progressive time and shape of a flood wave at successive points along a river. Also known as storage routing; streamflow routing.

Flood series A list of flood events, usually arranged in order of magnitude, which occurred during a specified period of time.

Flood stage The stage, on a fixed river gage, at which overflow of the natural banks of the stream begins to cause damage in any portion of the reach for which the gage is used as an index.

Flood tide (1) Tide during period between low water and the following high water. Also called as rising tide. (2) A tide at its highest point.

Flood volume The sum of the mean daily discharge passing at a particular site during a flood event.

Flood water retarding structure A structure providing for temporary storage of floodwater and for its controlled release.

Floodway A channel either natural, excavated, or bounded by dikes and levels, used to carry excessive flood flows to reduce flooding. Sometimes considered to be the transitional area between the active channel and the flood plain.

Floor (1) Forest floor. (2) All dead vegetable matter on the soil surface including litter and unincorporated humus.

Flora (1) The plants of an area or a period peculiar to that area or period as the case may be. (2) Bacteria in or on an plant, animal or soil.

Floral bud An unopened flower.

Floral diagram A diagram of a flower in cross-section showing the number and arrangement of floral parts.

Floral nectary Nectary in relation to flowers. Refers to glands which secretes nectar in a flower.

Floral receptacle See receptacle.

Florescence Bursting into bloom.

Floret (1) In grasses, the lemma and palea with included flower (stamens and pistil). In legumes, the individual flower, usually belong to a cluster. (2) A small flower, especially one of an inflorescence.

Floriculture A segment of horticulture concerned with commercial production, marketing and retail sale of cut flowers and potted plants, as well as the gardening and flower arrangement.

Florideophyceal A class of red algal, division Rhodophyta, having prominent pit connections between cells.

Floriferous Blooming freely, used principally of ornamental plants.

Florist Person dealing in cut flowers and plants usually grown by himself.

Floscular Having florets; composite flowered.

Floss (1) A soft downy or silky substance. (2) The loose pieces of silk in cocoon. (3) Lint of cotton and other plants resembling fine wool.

Flotation The raising of new capital by public subscription.

Flour beetle See long-headed flour beetle, saw-toother grain beetle; yellow meal worm; flat grain beetle; European meal worm.

Flour maize Zea mays var. Amylacea. Soft and of better quality than flint maize.

Flour moth See Indian meal moth.

Floury Pertains to a fine-textured soils which are smooth and dust-like when dry.

Flow A stream of water; movement of water, silt, sand, etc., discharge; total quantity carried by a stream; discharge from a pipe. The flow is usually expressed in cumecs. (cubic metre/sec), ha, cm, etc.

Flow chart When we try to tell a story of how a product is obtained as a result of series of processes, we can depict it in the form of a flow chart.

Flowage line A contour or line around a reservoir, pond, lake or along a stream, corresponding to some definite water level generally used in connection with the acquisition of rights to flood lands for storage purposes.

Flower (1) The characteristic reproductive structure of a seed plant, particularly if some or all of the parts are brightly coloured. (2) To develop blossoms.

Flower bud A bud when mature and open reveals a flower.

Flower head An inflorescence, containing usually small or medium-sized flowers clustered together.

Flower stalk Peduncle; pedicle.

Flowering stage The stage of the plant when it starts flowering.

Flow failure Failure in which a soil mass moves over relatively long distances in a fluid like manner.

Flow irrigation The system of irrigation of lands by gravity where the irrigation water is available to a higher level than the irrigated land. Also called as gravity irrigation.

Flow rate Amount of liquid passing through a gate, pipe or channel per unit time. Also called discharge.

Flow velocity The volume of water transferred per unit of time and per unit of area in the direction of the net flow of water in soil.

Flue The special passageways of heated air in a barn for curing tobacco.

Flue-curing Process of curing tobacco for cigarettes. Consists in yellowing, fixing the colour and drying the leaves at controlled temperatures. Takes about 5-6 days for curing.

Flueggea leucopyrus Securinga

leucopyrus. (Willd.) Muell. Arg. (Phyllanthus leucopyrus koen. ex. Roxb. A large shrub of family Euphorbiaceae which produces edible fruits.

Fluid film theory According to this theory reduction in friction due to application of lubricants is explained on the basis of rolling effect i.e. the lubricant is supposed to act like mass of globules, rolling inbetween two surfaces.

Fluidity The property of flowing possessed by gases and liquids.

Flume A complete and independently supported, artificial waterway used to carry water across depressions or over difficult terrain, or when other reasons make the construction of a normal conveyance channel or conduit impractical or uneconomical.

Fluming ratio The ratio of bed width of the contracted (flumed) work to the normal channel width downstream.

Fluorescence Property of absorbing light and re-radiating it in light of longer wavelength than the incident light.

Fluorine (F) A chemical which in excess causes fluorosis. In soils the critical limit is 100 ppm.

Fluorine poisoning A disease in cattles due to excessive uptake of fodder grown in soils with high fluorine content or water of such natures. Causes stunted growth, lameness, mottled and irregular teeth, thickening of jaws and bones of limbs. Also called fluorosis.

Fluorosis See fluorine poisoning.

Flush cut To prune a branch of a tree so that the cut is approximately even and parallel with the branch or trunk from which it was removed.

Flush irrigation Irrigation in which water flows over the surface of the ground in a continuous sheet of shallow depth, spreading slowly and soaking in. Normally, one such flush has to suffice for the whole growing period of the crops.

Flushing Removal of deposited sediment by means of a rush of water.

Flushing rate The rate at which the water of an estuary is replaced (usually expressed as the time for one complete replacement).

Fluted (1) Refers to an unevenness of the rind of citrus fruits in which there are easily distinguishable longitudinal ridges. (2) Also refers to similar rollers on which ridges are there.

Fluted scale See cottony-cushion scale.

Fluvial Of or pertaining to rivers; growing or living in streams or ponds; produced by river action, as a fluvial plain.

Fluvial deposits All sediments, past and present deposited by flowing water, including glaciofluvial deposits. Wage-worked deposits and deposits resulting from sheet erosion and mass wasting are not included.

Fluvioglacial See glaciofluvial deposits.

Fly A two winged insect belonging to the family Muscidae. Many blood-sucking, some vectors of disease; some predators, some scavengers.

Flying foxes Bats. Mammalia; order Chiroptera. A distinct group of mammals with membranous wings for flight. Typical flying fox Pteropus gigenteus Brunnich. Damages orchards and fruit plantations.

Fly weevil Angoumois grain moth.

Fly wheel It is heavy wheel attached to a rotating shaft of a engine, which because of its inertia makes the rate of rotation uniform.

Flywheel marking Marks on the face of a flywheel which serve as a guide for the proper timing of valve action in an engine.

Fodder Coarse grasses such as corn and sorghum harvested with the seeds and leaves and cured for animal feeding. Also the dry cured stalks and leaves of plants. Forage; roughage. Several crops are grown as fodder. Among the important ones are *bajra,* cluster bean, cowpea, Italian millet, *jowar*, maize, *methra*, *moth*, oats, *ragi*, *senji*, *saftal*, soyabean velvet bean etc.

Fodder oats Avena ludoviçiana, A. Sativa, A. Strigosa. See fodder.

Foehn A warm, dry wind on the lee side of a mountain range, the warmth and dryness being due to adiabatic compression as the air descends the mountain slopes.

Foehn air The warm, dry air associated with foehn winds.

Foehn cloud Any cloud form associated with a foehn, but usually signifying only these clouds of the lenticularis species formed in the lee wave parallel to the mountain ridge.

Foehn cyclone A cyclone formed (or at least enhanced) as a result of the foehn process on the lee side of a mountain range.

Foehn islands An isolated area where foehn air has not replaced cooler surface air.

Foehn period The duration of continuous foehn conditions at a given location.

Foehn storm A type of destructive storm which frequently occurs in October in the Bavarian Alps.

Foeniculum foeniculum See fennel.

— **officinale** See fennel.

— **vulgare** See fennel.

Fog Microscopic water drops suspended in air; finely atomised spray or smoke.

Fog climax A community that deviates from a climatic climax because of the persistent occurrence of a controlling fog blanket.

Fog dispersal Artificial dissipation of a fog by means such as seeding or heating.

Fog drip The moisture dripping to ground because of collection of fog on tree leaves or an shrubs.

Fog drop An elementary particle of fog, physically the same as a cloud drop. Also known as fog droplet.

Fog forest The dense, rich forest growth which is found at high or medium-high altitude on tropical mountain; occurs when the tropical rain forest penetrates altitude of cloud formation, and the climate is excessively moist and not too cold to prevent plant growth.

Fog generator A machine which disperses spray material in very fine droplets usually varying from 1 to 50 microns in dia and remains for quite sometime in the air. Used for spraying chemicals in dense foliage to equally spread the chemical. Also called as smoke generator.

Fog scale A classification of fog intensity based on its effectiveness in decreasing horizontal visibility.

Foggage English term meaning forage for winter pasture.

Foliaceous Consisting of or having the form or texture of a foliage leaf.

Foliage A name given to collective leaves of a plant.

Foliage plant A plant grown for the colour or shape of its foliage.

Foliar Of, pertaining to, or consisting of leaves.

Foliar diagnosis An estimation of the extent to which plants are getting certain necessary chemical elements from the soil, based on an examination of the colour and the growth habits of the foliage of the plants.

Foliar nutrition Supply of plant nutrients through leaves. The absorption of nutrients occurs through stomata and leaf cuticle or stem surface. Quick acting and is usually adopted to remove deficiency of micro-nutrients.

Foliar spray A spray of the fertiliser nutrients over the plant when leaves are present.

Foliation (1) The process of developing into a leaf. (2) The state of being in leaf.

Folic acid A constituent of vitamin B_2 complex. Vitamin M or B_{10} or B_{11} are the other names. Required for growth and blood formation. Present in green leaves of spinach, yeast and liver.

Folisol A great group of soils in the organic order. The soils are not usually saturated for more than a few days a year and consist of 10 cm (4 inches) or more of L-H horizons derived from leaf lifter, twigs, branches, and mosses. A lithic contact or fragmented material occurs at a depth of less than 160 cm. Minerals layers less than 10 cm thick may lie above the lithic contact.

Folkways Uniformly accepted ways of thinking and acting together are called as folkways.

Follicle (1) A simple, dry, dehiscent fruit, producing several to numerous seeds and composed of one carpel, which splits along one side. (2) Very small sac-like structure such as depression from which a hair or feather grows.

Fomentation The external application of moist heat; to get relief from pain.

Fomes noxius See cinchona brown root disease, coffee brown root disease, tea brown root disease, jack fruit brown root disease, rubber brown spot.

Food A material that can be ingested and utilised by the organisms as a source of nutrition and energy.

Food chain (1) The step by step transfer of food energy and materials, by consumption, from the primary source in plants through to increasingly higher forms of fauna. (2) The scheme of feeding relationships by trophic levels which unites the member species of a biological community.

Food crop A crop grown for the consumption by human being.

Food preservation To treat the food materials so as to keep them fresh, wholesome and free from decomposition during storage. Several treatments include refrigeration, canning, dehydration etc..

Food technology The science of food quantities, values, preparation, storage and marketing.

Food units Total digestible nutrients.

Food web The network of feeding relationship in a biological community; a system of interlocking food chains in a biological community.

Foot In mosses, liverworts, ferns, and other plant groups, a portion of young sporophyte which attaches the sporophyte to the gametophyte and often absorbs food from the latter.

Foot brake Brake operated by foot action as distinguished from the hand or parking brake.

Foot-candle A unit of illuminance equal to one lumen per square foot. This is the illuminance provided by a light source of one standards candle at a distance of one foot, hence the name.

Foothills The hills or sloping land at the base of mountains or a mountain range.

Foot rot Any disease that involves rotting of the stem or trunk of a plant.

Foot sprayer A sprayer operated by foot instead by the hand.

Foot valve Check valve used on the bottom of the suction pipe to retain the water in the pump when it is not in operation.

Forage Herbaceous plants or plant parts fed to domestic animals (generally, the term refers to such material as pasturage, hay, silage, and green chop in contrast to less digestible plant material known as "roughage"); to graze.

Forage crop A crop grown for forage.

Forage value (1) The relative value of plants for forage purpose. (2) The comparative value of the forage portion of an animals ration compared to the other feeds in the ration.

Forage quality Characteristics that make forage valuable to animals as a source of nutrients; the combination of chemical, biochemical, physical, and organioleptic characteristics of forage that determines its potential to produce animal meat, milk, wool, or work. Considered by some as synonymous with feeding value and nutritive value.

Forage testing Laboratory evaluation of a given forage of feed in terms of intended use.

Foragers (1) Animals that feed on consumers. (2) Foraging or worker bee which goes out in search of food.

Forb A weed nongrass like broadleaf herbaceous plant on which animals feed.

Forced outlet Basin or box outlet for a tile like in which the discharge will fill the basin and flow away over the ground surface. Used where a free fall outlet is not available.

Forced saving Saving achieved through an enforced reduction in the consumption of consumers goods by the community.

Forceps A plier like instrument used for wide ranging used by veterinary doctors and engineers.

Forcing A method by which plants are forced to mature earlier than their normal time by providing artificial light or withholding irrigation etc.

Ford crossing A paved crossing provided on small irrigation canals where it may by forced by traffic. Crossing used by animals is called 'cattle crossing', and that meant for carts is 'cart crossing'.

Forebay Reservoir or pond at the head of a penstock, pipeline, or pump.

Forecast A statement of expected future meteorological occurrences.

Forecast verification Any process for determining the accuracy of a weather forecast by comparing the predicted weather with the observed weather of the forecast period; used to test forecasting skills and methods.

Foreign matter Any substance which is unnatural to or not commonly developed in a product.

Foreign seed Seeds of weeds and crops other than the kind being tested.

Foreshore The portion of bank or shore, lying adjacent to water and slopping gradually to it, which is falling dry between high water and low water.

Forest An ecosystem consisting of plants and animals and their environment, with trees as the dominant form of vegetation. Broadly forest may be divided into tropical wet green, tropical moist deciduous forest, tropical dry deciduous, tropical thorn, subtropical pine, moist temperate, alpine.

Forest conservation Those measures concerned with the protection and preservation of forest lands and resources.

Forest cover All trees and associated plants which cover the ground in a forest.

Forest ecology The science that deals with the relationship of forest trees to their environment, to one another, and to other plants and to animals in the forest.

Forest fire Uncontrolled combustion of forest fuels.

Forest floor All dead vegetables and organic matter, including litter and unincorporated humus, covering the mineral soil surface, under forest vegetation.

Forest fly See tick fly.

Forest influence Influence of forests on water, soil, climate and health.

Forest litter Leaves, dead branches, bark, rotting wood etc. on the forest floor.

Forest management Measures concerned with the effective organisation of forest to ensure continued production of its goods and services.

Forest mapping The branch of forestry dealing with the preparation of maps showing the distribution and conformation of individual forest stands.

Forest measurements The branch of forestry concerned with the measurement of standing trees, cut wood, and lumber products.

Forest product Any material afforded by a forest for commercial use, such as tree products and forage.

Forest resources Forest land and the trees on it.

Forestry The management of forest lands for wood, forages, water, wildlife, and recreation.

Forestry multiple use The practice of forestry which combines two or more objectives, such as production of wood or wood-derivative products, forage and browse for domestic livestock, proper environmental conditions of wildlife, landscape effects, protection against floods and erosion, recreation, production and protection of water supplies, and national defence.

Forest soils (1) Soils developed under forest vegetation. (2) Soils formed in temperate climates under forest vegetation. (3) The natural medium for growth of tree roots and associated forest vegetation.

Forest stand The basic unit of forest mapping; a group of trees that are more or less homogeneous with regard to species composition, density, size, and sometimes habitat.

Forest utilisation The operation of harvesting and marketing of forest crop and other products of the forest.

Forest wind A light breeze which blows from forest toward open country on calm clear nights.

Forestation Afforestation.

Forester A person trained in forestry.

Fortran Formula translation; any of several specific procedure-oriented programming languages for digital computers.

Fork (1) The juncture of two tree-branches. (2) A tool or implement with a handle, a shank and two or more tines for digging soil or lifting hay or similar other works.

Forkert budding A modification of shield budding method in which the flap of the patch on the stock is not removed but is used to cover the bud. The flap is cut off at a later stage so that the bud may not dry out soon.

Form genus A name applied to plant

parts of certain fossils until a more complete identification is made.

Formaldehyde A gas obtained from methyl alcohol and soluble in water. Aqueous solution is powerful disinfectant and antiseptic used as insecticides, fungicide and fumigant.

Formal education Formal education is basically an institutional activity, uniform and subject-oriented full time, sequential, hierarchically structured, leading to certificates, degree and diplomas.

Formal groups Includes elected officers and committees. They are organised for the promotion of special interests.

Formalin A 40 per cent solution of commercial formaldehyde used as preservative of animal or plant tissues.

Fortified The products which have been enriched with nutrients not originally present in the product.

Fortunella japonica (Thumb.) Swingle (Citrus Japonica Thumb) A shrub or small tree of family Rutaceae grown for its edible fruit and as an ornamental plant.

Fossil An impression, a trace, or remains of a plant or animal from a formers geologic age.

Fouled Pasture spots or areas made unacceptable to the grazing animal by presence of urine or dung.

Foundation (1) The plants or animals used in developmental breeding. (2) The portion of the buildings below the ground surface. Supports the buildings.

Foundation bolts Bolts used for holding down or anchoring machinery or structural parts to the foundation on which they rest.

Foundation plate A plate to which a pump, an engine, or a motor, is bolted. A sole plate, A plate of cast iron placed in the bottom of a mold to receive the spindle to maintain a sweep.

Foundation seed stock Seed that has descended from a selection of recorded origin, under the direct control of the original breeder, a delegated representative, or a state or federal experiment station.

Foundation spur Spur in a fruit plant bearing one bud which remains after pruning; called foundation because of its providing shoots for next crop. Also called renewal spur.

Four-stroke cycle A power concluded in four strokes of the engine. In the auto engine, the first downstroke is the intake; the next upstroke allows compression; ignition and expansion on the third stroke, and exhaust on the fourth.

Four-way hybrid A hybrid resulting from mating two single crosses. Also called double cross.

Four-wheel drive The type of a automobile which has live axles both front and rear, driving power being delivered to all four wheels.

Foxglove See digitalis.

Foxgrape Vitis labrusca L. Family Vitaceae. A woody, shrubby wine, grown for its berries used for making wine.

Fox tail millet See Italian millet.

Fractus A cloud species in which the cloud elements are irregular but generally small in size, and which presents a ragged, shredded appearance, these characteristics change ceaselessly and often rapidly.

Fragaria chiloensis See garden strawberry.

— **indica** See Indian strawberry.

— **nilgerrensis** See Houtbois strawberry.

— **vesca** See perpetual strawberry.

Fragment A broken natural aggregate of sand, silt or clay.

Fragmentation Reproduction by a segment of the main plant, a form of sexual reproduction.

Frame working Method of rejuvenating old fruit trees or of one variety for converting to another desirable one. Method of grafting employed.

Frass The excrement of larvae and/or refuse left by boring insects.

Freak An unusual type of plant sometimes come across in breeding.

Freckle A dark-coloured lesion of a plant.

Free acidity The titratable acidity in the aqueous phase of a soil. It may be expressed in milliequivalents per unit mass of soil or in other suitable units.

Free-air temperature Temperature of the atmosphere obtained by a thermometre located so as to avoid as completely as practicable the effects of extraneous heating.

Freeboard The difference between the maximum flow line and the top of the bank or structure. It is the sum of the operational freeboard and the emergency freeboard. Also, called as total free board.

Free flow A condition of flow through or over a structure where such flow is not affected by submergence or the existence of tail water.

Free flowing seed Non chaffy; smooth seeds which will travel easily through sampling or mixing devices.

Free goods In economics, those attributed of the natural world which are valued by society but are not in individual ownership and do not enter into the processes of market exchange and the price system.

Free hold A land held for life.

Free oxides Oxides and hydroxides of iron, aluminium, silicon, manganese and titanium, usually of fine particle size, that occur uncombined with other elements and often as coatings on primary and secondary minerals.

Free ports Port at which imports may be landed without paying duties.

Free trade Trade which is unimpeded by tariffs, import and export quotas and other devices which obstruct the free movement of goods and services between countries.

Free water Gravitational, excess or drainage water.

Free weir A weir which is not submerged.

Free wheeling An overrunning clutch placed in second and high gear assembly, permitting easy engagement and full casting effect irrespective of engine speed.

Freeze-drying The evaporation of volatile material from a frozen product by high vacuum.

Freezing drizzle Drizzle that falls in liquid form but freezes upon impact with ground to form a coating of glaze.

Freezing level The lowest altitude in the atmosphere over a given location, at which the air temperature is 0° C; the height of the 0° C constant-temperature surface.

Freezing nucleus Any particle that, when present within a mass of super-cooled water, will initiate growth of an ice crystal about itself.

Freezing point The temperature at which a liquid solidifies.

Freezing precipitation Any form of liquid precipitation that freezes upon impact with the ground or exposed objects; that is, freezing rain or freezing drizzle.

Freezing rain Rain that falls in liquid form but freezes upon impact to form a coating of glaze upon the ground and exposed objects.

French bean Phaseolus vulgaris L. Family Papilionaceae. A tropical annual herb widely grown for its edible pod used as vegetable. Also called kidney bean, runner bean, pole bean.

French bean anthracnose Disease of French bean caused by Collectotrichum lindemuthianum (Sacc & Magn. Brir Cav). Characterised by brown or black necrotic spots on all green parts of plants, depressed in centre with raised margin.

French crab Crab species of apple.

French drain A small shallow trench filled with coarse rubble, clinker or similar material, with or without field drain pipes. Also called as rubble drain.

French tamarisk See jhan.

French wheat See buckwheat.

Frenching A non-parasitical disease of crops, characterised by a thickening, narrowing, and curling of leaves. Considered due to some nutritional deficiency.

Frequency A statistical expression of the presence or absence of individuals of a species in a series of subsamples, that is, the ratio between the number of sample area that contains a species and total number of sample area.

Frequency, or **Absolute frequency** The number of occurrences of a given type of event, or the number of members of a population falling into a specific class or between definite limits. Also called as absolute frequency.

Frequency curve The graphical representation of the frequency of occurrence of specific events, or a continuous frequency distribution, the variate being the abscissa and the frequency the ordinate.

Frequency distribution An arrangement or distribution of quantities pertaining to a single element in the order of their magnitude.

Frequency of irrigation The elapsed time in days from the start of one irrigation to the start of the next on the same field. Also called as irrigation interval.

Frequency table A table drawn up to show the distribution of the frequency of occurrence of a given characteristic according to some specific set of class intervals.

Fresh Refers to recently harvested or plucked food products.

Fresh ungerminated seed Seeds, other than hard seeds, which remain firm and apparently viable even after the appropriate treatment for breaking dormancy.

Fresh-water ecosystem The living organisms and nonliving materials of an inland aquatic environment.

Friable A consistence term pertaining to the ease of crumbling of soils i.e. ability to break easily into pieces, a desirable conditions in a soil. See also consistence.

Fringe water See capillary fringe.

Fritillary The common name for plants of the genus Fritillaria. The common name for butterflies in several genera of the subfamily Nymphalinae.

Frog It is a irregular piece of metal and is that part of the plough bottom to which the share, the mouldboard and the landside are attached rigidly.

Frogeye Any of various leaf diseases characterised by formation of concentric rings around the diseased spots.

Frog storm The first bad weather in spring after a warm period. Also known as whippoorwill storm.

Frond A very large leaf, especially a fern or fern like leaf.

Front A sloping surface of discontinuity in the troposphere, separating air masses of different density or temperature.

Front wheel drive A construction in which the live propelling axle is at the front end. The rear axle is dead.

Frontal (1) In the region of the forehead. (2) See front.

Frontal dune A shifting dune lying directly behind the berm, usually parallel to the shoreline formed by natural processes.

Frontal fog Fog associated with frontal zones and frontal passages.

Frontal lifting The forced ascent of the warmer, less-dense air and near a front, occurring whenever the relative velocities of the two air masses are such that they converge at the front.

Frontal precipitation Any precipitation attributable to the action of a front; used mainly to distinguish this type from air-mass precipitation and arographic precipitation.

Frontal thunderstorm A thunderstorm associated with a front; limited to thunderstorms resulting from the convection induced by frontal lifting.

Frost A consequence of low temperature of only a thin layer near the ground is chilled beyond freezing point it is called radiation frost. If the heavy cold air fills into the natural depression it is called pool or hill frost.

Frost action Freezing and thawing of moisture in materials and the resultant effects on these materials and on the structure of which they are a part.

Frost day An observational day on which frost occurs.

Frost heave The raising of a surface caused by ice in the underlying soil.

Frost lifting Lifting of seedlings above their normal soil as a result of the expansion of soil mass following the freezing of the soil water.

Frost point The temperature to which atmospheric moisture must be cooled to reach the point of saturation with respect to ice.

Frost smoke A rare type of fog formed in the same manner as a stem fog, but at colder temperatures so that it is composed of ice particles instead of water droplets.

Frosty mildew A leaf spot caused by fungi of the genus Cercasporella and characterised by pale to white lesions.

Fructescence The period of fruits maturation.

Fructification (1) Any fungal structure which contains or produces spores or the organ producing the spores or seeds. (2) The process of producing fruit. (3) A fruit and its appendages.

Fructosan Polysaccharide yielding primarily fructose on hydrolysis; the primary form of carbohydrate storage in certain forage crops.

Fructose Fruit sugar; $C_6H_{12}O_6$. Simple from of carbohydrate.

Fruit A fully matured plant ovary with or without other floral or shoot parts united with it at maturity, such as the pea pod, nut, tomato.

Fruit-body The spore-bearing structure, such as sporangiocarp, basidiocarp, conidiocarp.

Fruit bud A fertilised flower bud that matures into a fruit.

Fruit drop Any substantial dropping of immature fruit or of fruit as it approaches maturity.

Fruiting Body Fungus structure containing or bearing spores: mushroom, pycnidium, perithecium, apothecium, etc.

Fruiting spur Spur bearing three or more buds meant for bearing fruit in a fruit plant.

Fruit pit The endocarp and seeds of fruits of the genus Prunus, such as of prunes and peaches.

Fruit spot Any distinctive, small fruit caused by infection or physiological disturbances.

Fruit spur A short and stout branch of a fruit tree on which flowers and fruit set.

Fruit vegetables Groups of vegetables including tomatoes, brinjal, chilli etc.

Frustule The siliceous shell of a diatom, consisting of two valves, one overlapping the other.

Frutescent Becoming shrub-like in appearance or habit.

Fruticose Having the form of a shrub; branching freely, as certain lichens.

Fucales An order of brown algae composing the class Cyclosporeae.

Fucoid Resembling fucus in appearance; having a flat, dischotomously branched thallus.

Fucoxanthin The brown pigment of certain algae.

Fucus A genus of dischotomously branched brown algae; it is harvested in the help of industry as a source of algin.

Fuel oil A petroleum derivative most often used a fuel, with a higher flash point than kerosene.

Fugacious Quickly perishing; used principally to describe plant parts that fall soon after being formed.

Fulcrum A supporting organ such as a tendril or stipule.

Full pink Refers to the stage of blossom development in apple just before most of the buds expand to full bloom.

Fuller's earth A non-plastic clay consisting essentially of the mineral montmorillonite and similar in the respect to bentonite.

Fuller's teasel Dipsacus fullonum L. (D. sylvestris, Huds.). A herb of family Dipsacaceae.

Fulvic acid The alkali-soluble fraction of humus not precipitated by acid.

Fumigant A volatile disinfectant, destroying organisms by vapour. Carbondisulphide ethylene dibromide. etc.

Fumigation Exposure to the fumes of a gas that kills insect pests which damage plants or stored grains.

Functional disease A disorder which does not show any change in the tissues; when persisting may cause organic changes which may be recognised.

Functionaries Middleman between the producer and the consumer.

Fundus Boundary between underground and above ground portions of plant axis.

Fungi The allophytic plants that lack chlorophyll and are filamentous in structure; molds.

Fungi imperfecti Fungi which have not been connected with the perfect or sexual stage; most are imperfect states of Ascomycetes.

Fungicide Chemical or physical agent that kills or inhibits fungi.

Fungiform Shaped like fungus.

Fungistat An agent or chemical which prevents the growth and reproduction of fungi, but does not kill.

Fungistatic An agent preventing development of fungi without killing them.

Fungus A plant without chlorophyll which derives its nourishment from the organic matter of other plants.

Fungus gall A plant gall resulting from

an attack of a parasitic fungus.

Fungus spores The reproductive cells of fungus.

Funicle An ovule stalk; a small cord or band.

Funiculus The stalk which attaches an ovule to the inner portions of an ovary.

Funnel cloud The popular term for the tornado cloud, often shaped like a funnel with the small end nearest the ground.

Furred Having short decumbent hairs thickly covering the surface.

Furrow bank Furrow slice.

Furrow face The vertical side of the furrow made in the soil, away from which the turned soil is thrown.

Furrow irrigation A method of surface irrigation in which water is run in furrow between crop rows.

Furrow openers These are provided in a seed drill for opening the furrow before dropping the seeds.

Furrow slice The mass of the soil cut, lifted and thrown to one side is called 'Furrow slice'.

Furrow sole The bottom of the furrow on which the plough bottoms slides.

Furrow wall It is an undisturbed soil surface by the slide of a furrow.

Furrows Narrow ditches made by cultivation implements between rows of crops.

Fusarium A genus of fungi; most species saprophytes but many plant parasites.

Fusarium annum See chilli wilt.

— **coeruleum** See potato dry rot.

— **cucurbitae** See cucurbit fusarium root rot.

— **lini** See linseed wilt.

— **monoliniforme** See rice foot rot.

— **orthoceras var. ciceri** See gram wilt.

— **orthoceras var. entidise** See lentil wilt.

— **orthoceras var. lethyri** See lethyrus wilt.

— **oxysporum** See cotton wilt, tomato fusarium wilt.

— **oxysporum var. cubense** See banana panama wilt.

— **solani** See cucurbit fusarium root rot.

— **udum** See pigeon pea wilt.

— **udum vari crotalariae** See sanhemp wilt.

— **vasinfectum** See cotton wilt, sesamum wilt.

Fusiform Spindle-like narrowing towards the ends.

Fusoid Somewhat fusiform.

Futures (markets) Refers to a contract to deliver or receive commodities at some specific future month.

Fuzz Cotton fibre, thicker than lint and has no twists. It does not have much commercial use because it does not get separated from cotton seed at the time of ginning.

G

Gabion Rectangular or cylinderical wire mesh cage filled rock and used for protecting aprons, revetment etc. against erosion.

Gable Triangular upper part of wall at the end of ridged roof.

Gage or **Gauge** Device for registering precipitation, water level, discharge velocity, pressure, temperature, etc.

Gage Height (Hydraulics) The height of the water surface above some arbitrary datum, such as the bottom of the channel.

Gaging Station A selected section of a stream channel equipped with a gage, recorder, or other facilities for determining stream discharge.

Galactose Carbohydrate; isomer of dextrose and occurs in nature in combination as di- or poly-saccharide.

Gale A wind which has the velocity of 40 miles or more per hour at the height of 32 ft above the ground.

Galea A petal or other similar structure like helmet in shape, function or position.

Galerucella birmanica See water chestnut beetle.

Galileo See linear acceleration.

Gall Outgrowth or swelling, often more or less spherical, or unorganised plant cells as result of attack by bacteria, fungi, or other organisms.

Gall Abnormal tissue enlargement due chiefly to infectious micro-organisms or insects.

Galleria forest A modified tropical deciduous forest occurring along stream banks.

Galleria mellonella See greater wax-moth.

Gallery (1) Constructed chamber for collection of groundwater. (2) A relatively small, usually circular passage way made in the bark, or in the wood by an insect for feeding or for oviposition.

Gallnut A gall resembling a nut.

Gallon Volume occupied by a quantity of distilled water which weighs 10 lb in air against brass weights, with the water and air at 62° F and with the barometre at 30 inches, (76.2 cm). It is equal to 4.54596 litres.

Gallons per minute A unit used to express well discharge or pump discharge in irrigation and drainage.

Galvanotaxis Movement of a free-living organism in response to an electrical stimulus.

Galvanotropism Response of an organisms to electrical stimulation.

Galwang Albizia lucida Benth. Family Mimosaceae. A medium-sized tree nearly ever green with full crown of handsome dark green foliage. Lachost; raised from seed.

Gama grass See Gautemala grass.

Gamboge tree Garcinia cambogia Desr. Family Guttiferae. Medium sized tree producing edible fruit. Yellow gum resin from bark used for varnish; barks, leaves and fruits used as laxatives.

Gambrel roof A roof with two sloping sides stepped at different angles on each side of the centre ridge; the lower slope is steeper than the upper slope.

Gametangium (1) Gamete mother cell. (2) A structure which contains or bears gametes.

Gamete A reproductive cell, either male or female.

Gametogenesis The formation of gametes, or reproductive cells such as ova or sperm.

Gametophyte (1) The haploid generation producing gametes in plants exhibiting metagenesis. (2) An individual plant of this generation.

Gamma distribution A normal distribution whose frequency function involves a gamma function.

Gamma rays Rays sent out by certain radioactive substances at the same time as electrons; similar to X-rays but of shorter wavelengths.

Gamma-ray well logging Measurement of gamma ray intensity versus depth down the well bore; used to identify rock strata, their position, and their thicknesses.

Gammexane Trade name for the gamma arrangement of atoms in benzene hexachloride; used as insecticide.

Gamophyllous Having the leaves of the perianth united.

Gamosepalous Having sepals united at their edges. Also known as synsepalous.

Ganglion A mass of or enlargement or knot on nerve cell bodies from which nerve fibre radiate; a nerve centre. Mass of grey matter in central nervous system forming a nerve-nucleus.

Gang A combination of implements used simultaneously in cultivation of soil. Two gangs are generally used in field harrows.

Gangplow A plow with two or more cutters that turn parallel furrows.

Gangrene Death of tissues of the body on a massive scale due to the bacterial infection or by interference by injury, infection, freezing etc. With blood circulation or insufficient local nutrition.

Garner (1) A building or section of a building in which grain is stored. (2) A bin for weighing grain.

Ganoderma lucidum See coconut soot rot, arecanut foot-rot.

Garbin A sea breeze; in southwest France it refers to a south westerly sea breeze which sets in about 9 A.M., reaches its maximum towards 2 P.M., and ceases about 5 P.M.

Garcinia cambogia See gamboge tree.

— **cowa** See cowa.

— **indica** See Kokam butter tree.

— **mangostana** See mangosteen.

— **morella** See ceylone gamboge.

— **tinctoria** See egg tree.

— **xanthochymus** See egg tree.

Garden (1) A plot of land devoted to the growing of vegetables, flowers, fruits, herbs, ornamental plants etc. (2) Plants in a garden.

Garden asparagus See asparagus.

Garden bean Dolichos lablab var. typicus. A popular bean used as vegetable.

Garden beet See beet root.

Garden bond Consists of three stretchers in each course followed by a header, although this bond may have from two to five stretchers between headers.

Garden celery See celery.

Garden crop Any crop, flower or fruit, raised in a garden, in contrast to that raised in a field.

Garden cress Lepidium sativum L. A common cultivated herb of the Cruciferae family; young and ripened seeds used as vegetable and pulse.

Garden heliatrope See valerian.

Garden pea See pea.

Garden snail Any snail of the several species of genus Helix, which is destructive to garden crops.

Garden sorrel Rumex acetosa L. A herb of family Polygonaceae. The leaves are used as vegetable.

Garden strawberry Fragaria chiloensis (L.) Duchesne. Family Rosaceae. A herb cultivated for its fruits.

Gardenia campanulata Roxb. A large shrub or small tree of the family Rubiaceae, the leaves and fruits of which are edible.

— **resinifera** See dikamali.

Gardenland Land cropped with the help of water raised by lift irrigation to supplement rainfall to the extent necessary but not liberally as for wet land.

Gardenland crop A crop raised in a gardenland.

Garigul A low, open scrubland restricted to limestone sites in the Mediterranean area; characterised by small evergreen shrubs and low trees.

Garlic Allium sativum. A perennial plant of the order liliales and family Liliaceae grown for its pungent, edible bulbs. Important spices and condiment crop cultivated throughout India. Medicinal use in whooping cough and also for sore cyes. Garlic oil contains sulphur compounds. Bulb comprises a number of segments called cloves ranging from 6 to 50. Originated in central Asia and the Mediterranean region. Prefers modernate temperatures.

Garlic blight Disease of garlic caused by Alternaria palandui. Symptoms as in onion blight.

Garlic thrip See onion thrip.

Garuga pinnata Roxb. A common tree of the family Burseraceae which produces edible fruits. Leaf galls used for tanning and wood for making furniture.

Gas engine An internal combustion engine designed to operate on a mixture of gas and air as a fuel.

Gascoyne's Scarlet Blood red.

Gas pressure potential This potential component is to be considered only when external gas pressure differs from atmospheric pressure as in a pressure membrane apparatus. A specific term and definition are not available.

Gas vacuole A membrane-bound, gas-filled cavity in some algae and protozoans; thought to control buoyancy.

Gasket Paper, metal, rubber, or especially prepared material used between cylinder head and cylinder, or under similar conditions, to prevent leaking.

Gasoline engine An internal combustion engine designed to operate on a mixture of gasoline and air as a fuel.

Gasteropod A class of usually spiral-shelled molluscs with a flat creeping foot. Also gastropod.

Gastritis Acute chronic inflammation of the stomach lining.

Gastropod See gasteropod.

Gated pipe Portable pipe with small gates installed along one side for distributing water to corrugations or furrows.

Gate valve A sliding gate or slide which controls flow though the gate. One advantage is that it can afford full-bore opening.

Gather To collect; as to harvest a crop.

Gathetometre An instrument for measurement of small differences of heights. It consists of a small telescope, levelling apparatus which slides up or down a perpendicular standard fitted with a vernier scale.

Gauge (1) Device for registering water level, discharge velocity, pressure etc. (2) A unit of measuring thickness of wires and sheets.

Gauge height Elevation of water surface measured by a gauge.

Gauge wheel It is usually placed in hanging position and helps to maintain uniformity in respect of depth of ploughing in different soil conditions.

Gauging To determine, by measurement or other test, the size of an object, the capacity of a vessel, or the amount of its contents. To cut bricks or stones to make them uniform in size. To mix plaster of Paris with mortar to bring about quick setting.

Gauging station Selected section in the stream channel equipped with a gauge or facilities for obtaining stream flow data.

Gautemala gama grass See gautemala grass.

Gautemala grass Tripascum laxum Nash. Family Gramineae. A hardy, quick-growing grass with tussocky growth. Used as fodder and green manuring. Leaves used for mulching and thatching. Ratoons well. Propagated by stem cuttings. Also called Gautemala gama grass.

Gauze A thin, open-meshed muslin silk, linen or similar fabric used as a bandage after first making it sterile.

Geared pump A pump which is driven by an engine through the use of gearing.

Gear case A housing or metal box within which gears operate. Usually the gear case is filled with lubricant.

Gear metre A type of positive-displacement fluid quantity metre in which the rotating elements are two meshing gear wheels.

Gear ratio The relation between number of teeth on driving and driven gears.

Gear shift lever A lever by means of which the change-speed gears are shifted.

Gear shifting Throwing into mesh different sets of the change speed gears in order to vary the speed ratio and power between engine and rear axle.

Gecarcinucus jacquemontic A serious pest of paddy; crab.

Gel A colloidal solution in which the particles have precipitated.

Gelatin liquefaction Reduction of a gelatin culture medium to the liquid state by enzymes produced by bacteria in stab culture; used in identifying bacteria.

Gelatine A jelly-like substance obtained

from animal tissues. Used as food, as a medium for bacteria, as a stabiliser for ice-cream etc.

Gemma An asexual or vegetative outgrowth of a parent body, capable of growing into a new individual directly as in liverworts.

Gemmate Having buds, reproducing by gemmation.

Gemmation Act or manner of budding, arrangement of buds, reproduction by gemmae, formation of new individual by protrusion and separation of parts of the parents.

Gemmiferous Bearing buds.

Gemmiform Resembling a gemma or bud.

Gemmiparous Producing a bud or reproducing by a bud.

Gemmule The growing point of a plant embryo; reproductive cell of cryptogam; small gemma.

Gene (1) Specific region of a chromosome that is capable of determining development of a specific trait, composed wholly or in part of deoxyribonucleic acid (DNA). (2) The hypothetical unit of inheritance located in the chromosome, which by interaction with the other genes and the environment controls the development of a character. Genes are believed to be arranged linearly in the chromosomes. Chromogene, in contrast to cytogene.

Genealogy Pedigree. Record of descent from some one ancestry.

Genera Plural of genus.

General anaesthetic A substance that causes unconsciousness when inhaled; chloroform as an example.

General combining ability The average performance of a particular inbred in a series of hybrid combinations.

General purpose A tool which could be used for a variety of purposes.

General purpose tractor A multipurpose tractor used for almost all the farm operations.

Generation (1) Production of electricity etc. (2) Procreation, propagation of species, production by natural or artificial means; single step in descent or pedigree.

Generative Able to produce.

Generative organs See genital organs.

Generator A general term applied to machines which are used for the transformation of mechanical energy into electrical energy.

Generic Common to all species of a genus.

Generitype See genotype.

Genesis Origin, or evolutionary development as of a soil, plant or animal. Also kind of origin such as parthenogenesis etc.

Genetic Resulting from or produced by soil-forming processes, for example, a genetic soil profile or a genetic horizon.

Genetic complex A collective term for all the hereditary factors borne in an organism.

Genetic pan A natural subsurface soil layer of low or very low permeability having a high concentration of small particles and differing in certain physical and chemical properties from the soil immediately above or below the pan. See also claypan.

Genetic purity Trueness to type; varietal purity; plants/seeds conforming to characteristics of the variety as described by the breeder.

Genetic selection Selecting breeding stock primarily on the basis of qualities that are hidden from view, in control to physical selection.

Genetic shift Change in the genetic make up of varieties, if grown over a long period in areas outside their adaptation.

Genetics (1) Science which seeks to account for the resembalances and differences which are exhibited by organisms related by descent. (2) The study of the ancestry of some special organism or variety of plant or animal.

Genic sterility A type of male sterility conditioned by nuclear genes in contrast to cytoplasmic sterility. It may be transmitted by either the male or female parent.

Geniculate Having knee-like joints.

Genital organs External reproductive organs. Also called generative organs, genitals, genitalia.

Genitals See genital organs.

Genoa cyclone A cyclone, or low, which appears to have formed or developed in the vicinity of the Gulf of Genoa. Also known as Genoa low.

Genome A complete set of chromosomes (hence of genes), inherited as a unit from one parent.

Genotype (1) Group of organisms with the same genetic makeup. (2) The fundamental hereditary constitution of an organism.

Gensis Soils (1) The mode of origin of the soil, especially the processes or soil-forming factors responsible for the development of the solum, the true soil, from unconsolidated parent material. (2) The division of soil science dealing with soil genesis.

Gentianaceae A family of dicotyledonous herbaceous plants in the order Gentianales distinguished by lacking stipules and having parietal placentation.

Gentianales A family of dicotyledonous plants in the subclass Asteridae having well-developed internal phloem and opposite, simple, mostly entire leaves.

Genus A taxonomic category which includes one or more species. The genus name is the first name in the binomial. Similar genera are grouped in a family.

Geobiont An organism that spends its life in the soil.

Geobotany The study of plants as related to their geologic environment.

Geochemical and geothermal water surveys Surveys to determine the nature of and variations in the chemical character of ground water of the area and also the temperature of the ground water.

Geochemistry Science of the chemical composition of the earth crust.

Geographic environment The natural phenomena such as season, rainfall, temperature range, floods, droughts, soil, topography etc. of an area which may affect the origin, development and life of species.

Geography, Soil Geography dealing with the areal distribution of soil types.

Geologic map A representation of the geologic surface or subsurface features by means of signs and symbols and with an indicated means of orientation; includes nature and distribution of rock units, and the occurrence of structural features, mineral deposits, and fossil localities.

Geological Erosion See erosion.

Geology The study of science of the earth, its history and its life as recorded in the rocks; includes the study of geologic features of an area, such as the geometry of rock formations, weathering and erosion, and sedimentation.

Geometric similarity Similarity of form which exists between two objects or systems if the ratios of all corresponding linear dimensions are equal. This relationship is independent of motion of any kind.

Geophagus Pertaining to an organism feeding on soil.

Geophysical survey The exploration of an area in which geophysical properties and relationship unique to the area are mapped by one or more geophysical methods. In groundwater exploration relating to particular localities, this survey would give specific information as to the depth and extent of groundwater bodies as determined by recognisable lithological and stratigraphic changes.

Geophysics Science of the physical properties of the earth, its structure, composition and development.

Geophyte A plant with dormant parts such as tubers, bulbs, rhizomes underground; a land plant.

Geoponics The science of agriculture; the tillage or the ground.

Geosiridaceae A monotypic family of monocotyledonous plants in the order Orchdales characterised by regular flowers with three stamens that are opposite the sepals.

Geostrophic departure A vector representing the difference between the real wind and the geostrophic wind. Also known as ageostrophic wind; geostrophic deviation.

Geostrophic flux The transport of an atmospheric property by means of the geostrophic wind.

Geostrophic wind That horizontal wind velocity for which the coriolis acceleration exactly balances the horizontal pressure force.

Geostrophic-wind level The lowest level at which the wind becomes geostrophic in the theory of the Ekman spiral. Also known as gradient-wind level.

Geothermal A term relating to the heat of the earth.

Geotropism Response of a plant to the force of gravity. Positive by turning downwards as in the growth of a root; negative, when the direction is away from the source of stimulus as in the case of a normal stem.

Geographic-wind scale A graphical device used for the determination of the speed of the geostrophic wind from the isobar or contour-line spacing on a synoptic chart, it is a nomogram representing solutions of the geographic-wind equation.

Geraniaceal A family of dicotyledonous plants in the order Geraniales in which the fruit is beaked, styles are usually united, and the leaves have stipules.

Geraniales An order of dicotyledonous plants in the subclass Rosidae comprising herbs or soft shrubs with a superior ovary and with compound or deeply cleft leaves.

Geranium A genus of plants in the family Geraniaceal characterised by regular flowers, and glands alternating with the petals.

Geranium oil Oil distilled from several species of Pelargonium; used for adulterating or substituting otto of roses in the manufacture of perfumes and soaps.

Gerber test A volumetric measurement of the quantity of fat in a sample of milk; the European counterpart of the Babcock test.

Gerbillus indicus Common field rat.

Germ A primary source, especially one from which growth and development are expected. General designation for a unicellular microorganism.

German Reproductive glands.

Germ cell An egg or sperm or one of their antecedent cells. Multiplies to form a new individual, or shares in the reproduction of a new organism as contrasted with somatic or body cells.

Germicide An agent capable of killing germs, usually pathogenic microorganisms.

Germinal Of germs, of the nature of germs, in the earliest stage of development.

Germinal disc See germ spot.

Germinal spot The nucleus of an unfertilised egg.

Germination (1) The beginning or the process of development of a spore or seed. (2) Growth of the embryo in a seed; sprouting. (3) The resumption of growth by the embryo and development of a young plant from the seed. Germination, in a laboratory test, is the emergence and development from the seed embryo of those essential structures which, for the kind of seed being tested, indicate the ability to develop into a normal plant under favourable conditions in the soil.

Germination cabinet A type of seed germinator most commonly used in a seed laboratory.

Germination percentage The percentage of seeds which germinate under the given set of conditions.

Germination test The test conducted to find out the germination percentage in a given lot.

Germinative Having the ability to grow and develop.

Germinative energy The relative strength of sprouts and roots at the time of germination.

Germ nucleus An egg or sperm nucleus.

Germ plasm The material or physical basis of heredity.

Germ spot The germinal disc of an egg, seen as small, circular area on the surface of the yolk, which is the area of blasto-derm of a fertilised egg and point of which embryonic development starts.

Germ tube The tube formed by germination of a pollen grain through which the sperms are transferred. As Hypha produced by a germinated fungus spore.

Germling Early development of a plant from germination of the zygote in certain Thallophytes.

Gesneriaceae A family of dicotyledonous plants in the order Scrophulariales characterised by parietal placentation, mostly opposite or whorled leaves, and a well-developed embryo.

Gestalt A complex of weather elements occurring in a familiar form, and though not necessarily referring to basic hydrodynamical or thermodynamical quantities, may persist for an appreciable length of time and is often considered to be an entity in itself.

Gharbi A fresh westerly wind of oceanic origin in Morocco.

Gharra Hard squalls from the northeast in Libya and Africa that are sudden and frequent, and are accompanied by heavy rain and thunder.

Giant bee See rock bee.

Giant fescue Festuca gigantea (L.) Vill. Family Gramineae. Grass used as fodder.

Giant granadilla Passiflora quadrangularis L. Family Passifloraceae. A woody vine, native to tropical America, grown for its edible fruits.

Giant leaf hill mustard Brassica juncea (L.) Czern. & Cross. var. cunneifolia Roxb. (B. rugosa Prain. var. cunneifolia Prain.). Family Cruciferae. A leafy vegetable whose young leaves and tender shoots used as vegetable.

Giant mallee Eucalyptus obosa. Family Myrtaceae. Eucalyptus species introduced from Australia and is drought resistant. Rough bark at the base and ribbony smooth at the upper trunk and branches.

Giant rainers, Rainguns, or Extra large rainers A sprinkler having at least one nozzle greater than 1.25 mm. Sometimes, fitted with sectoring device.

Giant star grass Cynodon Plectostachyus Pilger. Family Gramineae. Extensive perennial grass with runners covering long distances. Grown as fodder and is adapted to wide variety of soils. Propagated by cuttings, root-stock or slips. Used for pasture hay. First cutting ready in $3^1/_2$ months. Also called East African star grass.

Giant taro (1) Alocasia Cucullata Schott. Family Araceae. A herb grown for its edible rhizomes. (2) Alocasia macrorhiza Schott. A herb whose corns are used for edible purposes. (3) Alococia indica (Roxb.) Schott. Family Araceae. Large-leaved caladium and a herb whose stem and root-stocks are edible.

Giantism See gigantism.

Gibberella fujikuroi See rice foot rot.

Gibberellic Acid, G.A. ($C_{19}H_{22}O_6$) these are the plant growth stimulating chemicals. These induce many different plant effects such as rapid stem growth, overcoming of dormancy, production of seedless fruits, and other responses. Tetra-cyclicdihydroxy-lactonic acid.

Gibberellins Group of substances isolated from certain organisms which act as growth stimulators.

Gibbsite Crystalline aluminium hydroxide, $Al(OH)_3$.

Giemsa stain Stain (azure and eosin dyes) used to demonstrate protozoa, viral inclusion bodies, and ricketisiae.

Gigantism The production of luxuriant vegetative growth, usually accompanied by a delay in flowering or fruiting; the appearance of giant forms of plants. Also called giantism. (2) An abnormal overgrowth of a part or the whole body of an animal.

Gilgai The microrelief of soils produced by expansion and contraction caused by changes in moisture. Gilgai is found in soils that contain large amounts of clay, which swells and shrinks noticeably with wetting and drying. It usually occurs as a succession of microbasins and microknolls in nearly level areas or as microvalleys and microridges parallel to the direction of the slope. See also microrelief.

Gill Lamella or hymenium-covered plate on underside of cap of a mushroom.

Gin A machine for separating cotton lint from cotton seed.

Ginger (1) Zingiber officinale Rosc. Herb cultivated for its aromatic rhizomes used as spice and in medicines. Grown from almost sea level to 1500 m or so above sea level. Thrives well under partial shade. Deep, well-drained friable loam, rich in humus ideal soil. Propagation by pieces of rhizome each with atleast one sound bud. Dry ginger a fifth of green by weight. (2) The dried or peeled rhizomes of ginger. The flavour is due to volatile oil, the pungency to gingerrol.

Ginger curing The process of preparing ginger for market. Two processes adopted in India. (1) Rhizomes after harvesting, sub-dried, cleaned of earth and soaked in water to peel of skin which is done by bamboo sticks or sea shells bits. Dried again in sun for 3-4 days and hand rubbed. Again soaked in water and rubbed with coarse cloth to remove remaining skin. (2) Peeled ginger soaked in thick lime water for sometime and fumigated with sulphur fumes for 12 hr and dried in sun for a day. Process repeated two or three times to obtain fully bleached white produce. Thoroughly dried and stored.

Ginger grass (1) Cymbopogon martinii (Roxb.) Wats. (Andropogon martinii Roxb.) Family Gramineae. A grass whose leaves yield an aromatic oil used in perfumery. Two varieties *motia* and *sofia*, oil from the former referred to a palmarosa oil, and from the latter which is inferior as ginger grass oil. Also called palman rosa. (2) Name also applied to Cymbopogon Caesius (Nees) Stapf. (Andropogon Caesius Nees). Family Gramineae. A perennial grass whose leaves yield an essential oil which is used in soap making.

Ginger grass oil See ginger grass.

Ginger rhizome rot See ginger soft rot.

Ginger soft rot Disease of ginger, caused by Pythium spp. Characterised by pale leaves turning yellow, withering and drying up, rhizomes becoming soft, watery and decayed in field and soft and crumbling on pressure in storage. Also called ginger rhizome rot.

Ginger tincture Tincture prepared by extracting parts of coarsely pounded ginger rhizomes with 90 per cent alcohol.

Ginning The separation of the lint from the seed of cotton.

Ginning outturn The percentage of lint cotton to seed cotton in the same batch or same variety.

Gir Breed of cattle, large and powerful bullocks but are slow and lethargic. Female best milkers.

Girdle A canker which surrounds stem, completely cutting off water supply, and thus causing death; girding roots also cause death.

Girdling To encircle the stem of a living tree with cuts that completely severe bark and cambium, and often are carried well into the cutter sapwood, for the purpose of killing the tree by preventing the passage of nutrients or by admitting toxic materials. Also see ringing.

Gironniera cuspidata (Bl.) Kurz. (G. reticulata Thw.) A large tree of the family Ulmaceae producing edible fruits.

— **reticulata** See G. cuspidata.

Girth The circumference of the trunk of a tree.

Gisekia pharnaceoides L. A herb of family Phytolocaceae, the leaves of which are edible.

Glabrous Having smooth surface, specifically, having the epidermis devoid of hair or down.

Glacial anticyclone A type of semi-permanent anticyclone which overlies the ice caps of Greenland and Antarctica. Also known as glacial high.

Glacial Drift All rock material carried by glacier ice and glacial meltwater, or rafted by icebergs. This term includes till, stratified drift, and scattered rock fragments.

Glacial soil Soil formed by the grinding action of ice and snow brought down by glaciers in the upper regions of Himalayas, found in Dehradun region and parts of Kashmir in India.

Glacial till See till.

Glaciation The geological action of glaciers.

Glacier wind A shallow gravity wind along the icy surface of a glacier, caused by the temperature difference between the air in contact with the glacier and free air at the same altitude.

Glaciofluvial deposits Material moved by glaciers and subsequently sorted and deposited by streams flowing from the melting ice. The deposits are stratified and may occur in the form of outwash plains, deltas, kames, eskers, and kame terraces. See also glacial drift and till.

Glacis A very gentle slope from a natural leavee or higher land along the bank of a large river to the back land of the river plain.

Glade A open grassy tract in a forest.

Gladiolus A genus of chiefly African plants in the family Iridaceae having erect, sword shaped leaves and spikes of brightly coloured irregular flowers.

Gland Organs that secretes substances which are used in the body or are eliminated from it; some stimulated to activity by herbs, others by hormones.

Gland The small bearing which closes the mouth of a stuffing box and takes the wear of the piston glare (paper). The reflection from a glossy paper.

Glandiferous Bearing acorns.

Glandiform Acorn shaped, like a gland.

Glandular Bearing glands.

Glandular responses (emotion) In a person experiencing a strong emotion, such as anger, the medulla of the adrenal gland pours and excessive amount of its secretion, adrenalin, into the blood. Travelling through the blood-stream to the various parts of the body, this secretion is responsible for many characteristics of strong emotional experience.

Glandular tissue Tissue of single or massed cells, parenchy matous and filled with, granular protoplasm adapted for secretion of aromatic substances in plants.

Glareous Growing in gravelly soil; refers specifically to plants.

Glass beads Beads — small rounded pieces of glass.

Glass-eye An animals eye in which the iris is pearly white.

Glauconite A dull green mineral found in minute quantities in soil; hydrated aluminium sillicate of iron and potassium.

Glaucoscent somewhat glaucous.

Glaucous Having a white or greyish powdery coating that gives a frosty appearance and rubs off easily.

Glaucous appearance Covered with a dull green grey waxy bloom which may be due to the difficulty of obtaining water, or excess of soluble salts, etc.

Glaves A foehn-like wind of the Faroe Islands. Also known as glave, glavis.

Glazed fruit or **vegetable** Essentially a candy, coated with thin, hard, smooth and shiny layer of heavy sugar syrup dried on the surface.

Glazed tile Tile with a vitreous coating.

Gleba Sporulating tissue in an angiocarpous fruit body.

Gleditschia triacanthos See honey locust.

Gleyed Soil Soil affected by gleysation.

Gleysation A soil forming process, operating under poor drainage conditions, which results in the reduction of iron and other elements and in gray colours, and mottles. See also Gleysolic and Gleysol.

Gleysol A great group of soils in the Gleysolic order. A thin (less than 8 cm, or 3 inches) Ah horizon is underlain by mottled gray or brownish gleyed material, or the soil has no Ah horizon. Up to 40 cm Ah horizon is underlain by mottled gray or brownish gleyed material, or the soil has no Ah horizon. Up to 40 cm of mixed peat (bulk density 0.1 or more) or 60 cm of fabric moss peat (bulk density less than 0.1) may occur on the surface.

Gleysolic An order of soils developed under wet conditions and permanent or periodic reduction. These soils have low chromas, or prominent mottling, or both, in some horizons. The great groups Gleysol, Humic Gleysol and Luvic Gleysol are included in the order.

Gliadin A substance interacting with glutenin to form gluten in cereals; the prolamine of wheat seed.

Gliocladium roseum See coconut leaf rot.

Gliricidia Gliricidia maculata H.B. & K. Family Papi lionaceae. Small tree native of tropical America whose leaves are used for green manuring; can stand frequent pollarding. Also called spotless gliricidia.

Gliricidia maculata See gliricidia.

Global radiation The total of direct solar radiation and diffuse sky radiation received by a unit horizontal surface.

Globate Globular; Globe-shaped.

Globe artichoke Cynara seolymus L. Family Compositae. Herbaceous perennial plant, native of Mediterranean region and Canary Isles cultivated throughout India. Flower stalks terminate into globular inflorescences with numerous substanding involucral bracts. The edible portion being the bracts and soft, fleshy, lower ends of receptacles of lower heads. Grown by vegetative propagation from suckers or off-shoots.

Globose Almost spherical.

Globular (1) Globe shaped; spherical. (2) Composed of globules.

Globule Small crops, pills spherical in shape like a small globe.

Globulin Group of simple proteins, insoluble in water but soluble in dilute salt solutions. Coagulate on boiling. Distributed like albumins, in plants and more in seeds.

Glochidiate Barbed at the tip.

Gloeosporium ampelophagum See grapevine anthracnose.

— **musarum** See banana anthracnose.

— **psidii** See guava anthracnose.

Glomerate Compactly clustered.

Glomerella lindemuthianum See anthracnose of bean, horse gram, swordbean and cowpea.

— **tucumanensis** See sugarcane red rot.

Glomerule Clustered flowerhead. Also clusters of small organisms, tissues or blood vessels etc.

Glucase A plant enzyme which produces grape sugar from maltose.

Glucocorticoids Hormones which when carried into the blood stream, increase rate of formation of glucose from proteins and fat. Produced by the tissues of the outer layers of the adrenal gland.

Gluco-proteins Group of proteins consisting of combination of a protein with carbohydrate group. Also called glyco-proteins.

Glucose ($C_6H_{12}O_6$) Most widely distributed monosaccharide hexose sugar; end product of hydrolysis of many polysaccharides; starch, maltose and lactose partially or wholly composed of glucose residues. Occurring in seeds and leaves of plants; ripe grapes contain it along with fructose white crystalline solid easily soluble in water. Also called Dextrose; grape sugar.

Glucosides Bodies of nitrogenous compounds combined with simple sugars; very common in vegetable kingdom but have little or no nutritive value.

Glue A viscous liquid having strong adhesive properties.

Glume One of the two empty chaffy bracts at the base of each spikelet in grasses.

Glut The supply of a commodity greatly in excess of the previous demand for it, leading in consequence to a considerable fall in price.

Glutinous Having a sticky surface.

Glutamin See amino acids.

Glutelins Group of simple proteins such as glutelin of wheat and oryzenin of rice. Insoluble in neutral solvents, soluble in acids alkalies.

Gluten The protein-rich constituent of wheat grain; represented by albuminoids. Carries from 8-26 per cent in soft wheat to hard wheats. Influences baking quality.

Glutenin A substance of cereals interacting with gliadin to form gluten.

Glycerine Clear, neutral, sweet and syrupy liquid obtained by saponification of animal or vegetable oils. Miscible in water and used as emollient, sweetening agent, solvent and as vehicle for drugs. Also called glycerol.

Glycerol See glycerine.

Glycine javanica Rhodesian kudzu vine. A climbing herb of family Polygonaceae.

— **hispida** See soyabean.

— **max** See soyabean.

— **soja** See soyabean.

Glycogen ($C_6H_{10}O_5$) Carbohydrate. Approximate counterpart in the animal body to dextrins in plants.

Glycophytes Nonhalophytic plants or plants that do not grow well when the osmotic pressure of the soil solution rises above 2 bars.

Glyco-proteins See gluco-proteins.

Glycoside Any of a group of organic compounds, occurring abundantly in plants, that produce sugars and related substances on hydrolysis.

Glycyrrhiza glabra See liquorice.

Glyptotermis dilatatus See rubber termite.

Gmelin's test A qualitative test for the pigments in bile; test solution is mixed with nitric acid containing nitrous acid; reaction is positive if colour appears at the acid-solution junction.

Gnat A small two-winged fly of which the female has blood-sucking mouthparts.

Gneiss Metamorphic rock derived from granite. Generally banded with light and dark stripes of varying width, the lighter band containing more of quartz and feldspars and the darker ones containing more of the biotite and hornblends. May also be schistose in character.

Gnorimoschema glapsigona See brinjal bud worm.

— **heliopa** See tobacco stem borer.

— **operculella** See potato tuber moth.

Goa bean Psophocarpus tetragonolobus (L.) DC. A twining herb of family Papilionaceae, native of India. Immature pods used as vegetable.

Goal It is the distance in any given direction one expects to go during a given period of time.

Gobar gas plant A simple plant for anaerobic digestion of cowdung and litter to produce a combustible gas useful for lighting and cooking purposes. In India large number of plants are being put up in villages. After the digestion, the residue is used as manure. Also bio-gas plant.

Go-devil A small, short sled (scoot) without a tongue, used in skidding logs.

Going long In the commodity markets, buying first and selling later.

Going short In the commodity markets, selling first and buying later.

Golden algae The common name for members of the class Chrysophyceae.

Golden-brown algae The common name for members of the division Chrysophyta.

Golden nematode Heterodera rostochinesis Woll. Cystforming root-knot eelworm. Most dreaded pest of potato. Also see root-knot eelworm.

Golden-shower Cassia fistula L. Family Caesalpiniaceae. Moderate sized tree indigenous to southern Madras, thrives under a variety of conditions. Drought resistant, frost tender, prefers moist soil with drainage. Propagated by direct sowing. Wood used for fuel, farm implements and building work. Also called Indian laburnum.

Golden wattle Acacia pycantha Benth. Family Mimosaceae. Tree native to Australia, now cultivated in Nilgiris. Extract of bark considered comparable to best Indian catechu.

Golden yellow sweet pea Crotalaria retusa L. Family Papilionaceae. Plant cultivated all over India for its stem bark used for obtaining fibre.

Golgi apparatus Organelle of cell cytoplasm made up of an irregular network of canals or solid strands.

Gomuti palm Arenga pinnata (Wurmb.) Merr. Family Palmaceae. A tree, sap drain from the tender portion of the stem used for making sugar and sugar products. Also called sugar palm.

Gonad A sexual gland, either ovary or testes, or ovotestes, which produces the reproductive cells.

Gonadectomy Excision of gonads.

Gonadin Active principle of sex glands controlling secondary sexual characteristics.

Gonadotropic hormone Any hormone which stimulates the gonads.

Gondi Cordia gharof (Forsk.). Ehrenb. & Asch. (Cordia rothii Roem & Schm.). Family Ehretiaceae. Tree whose fruits are edible.

Gondwana One of the rock formations in India. Occurring in plains of basin like depressions in the table land of peninsula. Has produced comparatively immature soils of less fertility.

Gonimoblast One of many short filaments arising from the fertilised carpogonium and giving rise to carpospores in the cystocarp.

Good A tangible commodity such as a house, car, refrigerator or foodstuff.

Good tilth A term referring to the tilth of the soils and is indicative of the physical condition favourable for germination and plant growth.

Goodwill The commercial benefit arising from a firm's reputation or business connections.

Gooseberry The plant and fruit of a bush, raised for its large, green, purplish black and red fruits. The species of Ribes and family Saxifragaceae.

Goose grass Eleusine indica(L) Gaertn. Herb of family Gramineae found throughout India and used as fodder.

Gopali Anisomeles indica(L) O. Kuntze (Anisomeles ovata R. Br.). Family

Labiatae. A woody herb used as fodder and for flavouring purposes.

Gorgan nut Euryale ferox Salisb. Family Nymphaeceae. A prickly aquatic herb whose seeds are roasted and eaten.

Goria Chrysopogon fulvus (Spreng.) Chiov. (Chrysopogon montanus Trin.) Family Gramineae. A perennial herb used as fodder.

Gormandizer sets The extremely stout and vigorous late tillers put out by certain sugarcane varieties (POS 2878) immature at the time of harvest, satisfactory as short crop sets. Also called water shoots.

Gosling blast A sudden squall of rain or sleet in England. Also known as gosling storm.

Gossypium arboreum See tree cotton.

— **arboreum var. nadam** See Egyptian cotton.

— **barbadense** See Egyptian cotton.

— **herbaceum** See Asiatic cotton.

— **hirsutum** See Upland cotton.

— **indicum** See tree cotton.

— **mexicanum** See upland cotton.

— **nanking** See tree cotton.

— **neglectum** See tree cotton.

Gossypol A bright yellow crystalline, toxic substance found in cotton seed.

Go to seed To mature beyond a usable condition.

Go to stalk To grow vegetatively at the cost of seed or fruit.

Gourd Any fleshy cucurbitous fruit of varying shape and size and of the family Cucurbitaceae.

Governor It is a mechanical device used on engines for maintaining a nearly constant speed of engine under different load conditions.

Governor hunting When the speed of the engine changes from too fast to too slow and again to low fast periodically it is called governor hunting.

Gowk storm In England, a storm or gale occurring at about the end of April or the beginning of May.

Grade (1) The slope of the land surface. (2) Any animal which has one pure bred parent and one of the unknown or mixed breeding. (3) The classification of products by standards of uniformity, size, trueness to type, freedom from blemish, fineness etc. (4) It is a measure of a plan angle and 100 grades is equal to one right angle or 90 degrees.

Graded contour furrow Furrow irrigation system in which the furrow grade deviates from the contour by a small, set amount.

Grades row farming Farming system where crop rows are established within prescribed slope limitations.

Graded Terrace A terrace having a constant or variable grade along its length.

Graded topocline A topocline having a wide range, or ranging into different kinds of environment, thus subjecting its members to differential selection so that divergence between local races may become sufficient to warrant creation of varietal, or even specific, names.

Gradeline In land levelling or irrigation preparations, the line usually marked by stakes to indicate the slope in any fixed direction of the field required after levelling.

Grader A machine with a centrally located blade that can be angled to a cast to either side, with independent hoist control on each side.

Grade-stablising structure A dam or drop structure constructed for the purpose of stabilising the grade of a gully or other watercourse, thereby preventing further head cutting or lowering of the channel grade.

Grade stake A stake upon which a grade has been marked. The top of the stake may be driven to the desired grade, or more commonly, horizontal crayon mark with arrow is used to mark the grade.

Grade up To improve a herd, flock etc. by breeding to a pure-bred sire.

Gradient (1) Slope. (2) Rate of change of elevation, pressure temperature, energy etc.

Gradient flow Horizontal frictionless flow in which isobars and streamlines coincide, or equivalently, in which the tangential acceleration is everywhere zero; the balance of normal forces (pressure force, cariolis force, centrifugal force) is then given by the gradient wind equation.

Gradient wind A wind for which Cariolis acceleration and the centripetal acceleration exactly balance horizontal pressure force.

Grading (1) System of out-breeding to bring uniformity and increased productivity in the off-springs. (2) Proper separation of produce into different grades. (3) To bring the land surface to a uniform grade.

Gradually varying unsteady flow Flow is gradually varied if the depth does not abruptly vary over a comparatively short distance.

Graft (1) An artificially induced vegetative fusion or union of parts from different individuals, the rooted part being called stock and the part or parts inserted to it the scion (or scions). (2) To make a graft. (3) The growth resulting from a graft.

Graftage Method of propagating plants vegetatively, in which the plants raised are composed of two parts, the root-stock and the shoot — the scion.

Graft chimera See graft hybrid.

Graft-hybrid An individual formed from graft and stock and showing characteristics of both. Also graft chimera.

Grafting The joining of two plant parts, usually stems, so that their tissues grow together.

Grafting by approach See inarching.

Grafting clay A specially prepared clay used for plastering graft joints, better and cheaper than graft wax. Prepared by mixing 2 parts of soil having a clay fraction of 75 per cent with 6.0 pH, 1 part of cowdung and an equal part of horse-dung soaked in water. When cowdung has fermented; the mud is kneaded into a fine dough.

Grafting knife A knife with one straight edge blade. It is specially designed for grafting.

Grafting wax A wax consisting of mixture of beeswax, tallow and resin used to seal a graft to make it air tight and water-proof.

Graft union The point or place on the root, stem or branch of a plant at which a graft has been made.

Grain (1) Fruit of the large-seeded grasses; a collective term for the cereals. (2) A descriptive term applied to the alignment of cells and tissues in wood. (3) Butter particles in butter making.

Grain (of rock) In the plane perpendicular to the rift along which the rock splits, but with more difficulty than along the rift.

Grain box A hopper in the seed-drill used for filling the seeds to be sown.

Grain density See particle density.

Grain drill See seed drill.

Grain dryer A mechanical device used to reduce moisture content in the harvested crop. Uses either heated air, or electrical energy or sun energy.

Grain sorghum Sorghum bicolor. A grass plant cultivated for its grain and to a lesser extent for forage. Also known as non-sachharine sorghum.

Grain-size Analysis (Mechanical analysis) See particle-size analysis.

Grain-size Distribution See particle-size distribution.

Grains or moisture The unit of measurement used to determine the amount of moisture in the air. 7000 grains equal one pound of water.

Gram (1) A pulse crop, see Bengal grain. (2) Unit of weight used in the metric system and is equal to .001 kg.

Gram blight Disease of gram caused by Mycosphaerella pinodes (B.&Bolx) Niessl. (Assochyta pisi Lib.). Characterised by the appearance of round or oval brownish spots with yellowish grey centre on leaves, stems and pods. Suitable crop rotation is one of the measures of control.

Gram caterpillar Heliothis armigera Hubn. Serious pest of gram, tomato and maize; also attacks cotton bolls and arhar pods. Green caterpillar with grey longitudinal streaks feeds on crop leaves and damages unripe seed, bores into podes. Also called gram pod borer.

Gram cutworm These belong to agrotis spp. and are one of the most serious pest of young rabi crops including pulses. Apart from these surface cut work, climbing cutworm are also found which harm the crop. The surface cut worm cut the tender stems of the plants just above the ground at night. The climbing cutworm damage by cutting the leaves from the top and edges leaving only the mid rib. Important climbing cut worm being Rhycia herculea corti and Draudt., Pseudelatia separata Walk. etc. Also called greasy gram cut worm.

Gram dry root rot Disease of gram caused by Macrophomina phaseoli, symptoms as in cotton dry root rot.

Gramgrass Any grass of the genus Bouteloria, pasture grass.

Gramineal The grasses, a family of monocotyledonous plants in the order Cyperales; characterised by distichously arranged flowers on the axis of the spikelet.

Gramineus Like or pertaining to the grass tribe.

Graminivorous Grass-eating.

Graminology Agrostology.

Gram negative Bacteria that are decolourised when stained by Gram's method but stain with the counterstain (red or brown).

Gram Panchayat Local administration unit at the village level in India.

Gram pod borer See gram caterpillar.

Gram pod fly Agromyza obtusa M. Pest of gram. A fly, laying eggs in pod shells. Maggots bore into and damage seeds.

Gram-positive Bacteria that are not decolourised by Gram's method: they retain the original violet colour of the Gram stain and are not stained by the counterstain.

Gram root rot Disease of gram caused by Opercullela padwickii. Characterised by plants drying up from top downwards, leaves becoming pale green and later yellow and finally drooping off; collar portion turning dark brown.

Gram rust The disease is not very common. It is caused by Uromyces cicerisarietini. The leaves become crowded with small round or oval brown to dark brown pustules which coalesce and form bigger pustules. They may be on both surface but major on lower surface of leaves. In severe cases pustules may also appear on petioles and stems.

Gram stain, Grams method Method of differential staining devised by Hans Christian Gram, a Danish bacteriologist.

Gram stem rot Disease caused by Sclerotinia sclerotiorum; symptoms as in cotton stem rot.

Gram wilt A physiological disease of gram and is quite common. Caused by Fusarium orthoceros var. ciceri or ozonium spp. The leaves in the initial stages become pale yellow and ultimately dry up. The disease is soil borne. Stunted growth, roots getting completely rotten.

Grana The minute, discrete, green bodies embedded in chloroplast.

Granadilla See giant granadilla.

Granary (1) A store house for husked or threshed grain. (2) A large food producing area.

Granite Igneous, crystalline, equigranular, coarse grained rocks consisting essentially of feldspar and quartz, with varying amount of mica, and other ferro-magnesiam minerals; hence granite stone.

Granular (1) See structure, soil. (2) Consisting of grains or appearing as if made of granules.

Granular fertiliser A fertiliser composed of particles of roughly the same composition and of about 2.54 mm in dia.

Grannulate To form into grains, granules or small masses.

Granulation (1) The cementation of particles into masses as grains, aggregates or clumps. (2) The formation of dextrose hydrate crystals in honey and crystals in other sugar syrup.

Granule Spheroidal soil aggregate, small particle of matter or grain.

Grape Vitis vinifera L. The common name for plants of the genus Vitis, family Vitaceae. Characterised by climbing stems with cylinderical-tapering tendrils and polygamodioecious flowers, grown for the edible, pulpy berries. Native of Armenia. Grows well in dry climate having short sharp winter and long dry summer. Tolerate frost during resting period in winter. Propagated by hard-wood cuttings. Also called wine grape, grapevine.

Grape sugar See glucose.

Grapevine See grape.

Grapefruit Citrus paradisi Macf. (Citrus decumana var. racemosa Roem.; C. decumana var. paradisi Nichols.; C. racemosa Marc.). Family Rutaceae. A fruit rich in ascorbic acid (vitamin C) and thiamine (Vitamin B). Propagated by budding.

Grapevine anthracnose Disease of grapevine caused by Gloeosporium ampelophagnum (Pass.) Sacc. Characterised by formation of depressed cankers on stems and twigs, arrested growth and drying of twigs. Development of brown spots with dark margins on berries, shrivelling and drying up of berries. Also called bird's eye of grape.

Grapevine black rot Disease caused by Guignardia bidwelli. Characterised by development of circular spots on half mature berries and formation of minute black dots within, shrinking and darkening of berries.

Grapevine brown leaf-spot Disease caused by Cercospora viticola. Dark brown angular spots on leaves and young shoots appear.

Grapevine chafer beetle Adoretus lasiophygus E. Beetle pest feeding on leaves of grapevine.

Grapevine downy mildew Disease caused by Plasmopara viticola Berl. and de Toni. Characterised by appearance of greenish-yellow spots which turn brown on leaves, dropping down of leaves, destruction of young branches.

Grapevine flea beetle Scelodonta strigicollis Mots. Insect pest which damages pruned grapevine, grub bores into roots and adult beetle feeds on leaves.

Grapevine girder beetle Sthenias grisator Fabr. Insect pest of grapevine. Adult make a ring around stem by girdling it.

Grapevine powdery mildew Fungus disease caused by Uncinula necato (Schw.) Burr. Characterised by appearance of white patches on leaf surfaces, distortion of leaves, blossoms, young berries.

Grapevine rust Fungus disease caused by Phakospora vitis. Characterised by appearance of numerous orange sori on lower leaf-surface.

Grapevine thrips Rhipiphorothrips cruentatus M. Pest infesting leaves and sucking sap from them.

Grapevine zinc deficiency disease Disease due to deficiency of zinc. Severe reduction of leaf-size, appearance of interveinal chlorosis, suppression of growth occurs. Die-back of leaders and extreme reduction in fruit yield results.

Graphs Getting a precise idea or drawing a comparison or conclusion from numerals or statistical data, needs presentation of data through graphs.

Graphidaceal A family of mosses formerly grouped with lichenised Hysteriales but now included in the order Lecanarales; individuals have true paraphyses.

Grass Botanically, any plant of the family Gramineae. Generally, in grassland agriculture the term does not include cereals when grown for grain but does include forage species of legumes often grown in association with grass.

Grass cover A good stand of grass of low herbaceous plants on soil; bind soils and help prevent erosion of soil.

Grass farming The cultivation of grass or legumes mainly for fodder purposes.

Grass minimum The minimum temperature shown by a minimum therometre exposed in an open situation with its bulb on the level of the tops of the grass blades or short turf.

Grass pea See lathyrus.

Grass silage Designating silage from grasses, legumes, or mixtures. Terms such as "alfalfa silage", "clover silage" are definite expressions that should be used where applicable.

Grass tetany (hypomaganesaemia) Condition of cattle and sheep marked by tetanic staggers, convulsions, coma, and frequently death; characterised by a low level of blood magnesium.

Grassed water way A natural or constructed water way, usually broad and shallow, covered with erosion-resistant grasses, used to conduct surface water from cropland.

Grasserie Disease of silkworm.

Grass crops Plants cultivated as forage and grain to be consumed by domestic livestock.

Grasshopper Any insect of the order orthoptera, family Acrididae (Locustidae). Have long legs for jumping; generally voracious, feeds on many kinds of plants.

Grassland Any area of herbaceous terrestrial vegetation dominated by grasses and graminoid species.

Grassland agriculture Farming system that emphasises the importance of grasses and legumes in livestock and land management.

Grassland farmer One who plans his row crop and livestock production around his grassland hectarage.

Grassland management Management of grassland with the aim of maximum utilisation of the forage with the minimum interference in the growth and vigour of the chief grasses. Brought about by following improved practices.

Grated inlet Direct inlet structure to a tile line protected with a grate.

Gravel Rock mineral particles fragments 2 mm to 7.5 cm in diametre.

Gravel envelope Selected aggregate placed around the screened pipe section of well casing or subsurface drain to facilitate the entry of water into the well or drain.

Gravel filter Graded sand and gravel aggregate placed around a drain or well screen to prevent the infiltration of fine materials from the aquifer into the drain or well.

Gravelly Containing appreciable or significant amounts of gravel. The term is used to describe soils or lands. See also coarse fragments.

Gravelly loam A loam which contains 25 to 50 per cent of gravel and more of sand and little of silt.

Gravimetric method A laborious but most accurate method of measuring soil moisture. The soil samples are taken in the field and are oven dried at a temperature of 105-110°C. The loss in moisture is expressed on dry weight basis or on volume basis. A standard for evaluating the accuracy of other methods of measuring soil moisture.

Gravitational potential The amount of work that must be done per unit quantity of pure water in order to transport reversibly and isothermally an infinitesimal quantity of water, identical in composition with the soil water, from a pool at a specified elevation and at atmospheric pressure, to a similar pool at the elevation of the point under consideration.

Gravitational water Water that moves into through or out of the soil by gravity. See also drainage water.

Gravity dam Dam which depends on its weight to resist overturning.

Gravity flow Flow of water due to gravity in the channels.

Gravity irrigation Irrigation in which water is not pumped but flows by gravity.

Gravity outlet Outlet for a drain line or system where the water flows by gravity.

Gravity separator It is a seed processing machine by which seed separations are made on the basis of differences in specific gravity.

Gray brown Luvisol A great group of soil in the Luvisolic order occurring in a moderate climate, higher than 5.5°C (42°F) mean annual temperature, and developed under deciduous and coniferous forest cover. These soils have a dark coloured mull-like surface (Ah) horizon, a light coloured eluviated (Ae) horizon, a brownish illuvial B (Bt) horizon, and a basic or calcareous C horizon. The solum is highly base saturated (NaCl extraction). This group includes soils formerly called Gray Brown Podzolic.

Gray Luvisol A great group of soils in the Luvisolic order occurring in moderately cool climates, where the mean annual temperature is usually lower than 5.5°C (42°F). The soils have developed under deciduous and coniferous forest cover, and have an eluviated light-coloured surface (Ae) horizon, a brownish illuvial B (Bt) horizon, and usually a calcareous C horizon. The solum is base saturated (NaCl extraction). The Ahe horizon, if presents is less than 5 cm (2 inches) thick. This group includes soils formerly called Gray Wooded.

Graze Partial defoliation of forage plants by the animal; to feed animals on growing grass or herbage; to forage.

Grazier (1) Animal that grazes on growing grass or herbage. (2) One who pastures (grazes) livestock, usually for commercial purpose.

Grazing, alternate Grazing of two or

more pastures in succession.

Grazing capacity The number of animals a specified pasture will carry for a specified period.

Grazing continuous The grazing of a specific range of pasture by livestock throughout a year or grazing season. The term is not necessarily synonymous with year long grazing.

Grazing deferred Delay of discontinuance of livestock grazing on an area for an adequate period of time to provide for plant reproduction, establishment of new plants, or restoration of vigor.

Grazing mixed Grazing two or more classes of livestock such as sheep and cattle on the same pasture.

Grazing pressure Number of animals per unit area of available forage.

Grazing, ration See grazing, strip.

Grazing, restrotation Intensive system of management whereby grazing is deferred on various parts of the range during succeeding years, allowing the deferred part complete rest for one year, two or more units are required.

Grazing, rotational System of pasture utilisation embracing periods of heavy stocking growth recovery during the same season.

Grazing strip (ration grazing) Confining animals to an area of forage to be consumed in a short period of time, usually a day.

Grease (1) Grease is a semifluid lubricant, used for slow moving parts and consists of a mixture of mineral oil, metallic soap and few other chemicals. (2) Term usually applied to inedible pig fat as distinguished from tallow which is fat of cattle and sheep. (3) Molted fat of dead animal.

Grease wood A low, stiff shrub similar in appearance to sagebush, but is indicative of strongly alkali soils, it is highly inflammable.

Greasy cut-worm See gram cut-worm.

Great diurnal range Difference in elevation between mean higher high waters and two low waters in each lunar day.

Great Group A category in the Canadian system of soil classification. It is a taxonomic group of soils having certain morphological features in common and a similar pedogenic environment. Examples are Black, Solonetz, Gray Brown Luvisol, Humic Podzol, Melanic Brunisol, Regosol, Gleysol, and Fibrisol.

Greater cardamom Amomum subulatum Roxb. Family Zingiberaceae. Seeds of the plant used in the preparation of sweet meat and for flavouring beverages. Also called Nepal cardamom.

Greater galangal Alpinia galanga(L.) Willd; (Languas galangal L.) Stuntz. Family Zingiberaceae. A perennial herb native of Indonesia and Malaysia. The rhizomes used as condiment and source of an essential oil. Flowers eaten raw and seeds as spice.

Greater wax-moth Galleria mellonella L. Serious pest on honey combs in bee-hives, the caterpillars feeding on combs and damaging them. Prevention by general cleanliness and care in storing empty combs.

Grecian foxglove Digitalis lanata Ehrh. Family Scorphulariaceae. A herb native of Europe. Leaves source of drug used on cardiac stimulant and tonic in heart disease.

Greco An Italian name for the northeast wind.

Green (1) A small piece of land covered with grass. (2) Unmatured; immature. (3) Uncured. (4) Pertaining to a plant whose foliage is green.

Green algae The common name for members of the plant division Chlorophyta.

Green almond Pistacia vera L. Family Anacardiaceae. Tree native of West Asia and cultivated for its edible seeds. Also called pistachio.

Green belt A strip of land maintained for the specific purpose of recreation or planting and kept free from buildings.

Green chop Mechanically harvested forage fed to animals while it is fresh and succulent. Preferred to "soiling", "zero grazing", or "green feed".

Green feed Uncure forage; rich in carotene and a good source of vitamin A.

Green gram Phaseolus aureus Roxb. (P. mungo auct; non L; P. radiatus and; non L.). Family Papilionaceae. Pulse cultivated throughout the country. Green pods as vegetable, grains whole or split as pulse and straw and husk as fodder for cattle. Does best on deep well-drained loam. Needs well distributed rainfall of 60-80 cm. Grown as kharif, rabi or summer crop.

Green hay (1) Uncured hay. (2) Hay which retains a green colour on being cured.

Green honey Honey stored in unsealed cells in a comb, not evaporated enough to be properly thick. Also called honey.

Green house Glass-enclosed, climate-controlled structure in which young or out-of season plants are cultivated and protected.

Greening The turning green of potato tuber due to exposure to sunlight. Also called sunburn.

Green leaf manuring The practice of applying leaves of plants and trees in the green stage to the soils for the sole purpose of green manuring.

Green house effect The effect that short wave radiation can pass easily through the atmosphere to the surface of the earth, while a proportion of the resulting heat is retained in the atmosphere, since the outgoing long wave reradiation cannot penetrate the atmosphere easily, especially when there is a cloud cover.

Green leafy vegetables Group of vegetables producing leaf. Easy to grow, nutritious and rich in vitamin and mineral salts. Propagated by seed sown direct in field. Nearly 20 leaf vegetables grown in India.

Green manure Plant material incorporated into the soil to improve it, while the plant material is still green.

Green manure crop A crop grown for use as green manure. See also green manuring.

Green manure tramper A light implement with four steel discs nearly 25 cm in dia and mounted on horizontal pipe shaft rotating in bearings. Used for burrying green manure in puddle.

Green manuring Method of growing a quick-maturing crop and ploughing it under to be incorporated into the soil. Supplies organic matter to improve physical condition and additional nitrogen particularly if the crop is leguminous.

Green mint Spearmint.

Green oats Oat crops before separating and grading.

Green onion An immature garden onion harvested before the bulb is of spherical shape, eaten fresh as salad.

Green peach aphid Myzus persicae Sulz. Insect pest of tobacco; tiny soft-bodies insect which suck sap from leaves, stems, flowers and capsules making plant look sickly. Tobacco leaves when attacked are unfit for curing. Also attacks potato and transmit virus disease like potato and tobacco leaf-curl.

Green pepper Green chillies.

Green peas The immature seeds of the garden pea (Pisum sativum) used as food.

Green revolution The establishment of new crop records through the adoption of new varieties of wheat and rice associated often with heavy inputs of fertilisers and improved irrigation.

Green scale Coccus viridis Green, a scale insect which is injurious to coffee and citrus crops.

Green sky A greenish tinge to part of the sky, supposed by seamen to herald wind or rain, or in some cases, a tropical cyclone.

Green snow A snow surface that has attained a greenish tint as a result of the growth within it of certain macroscopic algae.

Green sulfur bacteria A physiologic group of green photosynthetic bacteria of the Chloreceae that are capable of using H_2S and other inorganic electron donors.

Green tea See tea.

Green timber Freshly felled timber which has not lost much of its moisture or timber which still contains free water in its cell cavities.

Green vegetable Any vegetable whose edible portion is green. Also see green leafy vegetables.

Green wood cutting Soft wood cutting.

Green vitriol Ferrous sulphate.

Green wattle Acacia decurrens. Family Mimosaceae. Tree introduced into India from Australia. Extensively grown for its bark which is used for tanning. Wood used as fuel; useful for afforestation grasslands in the temperate zone, when rainfall is more than 150 cm; bark is not good for tan production. Frost resistance and does not grow below 1250 m above sea level.

Greenland anticyclone The glacial anticyclone which is supposed to overlie Greenland: analogous to the Antarctic anticyclone.

Gregale The Maltese and best-known variant of a term for a strong north-east wind in the central and western Mediterranean and adjacent European land areas, it occurs either with high pressure over Central Europe or the Balkans and low pressure over Libya, when it may continue for up to 5 days, or with the passage of a depression to the south or south-east, when it lasts only 1 or 2 days, it is most frequent in winter.

Gregarious Growing in clusters; habitually living and or reproducing in more or less pure groups; tending to herd together.

Gregarious flowering General flowering within one or a few years and over considerable areas, of all or most of the individual of certain species of trees that do not flower annually; as opposed to sporadic flowering.

Grevillea robusta See silver oak.

Grewia asiatica See phalsa.

— **elastica** See phalsa.

— **glabra** Blume (G. laevigata auct., plur. non Vahl.). A tree of the family Tiliaceae. Bark yield fibre used for cordage.

— **laevigata** See asar.

— **microcos** See asar.

— **oppositifolia** Buch. Ham. ex. Roxb. A small tree of family Tiliaceae. Bark as fibre and leaves and young tree as fodder.

— **populifolia** Grewia tenax (Forsk,) Fiori (G. Populifolia Vah.). A shrub of the family Tiliaceae which produces edible fruits and seeds.

— **tenax** See G. populifolia.

— **tillifolia** Vahl. A tree of family Tiliaceae which produces edible fruits.

— **villosa** Willd. A shrub of family Tiliaceae which produces edible fruits.

Grid A plane rectangular coordinate system imposed upon a geographical coordinate system.

Grid iron system A system in which parallel laterals enter into the branch of main drain only from one side.

Grinder A useful device for grinding various substances, as soil and plant samples.

Grinding compound A grease and abrasive or water-carried abrasive used in valve grinding operations.

Grindstone A revolving stone against which tools and materials are abraded by grinding. Grindstones are natural sandstone.

Grip A device for grasping or holding.

Grit Hard, gravelly substance such as small, rough edged stones or brick pieces.

Groin A structure perpendicular to the shore which serves to stabilise the shoreline by trapping littoral drift.

Groove A depressed or sunken channel, usually small; so designated, whether some other part does or does not fit into it.

Gross area The total area within the extreme limits set for irrigation by a project system, or a supply system, or a canal.

Gross-austausch The exchange of air-mass properties and the associated momentum and energy transport, produced on a worldwide scale by the migratory large scale disturbances of middle latitudes.

Gross change Any change that is observed by the normal human eye without the aid of a microscope.

Gross commanded area The portion of the gross irrigable area which can be commanded by gravity irrigation. In special cases, e.g. in India and Pakistan, gross commanded area also includes area irrigated by pumping or lifting the water by other devices.

Gross duty, or diversion duty Duty of water reckoned or measured at the source of diversion of irrigation supplies.

Gross energy The amount of energy given out when a substance is completely burnt to its final oxidation products; provides an indication of total heat value. Also called total heat, heat of combustion.

Gross irrigable area The gross area

less such areas as are excluded from the project by reasons of their being unsuitable for irrigation, either on account of the nature of the soil, or because the ground is too high to be irrigated by gravity flow or economically by pumps or other water lifting devices.

Gross lift area The portion of the gross irrigable area which can be irrigated only be pumping or lifting the water by other devices. Gross commanded area plus gross lift area equals gross irrigable area. The source of water may be subsoil water, reservoir, pumping from a river, pumping or lifting water from a gravity flow canal.

Gross pathological change A change visible by naked eye and is due to the disease in an organ or a tissue.

Ground anthrachose A disease caused by Collectrochum spp. The symptoms are found on the leaves and stipules in the form of necrotic spots. Acervuli are formed on the dying parts of the plant.

Ground clearance It is the height of the lowest point of the tractor from the ground surface, when the tractor is loaded to its permissible value of load.

Ground colour See surface colour.

Ground cover Prostrate or low plants that cover the ground instead of grass. All forest plants (Herbaceous and low growing shrubs) except trees.

Ground fire A fire that consumes all the organic material of the forest floor and also burns into the underlying soil, for example, a peat fire. It differs from a surface fire by being invulnerable to wind. In a surface fire the flames are visible and burning is accelerated by wind, whereas in a ground fire wind is not generally a serious factor.

Ground fog A fog that hides less than 0.6 of the sky and does not extend to the base of any clouds that may lie above it.

Ground frost In British usage, a freezing condition injurious to vegetation. Which is considered to have occurred when a minimum thermometre exposed to the sky at a point first above a grass surface records a temperature (grass temperature) of 30.4°F (–0.9°C) or below.

Ground moraine An unsorted mixture of rocks, boulders, sand, silt, and clay deposited by glacial ice. The predominant material is till, but some stratified drift is present. Most of the till is though to have accumulated under the ice by lodgment, but some till has been let down from the upper surface of the ice by ablation. Ground moraine is usually in the form of undulating plains having gently sloping swells, sags, and enclosed depressions.

Ground rainfall That portion of rainfall which reaches the soil surface.

Ground streamer An upward advancing column of high-ion density which rises from a point on the surface of the earth toward which a stepped leader descends at the start of a lightning discharge.

Ground sulphur Sulphur powder passing through 325 mesh and used as fungicide. The smaller the particle size, more effective the fungicide.

Ground survey, or **Field survey** A survey made by ground methods (as distinguished from the aerial survey) and includes control surveys, adastral surveys, topographic surveys and location surveys. A ground survey may or may not include the use of ground photographs but does include the use of aerial photographs for preliminary research check purposes.

Groundwater Water that stands or flows beneath the ground surface in soil or rock material which is thoroughly saturated. The upper surface of this saturated zone is called the watertable.

Groundwater accretion, depletion Increase or decrease of the amount of groundwater in storage.

Groundwater aquifer See aquifer.

Groundwater depletion curve See depletion curve.

Groundwater divide A line on a water table on each side of which the water table slopes downward in a direction away from the line.

Groundwater flow (1) The movement of ground water under hydraulic gradient. The movement could be observed by drawing contour maps of the groundwater levels. (2) That portion of the discharge of a stream which is derived from ground water.

Groundwater investigations or Geohydrological investigations Investigations of groundwater resources regarding their occurrence, movement, storage, discharge, quality, and their appraisal to determine the quantity they can yield.

Groundwater level See watertable.

Groundwater level indicator The instrument to obtain the water level below the ground surface. Electrical instrument could be used as an indicator. However, a simple device of a closed tube at the lower end with bottom clearance could be used well which makes sound when touches the ground water surface. The length of the cord is noted.

Groundwater runoff That part of the ground water which is discharged into a stream channel as spring or seepage water.

Groundnut Archis hypogaea L. Family Papilionaceae. Plant introduced in India in 16th century, originated in Brazil. Cultivated varieties vary in duration, habit of growth, size and shape of leaves, size and shape of pods, seeds, colour of seed-coat and number of kernels per pod etc. Broadly divided into bunch or erect and runner or spreading. Mainly a tropical crop but suited to substropical regions also, cannot stand frost, excessive drought or water stagnation. Grown on all soils but best is light sandy loams with moderate amount of organic matter and supplied with calcium. A legume, gets atmospheric nitrogen but needs potash and lime. Bunch type ripens early than spreading type.

Groundnut beetle grub The grub attack the root zone underneath the soil. They remain clinging to the rootlets and feed on them with the result the growth of the plants is stunted and in case of severe attack, crop suffers heavy loss.

Groundnut Bihar hairy caterpillar Diacrisia oblique Wlk. A pale buff-coloured medium-sized moth with black spots on the wings. Full grown caterpillars have orange coloured body covered with hairs. The caterpillar feeds on the epidermis of leaves and stems and as a result, the photosynthesis of the plant is adversely affected. The whole crop is defoliated in case of severe attack.

Groundnut cake A by-product highly nutritive substance obtained after oil has been taken out of the seeds. Have good amount of protein. When reduced to flour is palatable and balanced food for infants and adults. A suitable diet for diabetic patients. Blends well with wheat and other flours. Good feed for animals and milch cattles.

Groundnut collar rot Disease of groundnut caused by Pellicularia rolfsii. Characterised by plants turning yellow and drying up. It is also known as crown rot or seedling blight.

Groundnut dry root rot Disease caused by Macrophomina phaseoli and is very common in India. In the seedling stage affected plants quickly collapse developing a dry rot at or just below ground level. The decayed area is generally covered with a light red brown weft of mycelium being typical of this pathogen. In older plants a red brown lesion appears on the stem or on runners where they have been in contact with the soil.

Groundnut haulms Plants of groundnut after groundnut is removed. Fed green or after drying or conversion into silage to livestock. Rich in nutrients and slightly inferior to Lucerne hay. Applied also as green manure to paddy.

Groundnut irregular leaf-spot Disease of groundnut caused by Mycosphaerella arachidicola. Characterised by appearance of irregular circular spots, and bright yellow circular hollow on upper leaf surface.

Groundnut kernels Cheap and rich source of vegetable, protein content from 25-23 per cent, carbohydrate 10-20 per cent and fat 40-50 per cent, protein easily digestible and highly nutritive. Also a good source of phosphorus and lime. Contain vitamin A and B. Milk and curd manufactured from freshly harvested groundnut kernels.

Groundnut leaf miner Stomopteryx nerteria M. Insect pest of groundnut.

Small dark-headed greenish caterpillar that folds leaves and feeds on them. Moth small and dark brown. Also attacks soyabean and redgram. Infested fields show plants as if they are burnt.

Groundnut mosaic Disease caused by virus and is noted by mottling and malformation of leaves.

Groundnut oil Oil obtained from groundnut. Used as edible fat and used for making vegetable *ghee* by hydrogenation. Oil extracted in cold presses has pharmaceutical value similar to olive oil. Limited use in soaps, cosmetics and candles. Tests also indicate the possibility of its use as fuel in diesel engines.

Groundnut peg The elongating basal portion of the ovary after fertilisation into a stalk-iike structure, which reaches and enters the soil to form the pod.

Groundnut quality Quality of groundnut as judged by shelling percentage, natural test weight (weight-to-volume) size of kernels, colour of pods and kernels, oil content of kernels and free fatty acid content.

Groundnut root rot Disease caused by Pellicularia filamentosa. Characterised by red brown discolouration on stem which spreads and causes plant to collapse. Tissues blackened with sclerotia.

Groundnut rosette A virus disease of groundnut. In the beginning the diseased plants show faint mottling and vein necrosis of the young leaflet. Subsequently leaves may be completely chlorotic or show a mosaic pattern of dark green veins leading to light green leaflets. The petiole and rachis are shortened. The axis of the plants ceases to grow.

Groundnut shell The shell covering of groundnut; is 20-40 per cent by weight of the pods, is a good fuel and ash can be used as a manure containing 1.5 per cent potash and 3 per cent lime. Used in the manufacture of coarse boards, alcohol and acetone.

Groundnut stem borer Sphenoptera perroteti G. Insect pest; small shining dark-coloured beetle whose larva bores into stem and pupates in it.

Groundnut tikka disease Fungus disease caused by Cercospora personata (B & C) Ell. and Everth. and C. arachidicola Hori. In the beginning only leaf spots are common but later lesions also develop. The leaf spots due to C. personata are circular (1.6 mm in dia) and the necrotic lesions assume dark brown to black colour. The leaf spots due to C. arachidicola are irregular, larger in size surrounded by a bright yellow hollow which fades into green of the leaves.

Group A group is a unit or a collection of two or more individuals who are in social interaction over an appreciable period of time, who are in communication with each other, who have some common and agreed upon interests and values are stimulating each other psychologically for participation in common activities.

Group behaviour Group behaviour is the result of dynamic interaction between individuals in a social situation, groups dynamics can do much to make cooperative activities more democratic and more effective.

Grouped or nucleated villages In the grouped or nucleated villages, the residents dwell in a cluster. Their farms are spread around the villages dwelling. Their farm land may be in a block or the field may be scattered in many plots.

Grouping system A system similar to the natural system, except that a few short laterals may be provided in wet areas or ponds along the system.

Group variation Variation between different populations within a species, as distinct from individual variations which is the variation within a population.

Grout (1) A fluid or fluid mixture that can be poured or injected easily. (2) To fill the joints, hollows, cracks, seams, faults shattered zones or fracture zones, etc. with grout.

Grove Small wood, a group of trees, an orchard or a planting of trees such as a grove of mangoes.

Grow (1) To live and increase in stature, girth etc. (2) To cultivate (plants).

Grow out test Performed to determine the genuineness of seed as to species or variety, or freedom from seed borne infection.

Grow wild To grow uncultivated or untended.

Growing point Meristem.

Growing season Frost free period during which crops can be produced.

Growth (1) The increment in size of a living organism. (2) The development of an organism from its earliest stage to maturity. (3) Tumour gall etc.

Growth layer Annual layer.

Growth apposition Resulting in branching by displacement of a continued active terminal meristem by a more vigorous lateral i.e. no meristem abortion involved, of growth, subsituation resulting in branching by replacement of a terminal meristem which is determinate either by abortion or, most usually, by becoming a flower or inflorescence (See also plagiotropy by apposition or substitution).

Growth-habit The ultimate form of plant as expressed in its physiognomy, i.e. growth-habit is not defined by a precise differentiation sequence, but more or less by an overall appearance.

Growth promoting vitamin Vitamin A.

Growth Rate The annual growth, usually expressed as a percentage over the previous year, of productive capacity in a community.

Growth-regulating substances Plant regulators. Organic chemicals which modify or regulate physiological processes in an appreciable manner.

Growth stages Definite periods of growth in the life of a plant, such as tillering stage, milking stage etc.

Grub (1) Larva of a beetle. (2) Roots and crown remaining in the soil after a tree or shrub has been cut off.

Grubbers Implements or equipments to dig out roots.

Grubbing Digging out roots.

Gruel A food or feed prepared by mixing a ground feed in water cold or hot.

Gryllotalpa africana See mole cricket.

Gualtheria fragmentissima See wintergreen.

Guard cell One of the paired epidermal, chlorophyllous cells which enclose a stomate.

Guava Psidium guajava L. Family Myrtaceae. Fruit native of America and grown all over India. Medium sized tree growing as high as 10 m. Hardy, longlived and needs little attention. Fruit emits aroma, is pleasantly sweet and acidic. Source of vitamin C and minerals like calcium and phosphorous. Propagated by seed; anarching and layering popular.

Guava anthracnose Disease of guava caused by Colletotrichum psidii (Delacr.) Sheld. and Gloeosporium psidii. Delacr. characterised by appearance of small pin-point pustules, pink or black, on fruit surface, fruit having poor keeping quality.

Guava fruit canker Disease of guava caused by Pestalotiopsis psiddii. Characterised by appearance of brown or rust coloured circular and raised lesions on green fruit, older cankers showing depression in centre of spot.

Guava mealy scale Pulvinaria psiddii Maskell. Insect pest, colonies of mealy bugs cover leaves and suck juice, causing a sooty mould to develop.

Guazuma tomentosa See bastard cedar.

Gugal Commiphora roxburghii Arn. Engl. (Balsamodendron roxburghii Arn.) Family Burseraceae. Tree whose bark yields gum resin used as fixative in perfumery.

Guggula dhup Ailanthus tryphysa (Dennst.) Alston (Ailanthus malabarica DC) Family Simarubaceae. Tree whose resin from bark used as incense; leaves source of black dye; wood used for matchsticks.

Guide bund A training wall intended to lead water smoothly to, or away from, a barrage or weir.

Guignardia bidwelli See grapevine black, foot.

— **camelliae** See tea copper blight.

Guilandina moringa See drumstick tree.

Guinea grass Panicum maximum Jack. Family Gramineae. Perennial densely tufted grass, native of tropical Africa. Cultivated as fodder grass. Grows in varied conditions but not in waterlogged condition and cold climates. Propagated from slips. Gives first cutting in about $2^1/_2$ months. Good for stabilising bunds and make good silage.

Guinea squash See brinjal.

Gujhia weevil Tanymecus indicus Faust. Insect pest on wheat, poppy, barley, pea, gram, paddy and mustard. Also damages jowar, maize, cotton, sannhemp, jute, lentil, berseem, sunflower, potato, tobacco, cabbage, cauliflower. Damages young seedlings at night by cutting off roots or growing points or nibbling at roots.

Gujjar An Indian nomadic shepherd or cattle keeper.

Guizotia abyssinica See niger.

Gullet A tube-like structure leading to the interior of such flagellated organisms as Euglena.

Gulley See gully.

Gully A channel caused by erosion and the concentrated but intermittent flow of water during and immediately after heavy rains. It is deep enough to interfere with and not be removed by tillage operations. When small it is called rill.

Gully control The checking of gully erosion by mechanical, agronomic or vegetal means.

Gully control structure See grade stabilising structure.

Gully erosion See erosion.

Gully head The upper extremity of a gully which is usually the place where the water falls.

Gully plugging The planting of the entire ravine with stabilising vegetation. Considered best way of controlling gully erosion.

Gully-squall A nautical term for a violent squall of wind from mountain ravines on the Pacific side of Central America.

Gullying See gully erosion.

Gum Viscid secretion of some trees and shrubs that hardens on drying but is soluble in water.

— **Oleoresin** See gums.

Gumarabic acacia Acacia senegal (L.) Willd. (Acacia verek Guill. & Perr.). Family Mimosaceae. Tree of north Africa. Stem bark yields gum. Used in the textile, mucilage, paste, polish and confectionery industries.

Gum ghatti Anogeissus latifolia Wall. Family Combretaceae. A deciduous tree occurring in the drier parts of India. Gum from stem called ghatti-gum which is used in Calico-printing. Leaves for tanning and wood for making wheels.

Gum myrtle A shrub or tree of the genus Angophora family Myrtaceae.

Gum turpentine Essential oil obtained from the oleoresin exuded from living pine trees; the turpentine of commerce.

Gum tree See Eucalyptus.

Gum test This test consists of evaporating samples of fuel in special apparatus and is a measure of the presence of gum in the oil. Test in copper dishes over long period of times is also done to know the gum forming ability of the oil.

Gum vein Local accumulation of resin occurring as a wide streak in certain hardwoods.

Gumming See gummosis.

Gummosis Production of gummy exudates in diseased plants as a result of cell degeneration.

Gums Translucent amorphus substances which are degradation products of cell walls of woody species and which exude spontaneously from the tree. The process is accelerated when the bark or wood is wounded. Collected by making small and thin cut on the bark. Non-volatile colloidal products dissolving or swelling in contact with water.

Gun metal An alloy of copper and tin containing 8-10 per cent of the latter.

Gunnel It is the vertical face of the share and slides along the furrow wall.

Gunny bag or gunny sack Coarse sacking usually of burlap from jute fibre, used as container for agricultural produce or fertiliser. Also called burlap sack.

Gust A sudden, brief increase in the speed of the wind, it is of a more transient character than a squall and is followed by a lull or slackening in the wind speed.

Gustiness A quality of airflow characterised by gusts.

Gustiness components (1) The ratios, to the mean wind speed of the average magnitudes of the component fluctuations of the wind along three mutually perpendicular axes. (2)

The ratios of the root mean squares of the eddy velocities to the mean wind speed. Also known as intensity of turbulence.

Gustiness factor A measure of the intensity of wind gusts; it is the ratios of the total range of wind speeds between gusts and the intermediate periods of lighter wind to the mean wind speed, averaged over both gusts and lulls.

Gut A narrow, deep channel characterised by rapid currents.

Guti weather In Rhodesia, a dense stratocumulus overcast, frequently with drizzle, occurring mainly in early summer, and associated with easterly winds that invade the interior, bringing in cool and stable maritime air when an anticyclone moves eastward south of Africa.

Guttation The discharge of water from a plant surface, especially from a hydathode.

Gutter Ditch or channel.

Gutter-fly By-product in cotton mills; as good a manure as a cattle manure.

Guttiferae A family of discotyledonous plants in the order Theales characterised by extipulate leaves and conspicuous secretary canals or cavities in all organs.

Guttra In Iran, sudden squalls in May.

Guxen A cold wind of the Alps in Switzerland.

Guy Wire, rope, or chain used to steady, support, or hold in place mast, pole, chimney, or other similar piece of construction.

Guy rope Galvanised rope consisting of 6 strands, 7 wires each, and a hemp core.

Guzzle In the Shettand Islands, an angry blast of wind, dry and parching.

Gymnogynous Having a naked ovary.

Gymnosperm The common name for members of the division Pinophyta; seed plants having naked ovules at the time of pollination. Also naked seeds not enclosed in an ovary.

Gymnospermal The equivalent name for Pinophyta.

Gynaecium The carpels or pistils in a flower.

Gynaecomorph A male individuals which in appearance resembles a female.

Gynandromorph An individual with a part of the body exhibiting female characteristics and another part male.

Gynandrous With stamens and pistil united in one column.

Gynobase A gynoecium-bearing elongation of the receptacle in certain plants.

Gynodioecious Dioecious but with some perfect flowers on a plant bearing pistillate flowers.

Gynogenesis Development of a fertilised egg through the action of the egg nucleus, without participation of the sperm nucleus.

Gynomonoecious Having complete and pistillate flowers on the same plant.

Gynophore (1) A stalk that bears the gynoecium. (2) An elongation of the receptacle between pistil and stamens. (3) Pedical supporting ovary.

Gynostemium The column composed of the united gynoecia and androecium.

Gypsophilous Flourishing on a gypsum rich substratum.

Gypsum (Rock) An evaporite deposits of sedimentary rock from sea water containing the minerals gypsum. Also see calcium sulphate (gypsum).

Gyrate Arranged in rings or convolutions, go in circle or spiral.

Gyttja A Swedish word, Pronounced "Yuetya". A nutrient rich sedimentary peat consisting mainly of plankton, other plant and animal residues and mud. It is deposited in water in a fine divided condition.

H

Haar A wet sea fog or very fine drizzle which drifts in from the sea in coastal districts of eastern Scotland and north-eastern England; it occurs frequently in summer.

Habitant Any plant or animal that is permanently found in a place or section.

Habitat The external environment or a locality in which a plant or animal lives.

Habitat factor The influence of habitat on the growth of a plant or animal.

Hackberry (1) Celtis occidentalis L. A tree of the eastern United States characterised by corky or worty bark, and by alternate, long-pointed serrate leaves unequal at the base, produces small, sweet, edible drupaceous fruit. (2) Any of several other trees of the genus Celtis.

Hadromycesis Any plant disease resulting from infestation of the xylem by a fungus.

Haematic Of or containing blood.

Haematite (Fe_2O_3) Anhydrous sesquioxide of iron. Generally occurring as a iron ore and widely distributed in soils and is the cause of red colour in soils.

Hail Precipitation in the form of hard pallets of ice called hailstones.

Hail stage The thermodynamic process of freezing of suspended water drops in adiabatically rising air with temperature below the freezing point, under the assumption that release of latent heat of fusion maintains constant temperature until all water is frozen.

Hailstone A single unit of hail, ranging in size from that of a pea to that of a grape fruit, or from less than 6 millimetre to more than 13 centimetres diameter, may be spheroidal, conical, or generally irregular in shape.

Hail storm A storm during which hails fall.

Hair Any epidermal filamentous outgrowth consisting of one or more cells, varied in shape, a thread-like or filamentous outgrowth of epidermis of animals.

Hair weeds Algae submerged in water, though floating on surface when in profusion.

Hairworm Usually used for various nematodes.

Hairy caterpillar Diacrisia obliqua Wlk. Insect pest of jute, pulse crop etc. Defoliates plant completely.

Hairy root See malformation.

Half-breed (1) A cross between two botanical varieties of the same species. (2) A cross between two races.

Half life The time in which the radioactivity originally associated with a sample of isotopes will be reduced by one half through radioactive decay.

Half wrought timber Semi-finished timber on which further operations have to be done either by hand or by machine in order to complete the process.

Halimeda A genus of small, bushy-green algae in the family Codiaceae, composed of thick, leaf-like segments, important as a fossil and as a limestone builder.

Hallow stalk Any plant disease characterised by deterioration of the pith in the stalk.

Halloysite A 1:1 alumino-silicate mineral similar in structure to kaolinite. Due to variations in hydration, halloysite has variable interlayer spacing: 7.4 A up to 10 A (10 A = 1 mm) compared with 7.2 A for kaolinite.

Halminthes Group of parasitic worms.

Halo A narrow light-coloured band of diseased tissue sometimes found surrounding a lesion in a plant.

Halo blight A bacterial blight of beans and sometimes other legumes caused by Pseudomonas phaseolicola and characterised by water soaked lesions surrounded by a yellow ring on the leaves, stems, and pods.

Halocline The boundary between two masses of water, e.g. the surface waters of the Baltic which have a lower salinity q.v., due to the contribution of the fresh water of the rivers, than the corresponding current carrying the inflow of more saline water at a greater depth. Seen generally, the boundary between these masses the halocline is at a depth of about 60 metres. The layering is essentially stable since the bottom water, being more saline, is denser.

Halo length Maximum length of a fibre spread out in the form of a halo round a cotton seed and measured in millimetres.

Halomorphic soil A general term for saline and alkali soils.

Halophobe A plant which will not grow in a soil containing any appreciable amount of salt.

Halophytic vegetation Vegetation that grows naturally in soils having a high content of various salts. It usually has fleshy leaves or thorns and resembles desert vegetation.

Haloragaceae A family of dicotyledonous plants in the order Haloragales distinguished by an apical ovary of 2-4 loculi, small in florescence, and small, alternate as opposite or whorled, extipulate leaves.

Haloragules An order of dicotyledonous plants in the subclass Rosidae containing herbs with perfect or often unisexual, more or less reduced flowers, and a minute or vestigial perianth.

Halosere The series of communities succeeding one another, from the pioneer stage to the climax and commencing in salt water or on saline soil.

Hamamelidaceae A family of dicotyledonous trees or shrubs in the order Hamamelidales characterised by united carpels, alternate leaves, perfect or unisexual flowers, and free filaments.

Hamamelidae A small subclass of plants in the class Magnoliopside having strongly reduced, often unisexual flowers with poorly developed or on perianth.

Hamamelidales A small order of dicotyledonous plants in the subclass Hamamelidae characterised by vessels in the wood and a gynoecium consisting either by separate carpels or of united carpels that open at maturity.

Hamerophyte A cultivated plant.

Hamulus Small hook or hook-like process.

Hand duster or sprayer A manually operated device used for dusting or spraying insecticides and pesticides.

Hand hoe Garden hoe.

Hand pollination The pollination of flowers by hand, generally for producing hybrid seeds.

Hand pounding Manual pounding of rice for removing husk.

Hand weeding The removal of weeds manually.

Hank (1) Wool or cotton wound in a ring. (2) A measure of length for yarn. Normally 560 yards for worsted yarn and 840 yards for cotton yarn.

Hapaxanthy Development of an axis which is determinate by terminal, flowering, hence hapaxanthic shoot; a shoot which flowers terminally (of pleonanthy).

Haplant A plant with only haploid somatic cells, the zygote is diploid.

Haploid An organism or cell having only one complete set of chromosomes.

Haplosphase Haploid stage in the life cycle of an organism.

Haplostele A type of protostele with the care of xylem characterised by a smooth outline.

Hapteron A disk-like holdfast on the stem of certain algae.

Haptochlamydeous Having the sparophylls protected by rudimentary perianth leaves.

Haptophyceae A class of the phylum Chrysophyta that contains the Coccolithophorida.

Hard centres Term applied to jute fibre which is more or less clean and soft at both ends, but contains prominent dark strips of periderm in the middle portion.

Hard currency Currency which enjoys relatively stable value in international exchange and tends to maintain its value, at least in comparison with other currencies.

Hard dough stage A stage of ripening, intermediate between the milk and ripe stage.

Hard fibre A heavily lignified leaf fibre used in making cordage, twine, and textiles.

Hardness (of water) A measure of calcium and magnesium salts (carbonates, sulphates, chlorides, etc) expressed as parts of calcium carbonate per million parts of water.

Hard pan A horizon of the soil that is hardened or cemented soil particles with organic matter or with materials such as silica, sesquioxides or calcium carbonate.

Hard pan layer See hard pan.

Hard starch The compound starch cells that occupy the sides and back of the kernel of grain.

Hard rot (1) Any plant disease characterised by lesions with hard surfaces. (2) A fungus disease of gladiolus caused by Septoria gladioli which produces hard-surfaced lesions on the leaves and corms.

Hard seeds Seeds of Leguminosae and Malvaceae, which remain hard at the end of prescribed test period because they have not absorbed water owing to an impermeable seed coat, are classified as hard seeds.

Hard water Water which contains sulphates, chlorides and carbonates of calcium and magnesium. These when present in some higher quantities, get precipitated with soap.

Hard wheat A wheat producing hard, flinty kernels rich in proteins and suitable for making bread.

Hardwickia binata See anjun.

Hardwood Generally, one of the botanical group of trees that have broad leaves, in contrast to the needle-bearing conifers; also wood produced by broadleaved trees, regardless of texture or density.

Hardy A plant which is able to withstand unfavourable weather conditions.

Harness Gear of draught animals necessary for proper pulling of or moving a load.

Harrow An implement that is pulled over ploughed soil to break clods, level the surface, and destroy weeds. Secondary tillage implement.

Harrowing Cultivation of the soil with a harrow, a secondary tillage operation.

Hartho Securinega leucopyrus (Willd.) Muell-Arg. (Phyllanthus leucopyrus) Koen. ex Roxb.; Flueggea leucopyrus (Koen.) Willd. A large shrub or small tree of the family Euphorbiaceae which produces edible fruits.

Harvest To cut, reap, pick, gather any crop or product of value. Harvested crop. For forest trees the term commonly used is Harvest cut.

Harvester An implement power or bullock drawn to harvest the crop. Also see 'combine harvester'.

Hary-weinberg law The concept that both gene and genotype frequencies are in equilibrium from generation to generation in a large random mating population in the absence of mutation, migration, and selection.

Hashish Top leaves and tender parts of hemp, dried for smoking and chewing.

Hassock Tuft of matted grass.

Hastate Spear-shaped, more or less triangular with two based lobes divergent.

Haul distance, or **Haul** The distance measured along the centre line or most direct practical routes between the centre of the mass of excavation and the centre of mass of the fill as finally placed. It is the distance to which material is moved.

Haulm (1) The culm of grass. (2) The stem of a herb. (3) Stem stalks of peas and beans cured for fodder or hay.

Haustellum The proboscis meant for sucking (in insects).

Haustorium (1) An outgrowth of certain parasitic plants which serves to absorb food from the host. (2) Food absorbing cell of the embryo sac. in non-parasitic plants.

Hawk moth Achyrontia styx Wist. Insect pest of sesame caterpillar, feeds on leaves; also attacks brinjal, bean and potato.

Hay Entire herbage of forage plants, sometimes, including seed of grasses and legumes, that is harvested and cured dried for animal feed.

Hay crops A natural or cultivated crop which is used for hay-making. Principal grasses and field crops are Cenchrus ciliaris, Dichanthium annulatum, D. caricosum, Hetropogon contortus, Iseilema laxum, Iseilema wightii, Sehima nervosum and Themeda, berseem, Lucerne cowpea, soyabean, oats, sudan grass and Rhodesgrass.

Haylage Product resulting from ensiling forage with about 45 per cent moisture, in the absence of oxygen.

Haymaker A machine for curing hay.

Haze Fine dust or salt particles dispersed through a portion of the atmosphere; the particles are so small that they cannot be felt, or individually seen with the naked eye, they diminish horizontal visibility and give the atmosphere a characteristic opalescent.

Hazel nut Corylus avellana L. Family Betulaceae. Wild forms found in India. Grown for high nutritive value nuts. Also used for extracting oil. Raised from seeds; budding and grafting practised.

H-budding A double rectangular cutting resembling the letter H, made in the bark of the stock. The flaps are lifted in the centre for insertion of the bud.

Head A dense, short cluster of sessile or nearly sessile flowers on a very short axis or receptacle as in red clover or sunflower; also a fruit cluster. (2) A compact or orderly mass of leaves, as in lettuce. (3) The point region at which the trunk of a tree divides into branches.

Head (Hydraulics) (1) The height of water above any plane or reference. (2) The energy, either kinetic or potential, possessed by each unit weight of a liquid, expressed as the vertical to release the average energy possessed. Used in various compound terms such as pressure head, velocity head, and lost head.

Head ditch The main farm field channel.

Head of a canal The point where the supply is diverted into a canal from a reservoir, river, or parent canal.

Head of a drain The upstream end of a drain.

Head land Head land is the strip of unploughed land left at each end of the field for the tractor to turn while ploughing with a tractor.

Head mould False smut.

Head on To grow to a stock, as a bud in bud grafting.

Head out To form a flower head.

Head pruning The pruning of a plant in such a manner that the mature stem bears at its top a ring of arms spreading like an inverted cone.

Head reach The portion of principal or main canal, or branch canal from its source to the first off-take from it. In distributing canals, the head reach implies the portion of the canal from point of offtake to the first control point.

Head regulator A structure built at the intake end of an open channel to control the supplies of water and check trash and sediment entry into the channel.

Head row (1) A row drilled or sown at right angles to the regular rows across the end of a field. (2) A progeny row in a plant breeding nursery sown from the seed of a single panicle or head.

Head smut A fungus disease of maize and sorghum characterised by smut sori replacing the tassel and ear of panicle.

Head system of training A system of training grapevine in which top of the vine is cut to keep one or two buds at planting; one shoot only allowed to grow in first summer, trained straight and all side branches pinched back as they emerge. Lateral branches emerging below the cut continue to grow till vine is dormant in December. They are shortened at dormant pruning to spurs with 1 or 2 buds. Spurs produce a number of shoots during second summer, providing suitable arms for framework. At second dormant pruning 8 to 10 arms and one or two spurs on each arm retained cropping in 3rd year. Method low yielder.

Head-to-row Pedigree selection in grain crops where a number of heads are selected in the field. The seeds from each head are planted in a separate row with every tenth row being planted with a high-yielding variety for comparison. Each row is harvested and threshed separately and promising rows are planted in definite amounts and process repeated to see if a new promising variety is developed.

Headward erosion Lengthening of a valley at its upper end by gullying, produced by water which flows in at its head.

Head water Water level prevailing in the supply channel.

Heading back (1) Cutting back young trees in transplanting. (2) Also cutting tree tops to make them head at desired height and shape. Also see pruning.

Heading stage The stage of growth as in cereals when the heads start emerging from the sheath. Also called heading time.

Headworks, Diversion work, or **Diversion structure** A collective term for all works required at intakes of main or principal canals to divert and control river flows and to regulate water supplies into the main canal or canals.

Heart (1) A body organ. (2) Core or central portion of a tree or fruit.

Heart cam A cam, heart shaped in form used for the conversion of rotary motion into reciprocal motion.

Heart centre The wood immediately surrounding the pith which is characterised by discolouration, small checks, or pin-knots.

Heart rot (1) A rot involving disintegration of the heartwood of a tree. (2) A fungus disease of beets caused by Mycaspherela tabifica which results in decay of the central tissues of the plant. (3) A boron deficiency disease of sugarbeets that causes rot.

Heartwood The inner core of woody stem, wholly composed of non-living cells and usually differentiated from the outer enveloping layer (sapwood) by its darker colour.

Heat (1) A kind of energy manifested in various ways. (2) The period of sexual desire.

Heat capacity (volume) The amount of heat required to raise the temperature of a unit volume of soil by one degree. It may also be expressed in terms of weight. Also called volumetric specific heat.

Heat flux density The quantity of heat flowing per unit of time across a unit area.

Heather Calluna vulgaris. An evergreen health of northern and alpine regions distinguished by racemes of small purple-pink flowers.

Heat injury The injury caused to seed due to heat; drying at higher temperature may cause heat injury to seeds and may result in the lowering of seed germination.

Heat of combustion See gross energy.

Heat rigor A state of plants when they may survive for brief periods at temperature above the maximum.

Heavy A general term for soils containing high amount of clay. Heavy soil.

Heavy clay Soil material that contains more than 60 per cent clay, see also texture, soil.

Heavy metal toxicity Toxicity from metallic elements having high atomic weights, such as cadmium, cobalt, iron, lead, mercury, nickel, and zinc, may result in cellular injury, chromosomal damage, tumers, birth defects, as well as specified poisonings following ingestion or breathing of particles of such elements alone or often in compound form.

Heavy soil A soil having a high content of the fine separates, particularly clay, or a soil having a high drawbar pull and therefore, hard to cultivate. See also fine texture.

Heavy textured Fine textured soil in which clay and silt dominates. It has a dense structure and infirm to compact in consistency.

Hectare A land measurement unit equal to 100 × 100 m or 10,000 sq metres. Also equal to 2.47 acres.

Hedge Protective ring of plants around orchards, fields, buildings etc. Common hedge plants are: Inga dulcis, Parkinsonia aculeata L., Carissa carandas L., Casurina equisetifolia, Zizyphus spp., Duranta plumeiri Jacq., Cactus, Sesbania aegyptiaca, Acacia modesta Wall. Commiphera berryi Engl., Ipomea carnea, Punica granatum L., Thevetia nerifolia Schum.

Hedwigiaceae A family of mosses in the order Isobryales.

Heel The basal end of a plant stem cutting along with a piece of older stem.

Heeled in Trees or plants covered with moist earth in shallow trench or ditch.

Heel of the landside The rear end of the landside on a plough.

Height of head The distance of the lowest scaffold limb of a tree from the ground.

Height of hydraulic jump The difference between depth of water downstream and upstream of the jump.

Hekistotherm Plant adopted for conditions of minimal heat; can withstand long dark periods.

Helianthus annuus See sunflower.

— **tuberosus** See Jerusalem artichoke.

Helical groove A groove of uniform advance. The amount of advance in one revolution in a single helical groove is called the 'Pitch' in the case of a multiple helical groove, as double or triple, the amount of advance in one revolution is spoken of as the 'lead'.

Helical spring A compression type spring shaped like the frustum of a cone.

Helicoid cyme A type of determinate inflorescence having a coiled cluster, with flowers as only one side of the axis.

Heliconiaceae A family of monocotyledonous plants in the order Zingiberales, characterised by perfect flowers with a solitary ovule in each locule, schizocarpic fruit and capitate stigma.

Heliophyte Plants able to bear full sunlight.

Heliosis Production of discoloured spots on leaves through the concentration of sun. Also called solarization.

Heliothis armigera See gram caterpillar.

Heliotrope A plant whose flower or stem turns towards the sun. The process is known as heliotropism or phototropism.

Heliotrope tree Ehretia acuminata R. Br. Family Ehretiaceae. A tree whose fruit are edible.

Helix Spiral; genus including the common snail.

Hellula undalis See cabbage borer.

Helminthosporium avenae See oats leaf blotch.

— **blight** See rice brown eye spot.

— **frumentacei** See barnyard millet leaf spot.

— **gramineum** See barley stripe.

— **halodes** See coconut leaf rot.

— **heveae** See rubber bird's eye spot.

— **leucostylum** See ragi helminthosporium leaf blight.

— **nodulosum** See ragi helminthosporium leaf blight.

— **oryzae** See rice brown eye spot.

— **sacchari** See sugarcane leaf spot.

— **sativum** See barley foot rot and root rot.

— **sigmoideum var. irregularae** See rice irregular stem rot.

— **teres** See barley net blotch.

— **tetramera** See ragi helminthosporium leaf blight.

— **turcicum** See maize leaf blight.

Helminthosporium leaf spot Any leaf spot, commonly of grasses, caused by Helminthosporium species.

Helobiales An order embracing most of the Alismatidal in certain systems of classification.

Helopeltis antonii See tea mosquito.

— **theivora** See tea mosquito.

Helophyte A marsh plant, buds over winter under water.

Helophytia Differences in ecological control by fluctuations in water level such as in marshes.

Helotism Symbiosis in which one organism is a slave to the other, as between certain species of ants.

Hemeranthous Flowering by day.

Hemicellulose Heterogeneous polysaccharide fraction existing largely in the secondary cell wall of the plant.

Hemicryptophytes Perennial grasses or other plants whose upper parts die back to the ground surface during unfavourable season, leaving the reproduction buds still under the upper layers of soils such that these reproduce when the seasons is favourable.

Hemicydic Of flowers, having the floral leaves arranged partly in whorls and partly in spirals.

Hemidesmus indicus See Sarsaparilla.

Hemileia vastaraix See coffee rust.

Hemiparasite A parasite capable of a saprophytic existence, especially certain parasite plants containing some chlorophyll. Also known as semi-parasite.

Hemiptera Family of sucking insects which includes bugs, aphis, black flies.

Hemlock The common name for members of the genus Tsuga in the pine family characterised by two white lines beneath the flattened, needlelike leaves.

Hemp Cannabis sativa L. Family Cannabinaceae. Female flowering tops and leaves source of resin commonly known as *Charas*, *Bhang* obtained by drying leaves and flowering shoots of either of, or both the male and female plants. Fibre from plants used for rope making. Annual weed in cultivated fields in some areas.

Henna Lawsonia inermis L. (Lawsonia alba Lamk.) Family Lythraceae. Woody perennial shrub cultivated throughout India. Fresh and dry leaves provide dye and serve as fodder. Propagated from cuttings and seed. Flowers used for providing fragrance to cosmetic oils. Also called *Mehndi*.

Heptachlor A chlorinated hydrocarbon 1, 4, 5, 6, 7, 8, heptachloro 3a, 4, 5, 5a-tetrahydro-4, 7-endomethanoindene. It is more toxic than BHC. Used as contact and stomach insecticide and available as dust water dispersible powder and emulsifiable concentrate. Used against grasshoppers, locusts, thrips, beetles, caterpillars, white grub, termites and cut-worms.

Herb (1) A plant which does not develop much woody tissue. The tops die down to the ground each year. (2) A term used to designate certain aromatic and drug plants.

Herbaceous Plant growth that is relatively free of woody tissue.

Herbage (1) Leaves, stems and other succulent parts of forage plants upon which animals feed. (2) Non woody plants.

Herbarium (1) A collection of dried or otherwise preserved plant specimens used for reference and verification of names. (2) A building where herbarium is located.

Herbicide Phytotoxic chemical used to kill or inhibit plant growth.

Herb of grace Ruta graveolens L. Family Rutaceae. Small perennial under-shrub. Leaves have strong aromatic odour and bitter taste. Used as condiments and in medicine. Propagated from seed or cuttings. Also called *rue*, *sadab*.

Herbivore Animal that subsists principally or entirely on plants or plant products. Herbivorous animals.

Hereditable Capable of being passed by inheritance.

Heredity The transmission of parental qualities, expressed or latent to the progeny.

Heritability Portion of observed variability in progeny that is due to heredity, the remainder being due to environmental causes.

Hermaphrodite An individual with both male and female sex organs.

Hermetic Air-tight closure by fusion.

Hermetically sealed Pertaining to an air-tight closing of container.

Herringbone The name given to masonary or brickwork when laid up in a zig-zag pattern.

Herringbone drainage system A system in which the main drain and the laterals have a herringbone arrangement.

Herringbone furrow method A type of layout for furrow irrigation on the hillsides where soils are highly permeable. Concrete flumes of various sizes are laid down the slope. Water is intercepted and diverted into furrows at right angle, to the flumes. Water is run towards the centre of the field from two flumes bounding this field. By modifying the slope of the furrow water applied to each end of the furrow meets in the central portion.

Herringbone gear A gear in which the teeth slopes both ways from the centre line of the gear face, as would be the case if two spiral gears, one left hand and one right hand were fastened together, used for heavy work on mining machinery etc.

Herse convolvuli See sweet potato sphinx.

Hesperidium A modified berry, with few seeds, a leathery rind, and membranous extensions of the endocarp dividing the pulp into chambers; an example is the orange.

Hessian Light plain woven fabric made up of high quality white jute. Also called *burlap.* Use in packaging and drain tiles.

Heterandrous Having stamens differing from each other in length or form.

Heteroblastic With a morphological difference between juvenile and adult (e.g. in leaf shape and size).

Heterocarpus Producing two distinct types of fruit.

Heterochromous Of different colours.

Heterocyst An enlarged resting cell of certain members of the Cyanophyta. Usually it is transparent and thick walled.

Heteroecious Requiring two hosts to complete the life cycle as a parasite.

Heterogamous Bearing 2 or more kinds of flowers on the one inflorescence.

Heterogeneous Differing in kind; having unlike qualities; possessed of different characteristics; opposed to homogeneous.

Heterogen A variable group of plants or animals which arise as hybrids, sports, mutation etc., types of which may or may not breed true.

Heterogeneratae A class of brown algal distinguished by a heteromorphic alteration of generations.

Heterogenesis Birth of an organism otherwise than from parents of the same kind.

Heterogeneous parasite A parasite requiring two or more hosts for its complete development.

Heteromerous Of a flower, having one or more whorls made up of different number of members than the remaining whorls.

Heteromorphic Manifesting a type of alternation of generations in which the gametophytic plant is distinctly different from the sporophytic plant.

Heterophyllous Having more than one form of foliage leaves on the same plant or stem.

Heterophyte A plant that depends upon living or dead plants or their products for food materials.

Heteroploid See Aneuploid.

Heteropogon contortus See spear grass.

Heteroptera Order of insects to which bugs belong.

Heterosis Increased vigour or capacity of growth often resulting from crossing organisms genetically unlike. Negative heterosis also is recognised by some geneticists when decreased vigour results. Also called hybrid vigour.

Heterotic hybrid A hybrid which is an improvement over the parents.

Heterotic vigour See heterosis.

Heterosporous Producing spores of two types.

Heterostemony Presence of two or more different types of stamens in the same flower.

Heterostyly Condition or state of flowers having unequal styles.

Heterothallic Of a fungus, sexes separate in different mycelia.

Heterotrophic Capable of deriving energy for life processes only from the decomposition of organic compounds, and incapable of using inorganic compounds as sole sources of energy or for organic synthesis, See also autotrophic.

Heterotrophic bacteria Majority of soil bacteria which depend for their energy upon organic substances. Result in decomposition of cellulose and other carbohydrates, proteins, fats and waxes. Release nitrogen, phosphorus and other nutrients which are in a form in available to plants. An important class is nitrogen fixing bacteria.

Heterozygous The condition in which the homologous chromosomes of an individual possess different genes of the same allelic series. These plants do not breed true.

Heterozygote An individual to which two parents contributed unlike genes in regard to any given allelomorphic pair governing contrasted characters and which produces in turn two kinds of germ cells with respect to the characters. See Heterozygous.

Heterusia cingala see tea red slugs.

— **magnifica** See tea red slugs.

HE.TP. Hexaethyle tetraphosphate; an organo-phosphorus compound. Used in the control of aphids, thrips, scale insects, mealy bugs and mites. Available as dust and emulsifiable concentrate.

Hevea brasiliensis See rubber.

Hewn timber Timber converted to size by an axe. The ends are sometimes sawn.

Hexachlorophene A chemical compound used as insecticide.

Hexagonal system of planting A system of planting in which plants are grown at each corner of equilateral triangle by which six trees form a hexagon. The 7th tree is planted in the centre. More plants per hectare can be planted.

Hexonic acid See caproic acid.

Hexose Monosaccharide, ($C_6H_{12}O_6$). Comprises dextrose, levulose, galuctose and mannose.

Hibernate To pass unfavourable periods in a resting stage.

Hibiscus abelmoschus See musk mallow.

— **cannabinus** See mesta.

— **ficulneus** See ranbhindi.

— **furcatus** See kiasi.

— **esculantus** See lady's finger.

— **manihot** See ranbhindi.

— **Sabdariffa** See roselle.

— **Surattensis** See ranbhindi.

— **Syriacus** See Rose of Sheron.

— **tiliaceus** See Cuban bast.

— **vitifolius** See bankapas.

Hickory The common name for species of the genus Carya in the order Fagales; tall deciduous tree with pinnately compound leaves, solid pith, and terminal, scaly winter buds.

Hierarchy It is the rule of control of the higher over the lower. It is a graded organisation of several successive steps or levels, in which each of the lower levels is immediately subordinate to the next higher one, and through it to the higher steps right up to the top.

Hieroglyphus banian See rice grasshopper.

— **nigrorepletens** See phadka grasshopper.

Higgling The process by which price is determined in the market place where there is direct negotiation between buyer and seller.

High compression A state of compressibility of gases, usually in a combustion chamber, brought about by a small amount of compression space in the upper part of the chamber.

High finance Finance (q.v.) concerned with large sums of money.

High grade (1) Of good quality. (2) A fertiliser of high plant nutrition.

High grading The removal from the stand of only the best trees.

High water line of tide Intersection of plane of mean high water with the shore.

Hill bunt See wheat bunt.

Hill coconut Stercularia foetida, L. Family Sterculiaceae. A tree whose seeds are eaten. Also called wild almond.

Hill goose berry See hill guava.

Hill guava Rhodomyrtus tomentosa wight. A shrub of family Myrtaceae whose fruits are edible and bark used as tan.

Hill mango Commiphora caudata (Wight & Arn.) Engl. (Balsamodendron caudatum Wight & Arn.). Family Beuseraceae. A tree or shrub from which gum-resin is obtained and leaves used as medicine and as incense. Fruit edible.

Hilo grass Paspalum conjugatum, Berg. Family Gramineae. A perennial grass grown for fodder.

Hilum (1) A scar left where the seed stalk of funiculus breaks away, or where the seed was attached directly to the placenta when there is no seed stalk. (2) Eye of the seed. (3) Small notch, opening where vessel, nerves etc. enter.

Himalayan bamboo Arundinaria falcata Nees. (Bamboo falcata [Hort]). Family Gramineae. Bamboo found in western Himalayas and Sikkim.

Himalayan black cedar Alnus nitida Endl. Family Betulaceae. A large tree whose wood is used for match sticks, bark yields tan and dye, twigs used for matching baskets. Also call *Saroli.*

Himalayan rhubarb Rheum emodi wall. Family Polygonaceae. A herb whose roots and rhizomes are source of a drug used as a laxative, tonic and purgative.

Himalayan yellow raspberry Rubus ellipticus Smith. Family Rosaceae. A shrub producing edible fruits. Also called *lal anchu.*

Hinterland A land remote from population centres; an underdeveloped area.

Hippobasca capensis See tick fly.

— **masculata** See tick fly.

Hire purchase A procedure for purchasing goods under which the purchaser pays a deposit on receipt of the goods followed by a number of instalments until the debt is cleared.

Hispa armigera See rice hispa.

Histogen A clearly delimited region or primary tissues from which the specific parts of a plant organ are thought to be produced.

Histogenesis Production of organic tissues. Also called histogeny.

Histology The science dealing with microscopic structure of animals or plant tissues.

Hit and miss system It is system of governing the engine wherein the frequency of explosions or power strokes of the engine are regulated. At increased speeds, the exhaust valve gets opened which prevents fresh charge for entering the engine cylinder during suction strokes.

Hitch Connecting of an implement, machine to source of power.

Hitchenia caulina See Indian arrowroot.

H-layer See horizon, soil.

Hoary Having greyish or whitish colour, referring to leaves.

Hog An implement with a handle and a light cutting tool. Used for weeding, mulching of top soil. The process is known as hoeing.

Hog plum Spondias mangifera Willd. (Spondias pinnata (Koen.) Kurz.; Mangifera pinnata L.f.). Family Anacardiaceae. Tree producing fruits which makes good preserve. Barks used as a tan. Propagated by seeds.

Hollacarip (1) Having the entire thallus developed into a fruiting body or sporangium. (2) Lacking rhizoids and haustoria.

Holcus halepensis See Johnson grass.

— **sorghum** See sorghum.

Holdfast The basal attaching cell or organ of an algae or fungus, often composed of rhizoids.

Holding Amount of land held by a farmer or an institution.

Hollock Terminalia myriocarpa. Heurck & Muell-Arg. A tree of family Lombretaceae, evergreen whose wood is used as fuel or as building material and plywood. Raised by transplanting seedlings.

Holly The common name for the trees and shrubs composing the genus Ilex; distinguished by spiny leaves and small berries.

Holoptelea integrifolia See kanju.

Holostemma annulare See chhirvel.

— **reheedii** See chhirvel.

Holotrichia longipennis See white grubs.

Holocaenosis The nature of the action of the environment on living organisms.

Homeostasis The tendency of a system to maintain a dynamic equilibrium and, if disturbed to restore that equilibrium. In an ecological sense, a community in a steady state, with minimal free energy.

Homeostatic Maintenance of constancy or a high degree of uniformity in functions of an organisms or interactions of individuals in a population or community under changing conditions of the environment because of the capabilities of organisms to make adjustments.

Homeostatic forest A forest in a condition with minimum free energy.

Homestead A farm dwelling with surrounding ground.

Homoaeomerous Having algae distributed uniformly throughout the thallus of a lichen.

Homogamous Having all florets hermaphrodite of same sex.

Homogeneous Of the same kind, or nature, consisting of similar parts or of elements of like nature, opposed to heterogeneous.

Homogenetic Having common descent of origin. Having same characteristics due to this is thus called Homogeny.

Homogamy (1) Condition of having one type of flower, with stamens and pistils of uniform length. (2) Simultaneous maturation of male and female elements in the same flower.

Homoiochlamydeous Having perianth leaves alike, not differentiated into sepals and petals.

Homologous Chromosomes occur in somatic cells in pairs that are similar in size, shape and supposedly in function, one being derived from the male and one from the female parent. The two members of such a pair are spoken of as homologous chromosomes.

Homology Similarity of structure due to similarity of origin.

Homomorphism Having perfect flowers consisting of only one type.

Homomorphous Having similar or same form. Homomorphic.

Homonomous Having same law of growth.

Homoplastic Having similar structure.

Homoptera Order of insects to which jassids, pyrilla, white flies, aphids and scale insects belong.

Homospary Production of only one kind of asexual spore.

Homosphere The lower portion of a two-part division of the atmosphere (the upper portion is the heterosphere) according to the general homogeneity of atmospheric composition the region in which there is no gross change in atmospheric composition, that is, all of the atmosphere from the earth's surface to about 80-100 kilometres.

Homothalic Both sexes present in same mycelium.

Homothallism The condition of plants in which sexual reproduction takes place in a single thallus. The plants are self-compatible.

Homotype Part or organ, like any other in structure.

Homozygote An individual, the parent of which contributed to it a similar number of any given pair of genes and whose germ cells are all alike in regard to the genes for that character. See Homozygous.

Homozygous Having identical genes for any given pair or series of factors. Homozygous plants breed true for the character in question.

Honeycomb frost Ice in the soil in insufficient quantity to be continuous, thereby giving the soil an open, porous structure that readily permits water to enter.

Honey combing Internal cracks in timber due to drying stresses.

Honeycomb radiator Made from a number of small cells extending from front to back, laid one upon another, and soldered together. Water flows in a thin film around the cells or tubes and is cooled by the air flowing through them.

Honey extractor A device used for extracting honey by centrifugal force.

Honey locust Gleditschia triacanthos L.A spinous tree of the family Caesalpiniaceae. Ripe pods are used as fodder.

Hop Humulus lupulus L. Family Moraceae. A twining perennial herb grown for its cone-shaped fruits used by brewers, and in some medicines.

Hop clover Medicago lupulina L. A annual prostrate herb of family Papilinaceae grown as green manure and fodder crop.

Hopea parviflora See iron wood.

Hopper (1) The part of a seed drill or other machines in which seed or fertiliser is stocked. (2) One which hops; a hopping insect.

Hopperburn A disease of potato and peanut plants caused by a leaf hopper which secretes a toxic substance on the leaves, causing browning and shriveling.

Hordein See prolamines.

Hordeum sativum See barley.

— **vulgare** See barley.

Horizon (soil) A layer of soil, roughly parallel to the surface of the soil, with distinct characteristics which were produced by soil-forming processes. Horizons are classified as A_{00}, A_0, A_1, A_2, A_3, B_1, B_2, B_3, C_1, C_2, C_3 and D. Broadly as A, B, C and D.

Horizontal bench terrace A bench terrace which has no measurable slope from back to front of the bench.

Horizontal clevis It is a device to effect lateral adjustment of the plough relative to the line of pull.

Horizontal control Basic framework of points whose horizontal position azimuth and interrelationship have been determined accurately.

Horizontal eluviation A horizon.

Horizontal illuviation B Horizon.

Horizontal or flat storage A storage structure, the height of which is less than the width or diametre.

Horizontal photograph Photograph made with the camera axis horizontal.

Horizontal silo A silo built with its long sides parallel to the ground. Openings are left on both sides.

Horizontal suction This suction helps the plough to cut the proper width of furrow slice and is the maximum clearance between the land side and a horizontal plane touching point of share at its gunnel side and heel of the landside.

Horizontal training The training of plants to grow on a horizontal wire, trellis etc.

Hormogonium In certain blue green algae, a portion of the filament that becomes detached because of dead cells or heterocysts, but continues to grow into a new filament.

Hormonal weedicide Hormones which bend, contort or twist the stem and leaves of weeds and ultimate death; are applied as weedicides.

Hornbean Any tree of the genus Carpinus in the birch family distinguished by doubly serrate leaves and by small, pointed angular winter buds with scales in four rows.

Horned liverwort The common name for bryophytes of the class Anthocerotae. Also known as hornwart.

Horse chestnut Aesculus hippocastanum. An ornamental buckeye tree in the order Sapindales, usually with seven leaflets per leaf and resinous buds.

Horse-eye bean Mucuna prurita Hook. (Mucuna pruriens Baker non DC). A herb of family Papilionaceae, the fruits of which are used as vegetable.

Horse hair blight A fungus disease of tea and certain other tropical plants caused by Marasmius equierinis and characterised by black festoons of mycelia hanging from the branches.

Horse parslane Trianthema portulacastrum L. (T. monogyna L.) Family Aizoaceae. A common succulent herb, used as vegetable.

Horse dish Armoracia rsuticana. A perennial crucifer belonging to the order Capparales and grown for its pungent roots, used as a condiment.

Horse-radish tree See drumstick tree.

Horse tamarind See white popinac.

Horsegram Dolichos biflorus Roxb. Family Papilionaceae. A trailing herb. Most common pulse in south India grown mainly for feed and fodder of cattle and horses. Makes good hay and used as green manure crop. Leguminous and hence fixes nitrogen.

Horsegram anthracnose Disease caused by Glomerella lindemuthianum. Symptoms same as in bean anthracnose.

Horsegram die-back Disease of horsegram caused by Colletotrichum capsici. Symptoms same as in cowpea die back.

Horsegram dry root rot Disease caused by Macrophomina phaseoli. Symptoms same as in cotton dry root rot.

Horsegram Leaf spot Disease caused by Cercospora dolichi with symptoms similar to bean leaf rot.

Horsegram root rot Disease caused by Pellicularia filamentosa. Symptoms same as in cowpea root rot.

Horsegram rust Disease caused by Uromyces appendiculatis. Symptoms as in bean rust.

Horsetail The common name for plants of the genus Equisetum composing the order Equisetales.

Horticulture The art and science of growing plants. It includes growing of vegetables, fruits, flowers and ornamental shrubs and trees.

Horticultural crop Any fruit producing plant, A garden or orchard crop.

Hose Flexible pipe or tube, generally of rubber or plastic used for carrying and distributing liquids.

Hose clamp A clamp for making a tight joint between a hose and some less flexible object over which the end of the hose fits.

Hose coupling A device, generally a union, for joining together the ends of hose lengths.

Host Any plant attacked by a parasite.

Host range The range of plants and animals which are affected by a pathogen.

Host specific The parasite that lives only on one host.

Hot Designating a fruit such as chilli which has strong, pungent flavour.

Hot logging A logging operation in which logs go from the stump to the mill without pause.

Hot money Money which is transferred rapidly from one financial centre to another to take advantage of differences in shortterm interest rates or to escape the financial penalties of devaluation.

Hot water treatment Treatment of seed for controlling internal seed borne fungus. Seed is soaked for 4-5 hours in cold water and then dipped in hot water at 90°C for one minute and then in hot water at 54°C for 10 minutes. Seed is dried before use.

Household An economic unit consisting of a single person living alone, a married couple or a complete family, each unit with a source of income and responsibility for its disposal.

Household (group) It includes the members of the family and other persons who may be residing there to assist in farming and other household affair. There is possibility of a greater degree of interactions among the members of the same household and thus any change adopted or refused will involve the whole group.

House rat Rattus rattus. Omnivorous in habit. Feeds and damages all eatables.

Houtbois strawberry Fragaria nilgerrensis Schlecht ex. F. Gay. Family Rosaceae. A herb whose fruits are edible.

Hovenia dulcis See Japanese raisin tree.

Hubum clover Meliolotus alba var. annua Desr. Family Papilionaceae. Legume fodder. Good for semi-arid region and irrigated conditions. Commonly grown in north India.

Hue The aspect of colour that is determined by the wavelengths of light, and changes with the wavelength. Munsell hue notations indicate the visual relationship of a colour to red, yellow, green, blue, or purple, or an intermediate of these hues. See also Munsell colour system: chroma; and value, colour.

Hull The term applied to include the lemma and palea when they remain attached to the caryopsis after threshing. To remove hulls from seed. The outer, usually hard, covering of a fruit or seed.

Huller Any device which is used to remove hull from seeds. The process is commonly known as hulling and is generally employed in converting paddy into rice.

Humic acids A mixture of various dark-coloured organic substances precipitated by acidifying a dilute alkali extract from the soil. The term is used by some workers to designate only the alcohol-insoluble part of this precipitate.

Humic gleysol A great group of soils in the Gleysolic order. A dark-coloured A (Ah or Ap) horizon more than 8 cm (3 inches) thick is underlain by mottled grey or brownish gleyed mineral material. It may have up to 40 cm (16 inches) of mixed peat (bulk density 0.1 or more) or up to 60 cm (24 inches) of fibric moss peat (bulk density less than 0.1) on the surface. This group includes soils formerly classified as Dark Grey Gleysolic and meadow.

Humic layer A layer of highly decomposed organic soil material containing little fibre.

Humic mulch Decayed and disintegrated organic residue on the soil surface which acts as a mulch.

Humic podzol A great group of soils in the Podzolic order occurring in cool humid coastal regions, cool humid inland locations at higher altitudes, and some peaty depressions. The soils have a dark-brown to black Bh horizon at least 10 cm (4 inches) thick, having more than 1 per cent organic carbon, less than 0.3 per cent pyrophosphate-extractable Fe, a ratio of organic carbon to pyrophosphate-extractable Fe of 20 or more and a very low base saturation (NaCl) extraction. A thin iron pan or a series of very thin (totalling less than 2.5 cm) iron pans may be present.

Humicolous Of or pertaining to plant species inhabiting medium-dry ground.

Humid An area or climate that has more moisture than the actual agricultural requirement and where drainage facilities are generally essential to get rid of surplus moisture.

Humid climate A climate with sufficient precipitation to support forest vegetation, 500 mm or more in cool regions, up to 1500 mm in hot regions.

Humidity Atmospheric water vapour content, expressed in any of several measures, such as relative humidity.

Humidity, Relative The ratio of actual mass of water vapour per unit of volume to mass of water vapour that would saturate that volume at the same temperature and pressure, or roughly the per cent saturation of the space.

Humification The processes by which organic matter decomposes to form humus. In humus the initial structures or shapes can no longer be recognised. See also humus.

Humin The fraction of the soil organic matter that is not dissolved when the soil is treated with dilute alkali.

Humiriaceae A family of dicotyledonous plants in the order Linales characterised by exappendiculate petals, usually five petals, flowers with an intrastaminal disk, and leaves lacking stipules.

Humisol A great group of soils in the Organic order that are saturated for most of the year. The soils have dominantly humic middle tier, or middle and surface tiers if a terric, lithic, hydric, or cryic contact occurs in the middle tier.

Humivore An organism that feeds on humus.

Humo-ferric podzol A great group of soils in the Podzolic order. The upper 10 cm of the B horizon (Bf) contains between 0.5 per cent and 5 per cent organic carbon and 0.6 per cent or more pyrophosphate-extractable Al and Fe (0.4 per cent for sands). The ratio of organic carbon to pyrophosphate extractable Fe is less than 20. Most of the typical Podzols are classified as Humo-Ferric Podzols.

Humulus lupulus See hop.

Humus (1) The fraction of the soil organic matter that remains after most of the added plant and animal residues have decomposed. It is usually dark coloured. (2) Humus is also used in a broader sense to designate the humus forms referred to as forest humus. They include principally mor, moder, and mull. See also organic matter. Soil; mor, moder; mull; and horizon, soil. (3) All the organic material on and in the soil that undergoes continuous breakdown, change and synthesis.

Hundred-weight A unit of weight; 112 pounds avoirdupois.

Husbandry Art, science, processing and business of production of animals or poultry. Used in conjunction as animal husbandry or poultry husbandry.

Husk The outer coat of certain seeds, particularly if it is a dry, membranous structure.

Huskleberry The common name for shrubs of the genus Gaylussacia in the family Ericaceae distinguished by an ovary with 10 locules and 10 ovules, the dark-blue berries are edible.

Hyacinth bean See lablab bean.

Hyaline Colourless, or nearly transparent.

Hybrid Product of a cross between individuals of unlike genetic constitution or makeup. Species hybrid when it is the progeny of two parents much alike but differing in genes at one or more loci on one or more pairs of chromosomes. Intergeneric hybrid when it is the progeny of parents that differ greatly as of two species or of two genera. Numerical hybrid when it is a progeny of two parent that differ in chromosome numbers. Structural hybrid is the one whose parents differ in the structure of their chromosomes.

Hybrid Eucalyptus Also called Mysore hybrid. Result of cross pollination of a number of species of Eucalyptus found in the nandi hills of Mysore State. Fast growing, resistant to alkalinity, hardy, wood used as fuel, for pulp making and rayon.

Hybrid production The process of producing a hybrid from inbred lines in plants. The process consists of development of inbred lines by controlled self-pollination, determination of the best combining lines that go to make a cross productive, and commercial utilisation of the cross for seed production.

Hybrid vigour The phenomenon in which the cross of two stocks produces hybrids that show increased vigour. Also called Heterosis.

Hybridisation The process by which organisms of unlike hereditary constitution are crossed to produce F_1 progeny with genes for the characteristics of both parents.

Hybridise To create a hybrid.

Hydragogue Substance that drives water out from the body; especially a catharatic that causes watery purgation.

Hydrate Any compound that contain water which is expressed in molecules per molecules of the substance. Mono:1, di:2, tri:3, tetra:4, penta:5, hexa:6, hepta:7, etc.

Hydrated lime Calcium hydroxide.

Hydration Chemical combination of water with another substance.

Hydraulic Pertaining to fluids which are moving or at rest under forces of gravity or pressure.

Hydraulic cement A cement which will harden under water.

Hydraulic conductivity Hydraulic conductivity is the specific property designating the rate with which a porous medium transmit fluids under standard conditions regardless of the fluid transmitted. The term implies a measurement of both the properties of the fluid transmitted as well as the porous medium.

Hydraulic control systems It is a mechanism to raise or hold or lower the mounted or semi-mounted equipments by hydraulic means.

Hydraulic drive See fluid drive.

Hydraulic drop A local phenomenon in which a rapid change in the state of flow from subcritical to supercritical, resulting in a steep depression in the water surface, is caused by an abrupt change in the channel slope or cross section.

Hydraulic efficiency Efficiency of a pump or turbine to impart energy to or extract energy from water. The ability of a hydraulic structure or element to conduct water with minimum energy loss. The ratio of the capacity of a cross section of conveyance conduit and the one which is most efficient for the corresponding hydraulic shape.

Hydraulic elements Depth, area, wetted permetre, mean depth, hydraulic radius, velocity, energy, and other such quantities pertaining to hydraulics.

Hydraulic friction The combined effect of all sources of resistance to the flow of liquid.

Hydraulic gate A device consisting of a structure made of wood, metal or other material, for controlling the flow of water through a passageway or an opening in a structure.

Hydraulic grade line In a closed conduit a line joining the elevations to which water could stand in open gauge tubes. In an open conduit, the hydraulic grade line is the water surface.

Hydraulic gradients The slope of the hydraulic grade line. The slope of the free surface of water flowing in an open channel.

Hydraulic head See water, soil.

Hydraulic head charge The elevation with respect to specified reference level at which water stands in a piezometre connected to the point in question in the soil. Its definition can be extended to soil above the water table if the piezometre is replaced by a tensiometre. The hydraulic head in systems under atmospheric pressure may be identified with a potential expressed in terms of the height of a water column. More specifically, it can be identified with the sum of gravitational and matric-potentials and may be called the hydraulic potential.

Hydraulic jack A lifting jack, actuated by a small force pump enclosed within it, and operated by a lever from the outside.

Hydraulic jump The sudden and usually turbulent passage of water from low level below critical depth to high level above critical depth during which the velocity passes from supercritical to subcritical. It represents the limiting condition of the surface curve wherein it tends to become perpendicular to the stream bed.

Hydraulic jump basin, or Hydraulic jump dissipator A stilling basin in which energy is dissipated by hydraulic jump principle.

Hydraulic press (1) A machine used for crushing oil seeds, thereby extracting oil by hydraulic pressure. (2) A machine used to compact the material for binding. Reduces in volume and hence binding is made easy.

Hydraulic ram Device for lifting water by utilising energy of flowing water.

Hydraulic sprayer Power sprayer of

hydraulic type.

Hydric layer A layer of water in the control section of organic soils, extending from a depth of not less than 40 cm (16 inches) to a depth of more than 160 cm (64 inches).

Hydrobiology The study of the life of aquatic plants and animals.

Hydrocarbons A compound which contain two elements i.e. hydrogen and carbon.

Hyrdrochloric acid (HCl) A strong, mineral acid.

Hydroclimate The climate of water bodies such as lakes etc.

Hydrochory Dispersal of disseminules of water.

Hydrocyanic acid (HCN) Poison (also called prussic acid) produced as a glucoside by several plant species, especially sorghums. Hydrocyanic acid gas is deadly poisonous which is produced by mixing sodium cyanide with sulphuric acid.

Hydroflouric acid H.F. A strong, mineral acid.

Hydrogen (H) (1) an element. One of the essential elements for plant growth. (2) A colourless gas which is highly inflammable.

Hydrogen di-sulphide (H_2S) A compound of hydrogen and sulphur.

Hydrogenation It is the process of making vanaspati from oils. Pure hydrogen through a vegetable oil is bubbled under pressure in the presence of nickle in closed vessels at 160° to 180°C. Hydrogen thus added converts unsaturated compounds in the oil to saturated ones.

Hydrogen ion concentration The amount of hydrogen-ions present in a liquid or soil. The usual measurement is by the pH which is negative log of the hydrogen ion concentration. Determines whether the soil is acidic or alkaline.

Hydrogen peroxide A compound of hydrogen and oxygen used as bleaching agent and as an antiseptic.

Hydrogen swell An expansion of the can in canned goods due to formation of hydrogen as a result of corrosion of the tin plate.

Hydrogenic soil Soil developed under the influence of water standing within the profile for prolonged periods; it is formed mainly in cold, humid regions.

Hydrograph A graph showing the stage, discharge, velocity or some other features of water, with respect to time. For example, a graph showing the discharge of a stream against the time is called 'discharge hydrograph', A graph of sediment concentration or sediment charge with respect to time is 'sediment hydrograph', Other examples of hydrographs are 'flood hydrograph' 'overland flow hydrograph', stage hydrograph, 'well hydrograph'.

Hydrography Science dealing with certain aspects of all the waters of the earth's surface, particularly sounding, charting and mapping them, and analysing and describing their forms, positions and physical features of rivers, lakes, reservoirs, shallow, deeps, etc., and magnitudes and directions of winds, tides, currents, and the like.

Hydrologic equation The water inventory equation (inflow = outflow + storage) which expresses the basic principle that during given time interval, the total inflow to an area must equal the total outflow plus the net change in storage.

Hydrologic cycle The circulation of water from the sea, through the atmosphere, to the land, and thence, often with many delays, back to the sea or ocean through various stages and processes as precipitation, interception, runoff infiltration, percolation, groundwater storage, evaporation and transpiration, also the many short circuits of the water that is returned to the atmosphere without reaching the sea.

Hydrology Science dealing with waters of the earth in rivers, streams, lakes in or below the land surface, in the atmosphere (related aspects only) in all its states — their occurrence, distribution and circulation through the unending hydrologic cycle of precipitation, consequent runoff, stream flow, infiltration and ground water, eventual evaporation and reprecipitation. It is concerned with the physical, chemical and psysio-logical reactions of the water with the rest of the earth and its relation to the life of the earth.

Hydrolysis The process by which a substrate is split to form two end products by the intervention of a molecule of water.

Hydrometre An instrument to measure the specific gravity of liquids.

Hydromorphic soil A general term for soils that develop under conditions of poor drainage in marshes, swamps, seepage areas, or flats. See also Gleysolic.

Hydrophyllaceae A family of dicotyledonous plants in the order Polemoniales distinguished by two carpels, parietal placentation, and generally imbricate corolla lobes in the bud.

Hydrophyte (1) A plant that grows in a moist habitat. (2) A plant requiring large amounts of water for growth. Also known as hygrophyte.

Hydroponics Growing of plants in a nutrient solution with the mechanical support of an inert medium such as sand. Also called chemi-culture, aqua culture, tank farming, tray farming etc.

Hydroponics or nutriculture The technique or growing crops in an aqueous nutrient solution, without soil.

Hydrosere Community in which the pioneer plants invade open water, eventually forming same kind of soil such as peat or muck.

Hydrosols A group of natural soils in which water is the principal component.

Hydroxide of lime Hydrated lime.

Hydrosphere The water portion of the earth as distinguished from the solid part (lithosphere) and from the gaseous outer envelope (atmosphere).

Hydrous mica A term used in two different ways (i) the groups of clay-sized micas that have a higher lattice water content and lower potassium content than ideal mica (illite), (ii) interstratified montmorillonite- and vermiculite-mica minerals in which mica predominates.

Hyetal region A region in which the amount and seasonal variation of rainfall are of a given type.

Hyetography The study of the annual variation and geographic distribution of precipitation.

Hyetology The science which treats of the origin, structure, and various other features of all the forms of precipitation.

Hygrokinematics The descriptive study of the motion of water substances in the atmosphere.

Hygrology The study which deals with the water vapour content (humidity) of the atmosphere.

Hygrometre An instrument to measure relative humidity of air.

Hygroscopic Being sensitive to moisture, such as certain tissues. Possessing a marked ability to accelerate the condensation of water vapour; applied to condensation nuclei composed of salts which yield aqueous solutions of a very low equilibrium vapour pressure compared with that of pure water at the same temperature. Also called hydroscopic.

Hygroscopic coefficient (Obsolete). The weight percentage of water held by, or remaining in, the soil (i) after the soil has reached equilibrium with an unspecified environment of high relative humidity, usually near saturation, or with a specified relative humidity at a specified temperature. Moisture corresponding to pF of 6.0.

Hygroscopic depression The measure of a desiccant's capacity to take on water.

Hygroscopic water Water adsorbed by a dry soil from an atmosphere of high relative humidity, water lost from an air-dry soil when it is heated to 105°C; water held by the soil when it is at equilibrium with an atmosphere of a specified relative humidity at a specified temperature, usually 98 per cent relative humidity at 25°C, water is not available to vegetation. Also called hydroscopic water or hygroscopic moisture.

Hymatomelanic Acid The fraction of humus that is soluble in alcohol, after having been extracted with alkali and precipitated with acid, and which, when the alcohol is distilled, forms a brittle mass that is insoluble in alcohol.

Hymenium Spore-bearing layer of a fungus fruiting body.

Hymenoptera The group of insects such as bees and ants, which have two pairs of transparent, membrane like wings.

Hyperimmune Quality of possessing a degree of immunity greater than that found under similar circumstances.

Hypermetamorphosis A protracted and thorough-going metamorphosis; Metamorphosis involving two or more distinct types of larval instar in certain insects.

Hyperparasite An organism that is parasitic on or in a parasite.

Hyperplastic Term applied to a disease producing an abnormally large number of cells.

Hyperpycnal inflow A denser inflow that occurs when a sediment-laden fluid flows down the side of a basin and along the bottom as a turbidity current.

Hypersensitivity Immediate death of the tissue when attacked by a pathogen. The property is useful as it prevents further spread of infection.

Hypertrophy (1) The increase in size of part of a plant due to plant disease. (2) General increase in the bulk of an organ or a part of it, but not due to tumour formation.

Hypha Single thread of a fungus mycelium.

Hyphopodium More or less lobed appendage to a hypha.

Hypnineae A suborder of the Hypnobryales characterised by complanate, glossy plants with ecostate or costate leaves and paraphyllia rarely present.

Hypnobryales An order of mosses composed of procumbent and pleurocumbent plants with usually symmetrical leaves arranged in more than two rows.

Hypo (1) Sodium thiosulphate. (2) A prefix denoting under or less than normal.

Hypocarp A fleshy modified peduncle of certain fruits as cashew apple.

Hypochlorites Compounds containing oxygen, chlorine and a metal such as sodium hypochlorite (NaOCl). Used as disinfectant.

Hypocotyl The part of the embryo or seedling below the cotyledonary node and above the root, the transition region connecting the stem and root.

Hypodermic Sub-cutaneous; beneath the skin.

Hypodermis The outermost cell layer of the cortex of plants. Also known as exodermis, hypoderm, hypoderma.

Hypogeal Plants in which the cotyledons remain below the surface of the soil.

Hypogeal germination In which the cotyledon(s) remain within the seed coat and usually do not appear above ground as photosynthetic organs e.g. palms (cryptocotylar germination).

Hypogeous Growing or maturing under the soil surface.

Hypogynium Structure that supports the ovary in plants such as sedges.

Hypogynous Having all flower parts attached to the receptacle below the pistil and free from it.

Hypogyny The situation in a flower in which the ovary is above and free of the receptable and in which the sepals, petals, and stamens are attached below the ovary.

Hypoid gears A type of spiral bevel gear which permit the location of the pinion above or below the centre of the gear with which it meshes. Hypoid gears are now largely used for the rear axle drive in all passenger automobiles.

Hypomagnesaemia See grass tetany.

Hypoplastic Term applied to a disease with subnormal cell production. Also called hypoplasia.

Hypophysis The cell from which the root tip arises in dicotyledons.

Hypopodium The part of a branch which is localised between its origin and its first leaf; usually long in sylleptic branches, it remains very short in proleptic ones.

Hypotonic Solution having a lower osmotic pressure than that of a reference solution.

Hysteriales An order of lichnes in the class Ascolichenes including those species with an ascolocular structure.

Hysteresis The relation between matric potential and soil wetness can be obtained in desorption by subjecting a saturated sample to increasing suction and measuring the water content versus suction, or in sorption by wetting a dry sample while reducing suction. The equilibrium soil wetness at a given suction is greater in desorption than in sorption. This dependence of the equilibrium content and state of soil water upon the direction of the process leading up to it is called hysteresis. Lag in one of two associated processes or phenomena.

Hysterothecium Oblong or linear perithecium, sometimes considered an apothecium, opening by a cleft.

I

IPC Isopropyl-phenyl-carbamate. A weedicide applied as dust. Applied at pre-emergence for eradication of some annual grasses.

Icacinoceae A family of dicotyledonous plants in the order Celastrales, characterised by haplostemonous flowers, pendulous ovules, stipules wanting or vestigial, a polypetalous corolla. Velvet petals, and usually one (sometimes three) locules.

Ice accretion The process by which a layer of ice builds up on solid objects which are exposed to freezing precipitation or to supercooled fog or cloud droplets.

Ice age A major interval or geologic time during which extensive ice sheets (continental glaciers) formed over many parts of the world.

Ice fog A type of fog composed ot suspended particles of ice, partly ice crystals, 20-100 micrometres in diametre but chiefly, especially when dense, crystals 12-20 micrometres in diametre, occurs at very low temperatures and usually in clear, calm weather in high latitudes. Also known as frost flakes, frost fog, frozen fog, ice-crystal fog, rime fog.

Ice pellets A type of precipitation consisting of transparent or translucent pellets of ice five millimetres or less in diametre; may be spherical, irregular, or conical (rarely) in shape.

Icerya purchasi See citrus fluted scale.

Ice storm A storm characterised by a fall of freezing precipitation, forming a glaze on terrestrial objects that creates many hazards. Also known as silver storm.

Ideal irrigation interval The length of time the net amount applied by irrigation will last before the crop uses this moisture under peak use conditions and another irrigation is required.

Idiocerus atkinsoni See mango hopper.

— **clypealis** See mango hopper.

— **niveosparsus** See mango hopper.

Idiopathic A disease or symptom without any apparent cause.

Idle wheel or **idier** A gear transmitting motion between two active gears, or a pulley used against a belt to take up slack.

Idling Refers to the slow running of the engine when the car is not in motion.

Igneous rock Rock formed by the cooling and solidification of magma.It has not been changed appreciably since its formation. The two main classes are plutonic and volcanic. Granite, basalt, peridolite are other examples.

Ignition spark The high tension spark or arc induced in the secondary winding of the ignition coil and passed to the spark plugs where it explodes the gas in the engine cylinders.

Ilama Annona diversifolia Saff. Family Annonaceae. Fruit native of Mexicc.

Ileo-caecal valve Valve at the junction of small and large intestines.

Ileum Lower part of small intestine, anterior part of hind gut in insects.

Illicium verum See star anise.

Illite A group of clay minerals similar to montmorillonite but containing potassium, consisting of aggregates of small flakes, and possessing less swelling and expansive properties; A hydrous mica.

Illuvial horizon A soil horizon in which material carried from an overlying layer has been precipitated from solution or deposited from suspension as a layer of accumulation. See eluvial horizon. Usually called horizon B.

Illuviation The process of depositing soil material removed from one horizon in the soil to another, usually from an upper to a lower horizon in the soil profile. Illuviated substances include silicate clay, hydrous oxides of iron and aluminium, and organic matter.

Imago The last or adult stage in insect metamorphosis, the perfect insect.

Imbibition Absorption of moisture by a colloidal substance (organic substances) such as seed coats which is accompanied by swelling of tissues.

Imbibitional moisture Moisture over and above the hygroscopic moisture content of porous or soil particles, in case of soils available to plants.

Imbibitional water value The difference between the moisture equivalent and the xylene equivalent of a soil, which is a measure of the soils ability to imbibe water.

Imbricate Overlapping in a regular pattern, as the leaves of a moss or bracts of an involucre.

Immarginate Lacking a clearly defined margin.

Immature profile A still developing profile.

Immature soil A soil having indistinct or only slightly developed horizons. Also called juvenile soil.

Immature (plants or animals) Plants or animals which have not reached full development.

Immature seeds Not fully developed; not having parts developed.

Immature staple Refers to cotton which has been picked and baled before the fibre has been completely developed, weakened staple of inferior value.

Immersion cleaning Removing surface dirt from metal by dipping into a cleaning liquid.

Immobilisation (1) The conversion of an element from the inorganic to the organic form in microbial tissues so that the element is not readily available to other organisms or plants. (2) The action of rendering an object immobile.

Immune A condition of possessing immunity by an animal or a plant.

Immune serum A serum containing, immunising bodies.

Immunising agent A substance which when introduced into the body will produce antibodies in the blood that will resist or overcome an infection.

Immunity Exempt from disease, having qualities which do not permit infection. It may be natural or acquired.

Immunisation Process of increasing the resistance of a living organism.

Immunogenicity Ability to produce immunity or security against any particular disease or poison.

Immunology A branch of biological science concerned with the native or acquired resistance of higher animal forms and humans to infection with micro-organisms.

Impact assessment The evaluation of ecological effects to determine their impact on human needs.

Impacted crop See crop bound.

Impaction Condition of being firmly packed or lodged.

Impact loss (1) The loss of energy or pressure as a result of the impact. (2) The dispersal action exerted on the soil aggregates as a result of water falling on exposed soil.

Imparripinnate Unequally pinnate; pinnate with an odd terminal leaflet.

Impeded drainage A condition that hinders the movement of water by gravity through soils.

Impeller The rotating member in a pump, usually consisting of a number of vanes, which puts in motion the medium through which it travels.

Imperata arundinacea See cogon grass.

— **cylindrica** See cogon grass.

Imperfect flower A flower which bear stamens or carpels, but not both. Also called declinous flower.

Imperfect fungus One lacking any sexual reproductive stage.

Imperfect state State of life cycle in which asexual spores, or none, are produced.

Imperial bushel A dry measure used in Great Britain. It is equal to 1.0321 American bushels, which is equivalent to 32 quarts.

Impermeability The condition or property of a substance of not permitting water (liquids) or gases to move through it under the ordinary pressure.

Impermeable boundary A boundary which is resistant to penetration by water, air or roots.

Impermeable layer A layer of soil resistant to penetration by water, air or roots.

Impervious Resistant to penetration by

fluids or roots.

Impervious soils A soil through which water, air, or roots cannot penetrate. No soil is impervious to water and air all the time.

Impermeable Impenetrable, as when a seed coat allows no passage through to water or gases.

Implant A grafted portion of a tissue. To insert or to fix.

Implement A tool which helps in accomplishing a task more effectively and in a more systematic manner and relatively smaller time.

Implicit costs Costs in the form of lost opportunities to use resources, including time.

Impotence Temporary or permanent loss of reproductive power.

Impotent pollen A pollen which is unable to fertilise.

Impoverishment, soil The process or the result of making the soil less productive. Continuous cropping with poor management could be one of the process.

Improved breed Any group of animals or plants having a uniform breeding value in respect of certain qualities.

Improvement cutting To resort to cutting in a forest land to remove undesirable plant, or to remove plants which are not of good form and condition and also to increase the utility of forest for purposes other than wood production.

Improvement soil The processes for or the results of making the soil more productive for growing plants, by drainage, irrigation, addition of fertilisers and soil amendments, and so on.

Impulse As applied to impulse starter or impulse fuel pump, it refers to momentary acceleration given to speed of engine, shaft, or armature to produce starting increases of vacuum, or advance of timing position. Applied to auto engine and aviation engine.

Inactive See dormant.

Inactive front A front, or portion thereof, that produces very little cloudiness and no precipitation, as opposed to an active front. Also known as passive front.

Inadunate Not united. In crinoids, having the arms free from the calyx.

In-and-inbreeding See inbreeding. Also called line breeding.

Inarching A kind of repair grafting in which two plants growing on their own roots are grafted together and one plant is severed from its roots after the graft union is established. Also known as approach grafting.

Inarticulate Not segmented or not jointed.

Inborn Of or pertaining to a congenital or hereditary characteristics.

Inbred (1) Pertaining to off spring or progeny produced by self pollination in normally cross-pollinated plants. (2) A plant or animal produced by breeding between close relatives. (3) An individual with parent showing 50 per cent or more of common ancestory in their pedigree.

Inbred line A relatively homozygous line produced by inbreeding and selection.

Inbred-variety cross The F_1 cross of an inbred line with a variety, in corn, etc.

Inbreeding Breeding of closely related individuals; self-fertilization, as in some plants, is the most extreme form. Also called in-and-in breeding.

Incendiarism Malicious setting of fires.

Incidence Range of occurrence of influence, extent of effects. frequency.

Incineration A method of treating refuse q.v. to reduce the volume and weight and to leave an innocuous residue.

Incipient erosion Just start of the erosion process. Term specially used in case of gully erosion.

Incipient lethal level That level of a toxic substance beyond which 50 per cent of a population q.v. of organisms cannot live for an indefinite period of time. Incipient lethal levels denote threshold concentrations.

Incising Puncturing of the lateral surface of a resistant wood as an aid in securing penetration of the preservative. Also a cut.

Incitant Casual organism such as fungus, bacterium, nematode etc. which produces a particular disease.

Inclement weather When one or more factors comprising weather of an area are harsh to animals, man or crop.

Inclined separator See draper

separator.

Inclusion, bodies Round, oval, or irregularly shaped particles in cytoplasm or nucleus of cells parasitised by viruses; colonies of viruses.

Income The wealth, measured in money, which is at the disposal of an individual or a community, per year or other unit of time.

Income-tax, progressive A system of graded tax payments, individuals with small incomes paying a smaller proportion of their incomes than individuals with large incomes.

Income-tax, proportional A system of tax payments so designed that people pay in direct proportion to their incomes.

Incompatible Not capable of being mixed without undergoing distructive chemical changes or acting antagonistically.

Incomplete fertiliser A fertiliser lacking in one or more elements out of the three, essential for plant growth.

Incomplete flower A flower which lacks one or more of the four kinds (sepals, petal, stamens, carpels) of floral organs.

Incomplete metamorphosis Insect metamorphosis in which the young are hatched in general adult form and develop without the quiescent stage.

Incorrect confidence limits When the limits do not include the parametre.

Increaseate State of being swollen or thickened.

Increase To multiply a quantity of seed by planting it, rearing the plants that grow from it, and harvesting the seeds they produce. The seeds resulting from this process are called an increase.

Increaser Forage plant on the range that spreads under existing management.

Increment (1) The rate of growth of quantity of wood produced by a tree annually throughout its life. Also total increase in the wood of all the trees in a forest termed as annual increment of that forest. (2) Annual rise in pay of an employee.

Increment borer An instrument for measuring the growth of trees.

Incubation (1) The hatching of eggs by means of natural or artificial heat. (2) Period between infection and appearance of symptoms induced by parasitic organisms in a plant or animal. (3) The keeping of a culture of protozoa or bacteria in an incubator for development.

Incubation period (1) Period intervening between the time of infection and the appearance of the manifestations of a disease. (2) The period between the settling and hatching out of eggs.

Incubator A device for the artificial hatching of eggs. A small chamber with controlled oxygen, temperature, and humidity for newborn infants requiring special care. A laboratory cabinet with controlled temperature for the cultivation of bacteria, or for facilitating biologic tests.

Incubous The juxtaposition of leaves such that the anterior margins of older leaves overlap the posterior margins of younger leaves.

indarbela quadrinotata See bark-eating caterpillar.

Indefinite Not of fixed number etc. indeterminate or not limited.

Indehiscent (1) Remaining closed at maturity, as certain fruits. (2) Not splitting along regular lines. (3) with no special methods.

Indenated Hardened.

Indented cylinder separator The indented cylinder separator is a rotating, almost horizontal, cylinder with a movable horizontal separating trough mounted inside it. The length of the seed is the only factor affecting the separation made by the indented cylinder.

Independence Two events are independent if the probability that one of them will have a certain outcome is the same no matter what the outcome of the other, i.e., whether the probability of the other is given or not.

Indeterminate Pertaining to growth of plants, the flowers of which are borne on lateral branches, the central stem continuing vegetative growth, with blooming continued for a long period. Examples are alsike clover, alfalfa, and birdsfoot trefoil. (2) Group of tomato varieties producing a blossom cluster at every third internode being separated by three leaves. (3) In mathematics which cannot be determined such as division of zero by zero.

Indeterminate growth Growth of a plant in which the axis is not limited by development of a reproductive structure, and therefore growth continues indefinitely.

Index number A statistical indicator for reflecting the changes in magnitude of items in a series or groups of data, in terms of the value of a part of the respective data selected as a base.

Index of aridity A measure of the precipitation effectiveness or aridity of a region, given by the following relationship; index of aridity = *P/(T+10),* where *P* is the annual precipitation in centimetres, and *T* the annual mean temperature in degrees centigrade.

Index of association An indicator of the proper measurement of association of the properties of statistical data e.g. the index of association of column with row of a table.

Index of order association When both variables are quantitative, at least in the sense of being capable of being ranked, 'the index of order association' tells how much more likely it is that two individuals chosen at random from those in a table will have variables in the same order, than that they will have them in the opposite order.

Index species An adapted organism or plant to the habitat of a particular region such that it is unable to survive at other places.

Indexing Refers to the process used to test vegetatively reproduced plants or seeds stock for freedom from virus diseases before multiplying them. The process consists of growing a sample or grafting a portion of the plant to a susceptible variety under conditions suitable for symptoms to be shown.

Indexing host A susceptible host which is used for indexing, i.e., to show the presence or absence of a infection due to virus or insect.

Indian almond Terminalia catappa L. Family Combretaceae. Fruits used in tanning and also for increasing the potency of spirits. Also called tropical almond or Jangli badam.

Indian arrowroot (1) Hitchenia caulina Baker. A small shrub of Family Zingiberaceae; produces edible tubers. Leaves used in paper making. Also called *tikhur.* (2) Tacca leontopetaloides (L). O. Kuntze (Tacca pinnatifida Forst.). A herb of Family Taccaceae; produces edible rhizomes. Also called *diva.*

Indian bdellium tree Commiphora mukul (Hook ex stocks) Engl. (Balsamodendron mukul Hook ex Stocks). Family Burseraceae. Tree whose bark yields gum resin, which is used as an incense and as a fixative in perfumery.

Indian bean Polichos lablab L. var. Lignosus Prain. A perennial herb of family Papilionaceae. Seed is used. Also called Australian bean, field bean.

Indian black berry See jambolana.

Indian buckwheat Tartary buckwheat.

Indian butter bean Dolichos lablab L. var. typicus (Lablab niger Medic) Family Papilionaceae. Perennial twining herb, cultivated as an annual for pods, seeds and fodder. Also called *sem.*

Indian butter tree Diploknema butyracea (Roxb.) H.J. Lam. Bassia butyracea Roxb.; Madhuca butyracea (Roxb.) (Macb.) Family Sapotaceae. Flowers edible, seed oil used for cooking and bark for dyeing.

Indian cassia Cinnamomum tamala (Buch.-Ham.) T. Nees and Eberm. Family Lauraceae. A small tree cultivated in the hills of Assam the leaves of which are used as condiment.

Indian cherry See Sebestan.

Indian chestnut Castanopsis indica A. DC. A large tree, leaves of which are used for wrapping *bidis.* Tree of Family Fagaceae commonly found in Khasia hills and Bengal.

Indian clover Important winter fodder crop. Does well in alluvial soils with assured irrigation. Generally fed by mixing in *Bhusa* (Dry fodder of wheat stalks).

Indian colza See yellow *sarson.*

Indian crab apple Docynia indica Decne. (Pyrus indica wall.), Family Rosaceae. A tree found in the Eastern Himalayas and the Khasia hills the fruits of which are edible.

Indian Dill Anethum sowa Roxb. Family Umbelliferae. Leaves used for flavour and fruits as stimulant.

Indian frankincense Boswellia serrata Roxb. ex. Colebr. Family Burseraceae. Bark used for tanning, Oleo-gum resin derived from it is used as incense and rheumatism. Rosin used for varnishes. Flowers and seeds edible. Wood used for paper making.

Indian gooseberry See emblic.

Indian gutta percha tree Palaquium ellip. um (Dalz.) Engl. (Bassia elliptica Dalz.; Dichopsis elliptica Benth.). Family Sapotaceae. Tree found in the Western Ghats, Mysore and Madras. Oil extracted from seed used for soaps and as an illuminant.

Indian honey bee Apis indica F. The common honey be commonly reared in India. Lives in cavities of trees, hollows of rocks, hollowed out logs, boxes etc. Makes a series of parallel combs. Has gentle temperament and easy to handle. Gives 3-5 kg of honey per comb.

Indian horsechestnut Aesculus indica Colebr. ex Wall. Family Hippocastanaceae. A tree found in Kashmir, Punjab and Uttar Pradesh, whose fruits are edible and leaves and twigs of which are used as fodder.

Indian kino tree Pterocarpus marsupium Roxb. Family Papilionaceae. Tree whose wood is used for furniture. Gum has medicinal value and flowers and seeds edible. Found in Western peninsula and south India. Also called *Malabar kino.*

Indian laburnum Cassia fistula L. Family Caesalpiniaceae. Moderate sized tree, drought resistant, frost tender, does well in 50 to 75 cm rainfall. Propagated by direct sowing. Wood used as fuel, building work and for agricultural implements.

Indian leek See Chinese chive.

Indian mallow Abutilon theophrasti Medic. (Abutilon avicennae Gaertn.) A herb of family Malvaceae, whose stem produces fibre known as China jute or Tientsin jute used in rug-making. Also called China jute.

Indian meal moth Plodia interpunctella Hubn. Moth with fringed wings that lays eggs on stored food materials. Caterpillar prefers to feed on flour and other milled products and grain. Whole life cycle is about 5 weeks.

Indian Meteorological Department Commonly abbreviated as IMD. Located at Pune. Keep records of meteorological components and predicts rainfall and other adverse phenomena. Weather forecasts are issued by this department.

Indian mustard Brassica juncea (L) Czern. & Coss. Family Cruciferae. Oilseed crop commonly grown in Punjab, Uttar Pradesh, Bihar and Bengal, Oil edible and cake fed to cattle. Also called *Rai.*

Indian nettle tree See charcoal tree.

Indian oat Avena sterilis L. A herb grown for fodder. Family Gramineae.

Indian persimmon Diospyros peregrina (Gaertn.) Gurke (D. embryopteris Pers.), Family Ebenaceae. Tree whose fruits are edible. Dye from fruits used for colouring cotton and timber.

Indian poke Phytolacca acinosa Roxb. Family Phytolaccaceae. Broad leaved, perennial herb occurring wild in the Himalayan ranges and its leaves used as vegetable.

Indian rape Brassica campestris L var. toria Duth. & Full. Family Cruciferae. An oil-yielding crop grown under irrigated and rainfed conditions. Seeds contain pungent edible oil and cake is fed to cattle. Also called *toria.*

Indian rhododendron Melastoma malabathricum L. An ornamental large shrub of Family Melastomaceae, whose leaves are used for feeding silkworms and fruits are edible.

Indian rubber Ficus elastic Roxb. Tree of the family Moraceae native of India and Malaya. Latex from bark of branches used for making rubber. Also called Assam rubber.

Indian senna Cassia angustifolia Vahl. Family Caesalpiniaceae. A medicinal plant grown for its leaves. Grown largely in South India, used in tictures. Also called *senna.*

Indian shot Canna orientalis Rosc. (C. indica L. var. orientalis Rosc.) Family Cannaceae. A herb the tuberous rhizomes are used as vegetables.

Indian sorrel Oxalis corniculata. L.

(Xanthoxalis corniculata Small). Family Oxalidaceae. A common herb whose leaves and seeds are edible. Propagates by seed and vegetative parts.

Indian spinach Basella rubra L. (B. alba, L.). A vegetable crop of family Baselaceae (Leaves used as green vegetable).

Indian squash See squash melon.

Indian strawberry Duchesnea indica Focke (Fragaria indica Andr.). Family Rosaceae. A perennial herb, whose fruit are edible.

Indian tree cotton See tree cotton.

Indian valerian See valeriana wallichii.

Indian water lily Nymphaea esculenta Roxb. Family Nymphaeaceae. A floating aquatic herb whose tubers and seeds are edible.

Indian wheat Fagopyrum tataricum Gaertn. Family Polygonaceae. An annual plant, taller and coarser than buckwheat. The grains are inferior dark and bitter flour.

Indicated horse power The horse power delivered by an engine as calculated from the average pressure of the working fluid in the cylinders and the displacement. Abbreviated as 'ihp'.

Indicator Some thing that renders visible the completion of a reaction.

Indicator host A plant that shows symptoms of viruses that are latent in other hosts.

Indicator plant A plant used in geobotanical prospecting as an indicator of a certain geological phenomenon. Also plants that are characteristics of specific soil or site conditions.

Indicator species A species chosen to represent a particular environmental condition.

Indifferent plant A day neutral plant.

Indigenous Native to the locality, endemic, produced or living naturally in a specific environment.

Indigo Indigofera tinctoria L. Family Papilionaceae. A common small shrub grown as a cover and green manure crop. Plant is also of medicinal value.

Indigofera A very large genus of herbs, shrubs and undershrubs cultivated for the dye, indigo; many grown for fodder or for plant cover.

Indigo fera argentea var. caerulea See Egyptian indigo.

— **articulata** See Egyptian indigo.

— **endecaphylla** See trailing indigo.

— **teysmanni** Mig. Family Papilionaceae. A herb introduced from Ceylone and used as a herb.

— **tinctoria** See indigo.

Indigo psyllid Arytaina punctipennis Cr. A small light brown bug with spotted wings that sucks the sap from the indigo plant. Leaves curl up; knot formed at top of stem.

Indirect cell A closed circulation in a vertical plane in which the rising motion occurs at lower potential temperature than the descending motion, thus forming an energy sink.

Indispensable elements An essential element for the growth of a plant or an animal.

Individual selection Selection of individual animals or plant which have a desired character in breeding.

Indoculamus wightianus (Nees) Nakai (Arundinaria wightiana Nees.). Family Gramineae. A type of Bamboo used in making baskets and mats.

Indoplanorbis exustus See snail.

Induced resistance The resistance developed by a plant to the attack of pest or disease by farm practices and nutritional means. The degree of the resistance varies according to the conditions under which it is growing.

Induced resistant Resistance induced by management practices or nutrient supplies against pest, insects or diseases.

Induction flowmetre An instrument for measuring the flow of a conducting liquid passing through a tube, in which the tube is placed in a transverse magnetic field and the induced electromotive force between electrodes at opposite ends of a diametre of the tube perpendicular to the field is measured.

Induction period The number of times or periods of a given day-length to which a plant has to be exposed to cause flowering. Also see photoperiodism.

Indurated layer A soil layer that has become hardened, generally by cementation of soil particles. Does not loose the firmness on wetting.

Industrial crop Any crop that provides materials for industrial processes and products such as soyabeans, cotton (lint & seed), flax, and tobacco.

Industrial waste The liquid waste resulting from the processes employed in industrial establishments.

Industry In its widest sense, all the manifold activities of a country which provides employment, but more narrowly those activities concerned with the production of goods rather than services.

Inert A substance which is resistant to chemical action; not having active properties or values.

Inert matter Inert matter includes seeds and seed-like structure, namely pieces of broken or damaged seeds which are one half the original size or less; empty glumes, lemmas, paleas, unattached sterile florets and florets with a caryopsis less than minimum size; prescribed, and other matter, namely, soil, sand, stones, chaff, stems, leaves, cone scales, wings, pieces of bark, flowers nematode galls, fungus bodies, caryopses of graminae replaced by insect larvae, and all other matter not seeds.

Infant industry A young and growing industry, often considered not quite able to stand on its own feet in the face of foreign competition.

Infect To cause disease by introduction of germs, parasites, fungus etc.

Infection Process of beginning or producing disease, presence of multiplying parasites, bacteria, viruses etc. within the body of the host.

Infection court Place where an infection may take place, as leaf, fruit, petal, etc.

Infection stage The time or stage in the course of a disease during which the host responds, symptoms appear and disease develops.

Infections chlorosis A virus disease of plants characterised by yellowing of the green parts.

Infection disease Disease caused by infectious organisms that passes directly or indirectly from a diseased host to a healthy one.

Infectious organisms A organism capable of getting itself transferred from a diseased host to a healthy one resulting in making the healthy one infected with the same disease.

Inferior A product of poor quality; low grade.

Inferior ovary Ovary in the flower-stem located in such a way that it is below the petals.

Inferior result A result obtained in a subsequent test at another station, of a second submitted sample from the same seed lot, which indicates a lower quality then the first test, e.g. a lower purity percentage, a lower germination capacity; a higher count of certain seeds or a higher percentage of other seeds.

Infertile A male or female animal unable to reproduce, An unfertilised egg; a flower unable to produce seed; also refers to soils which are unable to support a normal crop growth.

Infest Attacks, annoy, overrun, disturb.

Infestation (1) Invasion of the body by microscopic parasites on the surface of the body. (2) Presence of a disease or of pathogens in soil, plant or animal, where it has the possibility of producing a disease.

Infilling See beating up.

Infiltration The downward entry of water into the soil.

Infiltration capacity See infiltration rate.

Infiltration head The head of water which makes it flow or infiltrate through the media.

Infiltration gallery A large, horizontal underground conduit of porous material or with openings on the sides for collecting percolating water by infiltration.

Infiltration index The average rainfall intensity over and above which the mass of rainfall equal the mass runoff.

Infiltration rate (1) A soil characteristic determining or describing the maximum rate at which water can enter the soil under specified conditions, including the presence of excess water. It has the dimensions of velocity (cm sec^{-1}). It was formerly called infiltration capacity. See also infiltration velocity. (2) The rate at which ground water enters an infiltration ditch, gallery, drain, sewer or other underground conduits.

Infiltration velocity The actual rate at which water is entering the soil at a given time. It may be less than the maximum (the infiltration rate), because of a limited supply of water (rainfall or irrigation). It is expressed in the same units that are used for the infiltration rate. See also infiltration rate.

Infiltrometre A device for measuring the rate of entry of fluid into a porous body, for example water into soil. Double rings is a commonly employed infiltrometre.

Infinite aquifer Water-bearing stratum that is very large or extensive in area.

Infinite soil Soil of essentially unlimited depth for plant root development.

Inflammation Protective reaction on the part of the tissues of the body brought about by the presence of an irritant — reaction to injury.

Inflate To fill air as inflating the tyres of an auto vehicle.

Inflation A condition in which the volume of purchasing power is constantly running ahead of the output of goods and services. With the result that as incomes and prices rise, the value of money falls.

Inflationary spiral An upward trend of prices, which is partly the result and partly the cause of increases in wages and salaries and other incomes such as profits, dividends, interest and rent.

Inflected Curved or bent sharply inward, downward, or towards the axis. Also known as inflexed.

Inflorescence Flowering part of a plant. A cluster of flowers and the manner in which flowers are arranged.

Influent An organism that disturbs the ecological balance of a community. An input stream of a fluid, as water into a reservoir or liquid into a process.

Influent stream A stream that contributes water to the zone of saturation of groundwater and develops bank storage. Also known as losing stream.

Infrafoliar Located below the leaves.

Infraorbital Beneath or below the orbit of eye.

Infraspecific Pertaining to a subdivision of a species, as sub-division and variety.

Infrastructural complement The population of latent meristems in a forest, representing potential growth.

Infrastructural set The population of functioning meristems in a forest, representing actual growth.

Infructescence An inflorescence's fruiting stage.

In full head Pertains to that stage of development of a cereal or grass crop when the inflorescence has grown to its fullest length.

Infusion (1) The introduction of a solution into the vein or a body opening. (2) An aqueous solution obtained by steeping vegetable drugs in cold or warm water.

Inga dulcis See Manila tamarind.

Inherent resistance A character of a plant which is inherited by it due to which it resists the attack of pest or disease.

Inheritance The passing on of the characters of the parents to the progeny by the germ cells.

Inherited character The character derived by an individual from one or the other of the parents.

Inhibition The prevention of growth or multiplication of organisms.

Inhibitor A chemical substance that acts to prevent a process from occurring; many chemicals both natural and artificial can act to prevent seeds germination.

Initial complex The visible complex of axes and active meristems produced by a tree when growing precisely according to its model.

Initial rainfall The rain that falls in the beginning of a strom before the depression storage is completely filled, is called the 'initial rainfall'; that falling near the end at rate less than infiltration capacity is called 'residual rain'; the intervening rainfall is the 'net supply rainfall'.

Initial soil storage The maximum amount of precipitation that is necessary to satisfy the moisture deficiency under specific conditions of soils before any runoff takes place.

Injury Results of transient operation of an adverse factor, as an insect bite, or action of chemicals.

Ink disease A fungus disease of the chestnut in Europe caused by Phytophthora cambrivora which produces black cankers and black exudate in the trunk. Also known as black canker.

Inlaying A graft consisting of removing the area of stock and preparing a scion which will fit it exactly.

Inlet Surface connection to a title drain, structure at the diversion end of a conduit, upstream end of any structure through which water may flow.

Inline valve, or one-line valve A valve installed in a pipeline to restrict or control flow of water going through the line.

Innate Pertaining to a natural or inborn character dependent on genetic constitution. Positioned at the apex of a supporting structure. Embedded in, especially of an organ such as the fruiting body embedded in the thallus of some fungi.

Inner bark The physiologically active layer of tissues between the cambium and the last formed periderm in woody plants.

Innovation In a general sense, the technical, industrial and commercial steps which lead to the marketing a new manufactured products and to the commercial use of new technical processes and equipment.

Inoculation Introduction of a disease agent into an animal or plant to produce a mild form of disease and render the individual immune. Treating a molten material with another material before casting in order to nucleate crystals. Introduction of microorganisms onto or into a culture medium. Addition of effective rhizobia to legume seed prior to planting for the purpose of promoting N-fixation.

Inoculum A small amount of substance containing bacteria from a pure culture which is used to start a new culture or to infect an experimental animal.

Inoperculate Not opening by a lid.

Inorganic A substance not of organic origin. A compound which does not contain carbon (except carbonates and cyanides).

Inorganic insecticide Insecticides of inorganic origin.

Inorganic nitrogen Nitrogen in combination with mineral elements, not in animal or vegetable form.

Inorganic soil A soil made up mainly of mineral particles. See, mineral soil.

Inositol A component of the vitamin B_2 complex; a factor which prevents baldness.

Insect An air-breathing animal of phylum with a distinct head, thorax and abdomen, has one pair of antennae, three parts of legs and usually two pairs of wings.

Insectarium See insectary.

Insectary A laboratory in which insects are bred in large numbers. Also called insectarium.

Insect control Regulation of insect populations by biological or chemical means.

Insect parasites Insects which lay eggs on the body of other insects, when the emerging larvae feed internally on them and kill them.

Insect predators Insects which feed on other insects.

Insect vector A insect which is a carrier of disease or virus etc.

Insecticide A chemical used for the purpose for killing insects. They may be inorganic or organic in nature.

Insecticide emulsion An insecticide preparation in which the active ingredient when insoluble in a cheap solvent is dissolved in a suitable solvent and emulsified in water to a required degree of concentration.

Insecticide solvent Substance in which an insecticide is dissolved for use as spray.

Insectifuge A chemical which does not kill insects but drives them away; an insect repellent.

Insectivorous Feeding on the diet of insects.

Insectology Science of insects and their relation to man.

Inseparable other crop plants Crop plants whose seed is difficult to separate from the main crop seed.

Insidious Gradual onset such that it is difficult to notice such as a disease is insidious if it develops slowly over a long period of time.

Insipid Lacking in favour and/or odour.

In situ measurements Measurements of a particular phenomena in place of occurrence.

Insolation Solar radiation received by the earth. Measured by the solar radiation per unit of horizontal surface.

Inspirator Device for drawing in of air or vapour.

Inspissate To thicken by evaporation or absorption of fluid.

Installment credit Credit to finance the purchase of consumer goods, repayable by series of payments over a period of time.

Instaminate Lacking stamens.

Instar Insect at a particular stage between moults.

Institution It can be defined as a system of relationships or a pattern for carrying out an idea or a desire which is regarded as necessary for the welfare of the group.

Instrumental error Systematic error arising from the imperfect condition of the instrument used.

Insuffation The blowing of a powder or vapour into a cavity for medication.

Insulation material Non-conducting material; may be for electricity or temperature. Used in insulation against electricity, heat and cold.

Insulin The anti-diabetic endocrine products of pancreas.

Insurance The elimination of risk of loss by the payment of premiums to an insurance company which undertakes to pay for specified losses.

Intake (1) The flow of fuel mixture in the intake manifold of a gasoline engine. (2) Also infiltration. (3) The head works of a conduit, the place of diversion of water.

Intangibles Costs or benefits that either cannot be quantified, or at least be priced.

Integrated logging A method of logging designed to make the best use of all timber products. It removes in one cutting all timber that should be cut, and distributes the various timber products to the industries that can use them to best advantage.

Integrated farming Production, processing and selling of agricultural crops as a unified business.

Integrifolius With entire leaves.

Integument A covering, investing or coating structure or layer as the skin of animals, the body wall of an insect and the shell, or outer layer of the ovules or plants.

Intensity (rainfall) The rate of occurrence of rainfall at a given time. It is obtained by data from recording type rain gauges or intensity recorders, directly. Expressed in cm/hour. Runoff occurs only if intensity at a given time is higher than the infiltration rate of the soil at that time.

— **(cropping)** Number of crops taken on a land surface in a full year. Usually expressed in per cent in which one crop denotes 100 per cent.

Intensive cropping Producing maximum crop-yields per unit area of the land. Includes soil and water management, crop treatment and farm management in a way that these help in attaining the objective namely maximum crop-yield.

Intensive farming Farming in which comparatively large amounts of capital and labour is utilised per hectare of cultivated land.

Interaction (communication) Interaction is the mutual and reciprocal influencing of each others behaviour.

Interaxillary Located within or between the axil of leaves.

Inter-breeding The mating of two cross-breds with each other or mating within a family or strain.

Intercalary Occurring between cells or regions of a different kind; as an intercalary meristem, opposed to apical meristem. Used in referring to growth occurring between the apex and the leaf.

Intercellular Between cells.

Interception The process by which precipitation is caught and held by foliage, twigs and branches of trees, shrubs and other vegetation and lost by evaporation, without reaching the surface of underground. Also the quantity thus evaporated. Also sometimes known as 'canopy interception'. Also see interception loss.

Interception loss The loss of precipitation due to interception by vegetation, plant canopy etc. This is considered as loss because it is not used effectively by plants or soil and is lost by evaporation again to the atmosphere.

Interceptor drain Surface or subsurface drain, or combination of both designed and installed to intercept flowing water.

Inter cropping A form of multiple cropping in which two or more crops simultaneously occupy the same field. Also called companion cropping.

Inter-crossing Cross-breeding.

Inter current Intervening. Refers to disease occurring during progress of another.

Interdigital Between fingers, toes or claws.

Interest The price paid by a borrower to a lender, usually on the basis of X per cent of the capital per annual. In extension it is defined as attention with a sense of concern focused upon some object.

Interest, neutral rate of (1) The rate of interest which preserves the *status quo* of an economy. (2) The rate of interest consistent with full employment, from this aspect it may be described as the 'Optimum rate of interest'.

Interference (1) The property by which the occurrence of one cross over reduces the chance of occurrence of another in its neighbourhood. (2) To interfere as open drainage channels in mechanical cultivation.

Interfertility The ability of a plant to produce viable seeds by cross pollination. Cross fertility.

Interflow That portion of rainfall which infiltrates into the soil and moves laterally through the upper soil horizons until intercepted by a stream channel or until it returns to the surface at some point down slope from its point of infiltration.

Interfluve The area of land between adjacent streams.

Interfoliaceous Between a pair of leaves, such as between those which are opposite or verticillate.

Interfruitfullness The ability of a variety of a fruit plant to mature fruit following pollination by another specific variety. Also called cross fruitfulness.

Intergeneric cross a cross between species belonging to different genera.

Intergeneric hybrid See hybrid.

Intergrade soil A soil that possesses moderately well-developed distinguishing characteristics of two or more genetically related taxa.

Interior dike Secondary dike in a multiple dike flood protection system.

Interior drain Drain installed in a dam or other earth structure as distinguished from peripheral or interceptor drain.

Intermediary Acting between the producer and the consumer.

Intermediate host A host in which a parasite develop only in part and goes to another host where it develops to its maturity.

Intermediate shelter belt A narrow shelter belt grown to supplement the basic shelter belt.

Intermediate stock A stock between scion and another stock.

Intermittent Not continuous. An action or phenomena occurring in intervals which may not be necessarily regular.

Intermittent attacks The return of symptoms after a period without symptoms of attack of a disease.

Intermittent grazing Grazing on two pasture blocks alternatively. This provides each block sufficient time to grow. Used in perennial grassland.

Intermittent growth Non-continuous growth without the predictable regularity or rhythmic growth, hence intermittent branching.

Intermittent leaching Removal or leaching of solutes from the root zone by ponding water on the surface intermittently as opposed to continuous leaching where ponding of the water is done continuously.

Intermittent parasite A parasite which goes to the host only when in need of food.

Intermittent stream Streams or channels in which water flows intermittently. Usually refers to stream which runs only in monsoon.

Internal drainage Drainage of the soil profile, may be natural or artificial.

Internal necrosis Death of tissues within the body tissues of a plant or an animal.

Internal parasite Forms of animal life which for the purpose of obtaining food and shelter live in the bodies of other larger animals.

Internal seed-borne disease Plant,

disease, where the mycelium of the disease producing fungus is present in the seed as deep-seated infection.

Internal seed-treatment Treatment of seed for internal seed-borne fungal disease.

International seed analysis certificate A form of certificate issued only by the International seed testing association and used for reporting the results of tests.

Internode The part of a plant stem between joints of nodes.

Inter planting Planting of seeds plants or crops in between rows of plants or crops already existing in a field.

Interpretation (germination) A seed shall be considered to have germinated when it has developed into a normal seedling. Broken seedlings and week, malformed, and obviously abnormal seedlings shall not be considered to have germinated.

Inter-row tillage Tillage between the rows. A cultural practice to remove weed and to improve soil physical properties. Done manually or mechanically.

Intersect To find a position by the triangulation method. To pass through or across.

Intersex Pertaining to an organisms which displays primary and secondary sexual characteristics intermediate between male and female.

Interspecific hybrid Plant or animal progeny resulting from crossing of two species.

Interspersion Intermingling of different vegetation types.

Intersterility The inability of a plant to produce viable seed if only the pollen of another variety is used. Also called cross sterility.

Interstices Voids, pores or spaces between soil particles or soil aggregates.

Interstratified clay mineral An aggregation composed of random or regular intergrowths of two or more clay minerals.

Intertidal area The area between high and low tide levels.

Interunfruitfulness The mobility of a variety of a plant to mature fruit if only the pollen of another variety is used.

Interveinal Between the veins of a leaf.

Intestinal worm Any internal parasite worm that inhabits the intestines.

Intolerance The in-capacity of a tree to develop and grow in the shade of and in competition with other trees.

Intracellular Within cells.

Intracellular fungus A fungus that occurs or grows within a cell.

Intractability The characteristics of a soil which causes it to be difficult to work.

Intracutaneous Within or into the substance of the skin. Also intradermic.

Intradermic See intracutaneous.

Intra fascicular Located or occurring within a vascular bundle.

Intrapetider (1) Enclosed by the base of petiole. (2) Between petiole and stem.

Intraspecific Within a species.

Intrazonal soil A soil having a morphology that show the influence of some local factor of relief, parent material, or age, rather than of climate and vegetation.

Intrinsic permeability The property of a porous material that is related to the ease with which gases or liquids can pass through it. The Darcy "K" multiplied by $\mu/\delta g$

μ is the viscosity of the fluid in poises,

δ is the density of the fluid in g cm^{-3} and

g is the acceleration of gravity in cm sec^{-2}.

See also permeability, soil and water soil.

Introduced Not native; introduced intentionally or accidentally in a new area; refers to plants, animals or diseases.

Introversion (adjustment) In this form, person give up outward struggle but carries a battle in thoughts and imaginations. This is a purely mental adjustment.

Intumescence Knob-like or pustule-like outgrowth of elongated cells on leaves, stems, etc., caused by environmental disturbances.

Inundation canal This term is ordinarily applied to a canal with or without some form of head regulator and dependent upon the surface level of the water in the river for its supplies.

Invasion To attack or invade; refers to

establishment of plants or animals in a new area in competition with the natives.

Invention It is a creation of things or an idea from previously existing things or ideas.

Inventory A classified list of all assets and liabilities, together with the value of each.

Inventory turnover The value of sales over say a year divided by the value of the inventory at the end of that year.

Inversion (1) A rearrangement of a group of genes in a chromosome in such a way that their order in the chromosome is reversed. (2) The process of conversion of cane sugar into dextrose and levulose.

Inversion (temperature) A warmer air layer overlying a colder one because of rapid heat loss by reradiation from the ground at night; the lowered temperature near the ground may cause sever damage to crops; especially in the spring or autumn, in urban areas, the air may accumulate a significant concentration of pollutants.

Inversion point, ecological The point in the forest where real vertical microclimatic gradient (e.g. of humidity and light intensity) cross the average vertical gradient (regression curve).

Inversion point, morphological The insertion of the lowest major limb of a tree above which all branch or reiterative complexes are progressively smaller.

Inversion surface, ecological Collectively microclimatic inversion points, as of light intensity and humidity.

Inversion surface, morphological Collective inversion points of individual trees in a forest, indicative of that forest's status.

Invert Lowest part of the internal cross-section of a lined channel or pipe.

Invertase A plant enzyme produced by yeasts which converts cane sugar into glucose and fructose.

Inverted siphon A pressure pipeline crossing a depression or passing under a high way; sometimes called a sag line from its U shape, consists of intake and outlet tanks connected together near the bottom by a pipeline of a suitable material. Used when a culvert cannot be used. The term is common but in appropriate as no siphon action is involved.

Inverted T Method of budding in which a incision on the stock is made which resembles the shape of the letter T.

Investment In economic science, capital expenditure on physical productivity assets.

Investment appraisal An assessment of whether investment in a proposed project will show a satisfactory rate of return.

Inviscate (1) To trap insect in a sticky substance or on a sticky surface. (2) To make a substance more sticky by mixing with a viscous matter.

In vitro In glass, in test tubes; outside the organism, as digestion in vitro.

In vivo In a living organisms such as in the animal or in the plant.

Involucre Plant-enveloping tissue, as a collective term, the floral bracts of a head-type inflorescence.

Involute The leaves which have their edge rolled inwards at each side.

Involution forms Abnormal forms assumed by micro-organisms growing under unfavourable conditions.

Iodine A trace element in soils raedily absorbed by plants from soil and water.

Iodine number A measure of the degree of saturation of fats and oils.

Ion Atom, group of atoms, or compound that is electrically charged as a result of the loss of electrons (cation) or the gain of elections (anions). Many of the salts in the soil solution exist as ions.

Ion activity The effective concentration of a particular ion in a solution or soil-water-system. It is expressed analogously to pH, as pCa, "pNa", and so forth.

Ion exchange The replacement of one ion by another having a similar charge.

Ions Acids, bases and salts (electrolytes) when dissolved in certain solvents are more or less dissociated into electrically charged units or parts of the molecules called ions. Ions carry charges of electricity and in consequence have different properties from the uncharged radicals.

Ipecac Any of several low, perennial, tropical South American shrubs or half shrubs in the genus Cephaelis of the Family Rubiaceae, the dried rhizome and root, containing emetine, cephaeline, and other alkaloids, is used as an emetic and expectorant. Also see Ipecacuanha.

Ipecacuanha Cephaelis ipecacuanha (Brot). A Rich. (Psychotria ipecacuanha Stokes). Family Rubiaceae. A low harbaceous plant, native of tropical America. Roots give commercial ipecac. Propagated by division of stem cuttings.

Ipomoea alba See moonflower.

— **aquatica** See swamp cabbage.

— **batatas** See sweet potato.

— **bona-nox** See moonflower.

— **digitala** L. A common climbing herb of Family Convolvulaceae. Used as fodder.

— **eriocarpa** See Ipomea hispida.

— **hispida** vahl. R&S (Ipomea eriocarpa R. Br.); Convolvulus hispidus Vahl.). Herb of Family Convolvulaceae. The leaves and stems used as vegetable.

— **maxima** (L.f.). G. Don ex. Sweet (Ipomoea sepieria Koening ex Roxb). A twining herb of Family Convolvulaceae used as vegetable.

— **muricata** (L.) Jacq. (Calonyction muricatum G. Don). Family Convolvulaceae. A herb whose flower pedicels are eaten as vegetable.

— **purga** Hayne (Exogonium purga (Hayne) Lindl.) A twining herb of family Convolvulaceae whose roots give resinous drug and propagated from cuttings.

— **reputans** See swamp cabbage.

— **sepiaria** See Ipomoea maxima.

Iridaceae A family of monocotyledonous herbs in the order Liliales distinguished by three stamens and an inferior ovary.

Iron (Fe.) An essential plant food element absorbed as ferrous or ferric ion. Also absorbed as complex organic compounds. Essential in the formation of cholorophyll, helps in enzyme systems which bring about reactions essential for plant development and respiration.

Iron bacteria Bacteria in the soil which secrete hydrated ferric oxide in badly drained soils whose water contains ferrous bicarbonate in solution, a condition made possible only in the absence of calcium carbonate.

Iron chlorosis Chlorosis due to the iron deficiency.

Iron deficiency (plants) Commonly appear in fruits plant and is characterised by cholorosis of leaves, scorching of leaf margins and tips in extreme conditions, restricted shoot growth and die back. Control by spraying Ferrous sulphate with Bordeaux mixture.

Iron minerals Found in rocks in non-silicate iron compounds which tend to give dark colour or shade to rocks and hence to soils when soils are formed from such rocks.

Iron pan A thin indurated soil horizon in which iron is a major constituent of the cementing material. Several kinds of cementing materials occur:
(i) iron organic matter complexes;
(ii) hydrous oxides of manganese and iron, and
(iii) hydrous iron oxides.

Iron phosphate ($FeSO_4$) Compound occurring in phosphate rock.

Ironwood Any of a number of hardwood trees in the United States, including the American harmbeam.

Irpex destruens See tea stump rot.

Irradiation Treatment of products with ultraviolet rays. The purpose is to increase the vitamin D content.

Irregular unsteady flow Flow in which the velocity or discharge at a point is not a determinate function of time.

Irrigable area Area capable of being irrigated, principally as regards quality and elevation of land.

Irrigated pasture An area of land irrigated for growing superior stands of edible grasses and clovers suitable for proper feeding of livestock.

Irrigating head See irrigating stream.

Irrigating stream Flow used for irrigation of a particular tract of land; flow of water distributed in single irrigation, or that in a single farm lateral.

Irrigation The artificial application of water to the soils for the benefit of growing crops.

Irrigation adequacy A combination of field moisture distribution factor and moisture storage capacity with a view to determining the adequacy of irrigation to meet crop needs.

Irrigation by surface flooding, or **surface flooding** A method of irrigation in which water is permitted to cover the surface of the land in a continuous sheet.

Irrigation ditch A small irrigation water conveying channel.

Irrigation efficiency The ratio of percentage of the irrigation water consumed by crops of an irrigated farm, field or project to the water diverted from the source of supply. When measured at the farm, headgate it is called 'farm irrigation efficiency' or 'farm delivery efficiency', when measured at the field or plot, it is designated as 'field irrigation efficiency', and when measured at the source of supply, it is called 'water conveyance and delivery efficiency or overall efficiency'.

Irrigation frequency Time interval between irrigations.

Irrigation methods Commonly used methods for applying irrigation water to crops. The method to be used depends upon soil, water and crop factors. Border strip, furrow, basin, contour are some of the surface irrigation methods. Modern and improved techniques include sprinkler and drip irrigation. It also include subsurface irrigation known as subirrigation. Buried drains are used to irrigate the land.

Irrigation module, or **hydro-module** Flow of water designed for or used in irrigating a unit area of land.

Irrigation potential Total possible area that has been brought under irrigation, plus that which can be planned for irrigation in a river basin, region or country, from available water resources, with designs based on what may be considered as good technical practice known at the time of assessment of the potential.

Irrigation project A project meant either solely or primarily for irrigation purposes including development and improvement of land, although in the latter case it may incidentally serve other purposes.

Irrigation requirement The quantity of water, exclusive of precipitation, i.e., quantity of irrigation water that is required for normal crop production. It includes surface evaporation and other unavoidable losses under the given conditions. It is usually expressed in water-depth units or depth-area units per unit area and may be stated as monthly, seasonal or for a crop period or annual. It is exactly the same as 'duty of water' when the latter is expressed in similar units. Irrigation requirement is also referred to as 'irrigation need' and/or 'grant-in-aid' and represents the difference between possible water expenditure and actual income.

Irrigation, or waterings Artificial water application expressed in depth-units for the requirements of the different stages of plant development or for restoring the soil to the desired moisture in the non-growing seasons.

Irrigation season The actual length of time in days between the dates of the first and last irrigation given to a crop during one season.

Irrigation water Water artificially applied to soils in the process of irrigation. It does not include precipitation.

Isanthous Having regular flowers.

Iseilema laxum Hack. Family Gramineae. A fairly drought-resistant perennial grass with prostrate habit providing a dense turf over the ground. High fodder value. Propagated from seed. Flourishes in loamy and black soils.

Isotropic surface A surface in space in which potential temperature is every where equal.

Isobar A line drawn through all points of equal atmospheric pressure along a given reference surface such as a constant height surface (notably mean sea level on surface charts), an isetropic surface, or the vertical plan of a synoptic cross-section. One of two or more nuclides having the same number of nucleons in their nuclie but differing in their atomic numbers and chemical properties.

Isobiochore A boundary line on a map connecting world environments that have similar floral and faunal constituents.

Isobryales An order of mosses in

which the plants are slender to robust and up to 90 centimetres in length.

Isocarpic Having the same number of carples and perianth divisions.

Isodiametric Having equal diametres or dimensions.

Isodyne Points of a cultivating implement having equal dynamometre pull; a line on a map of a cultivated field connecting points of equal dynamometre pull.

Isoetales A monotypic order of the class Isoetopsida containing the single genus Isoetes, characterised by long, narrow leaves with a spoon-like base, spirally arranged on an underground cormlike structure.

Isoetopsida A class of the division Lycopodiophyta; members are heterosporous and have a distinctive appendage, the ligule, on the upper side of the leaf near the base.

Isogamete A gamete with no apparent morphological difference from the gamete with which it unites; a like gamete.

Isogamous Reproducing by union of gametes of equal size and structure.

Isogamy Morphological similarity of fusing gametes.

Isogeneratae A class of brown algae distinguished by having an isomorphic alternation of generations.

Isogradient A line connecting points having the same horizontal gradient of atmospheric pressure, temperature, and so on.

Isosaline A line drawn on a chart or map of an area, indicating points of equal salinity q.v.

Isohel A line drawn through geographical points having the same duration of shine (or other function of solar radiation) during any special time period.

Isohyet A line drawn through geographic points recording equal amounts of precipitation for a special period or for a particular storm.

Isohyetal map or Isopluvial map A map showing isohyets.

Isolate (1) A breeding group restricted by isolation. (2) To keep a diseased animal or plant away from the rest which are healthy so as to avoid the spread of disease.

Isolation (seed crops) Separation from others; such as separation of seed fields from fields of other varieties of the same crop, same variety fields not conforming to varietal purity requirements; other related species, fields and fields affected by designated disease to prevent genetic and disease contaminant.

Isolation distance Distance to be maintained between the seed crop and the contaminant.

Isolation requirements Refers to the isolation required for maintaining the desired purity and health of a crop, such as the minimum specified isolation required for seed certification.

Isolation strip Area surrounding a sample plot given the same treatment as the plot.

Isomere A homologus structure or part.

Isomorphic Manifesting a type of alternation of generations in which gametophytic ad sporophytic plants are similar in appearance vegetatively.

Isomorphous substitution The replacement of one atom by another of similar size in a crystal lattice without disrupting or changing the crystal structure of the mineral.

Isopleth A line of equal value of a given quantity.

Isophyllous Having foliage leaves of similar form on a plant or stem.

Isopiestic line A line indicating on a map the piezometric surface of an aquifer.

Isopropyl-phenyl-carbonate A weedicide used as dust on some annual grasses. Used as pre-emergence weedicide. Does not affect broad leaved plants like cotton, crucifers and legumes but solanaceous and cucurbitous plants very susceptible. Barley; wheat susceptible, paddy, maize and sorghum etc. affected to a lesser extent. In short it is known as IPC.

Isoptera Order of insects to which termites belong.

Isotherm A line joining points of equal temperature on a chart or map.

Isotonic Solution having the same osmotic pressure as that of another solution taken as a standard reference.

Isotope A chemical having the same atomic number and position in the periodic elements as another but possessing a different atomic mass.

Isotope farm A carbon-14 growth chamber, or greenhouse, arranged as a closed system in which plants can be grown in a carbon-14 dioxide (CO_2^{14}) atmosphere and thus become labelled with C^{14}; isotope farms also can be used with other materials, such as heavy water (D_2O), phosphorus-35 (P^{35}), and so forth, to produce biochemically labelled compounds.

Isotropic aquifer Porous water-bearing media having essentially equal physical properties, particularly hydraulic conductivity in all directions.

Isotropic soil Soil having the same property (or properties) such as permeability, in all directions.

Isotype A animal or plant common to different regions.

Israeli babul Acacia tortilis. Exotic tree of medium size of Family Mimosaceae. Single stem and parasol-like crown; good fuel, timber and fodder tree.

Isthmus A passage or channel connecting two similar halves or cavities, as in the desmids.

Italian millet Setaria italica (L.) Beauv. (Panicum Italicum L.; Chaetochloa Italica Scribn.). Family Gramineae. Millet cultivated in Asia, African and American countries specially where the climatic hazard do not permit the cultivation of the other cereals. Annual grass attaining a height of 90-105 cm. The grains are used for food. The husked grains which are called rice are cooked and than utilised. The fodder is used as cattle feed. Raised rainfed and irrigated. Grown throughout the year specially as summer crop. Grown in soils ranging from light red loam to black cotton.

Italian millet blast Fungal disease of Italian millet caused by Piricularia oryzae; symptoms as in rice blast.

Italian millet downy mildew Fungal disease caused by Sclerospora graminicola. Characterised by dwarf growth of plants, production of excessive tillering from the crown and branches from axillary buds, turning of floral bracts into leaf like malformations, failure to form kernel, leaves showing necrosis and splitting and shedding.

Italian millet rust Fungal disease caused by Uromyces Setariae italicae (Diot.) Yoshino. Characterised by formation of small brown pustules on both surfaces of leaves, bursting of pustules early, exposing dusty mass of spores.

Italian millet smut Fungal disease caused by Ustilago Crameri. Characterised by affected grains replaced by black, loose spore mass which remains enclosed in a membrane till plant matures.

Italian millet udbatta disease Fungal disease caused by Balansia oryzae; symptoms same as in udbatta in rice.

Ixodidae Family of ticks comprising large number of genera and species. Often classified as continuous feeders or one-host ticks, and interrupted feeders or dropping-off ticks. The latter are further divided into three two-host ticks and three-host ticks.

J

Jack A mechanical device used for lifting heavy loads through short distances with a minimum expenditure of manual power.

Jack bean See sword bean.

Jack fruit Artocarpus heterophyllus Lank. (A. integrifolia, L.). Family Moraceae. Used as shade tree in coffee, areca, cardamom and pepper plantations. Fruit put to numerous culinary purposes; latex from bark used for plugging holes in earthenware or mhote buckets. Timber used for construction and furniture. Flourishes well under humid and warm climates up to 1600 m elevation. Prefers rich, deep and well-drained alluvial or open textured loamy lateritic soil. Propagated by seed.

Jack fruit black root rot Fungal disease of jack fruit caused by Rosellina sp. Symptoms as in tea black root rot.

Jack fruit brown root disease Fungal disease of jack fruit caused by Fomes noixus. Symptoms as in tea brown root disease.

Jack fruit bud weevil Ochyromera artocarpi M. Insect pest of jack fruit. Grub bores into tender buds and fruits.

Jack fruit charcoal rot Fungal disease of jack fruit caused by Ustulina zonata. Symptoms as in tea charcoal rot.

Jack fruit inflorescence rot Fungal disease of jack fruit caused by Rhizopus artocarpi. Characterised by rotting and shedding of young female inflorescence.

Jack fruit pink disease Fungal disease of jack fruit caused by Corticium salmonicolour, characterised by young woody branches losing their leaves; die-back; appearance of thin, pink incrustations on lower shaded side.

jack fruit rot Fungal disease caused by Phytophthora palmivora, characterised by fruit developing soft rot.

Jack fruit shoot borer Margaronia caesalis W.A. reddish brown caterpillar, turning green when old, boring into shoots, buds and fruits of jack fruit.

Jack tree See jack fruit.

Jackson Turbidity Unit (JTU) The standard unit used in measuring the turbidity of a water sample; defined in terms of the depth of water to which a candle flame can be clearly distinguished. The Jackson candle turbidimetre is the standard measuring instrument which compares the amount of light penetrating a given water sample with that penetrating a standard sample.

Jalap Ipomoea purga Hayne (Exogonium purga (Hayne) Lindl.) Family Convolvulaceae. A twining herb whose roots are source of resinous drug used as purgative. Propagated from cuttings.

Jalpai (1) Elaeocarpus floribundus Blume. Family Elaeocarpaceae. A tree cultivated for its edible fruit in Bengal and Assam. (2) See Ceylon olive.

Jam A preserved fruit food; mixture of fruit, sugar cooked to consistency of a gel, firm enough to hold fruit tissues. Contains fruit pulp and hence not clear.

Jamaica sorrel See roselle.

Jambolana Syzygium cumini (L.) Skeels. (Myrtus cumini, L.; Eugenia jambolana Lamk.) Family Myrtaceae. Important fruit considered indigenous to India or to West Indies. Cultivated and growing wild. Mostly planted as avenues in orchards or as wind break. Medium sized big tree and long lived. Fruits vary in size, and is of attractive purple colour and sub-acid to sweet in taste. Not exacting in soil requirement. Propagated from seed; inarching or approach grafting also practised. Also called jara plum.

Jamcha A piece of cloth used as tail cloth for the fly-catching net.

James Lange theory (emotions) According to this theory, emotional experience is nothing but the experiencing of the organic changes produced by the evoking stimulus. According to this theory, we feel sorry because we cry and we cry because we are sorry.

Jamrul Eugenia formosa wall. [Syzygium mappaceum (Korth.) Mansf.]. A tree of Family Myrtaceae. Fruit edible.

Jand Prosopis spicigera L. Family Mimosaceae. A moderate sized thorny tree with light foliage. Can grow in arid and dry regions. Prefers alluvial soils. Propagated by direct sowing. Wood used for many purposes. Thorny branches used for fencing.

Jangliechachinda Trichosanthes cucumerina L. A herb of Family Cucurbitaceae grown for its fruit used as vegetable. Also called *Rambel.*

Jangligailia Alysiacarpus longifolius Wight. and Arn. Family Papilionaceae. A herb used as fodder.

Jangli haldi Curcuma aromatica Salisb. (C. Zedoaria Roxb. non Rosc.) Family Zingiberaceae. Rhizomes used as carminative and tonic.

Japanese barnyard grass See Japanese barnyard millet.

Japanese barnyard millet Echinochloa frumentacea (Roxb.) Link. (Panicum crusgalli L. var. frumentacea Roxb.). Family Gramineae. A tufted annual grass cultivated as a kharif crop in India. Less important but quickest growing millets. Grain can be consumed but mostly by poor people. Resistant to drought and water logging. Straw of inferior quality.

Japanese clover Lespedeza striata (Thunb.) Hook, & Arn. Family Papilionaceae Introduced from Uruguay and grown for fodder. Also called lespedeza.

Japanese pedal thresher Appliance introduced from Japan for threshing paddy. Consists of a drum with nails or projections on the outside and revolved with a pedal arrangement and pinion and gears. The ears of the paddy are hold on revolving drum to separate out the grain. A easy and quick method comparative to the traditional method.

Japanese peppermint Mentha arvensis DC. var. piperascens Holms. Family Labiatae, cultivated for essential oil from leaves which is a source of menthol.

Japanese persimmon Diospyros kaki L.f. (D. chinensis Blume). Family Ebenaceae. Temperate fruit introduced in India in 1921. Grows well and fruits are oblate to slender, conical with thin orange pulp surrounding flattened seeds. Propagated by collar grafting by whip and tongue method.

Japanese raisin tree Hovenia dulcis Thunb. A large tree of Family Rhamnaceae, native of China. Grown for its edible, fleshy fruiting stalk.

Japonica One of the three sub-species of the sativa rice species. The other two are indica and javanica. Developed in Japan and widely used in sub-tropical and warm temperate regions of the world. Have been used in the cross-breeding with indica for evolving new varieties.

Jarewa Bothriochloa pertusa (L.). A. camus. [Andropogon pertusus (L.) Willd.] Family Gramineae. A perennial grass; a good fodder both for grazing or for stacking.

Java almond Canarium commune L. Family Burseraceae. A tree grown for its seed and oil which are edible. Oil also used for illumination.

Java black rot A fungus disease of stored sweet potatoes caused by Diplodia tubericola; the inside of the root becomes black and brittle.

Java cedar See bishop wood.

Java plum See jambolana.

Jelly Fruit product prepared by cooking essential or clear fruit extract strained free from insoluble matter and sugar as in jam. Sparkling transparent and attractive in colour.

Jerusalem artichoke Helianthus tuberosus L. A perennial herb of the family Compositae and native of North America. Edible tubers and also used for manufacture of industrial alcohol.

Jerusalem thorn Parkinsonia aculeata L. A small tree of family Caesalpiniaceae, native of tropical America and grown as hedge plant. Also called Christ's thorn.

Jetting A method of driving piles or well points into sand by using a jet of water to break the soil.

Jetty Dike built of piles, rock, or other material, extending into a stream or into the sea at the mouth of a river to induce scouring or deposition or for protection.

Jevons effect The effect upon the measurement of rainfall caused by the presence of the rain gauge; in 1861 W.S. Jevons pointed out that the rain gauge causes a disturbance in air flow past it, and this carries part of the rain past the gauge which would normally be captured.

Jews-mallow See olitorius tree.

Jharber Ziziphus rotundifolia Lamk. [Z. nummularia (Burm. f.) Wight and Arn.; Rhamnus nummularia Burm. f.]. Family Rhamnaceae. Wild species of ber. Good root stock for ber.

Jhau (1) Tamarix dioica Roxb. Family Tamaricaceae. The branches used for making baskets. (2) Tamarix troupii Hole. (Tamarix gallica auct. non L.). A bushy shrub of family Tamaricaceae. Galls formed on branches are used for Tanning. Also called French tamarix.

Jhingan Lannea coromandelica (Houtt.) Merr. [Lannea grandis (Dennst.) Engl.; Odina wodier Roxb.]. Family Anacardiaceae. Medium sized tree found in deciduous forest. Used for afforesting eroded areas. Young leaves are edible. Wood used for many purposes.

Jhum An Indian term for shifting cultivation commonly adopted in Assam.

Jimsonweed Datura stramonium. A tall, poisonous annual weed having large white or violet trumpet-shaped flowers and globose prickly fruits. Also known as apple of Peru.

Job description A careful description of a job and what it involves, enabling the qualifications and qualities required for it to be defined.

Job evaluation A systematic procedure for weighing up how much a job is worth in wages or salary compared with other jobs.

Job production The manufacture of an article to a customer's specification.

Job's tears Coix lacryma-jobi L. (C. lachryma L.) Family Gramineae. A perennial tropical herb, native of S.E. Asia and cultivated throughout India. Grains used as cereal. Produces white or grey bead-like seeds.

Joint (1) Body joints. (2) Joint between two single length of tiles.

Joint distribution The distribution of two or more variates.

Johnson grass Sorghum halepense (L.) Pers. (Holcus halepensis L.). Family Gramineae. A perennial grass resembling sorghum. A weed in cultivated fields. Infest crops such as sorghum and maize from which it cannot be distinguished in the early stages. Poisonous to stock in young stage. Used and grown as fodder. Suitable for hay-making. Heavy yielder.

Johnson grass ascochyta leaf spot Fungal disease of Johnson grass caused by Ascochyta sorghi. Characterised by indefinite spots of variable colour on older leaves and culms.

Johnson grass downy mildew Fungal disease caused by Sclerospora sorghi. Symptoms as in sorghum downy mildew.

Johnson grass grain smut Fungal disease caused by Sphacelotheca sorghi. Symptoms same as in sorghum grain smut.

Johnson grass head smut Fungal disease caused by Sphacelotheca reiliana. Symptoms same as in sorghum head smut.

Johnson grass loose smut Fungal disease caused by Sphacelotheca cruenta. Symptoms same as in sorghum loose smut.

Johnson grass red leaf spot Fungal disease caused by Collectotrichum graminicolum. Symptoms as in sorghum red leaf spot.

Johnson grass rust Fungal disease caused by Puccinia purpurea. Symptoms as in sorghum rust.

Joint family Collection of more than one primary family, on the basis of close blood ties and common residence, is called the joint family.

Joint oil See Synovia.

Joint space See crack width.

Joint wrapping Placement of a piece

of asphalt or coal tar impregnated paper, fibre glass or burlap over or around the junction of two tiles to help prevent inflow of sediment. See filter.

Jointer It is a small irregular piece of metal and is like a miniature plough. It turns over a small ribbon like furrow slice directly in front of the main plough bottom so that all trashes on the top of the soil is completely turned down and buried.

Jojoba Simmon dsia Californica. A shrub adapted to the arid portions of the south western United States and Mexico; liquid extracted from the seed may be used as a substitute for sperm oil.

Jonathan freckle A storage disease of apples characterised by small circular discolourations in the skin.

Jonathan spot A disease of apples characterised by circular depressed necrotic areas around the tenticels.

Juglandaceae A family of dicotyledonous plants in the order Juglandales having unisexual flowers, a solitary basal ovule in a unilocular inferior ovary, and pinnately compound, exstipulate leaves.

Juglandales An order of dicotyledonous plants in the subclass Hamamelidae distinguished by compound leaves; includes hickory, walnut, and butter nut.

Juglans regia See walnut.

Jugum One pair of opposite leaflets of a pinnate leaf.

Jujube Ziziphus mauritiana Lamk. (Z. jujube Lam. non Mill.) Family Rhamnaceae; one of the most common fruit trees of India. Fruit rich source of vitamin C and contains A and B. Dried fruit used as dessert, candied. Tree used for rearing lac insects. Hardy, grows well under varying soils and climatic conditions. Favours hot, dry climate. Propagated by seeds or by budding by ring or 'T' method, whip grafting.

Jumgermannials A leafly liverworts, an order of bryophytes characterised by cholorophyl-containing, ribbon-like or leaf-like bodies and an undifferentiated thallus.

Juncaceae A family of monocotyledonous plants in the order Juncales characterised by an inflorescence of diverse sorts, vascular bundles with abaxial phloem, and cells without silica bodies.

Juncales An order of monocotyledonous plants in the subclass Commelinidal marked by reduced flowers and capsular fruits with one too many anatropous ovules per carpel.

Jungle An impenetrable thicket of second-growth vegetation replacing tropical rain forest that has been disturbed, lower growth layers are dense.

Junk wind A south or southeast monsoon wind, favourable for the soiling of junks; the wind is blown in Thailand, China and Japan.

Junta A wind blowing through Andes Mountain passes, sometimes reaching hurricane force.

Juran A wind blowing from the Jura Mountains in Switzerland from the north west toward Lake Geneva; it is a cold and snowy wind and may be very turbulent, especially in spring. Also spelled Joran.

Jurassic The mid-Mesozoic geological period following the Triassic and preceding the Cretaceous; of this period.

Justicia adhatode See Malabar nut.

Jute Bast fibre, obtained from the jute plant. Exported as raw fibre or as manufactured goods. Fibres used as in the manufacture of packing cloth, hessian and bags for the transport and storage of cotton, food grains etc. and manufacture of rugs, carpets, curtain, lining, ropes and twines etc.

Jute apion Apion corchari Marshall. Serious insect pest of jute. Grub feeds inside bark, damages fibre bundles, as a result of mucilaginous substance oozes out, forming knots in the fibre.

Jute blackband disease Fungal disease of jute caused by Diplodia corchori. Characterised by dense black discoloured-band around stem, 60-90 cm, above ground level and wilting and dying of plants.

Jute board A fibre board made of Jute fibre.

Jute crop Corchorus capsularis L. and C. olitorius L. Shrub of the Family Tiliaceae from which best fibres of jute is extracted. A cash crop. Green leaves of C. olitorius edible; stalks stripped off fibre used as fuel and for making gunpowder charcoal. Requires warm and humid climate. It can withstand flooding in later stages to some extent. Cultivated on light sandy deltaic loams receiving periodic deposits of fertile silt from flood waters. Also called white jute, patsan.

Jute dry rot See jute stem rot.

Jute fibre The woody fibre of a plant native to India. The plant grows to a height of, from 3 to 4.5 m, and the prepared fibres are from 1.25 to 2.5 m long. Burlap webbing and rope are made from the fibres.

Jute fibre extraction The process of taking out fibre from jute stalks after retting. In this 8-10 stalks are taken together and their root ends are beat by a mallet. Wrapping the loosened fibres around the fingers is done and stems are agitated in the water till the fibre separates out. Rinsing, washing, cleaning, wringing and piling of the separated fibre is done. Finally strands are opened and hanged to dry in the sun. Fibre content is 4.5 to 7.5 per cent.

Jute paper A strong paper composed principally of jute fibre.

Jute powdery mildew Fungal disease caused by oidium sp. Symptoms same as in cotton powdery mildew.

Jute retting The steeping of stripped jute stems in water for stripping off the fibre, gently flowing, fairly deep, clear and soft and tepid water is considered best for the purpose. Traditional method of muddy stagnant water result in dark, discoloured fibre of low value. Complete but not too deep submersion desirable. Submergence in water in which jute has already been retted hastens retting process. Keeping bundles of stalks with the root end in water for 2-3 days before submerging ensures proper retting of root ends. Retting complete in 10-15 days during warmth and 25 or more days when water temperature is low.

Jute sclerotial disease Fungal disease of jute caused by Pellicularia rolfsii. Symptoms as in Barley sclerotial disease.

Jute semi-looper Anomis sabulifera Guen. Serious insect pest; a long greenish caterpillar feeds on leaves and later damages apical buds and top shoots; check plant growth. When attack is serious, plants get completely stripped off leaves.

Jute stem rot Fungal disease of jute caused by Macrophomina phaseoli (Maubl.) Ashby. Characterised by symptoms as damping off in seedling stage, and blackened discolouration and partial decay of tissue of basal portion of plant in older stage. Also called jute dry rot.

Junction Point of intersection of two open or closed drains; accessory used to effect a junction between two tile lines.

Junction box Box, manhole or other structure which serves to join one or more drains.

K

K selection Selective conditions which favour longevity of individuals at expanse of reproductive capacity.

K Strategist A species adapted to maximum survival of individuals at expense of reproductive capacity.

Kachnal Bauhinia racemosa Lamk. Family Caesalpiniacea. A small bushy tree, whose bark yields a fibre, flower buds and fruits edible.

Kadavanchi Momordica tuberosa Cogn. (M. cymbalaria Fenzl. ex Naud., Luffa tuberosa Roxb.). A trailing herb of Family Papilionaceae, which produces edible fruits.

Kaki Diospyros kaki. The Japanese persimmon; it provides a type of ebony wood that is black with gray, yellow and brown streaks, has a close, even grain, and is very hard.

Kal baisakhi A short-lived dusty squall at the onset of the south-west monsoon (April-June) in Bengal.

Kainit (Kainite) A potash salt containing potassium and sodium chloride and sometimes sulphate of magnesia, with more than 12 per cent potash. It is a fertiliser.

Kale Either of two biennial crucifers, Brassica oleracea var. acephala and B. fimbriata, in the order Capparales, grown for the nutritious curled green leaves.

Kallar soil See saline and alkali soils. Also called thur soils.

Kalo bilanni Antidesma acuminatum Wall. Family Euphorbiaceae. A tree which has edible fruit.

Kamala (1) Mallotus philippinensis Muell. Arg. Tree of the Family Euphorbiaceae whose capsule fruit bears hairs, used as vermifuge, with purgative action. Red dye obtained from surface of fruits used for dying silk. (2) Hairs of the capsule fruit of the tree. Also called rottlera.

Kanchari Alternanthera amoena Voss. (Telanthera amoena Regel). Family Amaranthaceae. A herb used as vegetable.

Kandla Bauhinia retusa Roxb. Family Caesalpiniaceae. Tree yields gum.

Kanghi (1) Abutilon asiaticum G. Don. Family Malvaceae. A shrub cultivated for its stem fibre, used in making ropes and cordage. (2) country mallow.

Kanju Holoptelea integrifolia (Roxb.) Planch. Family Ulmaceae. A large deciduous tree. Suitable for avenues on canal banks. Wood useful for bobbins, brush backs and agricultural implements.

Kanjura Commelina oblique Buch. Ham. Family Commelinaceae. Young shoots are eaten as vegetable.

Kankar Lime nodules found in alluvial soils, occurring either mixed with soil or as a impervious hard substratum at varying depth. It is due to precipitation of $CaCO_3$.

Kanta alu Dioscorea pentaphylla L. Family Dioscoreaceae. A herb found throughout India whose roots are edible.

Kanta-Katchu Lasia spinosa (L). Thw. (Lasia heterophylla Schott.). A perennial herb of Family Aracaceae, the young leaves of which are eaten as vegetable.

Kaolin A white non-plastic material consisting in the raw state, of the mineral kaolinite mingled with fragments of quartz, feldspar, mica, the residual of minerals of granite.

Kaolin, China clay, or Porcelain clay A white clay, essential hydrated aluminium silicate, resulting from the decomposition of feldspathic rocks.

Kaolin Group Hydrous aluminosilicates having a 1.1 phyllosilicate structure and no permanent charge.

Kapok Silky fibres that surround the seeds of the kapok or ceiba tree. Also known as ceiba; Java cotton; silk cotton.

Kapok tree Ceiba pentandra. A tree of the family Bombacaceae which produces pods containing seeds covered with silk cotton. Also known as silk cotton tree.

Karam sag Brassica oleracea L. var. acephala DC. Family Cruciferae. A type of mustard whose leaves and stems are used as vegetable.

Karanda Carrisa carandas L. Family Apocynaceae. Protective hedge plant yielding berry-like fruit used for making *chutney*, pickles and jelly. Rich in vitamin C.

Karer Caparis decidua (Forsk.) Pax (Capparis aphylla Roth.). Family Capparidaceae. A leafless shrub whose green and mature fruits edible. Wood hardy and heavy; resistant to termite attack.

Karmal (1) Dillenia pentagyna Roxb. Family Dilleniaceae. Tree found in western peninsula whose flowers are edible. Bark fibre used for cordage; old leave used as sand paper. (2) See Carambola.

Karnal bunt Fungal disease of wheat caused by Neovossia indica (Mitra) Mundkur. Characterised by diseased grains being partly converted into black sooty powder which smells like rotting fish.

Karst plain A plain on which karst features are developed.

Kartak Dicanthium caricosum (L.) A Camus. Family Gramineae. A grass used as fodder.

Kasaula An Indian term for an implement used for hoeing manually.

Karyotype chromosomal complement of an individual or a species.

Kassod tree Cassia siamea Lamk. Family Caesalpiniaceae. Medium sized ornamental tree. Flowers edible can be grown in dry areas and along canal banks.

Kasuri methi Trigonella corniculata. Family Papilionaceae. Fenugreek suitable for cooler areas.

Katabolism (catabolism) The destructive processes going on in an animal body.

Katafront A front (usually a cold front) at which warm air descends the frontal surface (except, presumably, in the lowest layers).

Katallobaric Of or pertaining to a decrease in atmospheric pressure. Also known as katabaric.

Kat ber Ziziphus xylocarpa Willd. A tree of Family Rhamnaceae. Serves as lac host and fruits used for tanning.

Kath bewal Grewia glabra Blume (G. Laevigata auct., plur. non Vahl.). A tree of Family Tiliaceae whose bark yields fibre for cordage.

Kathumbar Ficus hispida L. f. Shrub or small tree of Family Moraceae with edible fruits.

Kauki Manilkara kauki (L.) Dub. (Mimusops kauki L.). Tree of Family Sapotaceae whose fruits are edible.

Kavatta grass Setaria pallide-fusca (Schumach.) Stapt. and Hubbard. Family Gramineae. Herb used as fodder. Grains are also edible.

Keel (1) Boat-shaped structure formed by two anterior petals of Leguminoseae. (2) The ridge on blade or on other parts of grasses. (3) Longitudinal strip of metal attached at the centre bottom of the shoe or ditch bottom form of a trenching machine.

Kefir A low-alcohol-content beverage prepared from cow's milk treated with Lactobacillus casei and then fermented by the action of Sacchoromyces kefir or kefir grain.

Kelp The common name for brown seaweed belonging to the Laminariales and Fucales. Dried kelp contains 1.6 to 3.3 per cent nitrogen, 1.0 to 2.0 per cent phosphoric acid and 15-20 per cent potash. Can be used as a potash fertiliser.

Kentucky coffee tree Gymmocladus dioica. An extremely tall, dioecious tree of the order Rosales, readily recognised when in fruit by its leguminous pods containing heavy seeds, once used as a coffee substitute.

Keratin A complex nitrogenous compound.

Kerguelon faunal region A marine littoral faunal region comprising a large area surrounding Kerguelon Island in the southern Indian Ocean.

Kernal The inner portion of a seed within the integuments, also the whole grain of a cereal.

Kernal blight Any of several fungus disease of barley caused chiefly by Gibberella zeae, Helminthoporium sativum, and Alternaria species, shriveling and discolouring the grain.

Kernel spot A fungus disease of the pecan kernel caused by Conialhysium caryogenum and characterised by dull-brown roundish spots.

Ketchup See catchup.

Keto acid Formed in the liver of the animal from incomplete combustion of fat.

Ketone bodies Certain chemical substances such as diacetic acid, hydroxy butric acid, and acetone produced by the liver in animals. Excess production causes ketosis. When excessive amounts are eliminated in urine, it is called ketonuria.

Ketogenesis Production of ketone bodies or ketones.

Ketonuria See ketone bodies.

Ketsup See catchup.

Kettle Depression left after the melting of a detached mass of glacier ice burried in drift.

Key A wedge shaped strip of iron or steel used for preventing wheels from slipping around upon their axles. Keys are of various kinds and shapes.

Key-fruit Winged achenes hanging in clusters.

Key species Plant species that permits moderately close grazing and is fairly or potentially abundant.

Key stone The uppermost stone of an arch which looks its members together.

Key village A village defined under the key village scheme for particular upliftment or development programme in a village.

Khadar An Indian term for riverine soils formed by alluvial deposits.

Khantodi (1) Ceropegia hirsuta Wight & Arn. Family Asclepiadaceae. A climbing herb whose tubers are edible. (2) Ceropegia oculata Hook. A herb which produces edible tubers.

Khapra beetle Trogoderma granaria. Everts. Brown oval beetle whose grubs damage stored wheat, cereals, millets and pulses, Grub brownish white with bundles of reddish brown hair all over the body.

Khar batua Chenopodium murale L. Family Chenopodiaceae. A herb used as fodder. Also a annual weed in rabi crops.

Kharif crop (India and Pakistan) Summer crop or monsoon crop.

Kharpudi Ceropegia Lawii Hook. Family Asclepiadaceae. A succulent herb found in south India which produces edible tubers.

Khirni (1) Manilkara hexandra (Roxb.) Dub. (Mimusops hexandra Roxb.). A tree of family Sapotaceae; used as root stock for sapota; fruits edible. (2) See kauki.

Kiasi Hibiscus furcatus Roxb. A herb of Family Malvaceae whose leaves are edible and stem produces fibre for rope.

Kidney bean See French bean.

Kierpa Carallia brachiata (Lour.) Merr. Family Rhizophoraceae. A tree grown for its edible fruits.

Kiln An oven, furnace or large heated room for baking curing etc., such as brick kiln.

Kilogram Mass of the platinum-iridium cylinder deposited at the Bureau Internationale de Poids at Measures and declared international proto-type of the kilogram. It is equal to 1000 grams.

Kilogram-force Force which when applied to a body having a mass of 1 kg gives it an acceleration of 9.80665 m/sec^2.

Kikuyu grass Pennisetum clandestinum Hochst. ex. Chiov. Family Gramineae. Rhizomatous, stoloniferous perennial grass, with short internodes on stolons, forming dense growth and coverage. Suitable for moist hilly situations. Good for soil conservation.

Kimu Morus serrata Roxb. Tree of Family Moraceae. Edible fruits and leaves used as fodder.

Kinase An activator, specific but of unknown composition.

Kind Means one or more related species or subspecies of crop plants, each individually or collectively known by one common name e.g. wheat, paddy, etc.

Kinematic similarity Similarity of motion. When the ratios of the components of velocity at all homologies points in two geometrically similar system are equal, the two states of motion are kinematically similar, the paths of homologous particles will then also be geometrically similar.

Kinetic energy The energy that a body possesses as a consequence of its motion.

Kinetic flow factor A factor defined by twice the ratio of the kinetic energy head to the potential energy head.

Kineticity, or **Velocity-head ratio** Ratio of the kinetic energy head to the potential head.

Kingpin A steel pin or shaft, hardened and ground to size, used in fastening and supporting the steering knuckle to the axle beam, permitting free movement right and left of the front wheels.

Kinker Acacia jacquemontii Benth. Family Mimosaceae. Grown as hedge plant. Bark used as tan and branches as fodder.

Kino gum A brownish astringent substance or gum produced from Pterocarpus marsupium.

Kirni Canthium parviflora Lamk. Family Rubiaceae. A shrub or small tree whose wood is hardy and durable. Leaves and fruits edible.

Kitchen gardening Growing of vegetables near household to meet the internal requirement of vegetables.

Kite A bill often called an 'accommodation bill' drawn on a person who had received nothing in return. It is an expedient for obtaining credit for a short period.

Klendusity The ability of a susceptible plant to escape infection under conditions in which other susceptible varieties contract the disease. Such as a thick cuticle that prevents inoculation.

Knapsack duster The machine for applying insecticide or pesticides as dust. Air blast developed by billows to which handle is connected. Suitable for spot applications.

Knapsack sprayer A sprayer for spraying fertilisers, insecticides or pesticides in which pump operation is to be continuous to get the discharge of the liquid.

Knife harrow A type of harrow that consist of a frame holding a number of knives which scrap and partly invert the soil surface to smooth it and destroy small weeds.

Kniffen System of training in grapewine by which permanent arms of the vine are tied horizontally to wires and on the same level, above the ground.

Kniss Dioscorea deltoidea wall. Family Dioscoreaceae. A shrub whose roots are edible. Also yield a steroid hormone, cortisone used in medical field.

Knol khol brassica oleracea, L. var. gongylodes L. (var. caulorapa DC.). A biennial herb related to cabbage. Family Cruciferae. Culture like cabbage and cauliflower. Also called turnip rooted cabbage, turnip cabbage, stem cabbage.

Knot A branch base embedded in the tree or timber.

Knot cluster A number of knots with wood fibres deflected round the entire group.

Knot grass Paspalum distichum L. Family Gramineae. A grass with knotted stems, grown for fodder.

Knot hole A hole left as the result of the removal of a knot.

Knuckle joint One in which an eye at the end of a rod is engaged by the forked end of a second rod, the two being connected with a joint pin.

Kodo millet Paspalum scrobiculatum L. Family Gramineae. Hardy and remarkably drought resistance food grain. Grown on gravelly, stony, light upland, poor soils.

Kohlrabi A biennial crucifer, designated Brassica caulorapa and B. oleracea var. caulorapa, of the order Capparales grown for its edible turnip like, enlarged stem.

Kokam butter tree Garcinia indica (Thou.) Choisy. Tree of the family Guttiferae, the fruits of which are edible.

Kokko Albizia Lebbek (L.) Benth. (Mimosa Lebbek L.; Acacia Lebbek Willd.). Family Mimosaceae. A large deciduous tree with large white flowers and characteristically long and flat pods which remain on the tree long after ripening. Gum from the tree similar to gum arabic. Grown for shade in tea and coffee plantations. Wood useful for agricultural implements and furniture.

Kole cultivation A method of rice cultivation along the West Coast (India). The Sea water entry is checked by making embankments and rain water is pumped out in November. Levelling the soil without ploughing is done and sowing or transplanting is done.

Kovai fruit See little gourd.

Kreis test A test for rancidity in edible fats and oils.

Krummholz Stunted alpine forest vegetation. Also known as elfin-wood.

Kudaliya Desmodium triflorum (L.) DC. A herb of Family Papilionacea grown as a soil binder and fodder.

Kudzu Any of various perennial vine legumes of the genus Pueraria in the order Rosales cultivated principally as a forage crop. Used principally for soil conservation on slopes.

Kufri (1) Chrysopogon gryllus (L.) Trin. (Andropogon gryllus L.). A fodder of family Gramineae. (2) A variety of potato.

Kulai See chlorophytum tuberosum Dalz.

Kuleha Eucalyptus terminalis. Family Myrtaceae. Large tree with grey-white short, flaby bark, drought resistant.

Kumali Leea edgeworthii Santapu (L. aspera Edg. non Wall.). A shrub of Family Vitaceae, the fruits of which are edible.

Kumquat A citrus shrub or tree of the genus Fortunella in the order Sapindales grown for its small, flame-to orange-coloured edible fruit having three to five locules filled with an acid pulp, and a sweet, pulpy rind.

Kusa grass Desmostachya bipinnata (L.) Stapf. (Poa Cynosuroides Retz.). A perennial grass of Family Gramineae used for making rough ropes and thatching. Has medicinal value.

Kurtosis The extent to which a unimodal frequency curve is 'peaked' that is to say, the extent of the relative steepness of assent in the neighbourhood of the mode. Generally measured by the fourth moment.

Kutu Polygonum fagopyrum L. (Fagopyrum esculentum Moench). Family Polygonaceae. A buckwheat native of Central Asia, cultivated for its edible grains.

L

Labellum The lower petal.

Labial Of the lips, like serving as a lip; lip like part, or labium.

Labiate Refers to the form of a flower when calyx or the corolla are divided in two unequal parts, with one overhanging the other, like lips.

Labium The fused second maxillae of insects; labial.

Lablab bean Dolichos lablab L. var. typicus (Lablab niger Medic.) Family Papilionaceae. Perennial twining herb cultivated as an annual for its pods, seeds and fodder.

Lablab leaf spot Fungal disease of lablab bean caused by Cercospora dolichi. Characterised by circular to angular spots with grey centres and reddish border or leaves.

Lablab niger See lablab bean.

Lablab powdery mildew Fungal disease caused by Laveillula taurica var. macrospora. Characterised by grey white fungus growth on lower leaf surface.

Lablab rust Fungal disease caused by Uromyces appendiculatus. Characterised by minute, slightly raised spots on the leaves, developing into brownish powdery sori, which becomes dark brown or black.

Lablab yellow mosaic virus disease characterised by light and dark green patches on leaves and leaves turning yellow in advance stage.

Labour One of the 'factors of production'; the combination of all exertion by individuals, whether they be manual, physical or mental directed towards the production of wealth. Labour is quantitative, involving both efforts and time; it is also qualitative, both intelligence and skill determining its efficiency.

Labour income (farm) It is the amount of a farm operator receives for a year of labour and management after having paid all business expenses of the farm, including depreciation and making an allowance for a return on capital.

Labour force The proportion of population entering the labour market; employed plus unemployed.

Labour intensive Description of industries in which there is a considerable use of labour in relation to the amount of capital equipment, per unit of output.

Labour mix In a business enterprise, the proportion of salary-earners to wage-earners.

Labour, mobility of The case of movement of labour between areas and occupations. Movement from one area to another is known as geographic or lateral mobility. Occupational movement is described as vertical mobility.

Labour productive Productive labour is all efforts, by hand and brain, which contributes to the supply of goods and services which command a price.

Labour supply Supply of effort, the total number of hours of work that a population is willing to supply.

Labour turnover The number of workers who leave, or are replaced in a given period expressed as a percentage of the average number of workers employed during the period.

Lac Resin secreted by the glands of the body of lac insects. India a major producer of lac.

Laccifera lacca kerr Insect belonging to group of coccids (scale insects and mealy bugs) producing the lac resins through lac glands. Need 1,50,000 insects secretions to produce 450 gms of lac. Thrives on sap of host plants.

Lacerated Having the margin or apex deeply cut into irregular lobes.

Lacquer Gold-coloured varnish of shellac dissolved in alcohol; used for polishing wood; articles so coated.

Lactating Designating an animal that is actively producing and secreting milk. Preferred to "milking".

Lactometre An instrument to measure the specific gravity of milk.

Lactose ($C_{12}H_{22}O_{11}$) Milk sugar. Quickly fermented by micro-organisms into lactic acid.

Lac tree Schleichera oleosa (Lour.) Oken (Schleichera trijuga Willd.; Pistacia oleosa Lour.). Family Sapindaceae. Large deciduous tree raised from seed. A good lac host. Young fruits edible. Seed oil used for soap making and in perfumery.

Lactuca sative See lettuce.

— **sativa var. longifolia** See cos lettuce.

— **scariola** See wild lettuce.

— **serriola** See wild lettuce.

Lacustrine deposit Material deposited in lake water and later exposed either by lowering of the water level or by uplifting of the land. These sediments range in texture from sands or clays.

Lacustrine soil Soil formed in beds of lakes, and worked over by water.

Lady beetle Insect belonging to the Family Coccinellidae, beneficial because of its preying habit upon aphids, scale insects and mealy bugs. Used in biological control of insects.

Lady bird beetle See lady beetle.

Lady's finger Abelmoschus esculentus (L.) Moench. (Hibiscus esculentus L.). Family Malvaceae. Popular vegetable crop. Good source of vitamin A and B and also C. Rich in proteins and minerals. Good source of iodine. Thrives in most soils.

Lady's finger damping off Fungal disease caused by Pythium aphanidermatum and Phytophthora palmivora. Characterised by constriction at ground level and falling off of seedlings.

Lady's finger dry root rot Disease caused by Macrophomina phaseoli. Characterised by shredding of outer parts of roots.

Lady's finger grey leaf weevil Mylloceros sp. Insect pest of lady's finger. Adult weevil feeds on leaf tissue and makes holes in leaves.

Lady's finger leaf hopper bug Empoasca devastans. Insect pest of Lady's finger. Symptoms as in cotton leaf hopper.

Lady's finger leaf spot Disease caused by Cercospora hibisci. Characterised by appearance of brown spots on leaves.

Lady's finger powdery mildew Disease caused by Erysiphe cichoracearum DC. Formation of ash-coloured powdery growth on upper surface of old and new leaves. Chlorophyll content decreases.

Lady's finger rust Disease caused by Puccinia parri. Longitudinal lesions on plant appear.

Lady's finger yellow vein mosaic Virus disease in which veins spreading to main vein turn yellow. Entire leaf turning chlorotic. Formation of small size fruit. The disease is spread by white fly (Bemisia tabaci).

Laemophloeus minutus See flat grain beetle.

Laestadia theas See tea copper blight.

Laevulose (levulose) Fructose or fruit sugar.

Lag An event occuring in time (*t+k*) (*k* greater than zero) is said to lag behind an event occurring at time, *t*, the extent of lag being *k*. An event occurring *k* units before time '*t*' is regarded as bearing a negative lag, or a lead.

Lagenaria siceraria See bottle gourd.

— **vulgaris** See bottle gourd.

Lagg The depressed margin of a raised bog.

Lagoon A relatively shallow estuary with very restricted exchange with the sea and no significant freshwater inflow.

Lagtime (Hydrology) Watershed lag. (Flood Irrigation) The period between the time that the irrigation stream is turned off at the upper end of an irrigated area and the time that water disappears from the surface at the point or points of application.

Laissez faire or **non-directive leadership** Laissez faire leaders are those who perform their duties though patience and friendliness. They advise very little and offers no judgment. They are Yes-man. Production is not high and there is confusion in the group sometime.

Lajalu Neptunia oleracea Lour. (Neptunia prostrata Baill). A aquatic herb used as vegetable. Family Mimosaceae.

L Layer See horizon soil.

Lallei Albizia amara Bov. (Albizia wightii Grah, mimosa amara Roxb.) Family Mimosaceae. Wood used for fuel and tool handles and leaves as green fodder.

Lamella Gill.

Lamiales An order of dicotyledonous plants in the subclass Asteridae marked by its characteristic gynoecium, consisting of usually two biovulate carples, with each carpel divided between the ovules by a false partition, or with the two halves of the carpel seemingly wholly separate.

Lamina (1) A unit layer less than 1 cm thick in a stratified sequence. (2) A thin layer or scale; blade of leaf.

Laminal placentation Condition in which the ovules occur on the inner surface of the carpels.

Laminar flow That type of flow in which each particle moves in a direction parallel to every other particle, and in which the head loss is approximately proportional to the first power of the velocity. Streamlines remain distinct from one another over their entire lengths, except for the very minor effect of molecular mixing. It is sometimes designated 'streamline flow' or 'viscous flow'.

Laminariophyceae A class of algae belonging to the division Phaeophyta.

Laminated Generally refers to a soil, arranged in very thin layers or plates, horizontal to the soil surface.

Laminated wood An assembled product made up of layers of wood and adhesive in which the grains of adjacent layers are parallel.

Lamium album See white dead nettle.

Lammas shoot Late season flush of growth in a temperate tree which normally flushes annually.

Lampati Duabanga sonneratioides Buch.-Ham. (D. grandiflora Walp.). Family Sonneratiaceae. A large tree whose fruits are edible.

Lannate Wooly, covered with short hair like processes giving wooly appearance.

Lancedate Shaped like a lance or spearhead.

Land The solid part of the earth's surface or any part thereof. A tract of land is defined geographically as a specific area of the earth's surface. It's characteristics embrace all reasonably stable, or predictably cyclic, attributes of the biosphere vertically above and below this area, including those of the atmosphere, the soil, and the underlying geology, the hydrology, the plant and animal populations, and the results of past and present human activity, to the extent that these attributes exert a significant influence on the present and future uses of land by man. One of the factors of production.

Land blink A yellowish glow observed over snow-covered land in the polar regions.

Land breeze A coastal breeze blowing from land to sea, caused by the temperature difference when the sea surface is warmer than the adjacent land; therefore, the land breeze usually blows by night and alternates with sea breeze which blows in the opposite direction by day.

Land classification The arrangement of land units into various categories based on the properties of the land or its suitability for some particular purpose.

Land capability The suitability of land for a specific purpose. Physical land characteristics and climate mainly determine the risk and difficulties involve in the particular land-use.

Land drainage The disposal of storm water from rural areas, and the removal of water excess from agricultural lands to prevent or to relieve waterlogging, the accumulation of harmful amounts of salts and deterioration of soil structure. When specifically referred to agricultural land, it is also known as 'agricultural drainage'.

Land forming Process of changing the land configuration by manual or mechanical means for moisture control.

Landforms The various shapes of the land surface resulting from variety of actions such as deposition or sedimentation (eskers; lacustrine basins) m erosion (gullies, canyons) and earth crust movements (mountains).

Land irrigability classification Classification of land on the basis of some parametres to indicate whether the land is suitable for irrigated agriculture.

Landlord One who owns the land but does not cultivate himself and rents out to others for cultivation.

Land reclamation Making land capable of production, more intensive land use and more profitability. This may be achieved by changing its character through application of chemical amendments, clearing, erosion control etc.

Land requirement Refers to the requirements as to preceding crop for seed certification.

Land resource area An area of land reasonably alike in its relationship to agriculture with emphasis on combinations and/or intensities of problems in soil and water conservation, ordinarily larger than a land resource unit and smaller than a land resource region.

Land resource unit A subdivision of land resource area with emphasis on a specialised type of agriculture, intensities, or problems in soil and water conservation, it has a narrower range in relationship to agriculture with emphasis on soil and water conservation.

Landside Landside is that part of the plough that slides along the face of the furrow wall, helps to resist the side pressure exerted by the furrow slice on the mould board and also helps in stabilising the plough while it is in operation.

Land sky The relatively dark appearance of the underside of a cloud layer when it is overland that is not snow-covered, used largely in polar regions with reference to the sky map; it is brighter than water sky, but much darker than ice-blink or snow-blink.

Landscape (1) All the natural features such as fields, hills, forests, and water that distinguish one part of the earth's surface from another part. Usually it is the portion of land or territory that the eye can see in a single view, including all its natural characteristics. (2) To beautify a terrain for human enjoyment.

Land smoothing or **land grading** The operation of producing a plane land surface with a continuous slope. Th process may in general be included in land forming.

Land surveyor A specialist who measures land and its natural features and any man-made features such as buildings or roads for drawing to scale as plans or maps.

Landslide (landslip) (1) A mass of material that has slipped downhill by gravity, often assisted by water when the material is saturated. (2) Rapid movement of a mass of soil rock, or debris down a slope.

Land topography, or **relief** The surface configuration of the land.

Land type See association soil.

Land-use map A map showing land-use classes as well as other earth surface features such as roads, manufacturing plants, and harbours.

Land-use pattern The areal design or arrangement of land uses, major and minor and of operation units convenient for cultivation.

Land-use planning The development of plans for the uses of land that will, over a long period, best serve the general welfare, together with the formulation of ways and means of achieving such uses.

Langley A unit of radiation; equal to one gram-calorie per square centimetre.

Langsat Lansium domesticum Jack. Family Meliaceae. A Malayan fruit. Propagated by seed.

Lannea coromandelica See jhingam.

— **grandis** See jhingam.

Lanolin a yellowish viscous mass of wax-like consistency, very resistant to acids and alkalis, emulsifies with water and used in ointments. A wool-fat.

Lansium domesticum See langsat.

Lantana Lantana camara L. var. aculeata (L.) Moldenke (L. Camara auct., non L.). Family Verbenaceae. A perennial shrub with rough haired leaves and rough stem. Forms dense undergrowth, especially in teak forests. Produces plenty of seed, forms underground suckers.

Lantana camara See lantana.

Lap An accurately finished tool having its surface charged with an abrasive substance.

Lap joint A joint produced by the overlapping of contiguous faces of wood or metal etc.

Laphygma exigna See lucerne caterpillar; tobacco leaf caterpillars.

Lapidicolous Living under a stone.

Lappa clover Trifolium lappaceum L. Family Papilionaceae. Herb grown for fodder. Also called Burlock clover.

Lapse line A curve showing the variation of temperature with height in the free air.

Lapse rate (1) The rate of decrease of temperature in the atmosphere with height. (2) Sometimes, the rate of change of any meteorological element with height.

Larch The common name for members of the genus Larix of Thepine family, having deciduous needles and short, spur-like branches which annually bear a crown of needles.

Larch canker A destructive fungus disease of larch and sometimes pine and fir caused by Dasyscypha willkommii and characterised by flat, depressed cankers on the twigs and branches.

Large Indian bamboo Dendrocalamus giganteus Munro. Family Gramineae. Largest of Indian bamboo; produces clums up to 25 cm in dia. Cultivated in Assam, Bengal and Kerala.

Large-leaved caladium Alocasia indica (Roxb.) Schott. Family Araceae. Herb grown for its edible stems and rootstocks.

Laria affinis See pea beetle.

Larva The name of that stage of insect's life hatching from the egg which have complete metamorphosis. Caterpillars maggots and grubs are larve.

Larval flukes Immature stages of trematodes which may occur in water, in snails and insects or on vegetation.

Lorvicide A insecticide which destroys insect larve.

Larviparous Producing live larve.

Lasia spinosa See kanta katchu.

— **heterophylla** See kanta katchu.

Lasioderma serricorne See cigarette beetle.

Late (1) Refers to a variety which flowers or fruits later than normal. (2) An event such as sowing which is done later than normal.

Latent Lying dormant but capable of development under favourable circumstances.

Latent bud An auxillary bud whose development is inhibited, sometimes for many years, due to the influence of apical and other buds. Also known as dormant bud.

Latent heat The amount of heat energy expanded in changing the state or phase of a body without raising its temperature.

Latent infection Term usually applied to fungus infection in fruits while still on trees and nearing maturity.

Latent period Preinfective period.

Latent virus The virus which is present in a plant without the production of symptoms.

Latent instability The state of that portion of a conditionally unstable air column lying above the level of free convection; latent instability is released only if an initial impulse on a parcel gives it sufficient kinetic energy to carry it through the layer below the level of free convention, within which the environment is warmer than the parcel.

Lateral Relating to the side or crosswise of the length.

Lateral bud Any bud that develops on the side of stem.

Lateral bund In soil conservation, an embankment constructed at intervals along the contour bund, and constructed along the slope of the land and at right lines to the contour bund, to prevent flow towards and accumulation of water in a depression especially when there is deviation from contour.

Lateral drain A single line which collects the water from the soil and empties into a main or sub-main.

Laternal erosion Sheet erosion.

Lateral motion Motion in side-wise direction.

Lateral root A root branch arising from the main axis.

Laterals Diagonal braces between two members to increase rigidity.

Lateral stability Stability with reference

to disturbances involving rolling, yawing, or side slipping, i.e., disturbances in which the position of the plane of symmetry of the aircraft is affected.

Lateral thrust The pressure of a load which extends to the sides.

Laterite soils Soils formed *in situ* by the leaching out of the bases from the parent rock under monsoon conditions of dry and wet cycles. Well drained but poor in plant nutrients. Broadly divided into high and level laterites.

Laterisation Process of leaching out of calcium at or near the surface of the soil in heavy rainfall areas under tropical conditions due to intense weathering taking place to a great depth. Iron and Aluminium concentrate at or near the surface of soil.

Latex A milky or clear, sometime coloured juice or emulsion found in some plants in special secretory cells, usually the phloem and the leaves. Plants of the families Apocynaceae, Euphorbeaceae and Moreceae produce latex which is source of rubber.

Latexosis The production of latex by plants.

Latheticus oryae Waterh. A pale yellow beetle that infects flours and meals of cereals and dried vegetable matter. The grub, yellow and with numerous hairs on its body, also feeds on flours and meals. Also called long-headed flour beetle.

Lathyrism Paralytic condition of animals caused by consumption of seed of certain plants of the genus Lathyrus.

Lathyrus Lathyrus sativus L. Family Papilionaceae. A pulse crop grown for fodder and green manure. Low grade pulse but fodder of high nutritive value. Grain on excessive consumption is harmful to man and animal. Hardy, cold weather crop, grows best in water retentive low lying heavy soils.

Lathyrus aphaca See yellow vetchling.
— **sativus** See lathyrus.
— **ochrus** DC. A herb of Family Papilionaceae grown for fodder.

Lathyrus downy mildew Disease caused by Peronospora lathyripalustris. Characterised by leaves and twigs showing greyish, violet downy growth chiefly on under surface with affected part turning yellow or brown.

Lathyrus powdery mildew Disease caused by Erysiphe polygoni. Symptoms as in bean powdery mildew.

Lathyrus rust Disease caused by Uromyces fabae. Symptoms as in broad bean rust.

Lathyrus wilt Disease caused by Fusarium orthoceras var. lathyri. Characterised by chlorosis of leaves and discolouration of vascular tissues of roots; plants withering and drying up.

Laticifer A latex duct found in the midcortex of certain plants.

Latiferous Containing or secreting latex.

Latin squares An experimental design. This type of design is similar in principle to a randomised block arrangement, but in a 'Latin square' two cross-groupings of the experimental units are carried out, corresponding to the rows and columns of a square. The treatments are subject to the double restriction that each treatment occurs once and once only in each row and in each column. Thus, the differences between the rows and columns can be eliminated from the experimental comparisons. Especially adapted for accurate comparisons with a small number of treatments.

Latitude zonation Latitudinal (normal) Zonation of soils. See also vertical zonation.

Latmahuria Digera muricata (L.) Mart. (Digera arvensis Fork.; Achyranthes muricata L.). Family Amaranthaceae. A herb whose tender twigs and inflorescence are used as vegetable.

Lattice designs Designs developed for testing a large number of treatments, in which the number of blocks exceed the number of complete replications.

Lattice energy The energy required to separate the ions of a crystal to an infinite distance from each other.

Lattice structure The orderly arrangement of atoms in a crystalline material.

Lauraceae The laurel family of the order Magnoliales distinguished by definite stamens in series of three, a single pistil, and the lack of petals.

Lauter tub A tank with a perforated false bottom in which the extract (wort) is separated from the insoluble grain residue during mashing of barley malt.

Laveillula taurica See castor powdery mildew.

— **taurica** var. macrospora. See lablab powdery mildew.

Law of diminishing returns An economic law which states that as an increasing amount of a variable factor of production is applied to a fixed factor of production, the result is a decrease in average returns per variable unit applied.

Lawn mower A mechanical machine used for cutting grass to even height.

Lawsonia alba See henna.

— **inermis** See henna.

Lax Not campact.

Layering A propagation method by which root formation is induced on a branch or a shoot attached to the parent stem by covering the part with soil.

Leaching Removal from the soil of excess salts in solution by applying excess water of good quality to keep the water flux downward. The operation is generally carried out to reclaim salt affected soils.

Lead arsenate [$Pb_3 (AsO_4)_2$] A white crystalline compound stomach poison, insecticide applied as dust or spray. Basic form more stable than acid form, has less phytotoxic effect on crop.

Lead time The elapsed time between the placing of an order and receipt of the goods.

Leader (1) The terminal shoot of a tree or plant; main or dominant stem. (2) A person who takes lead in adopting new technology.

Leadership It is an activity of influencing people to co-operate towards some goal which they come to find desirable.

Leaf A modified aerial appendage which develops from a plant stem at a node, usually contains chlorophyll, and is the principal organ in which photosynthesis and transpiration occur. In grasses it consists of sheath and blade.

Leaf area index The leaf substended per unit area of land. Commonly denoted as LAI.

Leaf blight Any of various blight disease which cause browning, death, and falling of the leaves.

Leaf blotch A plant disease characterised by discoloured areas in the leaves with indistinct or diffuse margins.

Leaf bud A bud that produces a leafy shoot.

Leaf cist Any of several diseases of conifers characterised by falling of the needles.

Leaf curl A fungus or viral disease of plants marked by the curling of leaves.

Leaf cushion The small part of the thickened leaf base that remains after abscission in various conifers, and also in some extinct plants.

Leaf drop pre-mature falling of leaves, associated with disease.

Leaf fibre A long multiple celled fibre extracted from the leaves of many plants that is used for cordage, such as sisal for binder, and abaca for Manila hemp.

Leaf gap A region in the parenchyme of the vascular cylinder associated with the branching of vascular tissue (leaf trace) from the cylinder into a leaf.

Leaf lettuce Lettuce types; does not form heads; producing large, crumpled leaves, with wavy or frilled margins.

Leaf miner An insect that burrows into and feeds between the epidermal layers of leaves when in larval stage.

Leaf mottle A fungus disease characterised by chlorotic mottling of the leaves; for example, caused by verticillium dahliae in sunflower.

Leaf mould (1) The spongy humus developed under cover of forest vegetation especially of deciduous trees, in a temperate, humid climate. (2) Decomposed leaf compost used in preparing pot mixtures for raising seedlings in pots.

Leaf roll Any of several virus diseases characterised by upward or inward rolling of the leaf margins.

Leaf rot Any plant disease characterised by break-down of leaf tissues, for example, caused by Pellicularia koleroga in coffee.

Leaf rust Any rust disease primarily affecting leaves; common in coffee, alfalfa, and wheat, barley and other cereals.

Leaf scald A bacterial disease of sugarcane caused by Bacterium albilineans which invades vascular tissues, causing creamy or greyish streaking and withering of the leaves.

Leaf scar A mark on a stem, formed by secretion of suberin and a gum-like substance, showing where a leaf has abscised.

Leaf scorch Any of several disorders and fungus diseases marked by a burned appearance of the leaves, for example, caused by the fungus Diplocarpon earliana in strawberry.

Leaf sheath The lower part of the leaf which encircles the stem in grasses and grass like plants.

Leaf spot Any of various diseases or disorders characterised by the appearance of well-defined discoloured spots on the leaves.

Leaf spring A spring made up of several flat plates superimposed one upon another, such as used on automobiles, wagons, cars, etc.

Leaf stalk Petiole.

Leaf stem Petiole.

Leaf stripe Any of various plant disease characterised by striped discolourations on the foliage.

Leaf trace A branch of vascular tissues of a stem, extending out into a leaf.

Leaflet (1) One part of a compound leaf, also a small or young foliage leaf. (2) A small, immature leaf.

Leafy Refers to a plant, usually herbaceous with numerous or large leaves such as in leafy vegetables.

Leak A watery rot of fruits and vegetables caused by various fungi, such as Rhizopus nigricons in strawberry.

Leaky aquifers A confined aquifers covered with semi-impermeable layer at the top from which water from the unconfined aquifer above leaks into the lower aquifer particularly while pumping.

Lean cement concrete The mixture of cement concrete in which sand proportion is much higher than cement.

Lean mixture A fuel mixture which contains too much air in proportion to gasoline.

Lean to A shed or building built against or alongside of another structure.

Learning It is the process by which an individual through his own activity becomes changed in his behaviour.

Lease An arrangement whereby the use of land or buildings is granted by a landlord, or lessor to a tenant, for a specified term of years at an agreed annual rental.

Least square estimator An estimator obtained by the method of least squares.

Least square fit The least squares method gives rise to a straight or curvilinear trend from which the mean of the squared vertical deviations is a minimum. Hence, the 'least square line' of best fit.

Least square line The least square method produces a 'least square line' (also called 'a line of best fit' by minimising the mean squared vertical deviations from the 'regression line' where the regression between the variates is linear.

Least square method It is a method of determining the arbitrary constants in the equation of a curve, based upon the principle that the most probable value of a quantity to be deduced from a set of measurements is obtained by making the sum of the squares of the residuals a minimum.

Leather The cured, tanned skin of animals, particularly of bovines.

Leather rot A hard rot of strawberry caused by the fungus Phytophthora cactorum.

Labbeck-tree See kokko.

Lecanoraceae A Family of lichens in the order Lecanorales characterised by a crustose thallus and a distinct thalloid rein on the apothecia.

Lecanorales An order of the Ascolichenes having open discoid apothecia with a typical hymenium and hypothocium.

Le-Chateliers principle If conditions of a system, initially at equilibrium, are changed, the equilibrium will shift in such a direction as to tend to restore the original conditions.

Lecideaceae A temperate and boreal family of lichens in the order Lecanorales; members lack a thalloid rim around the apothecia.

Lecithin A phosphorised fat or phospholipide of cell protoplasm.

Lectins Group of protein substances, natural antibodies. May be obtained from lima beans.

Lecythidaceae The single family of the order Lecythidales.

Lecythidales A monofamilial order of dicotyledonous tropical woody plants in the subclass Dilleniidae, distinguished by entire leaves, velvate sepals, separate petals, numerous centrifugal stamens, and a syncarpous, inferior ovary with exile placentation.

Ledger A book for keeping accounts. All credits and debits are listed under appropriate heads.

Leea aspera See kumali.

— **edgeworthii** See kumali.

— **macrophylla** See dholsamudra.

Leek Allium porrum L. (Allium ampeloprasum L.). Family Liliaceae. Biennial plant grown as annual; blenched stems and leaves used as vegetable. Base of plant slightly swollen.

Leersia hexandra See rice grass.

Left hand tool Side tool ground to an angle on the right-hand side, and which, therefore, cuts from left to right.

Leggy A plant which has unusually long stem.

Leghaemoglobin From "leg" (leguminous) and "haemoglobin", a complex respiratory pigment of red corpuscles; a hemoprotein similar to blood haemoglobin.

Legume Plant member of the Family Leguminosae, with the characteristic of forming nitrogen-fixing nodules in its roots, in this way making use of atmospheric N possible; the pod of a leguminous plant.

Legume forage Any plant of the legume family (Leguminosae) used for livestock feed, grazing, hay, or silage.

Legume inoculation The application of a bacterial culture to the soil or seed of legume prior to planting so that nodules containing the bacterium may form on the roots and plant may thrive.

Legumin vegetable casein; a globulin in seeds of leguminosae.

Leguminosae The legume family of the plant order Rosales characterised by stipulate, compound leaves, 10 or more stamens, and a single carpel, many members harbour symbiotic nitrogen-fixing bacteria in their roots.

Lehmann love grass Eragrostis lehmanniana Nees. Family Gramineae. Grass Grown for erosion control and fodder. Introduced from U.S.A.

Leitneriales A monofamilial order of flowering plants in the subclass Hamameliobe; members are simple-leaved, dioecious shrubs with flowers in catkins, and have a superior, pseudomonomerous ovary with a single ovule.

Lemma The lower bract of the grass spikelet which encloses the caryopsis. Formerly called "Flowering glume".

Lemnaceae The duckweeds, a family of monocotyledonous plants in the order Arales; members are small, free floating thalloid aquatics with much reduced flowers that form a miniature spadix.

Lemon Citrus limon (L.) Burm, f. (C. medica L. var. limon L.; C. limonum Risso; C. medica var. limonum Wight & Arn.). A small evergreen tree belonging to the order Sapindales cultivated for its acid citrus fruits which is a modified berry called a hesperidium.

Lemon butterfly Papilio demoleus L. Insect pest of citrus; caterpillar feeds on leaves; in severe cases all leaves are eaten up.

Lemon grass See West Indian lemon grass and East Indian lemon grass.

Lemon grass oil Oil extracted from lemon grass. One hectare of grass yields 60-80 bottles of oil.

Lemon-scented Spotted gum Eucalyptus citriodora Hook. Exotic Eucalyptus tree with tall straight clean bole and smooth, bluish grey bark. Does best on soils with considerable moisture. Plants susceptible to termite attack. Oil extracted from leaves.

Lemon-vine See West Indian gooseberry.

Length of hydraulic jump The distance measured from the front face of the jump to a point on the surface immediately downstream from the roller.

Length-of-run (1) The distance water must run in furrows or over the surface of a field from one head ditch to another or to the end of a field. (2) The period of time that one irrigator is allowed at the irrigating head, in the system of rotation deliveries.

Lens culinaris See lentil.

— **esculenta** See lentil.

— **phaseoloides** See nicker bean.

Lentic Of or pertaining to still waters such as lakes, reservoirs, ponds, and bogs. Also spelled lenitic.

Lentical A loose structured opening in the periderm beneath the stomata in the stem of many woody plants that facilitates gas transport.

Lenticularis A cloud species, the elements of which have the form of more or less isolated, generally smooth lenses; the outlines are sharp. Also known as lenticular cloud.

Lentil Lens exulenta Moench, Lens culinaris Medic.; Ervum lens L. An annual legume having pinnately compound, vetch-like leaves; cultivated for its thin, lens-shaped, edible seed.

Lentil dry root rot Disease caused by Macrophomina phaseoli; Symptoms as in cotton dry root rot.

Lentil rust Disease caused by Uromyces fabae; Symptoms as in broad beans rust.

Lentil wilt Disease caused by Fusarium orthoceras var. lentidis; Symptoms as in lathyrus wilt.

Lepidium sativum See garden cress.

Lepidophyllous Having scaly leaves.

Lepidoptera Order of insects with four membranous scale-covered wings to which butterflies and moths belong.

Leptadenia Leptadenia pyrotechnica (Forsk.) Decne. (L. Spartium Wight.). A shrub of Family Asclepiadaceae. The slimy fruits and twigs are used as food.

Leptadenia pyrotechnica See leptadenia.

— **spartium** See leptadenia.

Leptocaul A thin-twigged and usually much branched tree.

Leptocorisa acuta See rice bug.

— **varicornis** See rice bug.

Leptophyll A growth-form class of plants having a leaf surface area of 15 square millimetres or less; common in alpine and desert habitats.

Leptothyrium pomi See apple sooty-blotch and fly-speck.

Leren Calathea allouia. A tuber food crop of the Family Marautaceae cultivated in the Caribbean; Central and South America.

Lesion Localised spot of diseased tissue.

Leskeineae A suborder of mosses in the order Hypnobryales; plants are not complanate; paraphyllia are frequently present; leaves are costate, and cells are not generally differentiated.

Lespedeza Any of various legumes of the genus Lespedeza having trifoliate leaves, small purple pea-shaped blossoms, and one seed per pod. Also Japanese clover.

Lesser grain borer Rhizopertha dominica Fabr. Destructive pest of all cereal grains in store. Also breeds in flour and dry fruits, larva bores into grain and feeds inside.

Lesser wax-moth Achroia grisella F. Insect pest in bee colonies.

Lesser yam Dioscorea esculenta (Lour.) Burkill. (D. aculeata L., D. fasciculata Roxb.). Family Dioscoreaceae; a shrub whose roots are edible.

Lethal Deadly.

Lethal gene A gene that renders inviable an organism or a cell possessing it.

Lettuce Lactuca sativa L. An annual plant of the order Asterales cultivated for its succulent leaves; common varieties are head lettuce; leaf or curled lettuce, romaine lettuce, and iceberg lettuce.

Leucaena glauca See white popinac.

— **leucocephala** See white popinac.

Leucinodes orbonalis See brinjal shoot and fruit borer.

Leucodontineae A family of mosses in the order Isobryales with foliated branches, often bearing catkins.

Leucoplast A non-pigmented plastid; capable of developing into a chromoplast.

Leucosin A protein-like food reserve in

certain algae.

Levee A natural or artificial embankment along a river or stream to confine its spread.

Leveillula taurica See sunhemp powdery mildew; tomato powdery mildew, cotton powdery mildew; clusterbean powdery mildew.

Level (1) A surveying instrument. (2) To make a filed level. (3) The amount of a component in a mixture. (4) Level of the water in streams or underground.

Leveller A scraper, drag, or any other form of device, implement or machine for smoothing land, dumping and excavating, carrying, spreading earth during irrigation preparations.

Levelling or grading The moving of soil in preparing land for irrigation to produce a level or uniformly sloping surface.

Level of significance, or significance level Many statistical tests or hypothesis depend on the use of the probability distributions of the statistic chosen for the particular test to see whether chance has been reduced to a low enough level for the particular problem at hand. If the occurrence by chance is reduced to about 1 out of 100, this is referred to as the '0.01 or 1 per cent level of significance'. More generally, any difference that chance would cause to be exceeded with a probability of P is said to lie on the P 'level of significance', and P may be quoted as a decimal fraction or percentage.

Level terrace (1) A contour terrace. (2) A terrace that follows the absolute contour as contrasted with a graded terrace.

Levulose Fructose.

Levy Collecting of assessment.

Ley An English term designating the biennial or perennial hay or pasture portion of a rotation that includes cultivated crops.

Ley farming System of growing grass in rotation with agricultural crops.

Libration cutting An improvement cutting. In this process young trees are saved off by removal of overhead shade.

Liberian coffee Coffea liberica Hiern. Family Rubiaceae. A shrub or small tree of coffee. One of the variations grown in India.

Lichen A symbiotic mutualistic association of an algae type and fungal type.

Lien The right to retain the property of another person until debts due from that person have been paid.

Liebig's law of the minimum The law states the minimal proportions of plant nutrients necessary for plant growth. Law given by German scientist Baron van Liebig.

Life cycle The various phases through which an individual species passes to maturity.

Life irrigation A irrigation necessary to save the crop or atleast to improve the yield substantially over the yield when that particular irrigation is skipped.

Life zone A portion of the earth's land area having a generally uniform climate and soil, and a biota showing a high degree of uniformity in species composition and adaptation.

Lift irrigation When the water supply is at low level to run by gravitation on to the land, the water is lifted by pumps or other water lifting devices and then made to flow on land under irrigation by the force of gravity.

Lift irrigation area, or **lift area** Area interspersed between gravity flow areas of which the levels are too high to permit irrigation by gravity flow from the source, but which can be irrigated economically water raised by pumps or other water lifting devices to the necessary level at some point in the supply system.

Light Form of radiant energy necessary for plant growth. Insufficiency may cause lack of chlorophyll production. Colouration of fruit depends on light.

Light demander A species of plants that demand abundant light for its best development.

Light period See photoperiodism.

Light soil A soil such as sands and sandy loams which are easy to work with.

Ligneous Woody, of the nature of wood.

Lignification The impregnation of cell walls with lignin, most commonly in cells of woody tissue.

Lignify To convert cell wall constituents into wood or woody tissue by chemical and physical changes.

Lignin Complex non-carbohydrate strengthening material in the thickened cell walls of plants; practically indigestible.

Lignite A variety of coal intermediate between peat and bituminous coal, especially one in which the texture of the original wood is distinct.

Lignocellulose The substance of which woody tissue is mainly composed.

Lignosa Woody vegetation.

Lignumvitae Guaiacum sanctum. A medium sized evergreen tree in the order Sapindales that yields a resin or gum known as gum guaiac or resin of guaiac. Also known as Hollywood lignumvitae.

Ligule (1) A small outgrowth in the axis of the leaves in Selaginellales. (2) A thin outgrowth of a foliage leaf or leaf sheath. A small lobe on the parapodium of certain polychaetes.

Liliaceae A family of the order liliales distinguished by six stamens, typically narrow, parallel-veined leaves, and a superior ovary.

Liliales An order of monocotyledonous plants in the subclass Liliidae having the typical characteristics of the subclass.

Liliidae A sub-class of the Liliopsida, all plants are syncarpous and have a six-membered perianth, with all members petaloid.

Liliopsida The monocoty ledons, making up a class of the Magnoliophyta, characterised generally by a single cotyledon, parallel-veined leaves, and stems and roots lacking a well-defined pith and cortex.

Lily Any of the perennial bulbous herbs with showy unscented flowers constituting the genus Lilium.

Lima bean Phaseolus limensis Macf. Family Papilionaceae. Tall climbing plant producing large flat seeds. Young pods used as vegetable; seeds used as pulse. Also called sieva bean.

Lima bean anthracnose Disease caused by Colletotrichum lindemuthianum. Characterised by elongated dark brown lesions on plant, canker on cotyledons in seedlings, lesions on pods.

Lima bean ashy stem blight Disease of lime bean caused by Macrophomina phaseoli. Characterised by black sunken cankers just above cotyledonary node, lesions extending up the stem and killing growing point, yellowing of leaves, stunting and appearance of black sclerotia.

Lima bean common and yellow mosaic Virus disease characterised by vein clearing and mottling on leaves with extreme malformation in some cases. Symptoms of common mosaic overlapping those of yellow mosaic.

Lima bean leaf spot Disease caused by Cercospora cruenta. Characterised by circular to angular spots, grey in centre and often surrounded by reddish border on older leaves and on pods.

Lima bean powdery mildew Disease caused by Erysiphe polygoni. Characterised by faint, slightly discoloured specks from which white powdery spots spread and overrun leaf, stem and pod.

Lima bean rust Disease caused by Uromyces appendiculatus.

Lime (plant) Citrus aurantifolia. A tropical tree with elliptic oblong leaves cultivated for its acid citrus fruit which is a hesperidium.

Lime *(Agricultural)* A soil amendment consisting principally of calcium carbonate, and including magnesium carbonate and perhaps other materials. It is used to supply calcium and magnesium as essential elements for growth of plants and to neutralise soil acidity.

Lime berry Triphasia trifolia (Burm. f.) P. Wilson (Triphasia trifoliata DC., Limonia trifolia Burm. f.). A common shrub of Family Rutaceae with edible fruits.

Lime concretion An aggregate of precipitated calcium carbonate or other material cemented by precipitated calcium carbonate.

Lime pan An aggregate cemented by precipitation of calcium carbonate.

Lime requirement The amount of agricultural limestone, or the equivalent of another liming material, required per hectare to a soil depth of 15 cm (of for 2,240 t of soil) to raise the pH of the soil to a specific value under field conditions.

Limestone Rock composed mainly of calcium carbonate.

Lime Sulphur Fungicide used for the control of mildews. Also used against thrips, mites and as dip for sheep.

Limestone Rock composed mainly of calcium carbonate.

Liming materials Materials used for liming. Common materials are finely ground limestone; the fine limestone; burnt lime; refuse from paper mills, sugar factories and fertiliser plants; also basic slag.

Limiting factor A factor in the activities of an undertaking which at particular point in time or over a period will tend to limit the volume of output.

Limnetic Of, pertaining to, or inhabiting the pelagic region of a body of fresh water.

Limnocharitaceae A family of monocotyledonous plants in the order Alismatales characterised by schizogenous secretary canals, multiaperturate pollen, several of many ovules, and a horseshoe-shaped embryo.

Limno layer In organic soil, a layer at-least 5 cm thick composed of marl, di-atomaceous earth, or copprogenous earth (sedimentary peat).

Limnology The scientific study of the physical, chemical and biological condition of lakes, ponds, and streams.

Limonia acidisima See sour lime.

— **trifolia** See lime-berry.

Limonite (2 Fe_2O_3 $3H_2O$). Hydrous sesquioxide of iron; widely distributed in rocks. Main source of yellow, red and brown colour in soils.

Linaceae A family of herbaceous or shrubby dicotyledonous plants in the order Linales characterised by mostly capsular fruit, stipulate leaves, and exappendiculate petals.

Linales An order of dicotyledonous plants in the subclass Orsidae containing simple-leaved herbs or woody plants with hypogynous, regular, syncarpous flowers having five to many stamens which are connate at the base.

Lindane A 99 per cent pure gamma isomer preparation of benzene hexachloride; acts as stomach and contact insecticide. Also used as fumigant. Available as dust, water dispersible powder and emusifiable concentrate.

Line Group of individuals from a common ancestry; more narrowly defined group than a strain or cultivar.

Line breeding Systematic inbreeding which involves the repetition of certain desirable individuals and their offsprings; makes it possible to fix definite characteristics in offsprings. Also see in and in breeding.

Line chart Akin to the 'scatter diagram' is the line chart when successive plots of two varieties are connected by lines.

Linear velocity It is the rate of change of position or displacement with respect to time. For its complete specifications, direction of motion as well as magnitude has to be specified. It is expressed in cm/sec or equivalent units.

Linear sympodium A single stem made up of successive, usually determinate, mono-axial units.

Line drains Drains installed at the lower points in the burried pipelines of a sprinkler irrigation system.

Line graph It can show the growth trend or frequency of a factor. This is a graphic transaction of the numerical or statistical data. Many things can be shown by lines or dotted lines or lines of different colours or types.

Line of pull It is the straight line passing through the centre of the pull, the point of hitch and the centre of resistance.

Line of seepage The free water surface of the zone of seepage.

Line sowing A method of sowing seeds in line. Helps in the use of implements for interculture.

Line valve Valve in a pipeline for the purpose of controlling flow.

Lined channel An irrigation channel lined with materials like brick masonary, stone, cement concrete or bituminous materials mainly to reduce seepage losses. It is expected that by lining seepage can be reduced to 1/4th of the volume in a unlined channel.

Lined section Section of an irrigation channel protected with lining.

Lining A protective covering over the entire, or a portion, of the perimetre of a conduit, or reservoir, to prevent seepage losses, to withstand pressure, or to resist erosion. Linings are sometimes provided to reduce friction or otherwise improve conditions of flow.

Linkage Association of characters in inheritance, due to the fact that the genes determining them are physically located in the same chromosomes. Such a group of linked genes is called a linkage group.

Linseed Linum usitatissimum L. Family Linaceae. An annual herbaceous crop producing seeds whose oil is of commercial importance. Stalks used as fuel and sometimes for extraction of coarse fibre. Grows well in deep clayey black soils and alluvial loams. Also called *Alsi*.

Linseed aphid Myzus persicae Sulz. Pest congregates in small colonies, causing stunted plant growth.

Linseed blight Disease caused by Alternaria lini. Characterised by appearance of dark brown irregular patches on leaves which curl and dry up, diseased buds covered with dark brown spots and fail to open.

Linseed cut worm Agrotis spp. Insect whose caterpillar cuts plants near roots under the soil.

Linseed gall fly Dasyneura lini Barnes. Maggot feeds on parts of bud which then does not develop seed capsules.

Linseed oilcake Residue cake left after extracting oil; used for feeding cattle and as manure.

Linseed rust Disease caused by Melampsora lini (Pers.) Lev.; characterised by formation of yellowish orange pustules on both surfaces of leaves, stems, flower stalks, shedding of leaves and later appearance of reddish brown to black, elongated, crust-like spots on flower-stalks and stem.

Lint (1) Cotton fibre; the epidermal hair of the cotton seed. (2) Cotton freed of seed.

Lint frequency The weight in grams of cotton fibre of uniform length produced per square centimetre of seed surface.

Lint index The weight of lint cotton produced by 100 seeds.

Lint percentage Ginning outturn.

Linters Cotton fibres which remain on cotton seed after ginning and removed in cotton oil mills. Important source of cellulose.

Linum usitatissimum See linseed.

Lipaphis erysimi See mustard aphid.

Lipase An enzyme that digests fats into fatty acids and glycerol.

Lip cell One of the cells (usually two) formed along the line of dehiscence in a fern sporangium.

Lipide One of the various fat-soluble substance occurring in plants and animals, as a sterol, carotene, terpine.

Lipoid Concerning fat or fatty tissue.

Lipolysis The hydrolysis of fats by enzymes, acids, alkalies or other means, to yield glycerol and fatty acids.

Lipolytic organism Organism which effect chemical breakdown of fat.

Liquefaction (spontaneous liquefaction) The sudden large decrease of the shearing resistance of a cohesionless soil. It is caused by a collapse of the structure by shock or other strain and is associated with a sudden, temporary increase of the interstitial water pressure. It involves a temporary transformation of the material into a fluid mass. See also quick clay.

Liquid assets Assets either in the form of money, or which can be quickly converted into money.

Liquidation The conversion of assets into cash, when they are not already in the form. In terminating or winding up a business 'going into liquidation' means not only converting assets into cash but also the paying of all business debts and the distribution of the balance of cash, if any, among the owners of the business.

Liquidity The ability to meet current financial liabilities in cash.

Liquid limit Water content at which a cohesive soil reach a state when the attractive forces cease to exist and soil possesses no cohesion is known as the liquid limit.

Liquorice Glycyrrhiza glabra L. Family Papilionaceae. A herb whose extract from roots is used in confectionery and as flavouring material in tobacco industry. Roots used during sore throat. Also called *mulhatti*. Also spelled as licorice.

Lirella A long, narrow apothecium with a medial longitudinal furrow, occurring in certain lichens.

Listed security A security recognised by a stock exchange for trading on that exchange.

Lister An implement used for furrowing land. It frequently has a planting attachment.

Litchi Litchi chinensis Sonn. (Nephelium litchi Camb; Dimocarpus litchi Lour.). Family Sapindaceae. Popular fruit. Exacting in climate; needs freedom from frost, high humidity, rich deep soil and plenty of moisture.

Litchi chinensis See Litchi.

Litchi fruitstone borer Argyroploce illepida But. Serious insect pest, caterpillar eats funicle or stalk of seed and bores into seed or pulp. Attacked fruit unfit for consumption. Also attacks bael (Aegle marmelos); wood-apple, tamarind, babul, amaltas and sunhemp.

Litchi leaf-curl mite Eriophyes sp.; pest, white mite curls up leaves, which turn brownish on lower surface and fall off.

Lithic Layer Bedrock under the control section of a soil. In organic soils, bedrock occurring within a depth of between 10 cm (4 inches) and 160 cm (64 inches) from the surface.

Lithochromic soil A soil which owes its colour to parent material or rock from which it was derived.

Lithophyte A plant that grows on rock.

Lithosequence A group of related soils that differ from each other in certain properties primarily because of differences in the parent rock.

Lithosere A succession of plant communities that originate on rock.

Lithosols An azonal group of soils having an incomplete solum or not clearly expressed soil morphology and consisting of freshly and imperfectly weathered mass of rocks.

Litmus paper A paper when saturated with a violet blue colouring matter which turns red in acids. The colour is restored by alkali. A red coloured paper turns blue in alkali. Thus it is used to test acidity/alkalinity.

Litre Volume occupied by a mass of one kilogram of pure air-free water at the temperature of its maximum density (4° C) and under normal atmospheric pressure. Under normal circumstances it is taken as 1000 cc.

Litsea citrata See siltimber.

— **monopetala** See meda.

— **polyantha** See meda.

Litter (1) Waste material such as plant refuse, straws of cereal grains, hay, leaves, stalks, peat moss etc. (2) Group of young animals brought forth at a single birth.

Little gourd Coccinia cordifolia (L.) Cogn. (Coccinia indica Wight and Arn., Cephalandra indica Naud.). Family Cucurbitaceae. A twining or spreading shrub cultivated as a vegetable crop.

Little leaf disease Any of various plant diseases and disorders characterised by chloratic, under-developed and sometimes distorted leaves. May be due to zinc-deficiency or pests or viruses.

Little millet Panicum miliare Lamk. Family Gramineae. Grown as a poor man's crop capable of withstanding drought and waterlogging. Used similarly as barnyard millet.

Little peach disease A virus disease of the peach tree in which the fruit is dwarfed and delayed in ripening, the leaves yellow, and the tree dies.

Littoral Of or pertaining to the shore, especially of the sea; coastal.

Littoral current Current that moves along the seashore parallel to shore line.

Littoral drift The movement of sand by littoral (longshore) currents in a direction parallel to the beach along the shore.

Littoral zone The part of the ocean immediately adjacent to the shore.

Live fence Fence made by closely planted thorny or other plants, maintained in live condition.

Live knot, or sound knot A knot free from decay and other defects, firmly intergrown with surrounding wood.

Liverwort a member of the class Hepaticae, having a flat, lobed thallus.

Livestock Term used to denote animals kept on farm. Also called stock. The term does not include poultry birds.

Live virus A virus whose ability to infect is still intact.

Live weight The gross weight of a live animal.

Livistona australis see cabbage palm.

Lixiviated soil Leached soil, lixiviation is a natural process in the formation of soil, particularly in humid region. Lixiviation is the process of separating soluble and insoluble components by percolation of water.

Livingston atmometre A clay atmometre in the form of a sphere.

Load factor The total number of units actually sent out from a plant during a period (usually a year) expressed as percentage of what would have been sent out if the plant had been worked at maximum capacity throughout the period.

Loam A soil made up of sand, silt, and clay particles in such amounts as to exhibit properties of sand and clay in about equal proportions. Soil material usually contains 7 to 27 per cent clay, 28 to 50 per cent silt and less than 52 per cent sand. See also texture soil.

Loamy Intermediate in texture and properties between fine textured and coarse textured soils. It includes all textural classes having 'loam' or 'loamy' as a part of the class name, such as clay loam or loamy sand. See also loam and texture soil.

Loamy coarse sand See texture soil.

Loamy very fine sand Fifty per cent or more very fine sand. See texture, soil.

Loamy fine sand Fifty per cent or more fine sand or less than 25 per cent very coarse, coarse, and medium sand and less than 50 per cent very fine sand.

Loamy sand Soil material that contains at the upper limit 85 to 90 per cent sand, and the percentage of silt plus 1.5 times the percentage of clay is not less than 15, at the lower limit it contains not less than 70 to 85 per cent sand, and the percentage of silt plus twice the percentage of clay does not exceed 30 per cent.

Lobate Divided into lobes. Also called lobose.

Lobe Rounded and flattish, projecting or pendulous sub-division of an organ, especially a sub-division marked off by a fissure.

Lobed Divided by clefts, as a maple leaf.

Loblolly pine Pinus teada. A hard yellow pine of the central and southeastern United States having a reddish brown fissured bark, needles in groups of three, and a full bushy top.

Lobule A small lobe; one of the subdivisions of a lobe.

Local infection An infection in which only a part of plant or animal is affected.

Locality factors Climate, soil, topography and biotic factors on which the vegetation of a particular area is determined.

Locality groups Are composed of people in a specific geographic area who have developed a feeling of belonging and who tend to associate with each other more than with outsiders.

Localised disease A disease which affects only one part instead of the whole body.

Locellus A small compartment of the ovary.

Lock (1) A small tuft of cotton or wool fibres. (2) A locule or the cotton in a locule. (3) A single cavity in the ovary.

Lock out The refusal by an employer to admit his employees to their place of work. Opposite of strike.

Locomotory Pertaining to locomotion, i.e., move from place to place.

Loco weed Any species of Astragalus containing selinium taken up from the soil.

Locule A cavity in a sporangium or ovary.

Locust Either of two species of commercially important trees, black locust (Robinia pseudoacacia) and honey locust (Gladissia triacanthos), in the Family Leguminosae. The common name for various migratory grass hoppers of the family Locustidae.

Lodging The inelastic displacement of the plant from the vertical or initial position.

Loess Material transported and deposited by wind and consisting of predominantly silt sized particles.

Loganiaceae A family of mostly woody dicotyledonous plants in the order Gentianalies; members lack a latex system and have fully united carpels and axile placentation.

Log To cut and remove logs from an area.

Log rule (1) A table indicating the amount of lumber which can be sawed from logs of given sizes. (2) A log-scaling stick.

Log scale A scaling stick for measuring the contents of logs in terms of board feet.

Logarithmic normal or **lognormal distribution** If the logarithms of a set of variate values are distributed according to the normal distribution, the variate is said to have a logarithmic normal distribution, or be distributed lognormally.

Logged-off land Land cleared of trees but from which stumps are yet to be removed.

Logger (1) A person who is engaged in logging operations. (2) Locally, a person who hauls logs to landings and skidways.

Logging (1) The cutting of trees for lumber. (2) Logging a bore (tubewell) useful for determining availability of aquifer.

Logistic curve An S-shaped curve of population growth which is initially slow, then steepens, and finally flattens out at an asymptote determined by the carrying capacity q.v. of the environment q.v.

Lolium perenne See perennial rye grass.

— **temulentum** See rye grass.

Loment A dry, indehiscent single-called fruit that is formed from a single superior ovary; splits transversely in numerous segments at maturity.

Longan Euphoria longan (Lour.) Steud. (E. Longana Lamk., Nephelium longana Camb.) Family Sapindaceae. A tree cultivated for its edible fruit.

Long crop The term is used generally to denote a crop that takes more than four months to mature. As a relative term; it denotes longer of the two crop-being called 'short crop' when, however, two crops are sown simultaneously in the same field, the one that stays longer in the field is called 'long crop'.

Long fruited blood wood See kuleha.

Long-headed flour beetle Latheticus oryae Waterh. A pale yellow beetle that infects flour and meals of cereals and dried vegetable matter in storage. The grub, yellow and with numerous hairs on its body, feeds on flours and meals.

Long day plant A plant that flowers in response to a long photoperiod.

Longicorn Having long antennae as certain beetles.

Longitarsus nigripennis See pepper flea beetle.

Longitude Angular distance east or west from a standard meridian, as that of greenwich, to the meridian of any place, reckoned to 180° E or W.

Longitudinal section, L. Section, or Profile map A vertical section of a canal along its centre line.

Longitudinal smoothing Land smoothing operation where all soil movement is done parallel to crop row direction for the purpose of obtaining a uniform row grade.

Long melon Cucumis melo L. var. utilissimus Duth. & Full. Family Cucurbitaceae. A warm season crop grown for its fruit eaten raw and used in *Salad.* Light green and green varieties. Also called *kakri.*

Long pepper Piper longum L. Family Piperaceae. A climber of the pepper family, produces berries of climber of the pepper family, produces berries of value in Indian medicines. Grows wild in many places.

Long-range forecast A forecast for a period greater than 5 days (or a week) in advance. Also known as extended range forecast.

Long shoot Where there is shoot dimorphism, and extended shoot contributing to the architecture of a tree.

Long throat Throat having significant linear dimension in the direction of flow.

Long staple Cotton fibres more than 30 mm long.

Longshore current A current, created by waves which moves parallel to and against the shore, particularly in shallow water, and which is most noticeable in the surf or breaker zone; littoral current.

Lonia See alkali seepweed.

Long whip grafting A method of grafting in which the stock is cut at an angle of 15° to 25° and the scion at an angle of 65° to 75°.

Lookout. A station or post used primarily in the detection of fires. A permanent lookout is generally equipped with a lookout tower or structure.

Loopers Certain caterpillars, which have a looping motion while crawling.

Loose A soil consistence term. See also consistence.

Loose housing System of cattle management in which animals are given free access to feeding troughs, water, resting area. Not tied down at a place.

Lopping Pruning or cutting away parts of shrubs or trees.

Loquat Eriobotrya japonica (Thunb.) Lindl. Family Rosaceae. Ever green, hardy, low spreading tree growing to about 7 m. Edible fruits used in making jelly, jam and other preserves. Succeeds in sub-tropical climate. Propagated by seeds, budding, cutting, inarching or grafting.

Lorantheceae A family of dicotyledonous plants in the order Santalales in which the ovules have no integument and are embedded in a large, central placenta.

Loranthus Longiflorus. Plant parasite which generally grows wild on branches of mango tree absorbing their sap. Has thick, succulent light green leaves, pale flowers and bright red berries, which attracts birds, seed sticky and is spread by birds. If not removed in time, the branch is killed.

Lot A lot is a specified quantity of seed which is physically identifiable.

Lotus corniculatus See bird's foot trefoil.

— **hispidus** Desf. A herb of the Family Papilionaceae grown for fodder.

— **tetragonolobus** See winged pea.

— **uliginosus** Schkuhr. A herb of Family Papilionaceae grown for fodder.

Low head high volume pumps These are the pumps which lift large volumes of water but lift is much lower. Used for pumping from open surface bodies and commonly for drainage.

Low land A term designating a land which has lower elevation than adjacent lands and therefore is poorly drained.

Lower plastic limit See plastic limit.

Luban See Indian frankincense.

Lubricant Substance used in machines to reduce friction between moving parts. Grease and oils derived from petroleum are common lubricants.

Lubrication The act of applying lubricants.

Lubricity Oiliness, smoothness.

Lucerne Medicago sativa L. Family Papilionaceae. Perennial legume, grown for fodder. Has high water requirement. First cutting $2^1/_2$ to 3 months after sowing. Fodder highly nutritious. Also called alfalfa.

Lucerne bacterial disease Disease caused by Xanthomonas alfalfa. Characterised by formation of water-soaked spots surrounded by straw coloured halo on leaves.

Lucerne caterpillar Laphygma exigua Hubn. Insect pest of lucerne which also attacks linseed, safflower, indigo, lentil, cabbage, onion, maize, coriander, jute and cotton. In severe cases plant is completely defoliated.

Lucerne common leaf spot Disease caused by Pseudopeziza medicaginea. Characterised by small, circular, brown spots with black raised discs of leaves.

Lucerne crown wart Disease caused by Physoderma alfalfae. Characterised by formation of dull-coloured large galls on roots at soil level.

Lucerne downy mildew Disease caused by Peronospora aestivalis. Characterised by greyish white powdery patches on leaf under-surfaces, with corresponding yellowish and necrotic portions on upper surface.

Lucerne dry rot Disease caused by Macrophomina phaseoli-Symptoms as in cotton dry rot.

Lucerne rust Disease caused by Uromyces striatus. Characterised by appearance of small, round or oblong dark brown pustules on leaves and petioles.

Lucifugous Shunning day light.

Lucuma mammosa See marmalade plum.

Luffa acutangula See ridged sponge gourd.

— **aegyptiaca** See smooth sponge gourd.

— **cylindrica** See smooth sponge gourd.

Lumber (1) Converted timber. (2) To cut trees for logs.

Lumber jack One who works on logging operations; colloquil for logger.

Lumen Central part of plant cell.

Luminiferous Producing or transmitting light.

Lump Compact, shapeless mass.

Lunar day The interval between two successive transits of the moon's centre over the same meridian.

Lunate Crescent shaped.

Liquid limit (Upper plastic limit; Atterberg limit) (i) The water content corresponding to an arbitrary limit between the liquid and plastic states of consistence of a soil (ii) The water content at which a paste of soil, cut by a standard-sized groove, will flow together from a distance of 12 mm under the impact of 25 blows in a standard liquid limit apparatus.

Lupine A leguminous plant of the genus Lupinus with an upright stem, leaves divided into several digitate leaflets, and pea-shaped blossoms.

Lupinus albus See white lupine.

— **luteus** See yellow lupine.

Luscious Richly sweet in taste or smell.

Lush growth Luxuriant growth.

Luvic gleysol A great group of soils in the Gleysolic order developed under wet conditions, under grass or forest or both. The soils have Aeg and Btg horizons.

Luvisolic An order of soils that have eluvial (Ae) horizons, and illuvial (Bt) horizons in which silicate clay is the main accumulation product. The soils developed under forest or forest-grassland transition in a moderate to cool climate.

Luxury A commodity or service which satisfies a secondary need in respect of food, clothing, shelter or entertainment.

Luxury consumption The consumption of an element above the point at which the element increases growth, resulting in an increase in content of that element in the plant.

Lycaenid butterflies Insect pest on pulses. Larve feed on flower buds and petals.

Lychee A tree of the genus litchi in the Family Sapindaceae, especially L. chinensis which is cultivated for its edible fruit, a one seeded berry distinguished by the thin, leathery, rough pericarp that is red in most varieties. Also spelled litchi.

Lycopene ($C_{40}H_{56}$) A carotenoid, responsible for the red colour of tomatoes.

Lycopersicum esculantum See tomato.

Lye A crude caustic soda solution. Used in soap making and for peeling of fruits and vegetables in industries.

Lycopodiophyta A division of the subkingdom Embryobionta characterised by a dominant independent sporophyte, dichotomously branching roots and stems, a single vascular bundle, and small, simple, spirally arranged leaves.

Lyon bean Mucuna cochinchinensis (Lour.). A herb of Family Papilionaceae grown as a cover crop and for green manure. Pods used as vegetable. Also called *tohar sem*.

Lyophobic Referring to a substance in a colloidal state that has a tendency to repel liquids.

Lysimetre (1) A device for measuring percolation and leaching losses of water and solutes from a column of soil under controlled conditions. (2) A device for measuring gains (precipitation and condensation) and losses (evapotranspiration) of water by a column of soil.

Lysine An amino acids found in growing grass. An important content of cattle nutrition as it affects milk.

Lysis (1) The gradual disappearance of the symptoms of disease. (2) The destruction or dissolving of cells.

Lysols Cresylic acid-containing disinfectant.

M

Maba buxifolia See angaru.

Macadamia ternifolia F. Muell. A nut tree of family Proteaceae. Also known as Queensland nut.

Macaroni-wheat Triticum durum. One of the important species of wheat grown in India. Also see durum wheat.

Macassur ebony Diospyros ebenum Koenig (D. assimilis Bedd.; D. Sapota Roxb.) Family Ebenaceae. Tree whose fruits are edible and wood used for various purposes.

Macerate To soften or separate constituents, as of food, by crushing, soaking etc. as depulp.

Machilus bombycina King ex Hook f. A middle sized tree of family Lauraceae. Leaves are fed to silkworms.

Machinable Material capable of being finished by tools or cutters in or on a machine tool.

Machine Apparatus for applying mechanical power. Each machine has several parts having defined function to perform.

Machine rating The amount of power a machine can deliver without overheating.

Machine shop A workshop in which work, metal or other material, is machined to specified size and assembled.

Macro-climate The climate of a large geographic region.

Macro-consumer A large consumer which ingests other organisms or particulate organic matter. Also known as biophage.

Macro-conidia Large conidia.

Macro-distribution The distribution of national income into such broad aggregates as wages, rent, interest and profits.

Macro-economics A branch of economics concerned with the analysis of the economy in the large, i.e., with such large aggregates as the volume of employment, saving and investment, the national income, and so on.

Macro-element See Macro-nutrient.

Macro-fauna (1) Widely distributed fauna. (2) Fauna of a macro-habitat.

Micro-flora Plants visible to the naked eye. (i) Widely distributed flora. (ii) Flora of a macro-habitat.

Macro-habitat An extensive habitat presenting considerable variation of the environment, containing variety of ecological niches, and supporting a large number and variety of complex flora and fauna.

Macro-meteorology The study of the largest scale aspects of the atmosphere, such as the general circulation, and weather types.

Macro-nutrient A chemical element necessary in large amounts usually greater than one ppm in the plant, for the growth of plants and usually applied artificially in fertiliser or liming materials. Macro refers to the quantity and not to the essentiality of the element to the plant. See also micronutrient. Also called macro-elements.

Macro-phyllous Having large or long leaves.

Macropes excavatus See sugarcane black bug.

Macrophoma musae See banana black finger.

Macrophomina phaseoli A dry root rot of brinjal, cotton, cowpea, clusterbean, lentil and several others.

Macropodus (1) Having a large or long hypocotyl. (2) Having a long stem or stalk.

Macroscopic Large enough to be seen with the naked eye.

Macroscopic velocity See flow velocity.

Macrospore The larger of two spore types produced by heterosporous plants, the female gamete. Also known as megaspore.

Macropores Soil pores big enough to facilitate rapid downward movement of water by force of gravity, as in sands and well aggregated soils.

Macrostylous (1) Having long styles. (2) Having long style and short stamens.

Maculate Marked with speckles or spots.

Madagascar potato Coleus rotundifolius (Poir). A Cheval. (C. parviflorus) Benth.; Plectranthus tuberasus Blume. Family Labiatae. A herb or shrub cultivated for its underground stem used as vegetable.

Madar Calotropis gigantea (L.) R. Br. ex Ak. Family Ascelepiadaceae. A large perennial undershrub found as a weed. Floss obtained from seeds used for stuffing purposes. Also called *Ak*.

Madeira vine Boussingaultia beselloides H.B. and K. Family Basellaceae. A tropical ornamental climber, with thick, shiny leaves and white fragrant flowers. Leaves and tubers edible.

Made land Areas filled with earth, or earth and trash mixed usually by or under the control of man. A miscellaneous land type.

Maduca butyracea See Indian butter tree.

— **indica** See mahua tree.

— **latifolia** See mahua tree.

— **longifolia** See mowra fat.

Maggot Legless larva as of fly.

Magmatic water Water derived from or existing in molten igneous rock or magma. Also known as juvenile water.

Mangesia (Mg) Magnesium oxide, hydrated magnesium carbonate. While powder used as antacid and cathartic.

Magnesia magma Milk of magnesia.

Magnesium (Mg) A white metal essential for animal and plant growth. Absorbed as Mg^{++} by the plant. Important constituent of chlorophyll. Excess in soil is not useful as it reduces hydraulic conductivity. See Sodium adsorption ratio.

Magnesium deficiency Condition brought about in plant by lack of magnesium particularly in light sandy soils. Spraying 0.1 to 0.2 per cent of magnesium sulphate helps in overcoming the deficiency.

Magnetic separator The separation of seeds in this machine is done on the basis of magnetic properties. Seeds are selectively pre-treated with magnetic material e.g., finely ground iron powder.

Magnetite See iron minerals.

Magneto anemometre A cup anemometre with its shaft mechanically coupled to a magnet; both the frequency and amplitude of the voltage generated are proportional to the wind speed, and may be indicated or recorded by suitable electrical instruments.

Magnolia A genus of trees, the type genus of the Magnotiaceae. With large, chiefly white, yellow or pinkish flowers, and simple, entire, usually large evergreen or deciduous alternate leaves.

Magnoliaceae A family or dicotyledonous plants of the order Magnoliales characterised by hypogynous flowers with few to numerous stamens, stipulate leaves, and uniaperturate pollen.

Magnoliales The type order of the sub-class Magnolidae, members are woody plants distinguished by the presence of spherical ethereal oil cells and by well-developed perianth of separate sepals.

Magnolidae A primitive sub-class of flowering plants in the class Magnoliopsida generally having a well-developed perianth, numerous centripelal stamens, and bitegmic, crassinucellate ovules.

Magnoliophyta The angiosperms, a division of vascular seed plants having the ovules enclosed in an ovary and well-developed vessels in the xylem.

Magnoliopsida The dicotyledons, a class of flowering plants in the division Magnoliophyta; generally characterised by having two cotyledons and net-veined leaves, with vascular bundles borne in a ring enclosing pith.

Mahogany Any of several tropical trees in the Family Meliaceae of the order Geraniales. The hard wood of these trees, especially the red or yellow-brown wood of the West Indies mahogany tree is good as timber.

Mahua refuse By-product of mahua distillation. A good source of manure, equivalent to cattle manure.

Mahua tree Madhuca indica (Roxb.) Macb. Madhuca latifolia (Roxb). Family Sapotaceae. A large tree cultivated for its many uses. Flowers used as vegetable and are source of alcohol. Seed oil used in soap making and cooking. Oil cake used as manure.

Mahubarana fat A pale-yellow solid oil, melting point 40-44°C, obtained from seeds of trees of the genus Boldoa; used for soaps and candles.

Maillard product Lignin artifact that is an artificial indigestible polymer between proteins and amino acids and degradation products of sugar and other carbohydrates.

Main A duct or pipe that supplies or drains ancillary branches.

Main axis The central or major line of growth, as the primary rootstem axis.

Main drain, or **conducting drain** A principal drain which serves as an outlet to all subsidiary drains which discharge into it.

Main line In a sprinkler irrigation system, the conduit conveying water under pressure from the source of supply to feed sub-mains or sprinkler laterals. The supply line, feeding the giant sprinkler system, is often referred to as a 'lateral'.

Main line riser outlet valve Valve attached to the main line at intervals for the purpose of connecting laterals.

Main stream The principal or largest stream of a given area or drainage system. Also known as master stream; trunk stream.

Mainphal Vangueria pubescens. Kurz. A large shrub or small tree. Young leaves eaten as vegetable also called *pundrika*.

Maintainer line The 'B' line.

Maize Zea mays L. Family Gramineae. An important crop of India grown as fodder and for seeds. Starch extracted from grain. Needs moisture in early stages of growth but cannot tolerate waterlogging. Crop for fodder cut when tassels appear.

Maize borer Chilo zonellus Swinh. Insect pest destructive to maize, sorghum and pearlmillet. Also attacks paddy and sugarcane. Caterpillars bore into stem, feeds on inside, attacked plants show poor growth and lodges easily and producing dead hearts. Caterpillars also bore into maize cobs and destroy grain.

Maize brown spot Fungal disease caused by Physoderma zeamaydis. Characterised by light green oval, slightly water-soaked lesions on leaves and leaf sheaths, later turning reddish brown, stalks often break at infected centres.

Maize downy mildew Fungal disease caused by Sclerospora philippinensis Weston. Characterised by yellowish streaks with whitish powdery growth appearing on leaves which later turn brown. Affected plants pale, stunted and bunchy.

Maize head smut Disease caused by Sphacelotheca reiliana. Characterised by smut sori replacing tassel and ear, forming hard, black, sooty mass, with affected flowers growing into leafy structures. Affected plants remain stunted, brittle and green, longer than others.

Maize leaf blight Disease caused by Helminthosporium turcicum Pass. Characterised by long, elliptical, greenish brown, thin and semi-transparent lesions on leaves. Diseased plants look burnt or frost-bitten.

Maize leaf rust Disease of maize caused by Puccinia sorghi. Characterised by reddish brown, open pustules on leaves which sometimes dry up.

Maize leaf spot Disease caused by Cercospora sorghi. Characterised by formation of circular to somewhat irregular brown spots on leaves, which later turn olivaceous.

Maize smut Fungal disease caused by Ustilago zeae. Characterised by soft tumours on all parts of plant, mostly on cob. Tumours when mature rupture and release black dusty mass of spores.

Make and break The term may be applied to several electrical devices. Primarily, there are a pair of contact points, one stationary, and the other operated by a cam which makes the break in a circuit between these points.

Makedshingi Caralluma fimbriata Wall. Family Asclepiadaceae. A fleshy herb which is eaten as vegetable.

Malabar mountain ebony Bauhinia malabarica Roxb. Family Caesalpiniaceae. A tree whose leaves are used for flavouring foodstuffs and bark used as tan. Tender seeds and leafy tips edible.

Malabar nut Adhatoda vasica Nees. (Justicia adhadota L.) An evergreen shrub whose twigs and leaves are used as green manure. Leaves also source of medicinal preparations.

Malathion O, O-dimethyl S-(1, 2-dicarbethoxyethyl) phosphordithioate. An organo-phosphorus compound. Slightly soluble in water, can be mixed with many organic solvents like alcohols, esters, ethers, ketones and vegetable oils. Mostly used as contact insecticides but also acts as stomach poison. Has some fumigation action. Used to control storage pests, mites, aphids, mangohoppers, young stages of mealy bugs, thrips and scaly insects.

Malayan apple Syzygium malaccense (L.) Merr. & Perry. Eugenia malaccensis L. Family Myrtaceae. Ornamental fruit tree from Malacca. Produces oval fruits with crispy and juicy flesh with good sub-acid flavour. Propagated by seed, inarching.

Male A staminate plant.

Male sterile Producing no functional pollen.

Malestomataceae A large family of dicotyledonous plants in the order Myrtales characterised by an inferior ovary, axile placentation, up to twice as many stamens as petals (as sepals), anthers opening by terminal pores, and leaves with prominent, subparallel longitudinal ribs.

Malformation Symptom found in plants due to disease whereby plant forms show hypertrophy or atrophy, but when more localised, shows wart, gall, blisters, tumour, leaf-curt, hairy root, witches broom, canker and rosette. Assists in the identification of pathogen causing diseases.

Malignant (1) Virulent. (2) Threatening to produce death. (3) Resistant to treatment. (4) Tending to grow worse or recur after removal.

Malling root stock Root stock of apple from Malling Research station, England. Classified into four groups very dwarfing, semi-dwarfing, vigorous and very vigorous.

Mallophaga A suborder of the order Anoplura, to which the common biting lice belongs.

Mallotus philippinensis See kamala.

Malnutrition An unhealthy condition due to poor nutrition.

Malpighiaceae A family of dicotyledonous plants in the order Polygalales distinguished by having three carpels, or toothed, and indehiscent fruit.

Malpighia glabra See Barbados cherry.

— **punicifolia** see West Indian cherry.

— **urens** A tree of family Malphigniaceae which produces edible fruits.

Malt A nutrient material made from grain, commonly barley, which has been soaked, allowed to germinate and dried.

Maltase An enzyme that catalyzes the conversion of maltose to dextrose.

Malt beverage Any of various fermented alcoholic beverages, including bear, ale, stout, and porter, barley malt is the principal ingredient.

Malt extract A sugar obtained from the filtrate of a mixture of malted barley and grain.

Malthusian theory A theory expounded by T.R. Malthus (1766-1834) according to which human population tends to increase at a greater rate than food supply, unless birth rate is controlled, threat of poverty and famine are inevitable.

Maltose Malt sugar, $C_{12}H_{22}O_{11}H_2O$. Disaccharide.

Malt whiskey Whiskey produced in a pot still from malted barley.

Malus baccata See Siberian crab apple.

— **pumila** See apple.

— **sylvestris** See apple.

Malvaceae A family of herbaceous dicotyledons in the order Malvales characterised by imbricate or contorted petals, mostly uni-locular anthers, and minutely spiny, multiporate pollen.

Malvales An order of flowering plants in the subclass Dilleniidae having hypogynous flowers with valvate calyx, mostly separate petals, numerous centrifugal stamens, and a syncarpous pistil.

Mammea americana See mammee-apple.

— **longiflora** See Alexandrian laurel.

Mammee-apple Mammea-americana L. A tree of family Guttiferae whose fruits are edible.

Management The human factor within a production process which delimits production process, accumulates information on their solutions, analysis information, reaches decisions, acts on them and bears responsibility for such actions.

Management soil (1) The total of all tillage operations, cropping practices, fertiliser, lime and other treatments conducted on or applied to a soil for the production of plants. (2) The division of soil science dealing with the items listed in (1).

Management groups, soil group of soil units having similar adaptations or management requirements for one or more specific purposes, such as adapted crops or crop rotations, drainage practices, fertilisation, forestry, and highway engineering.

Mandarin A large and variable group of citrus fruits in the species Citrus reticulata and some of its hybrids, many varieties of the trees are compact with willowy twigs and small, narrow, pointed leaves; includes tangerines, king oranges, temple oranges, and tengelos.

Mandarin orange Citrus reticulata Balanco. (Citrus nobilis Andrew). Family Rutaceae. One of the major species grown in India. Grows successfully in all tropical and subtropical parts of India. Peel used for marmalades; essential oil used in confectionery, toilet and pharmaceutical preparations.

Manganese Trace food element absorbed as Manganese (Mn^{+++}) ion and functions closely with iron. Helps in chlorophyll formation and acts as catalyst in oxidation reduction reactions within plant tissues.

Manganese chlorosis A discolouring disease plants due to manganese deficiency.

Manganese deficiency Condition of plant due to deficiency of manganese. Common in fruit trees. Control by spraying 450 gm of manganese sulphate and 225 gm quick lime in 100 gallons of water.

Mangel Beta vulgaris var. macrorhiza. A biennial herb of Family Chenopodiaceae grown for fodder.

Mangifera indica See mango.

— **pinnata** See hog plum.

Mango Mangifera indica L. Family Anacardiaceae. Most important fruit crop of India. Grows in humid as well as dry climates. When young needs shelter from frost. Fruit rich source of vitamin A and C.

Mango anthracnose Disease caused by Collectotrichum gloeosporides Hayes. Characterised by appearance of dark blister-like spots on young twigs, flowers and fruits; blackening, drying up and falling of blossoms; shedding of young fruits; development of black spots on older fruits.

Mango bacterial leaf spot Disease caused by Pseudomonas mangiferae-indicae. Characterised by formation of minute water soaked lesions bound by veins on leaf blade lesions turning dark brown with age and surrounded by chlorotic halo.

Mango bark eating caterpillar Inderbela quadrinotata walker. Dirty brown caterpillar feeding on bark of mango tree, producing long, ribbon like, loose dark mess on stems and branches; bores into wood and makes tunnels.

Mango canker and die back Disease of mango characterised by leaves and twigs developing dieback.

Mango fruit fly Dacus dorsalis Hendel. Pest of mango. Maggot feeds on pulp of fruit causing fruit to rot; attacked fruits fall off. Also attacks guava, loquat, pomegranate, apricot, plum, peach, fig, apple, persimmon, banana and citrus.

Mango ginger Curcuma amada Roxb. Family Zingiberaceae. Species of turmeric whose rhizomes resemble those of ginger and have a mango-like fragrance. Also called *aam haldi.*

Mango hopper Idiocerus clypealis Leth., I. atkinsoni Leth., I. niveosparsus Leth. Causes severe damage by sucking sap from inflorescene resulting in fall of flowers, Also secretes honey-dew which helps development of sooty mould.

Mango leaf-cutting weevil Deporaus marginatus Pasc. Small greyish beetle, pest of mango. Cuts new mango leaves near base, causing heavy damage to young seedlings.

Mango leaf gall makers Procontarinia matteiana Kieff. Larvae feed on leaves and produce galls.

Mango mealy bug Drosicha mangiferae Gr. Serious insect pest of mango. Nymph and adult cluster around tender stem and twigs and suck sap; later attacks inflorescence which does not set fruit; also secretes honey dew giving rise to fungus diseases. Also attacks guava, Jack fruit, litchi, papaya and citrus.

Mango mites Eriphyes and Typhlodromus sps. or Tyrophagus asiaticus Evans. Mites attack mango shoots and inflorescence, causing malformation.

Mango nut weevil Cryptorhychus mangiferae Fabr. C. gravis Fabr. Insect pest whose grub bores into stone.

Mango pink disease Disease caused by Corticium salmonicolor. Symptoms as in tea pink disease.

Mango powdery mildew Fungal disease caused by Oidium mangiferae Berthet. Characterised by shedding of flowers and young fruits.

Mango shoot gall psyllid Apsylla cistellata Buckton. Makes cone-shaped galls in growing buds of mango shoots, checking normal formation of inflorescence or leaves; makes shoots dry up.

Mango sooty mould Disease caused by Capnodium ramosum. Symptoms as in sapodilla sooty mould.

Mangosteen Garcinia mangostana L. Family Guttiferae. Delicious tropical fruit used mainly as dessert. Rind has 1 to 13 per cent tannin. Propagated by seed; rootings by hard-wood cuttings and inarching also practised.

Mangosteen oil tree See kokum butter tree.

Mango stem borer Batocera rufomaculata De Geer. (B. rugra L.). Insect pest of mango; also attacks fig, mulberry and jack fruit. Grub tunnels into branches and trunk of tree and fees on wood inside.

Mangrove A tropical tree or shrub of the genus Rhizophora characterised by an extensive, impenetrable system of prop roots which contribute to land building.

Mangrove swamp A tropical or subtropical marine swamp distinguished by the abundance of low to tall trees, especially mangrove trees.

Mangum terrace First terrace constructed in the United States to such shape that the channel and embankment could be farmed.

Manhole An opening through which a workman may gain access to a drain, boiler, sewer, tunnel, etc. or other underground area.

Man-hour of labour The amount of accomplishment expected of a farm worker in a 10 hour day, using adequate equipment and operating at the average rate of performance.

Manifold Usually a casting or connection containing several branches or parts, as an exhaust manifold on an automobile engine, for conducting the exhaust from each cylinder into a single exhaust pipe.

Manifold vacuum State of atmospheric tension existing in a manifold when the engine is in operation.

Manihot esculenta See tapioca.

— **utilissima** See tapioca.

Manila grass Zoisia matrella (L.) Merr. Family Gramineae. A perennial grass used for fixation and reclamation of sand dunes.

Manila hemp Musa textilis Nee. Family Musaceae. A banana-like plant.

Manila maguey See cantala.

Manila tamarind Pithecellobium dulce (Roxb.) Benth. (Mimosa dulcis Roxb., Inga dulcis Willd.) Family Mimosaceae. Tree whose wood is of multipurpose use, seeds edible, leaves used as fodder. Suitable for afforesting arid areas; can grow on poor soils and waste lands.

Manilkara zapota See sapodilla.

— **hexandra** See khirni.

— **kauki** See kauki.

Manioc Tapioca; meal made from tapioca.

Manna Sweet exudate of certain plants; honey dew.

Mannan Simple ploysaccharide found as reserve carbohydrate in seeds and wood pulp.

Mannitol A white sweetish crystalline carbohydrate alcohol derived from the sugar mannose.

Mannose carbohydrate; isomer of dextrose and occurs in nature in combination as di-or-polysaccharide.

Mantis Insect of Family Mantidae, predaceous on other insects, capturing the prey with its unusually long front legs.

Manual switch A hand or foot operated switch as opposed to one operated in any other manner.

Manubrium The inner prismatic cell from which the antheridial filaments arise in the antheridium of the Charales.

Manure The excreta of animals, with or without the admixture of bedding or litter, in varying stages of decomposition. It is also called barnyard manure or stable manure. This is the usual meaning in North America. In some countries, manure is used to refer to any fertiliser.

Maple Any of various broad-leaved, deciduous trees of the genus Acer in the order Sapindales characterised by simple, opposite, usually palmately lobed leaves and a fruit consisting of two long-winged samarus.

Maple sugar Sugar obtained from boiled maple syrup.

Maple syrup Syrup made from the sap of certain maple trees, particularly the sugar maple.

Maquis A type of vegetation composed of shrubs, or scrub, usually not exceeding three metres in height, the majority having small, hard, leathery often spiny or needlelike drought-resistant leaves and occurring in areas with a Mediterranean climate.

Mar An Indian term for imperfectly drained clay loams wet lands with sufficient organic matter.

Maranta arundinacea See west Indian arrowroot.

Marantaceae A family of monocotyledonous plants in the order Zingiberales characterised by one functional stamen with a single functional pollen sac, solitary ovules in each locule, and mostly arillate seeds.

Marattiaceae A family of ferns coextensive with the order Marattiales.

Marattiales An ancient order of ferns having massive eusporangiate sporangia in sori on the lower side of the circinate leaves.

Marbled Mottled or streaked, like certain kinds of marble.

Marc The insoluble residue left after processing fruits, sugarcane, sugarbeet etc.

Marcescent Designating leaves which wither on the plant without falling off.

Marcgraviaceae A family of dicotyledonous shrubs or vines in the order Theales having extipulate leaves with scanty or no endosperm, two integuments, and highly modified bracts.

Marchantiales The thallose liverworts, an order of the class Marchantiopsida having a flat body composed of several distinct tissue layers, smooth-walled and tuberculata-walled rhizoids, and male and female sex organs borne on stalks on separate plants.

Marcottage Artificial propagation by stimulation of aerial root formation in woody axis.

Marchantiopsida The liverworts, a class of lower green plants, the plant body is usually a thin, prostrate thallus with rhizoids on the lower surface.

Margarine A plastic food, fat product composed of processed vegetable oils or animal fats or both, cultured milk, salt; and emulsifiers.

Margarine indica See pumpkin caterpillar.

Margaronia caesalis See jack-fruit shoot borer.

Marginal (1) Lands not productive enough to be called profitable. (2) At or near the margin, edge or border. (3) A term used in economics in calculating the profit increase as compared to cost of the input resulting in the increased profit.

Marginal blight A bacterial disease of lettuce caused by Pseudomonas marginalis; characterised by brownish marginal discolouration of the foliage.

Marginal bund In soil conservation, an embankment which is constructed at the lowest part of the catchment without any reference to contour.

Marginal chlorosis A virus disease characterised by yellowing or blanching of leaf margins; common disease of peanut plants.

Marginal Consumer A consumer who would increase or reduce the consumption of something by one unit if the price of that something fell or rose by one unit.

Marginal cost The additional cost of producing each successive increment of output; the cost of producing M+1 units minus the cost of producing M units.

Marginal firm A firm which is the last to enter, or the first to leave a particular line of production. It is not necessarily the least efficient firm and may, indeed be the most efficient.

Marginal land Land which is the last to be brought into, or the first to be taken out of, a particular line of production. This will not necessarily be a land which is just worth cultivating.

Marginal placentation Arrangement of ovules near the margins of carpels.

Marginal producer A producer who would be just induced to produce a certain commodity at the prevailing level of profits being obtained from it, but if profits fall slightly, would withdraw from production.

Marginal product The amount added to the total product by the addition of one more unit of capital and labour.

Marginal productivity The productivity of the last unit of a factor of production.

Marginal profit The difference between the selling price and the marginal cost of an article.

Marginal purchaser A person who is on the margin of doubt as to whether to buy an article or spend the money on something else, and who just buys.

Marginal unit The marginal unit is the last to be added, or the first to leave a supply.

Marginal utility The extent to which a consumer's satisfaction would be increased or decreased if he had one more or one less unit of a commodity.

Marginal utility of money The increase of decrease in total utility resulting from the addition or subtraction of one unit of currency to or from the total supply of money of a consumer.

Marginal value The price at which consumers will just be induced to buy the last increment of the supply of a commodity.

Marginal variegation Narrow or broad zones of pale tissue which mark the edge of leaves or leaflets.

Marginal weeds Rooted higher plants occurring in marginal areas of ponds and in swamps.

Maricolous Living in the sea.

Mariculture The cultivation of marine organisms, plant and animal, for purposes of human consumption.

Marijuana The Spanish name for the dried leaves and flowering tops of the hemp plant (Cannabis sativa), which have narcotic ingredients and are smoked in cigarettes. Also spelled as marihuana.

Marine climate A regional climate which is under the predominant influence of the sea, that is, a climate characterised by oceanity; the antithesis of a continental climate. Also known as meritime climate, oceanic climate.

Marine forecast A forecast, for a specified oceanic or coastal area, of weather elements of particular interest to maritime transportation, including wind, visibility, the general state of the weather, and storm warnings.

Marine meteorology That part of meteorology which deals mainly with the study of oceanic areas, including island and coastal regions; in particular, it serves the partial needs of surface and air navigation over the oceans.

Marine soil Soils made of materials carried by rivers and deposited in the ocean and subsequently raised by the elevation of sea bed.

Maritime air A type of air whose characteristics are developed over an extensive water surface and which therefore, has the basic maritime quality of high moisture content in at least its lower levels.

Maritime tropical air The principal type of tropical air, produced over the tropical and subtropical seas; it is very warm and humid, and is frequently carried poleward on the western flanks of the subtropical highs.

Marjoram Any of several perennial plants of the genera Origanum and Majorama in the mint family, Labiatae; the leaves are used as a food seasoning.

Market An area, however, large or small, where buyers and sellers are in sufficiently close contact with each other to ensure that the price of a commodity tends to be the same in all parts of the market, allowance being made for transportation costs, tariff barriers, and other obstacles.

Marketing board An organisation set up to assist producers in the marketing of their product.

Market supply curve A curve on a graph showing the relationship between the prices of a commodity and the quantities which traders are willing to sell at each of those prices.

Marking gauge A woodworker's gauge, used for scribing lines parallel to the edge of a board. It consists of a bar with an inserted pin or scribe, and a sliding head which may be adjusted by means of a thumb screw.

Marking nut-tree Semecarpus anacardium L. f. (Anacardium Orientale L.) Family Anacardiaceae. A tree whose fruits give black resin which is used in marking-ink to write on cloth. Seed oil used for protection against white ants and for floor dressing. Nuts used as tan.

Marking, Timber Selecting and indicating, usually by blaze or paid spot, trees to be cut or retained in a cutting operation; spotting.

Mark-up The percentage of profit added to the cost price of goods by traders in order to determine their selling prices.

Marl A soft, unconsolidated-earthy deposit consisting of calcium carbonate or magnesium carbonate, or both and often shells, usually mixed with varying amounts for clay or other impurities.

Marmalade plum Calocarpum sapota (Jacq.). Merr. (Lucuma mammosa Gaertn. F.) Family Sapotaceae. A tree which produces edible fruits.

Marmandai Pluchea lanceolata Oliver & Hiern. Family Compositae. A small perennial broad-leaved herb grown for fodder and also found as a weed.

Marrow (1) Pith of certain plants. (2) Vegetable marrow.

Marrubium Marrubium vulgari. An aromatic plant of the mint family, Labiatae; leaves have a bitter taste and are used as a tonic and anthelmintic. Also known as hoarhound; horehound.

Marsh A transitional land-water area, covered at least part of the time by estuarine or coastal waters, and characterised by aquatic and grass-like vegetation, especially without peat-like accumulation. Periodically flooded or continually wet areas having the surface not deeply submerged. Subclasses include fresh water and salt water marshes. See also Swamp.

Marshy soils Soils found in the alluvial and coastal belts, low lying areas formed by extinct and dried rivers and lakes giving rise to water-logged conditions and marshy patches. Generally bluish grey due to reduction of iron to ferrous state. Have varying amounts of organic matter.

Marvel grass Dichanthium annulatum (Forsk.) Stapf. Family Gramineae. A perennial tufted grass with a creeping rhizomatous stem. One of the best fodder grasses.

Masel Iseilema laxum Hack. Family Gramineae. A fairly drought resistant perennial grass providing a dense turf over the ground. High fodder value. One of the best fodder grass in India.

Mashoni Teramnus lobialis Spreng. Herb of Family Papilionaceae used as cover and pasture crop.

Mask A human, animal or grotesque face without the rest of the body, used as an ornament.

Masked Refers to disease symptoms that are hidden.

Maslin (Meslin) A mixture of growing grains, such as oats and barley, or the product milled from it.

Masonry A construction of stone or similar materials such as concrete or brick.

Masonry dam A dam constructed of stone or concrete blocks set in mortar.

Masonry lining A type of canal lining involving the use of bricks, tiles, concrete blocks and stones. According to the construction used, they are separately called 'brick lining', 'tile lining', 'block lining' and 'stone lining'.

Mass concrete Concrete set without structural reinforcement.

Mass centre The centre of gravity of the mass curve.

Mass curve or **Mass diagram** A curve in which values of cumulative rainfall or runoff are plotted against time. Also sometimes referred to as 'summation graph' or summation curve.

Massive soil structure Large, uniform masses of cohesive soil which sometimes have irregular cleavage, as the C horizons of many class soils.

Mass production In large scale manufacture of a standardised product by the use of specialised labour and capital equipment.

Mass rainfall curve or Mass precipitation curve A plot of accumulated precipitation against time.

Mass selection Selection of some desired character or characters, where progeny of the plants or heads selected are grown in bulk.

Massulae Tissue surrounding the microspores in Azolla.

Mass vasting A general term for a variety of processes by which large masses of earth material are moved by gravity from one place to another.

Master cylinder The fluid cylinder in the brake system; the brakes are applied through pressure on the foot pedal.

Mate Cross-pollination.

Matinal The morning winds, that is, an east wind.

Matricaria A genus of weedy herbs having a strong odour and white and yellow disk flowers; chamomile oil is obtained from certain species.

Matric potential The amount of work that must be done per unit quantity of pure water in order to transport reversibly and isothermally an infinitesimal quantity of water, identical in composition with the soil water, from a pool at the elevation and the external gas pressure of the point under consideration, to the soil water. See water, soil.

Matri-lineal When it is along female line.

Matri-local joint families where the husband goes to live with the family of his wife, and the family line goes with the mother.

Matrix, soil The main soil constituent or material that encloses other soil features, for example, concretions embedded in a fine-grained matrix.

Matroclinous Resembling the female parent.

Matronymic If the name goes with the mother's name.

Matted growth Dense growth.

Mattocking A cultural operation in banana consisting of gradually removing the pseudostem of a parent plant of banana after the harvest of bunches. Helps follower to get interrupted growth.

Maturation Plant process of coming into full development as to mature or ripen.

Mature forest A forest that has reached the stage of utilisation.

Mature green stage Stage at which tomato fruit is still green but fully developed. The tomato is harvested at this stage when it is to be transported long distances.

Mature-plant resistance A term applied particularly to resistance to stem rust in the stages from heading to maturity where this resistance is not correlated with seedling reaction.

Mature profile A soil profile that has attained full development; occurs in moderately level countries which have been occupied by natural vegetation for a long time and is not interfered by human factors.

Mature, soil A soil having well-developed soil horizon produced by the natural processes of soil formation.

Maturity For a given species or stand, the approximate age beyond which growth declines or decay begins to increase at a rate likely to assume economic importance.

Maund A basic weight unit used in India in the past . Now with the use of M.K.S. System, it is not used.

However 1 Maund = Approximately 36 kg.

Mauritius grass See para grass.

Mauritius hemp A hard fibre obtained from the leaves of the cabuya, grown on the island of Mauritius; not a true hemp.

Mauritius raspberry Rubus rosaefolius Sm. Family Rosaceae; shrub cultivated for its edible fruits.

Maximum available water The difference between water contained at field capacity to that at the wilting point.

Maximum effective rainfall Part of rainfall producing the maximum flow (according to a hydrograph).

Maximum level of emergency operation The maximum level to which water surface is expected to rise when operating a canal system under emergency conditions.

Maximum normal flow The maximum flow or discharge assumed as normal for design purposes.

Maximum possible precipitation The largest precipitation (duration and amount) that theoretically could occur in a drainage area or başin during a certain climatic era.

Maximum probable precipitation Precipitation of a given amount and duration that can reasonably be expected to occur on a basin with a known frequency.

Maximum sustained annual yield Term used in forestry for securing to the fullest possible extent a sustained annual or periodic return from a forest.

Maximum thermometre A thermometre that registers the maximum temperature attained during an interval of time.

Maximum-wind level The height at which the maximum wind speed occurs, determined in winds, aloft observation. Also known as maximum wind level.

Maximum-wind topography The topography of the surface of maximum wind speed. Also known as maximum wind topography.

Mazaedium The fruiting body of certain lichens, with the spores lying in a powdery mass in the capitulum.

Mazari palm Nannorrhops ritchieana H. Wendl. (Chamaerops ritchieana Griff.). Family Palmaceae. A low gregarious palm whose leaves yield fibre. Inflorescence and fruits eaten as vegetable.

M.B.F. Thousand board feet.

M.B.M. Thousand (feet) board measure.

MCPA 2 methyle-4-chlorophenoxyacetic acid; a hormone used as weedicide. Available mainly as sodium salt, and closely resembles 2, 4-D in its weed killing properties.

Meadow An area of land devoted to the production of forage, mainly perennial, that is harvested for hay. Also known as pelouse.

Meadow fescue Festuca elatior L. Family Gramineae. An upright, tufted perennial shrub used as fodder.

Meadow foxtail Alopecurus pratensis L. Family Gramineae. A perennial fodder grass.

Mealy bug Species of insects of genus Pseudococcus, family Pseudococcidae, widely distributed; small, elliptical infesting fruit trees and other plants.

Mean The arithmetic average.

Mean annual evaporation The mean value in depth units of evaporation, the period of observation being of adequate duration to secure approximate constancy.

Mean annual increment The total growth of a stand of trees divided by the total age.

Mean annual rainfall The mean of annual rainfall observed over a period which is sufficiently long to produce a fairly constant mean value. Also known as mean annual precipitation.

Mean annual runoff The value of the annual volume of water discharged by the stream draining the area, the period of observation being sufficiently long to secure a fair mean; similarly 'mean monthly runoff'.

Mean annual temperature The mean of the maximum and minimum temperatures recorded over a period of years in a given locality.

Mean depth Average water depth in a stream channel or conduit computed by dividing the cross-sectional area by the surface width.

Meandering stream A stream having a

pattern of successive meanders. Also known as snaking stream.

Mean range The arithmetic mean of the ranges of a set of samples of the same size.

Mean river level The average height of the surface of a river at any point for all stages of the tide over a 19-year period.

Mean sea level The mean level of sea at a given place.

Mean square In general, the mean square of a set of values is the arithmetic mean of the squares of their differences from some given value, namely their second moment about that value.

Mean square deviation The second moment of a set of observations about some arbitrary origin. If that origin is the mean of the observations, the mean square deviation is equivalent to variance.

Mean square error of estimate The square root of the arithmetic mean of the squared deviations between the estimates for different samples and the population parametre under estimation. In the balance, it is best to have as small a mean square error as possible, even if this entails a biased estimator.

Mean temperature The average temperature of the air as indicated by a properly exposed thermometre during a given time period, usually a day, month or year.

Mean tidal range Difference in elevation between mean high water and mean low water; 'spring range' and 'neap range' are the near ranges of spring tides and neap tides respectively.

Mean tide level, or Half tide level The plane, or surface, that is exactly half way between high water and mean low water. Because of the lack of symmetry of the tidal curve, this is not exactly the same as mean sea level.

Mean velocity The average velocity at a given section of the stream. It is calculated by dividing the total discharge by cross-sectional area at that section.

Measurable precipitation The precipitation more than .127 mm at a place. Less than this considered as trace.

Measuring weir Weirs usually used for measuring discharge. See weir.

Mechanical ability Persons who understand things as in skilled trades, has much of science, manual skills combination. Individual and building traders are higher in mechanical ability.

Mechanical analysis See particle-size analysis and particle-size distribution.

Mechanical control The term generally used in erosion control when it is done through structures such as earth dams, terraces etc. in contrast to vegetative control.

Mechanical eluviation Eluviation in which apart from chemical substances finer particles of the clay soils are leached resulting in the development of a profile characterised by light texture.

Mechanical practices Soil and water conservation practices that primarily change the surface of the land or that store, convey, regulate, or dispose of runoff water without excessive erosion.

Mechanical seed drill See seed drill.

Mechanical soil analysis The separation of particles groups (separates) by mechanical means and determination of percentage of each group in a given soil sample.

Mechanical tissue Supporting tissue, usually fibrous or woody.

Mechanical ventilation Process of supplying or removing air with a mechanical device to or from any space, such as the interstitial space between products.

Mechanics and engineering, Soil The sub specialisation of soil science dealing with the effect of forces on the soil and the application of engineering principles to problems involving the soil.

Meconic acid White crystalline acid obtained from opium.

Meconium (1) Waste product of pupa or other embryonic form. (2) The juice of opium poppy.

Mecron The unhumified plant matter on a forest soil.

Mediacid Soils having pH in the range of 4.5 to 5.0.

Median The median is that value of the

variate which divides the total frequency into two halves, that is above and below which half the values lie.

Medicago denticulata See toothed bur clover.

— **hispida** See bur clover.

— **lupulina** See hop clover.

— **obscura** Retz. A herb of Family Papilionaceae grown as fodder crop.

— **sativa** See lucerne.

— **turbinata** Willd. A herb of Family Papilionaceae. Used as fodder.

Medication The introduction of chemical into the sapstream of a living plant to kill or make it repellent to insect attack.

Mediterranean climate A type of climate characterised by hot day sunny summers and a winter rainy season; basically, this is the opposite of a monsoon climate. Also known as etesian climate.

Mediterranean faunal region A marine littoral faunal region including that offshore portion of the Atlantic Ocean from northern France to near the Equator.

Medium The supporting substance on or in which plants, fungi and bacteria are cultured or grown. The term usually includes the nutrients as well.

Medium cane Saccharum sinense. Group of sugar canes thin and medium like *panshahi, hagari, mungo* etc.

Medium sands A soil separate that includes particles ranging from 0.5 to .25 mm.

Medium sandy loam A soil class which contains 30 per cent or more of very coarse, coarse and medium, sand combined but less than 25 per cent very coarse sand, and less than 30 per cent of very fine or fine sand. The clay is less than or equal to 20 per cent and silt less than or equal to 50 per cent.

Medium term loan A loan repayable in 3 to 5 years.

Medium texture Intermediate between fine textured and coarse-textured soils. It includes the following textural classes: very fine sandy loam, loam, silt loam and silt.

Medulla Loose layer of hyphae inside a thallus; body of a sclerotium.

Medulla cell A layer of globular cells in the centre of a wool fibre.

Medullary rays Groups of parenchymatous cells, extending radially between the vascular bundles, which provide for radial transfer of the food and water in a plant.

Megagametophyte A female gametophyte resulting from the growth of a megaspore and producing female gametes or eggs, the embryo sac of angiosperms.

Megaphyllous Having relatively large leaves.

Megasorus A sorus containing only megasporangia, and therefore producing only megaspores.

Megaspore (macrospore) A spore having the property of giving rise to a gametophyte (embryo sac) bearing only a female gamete. One of the four cells produced by two meiotic divisions of the megaspore mother cell (megasporocyte).

Megasporophyll A leaf bearing megasporangia.

Megasporangium A sporangium containing only megaspores.

Megaspore mother cell The cell which undergoes two meiotic divisions to produce four megaspores.

Megathermophyte A plant that requires great heat and abundant moisture for normal growth. Also known as macrothermophyte.

Mehal Prunus pashtia Buch. - Ham. Small tree of Family Rosaceae used as root stock for pear.

Megastrobilus A cone or strobilus bearing megasporangia.

Meiosis The process by which the chromatin material becomes reduced qualitatively and quantitatively to half the somatic number of chromosomes. It is completed in the two division, meiotic mitosis, which procede the formation of gametes in animals or of spores in plants.

Meiwa kumquat Fortunella crassifolia. An evergreen tree or shrub of Family Rutaceae grown for its very small highly aromatic fruit chiefly used for preserves.

Melampsora lini See linseed rust.

— **ricini** See castor rust.

Melanic brunisol A great group of soils in the Brunisolic order. The soils have mull Ah horizons thicker than 5 cm and base-saturated Bm horizons. This group includes soils formerly classified as Brown Forest.

Melanised soil A soil which has a very dark colour to a considerable depth in the surface layer due to incorporation of humus.

Melanopsichium eleusinis See ragi smut.

Melanose (1) A fungus disease of grapevine caused by Septoria ampelina; leaves are infected and fall off. (2) A fungus disease of citrus trees and fruits caused by Diaporthe citri, characterised by hard, brown, usually gummy elevations on the rind, twigs, and leaves.

Melassic acid An acid generated in molasses and glucose by the action of alkalies.

Melastoma Malabathricum See Indian rhododendron.

Melia azadirachta See neem tree.

Meliaceae A family of dicotyledonous plants in the order Sapindales characterised by mostly extipulate, alternate leaves, stemans mostly cannate by their filaments, and syncarpous flowers.

Melilotus alba See hubum clover.

— **indica** See sweet clover.

— **officinalis** See yellow sweet clover.

— **parviflora** See sweet clover.

Melinis Melinis minutiflora P. Beauv. Family Gramineae. A grass which secretes a viscous oil with the odour of cumin. Good fodder; also an insecticide and repellent for mosquitoes.

Meliola butleri See citrus sooty mould.

Melitology Branch of entomology concerned with bees.

Melliferous Honey-bearing.

Mellow Soft tender, juicy, ripe fruit.

Mellow soil A very soft, very friable, porous soil having no tendency towards hardness or harshness. Also consistence.

Melocanna bambusoides See tarai bamboo.

Melochia corchorifolia L. An undershrub of Family Sterculiaceae. Leaves eaten as vegetable; stem yield fibre.

Melolontha nepalensis See apple leaf and fruit-eating beetles.

Melon Either of two soft-fleshed edible fruits, muskmelon or watermelon, or varieties of these.

Melon fly Dacus cucurbitae Coq., D. ciliatus Loew. Pest on fruit of melon and other cucurbits. Maggot bores into fruit, making it rot.

Melothria heterophylla See amantmul.

Meltwater Water derived from melting ice or snow, especially glacier ice.

Membrane (1) A thin film, skin or layer of tissues covering a part of animal or plant. (2) A thin covering of cells or of unicellular organisms. (3) A layer of material of low permeability placed at the bottom of ponds to reduce seepage.

Memecylon edule See ironwood tree.

— **tinctorium** See ironwood tree.

Mendelian genetics Genetic principles which follow Mendel's law.

Mendel's Law one of the two laws concerning hereditary transmission of genes or character determinates, discovered by Gregor Medel in 1886. The two laws, known as principle of segregation and the principle of independent assortment, state that inheritance is subject to definite rules or laws; offspring of parents with contrasting characters will resemble the dominant parent, the second generation resulting from crossing will produce, 25 per cent resembling the dominant grand parent, 50 per cent resembling the parents and 25 per cent resembling the recessive grandparent.

Mensuration, Forest A science dealing with the measurement of volume, growth, and development of individual trees and stands, and the determination of various products obtainable from them.

Mentha arvensis See field mint.

— **arvensis** var. piperascens. See Japanese peppermint.

— **longifolia** See mint.

— **pulegium** See European pennyroyal.

— **Spicata** See spearmint.

— **Sylvestris** See mint.

— **viridis** See spearmint.

Menthol White crystalline substance obtained from peppermint oil, useful in treatment of cold; antiseptic.

Mercerize Prepare cotton goods for

dyeing by use of chemicals.

Merchantable Trees or stands of a size and quality suitable for marketing and utilisation. They may or may not be so located as to be accessible for logging. Also, a specific grade of southern yellow pine timbers.

Mercuric chloride Corrosive sublimate used as disinfectant and in control of fungi, gnats, earthworms etc.

Mercurous chloride Calomel.

Meriandra benghalensis See Bengal sage.

Mericarp One of the two carpels of the fruit of an umbeliferous plant.

Merismatic Dividing or separating into cells or segments.

Meristem Formative plant tissue composed of undifferentiated cells capable of dividing and giving rise to other meristematic cells as well as to specialised cell types, found in growth areas.

Mesa A rather flat topped steep-sided hill or mountain that is usually composed of nearly horizontal strata of bedrock.

Mesarch Characterised by primary xylem in which maturation of the metaxylem is both centripetal and centrifugal, that is, both towards the centre and the outside of a stem or other organ.

Mesic layer A layer of organic material at a stage of decomposition between that of the fibric and humic layers.

Meso Intermediate, middle.

Mesocarp The middle layer of pericarp, thick, fleshy and edible in the stone fruits.

Mesoclimate The climate of small areas of the Earth's surface that may not be representative of the general climate of the district. The mesoclimate is intermediate in scale between the macroclimate and the microclimate.

Mesocotyl In some monocotyledons, the part of the seedling stem which elongates below the shoot.

Mesoderm The embroyonic layer lying between ectoderm and endoderm.

Mesome In the telome theory, the internodes below the telomes.

Mesophile An organism growing best at moderate temperature of 25 to 40°C.

Mesophyll The chlorophyllous leaf tissues between the epidermal layers.

Mesophytes Plants which are intermediate between hydrophytes and xerophytes, that is, which grow in soil containing moderate amounts of available moisture.

Mesotherm (1) A plant that grows in intermediate condition of temperature along with plenty of water. (2) A plant that grows under conditions of alternating high and low temperatures.

Mesotrophic Refers to soil that has suffered moderate leaching, such that the sum of the exchangeable Ca, Mg, K and Na, expressed in me/100g clay, is 5-15. The figure is calculated from the S-value and the clay content. Such soils are said to have a medium base status.

Mesozoic The geological era between the Paleozoic and Caenozoic.

Mesquite Prosopis juliflora (Sw.) DC. (Mimosa juliflora Sw.) Family Mimosaceae. A small or medium sized tree which does not stand waterlogging. Coppices well. Wood strong, hard and makes good fuel. Pods excellent fodder. Used for halting desert from spreading. Also called *kabuli kikar*.

Messisol A great group of soils in the Organic order that are saturated for most of the year. The soils have a dominantly mesic middle tier, or middle and surface tiers, if a terric, lithic, hydric, or cryic contact occurs in the middle tier.

Mesta Hibiscus cannabinus L. Family Malvaceae. A fibre crop whose fibre is used in cordage, fishing nets and canvas. Also mixed with jute in manufacture of sugar bags. Tops and branches fed to cattle. Seeds also used for feeding milch cattle. Fibre forms about 16 per cent of the weight of dry stalks.

Metabolic body size Weight of animal raised to the three-fourths power.

Metabolism Complex of physical and chemical processes involved in the maintenance of life.

Metabolite Anything produced by metabolism.

Metabolizable energy (ME) Food intake gross energy minus faecal energy, minus in the gaseous products of digestion, minus urinary energy. Also called available energy.

Metagenesis Alternation of generations.

Metaldehyde A crystalline polymer of acetaldehyde used at 2 to 3 per cent in poison baits for snails and slugs.

Metamere One of the several similar segments of the body.

Metamorphic rock Rock derived from pre-existing rocks, but differing from them in physical, chemical and mineralogical properties as a result of natural geological processes, principally heat and pressure, originating within the earth. The pre-existing rocks may have been igneous, sedimentary, or another form of metamorphic rock.

Metamorphosis Change in form and structure undergone by an animal from embryo to adult stage, as in insects; transformation of one structure into another.

Metaphase The stage of cell division in which the chromosomes are arranged in an equatorial plate or plane; precedes the anaphase stage.

Metaphosphate An acid derived from phosphorus pentoxide. It contains the least water of the phosphoric acids.

Metaphosphate fertiliser A fertiliser which contains higher percentages of P_2O_5 than ordinary acid phosphate and superphosphate.

Metaplasm Part of protoplasm that contains formative material.

Metastasis (1) Transference of bodily function, disease from one part of organ to another. (2) Transformation of chemical compounds into others in the process of assimilation by an organism.

Metasystox 0, 0-dimethyl-0-2-ethyl-mercaptoethyle thionophosphate. Organo-phosphorus compound, far less toxic than systox; systematic insecticide for white flies, aphids and mites. Also called methyldemeton.

Metatarsus Part of foot between tarsus and toes; first joint of tarsus in insects.

Metaxenia Immediate effects of the pollen upon the development of the mother tissue of the fruit.

Meteoric water Water which is derived from the atmosphere.

Meteorology The branch of science that deals with atmospheric phenomena and the basic laws that produce and control such phenomena.

Methallyl chloride ($CH = C(CH_3) = CH_2Cl$) A chemical compound used as fumigant for soil and grain.

Methanesulphonyl fluoride (CH_2SO_2F) A greenhouse fumigant.

Methionine An essential aminoacid containing sulphur.

Methoxychlor Dichloromethoxyphenyl trichloroethane. A chlorinated hydrocarbon insecticide; related to DDT.

Methyl bromide (CH_3Br) Highly toxic insect fumigant. Gas 3.3 times heavier than air.

Methyl formate An insect fumigant.

Methyl red A chemical dye stain used in determining the classification of taxonomy of certain bacteria, such as coliform bacilli.

Metre Distance, at zero degree centrigrade and under normal atmospheric pressure, between the axes of two median lines traced on the platinum iridium bar deposited at the Bureau International des poids et Mesures and declared International proto-type of the metre. A linear unit of measurement is equal to 100 cm. 1000 metres constitute one kilometre.

Metric system A decimal system of units based upon the metre as the unit of length, the gram as the unit of weight or mass, and the second as the unit of time.

Metric technical unit of mass Mass that acquires an acceleration of one metre per second per second under the influence of a force equal to one kilogram force.

Metriona circumdata See sweet potato leaf beetles.

Metroxylon rumphii See sago palm.

— **Sagu** See sago palm.

Mexican linaloe Bursera penicillata (Sesse & Moc. & DC). Engl. (Bursera delpechianum Poiss). Family Burseraceae. A tree producing aromatic fruits from which oil is extracted which is used in perfumery and for flavouring food and beverages.

Mexican poppy Argemone mexicana L. Family Papaveraceae. An annual herb spread throughout India. Bears ash-grey spiny leaves with ragged edges and bright yellow leaves. Prickly capsules bear numerous seeds, easily scattered. Seed contains the alkaloid argemonin, which cause dropsy. Seeds source of semi-dry oil used as lubricant and illuminant. Plant used to reclaim *usar* land.

Meyna laxiflora See pundrika.

Mhote A water lifting device. Leather bucket is used for water lifting with one or two pairs of bullocks takes about 7 to 10 days in irrigating a hectare of land.

Miasma Infectious or noxious emanation.

Mica A mineral group consisting of phyllosilicates having sheet like 2:1 lattice structures, generally with potassium in the interlayer position. Than called Potash mica or muscovite mica, white, clear and transparent. Magnesium mica or biotite mica, black. Soft; hardness 2 to 2.5 and specific gravity 2.8 to 3.0; when present, gives a greasy or soapy feel to soil.

Micell (1) A sub-microscopic, crystalline, multimolecular unit of structure built from complete molecules which forms the foundation walls of cells. (2) The nucleus of a colloid which is negatively charged. Also called micella, micelle.

Michai Ipomoea muricata (L) Jacq. (Calonyction muricatum G. Don.). Family Convolvulaceae. A herb, the flower pedicels of which are eaten as vegetable.

Michelia champaca See champac.

Micorrhiza Means 'fungus root', name given to symbiotic association of fungal mycelium and roots of certain trees and shrubs; aids host plant in the absorption of certain nutrients.

Microaerophilic Term applied to micro-organism that require free oxygen for their growth but in an amount less than that of the oxygen of the atmosphere.

Microbe A microscopic organism.

Microbial insecticide Pathogen used for killing insects. Harmless to beneficial insects.

Microbial processes The activities and influence of micro-organisms in modifying soils or other substances.

Microbicide An agent that kills microbes.

Microbiology, soil The sub-specialisation of soil science dealing with soil-inhabiting micro-organisms and their relationship to agriculture, including both plant and animal growth.

Microbispora A genus of bacteria in the Family Micromonosporaceae; spores are produced in longitudinal pairs on aerial mycelia only.

Microclimatology The study of a microclimate, including the study of profiles of temperature, moisture and wind in the lowest stratum of air, the effect of the vegetation and of shelter belts and the modifying effect of towns and buildings.

Microclimate (1) The climate of a small area resulting from the modification of the general climate by local differences in elevation or exposure. (2) The sequence of atmospheric changes within a very small region. (3) Atmospheric environment conditions in the immediate vicinity of the plant including interchanges of energy, gases and water between atmosphere soil.

Micrococcus A specific genus of small round, bacterial cells of the Family Coccaceae.

Microconidia Very small spores, now considered spermatia of a fungus also having larger conidia.

Microenvironment (1) Evolutionary processes resulting from the accumulation of minor changes over a relatively short period of time; evolutionary changes due to genemutation. (2) Evolution of species.

Microfauna The part of the animal population consisting of individuals that are too small to be clearly distinguished without the use of a microscope. It includes protozoa and nematodes.

Microfibril (1) A sub-microscopic thread like unit of the cellulose matrix of the cell-wall in wood. (2) Small fibre, a number of which cemented together forms the spindle cell of wood.

Microflora (1) The part of the plant population consisting of individuals that are too small to be clearly distinguished without the use of a microscope. It includes algae, bacteria and fungi. (2) Vegetation of a small area in contrast to that of a region.

Microgametophyte A male gametophyte resulting from the growth and development of a microspore.

Micrometeorology That portion of the science of meteorology that deals with the observation and explanation of the smallest scale physical and dynamic occurrences within the atmosphere, studies are confined to the surface boundary layer of the atmosphere, that is, from the earth's surface to an altitude where the effects of the immediate underlying surface upon air motion and composition become negligible.

Micrometre Instrument for measuring width, length etc. with precision particularly of small objects.

Micron One millionth part of a metre. Unit of measurement of small bodies.

Micronised sulphur Sulphur in an extremely fine condition, with a particle size less than 10 microns. Used for dusting or spraying against plant diseases.

Micronutrient A chemical element necessary in only small amounts, usually less than 1 ppm in the plant, for the growing of the plants and the health of animals. Examples of these elements are boron, molybdenum, copper, iron, manganese and zinc. "Micro" refers to the amount, not the essentiality of the element to the organism. See also macronutrient. Also called microelements or trace elements.

Micro-organism A form of life of microscopic size; such as bacteria, fungi or protozoa.

Micro parasite A microscopic parasite.

Micropedology A branch of pedology dealing with microscopic structures, arrangements and minerological composition of soil mass.

Microphyllous (1) Having small leaves. (2) Having leaves with a single, unbranched vein.

Microphyte (1) A microscopic plant. (2) A dwarfed plant due to unfavourable environmental conditions.

Micro-pores Soil pores which are so minute that they impede the liquid flow of water in soil. It is because of this that in spite of total pore space being high in clays, hydraulic conductivity is less than coarse texture soils such as sands.

Micropyle A minute opening in the integument at the tip of an ovule through which the pollen tube commonly enters; persists in the seed as an opening of a scar between the hilum and point of radicle.

Microrelief Small-scale, local differences in relief including mounds, swales, or hollows. See also cradle knoll and gilgai.

Microscope An instrument which enables to see things which are not visible with naked eye. Single or number of lenses magnify the object.

Microscopic Too small to be seen except with the aid of a microscope; true of most of the fungus structures.

Microsorus A sorus containing only microsporangia and therefore producing only microspores.

Micro-sporangium A sporangium producing only microspores.

Microspore One of the four cells produced by the two meiotic divisions (mitoses) of the microspore, mother cell (microsporocyte). A spore having the property of giving rise to a gametophyte bearing only male gametes. Also called pollen grain.

Microspore-mother-cell A cell from which other cells are formed. Also called pollen-mother-cell.

Microsymbiont One of the organisms (bacteria) in a symbiotic relationship.

Microtermes obesi See sugarcane termites.

Microthermic tree A tree which tolerates a cold climate and a short growing season.

Mid-bloom The middle period of blooming.

Midden A refuse heap marking the site of previous habitation.

Middleman A trader in between producer and the consumer.

Middle tier The tier just below the surface tier. It is 80 cm thick or extends to a lithic or hydric contact. The great group classification of Organic soils is usually based on this tier.

Middlings Coarse particles separated

in the milling of wheat; contains germ of wheat grain and fine bran. Nutritious, more rich in proteins and digestible nutrients than bran. Used as cattle and poultry feed.

Midget Extremely small individual.

Mid-rib A large central vein of a leaf, continuation of the petiole.

Mid-range For a set of values arranged in order of magnitude X_1, $X_2 ... X_n$, the mid range is defined as $^1/_2(X_n+X_1)$.

Mignonette tree See henna.

Mignonette vine See Madeira-vine.

Migratory locusts Locusta migratoria L.; yellowish brown locust, found generally scattered in several states. Also see locusts.

Migratory villages In such villages, people live for months, arrange food supplies from natural sources in the form of wild fruits, animals etc. When the food supply of a place is exhausted, they shift to another place, where they can find better supplies.

Mild steel A steel very low (0.15-0.25%) in carbon. It welds, but does not tember.

Mildew Plant disease in which the pathogen is a growth on the surface.

Miliary Small, resembling a millet seed in size.

Milliequivalent per litre A unit of measurement of fonic concentration in solutions.

Milk of lime A whitish suspension made by slaking calcium oxide (quicklime) in water. Disinfectant and used for whitewash.

Milk weed Euphorbia prostrata. Family Euphorbiaceae. Most common annual of prostrate habit characterised by milky juice occurring in crop and garden lawns. Produces numerous tiny seeds.

Milled rice Hulled rice grain, rice grain with bran coat removed.

Miller law A law administered by the Food and Drug Administration that regulates the production and use of agricultural fungicides in the United States, and will not allow materials to leave poisonous residues on edible crops.

Millets Annual grasses and cereals of the Family Gramineae producing small seeds. Grown for grain and forage. Some of the millets are barnyard millet, greatmillet, common millet, finger millet, maize etc.

Millettia auriculata See ganj.

Milligram equivalent Of a salt constituent is the number of milligrams represented by its combining weight.

Milligram equivalent per million Parts per million divided by its combining weight.

Millimetre A unit of measurement of length. 1/100th of a metre or 1/10th of a centimetre.

Milling quality The percentage of whole kernels and total milled rice obtained from a given quantity of unmilled rice.

Milo Sorghum vulgare var. subglabrescens. Family Gramineae. A drought resistant annual grass. Grown for livestock feed.

Mimosa amara See lallei.

— **dulcis** See manila tamarind.

— **juliflora** See mesquite.

— **lebbeck** See kokko.

— **leucophloea** See safed kiker.

— **odoratissima** See black siris.

— **pudica** See touch-me-not.

Mimusops elengi See Spanish cherry.

— **hexandra** See khirni.

— **kauki** See kauki.

Mine (1) Mine of minerals as coal mines etc. (2) To mine minerals. (3) To bore and eat into a gallery or passageways.

Mine dump Area covered with overburden and other waste materials from ore and coal mines, quarries, and smelters, and usually having little or no vegetative cover. A miscellaneous land type.

Mineral A homogeneous naturally occurring phase, sometimes restricted to inorganic, crystalline phases.

Mineral elements The inorganic constituents of soil, plant and plant products.

Mineral mixture A feed containing salt, limestone, phosphate, mineral elements etc.

Mineral oil Oil obtained from petroleum and other hydrocarbon minerals.

Mineral, Soil (1) Any mineral occurring as a part of or in the soil. (2) A natural inorganic compound with definite physical chemical, and crystalline properties (within the limits of isomorphism) occurring in the soil. See also clay mineral. (3) A soil consisting predominantly of, and having its properties determined predominantly by, mineral matter. It contains less than 17 per cent organic carbon except for an organic surface layer that may be up to 40 cm (16 inches) thick if formed of mixed peat (bulk density 0.1 or more) or 60 cm if of fibric moss peat (bulk density less than 0.1).

Mineral water Water containing naturally or artificially supplied minerals or gases.

Mineralogical analysis The estimation or determination of the kinds or amounts of minerals present in a rock or a soil.

Mineralogy The science of minerals.

Mineralogy, Soil The subspecialisation of soil science dealing with the homogeneous inorganic materials found in the earth's crust to the depth of weathering or sedimentation.

Mineralisation The conversion of an element from an organic form to an inorganic state as a result of microbial decomposition.

Mine wash Water deposited accumulations of sandy, silty, or clayey material recently eroded in mining operations. It may clog streams and channels and damage land on which it is deposited. A miscellaneous land type.

Minim A drop; Sixtieth part of a fluid drachm.

Minimum tillage The least amount of tillage needed for quick germination and a good stand of crop.

Minor element See micro-nutrient.

Minor forest products Economic forest products other than timber. Include tan substances, fibres and flosses, bamboos and canes, fodder and fodder grasses, gums, resins, honey, lac and perfumed wood.

Minors, distributary minors, or Tertiary canals Canals taking off from distributaries and supplying water to subminors or outlets or turnouts.

Mint Mentha longifolia (L.) Huds. (Mentha sylevestris L.) Family Labiatae. An aromatic herb, used as a flavouring agent. Dried leaves used as carminative and stimulant. Also called *pudina.*

Minute (1) A unit of time and is equal to 60 seconds. (2) A unit of plane angle and 60 minutes is equal to 1/180 radians. Sixty minutes also equal to one degree

Miracidium The ciliated embryo or larva which hatches from the egg of a trematode (fluke).

Mirtenga Protium serratum (Wall. ex Colebr.) Engl. (Bursera serrata Wall. ex. Colebr.). Family Burseraceae. A tree whose fruits are edible.

Miscellaneous land type A mapping unit for areas of land that have little or no natural soil, or that are inaccessible for orderly examination, or where, for some reason, it is not feasible to classify the soil, for example rough mountainous land, eroded slopes, and marshes.

Miscible Refers to two or more substances, when mixed together, forming a uniformly homogeneous solution.

Misfit stream A stream whose meanders are either too large or too small to have eroded the valley in which it flows.

Mist (1) fine liquid droplets suspended in or falling through a moving or stationary gas atmosphere. (2) Liquid particles in a fine state of division but of size which make up spray material.

Mist droplet A particle of mist, intermediate between a haze droplet and a fog drop.

Mist-O-Matic treater A type of machine used for treating seeds. It applies treatment as a mist directly to the seed.

Milter gate Either of a pair of canal lock gates that comes out from the side walls and meet at an angle pointing toward the upper level.

Mist sprayer Essentially a low volume sprayer used for applying fungicide or insecticides in concentrated form. Break up and atomizes insecticide and disperses them into an air current of high velocity. Dilution with water reduced by 20 to 80 per cent or more. Used for treating large areas of fruit trees, high and dense crops. Also called mist blower, drift sprayer.

Mites Arachnids like the tick; belong to the Order Acarena. Together with insects make up the group Arthropodes; have 4 pairs of legs, body not divided into distinct regions. Parasite on plants and animals. Cause mange and allied skin diseases in animals.

Mitha karela Cyclanthera pedata Schrad. Family Cucurbitaceae. A cucurbit vegetable.

Miticide Any chemical or compound such as parathion used for killing mites.

Mitochondrion (pl. mitochondria) Small, spherical or rod shaped cytoplasmic organelle(s).

Mitosis The process by which the nucleus is divided into two daughter nuclei following equational splitting of chromosomes. Somatic mitosis. The process by which the daughter nuclei are identical, quantitatively and qualitatively. Meiotic mitosis. Two nuclear divisions that result in sexual spores in higher plants and in gametes in animals. Both divisions are necessary to complete reduction.

Mitotic cycle Entire sequence of processes of cell division in which the diploid number of chromosomes is retained in both daughter cells.

Mix To cross or interbreed plants or animals more or less by chance.

Mixed cloud A cloud containing both water drops and ice crystals, hence a cloud whose composition is intermediate between that of a water cloud and that of an ice-crystal cloud.

Mixed cropping system of sowing two or three crops on the same land one of them being the main crop. This is done particularly in rainfed farming to avoid risk of failure of the main crop.

Mixed culture Culture containing two or more kind of organisms.

Mixed fertilisers A mixture of fertilisers containing more than one material in fixed proportions. The fertilisers are mixed to apply balanced nutrients and to reduce labour requirement of applying fertilisers separately. A fertiliser containing N, P, K, is also know as complete fertiliser.

Mixed flow pump Pump in which pressure is developed partly by centrifugal force and partly by the lifting action of the vanes on water.

Mixed forest Forest composed of trees of two or more species.

Mixed tides Those which are preponderantly semi-diurnal at sometimes and diurnal at others, according to the declination of the moon.

Mixing ratio In a system of moist air, the dimensionless ratio of the mass of water vapour to the mass of dry air, for many purposes, the mixing ratio may be approximated by the specific humidity.

Mobile soil colloids Soil colloids so dispersed as to move with percolating water.

Model length The length of the maximum number of fibres in a given sample of lint. Also called staple length.

Mode The class of greatest frequency in a frequency distribution.

Model (botanical) Short for architectural tree model, generalised growth programme which determines successive architectural phases of a tree.

Moder A zoogenous forest humus made up of plant remains partly disintegrated by the soil fauna (F layer), but not matted as in raw humus. It is transitional to a zone of spherical or cylindrical microdejections on arthropods that is permeated by loose mineral particles in its lower part and often throughout. Although incorporation of organic matter is intense, it is shallow, because none of the organisms concerned with moder formation have important burrowing activity. The mixing of organic and mineral particles is purely mechanical. Organic carbon under the F layer varies from 23 per cent to 29 per cent, but may exceed 35 per cent. The C : N ratio is 20 to 25 and sometimes lower. Various subgroups can be recognised by their morphology and chemical characteristics.

Moderately coarse texture Consisting predominantly of coarse particles. In soil textural classification, it includes all the sandy loams except the very fine sandy loam. See also coarse texture.

Moderately fine texture Consisting predominantly of intermediate sized soil particles with or without small amounts of fine or coarse particles. In soil textural classification, it includes clay loam, sandy clay loam, and silty clay loam. See also fine texture.

Moderately sinuous canals Canals

having a degree of sinuousity in alignment which is typical of moderately rolling topography.

Modified leader System of training fruit trees. Developed by first training tree to the leader type by allowing the central axis to grow for four to five years as in leader type and then heading the central stem to allow lateral branches to grow as in open centre system. A intermediate system.

Modified loess Loess that has lost its typical characteristics by secondary processes, including temporary immersion, erosion and subsequent deposition, chemical changes involving the destruction of the bond between the particles; or chemical decomposition of the more perishable constituents, such as feldspar.

Modified stem A plant stem like a tuber, rhizome, stolon or runner, which differs from the normal upright stem.

Modifier or modifying gene A gene that affects the expression of another nonallelic gene.

Modular construction Way of building trees or branch complexes out of modules.

Modulation (Of humidity and light gradients), progressive but non-uniform change in micro-climatic conditions throughout the forest canopy.

Modulators the variable factors that limit the carrying capacity of the coastal ecosystem on a short term (often day to day) basis. These include temperature, available mineral nutrients, dissolved gas concentration, and presence of toxic chemicals.

Module Shoot unit with determinate growth, either by apical abortion or conversion of apex to an inflorescence

Mohwa butter Oil extracted from seeds of tree Bassia latifolia; used in soap making. Oil seed cake used as manure.

Moisture content The moisture content of a sample is either the loss in weight when it is dried, or the quantity of water collected when it is distilled. It is expressed as a percentage of the weight of the original samples.

Moisture, dry basis A basis for representing moisture content of a product, as parts of water per part of dry matter (DM), that is equal to the weight of the DM and the answer multiplied by 100 = per cent. It may be more than 100 per cent.

Moisture equivalent The weight percentage of water retained by previously saturated samples of soil 1 cm thick, after it has been subjected to a centrifugal force of 1000 times gravity for 30 min.

Moisture release curve See moisture retention curve.

Moisture retention curve A graph showing the soil moisture percentage (by weight or by volume) versus applied tension or pressure over a specified range.

Moisture, Soil Water contained in the soil.

Moisture storage capacity factor The ratio of the average available moisture contained in the plant root depth at the time of sampling to the average available moisture capacity of that particular soil to the root depth.

Moisture tension The equivalent pressure that must be applied to the soil water to bring it to hydraulic equilibrium, through a porous permeable wall or membrane, with a pool of water of the same composition. 1000 cm being equal to one atmospheric tension.

Moisture-utilisation efficiency, Moisture-use efficiency or Water-utilisation efficiency The units of marketable crop produced for a unit depth of water used in evapotranspiration.

Moisture volume percentage The ratio of the volume of water in a soil to the total bulk volume of the soil.

Moisture weight percentage The moisture content expressed as a percentage of the oven dry weight of soil. Also dry-weight percentage.

Moisture, wet basis Used for commercial designation. Obtained by dividing the weight of water present in the material by the total weight of material, including water and dry matter.

Moke disease Bacterial wilt of banana caused by Pseudomonas solanacearum.

Molascake Blocks made out of a mixture of molasses and bagasse screenings and groundnut cake, mineral mixtures also being added at times; used for feeding animals.

Molasses (1) A thick brown viscid syrum that drains from raw sugar during sugar manufacturing process and sometimes used for reclamation of soils. (2) A syrup obtained by evaporation of sweet juice of any plant.

Mold board plow A plow equipped with a curved iron plate (moldboard) that lifts and turns the soil. Also known as turnplow, or mould board plough.

Molds Fungi with conspicuous mycelium or spore masses, often saprophytes. Multicellular fungi.

Mole (mol) A unit of mass numerically equal to the molecular weight of the substance. The gram mole or gram molecule is the mass in grams numerically equal to the molecular weight.

Mole, or **Mole ball** An egg-shaped device pulled behind the tooth of a subsoil plough to open drainage passages.

Mole cricket Gryllotalpa africana Pal Beaur. A major insect pest of potato. Burrows in field, bores into tuber, usually infests damper parts of field; both nymph and adult harmful.

Molecular One of the minute groups of atoms of which material substances are conceived to consist. The smallest portion to which substances can be divided without losing chemical identity.

Molecular conductivity The conductivity of a volume of electrolyte containing 1 mole of dissolved substance.

Molecular diffusion The transfer of mass between adjacent layers of fluid in laminar flow.

Molecular effusion Mass transfer flow mechanism of free-molecule transfer through pores or orifices.

Molecular genetics The approach which deals with the physics and chemistry of the processes of inheritance.

Mole drainage The method of draining soil by means of mole drains.

Mole drains Cylindrical drains formed in the subsoil by pulling a steel ball through it.

Mole plough, or Mole plow A plough by which a steel ball following a pointed-iron shoe rnakes an underground channel, resembling the track of a mole, to serve as a drain.

Molybdenosis Molybdenum poisoning.

Molybdenum (Mo) Trace food element of plants absorbed as MoO_4 ion. Utilised in enzyme systems which brings about oxidation-reduction-reactions in plants. Essential for symbiotic and non-symbiotic nitrogen fixation.

Molybdenum deficiency condition in plants brought about by the deficiency of molybdenum. More evident in legumes and more so in clovers. Stunted growth, yellowing of plants, drying up of leaf tissues, distortion and dying of growing tips in cauliflower, cabbage, potato and lettuce occur.

Molybdenum poisoning The poisoning of farm animals as a result of presence of molybdenum in higher amounts in animals feed.

Moment The product of a force, and the perpendicular of its line of action from the point on which it acts.

Moment of a force The measure of the tendency of the force to cause rotation about the axis. The product of the force and the perpendicular distance from the axis to its line of action. It is measured in m-kg or cm-gms. Also known as torque.

Moment of inertia The sum of the products of each particle of a moving body multiplied by the square of the distances of the particles from their neutral axis.

Momentum the momentum of a particle is defined as the product of its mass and velocity.

Momentum-transport hypothesis The hypothesis that the principle of conservation of momentum is valid in turbulent eddy transfer.

Momordica balsamina See balsam apple.

— **charantia** See bitter gourd.

— **cochinchinensis** See *bhat karela*.

— **cylindrica** See smooth sponge gourd.

— **cymbalaria** See *kadavanchi.*
— **dioica** See small bitter gourd.
— **mixta** see *bhat karela.*
— **tuberosa** See *kadavanchi.*

Monadelphous Having the filaments of the stamens united into one set.

Monandrous Having a single stamen.

Monarch A protostele in which the central xylem is circular and without arms.

Monarchic theory (intelligence) The theory believed in one great power of mind dominating over others. Spearman defined this to be an 'all pervading mental power' or to put it in the words of Ross "conscious adaptation to new situations" or "inborn all-round mental efficiency". According to this view, a person who can perform an intellectual task very well, can also perform another task equally well.

Money Any generally accepted medium of exchange, something that will be accepted by most people in exchange for goods and services and for the payment of debts.

Mongrel seed Seed for planting which is a mixture of seeds of several strains or varieties; not pure or of a single variety. Plants thus produced are called mongrel.

Moniliform Shape or structure resembling a string of beads.

Monkey bread tree Adansonia digitata L. Family Bombacaceae. Tree dried fruits of which are used as floats for fishing nets, pulp makes a cooling drink. Bark yields strong fibre useful for cordage. considered one of the most long-lived plant in the world.

Mono-Ammonium phosphate A chemical containing 12.17 per cent phosphoric acid when in its pure state. Used in fertilisers. Compound fertiliser containing 11 per cent nitrogen and 48 per cent phosphoric acid.

Mono-basic Having one base or atom of a base.

Monocarpic Bearing fruit once and then dying.

Monocarpous Having a single ovary.

Monocephalous A flower with one head or umbel.

Monochlamydous Referring to flowers having only one set of floral envelopes, that is, either a calyx or a corolla.

Monocaul With a single trunk.

Monocecious With separate male and female flowers on the same plant.

Monoclimax A climax community controlled primarily by one factor, as climate.

Monoclinic Having both stamens and pistils in the same flower.

Monocolpate pollen Pollen grains having a single furrow.

Monocotyledon Any plant of the class Liliopsida; all have a single cotyledons in corn.

Monoculture A system in which same crop is grown year after year. Also called single cropping.

Monocytic Having the nature of moncytes, large, white blood cells with an oval or indented nucleus.

Monoecious (1) Having both staminate and pistillate flowers on the same plant. (2) Having archegonia and antheridia on different branches. (3) In rusts, all stages of life cycle on single species of plant.

Moneglucoside of delphinidin dimethyl ether A complex organic chemical which is supposed to be responsible for the colour in some fruits as in grapes.

Monogynous Having only one pistil or stigma.

Monohybrid A hybrid whose parents differ in only single character.

Monolayer With leaves spread in a single horizontal layer, i.e., without overlap; hence; multilayer, with leaves distributed vertically and some degree of overlap.

Monolepta signata See Prodenia litura; colocasia leaf-eaters.

Monolith A vertical section of a soil profile removed from the soil and mounted for display or study.

Monomorphic Not changing form during development. Also monomorphous.

Monopetalous Having corolla in one piece, or petals united into tube.

Monophagous parasite A parasite restricted to only one host.

Monophyllus A plant part which is made up of a single leaf sometimes plant bearing only one leaflet.

Monoploid Possessing only one chromosome number. Similar to haploid but used to avoid implication of half of diploid number as for plants possibly polyploid.

Monopodial Stem branching in which there are lateral shoots on a primary axis.

Monopterous Pertains to single-winged seed of some plants.

Monopodium An axis established by a single indeterminate meristem, i.e, branches remain subordinate; hence monopodial growth, growth by continued activity of a single meristem.

Monopoly A firm which produces a sufficiently large proportion of the total output of a commodity to enable it to influence the price of the commodity by variations in output.

Monosaccharides Simple sugars and include pentoses (C_6-H_{10}-O_5) and hexoses (C_6-H_{12}-O_6).

Monosigales A botanical order equivalent to the Choanoflagellida in some systems of classification.

Monosiphonous Consisting of as single filament.

Monosomic Pertaining to a diploid organism which lacks one chromosome of its proper complement.

Monospermous Having or producing one seed.

Monosporangiate With microsporangia and megasporangia borne in separate flowers.

Monosporangium A sporangium producing monospores.

Monospore A simple or undivided non-motile asexual spore; produced by the diploid generation by some algae.

Monostichous Arranged in, consisting of one layer or row.

Monotheca muscatensis See gurgura.

Monotropaceae A family of dicotyledonous herbs or half shrubs in the order Ericales distinguished by a small, scarcely differentiated embryo without cotyledons, lack of chlorophyll, leaves reduced to scales, and anthers opening by longitudinal slits.

Monotrophic Refers to bees which visit only one kind of flowers.

Monoxenous parasite A parasite which requires only one host for its complete development.

Monsoon climate The type of climate which is found in regions subject to monsoons.

Monsoon fog An advection type of fog occurring along a coast where monsoon winds are blowing, when the air has a high specific humidity and there is a large difference in the temperature of adjacent land and sea.

Monsoon low A seasonal low found over a continent in the summer and over the adjacent sea in the winter.

Monstera Monstera deliciosa Liebm. (Monstera lennea Koch.). Family Araceae. A attractive ornamental vigorous scrambling climber. Develops large fruits resembling maize, cob, edible. propagated by stem cutting or by seeds. Also called *Amarphal*.

Monstera deliciosa See monstera.

— **lennea** See monstera.

Monticule (1) Small hill. (2) Small mound caused by volcanic eruption.

Montmorillonite A specific aluminous, members of the specific group. A group of clay minerals with the chemical formula $(OH)_4$ I-Al (FeMg) Si_8O_{20} occurring as minute equidimensional flakes, and possessing swelling or expansive properties in the presence of water, is the chief constituent of bentonite and Fuller's earth.

Montmorillonite group Clay minerals having a 2 : 1 expanding crystal lattice. Isomorphous substitution gives the various types and causes a net permanent charge balanced by cations in such a manner that water may move between the sheets, giving reversible cation exchange and very plastic properties. Synonymous with smectite.

Moonflower Ipomoea alba L. (Ipomoea bonanox L.). An ornamental herb of Family Convolvulaceae, the leaves and fleshy calyx of which are eaten as vegetable.

Moor A constantly wet, poorly drained, elevated land covered with grasses and low shrubs, often having peat deposits.

Mor (or Raw humus) A non-zoogenous forest humus form distinguished by a matted F layer and a holorganic H layer with a sharp delineation from the A horizon. It is generally acid, having high organic carbon content

(52% or more) and a high C : N ratio (25:35, sometimes higher), and chemical and biological properties.

Moraceae A family of dicotyledonous woody plants in the order Urticales characterised by two styles or style branches, anthers inflexed in the bud, and secretion of a milky juice.

Moraine An accumulation of earth, generally with stones, carried and finally deposited by a glacier, several kinds of moraines are distinguished, such as ground moraine and end moraine.

Morbidity (1) Being diseased. (2) The extent to which disease exist.

Mordant Chemical added to a dye to make it stain more intensely.

Moringa oleifera See drumstick tree.

— **pterygosperma** See drumstick tree.

Morphia An alkaloid narcotic principle of opium; largely used for alleviating pain. Morphine.

Morphogenesis Study of processes of origin, differentiations, and development of plant tissues and organs.

Morphological inversion point See inversion point morphological.

Morphological inversion surface See inversion surface morphological.

Morphology (1) A branch of biology that deals with structure and form of an organism at any stage of its life history. (2) Refers to pedology or soil science, the study of soil horizons and their arrangement in the profile. (3) Refers to form or shape of physical features of land.

Morphology, soil (1) Physical constitution, particularly the structural properties, of a soil profile as exhibited by the kinds, thickness, and arrangement of the horizons in the profile, and by the texture, structure, consistence and porosity of each horizon. (2) The structural characteristics of the soil or any of its parts.

Mortality Death. Destruction of plants and animals due to pest and disease attack, generally on large scale. (2) Death rate or total deaths in a given period.

Mortgage The conveyance of property by a debtor to a creditor as security for a debt, with a condition that the property shall be reconveyed on payment of the debt.

Morus acidosa See common mulberry.

— **alba** See white mulberry.

— **australis** See common mulberry.

— **indica** See common mulberry.

— **nigra** See black mulberry.

— **rubra** See red mulberry.

— **serrata** See *kimu*.

Mosaic An assemblage of vertical serial photographs to represent a composite picture of the land features.

Mosaic disease A diseased condition in plants, often caused by viruses, in which different parts of the leaf show unequal development of chlorophyll, resulting in a mottled appearance.

Moss Any plant of the class Cryatae, occurring in nearly all damp habitats, except the ocean. (2) Wet spongy soil.

Moss peat Species of moss found in marshes with the structure of leaves suitable for absorbing and retaining large amounts of moisture. Useful to nursery man and florist for various purposes. Also called sphagum.

Mossy Term applied to jute fibre which shows moss-like material consisting of fine adventitious roots and extraneous vegetable matter.

Most economic section Cross section which, with a given surface slope and given amount of excavation, has the maximum capacity.

Most efficient section Cross-section which, with a given surface slope and area, has the maximum capacity.

Motes Immature seed and particles of trash in cotton ginning.

Moth A small nocturnal insects belonging to the order Lepidoptera. The adults have wings covered with scales. Distinguished from butterflies in not having clubbed antenne. Many destructive to crops and stored products in caterpillar stage.

Moth bean Phaseolus aconitifolius Jacq. (Phaseolus trilobus Wall, Dolichos dissectus Lamk.) Family Papilionaceae. A trailing annual herb with yellow flowers cultivated for edible seeds and forage. Most drought resistant of *kharif* pulses. Useful against erosion. Green seed and black seed crops.

Mother (1) Female parent. (2) Mucilaginous substance produced in vinegar during fermentation by mould-fungus.

Mother block A group of plants known to be disease free or true to type from which cuttings are made for commercial propagation.

Mother bulb A bulb which has several points of division and from which they are separated for propagation.

Mother cane A cane or branch used in propagation in continuous or modified continuous layerage, as in grape.

Mother plant Any plant whose parts are used for propagation.

Mother root Root bedded for propagation of a crop as in sweet potato.

Mother tree The original tree from which varieties are developed.

Motivation It is a process of creating a strong desire for appealing to the fundamental wants of the people, so that they are influenced to change.

Motor (1) A machine for transforming electrical energy into mechanical energy. (2) An internal combustion engine, as the gasoline engine of an automobile.

Motor analyser A group of instruments, usually assembled in a single case or on a single panel, by which such items are cylinder compression, air-fuel ratio, and ignition timing can be checked.

Mottled zone A layer that is marked with spots or blotches of different colour or shades of colour. The pattern of mottling and the size, abundance, and colour contrast of the mottles may vary markedly and should be specified in the soil description.

Mottle-leaf (1) A virus disease characterised by chlorotic mottling and wrinkling of leaves. (2) A zinc-deficiency disease characterised by partial chlorosis of the leaves and stunting of the plant.

Mottles Spots or blotches of different colour or shades of colour interspersed with the dominant colour.

Mottling Formation or presence of mottles in the soil.

Mould Minute living organism visible to naked eye. Recognised in later stage of development as white, grey, brown, soft, fluffy patches growing upon foods kept in damp places and at moderate temperatures. Colour depend upon mould type. Some useful grow readily in presence of acids. Easily destroyed by heat (boiling).

Mouldboard The wing like part of a mouldboard plough. Shape assists in turning the soil.

Mould board plough See moldboard plough.

Mounted type Implements A mounted implement is one which is attached to the tractor, such that it can be controlled directly by the tractor steering unit. The implement is carried fully by the tractor when out of work.

Mound building termite A termite group which makes mounds of different shapes and sizes, depending upon species, with small chambers connected by numerous vertical and transverse galleries.

Mould layering Method used in the propagation of apple root-stock. Also called stooling.

Mountain ebony Bauhinia variegata. L. (B. candida Roxb.) Family Caesalpiniaceae. Medium sized tree whose wood used for agricultural implements and bark for tanning. Leaves and flower buds eaten as vegetables. Also called *Kachnar.*

Mountain papaya Carica candamarcensis Hook. f. Family Caricaceae. A small tree whose fruits are edible.

Mouse trap, or Poli scoop An adjustable metal sheet pretruding into the water stream in the aluminium flume, which intercepts and deflects water into the furrow.

Movable dam A water barrier that may be opened wholly or partially. Movable parts are gates, stop logs etc. These are used to control the opening area of the dam.

Moving average method fit The averaging at the central year of a series of overlapping means to smooth out, or damp down, undue high and low short-term values. Hence the 'moving average fit'.

Mow (1) To cut with a mower, sickle, or scythe. (2) storage space for hay.

Mower The machine for mowing hay, weeds, grass etc.

Mowra fat Madhuca longifolia (L.) Macb. (Bassia longifolia. L.). Family Sapotaceae. Tree whose flowers are used in the manufacture of alcohol, oil from seeds used for cooking; bark used as tan. Also called *Mohua*.

Mucilage A substance of varying composition, hard when dry, swelling and slimy when moist, produced in cell walls of certain plants. A Viscid solution or paste consisting of mucilaginous principles of vegetable substances.

Mucin compound Any of a group of muco-proteins found in various secretions and tissues of man and animals, such as saliva, lining of the 'stomach', and skin.

Muck soil An organic soil consisting of highly decomposed materials, Mucky peat and peaty muck are terms used to describe increasing stages of decomposition between peat and muck.

Muck Decomposed organic material, usually 20 to 50 per cent organic matter mixed with mineral matter, dark in colour, and accumulated under poor drainage conditions.

Mucronate Tipped with an abrupt point.

Mucuna cochinchinensis See Lyon bean.

— **deeringiana** See florida velvet bean.

— **gigantea** DC. A herb of Family Papilionaceae. Whose Seeds are edible.

— **nivea** See Lyon bean.

— **prurita** See horse-eye bean.

— **pruriens** See horse-eye bean.

Mud (1) Freshly deposited alluvium of flood waters. (2) Fine textured soil so saturated with water as to be semi-fluid or sticky or plastic in condition.

Mudflow slide Movement of large masses of "quick clay".

Mulberry Temperate tree, grows well in worm regions too. Hardy, belongs to Family Moraceae whose fruit is consumed fresh or as preserved. Used for producing silk worm.

Mulberry silkworm Bombyx Mori. Silk worm feeding on mulberry.

Mulch (1) Any material such as straws, dust, leaves, plastic film, or loose soil that is spread on the surface of the soil to protect the soil and the plant roots from the effects of raindrops, soil crusting, freezing, and evaporation. (2) To modify soil so as to serve as mulch.

Mulch farming A system of farming in which the organic residues are not ploughed into or mixed with the soil, but are left on the surface as a mulch.

Mulch to To apply mulch to the surface of the soil.

Mulga acacia Acacia aneura F. Muell. Family Mimosaceae. A tree whose leaves are used as fodder. Also called mulga wattle.

Mulga wattle See mulga acacia.

Mull A zoogenous forest humus form consisting of an intimate mixture of well-humified organic matter and mineral soil that makes a gradual transition to the horizon underneath. It is distinguished by its crumb or granular structure, and because of the activity of the burrowing microfauna (most earthworms), partly decomposed organic debris does not accumulate as a distinct layer (F layer) as in mor and moder. The organic matter content is 5-20 per cent and the C:N ratio is 10-15. Various sub-groups can be distinguished by the morphology and chemical characteristics. Ah horizon.

Multiflorus Bearing more than three flowers.

Multilateral Having many sides.

Multilayer See monolayer.

Multimodal distribution A frequency distribution with more than one model value.

Multinomial distribution The discrete distribution associated with events which can have more than two outcomes; it is a generalisation of the binomial distribution.

Multinucleate Several nuclei in same cell.

Multi-parasitism A single host suffering from simultaneous attack of more than one primary parasites.

Multiparous (1) Bearing several individuals at a single birth. (2) Having had several offsprings from more than one progeny.

Multiple annual ring An annual ring of wood in trees consisting of two or more false rings.

Multiple bar graph It compares two or more bars with a numbered scale to show the desired information. The bars can either be vertical or horizontal depending upon the size of the graph paper and contents of the data to be graphed.

Multiple cropping A system of growing several food crops on the same field in one year.

Multiple cross A cross between two parents of different origin.

Multiple factor hypothesis The type of inheritance in which a character is dependent on many different genes or factors.

Multiple factor inheritance Inheritance in which a character is determined by many genes with similar effects.

Multiple factors Two or more factors which have similar effects on a particular character.

Multiple fruit A cluster of matured ovaries produced by separate flowers; a compound fruit.

Multiple plough A plough with two or more bottoms. Also called gang plough.

Multiplication plot A field for growing a crop to produce seed in larger quantities from a comparatively small lot of selected seeds.

Multivoltine Type of silk worm which gives more than two crops a year.

Mummy Dried, shriveled fruit, result of disease.

Munsell color system A colour designation system specifying the relative degree of the three simple variable of colour; hue, value, and chroma. For example; 10 YR 6/4 is the colour of a soil having a hue of 10 YR, value of 6, and chroma of 4. These notations can be translated into several different systems of colour names.

Munj grass Erianthus munja (Roxb.) Jesw. (Saccharum munja Roxb.) Family Gramineae. Grass whose stem fibre used for making baskets, mats and cordage; leaves used for thatching, also source of paper. Grass used as soil binder. Also called *Munja*.

Muntingia calabura See chinese cherry.

Muntz metal An alloy of copper and zinc with 60 : 40 composition; it contains enough brass, which makes it harder, stronger and less ductile.

Muriform Having cross and longitudinal septa.

Muriate of ammonia Ammonium chloride.

Muriate of potash (KCl) A potash salt containing 48 to 62 per cent potash (K_2O). It is readily available to plants. Used as fertiliser.

Murraya koenigii See curry leaf tree.

Murrum Gravelly soil.

Musa paradisiaca See banana.

— **sapientum** See banana.

— **textilis** See manila hemp.

Muscology A study of mosses.

Muscovado Unrefined sugar obtained from sugarcane juice by evaporation and draining of molasses.

Muscovite Mica.

Mushroom (1) General term for the fruiting body of fungi or basidiomycites class. (2) The fleshy, edible fruiting bodies of fungi. (3) The cultivated or wild edible mushrooms.

Muskeg A peat bog or tussock meadow, with variably woody vegetation.

Muskmallow Abelmoschus moschatus medic. (Hibiscus abelmoschus L.). Family Malvaceae. A tropical annual or biennial shrub whose seed give oil used as exalting agent in fine perfumes. Source of fibre from which cloth is made.

Muskmelon Cucumis melo. The edible, fleshy, globular to long tapered fruit of a trailing annual plant of order Violales; surface is uniform to broadly matured to wrinkled, and smooth to heavily netted, and flesh is pale green to orange; varieties include cantaloupe, Honey dew, Casaba, and Persian melons.

Mustard Common name for some of the herbs of the genus Brassica Family Cruciferae. Grown for seed (mustard) see Indian mustard.

Mustard aphid Lipahis erysimi Davis. Pest of mustard which pierces leaves, flowers or capsules and sucks sap, curls up leaves and makes plants wither and die, secretes honeydew which helps mouldy growth. In severe attack crop turns black. Also attacks raddish, spinach, turnip, knol-khol, cabbage, cauliflower, rai, toria and other cruciferous crops. Locally called *Chepa*.

Mustard blight Disease caused by Alternaria brassicae (Berk.) Sacc.

Characterised by appearance of small circular spots on leaves, stem and pods, with plant looking pale and withered. Also attack cruciferous vegetables such as cauliflower and Turnip.

Mustard downy mildew Disease of mustard and cruciferous vegetables. Caused by Peronospora parasitica (Fr.) Tul. It appears along mustard white rust. Characterised by appearance of scattered, irregular patches with greyish white downy growth on leaves and sometimes on inflorescence. Withering of leaves occur when attack is severe.

Mustard saw fly Athalia proxima Klug. Pest whose larva feeds on leaves, cutting irregular holes. Also attacks raddish, cabbage and turnip.

Mustard tree Salvadora persica L. A large shrub of Family Salvadoraceae; leaves of which are eaten as vegetable.

Mustard white rust Disease caused by Cystopus candidus (Pars. Exchev.) It occurs along with downy mildew. Characterised by appearance of white, chalky, erupted pustules on the green parts of plants, swelling and distortions on inflorescence and swollen and fleshy floral organs, which turn greenish or violet.

Mustiness An offensive, punget odour caused by moulds.

Mutagen A substance or treatment which under suitable conditions may cause mutation.

Mutant An individual which has suddenly acquired a heritable variation not present in the parent form.

Mutation A sudden new variation that is inherited. Variously used to include individual gene changes and chromosomal changes.

Mycelia sterilia Fungi Imperfecti where spores, except for chlamydospores, are not present.

Mycelium A mass of threadlike filaments, branched or composing a network, that constitutes the vegetative structure of a fungus.

Mycobacterium Slender, aerobic, rod-shaped micro-organism.

Mycoderma The membrane or mother formed on fermenting wine, vinegar etc. The genus of fermentation fungi found in yeast which produces this membrane.

Mycology The branch of botany that deals with the study of fungi.

Mycoplasma The single genus of the Family Mycoplasmataceae.

Mycorrhiza The association, usually symbiotic, of fungi with the roots of seed plants. See also ectotrophic mycorrhiza and endotrophic mycorrhiza.

Mycosis Any infection caused in animals or plants by fungi.

Mycosphaerella arachidicola See groundnut irregular leaf spot.

— **gossypina** See cotton leaf spot.

— **musicola** See banana leaf spot.

— **pinodes** See gram blight.

Myldocerus spp See cotton grey weevil.

Myriangiules An order of parasitic fungi of the class Ascomycetes which produce asci at various levels in uniascal locules within stromata.

Myriapod Animal with many legs as centipedes and millipedes.

Myricaceae The single family of the plant order Myricales.

Myrrica nagi See bay-berry.

— **rubra** See bay-berry.

Myristica beddomeii A wild species of nutmeg; used as rootstock for approach grafting of nutmeg.

— **fragrans** See nutmeg.

— **malabarica** See Bombay mace.

Myricales An order of dicotyledonous plants, in the subclass Hamamelidae, marked by its simple, resionous dotted, aromatic leaves, and a unilocular ovary with two styles and a single ovule.

Myrmecophile An organism usually a beetle, that habitually inhabits the nest of ants.

Myrmecophyte A plant that houses and benefits from the habitation of ants.

Myrobalan Fruits of forest trees species used in the tanning industries.

Myrobalan bellerica See belleric myrobalan.

Myrobalanus bellerica See belleric myrobalan.

Myrrh An aromatic gum resin produced by trees or shrubs of the genus Commifera, Family Burseraceae, used in the manufacture of perfumes.

Myrsinaceae A family of mostly woody dicotyledonous plants in the order Primulales; characterised by flowers without staminodes, a schizogenous secretary system, and gland dotted leaves.

Myrtaceae A family of dicotyledonous plants in the order Myrtales, characterised by an inferior ovary, numerous stamens, anthers usually opening by slits, and fruit in the form of a berry, drupe, or capsule.

Myrtales An order of dicotyledonous plants in the subclass Rosidae, characterised by opposite simple, entire leaves and perigynous to epigynous flowers with a compound pistil.

Myrtus cumini See Java plum.

Myxamoeba A uninucleate, amoeboid, reproductive cell in the Myxomycetes.

Myxomycophyta An order of microorganisms, equivalent to the Mycetozoia of zoological classification.

Myxophyceae An equivalent name for the Cyanophyceae.

Myzus persicae See linseed aphid, potato aphid.

N

Nailhead spot A fungus rot of tomato caused by Alternaria tomato and marked by small brown to black sunken spots on the fruit.

Najadaceal A family of nanocotyledonous, submerged aquatic plants in the order Najadales distinguished by branching stems and opposite or whorled leaves.

Naked Unsheathed; plain, undisguised; devoid of; treeless, leafless; barren; without pericarp, leaves, hairs, scales, shell etc.

Naked bud (1) A resting meristem without obvious morphologically specialised protective organs. (2) A bud covered only by rudimentary foliage leaves.

Naked flower (1) A flower without petals or sepals. (2) A flower without a perianth.

Naked seed Achene.

Nanism Dwarf growth.

Nannorrhops ritchieana See mazari palm.

Naphthalene (C_{10} H_8) A chemical obtained from coaltar.

Naphthalene acetic acid An organic acid. Commonly employed in mixtures of hormones used to stimulate root growth.

Napiform Turnip-shaped, referring to roots.

Nappe A sheet or curtain of water overflowing a weir, fall, etc. The nappe has an upper and a lower surface.

Napped Term used for the matting or clumping of cotton fibres.

Napunyah Eucalyptus ochrophloia. Family Myrtaceae. Has drooping, smooth branches and yellowish, hard, roughbark, a shade and timber tree of Eucalyptus species introduced from Australia.

Nappy Coarse and sticky lint cotton.

Narceine An alkaloid obtained form opium; sometimes used instead of morphia.

Narcosis State of unconsciousness.

Narcotic A drug or substance inducing drowsiness, sleep or even producing unconsciousness.

Narrow-based ridge terrace Type of contour terrace also known as bund. A narrow based embankment made across slope on a level or graded longitudinally, if water removal is required. Vertical spacing 90 to 180 cm and used on lands up to 10 per cent slope. A soil conservation measure.

Narrow base terrace A terrace similar to a broad base terrace in all respects excepting the width of the ridge and channel.

Narthex asafetida Ferula narthex Boiss. Herb of Family Umbelliferae; produces an oleo-gum-resin, which is used for flavouring food products.

Nasturtium fontanum See water-cress.

— **officinale** See water-cress.

Natal grass Pennisetum pedicellatum Trin. Family Gramineae. Grass used for fodder.

Natal plum (1) Carissa arduina Lamk. (Carissa bispinosa Desf., Arduina bispinosa L.) A small thorny bush of Family Apocynaceae which produces edible fruit. A species of *Caronda*. (2) Carrisa grandiflora A. DC. (Arduina gradiflora E. Mey.). A large spiny shrub of Family Apocynaceae. Cultivated for its edible fruits.

Natant Floating on water-suriace.

National income A measure of the money value of the goods and services becoming available to a nation from economic activity during a prescribed period.

Native grass Grass species indigenous to an area; not introduced from another environment of area.

Native pasture One composed of native plants or naturalised exotic plants.

Natural cross A hybridisation or interbreeding without man's assistance and occurring naturally.

Natural drainage As applied to a given area, refers to the natural condition of that area for disposing of surface water from it.

Natural enemy Any natural organism available or introduced in an area which preys or feeds upon another.

Natural habitat The place which a plant or animal naturally occupies; natural environment.

Natural immunity Immunity from diseases resulting from qualities inherent in a plant or animal.

Natural indigo Blue dye extracted form various species of indigofera.

Natural reproduction Renewal of plants by self sown seeds, sprouts or rhizomes etc. In case of forest it is called natural regeneration.

Natural resins Include all the naturally occurring resins which are used in varnish manufacture.

Natural resources Resources natural to an area. These are soil, water and mineral deposits.

Natural runoff From an area is that part of the runoff from it which it will discharge without artificial works intended to collect the waters which fall upon it.

Natural selection The automatic selection which takes place under natural conditions by reasons of death or partial inhibition of individuals less fitted to thrive under the conditions obtaining.

Natural systems The drains which follow the natural depression in the land, and diverge like the frame of a tree. Also called as random systems.

Natural ventilation Process of supplying or removing air using the natural forces of wind and temperature difference to or from any space, such as interstitial space between products.

Naturalised disease A disease caused by organism of some foreign origin but has established in the new habitat.

Naturalised plant Plant introduced from another environment; has become established in and more or less adapted to a given region by being grown for many generations.

Navel Umbilicus. Place of attachment of umbilical card to body of embroy. Seen as depression in front of belly.

Navicular Boat-shaped.

Neap tide High water is lower than normal; it occurs usually when the moon is in the first and third quarters. At these times the tidal forces of the sun and moon are opposed.

Neat cement A pure cement mortar, i.e., not out down by the admixture of sand.

Neck The tapering portion of an archegonium. A sperm enters an archegonium through a canal extending lengthwise through the neck.

Neck canal cell A cell or cells within the archegonial neck.

Neckeraceal A family of mosses in the order Isobryales distinguished by undulate leaves.

Neck rot A fungus disease of onions caused by species of Botrytis and characterised by rotting of the leaves first above the bulb.

Necrosis Death of organs of a plant, either as blight or death of tissues in localised areas, usually inside fruits and stems, or die back or blast or anthracnose or limited lesions. Characteristic symptoms in certain plant diseases.

Necrotic centre The degenerating and dying cells in the centre of a lesion.

Necrotic ring spot A virus leaf spot of cherries marked by small, dark water soaked rings which may drop out, giving the leaf a tattered appearance.

Nectar (1) Sweet substance secreted by nectaries or special glands or flowers or certain leaves. (2) Substance containing spores and attracting insects produced by certain fungi.

Nectarine A smooth, fuzzless fruit originating as a spontaneous somatic mutation of the peach, Prunus perxica and P. perssica var. nectarina.

Nectary A secretary organ or surface modification of a floral organ in many flowers, occurring on the receptacle, in and around varies, on stamens, or on the perianth; secretes nectar.

Nectria cinnabarina See tea die back.

Neem tree Azadirachta indica A. Juss. (Melia azadirachta L.) Family Meliaceae. Useful tree with hard, close grained wood; bark, leaves, flowers, seeds have medicinal value, oil cake from seed used as manure. Also called *Nim*.

Negative rain Rain which exhibits a net negative electric charge.

Negative staining Staining the back ground, not the organism.

Negative tropism Tendency to move away from the source of stimulus. As an example roots of plants moving away from light source.

Nelumbium speciosum See East Indian Lotus.

Nelumbo nucifera See East Indian Lotus.

Nemagon 1, 2-dibromo-3-chloropropane. Soil fumigant and nematocide. Can be applied around roots without any harm. Available as granules and emulsion. Not to be used in potato, tobacco, onion and few others.

Nemaral Pertaining to or inhabiting a grove or wooded area.

Nemas Nematode.

Nemataceal A Family of masses in the order Hooksiales, distinguished by having perichaetial leaves only.

Nematocide Chemical or physical agent killing nematodes.

Nematode root galls See root knot.

Nematodonteal A group of masses included in the sub class Eubya in which there may be faint transverse bars on the peristome teeth.

Nematoid Thread-like; filamentous.

Nematology A branch of zoology dealing with the study of nematodes and their relationship with plants and animals.

Neocarpy Production of fruit by an otherwise immature plant.

Neoluvium The soil or residue of weathering which results form the eluviation of recent deposition, such as alluvium and glacial drift.

Neoformation Formation of an organ or part of a shoot system immediately prior to its expansion.

Neopalynology A field of palynology concerned with extinct micro-organisms and dissociated microscopic parts of megaorganisms.

Neoplasm Growth of a tumour. Any abnormal growth.

Neovossia indica See Karnal bunt.

Nepal ebony persimmon Diospyros tomentosa Roxb. Family Ebenaceae. Fruits edible and leaves used for wrapping bidis.

Nephantis serinopa See coconut black headed caterpillar.

Nephelium lappaceum See rambuttan.

— **litchi** See litchi.

— **longana** See longan.

— **mutabile** See pulasan.

Nephology The study of the clouds.

Nephotettix apicalis See paddy green jassids.

— **bipunctatus** See paddy green jassids.

Nephritis Inflammation of kidneys.

Neps The small tangles of fibres visible as dots or specks in a thin web of fibres held to the light or against a dark background.

Neptunia oleracea See Lajalu.

— **prostrata** See Lajalu.

Neptunian Maintaining the aqueous origin of certain rocks; produced by the action of water.

Nervate Referring to leaves having ribs.

Nervation The disposition of nerves or veins in a leaf or membranous wing.

Nerve vein of insect wing; a vain of leaf.

Nervine Term designating medicine or drug relieving nerve disorders.

Nervure One of the tubes framing an insects wing; principal vein of the leaf.

Net blotch A fungus disease of barley caused by Helminthasparium teres and marked by spotting of the foliage.

Net cash income The difference between total receipts and total expenses over a year.

Net duty of water Duty of water reckoned or measured at the farm. Also called as farm duty.

Net energy Difference between metabolizable energy and heat increment; includes the amount of energy used either for maintenance only or for maintenance plus production.

Net increment The addition of crop or tree growth in a field or forest. It represents the increase in grain yield or usable timber.

Net radiation The difference between upward and downward radiation flux, or, alternatively, the difference between the net short wave and the net long wave radiation.

Net sown area Under revenue classification total area sown in an area. If an area is cropped twice it is added in the total area calculation.

Net venation A scheme of vein arrangement in leaves, in which the veins branch frequently, forming a network.

Net weight Weight of the actual product. The total weight minus the weight of the container or covering.

Nettle tree See hackberry.

Neural Of the nerves, of the central nervous system.

Neural response For every emotional response, there is a neural response. The brain gets disturbed. All parts of the nervous system get involved in the emotions.

Neuroptera Order of insects having four naked membranous transparent wings with reticulate neuration, to which aphidlions and antlions belong.

Neurotic Term describing drug affecting nervous system.

Neurotoxicity The state or condition of being poisonous to the brain or nerves.

Neuter Asexual; having neither stamen nor pistil; sexually undeveloped female insect; castrated animal.

Neutral day plant See day-neutral plant.

Neutraliser An alkali used to neutralise acidity.

Neutral stability The state of an unsaturated column of air in the atmosphere when its environmental lapse rate of temperature is equal to the dry adiabatic lapse rate.

Neutral soil A soil having pH in the range of 6.6 to 7.3. In this range soil is called neither alkali nor acidic.

Neutral stress The stress transmitted through the interstitial fluid of a soil or rock mass. Also known as neutral pressure; pore pressure; pore-water pressure.

Newton Force which, when applied to a body having a mass of one metre per sec per sec. It is equal to 10^{-5} dynes.

Newtonian flow Flow system in which the fluid performs as a Newtonian fluid, that is, shear stress is proportional to shear rate.

New wood The current year growth in plants.

New Zealand flax Phormium tenax Forst. Tree of Family Liliaceae whose leaves yield fibre of high lustre, soft and flexible. Also called New Zealand hemp.

New Zealand spinach Tetragonia tetragonoides Kuntze. (T. expanse Murr.). Family Aizoaceae. A herb whose leaves are used as vegetable. Drought resistant.

Niacin See nicotinic acid.

Nibs Fragments of crushed cocoa beans.

Nibble To take small bites at.

Niche The energetic position of a species in a community.

Nick The two parents for producing hybrid seed are said to nick when they produce high yields of seed of a highly productive and desirable hybrid.

Nickel Ni. A metal which is found in traces in soils. It is deleterious to plant growth.

Nicker bean Entada phaseoloides (L.) Merr. (Entada scandens Benth.), Lens phaseoloides L.) Family Mimosaceae. A huge climbing shrub whose seeds are edible.

Nicotiana rustica See tobacco.

— **tabacum** See tobacco.

Nicotine ($C_{10}H_{14}N_2$) Alkaloid found in tobacco leaves and stem. A colourless liquid in pure form. The per cent nicotine is in the order of bidi tobacco > snuff tobacco > cigarette tobacco > cigar tobacco > *hookah*. Nerve stimulant.

Nicotine sulphate An insecticide which acts as a contact insecticide with marked fumigant action. Used against aphids.

Nicotinic acid ($C_5H_5NO_2$) Member of the vitamin B complex. Also called niacin.

Nidus (1) Place in which insect deposits eggs. (2) Place in which spores or seeds develop. (3) Place of origin or development of disease.

Nigella sativa See black cumin.

Niger Guizotia abyssinica (L.f.) Cass. Family Compositae. A oil seed crop whose seeds contain 38 to 50 per cent of pale, yellow, odourless, edible oil with a pleasant nutty taste. Oil used for many purposes, and oil seed cake used for cattle feed.

Niger seed See niger.

Night-soil Human excrement or faecal matter and human urine. Can be used as manure for cropland.

Nigrescence Turning black.

Nimbus Rain cloud.

Nip (1) A knot in cotton fibre. (2) A sudden set back in the plant or animal due to severe cold or frost.

Nipa fruticans See water coconut.

— **palm** See water coconut.

Nitrate poisoning Condition sometimes resulting when ruminants ingest nitrates (NO_3) that rumen bacteria convert to nitrite (NO_2); the nitrites compete with oxygen, typing up the oxygen-carrying mechanism in the blood and causing the animal to suffocate.

Nitrate of potash (KNO_3) A fertiliser. Contains 46.6 per cent of potash.

Nitrate of soda ($NaNO_3$) One of the chief sources of nitrogen for commercial fertilisers. Pure form contains 16.48 per cent of nitrogen.

Nitrate reduction The biological reduction of nitrate to nitrite form.

Nitrates Compounds of nitrogen and oxygen with some metallic elements such as sodium, potassium, calcium. Principal source of nitrogen for commercial fertilisers.

Nitre (1) Nitrate of potash (salt peter). (2) Nitrate of sodium or Chile Salt peter.

Nitrification Biological conversion of NH_4 into nitrite and then to nitrate which is available to plants.

Nitrifiers See nitrifying bacteria.

Nitrifying bacteria Bacteria in the soil which converts ammonia into nitrites and are called Nitrosomonas. The bacteria which converts nitrites into nitrate are called nitrobacter. These are found in top 25-30 cm of the soil surface and perform best in the temperature range of 25° to 38°. Cease activity at pH below 5.0.

Nitrobacter See nitrifying bacteria. Groups of bacteria which converts nitrites into nitrates.

Nitrobacteria Minute, rod-like bacilli which convert nitrogen into soluble nitrates in soils.

Nitrogen (1) N. A major nutrient of plant which is assimilated by plants either in NH^+_4 or NO^-_3 ionic forms. Imparts, healthy dark green colour to the leaves and promotes rapid growth of stem and leaves. Protein content of fodder and food products is increased. If not available to plants in appropriate quantities, undesirable symptoms appear and the situation is termed as nitrogen deficiency. Symptoms of deficiency appear first in older leaves which become pale yellow and result in premature defoliation. Growth of tops and roots is restricted. (2) A colourless, tasteless, odourless gas constituting about 78 per cent of the atmospheric air by volume.

Nitrogen assimilation The incorporation of nitrogen into organic cell substances by living organisms.

Nitrogen carrier Any organic or inorganic nitrogen containing fertiliser.

Nitrogen cycle The circular manner in which nitrogen is first taken up by plants and then returned to soil. The simple compound like ammonia and nitrate are assimilated by plants and converted to protein or other organic introgenous compounds. The protein is taken up from plant tissues by animals as food and returned to soil as manure. Thus nitrogen taken from the soil is returned to the soil.

Nitrogen excess Injury to plants due to excess of nitrogen. Quality of crop is impaired, maturity is delayed and susceptibility to diseases is increased. It causes lodging of cereal crop as stem lengthening occurs.

Nitrogen fixation The conversion of elemental nitrogen to organic combinations or to forms readily utilisable in biological processes. Symbiotic fixation is the term used when brought about in root nodules of leguminous plants and non-symbiotic when brought about by free living micro-organisms.

Nitrogen fixing bacteria Species of the genus Rhizobium, which live symbiotically in root nodules of the leguminous plant. Species of the genus Azotobacter live non-symbiotically and act independently.

Nitrogen fixing plant Any plant which can extract the nitrogen of the air and assimilate and fix it for use by plants with the help of bacteria living in the root nodules.

Nitrogen free extract Soluble and digestible carbohydrate; estimated in a food stuff by subtracting the sum of crude fibre, fat, crude protein, ash and water from the whole.

Nitrogenous ratio Ratio obtained by multiplying the fat content of a feed stuff by 2.25 to convert it into terms of starch, and balancing the sum of proteins nutrients against non-protein. The ratio gives the energy value of a ratio against its body building power. Also called nutritive ratio or albuminoid ratio.

Nitrojection The application of agricultural ammonia by injection into the soil.

Nitrophilous Refers to plants which are tolerant to alkali.

Nitrosomonas A specialised group of bacteria in soils which oxidise ammonia into nitrites. See nitrifying bacteria.

Nobbe trier It is a pointed tube, long, enough to reach the centre of the bag, with an oval hole near the pointed end. It is made in different dimensions to suit various kinds of seeds.

Noctiflorous Flowering at night.

Noctuida Family name of one group of cut worms.

Nocturnal Of the or occurring at night.

Node (1) The joint on a stem from which roots, leaves, and branches have their origin. (2) A lymph gland. (3) Hard tumour.

Nodical Central point in system.

Nodosity Knottiness; knot, protuberance.

Nodostoma subcostatum See banana beetle.

Nodulation The production of nodules on roots of legumes by specialised legume bacteria.

Nodule (1) A small knob-like structure as root nodule, lymphatic nodule. (2) Small knotty tumour, ganglion or any roundish mass of abnormal tissue.

Nodule bacteria See rhizobia.

Nodule soil A rounded unit within the soil matrix that differs from the surrounding material because of the concentration of some constituent or a change in fabric.

Nodus A knotty point.

Nomad Member of the tribe roaming from place to place in search of grazing for their livestock or food for themselves.

Nomad biological A pioneer tree on a disturbed site (synonyms : scar species, weed tree).

Nomad heliophilous Frequently occurring but short-living tree species; favourable regions of high light intensity in a forest community and generally without reiteration.

Nomad sciaphilous Frequently occurring somewhat longer-living tree species tolerant of low light intensity in a forest community, and generally without reiteration.

Nomenclature System of naming plants, animals, organ etc.

Nomograph A diagram for the graphical solution of problem that involve formulas in two or more variables.

Nonarticulated growth Continuous growth resulting in a nonarticulated shoot.

Non-acid forming fertiliser The fertiliser which when applied to soils do not increase the residual acidity.

Non-beneficial consumptive use The water consumed by native vegetation, evaporated from bare and idle land surfaces and from water surfaces.

Non-contributing area An area with closed drainage.

Non-capillary porosity See air porosity.

Non-drying oil An oil which does not change its consistency when exposed to air. This is so with oils having iodine value less than 90.

Non-lignified Refers to non-woody type of crude fibre in feeds; contains little lignin.

Non-linear correlation The correlation between variates where the regression is not linear.

Non-motile Unable of locomotion.

Non-orthogonal Data are said to be non-othogonal, if they lead to estimates of various effects which are not independent of one another.

Non-palatable Designating materials not relished.

Non-parametric A method, for example, of testing hypothesis or setting up a confidence interval, which does not depend upon the form of the underlying distribution.

Non-pathogenic Not capable of producing disease.

Non-persistent virus Virus whose vectors lose their ability to transmit a disease within a day of leaving the infective plant.

Non-protein nitrogen Broad class of nitrogenous substances not comprising protein such as glutamine, glutamic acid, asparagin aspartic acid, and gamma-amino butyric acid.

Non-recoverable deep percolation losses A form of outlet from a valley and is independent of the factors influencing transpiration and evaporation.

Non-recurrent parent Used in backcrosses to refer to the original parent not used in backcross generations.

Non-specific disease A disease not related to any infection, parasitism, nutritional or genetic factors.

Non-toxic Not harmful to plants or animals.

Non-verbal (communication) Non-verbal communications includes the gestures, facial expressions, movements of arms, etc., which make our communications more effective.

Norm Standard, custom, code, law or institution.

Normal (1) A mean of average value established from a series of observations for purpose of comparison of some meteorological or hydrological events. (2) Conforming to standards.

Normal depth The depth of water in an open conduit that corresponds to uniform velocity for a given flow.

Normal deviate The value of a deviate of the normal distribution.

Normal erosion The slow and steady erosion of soil normally taking place in any landscape. Also called geologic erosion.

Normal forest One in which growing stock is so distributed by size and age classes as to provide a sustained yield of nearly equal annual volume through growth.

Normal growing stock The maximum volume which any given site is capable of maintaining in relation to economic conditions and the desires of the operator.

Normal increment The best average increment which is attainable by a given species on a given site as represented in normal yield tables.

Normal moisture capacity See field capacity.

Normal seedlings The seedlings which show the capacity for continued development into normal plants when grown in good quality soil, and under favourable conditions of water supply, temperature and light.

Norm of irrigation (1) Depth of irrigation. (2) Quantity of water designed for or used in irrigating a unit area of land.

North-east monsoon This part of the monsoon accounts only for 2 per cent of the annual rainfall in the Indian sub-continent. The north-east early air current that blows over the sub-continent and the seas from Nov. to Jan.

Nose The point of a plough share.

Nosema bambycis Causes dangerous fungus disease of silk work, i.e., pebrine.

Notch (1) One of the water measuring device during flow in a channel as V notch. (2) The opening in a dam or spillway for water to pass.

Notch graft A plant graft in which the sion is inserted in a narrow slit in the stock.

Notified kind/variety In relation to any seed this means any kind or variety thereof notified under section 5 of Seeds Act, 1966.

Noxious A term applied to seriously harmful weeds.

Noxious weed seed Seeds from any plant of those harmful to agriculture. These species are designated by law for seed law enforcement called as primary which are seeds of noxious weed species restricted by law from being present in any quantity in agricultural seeds or secondary which are seeds of noxious weeds species permitted by law to be present in limited numbers in agricultural seeds.

Nozzle Mouthpiece of a spout through which liquids are discharged in a stream or as a spray as in sprinkler irrigation. The discharge can be controlled by the pressure head of the water.

N P K Short form of three major nutrients for plants. Each stands for nitrogen, phosphorous and potassium.

Nucellus The oval central mass of

tissue in the world ovule; contains the embryo sac.

Nuciferous Nut-bearing.

Nucleole Nucleus of or within nucleus.

Nucleo-proteins Group of proteins consisting of histone or protamine and are special characteristics of both plant and animal cell nuclei.

Nucleus (seed) A group of progenies of individual plants taken at random from a variety for the purpose of purifying that variety of mixtures and off type.

Nuclide The nucleus of an isotope.

Nucule Formerly, the cogonium of the charales.

Numerical hybrid See hybrid.

Nuptial flight Flight taken by queen bee with drones during which she is fertilised.

Nurse crop A companion crop.

Nurse graft A plant graft in which the scion remains united with the stock only until roots develop on the scion.

Nursery (1) A group of young trees in a plantation. (2) A place where the young plants are grown for transplanting elsewhere.

Nurture Feeding to promote growth; to take every care; to rear.

Nut (1) A hard indehiscent one-seeded fruit enclosing edible kernel. (2) Term commonly used for such fruit such as coconut, almonds, groundnut etc. which are botanically not nuts.

Nutate Droop down.

Nutation Curvature in the stem of a growing plant.

Nut grass Cyperus rotundus L. (Cyperus hexastachyos Rottb.) Family Cyperaceae. Noxious perennial weed propagated by roots or nuts.

Nutlet The stone formed in drupacious fruits.

Nutmeg Myristica fragrans Houtt. Family Myristicaceae. Used as spice and condiments. Oil used as medicine, in soap and perfume industry. Essential oil from leaves toxic to weeds. Used in chewing gum, flavouring essences and cosmetics.

Nutramins See vitamins.

Nutrient (1) A chemical element or compound assimilated by a plant or animal as food or which can be united with other substances to form food. (2) Also a element or compound in the soil present which is essential to the growth of plants.

Nutrient, animal Food constituent or group of food constituents of the same general chemical composition required for support of animal life.

Nutrient level The amounts of nutrients which are present in the soil and are available to plants for assimilation. Sometimes proportions of these elements are also expressed as levels of available nutrients.

Nutrient, plant Element essential to plant growth used in the elaboration of food and tissue.

Nutrient solution The solution of all the required nutrients for the plant growth without soil and used in hydroponics. Also called culture solution.

Nutrient toxicity Toxic effect of nutrients when in excess. Particularly used in case of micro-nutrients toxicity.

Nutrients The elements of substances that promote growth and development of plant or animal life.

Nutritive ratio See nitrogenous ratio.

Nutritive value The relative capacity of a feed to provide nutrition. Usually termed as high, low etc.

Nutritive value index (NVI) Daily digestible amount of forage eaten per unit of metabolic body size relative to a standard forage.

Nut structure The aggregation of fine particles of soil into small, subangular, blocky or roughly rounded, masses of definite shape or size.

Nux vomica Strychnos nuxvomica L. Family Loganiaceae. A tree. Also dried seeds of nux vomica plant. Contains strychnine and is very poisonous.

Nyctinarty A nastic movement in higher plants associated with diurnal light and temperature changes.

Nymph Stage in the life of insects and related forms like the ticks. A stage immediately preceding the adult one.

Nymphaea alba See European white water lily.

— **esculenta** See Indian water lily.

— **lotus** See white lotus.

— **lotus var. pubescens** See white lotus.

— **nelumbo** See East Indian lotus.
— **nouchali** See white lotus.
— **stellata** See blue lotus of India.

Nymphaeaceal A family of dicotyledonous plants in the order Nymphaeales distinguished by the presence of roots, perfect flowers, alternate leaves, and uniaperturate pollen.

Nymphaeales An order of flowering aquatic herbs in the subclass Magnoliidae; all lack combium and vessels and have laminar placentation.

Nymphula depunctalis See rice caseworm.

Nyssaceal A family of dicotyledonous plants in the order Carnales characterised by perfect or unisexual flowers with inbricate petals, a solitary ovule in each locule, a unilocular ovary, and more stamens than petals.

O

Oak Any tree of the genus Quercus in the Order Fegales, characterised by simple, usually lobed leaves, scaly winter buds, a star-shaped pith, and its fruit the acon, which is a nut, the wood is tough, hard and durable generally having a distinct pattern.

Oak wilt A fungus disease of oak trees caused by Chalora quercine, characterised by wilting and yellow and red discolouration of the leaves progressing from the top downward and inward.

Oasis effect Vertical energy transfer from the air to the crop.

Oats Avena sativa L.; Avena sterilis L. var. culta. Family Gramineae. Mainly grown as fodder in India but sometimes also for grain. A useful cereal owing to its oil and mineral content but its higher fibre content reduces its value. Oat meal taken by man. Considered good for horses and milch cattle.

Oats covered smut Disease caused by Ustilago kolleri Wille. Characterised by replacement of grains of affected plants by black mass of smut spores.

Oats leaf blotch Disease caused by Helminthosporium avenae. Characterised by appearance of oblong to linear blotches with irregular margin on leaf blades, light reddish brown in colour, frequently with sunken centres. Infected leaves turn yellow and dry up.

Oats loose smut Disease caused by Ustilago avenae. Characterised by the transformation of grain into black powdery mass of smut spores.

Oats stem rust Disease caused by Puccinia graminis avenae. Characterised by the appearance of reddish brown to dark brown oblong pustules on the culms and leaf sheaths which later turn black; shrivelling up of grains.

Oberea brevis See soyabean stem borer beetle.

Objectionable weed Weed plant whose seed is difficult to separate once mixed with crop seed and which is poisonous or injurious or has a smothering effect on the main crop. It is difficult to eradicate once established, has a high multiplication ratio thus making its spread quick and serves as an alternate host for crop diseases and pests.

Objective It is direction of movement; goal.

Obligate parasite An organism which can only obtain its food from living protoplasm.

Obligate saprophytes Micro-organisms not related to living cells which get their nutrients from dead organic tissues or available inorganic materials.

Oblique photograph Photograph taken with a single-lens camera with the axis at an appreciable angle to the vertical, generally at 45° or more. A high oblique photograph is one that includes the horizon, a low oblique photograph does not include the horizon.

Obnoxious weed Any weed when eaten by milch animal imparts obnoxious flavour or odour to milk.

Obovate Inversely ovate, narrowest at base.

Observed value A value of a quantity that is obtained by instrumental measurement of the quantity.

Obsolescence The loss in value of physical assets due to technological changes which have led to more efficient assets, rather than due to physical deterioration.

Obtuse (1) Of a leaf having a free end; rounded or blunt. (2) Greater than a right angle.

Occlusion (1) The process of healing over of the wound caused by cutting or breaking of a limb of a tree. (2) The absorption of gases by solids.

Ocellus A large cell of leaf epidermis, specialised for reception of light.

Ochnaceae A family of simple, stipulate leaves a mostly gynobasic style, and anthers that generally open by terminal pores.

Ochrocarpus longifolius See Alexandrian laurel.

Ochyromera artocarpi See jack bud weevil.

Ocimum Kilimandscharicum See camhor basil.

— **tectorius** See salza.

Octane rating A measure of the antiknock qualities of gasoline. Iso octane produces the least knock and is rated at 100. Normal heptane produces the most knock and is rated at zero. A fifty-fifty mixture of these two would produce a knock rated at 50.

Octane selector A scheme for adjusting the timing in order to get maximum efficiency from various grades of gasoline.

Odd-pinnate Of a compound leaf, having a single leaflet at the tip of the petiole with leaflets on both sides of the petiole. Also known as imparipinnate.

Odina wodier See jhingan.

Odoiporus longicollis See banana stem weevil.

Odometre An instrument for measuring distance traversed, as of a vehicle.

Odonata Order of insects to which the dragon flies belong.

Odontotermes assumathi See sugarcane termites.

— **obesus** See sugarcane termites.

— **taprobanes** See sugarcane termites.

Oecology Branch of biology dealing with habits, mode of life, relation to their surroundings etc. of living organisms.

Oedema Abnormal swelling of plant parts due to excessive intake of water.

Oenanthe javanica See water dropwart.

— **stolonifera** See water dropwart.

Oestrus ovis See headmaggot fly.

Oedogoniales An order of fresh-water algae in the division Chlorophyta; characterised as branched or unbranched microscopic filaments with a basal holdfast cell.

Offal The by-product of grain milling, such as bran, shorts etc.

Offer the price at which a person is ready to sell; purchase price. Hence to 'buy at best offer'.

Off-season The crop in a season of the year when that particular crop is not generally grown.

Off-shoot A lateral shoot or branch which arises from one of the main stems of a plant.

Off type Plant or seed deviating significantly from the characteristics of a variety as described by the breeder in any observable respect such as off type strain.

Offtake canal A canal for taking away water from another canal.

Ogee Profile of an overflow dam or spillway shaped in form of an "S".

Oidium erysiphoides See sanhemp powdery mildew.

— **heveae** See rubber powdery mildew.

— **mangiferae** See mango powdery mildew.

— **tingitanium** See citrus powdery mildew.

Oils A greasy or unctuous liquid of animal, vegetable, or mineral origin; in addition to its use as a lubricant, oil has a very wide use in the industries.

Oil base Sprays of poison or insecticide in which oil is used as a base or carrier.

Oil cake Leftout of seeds from which oil has been extracted. Used as fertiliser or cattle feed.

Oil cell A specialised and enlarged parenchyma cell or ray cell which secretes an essential oil.

Oil and fats The esters of glycerol and fatty acids, in food and foodstuffs. Oils are those lower members which are liquid at room temperature.

Oil filter The cylinder containing screens or filtering material, attached to the motor in such a manner that when the motor is running, the lubricating oil is constantly passing through the filter, thus removing grit and foreign matter. The filtering unit should be renewed at every 8,000 miles.

Oil line The tubing and devices connected therewith which comprise the circuit for lubricating oil.

Oil pan The lower half of the crank-case, made either of pressed steel or cast aluminium.

Oil tight Constructed or packed so as to prevent oil leakage.

Oil wasteland Land on which oily wastes have accumulated, including slush pits and adjacent areas affected by oil waste. A miscellaneous land type.

Oilless type bearings (1) Bearings which require oiling only at long intervals. Usually made of some oil absorbing porous metal. The warmer the bearing, the more actively it lubricates. (2) Bronze bearings with graphite inserts.

Okra Hibiscus esculentus. A tall annual plant grown for its edible immature pods. Also known as gumbo.

Olacaceae A family of dicotyledonous plants in the order Santaliales characterised by dry or fleshy indehiscent fruit, the presence of petals, stamens, and chlorophyll, and a 2.5 celled ovary.

Old growth Timber stands in which no cutting has been done. Synonyms: first-growth timber, virgin timber.

Oleaceae A family of dicotyledonous plants in the order Scrophulariales characterised generally by perfect flowers, two stamens, axile to parietal or apical placentation, a four-lobed corolla, and two ovules in each locule.

Olea europaea See olive.

— **ferruginea** See Indian olive.

Oleaginous Having properties of or producing oil; greasy.

Oleaster Elaeagnus angustifolia L. (E. hortensis Bieb.). Family Elaeagnaceae. A shrub or small tree whose fruits are edible.

Olein A colourless oily compound of glycerine and oleic acid found in most fats and mixed oils.

Olenecambtus biolobus See fig beetle.

Oleoresins Mixture of essential oils and resins; the exudations produced by the combination of volatile odoriferous substances found secreted in the intercellular spaces with resin formations. Important ones are Dipterocarpus spp. of which Dipterocarpus turbinatus is the best.

Olericulture Vegetable gardening.

Oligarchic theory (intelligence) It holds that cognitive abilities are manifestations not of a single commanding faculty, but of a few main intellectual powers of groups of abilities. In related subjects he does fairly well but fails in unrelated subjects.

Oligocarpous Having few fruits.

Oligoirophic Of a lake, lacking plant nutrients and usually containing plentiful amounts of dissolved oxygen without marked stratification.

Oligomerous Having one or more whorls with fewer members than whorlo of the flower.

Oligonychus indicus See sugarcane mite.

— **coffeae** See tea red spiders.

Oligophagous parasite A parasite which can develop upon a few closely related host species.

Olitorius jute Corchorus olitorius L. Family Tiliaceae. Commercial variety of jute cannot withstand water stagnation. Has edible leaves, long cylindrical pods, strong but somewhat dull coloured fibre.

Olive Any plant of the genus Olea in the order Schrophulariales, especially the over green/olive tree (O. europeae) cultivated for its drupaceous fruit, which is eaten ripe (black olives) and unripe (green), and is of high oil content. Also called *Zaitun*.

Olive knot A bacterial disease of the olive caused by Pseudomonas sevastonoi and characterised by excersecnees on the foliage and branches, and sometimes, on the trunk. Also known as olive taberde.

Olive oil The edible oil obtained from the fruit of olive; chemically consists of olein, palmitin, some stearin and a small amount of linolein.

Olivine A ferromagnesium silicate, which together with serpentine forms about 1 per cent of the minerals found on earth's crust.

Ombrophilous Able to throng in areas of abundant rainfall.

Ombrophobous Unable to live in the presence of long, continuous rain.

Omnivore An organism which obtains energy from the consumption of both plants and animals.

Omum Trachyspermum ammi (L.)

Sprague (Carum copticum Benth. and Hook. f., Sison ammi L.) Family Umbelliferae. Annual plant with predominant branching habit. Thymol prepared from oil used in place of menthol. Fruit used as spice, stimulant and tonic. Also called *ajwan.*

Onagraceae A family of dicotyledonous plants in the order Myrtales characterised generally by an inferior ovary, axile placentation, twice as many stamens as petals, a four nucleate embryo sac, and many ovules.

Onchosphere The first larva of the tapeworm enclosed in the egg-shell and with six hooks.

On cost A charge made on the direct cost of production of a commodity as a contribution to the recovery of indirect expenses.

One off' production The production of an article to suit the particular requirements of a customer. Opposite of mass production.

One-third atmosphere percentage See moisture tension soil.

Onion (1) Allium Cepa L. A biennial plant in the order liliales cultivated for its edible bulb. (2) Any plant of the genus allium.

Onion leaf blight Disease caused by Alternaria palanduii Ayyangar and A Porri. Characterised by appearance of brownish oval spots on leaves of young plants, and dying of leaves.

Onion smudge A fungus disease of the onion caused by Colletotrichum circinans and characterised by black concentric integral rings or smutty spots on the bulb scales. Also known as onion scale.

Onion smut A fungus disease of onion especially seedlings, caused by Urocystis cepulae and characterised by elongate black blisters on the scales and foliage.

Onion thrip Thrips tabaci L. pest of Family Thripidae; nymph and adult suck sap from lower surface of leaves, distorting and twisting them. Also attack cabbage, cauliflower, turnip and cotton.

Onshore wind Wind blowing from the sea toward the land.

On the cob Unshelled corn cob.

Ontogeny The life history, or development of an individual, as opposed to that of the race.

On year Refers to the year in which biennial bearing fruit tree produces fruits.

Onyx Kind of quartz allied to agate with different colours in layers.

Ooecium Bud-like sac for receiving and fertilising ova in polyzoa.

Oogamus Characterised by a type of sexual reproduction involving a large non-flagellated egg and a small sprem.

Oogenesis The process by which germ cells are produced by the female; production or development of ovum.

Oogonium The unisexual female sex organ in oogamous algae and fungi.

Oolite Granular limestone, each grain being a particle of sand in carbonate of lime.

Oomycetes Sub class of the Phycomycetes, gametangia of unequal size.

Oosperm A spore which is produced by heterogamous fertilisation, and from which the sparophytic generation develops.

Oospore Resting spore formed in a fertilised ooganium.

Open centre A system of training fruit trees in which main stem is cut after it has achieved certain height and subsequently vegetative growth is induce in lateral branches. It results in bulk of the crop grown near to the ground.

Opegraphaceae A family of the Hysteriales characterised by elongated ascocarps; members are crustose on bark and rocks.

Open channel Any natural or artificial covered or uncovered conduit in which liquid (usually water) flows with its top surface bounded by the atmosphere.

Open-channel drainage Drainage accomplished by open channels or ditches.

Open channel flow Flow in a channel with a free surface in contact with the atmosphere and includes flow in pipes and closed conduits flowing partly full.

Open conduit Any conduit in which water flows with a free surface.

Open ditches, or open drains Drains open to the sky and which provide outlets for tile drains and tributary ditches and remove surface water directly.

Open ended Descriptive of a decision or policy which is not restrictive, allowing the broadest possible interpretation.

Open pollination Open pollinated seed is seed produced as a result of natural pollination as opposed to hybrid seed produced as a result of controlled-pollination.

Open traverse A surveying traverse in which the last leg, because of error, does not terminate at the origin of the first leg.

Operating gear, or lifting gear Any mechanical device used for adjusting the position of the gates, wickets, shutters, etc., of a moveble weir.

Operational freeboard The difference between the maximum flow line and the maximum level of normal operation.

Operational research The application of scientific and mathematical techniques to complex problems arising in the direction and control of systems of man, machines materials and money, to improve the speed, quality and cost of performance.

Operational wastes, or **Delivery losses** Losses due to lack of efficiency in management and breaks in the conduits.

Operculate With a cover of lid, as in some asci.

Opercullela padwickii See gram root rot.

Operculum The lid covering the apex of a moss capsule.

Ophioglossales An order of ferns in the subclass Ophioglossidae.

Ophioglossidae A small subclass of the class Polypodiopsida; the plants are homosparous and eusporangiate and are distinguished by the arrangement of the sporogenous tissue in the characteristic fertile spike of the sporophyte.

Opiate Medicine containing or made of opium.

Opisometre An instrument for measuring the length of curved lines, such as those on a map; a wheel on the instrument is traced over the line.

Opium Reddish brown heavy scented bitter drug produced from juice of opium poppy. Used as sedative in medicines.

Opium poppy Papaver somniferum L. Family Papaveraceae. An annual herb from which latex is obtained which is source of opium. Yields numerous alkaloids, such as morphine, codeine, narcotine.

Oppilate Block up, obstruct.

Opportunity cost The real cost of satisfying a want, expressed in terms of the cost of the sacrifice of alternative activities.

Optical air mass A measure of length of the path through the atmosphere to sea level traversed by light rays from a celestial body, expressed as a multiple of the path length for a light source at the zenith.

Optimal estimate An estimate will be called 'optimal' if there does not exist another estimate whose mean square error is always smaller than the mean square error of the former all values of the parametre.

Optimum consumptive use Consumptive use which produces a maximum crop yield.

Optimum output Output at which factors of production are combined in the most efficient way.

Optimum water content The amount of water in the soil which is optimum to carry out certain operations such as levelling or for the growth of plants/crops.

Optimum water requirements The seasonal depths of beneficial water use that result in maximum yields of different crops, where the depths include soil moisture supplied by precipitation as well as water delivered by irrigation.

Option demand Demand characterised by a willingness to pay for retaining an option to use an area or facility that would be difficult or impossible to replace and for which no close substitute is available.

Optional capital The difference in capital cost between two projects or schemes, which are true alternatives.

Orach Atriplex hortensis L. Family Chenopodiaceae. A Leafy vegetable with tender stems and soft attractive green leaves; hardier than spinach.

Orange Any of various evergreen trees

of the genus Citrus, cultivated for the edible fruit, which is a berry with an aromatic, leathery rind containing numerous oil glands.

Orange oil An edible oil produced by crushing or grinding whole oranges and extracting it centrifugally, or by rasping the peel.

Orange stem borer Stromatium barbatum Fabr. Major pest which causes great damage to trees. Grub bores into branches of tree and tunnels inside causing branch to wither and tree to dry up. Damages orange, pomelo, grapefruit and malta.

Orbicular Circular, discoid, ring shaped, spherical, globular or rounded.

Orchard A group of fruit-bearing, nut-bearing, or sugar maple trees under cultivation.

Orchard grass Dactylis glomerata L. Family Gramineae. A perennial bund grass used as fodder.

Orchid Any member of the Family Orchidaceae; plants have complex, specialised irregular flowers usually with only one or two stamens.

Orchidaceae A family of monocotyledonous plants in the subclass Liliidae; plants are mycotropic and sometimes nongreen with numerous tiny seeds that have an undifferentiated embryo and little or one endosperm.

Orchis Orchis latifolia L. A perennial herb of Family Orchidaceae, the tuberous crop of which is eaten.

Orchis latifolia See Orchis.

Order, soil A category in the Canadian system of soil classification. All the soils of Canada have been divided into eight orders: Chernozemic; Solonetzic, Luvisolic; Podzolic; Brunisolic; Regosolic; Gleysolic; and Organic. All the soils within an order have one or more characteristics in common.

Ore Native mineral from which useful metals are extracted.

Oregano A spice prepared from leaves of various aromatic mints; especially wild marjoram.

Organ complex The total interacting complex of organs forming an individual tree or part of a tree; such as a branch system (e.g. plagiotropic complex).

Organic (1) Of plant or animal origin. (2) Chemical compounds containing carbon.

Organic disease Disease due to visible structural changes in some tissues or organs of the body.

Organic insecticide Insecticide of organic origin, containing chlorinated hydrocarbons and organophosphorus compound or extracted from plants such as pyrethrin or nicotines.

Organic manure (1) Farmyard manure. (2) Any manure such as oilseed cakes, treated sewage etc.; of organic in composition.

Organic evaluation The processes of change in organisms by which descendants come to differ from their ancestors; and a history of the sequence of such changes.

Organic matter, soil The organic fraction of the soil; includes plant and animal residues at various stages of decomposition, cells and tissues of soil organisms, and substances synthesised by the soil population. It is usually determined on soils that have been sieved through a 2.0 mm sieve.

Organic phosphorus Phosphorus present as a constituent of an organic compound; or a group of organic compounds such as glycerophosphoric acid, inositol phosphoric acid, and cytidylic acid.

Organic soil A soil whose solid part mainly consists of organic matter.

Organisation It is a set of two or more parts which are so related that they function as a whole.

Organisational chart This shows the structure of a organisation, say State Department, Agricultural University, Development Department.

Organogen Any chemical element, such as nitrogen, carbon, oxygen, phosphorus etc., without which an organism cannot exist.

Organoleptic Affecting one or more of the organs of special sense (sight, touch, hearing, smell, and taste).

Orientation The term 'orientation' is used as equivalent to 'attitude' in its more inclusive sense of referring to both cannotive and cognitive tendencies.

Orifice Mouth or aperture used for

measuring water discharge in hydraulics.

Ornamental tree A tree which is used to beautify places. The tree itself is beautiful due to form, flowers, foliage or decorative fruit.

Orobanche Orobanche cernua Loefl and O. indica. Family Orobanchaceae. A fleshy annual plant parasite causes severe damage to tobacco, brinjal, mustard, rape and other garden crops.

Orobanche cernua See orobanche.

— **indica** See orobanche.

Oromorphic A soil which retains mainly the characteristics of the parent material.

Orophyte Any plant that grows in the subalpine region.

Ortet The original single plant from which a clone is obtained.

Orthoeluvium Primary parent material of soil.

Orthogonal Two varieties, or two linear functions, are said to be orthogonal, if they are statistically independent.

Orthoptera Order of insects to which crickets, grasshoppers and locusts belong.

Orthostichy Vertical series of leaves or leaf scars.

Orthotropism The tendency of a plant to grow with the longer axis oriented vertically.

Orthotropous Straight and symmetrical, with the chalaza at the base and the micropyle at the opposite end, as an ovary.

Orthotropy Gravitational response which produces a vertical axis, hence orthotropic shoot with the complex of characters resulting from this response, i.e., radial symmetry and vertical orientation (cf. plagiotropy).

Orts Rejected feedstuffs left under condition of *ad libitum* stall feeding.

Ortstein (1) An indurated layer in the B horizon of Podzols in which the cementing material consists of illuviated sesquioxides and organic matter. (2) As a sub-group of Podzolic soils. Ortstein indicates a Bjfc of Bfc. horizon that is strongly cemented, occurs over atleast one-third of the exposure, and is atleast 2.5 cm thick.

Oryctes rhinocerus See coconut beetle.

Oryza rufipogon See wild rice.

— **sativa** See rice.

Oryzaelphilus surinamensis See saw-toothed grain beetle.

Oryzenin See glutenins.

Osage orange Maclure pomifera. A tree in the mulberry family of the Urticales characterised by yellowish bark, milky sap, simple entire leaves, strong axillary thorns, and aggregate green fruit about the size and shape of an orange.

Oscillating flow, or **Periodic unsteady flow** Flow in which the velocity and discharge are also functions of time, but the flow is periodic with a complete reversal of direction of flow as in tide phenomena.

Oscillating pipeline An applicator used in overhead irrigation. The pipeline is oscillated by an oscillator which is mechanical device worked by hydraulic pressure.

Oscillation An oscillation in a time series (or more generally, in a series ordered in time or space) is a more or less regular fluctuation about the mean value of the series.

Oscillator, Oscillating arm A driving mechanism that rotates the sprinkler head, it is actuated outward by the escaping jet of water and brought back by means of a torsional spring which propels the rotating member forward by regularly controlled impacts.

Oscillatoriales An order of blue-green algae (Cyanophyceae) which are filamentous and truly multicellular.

Oscular Osmosis.

Osmometre An apparatus for the measurement of osmotic pressure.

Osmosis Passage of fluids or other substances (either a solute or a solvent) through a membrane.

Osmotic potential See water soil.

Osmotic pressure See water soil.

Ostiole The pore or opening of a fruiting body of algae or fungi.

Other crop seed Seeds of plants which are grown as crops other than main crop.

Otaheite gooseberry Phyllanthus acidus (L.) Skeels. (Averrhoa acida L., Cicca disticha L., Phyllanthus distichus Muell-arg.). Family Euphorbiaceae. Plant of *aonla* family grown for ornamental foliage. Fruits used for pickles. Leaves also edible. Bark used for tan. Also called otaheite gooseberry leaf-flower, tahiti gooseberry, West India gooseberry etc.

Otaheite gooseberry leaf flower See otaheite gooseberry.

Other fallows A class of land under revenue classification. Fallow for 1-5 years but not more.

Othreis ancillia See citrus fruit-sucking moths.

— **fullonia** See citrus fruit-sucking moths.

— **materna** See citrus fruit-sucking moth.

Other seeds Other seeds include seed and seed-like structures of any plant species other than that of pure seed.

Ounce Unit of weight; 1/16 pounds in avoirdupois.

Outbreeding The mating of unrelated plants or animals which have no common ancestors on both sides of their pedigrees for atleast up to 4-6 generations. Grading, cross-breeding and out crossing are 3 kinds of out breeding.

Outcrop Emergence of stratum, vein or rock at the surface.

Out cross A cross usually natural to be a plant of a different genotype; mating of hybrid with a third parent, an off type plant resulting from pollen of a different sort.

Outer bark The layer of dead tissue; generally outside the last formed periderm of a tree or woody plant.

Outfall The point where water is discharged.

Outlet An arrangement to dispose off the surplus water. Also a waste weir.

Outlet area, or **turnout area** The block of land served by an outlet or turnout. It may comprise land of one farmer or a number of farmers. Commonly known as '*chak*' in some states of India, and in West Pakistan.

Outlet ditches, or **Outlet channels** Large open ditches whose primary function is to collect from large areas all free surface flow resulting from any cause, and the discharge from tile drainage systems, and to carry it to final outlets (rivers, lakes, seas, etc.).

Outwash Sediments washed out by flowing water beyond the glacier and laid down as stratified drift in thin forest bends. The particle size may vary from boulders to silt.

Oval Broadly elliptical, egg shaped.

Ovary (1) The swollen (enlarged) part of the pistil that contains the ovules. (2) One of the two female reproductive glands in animals.

Ovate Shaped like an egg, the broader end at the base.

Oven dry soil Soil that has been dried at 105°C, until it has reached constant weight.

Over running clutch A clutch designed with inner and outer rings; the inner is milled with several triangular slots in which hardened steel rollers are inserted. By this construction, the inner ring will drive the outer, but if the outer ring moves faster than the inner, it will slide or overrun the inner ring.

Overcapacity Excess productive capacity likely to arise in an industry subject to constant fluctuations in the demand for its products. Overcapacity or surplus capacity may also arise from long term changes in demand.

Overconsolidated soil deposit A soil deposit that has been subjected to an effective pressure greater than the present over burden pressure.

Overdrive An arrangement of gearing which produces more revolutions of the driven than of the driving shaft.

Over grazing Grazing so heavy that it impairs future forage production and causes deterioration through damage to plants or soil or both.

Over grazed range A range deteriorated from its productive potential due to continued over use.

Over heads Costs not directly chargeable to any unit produced. They include rent of premises, salaries of salesman and research salaries of directors, etc. and all other general expenses, as opposed to costs of raw materials, wages, etc., the proportion of which that goes into each units can be determined.

Over irrigation Applying water to the field in excess of the water requirements of the crops.

Over land flow The flow of storm water over the ground before becoming channelised.

Overland runoff Runoff water flowing over the land surface before it reaches a definite stream channel.

Overload Running water in excess of the designed discharge in irrigation canals.

Overrun The excess of the amount of lumber actually saved from logs over the estimated volume or log scale, usually expressed in per cent of log scale.

Over story Upper crown canopy of a forest.

Over time It is the time required to adopt in innovation from its origin.

Over trading A situation in which a firm has little working capital.

Overtree sprinkler method The sprinkler method used in orchards to distribute water over the top of the trees.

Overuse Excessive use of the current year's growth which will result in range deterioration or over-grazing, if continued; over-utilisation.

Oviside An agent which kills eggs of insects and other parasites.

Ovule The macrosporangium of flowering plants, consisting of the nucellus plus the integuments; develops into a seed following fertilisation.

Ox Bos indicus. One of the two species of cattles in India. Any bovine animal usually horned, cloven footed ruminant quadruped used for draught purposes. Used in carrying out farming operations.

Oxalacetate An intermediate compound formed from sugars and sugar producing aminoacids essential for the oxidation of carbohydrates, fats and proteins for energy.

Oxalic acid An organic acid occurring in many plants.

Oxalis corniculata See Indian sorrel.

— **cernua** See Bermuda butter-cup.

— **pes-caprae** See Bermuda butter-cup.

Oxbao A closely looping, U-shaped stream meander whose curvature is so extreme that only a neck of land remains between two parts of the stream. Also known as horseshoe bend. The abandoned, horseshoe-shaped channel of a former stream meander after the stream formed a neck cutoff. Also known as abandoned channel.

Oxidation Combining with oxygen. A chemical change.

Oxolidaceae A family of dicotyledonous plants in the order Geraniales, generally characterised by regular flowers, two or three times as many stamens as sepals or petals, a style which is not gynobasic, and the fruit which is a beakless, loculicidal capsule.

Oxya velox See rice grasshopper.

Oxycarenus laetus See dusky cotton bug.

Oxygen A colourless, odourless gas. Essential for all living things. Chemically denoted by O. Helps in burning but it does not burn itself.

Oxygen absorbed (OA) The amount of oxygen in parts per million absorbed by a sample of water from acidic permanganate ($KMNO_4$) in 4 hours at 27°C.

Oxygen sag curves Graphical curves which relate the dissolved oxygen content of water against time flow, within the context of the process of self-purification, following the discharge of pollutants into a stream.

Oxygenate Supply, treat or mix with oxygen; oxidize; charge blood with oxygen.

Oxylophyte A plant that thrives in or is restricted to acid soil.

Oxypotalous Having sharp-pointed petals.

Oxyphytia Discardant habitat control due to an excessively acidic substratum.

Ozonium texanum var. parasiticum See cluster bean ozonium wilt.

Ozonosphere A layer of the atmosphere between about 20 kilometres and 50 kilometres above the surface of the earth containing ozone.

P

P_1, P_2 etc. The first, second etc., parental generations from a parent.

Pachycaul A thick-twigged, little-branched tree.

Pachydiplosis oryzae See paddy gall fly.

Pachyrhizus angulatus See potato bean.

— **erosas** See potato bean.

Packaging The process of filling, weighing and sewing bags.

Paddock Small fenced field used for grazing purposes.

Paddy Rice with husk.

Paddy cutworm Pseudaletia (Cirphis), unipunctata Haw., P. albistigma M. Insect pest of paddy; caterpillar feeds on leaves and cuts stem and unripe years after crawling up plants. Also called paddy climbing cutworm, army worm.

Paddy eelworm Ditylenchus angustus Butler. Nematode pest of paddy. Plants supposed to be suffering from *Ufra disease*. Turns leaves and leaf sheaths yellow, later brown which dry up. Plants get stunted, ears turn dark brown, twisted and distorted: Grains either are not formed or are partially formed.

Paddy gall disease See paddy gall fly.

Paddy gal fly Pachydiplosis oryzae. Wood Mason. It is commonly referred to as *gall diseases*. Serious pest of paddy. Maggot burrows into growing point of paddy plant at the base and forms gall causing pale white or slightly bluish, silvery shoot in place of main shoot, which forms up grain.

Paddy green jassids Nephotettix bipunctatus Fabr.; N. apicalis Mots., Tettigoniella spectrar. Pest of paddy. Nymphs and adults suck sap of young paddy leaves. It results in yellowing and drying up of plants.

Paedomorphism Persistence and/or elaboration of juvenile characters (may be ontogenetic or phylogenetic).

Pagoda habit Tree physiognomy with regular tiers of horizontal branches, the tiers progressively shorter distally, and the end of each branch upright.

Paired comparison The comparison of two samples of equal size where the members of one can be paired off against the members of the other.

Paleoniaceae A monogeneric family of dicotyledonous plants in the order Dilleniales; members are mesophyllic shrubs characterised by cleft leaves, flowers with an intra-staminal disk, and seeds having copious endospern.

Painted bug Bagrada cruciferarum Kirk. An insect pest of several crops such as mustard, cabbage, cauliflower and radish.

Palaeo Ancient also paleo.

Palaesol A soil formed in past climatic or geologic period.

Palatability Plant characteristics eliciting a choice between two or more forages or parts of the same foragees or parts of the same forage, conditioned by the animal ánd environmental factors that stimulate a selective intake response. Acceptability.

Pale Pointed piece of wood for fence; stake.

Palea (Palet) (1) The upper, enclosing bract of grass flowers. (2) A chaffy scale found on receptacle of the disk flowers of same, composite plants.

Paleate Having a covering of chaffy scales, as some rhizomes.

Paleobioclimatology The study of climatological events affecting living organisms for millenonia or longer.

Paleobotany The division of paleontology concerned with the study of fossil plants.

Paleozoic The oldest geological era having abundant fossils. It includes divisions from the Cambrian to the Permian.

Palisade cell One of the columnar cells of the palisade mesophyll which contain numerous chloroplasts.

Palisade mesophyll A tissue system of the chlorenchyma in well-differentiated broad leaves composed by closely spaced palisade cells oriented parallel to one another, but with their long axes perpendicular to the surface of the blade.

Palisade tissue A leaf tissue composed of long cylindrical chlorophyllous cells, as the portion of a mesophyll lying directly below and at right angles to the upper epidermis.

Pallets Wood frame structure which are used to support a stack of bags, to prevent them from touching the floor and absorbing moisture; mattress.

Palliate alleviate disease without curing. Palliative is used for the agent to palliate.

Palm A tropical or sub-tropical tree or shrub of the family Palmaceae. Usually with upright, unbranched stem and head of large pinnate or fanshaped leaves. Grown for wood, fruits, resins, fibre or as a source of drug.

Palmate Having the lobes, divisions, or leaflets radiating from a common centre.

Palmately veined A type of net venation in which the main veins of a leaf blade branch out from the apex of the petiole like the finger of a hand.

Palmyra palm Borassus flabelli fer L. (B. flabelliformis Roxb.) A tall palm of family Palmaceae.

Palmetto fibre Brush or brooms fibre obtained form young leaf stalks of the cabbage palm tree (Sobal palmetto).

Palm-grass Setaria palmifolia (Koenig) Stapf. A perennial grass used for fodder and belonging to family Gramineae.

Palmitic acid One of the fatty acids of butter-fat.

Palmitin A fat occurring in adipose tissue, milk, palm oil.

Palm nut The edible seed of the African oil palm.

Palm oil Oil extracted from African palm oil.

Palp Joint sense organ in insects. Also known as palpus.

Palpus See palp.

Paludal Pertaining to wet or swampy land.

Palustrine Being, living, or thriving in a marsh.

P-aminobenzoic acid A member of vitamin B complex which promotes growth in chicks.

Pampa An extensive plain in South America, usually covered with grass.

Pan-hard Horizon or layer in soil which has a very high clay content, is strongly compacted and restricts the flow of water.

Panama disease A fungus disease of banana caused by invasion of the vascular system by Fusarium oxysporum cubense, resulting in yellowing and wilting of the foliage and ultimate death of the shoots.

Pan coefficient The ratio of the amount evaporated from a large body of water to that measured in an evaporation pan.

Pandemic Very widespread epidemic, even of worldwide extent.

Pandonaceae The single, pantropical family of the plant order pandanales.

Pandanales A mono-familial order of monocotyledonous plants, members are more or less arborescent and sparingly branched, with numerous long, firm, narrow, parallel-veined leaves that usually spring margins.

Pandanus latifolia Pandanus latifolius Perr. A small tree of family Pandanaceae used for flavouring cooked rice.

Pandanus odoratissimus P. tectorius Soland ex Parkinson (P. odoratissimus Roxb.) A shrub or small tree yielding essential oil used in perfumeries. Leaves yield fibre. Family Pandanaceae. Also known as screw pine.

Pandanus tectorium See Pandanus odoratissimus.

Pandurate Of a leaf, having the outline of a fiddle.

Panic grass Any grass of the genus Panicum, family Gramineae. For example Guinea grass, para grass, foxtail millet.

Panicle A compound inflorescence with pedicled flowers, usually loose and irregular, as in oats, rice etc. In grasses, the panicle may be compact and specialized (timothy) or open (smooth bromegrass).

Panicum antidotale See Blue Panic grass.

— **colonum** See Shama millet
— **crus-galli** See Japanese barnyard millet
— **glaucum** See Cat tail millet
— **italicum** See Italian millet
— **maximum** See Guinea grass
— **miliaceum** See Common millet
— **milare** See Little millet
— **repens** See Torpedo grass.

Panphytotic Designating a worldwide distribution of a plant disease.

Pans Horizons or layers in soils that are strongly compacted, indurated, or very high in clay content. Also caliche, claypan, fragipan, genetic pan, iron pan, and pressure or induced pan.

Pantandrous Of a flower having five stamens.

Pantropic Turning to any direction, invading many tissues.

Papain Milk of raw papaya fruit in dried condition. Used as a digestant.

Papaver somniferum See opium poppy.

Papaveraceae A family of dicotyledonous plants in the order Papaverales, with regular flowers, numerous stamens, and a well-developed latex system.

Papaya Carica papaya L. Family Caricaceae. Quick and heavy yielding fruit rich in vitamins and minerals.

Papaya collar rot Disease of papaya characterised by rotting of tree trunk at collar and wilting of leaves. Caused by Pythium aphanidermatum.

Papaya foot rot Disease in which water-soaked lesion on tree trunk near ground appear. Peeling off of bark. It is caused by Phytophthora parasitica.

Papaya mosaic A virus disease characterised by leaves getting small, curled or crinkled. Appearance of green blister like patches on leaves and dying of plants.

Paper mulberry Broussonetia papyrifera (L.). Family Moraceae. A small tree source of strong lustrous fibre.

Papilionaceae Family name of the group to which most of the cultivated legumes belong.

Papaverales An order of dicotyledonous plants in the subclass Magnoliidae, marked by a syncarpous gynoecium, parietial placentation, and only two sepals.

Papilionoideae A sub-family of the family Leguminosae with characteristics irregular flowers that have a banner, two wing petals, and two lower petals united to form a boat shaped keel.

Papilla Small, nipple-like projection. Manifesting small nipple-like projection is papillate.

Papilionaceous Butterfly-shaped. A term referring to legume flowers.

Pappus An appendage or group of appendages consisting of a modified perianth on the ovary or fruit of various seed plants, adopted to dispersal by wind and other means.

Paprika Capricum anuum. A type of pepper with non-punnent flesh, grown for its long red fruit from which a dried, ground condiment is prepared.

Papule Small fleshy projection on plant.

Parabolic weir A weir with parabolic shaped notch. It can be so proportioned so as to give a parabolic discharge equation.

Parachute A special structure of seeds which assists in dispersal.

Paraeluvium Residual product which results from the eluviation of the weathering products of sedimentary rocks.

Paraffin A product obtained from dry distillation of wood, coal, petroleum, colourless, odourless, tasteless fatty substance.

Para grass Brachiaria mutica (Forsk) Stapf. (Panicum purpurascens Raddi.) Family Gramineae. A coarse perennial grass with rhizomes freely rooting at nodes and growing rapidly. Grown for fodder. Palatable at all stages and propagated by cuttings. Also known as buffalo grass.

Paraleliotropism The tendency in plants to turn leaves parallel to incidence of light rays.

Parallel Lying side by side; extending in the same direction and equidistant at all points, having the same direction, similar.

Parallel connected transformers When two or more transformers have their primary windings connected to the same source of supply in such a manner that the impressed voltage in each case is the same as that of the line.

Parallel drainage pattern A drainage pattern characterised by regularly spaced streams flowing parallel or nearly parallel to one another over a large area.

Parallel canal, or **side canal** A canal built by the side of, and parallel to the parent canal to irrigate areas not permitted to be irrigated from the parent canal. Called 'ditch canal' in India and Pakistan and 'gannabia' in U.A.R.

Parallel evolution Evolution of similar characteristics in different groups of organisms.

Parallel field ditch system A system similar to bedding, except that the drains are spaced farther apart and have a greater capacity than the dead furrows. The minimum depth of the ditch is 75 cm and minimum cross-section areas 150 square cm.

Parallel system A system of drains with long parallel laterals emptying into a single main drain.

Parallel veined Of a leaf, having the veins parallel, or nearly parallel, to each other.

Parallel venation A type of leaf venation, in which principal veins are parallel to each other and to the longitudinal axis of the leaf.

Paramo A biological community, essentially a grassland, covering extensive high areas in equatorial mountains of the Western Hemisphere.

Paramylum The starchlike reserve food in the Euglenophyta.

Paranthropophytia Discrepant control of regions or areas due to immediate and continuous or periodic interference, as a certain cultivation practices.

Paraphysis A sterile hyphal element in the hymenium, especially in the Ascomycetes, usually clavate or filiform.

Paraphysoids Threads of hyphal tissues between asci, like delicate paraphysis but without free ends.

Parasa lepida See Wood-apple slug caterpillar.

Parasite An organism that lives on or in a second organism, usually causing disease in the latter.

Parasitic Pertains to disease caused by parasite. A carrier.

Parastichy Regular oblique series of leaves or leaf scars.

Paratetraanychus indicus See sugarcane insect pests.

Parathion O=O-diethyl O-P-nitrophenyl phosphorothioate. An organic phosporous compound which acts as a contact and stomach insecticide applied as dust, suspension or emulsion. Highly toxic to humans.

Parboiling A process in rice milling to improve milling quality and to obtain higher outturn of whole rice.

Parenchyma A tissue of higher plants consisting of living cells with thin walls that are agents of photosynthesis and storage; abundant in leaves, roots and the pulp of fruit, and found also in leaves and stems.

Parent canal With reference to any canal, the canal which feeds it with water supply.

Parent company A firm that controls one or more other firms through stock ownership.

Parent material The unconsolidated and more or less chemically weathered mineral or organic matter from which the solum of a soil has developed by pedogenic processes.

Parent rock The rock from which the parent materials of soils are formed.

Parent seed (1) The single cross seeds from which hybrid seed is obtained by crossing. (2) Any seed which is to be multiplied.

Parging A thin coating of mortor or plaster on a brick or stone surface.

Parietal placentation Having the placenta and therefore, the seed developing along the inner walls of the ovary.

Parietales An order of plants in the Englerian system; families are placed in the order Violales in other systems.

Paripinnate Pinnate without a terminal leaflet.

Paris green Acetoarsenite of copper. One of the early arsenical compounds to be used as stomach poison for leaf eating insects. Used as dust/spray. Highly poisonous.

Parkia roxburghii Parkia roxburghii G. Don. A small tree of family Mimosaceae whose fruits are edible.

Parkinsonia aculeata See Jerusalem thorn.

Parrot Psittacula cyanocephala, a bird pest.

Parsely Petroselinum crispum. A biennial herb of European origin belonging to the order Umbellales grown for its edible foilage.

Parsnip Pastinaca sativa L. (Peucedamum sativum (L.) Hook. f.) A bienniol herb of Mediterranean origin belonging to the order Umbellales; grown for its edible thickened tap root.

Parthenocarpy Production of fruit without fertilisation.

Parthenogamy The fusion of two gametes.

Parthenogenesis The development of an egg into a new individual without fertilisation by a sperm.

Partial association The measures of association in subpopulations of qualitative characteristics analogous to the partial correlation between quantitative variates.

Partial correlation The correlation between two varieties in a conditional distribution for which one or more other variables are held fixed. Specifically, the product-moment correlation coefficient in the conditional distribution.

Partial etiolation A low chlorophyll content of a plant or plant part resulting from insufficient light.

Partial-duration series A statistical grouping or graph involving all rates or volumes for a period or record that are greater than some selected minimum value. In contrast, the Annual Series is a graph or tabulation in which only the maximum value for each year is included in the series.

Partially developed underground water basin An underground water basin wherefrom the use of underground water does not exceed the naturally occurring recharge but a considerable development, either agricultural or urban, dependent on the use of underground water has taken place on the land above.

Partial sterilisation The elimination of a portion of a population of micro-organisms, usually by heat or chemical treatment. The process is selective and certain organisms or group of organisms are destroyed to a greater extent than others.

Partical density The mass per unit volume of the soil particles. It is usually expressed in grams per cubic centimetre. Has been called grain density. See also bulk density, soil.

Particle size The effective diametre of a particle measured by sedimentation, sieving, or micrometric methods. Has been called grain size.

Particle size distribution The amounts of the various soils separates in soil sample, usually expressed as weight percentages. Has been called grain-size distribution.

Particle-size analysis The determination of the various amounts of the different separates in a soil sample, usually by sedimentation, sieving, micrometry, or combinations of these methods. Has been called grain-size analysis or mechanical analysis.

Partnership An association of persons carrying on a business in common with a view to profit.

Partnership limited Partnership which includes limited partners whose financial obligations to the firm are limited to the amount of capital they have invested.

Parts per million (ppm) Weight units of any given substances per one million equivalent weight units of oven dry soil; for soil or other solutions, the weight unit or solute million weight units of solution.

Parmo A destructive fungus disease of flax caused by Mycosphaerella linorum.

Paspalum Any grass of the genus paspalum, family Gramineae.

Paspalum conjugatum See Hilo grass.
— **dilatatum** See Dallis-grass.
— **distichum** See Knot grass.
— **notatum** See Bahia grass.
— **scrobiculatum** See Kodo millet.
— **vaginatum** See Seashore paspalum.

Passifloraceae A family of dicotyledonous, often climbing plants in the order Violales, flowers are polypetalous and hypogynous with a corona, and seeds are arillate with an oily endosperm.

Passiflora edulis See Passion fruit.
— **incarnata** See Wild passion flower.

— **laurifolia** See Water lemon.

— **quadrangularis** See Giant granadilla.

Passion fruit Passiflora edulis sims. Family Passifloraceae. Fruit good for squash. Fruit fairly acidic.

Pasteurism Prevention or cure of disease by successive inoculations.

Paste mixer Device for the blending together of solid particles and a liquid, with the final formation of a single paste phase.

Pasteurella A genus of gram-negative, avoid or rod-shaped bacteria of uncertain affiliation; distinguished by bipolar straining; parasites on mammals; including humans, and birds.

Pasteuria A genus of budding bacteria, cells are spherical to pear-shaped, reproduce by budding at the rounded end.

Pasteurization Heating of milk, etc., for a short time to a temperature that will destroy pathogenic bacteria but will not affect the food properties and flavour of the material.

Pasteurizer An apparatus used for pasteurization of fluids.

Paste wood filler Material composed of ground silica and quick drying varnish, thinned to brushing consistency with mineral spirits or turpentine prior to application, worked into the pores of the wood to obtain a smooth finish.

Pastoral Kind of livelihood dependent entirely on pasturing.

Pasturage Vegetation on which animals graze, including grasses or grasslike plants, legumes, forbs, and shrubs.

Pasture An area devoted to the production of forage (introduced or native) and harvested by grazing.

Pasture carrying capacity Number of animals a given pasture will support at a given time or for a given period of time.

Pasture, crop-land Includes rotation pasture and other land used only for pasture but that could be used for crops without additional improvements. Preferred to "rotation pasture".

Pasture land Land used primarily for the production of adapted domesticated forage plants to be grazed by livestock.

Pasture, permanent Pasture of perennial or self-seeding annual plants maintained through several years for grazing.

Pasture renovation Improvement of a pasture by the partial or complete destruction of the sod, plus liming, fertilizing, seedling, and weed control as may be required to establish desirable forage plants.

Pasture rotation See pasture cropland.

Patanga succincta Bombay locust greenish pale locust and highly damaging.

Patch bolt A flat-head bolt threaded the full length of the body. The square shank, which projects from the head to permit setting the bold with a wrench, is removed after the-bolt is in position. Patch bolts are used in making repairs to boiler plate, structural parts, etc.

Patch budding Budding in which a small rectangular patch of bark bearing the scion (a bud) is fitted into a corresponding opening in the bark of the stock. Also called flap and patch budding.

Patent A grant from a government to a person or persons conferring for a specified time the exclusive privilege of making, using or selling a new invention.

Pathobiology The study of biology of diseases or the disease processes.

Pathogen Any organism or factor causing disease. Capable of causing diseases is pathogenic.

Pathogenesis The development of morbid conditions of disease.

Pathology Science dealing with all relations of diseases. Its nature, structural and functional changes brought out by it.

Patri-local joint families When the married couple and their off spring put up with the husband's family or in a new household, which he sets up on his own.

Patronymic If offsprings inherit the father's name.

Pauperization Reduction of vigor and size such that architecture is minimally expressed; may also result in differentiation and reiteration.

Pavna Sehima nervosum (Rottl.) Stapf. (Andropogon hervosus (Rottl.). A perennial grass family Gramineae used as fodder.

Payment by results A technique for increasing productivity; it includes piece-work methods of payment; incentive bonuses, and other schemes relating earnings to output.

PDB (Paradichlorobenzene) A useful soil fumigant. Usually applied in apple trees for control of wooly aphis.

Pea An annual or perennial herb of the genus Pisum. Family Leguminosae. An important vegetable.

Pea-diseases Pea damping off caused by Pythium spp.; pea downy mildew caused by Peronospora pisi; pea foot rot and blight caused by Ascochyta pinodella, A. pisi, and A. pinodes; pea mosaic a virus disease and pea rust caused by Uromyces fabae and U. pisi are important pea diseases.

Pea-insect pests Pea beetle (Laria affinis Frol) attacking pea in field and store, pea leaf miner (Phytomyza atricornis Meign) maggots of which make shining white zig-zag galleries on leaves are important insect pests.

Peach Prunus persica (L). A low, spreading, freely branching tree of the order Rosales, cultivated in less rigorous parts of the temperate zone for its edible fruit, a juicy drupe with a single large seed, a pulpy yellow or white mesocarp, and a thin firm epicarp.

Peach-diseases Peach leaf curl caused by Taphrina deformans Terl; peach sun scald, a physiological diseases are two important diseases.

Peach-insects pests Peach fruit fly (Dacus zonatus Saunders) a pest which causes dark-coloured punctures on fruits; peach leaf curling aphid (Anuraphis helichrysi Kalt) which suck sap from young leaves, leaves get curl up and peach scale (Pseudaulacaspis pentagona Targ.) which sucks juice from twigs are the few insect pests of peach.

Peak A maximum quantity or height which occurs over a relatively short period of time, e.g. peak discharge.

Peanut Arachis hypogaea. A low, branching, self-pollinated annual legume cultivated for its edible seed, which is a one-loculed legume formed beneath the soil in a pod. Also groundnut.

Pear Any of several tree species of the genus pyrus in the order Rosales, cultivated for their fruit, a pome that is wider at the apical end and has stone cells throughout the flesh.

Pear-diseases Pear pink disease caused by Pellicularia salmonicolor (B & Br.); pear sooty blotch and fly speck caused by Leptothyzium pomi (Mont. Fr.) Sacc; pear stem black caused by Coniothecium chomatosporum corda are the diseases in pear.

Pearl millet See *Bajra*.

Peat Unconsolidated soil material consisting largely of undecomposed; or only slightly decomposed, organic matter, that has accumulated under excessive moisture.

Peat moss Moss, especially sphagnum moss, from which peat has been produced.

Peat soil An organic soil having no, or only slight decomposition.

Peavey A stout wooden lever for rolling logs. A curvey metal hook is hinged to the lower part of the handle, and the tip is armed with a sharp steel spike.

Pebble Small stone worn and rounded by action of water.

Pebrine Dangerous disease of silk worm caused by Nosema bombycis. worms show high mortality.

Pecan Carya illinoensis. A large deciduous hickory tree in the order Fagales which produces an edible, oblong, thin shelled nut.

Pectase An enzyme of plants which forms vegetable jelly. Also called pectinase.

Pectin Polysaccharide from the middle lamella of the plant cell wall; jelly-forming substance found in fruit.

Pecuniary culture A social system in which the accumulation of physical wealth is both acceptable to that society, and valued highly by that society.

Pecuniary effects Effects arising from changes in relative prices which only redistribute goods and services among people without affecting total production or consumption opportunities. Such effects involve improvements in the economic position of some persons but correspondingly impaired positions for others.

Ped A unit of soil structure such as a

prism, block or granule, which is formed by natural processes, in contrast with a clod, which is formed artificially; a natural aggregate.

Pedators Animals that feed primarily on forages.

Pedicel (1) The stem of a fruiting on spore-bearing organ. (2) The stem of a single flower.

Pediculus Infested with lice.

Pedigree Having a known line of descent; genealogical table.

Pedigree method of breeding A system of individual plant selection during the segregating generations of a cross where the progeny plants usually are separately spaced and the pedigree of particular selections is known.

Pedocal A mature soil with a horizon of accumulated carbonates in the soil profile.

Pedogenesis See genesis, soil.

Pedology The aspects of the soil science dealing with the origin, morphology, genesis, distribution, mapping and taxonomy of soils, and classification in terms of their use.

Pedosphere The outer soil layer of earth in and on which organic life exists.

Peduncle The primary stalk supporting either an inflorescence or a solitary flower. In grasses, the uppermost internode of the culm.

Peel Rind; the outer coverage or skin of fruits and vegetables.

Peeler (1) Usually one who removes bark from timber cut in the spring months when bark slips. (2) A log used in the manufacture of rotary-cut veneer.

Peg The plant mechanism by which the fertilised peduncle forces the ovary under the ground where it develops into a pod as in ground nut.

Peg graft A graft made by driving a scion of leafless dormant wood with wedge-shaped base into an opening in the stock and sealing with wax or other material.

Pelite Rock composed of fine particles such as clay, which yields silty and clayey soils on weathering.

Pellet A small, cylindrical, compressed mass of livestock feed.

Pellicle The thin, firm outer membrane of certain unicellular organisms such as Euglena.

Pellicular water Films of groundwater adhering to particles or cavities above the watertable.

Pelophyte A plant growing in clayey soil.

Peltate Of leaves, having the petiole attached to the lower surface instead of the base.

Pempheres offinis See Cotton stem weevil.

Peneplain A rugged area that was high at one time, but has been reduced by erosion to a low, gently rolling surface resembling a plain.

Penetrability The ease with which a probe can be pushed into the soil. It may be expressed in units of distance, speed, force or work depending on the type of penetrometre used.

Penetration test A test to determine the relative values of density of non-cohesive sand or silt at the bottom of boreholes.

Penetrometre Device for measuring the force required to push a probe rod into the soil.

Penicillin A powerful, nontoxic, anti-infective substance found in the mold fungus Penicillium.

Pennales An order of diatoms (Bacillariophyceae) in which the form is often circular, and the markings on the valves are radial.

Pennisetum ciliare See Anjan grass.
— **clandestinum** See Kikuyu grass.
— **nervosum** See Bent spike Pennisetum.
— **pedicellatum** See Natal grass.
— **purpureum** See Napier grass.
— **typhoideum** See Bulrush millet.
— **typhoides** See Bulrush millet.

Pentadelphous Having the stamens in five sets with the filaments more or less united within each set.

Pentagynous Of a flower, having five styles.

Pentalonia nigronervosa See black banana aphid or cardamom marble disease.

Pentamerous Having each whorl of the flower consisting of five members or a multiple of five.

Pentandrous Having five stamens.

Pentavalent With combining power of five atoms of hydrogen.

Pepo A fleshy indehiscent berry with many seeds and a hard rind, characteristics of the cucurbitaceae.

Pepper Any of several warm-season perennials of the genus Capsicum in the order Polemoniales, especially C. annum which is cultivated for its fruit, a many-seeded berry with a thickened integument.

Pepper-diseases The diseases are pepper black root rot caused by Rosellinia bunodes; pepper pollu or anthracnose caused by Colletotrichum piperi Roy; Pepper wilt caused by sclerotium rolfsii Sacc. & P.

Pepper-insect pests Pepper flee beetle (Longitarsus nigripennis M.), whose grub bores into ripening berry is the common insect pest.

Peppermint Any of various aromatic herbs of the genus mentha in the family Labiatae especially M. piperita.

Pepsin Digestive enzyme, occurring in the gastric contents and acting on proteins.

Peptones Simple, water soluble proteins which do not co-agulate when heated. Used as digestible nutrients or as culture media.

Percentage distribution A frequency distribution with the total frequency equated to 100 and the individual class frequencies in proportion to that figure.

Percentage germination The percentage germination on the seed analysis certificate indicates the proportion of the number of seeds which have produced seedings classified as normal, under the conditions and within the period specified.

Percentage regulation of speed It is the variation in engine speed between full load and no load conditions. It is usually expressed as a percentage of rated speed.

Percentage runoff The amount of runoff expressed as a percentage of total rainfall on a given area.

Percentiles The set of partition values which divide the total frequency into 100 equal parts.

Perception It is how one feels about a thing and interprets in one's neural mechanism. The farmers and the extension educationists should perceive in the same way to bring out an effective change about an innovation.

Perched stream A stream whose surface level is above that of the water table and that is separated from underlying groundwater by an impermeable bed in the zone of aeration.

Perched water Groundwater that is unconfined and separated from an underlying main body of groundwater by an unsaturated zone. Also known as perched groundwater.

Perched watertable A watertable due to the "perching" of water on a relatively impermeable layer at some depth within the soil. The soil within and below the impermeable layer is not saturated with water. Also known as apparent watertable.

Percolation, soil water The downward movement of water through soil, especially the downward flow of water in saturated or nearly saturated soil at hydraulic gradients of the order of 1.0 or less.

Perennial (1) A plant that lives for an indefinite period (atleast more than two years) dying back seasonally and producing new growth from a perennating part. (2) Flowing during all the year as perennial stream, perennial river or canal.

Perennial system of irrigation, or **perennial irrigation** System of irrigation is said to be perennial when the land under the crops receives assured supplies from the rivers or other sources of supply throughout the year.

Perfect flower A flower having both stamens and pistils.

Perfect state Stage of life-cycle in which spores are formed after nuclear fission.

Perfoliate Pertaining to the form of a leaf having its base united around the stem. Pertaining to the form of certain insect antennae having the terminal joints expanded and flattened to form plates which encircle the stalk.

Perforated pipe method A sprinkler method consisting of portable, lightweight pipes in which small holes (perforations) have been drilled in such an arrangement as to cause the water to be distributed fairly evenly over strips of land on both sides of the pipe.

Performance standard A specific measure for control of human activity.

Perianth The floral envelope, i.e., calyx and corolla.

Periblem A group of embryonic cells in root or stem tip that gives rise to the cortex.

Pericarp The wall of a fruit, developed by ripening and modification of the ovarian wall.

Pericentral cell In the red algae, the cell which ultimately gives rise to the carpogonium and the auxiliary cells.

Periclinal Parallel to the outer surface, as the periclinal walls in cell division.

Pericycle The outer boundary of the stele of plants, may not be present as a distinct layer of cells.

Pericyclic fibres Thick walled, elongated cells arising in the pericycle.

Peridium the outer coat or wall of the fruiting body of certain fungi; the outer covering of the basidiocarp of pufballs.

Peridolite Igneous rocks.

Peridorm A group of secondary tissues forming a protective layer which replaces the epidermis of many plant stems, roots, and other parts, composed of cork cambium, phelloderm, and cork. The superficial transient layer of epithelial cells of the embryonic epidermis.

Perigean tide Tides with increased range which occur when the moon is in the perigee.

Periglacial Indicative of all cold climate processes, whether or not they occur in the immediate vicinity of glaciers.

Perigynium A fleshy cup or tube like structure surrounding the archegonium of various bryophytes.

Perigynous Having a flower structure in which the stamens and perianth parts are borne on a cup or disk formed by the receptacle surrounding the ovary.

Perimedullary Surrounding a central core, as the tissues around pith.

Periodic block The part or parts of a forest set aside to be regenerated or otherwise treated during a specified period.

Periodic increment The growth of a stand of trees for any specified period. It is annual if the stand growth is divided by number of years in the period.

Periodic variations The recurrent patterns of change within the period that results from the operation of forces connected with the period of custom at different times of the period. When connected with the climate or season, these variations are referred to as 'seasonal variations'.

Period of concentration or **Time of concentration** The time taken by the runoff from the farthest point of the catchment to reach the point of concentration.

Peripheral Pertaining to the surface of the body or a plant.

Periphysis The organizational status of a tree meristem influenced by environment.

Periplast The outer cell membrane, especially of such organisms as Euglena.

Perisperm The nutritive tissue in a seed formed outside the embryo sac. It usually comes from the nucleus.

Peristome A ring of teeth surrounding the opening of a moss capsule and by its hygroscopic movements scattering the spores.

Perithecium A spherical or pear shaped ascocarp opening by means of a terminal pore (of the pyrenomycetes).

Permafrost Any part of the earth's mantle maintaining a temperature of 0°C or lower continuously for two or more years; of high latitudes in both hemispheres and of high altitudes of major mountain systems. Also perennially frozen soil horizon.

Permafrost table The upper boundary of permafrost usually coincident with the lower limit of seasonal thaw. See also permafrost.

Permanent charge The net negative or positive charge of clay particles inherent in the crystal lattice of the particle. It is not affected by changes in pH or by ion-exchange reactions.

Permanent parasite A parasite which spends all its life stages in or on the body of the host.

Permanent pasture One that is used indefinitely for grazing by livestock.

Permanent sprinkler method, or **stationary sprinkler method** The method made up of permanently located pipeline system, usually buried, conveying the water from the source of supply to sprinklers.

Permanent villages In such villages, the settled human aggregates live for generations. They develop their farm practices, village organisations and social relations with the neighbouring villages. After some time, they organise a local or regional government.

Permanent wilting percentage See wilting coefficient.

Permanent wilting point The percentage of moisture at which plants wilt and fail to recover when placed in an atmosphere saturated with water vapour. Also known as wilting percentage.

Permeability, soil (1) The ease with which gases and liquids penetrate or pass through a bulk mass of soil or a layer of soil. Because different soil horizons vary in permeability, the specific horizon should be designated (2) The property of a porous medium that relates to the ease with which gasses or liquids can pass through it. Previously, it was considered to be the 'K' in Darcy's law. It is the 'K' in intrinsic permeability.

Permanganate of potash A chemical used as disinfectant, fungicide and oxidizing agent. Chemically it is $KMnO_4$.

Permatode Roundworms living in soil or water.

Permian The most recent period of the Paleozoic. The permian follows thee Carboniferous and precedes the Triassic.

Permissible area That portion of area served by an outlet which is intended to be irrigated in a given period is known as the permissible area for that period, e.g. annual 'permissible area'.

Permissible velocity The maximum velocity of water at which it flows in a channel without scouring the sides of the channel.

Peroxyacetyl nitrate (Pan) A reactive gas formed in photochemical smog, causing eye irritation as well as plant damage.

Perpelic Plants which grow on clay soils.

Persea americana See Avocado.

— **drymifolia** See Avocado.

— **gratissima** See Avocado.

Persian clover Trifolium resupinatum L. Family Papilionaceae. Annual fodder legume.

Persimmon Tree of genus Diospyros, family Ebenaceae. Some produce edible fruits.

Persistent Of a leaf, withering but remaining attached to the plant during the winter.

Pervious soil A soils permitting water flow easily; a light texture soil.

Pest A destructive insect.

Pesticides Chemicals that kill organisms that are injurious to man or to the crops and animals upon which he depends for food, fibre, and shelter. These organisms include insects. mites, micro-organisms and rodents. Pesticides include insecticides, fungicides, herbicides, and others.

Petal One of the sterile, leaf-shaped flowers part that make up the corolla.

Petalody A morphological change in which other parts of a flower are changed to petals.

Petiolar Growing on a small stalk.

Petiole A leafstalk. The slender stalk by which the blade of a leaf is attached to the stem.

Petrosaviaceae A small family of monocotyledonous plants in the order Triuridales; characterised by perfect flowers, three carples, and-numerous seeds.

Petri dish Round glass dish with cover used for growing bacterial cultures.

Petrography Scientific description of formation and composition of rocks.

Petrol European name for gasoline. The term is little used in the United States.

Petroleum A natural oil taken from the earth, used extensively for heating and lighting. By distillation, it yields many products valuable in the industries, such as gasoline, kerosene, paraffin, etc.

Petrology Study pertaining to rocks.

pF The logarithm of the soil moisture tension measured by the height in centimetres of a column of water that produces a tension of equal force.

Phaeophyta The brown alagae, constituting a division of plants; the plant body is multicellular varying from a simple filamentous form to a complex, sometimes branched body having a basal attachment.

Phage An agent causing destruction or lysis of micro-organism.

Phage typing Use of bacteriophages and their lytic properties to classify bacteria.

Phakospora vitis See Grapevine rust.

Phalanx Segment of digits of vertebrates; bundles of stamens united by filaments.

Phalloid Resembling a mushroom of the genus phallus, named for a fancied likeness to the human penis.

Phanerogam A plant whose flowers develop pistils and stamens.

Phanerocotylar See epigeal germination.

Phase converters A device which enables a three phase motor to operate on single phase power.

Phase, Soil A sub-division of a soil type or other unit of classification having characteristics that affect the use and management of the soil, but that do not vary sufficiently to differentiate it as a separate type. A variation in a property or characteristic such as degree of slope, degree of erosion, or content of stones.

pH See pH, Soil

pH dependent cation exchange capacity The difference between the effective cation exchange capacity and the cation exchange capacity of a soil measured at a pH higher than that of its natural value.

pH measurement Determination of the hydrogen-ion concentration in an ionized solution by means of an indicator solution (such as phenolphthalein) or a pH metre.

pH metre An electronic voltmetre using a pH responsive electrode that gives direct conversion of voltage differences to differences of pH at the temperature of the measurement.

pH, soil The negative logarithm of the hydrogen ion activity of a soil. The degree of acidity or alkalinity of a soil as determined by means of a glass, quinhydrone, or other suitable electrode or indicator at a specified moisture content or soil water ratio, and expressed in terms of the pH scale.

Phellem Cork or cells of the bark.

Phelloderm Secondary tissue formed by cork cambium or phellogen on its inner surface.

Phellogen The meristematic portion of the periderm, consisting of one layer of cells that initiate formation of the cork and secondary cortex tissue.

Phenol Coal tar derivative with a characteristic aromatic odour. Antiseptic and gemicide. Chemically C_6H_5OH. Also called carbolic or phenic acid.

Phenol coefficient Disinfecting property of a chemical as compared with that of phenol (carbolic acid).

Phenospermy The production of seed which can not reproduce.

Phenothiazine A organic stomach poison.

Phenotype An organism exemplified by its observed characters without reference to its genetic constitution; individuals of a phenotype may look alike without breeding alike. See Genotype.

Phialide A cell which develops one or more open ends from which a basipetal succession of conidia develops without an increase in length of the Phialide itself.

Philosophy A body of principles underlying a given branch of learning.

Phloem A complex, food-conducting vascular tissue in higher plants; principal conducting cells are sieve elements. Also known as sieve tissue.

Phloem necrosis A pathological state in a plant in which the phloem undergoes brown discolouration and disintegration.

Phony disease A virus disease of the peach characterised by dwarfing, abnormal darkening of leaves, and a light crop of small, highly coloured fruit, tree stops bearing fruit after a few years.

Phoresy A type of commensation which involves the transpiration of one organism (the guest) by a larger organism (the host) of a different species.

Phosphate rock A natural rock containing one or more calcium phosphate minerals and used in manufacturing of phosphorous fertilisers.

Phosphorescene The emission of light without appreciable heat; the property of continuing to be luminous in the dark after exposure to light or other radiation.

Phosphorus Essential plant food element absorbed in the form of $H_2PO_4^-$.

Photic zone The uppermost layer of a body of water (approximately in the upper 100 metres) that receives enough sunlight to permit the occurrence of photosynthesis.

Photoautotrophic Able to synthesise its own food in the presence of light, as a plant.

Photochemical Pertaining to a chemical reaction activated by light.

Photoecology The application of air photography to ecology integrated land resources studies, and forestry.

Photogrammetry The science or art of obtaining surveys or marking maps by means of photography.

Photo-index map An index map by assembling the individual photographs into their proper relative position and photographically copying the assembly at a reduced scale.

Photomap A mosic map made from aerial photographs showing physical and cultural features as on a planimetric map. See also planisaic.

Photolysis Chemical changes brought about by the absorption of light.

Photoperiod Light period or length of day to which plants are exposed

Photoperiodic adaptation The adjustment of plants to a definite length of daily exposure to light.

Photoperiodism Response of a plant organism to the relative lengths of day and night (light and dark), particularly in respect to floral initiation.

Photoreaction A reaction that is initiated or hastened by light.

Photorespiration The respiration of plant cells that occurs in the presence of light and while the cells are concurrently carrying on photo-synthesis, especially active in C_3 plants and nearly or completely absent in C_4 plants. Plants also respire in the dark at the expense of substrates generated by photosynthesis in earlier light periods.

Photosynthesis The formation of carbohydrates from water and the carbon dioxide of the air in the chlorophyll-containing tissues of plants exposed to light. Also called assimilation.

Photothermal Pertaining to combined effects of light and temperature. Also called heliotropism.

Phototropism A growth-mediated response of a plant to stimulation by visible light.

Phragmoid Having septae perpendicular to the long axis, as the conidia of certain fungi.

Phreatic water Groundwater in the zone of saturation.

Phreatophyte A plant with a deep root system which obtains water from the ground water or the capillary fringe above the water table.

Phthorimaea glapsigona See Brinjal bud worm.

Phycocyanin A blue green pigment associated with chlorophyll in the Cyanophyta.

Phycoerythrin A red pigment associated with chlorophyll in the Rhodophyta.

Phycology The study of algae. Also algology.

Phyctia clientella See Brinjal leaf caterpillars.

Phyllary A bract of the involucre of a composite plant.

Phylloclade A flattened stem that fulfills the same functions as a leaf.

Phyllocnistis citrella See citrus leaf miner.

Phyllode A broad, flat petiole that replaces the blade of a foliage leaf.

Phyllody Metamorphosis of an organ into a foliage leaf.

Phylogeny The evolutionary or ancestral history of organisms.

Phyllogoniaceae A family of mosses in the order Isobryales in which the leaves are equitant.

Phyllotaxy The arrangement of leaves on a stem.

Phyllomorphic branch Determinate short-lived and strongly plagiotropic branch axis with some or many of the biological characters of a compound leaf.

Phyllophaga longipennis See White grubs.

Phyllophagan Those animals who feed on leaves.

Phyllosilicates Silicate structures in which the SiO_4 tetrhedra are linked together in infinite two-dimensional sheets and are condensed with layers of AlO or MgO octahedra in the ratio 2:1 or 1:1 isomorphous substitution of certain elements often occurs.

Phyllotaxis The arrangement of leaves on a stem or axis.

Phyllotreta cruciferae See cucurbit flea beetle.

Phylum A large taxonomic unit, subdivided into closely related classes.

Physical climate The actual climate of a place, as distinguished from a hypothetical climate, such as the solar climate or mathematical climate.

Physical poison A poison having physical rather than bio-chemical effect.

Physical properties of soils The characteristics, processes or reactions of a soil that are caused by physical forces, and are described by, or expressed in physical terms or equations. Sometimes physical properties are confused with and hard to separate from chemical properties hence the terms "physical-chemical" or "physicochemical". Examples of physical properties are bulk density, water holding capacity, hydraulic conductivity, porosity, and pore-size distribution.

Physical weathering The breakdown of rock and mineral particles into smaller particles by physical forces such as frost action and wind. See also weathering.

Physiogenic disease Caused by unfavourable environmental factor.

Physiognomy The general external appearance of a community of plants as determined by the dominant species.

Physiography Description of nature of natural phenomena.

Physiological ecology Study of the physiological mechanisms or correlates of behaviour.

Physiology (1) (seed) The study of the functions, and activities of seed. (2) Science that deals with life, living organisms, their functions and activities.

Physiologic races Biotypes or groups of biotypes within species that behave more or less consistently in pathogenicity on certain differential varieties or host plants. Physiologic races sometimes are differentiated on the basis of cultural or physiochemical characters.

Physiosorption The process of attachment of nonionic substances such as polar water molecules, acetic acid molecules, or nucleic acids to clays or other solid phase surfaces. The attachment of large molecules to clay particles by ionic processes is not physiosorption.

Physoderma alfalfae See Lucerne crown wart.

— **trifolii** See Clover crown wart.

— **zeamaydi** See Maize brown spot.

Phyte Denoting a vegetable organism. For example saprophyte.

Phyto Denoting a plant.

Phytoclimatology The study of the micro-climate in the air space occupied by plant communities, on the surface of the plants themselves and in some cases, the air spaces within the plants.

Phytoedaphon The microflora of the soil.

Phytogenesis The evolution and development of plants.

Phytohormone Organic chemical compounds which modify physiological processes in an appreciable manner when used in small quantities.

Phytogenic soils (Obsolete) Soils developed under the dominant influence of natural vegetation, mainly in temperate regions.

Phytome The vegetable substance composing a plant.

Phytometre (1) A plant or group of plants used to measures the physical factors of the habitat in terms of physiological activities. (2) A device for measuring transpiration consisting of a vessel containing soil in which one or more plants are rooted and sealed so that water can escape only by transpiration from the plant.

Phytomorphic soils Well drained soils of an association that have developed under the dominant influence of the natural vegetation characteristic of a region. The zonal soils of an area.

Phytomyza atricornis See pea-insect pests.

Phyton A rudimentary plant's smallest

detached portion which can form another plant.

Phytopathogen An organism that causes a disease in a plant.

Phytopathology Plant pathology, science of plant diseases.

Phytophagus Herbivorous.

Phytoplankton The plant component of the plankton.

Phytoplasm Protoplasm.

Phytosis Production of disease such as by fungi or any other vegetable parasites.

Phytotoxic Poisonous to plants.

Phytotoxicant Poisonous to plants.

Phytotoxin Any toxin originating in plants.

Pierce's disease A virus disease of grapes in which there is mottling between the veins of leaves, early defoliation, ripening and withering in the fruit.

Piezometre An instrument for measuring the pressure head of liquids.

Piezometric surface The surface at which water will stand in a series of piezometres.

Pigeon pea Cajanus cajan (L.) Milsp. (C. indicus Spreng.). Family Papilionaceae. A pulse crop whose green pods are used as vegetable and seed as pulse. A *kharif* crop grown in rabi as well. Called *arhar* (Hindi).

Pigeon pea pod fly Agromyza obtusa Mall. Insect pest whose larvae enter pods and feed on seed.

Pigeon pea wilt Disease caused by Fusarium udam Butler. Leaves turn pale and yellow and wither. Plants die in patches till they are mature.

Pigment Substances that appear coloured by virtue of differential absorption of radiant energy. These substances impart colour to tissue of plants. Green colour is a result of chlorophyll; orange and some red colours are due to many carotenoids; many red to blue colours are antho cynanins, light yellow colours are flavones.

Pile A long, straight, stiff structural element, usually of circular or square cross section, which is driven, or otherwise introduced into the soil, usually for the purpose of providing vertical or lateral support.

Pileus Hymenium-supporting part of fruit body of a higher fungus; the cap of a mushroom.

Pimento Capsicum annuum. A type of pepper in the order Polemonials grown for its thick, sweet-fleshed red fruit.

Pina A fibre obtained from the large leaves of the pineapple plant. Also known as pineapple fibre.

Pinching Removing of terminal shoots of plants to encourage bushy growth or to prevent flower production in early stages.

Pine Any of the cone-bearing trees composing the genus pinus; characterised by evergreen leaves (needles), usually in tight clusters of two to five.

Pine apple Ananas sativus. A perennial plant of the order Bromeliales of family Bromeliaceae, with long, sword like, usually rough-edged leaves and a dense head of small abortive flowers; the fruit is a sorosis that develops from the fleshy inflorescence and ripens into a solid mass, covered by the persistent bracts and crowned by a tuft of leaves.

Pineapple crown Cluster of short leaves on the head of pineapple fruit; used in propagation.

Pineapple-diseases Pineapple soft rot or pineapple stem end rot caused by cerotostomella paradoxa (de Seynes) Dade; and pineapple storage rot caused by ceratostomella paradoxa are important.

Pineapple mealy bug Pseudococcus bromeliae B. Small mealy bug which infests shoots and fruits and sucks saps of reddish colour.

Pineapple slip Shoot borne on the fruiting stem of pineapple plant; also used in propagation. Also called pineapple gill.

Pine needle oil An essential oil derived from various pines; colourless to yellowish oil with balsamic aroma; soluble in alcohol; used in perfumes and medicines. Also known as Douglas fir oil; fir wood oil.

Pine nut The edible seed borne in the cone of various species of pine (Pinus), such as stone pine (P. Pinea) and pinion pine (P. cembroides var. edulis).

Pinepsida A class of gymnospermous plants in the sub-division Pinicae characterised by entire-margined or slightly toothed, narrow leaves.

Pinicae A large sub-division of the Pinophyta. Comprising woody plants with a simple trunk and excurrent branches, simple, usually alternative, needle like or scale like leaves, and wood that lack vessels and usually has resin canals.

Pinion The smaller gear of a pair, either bevel or spur, regardless of size.

Pinion gear and shaft The pinion gear and shaft, made as a unit from one piece of steel, carries the power from the propeller shaft to the rear axle. The unit is supported by two tapered roller bearings; all are contained in a housing attached to the front of the rear axle housing.

Pink bollworm Platyedra gossypiella Saund. Pest of cotton whose caterpillar attacks buds, flowers, bolls and seeds. Also cuts lint, discolours and soils it.

Pink borer Sesamia inferens Walk. Insect pest of many cereals and sugarcane. Boses into stems. Also called stem borer.

Pink disease A fungus disease of the bark of rubber, cacao, citrus, coffee, and other trees caused by Corticium salmonicolour and characterised by pink covering of hyphae on the stems and branches.

Pink root A fungus disease of onion and garlic caused by various organisms, especially species of Phome and Fusarium, marked by red discolouration of the roots.

Pink rot (1) A fungus disease of potato tubers caused by Phytophtora erythroseptica and characterised by wet rot and pink colour of the cut surfaces of the tuber upon exposure to air. (2) A rot disease of apples caused by the fungus Tricothecium roseum. (3) A watery-soft rot of celery caused by the fungus sclerotinia sclerotiorum.

Pink stage Term applied to tomato fruit indicating the time to pick it for marketing. Pink colour varies from trace at the blossom end to considerable amount covering the surface.

Pinophyta The gymnosperms, a division of seed plants characterised by vascular plants with roots, stems, and leaves, and with seeds that are not enclosed in an ovary but are borne on cone scales or exposed at the end of a stalk.

Pinna A leaflet of a fern leaf; the primary division of pinnate leaf.

Pinnate Feather-like; having the leaflets of the primary divisions of a leaf arranged on each side of an axis.

Pinnate leaf Leaflets arranged along both sides of a common petiole, as in peas.

Pinnately veined Manifesting a net venation in which the secondary veins branch out in parallel fashion from the single midrib of a leaf blade.

Pinnule One of the ultimate divisions of a bi-pinnate leaf.

Pint A unit of measuring volume of liquids. One-eighth of a gallon.

Pinus The genus of the family Pinaceae; to true pines, coniferous trees of north temperate regions having early deciduous primary leaves, needle-like secondary leaves, and cones with woody scales.

Pioneer plant A plant species capable of invading bare areas.

Pip (1) Single blossom of clustered inflorescence. (2) Seed of apple, pear, orange. (3) The raised crown or individual root stock.

Piperaceae A family of dicotyledonous plants in the order Piperales characterised by alternate leaves, a solitary ovule, copious perisperm, and scanty endosperm.

Piperales An order of dicotyledonous herbaceous plants marked by ethereal oil cells, uniaperturate pollen, and reduced crowded flowers with orthotropous ovules.

Pine culvert A buried pipe for carrying a water course below ground level.

Pipe coupling A threaded sleeve used to connect two pipes.

Pipe die A screw plate used for cutting threads on pipe.

Pipe fittings A general term referring to ells, tees, various branch connectors, etc., used in connecting pipes.

Pipe model theory Theory which relates units of crown structure to vascular units, or 'pipes' in the axes supporting them (cf. unit pipe).

Piricularia oryzae See rice blast and Italian millet blast.

Pisciculture Cultivation of fish under artificial conditions.

Pistachio Pistacia vera. A small, spreading dioecious evergreen tree, with leaves that have three to five broad leaflets, and with large drupaceous fruit, the edible seed consists of a single green kernel covered by a brown coat and enclosed in a tough shell.

Pistill The ovule-bearing organ of angiosperms; consists of an ovary, a style, and a stigma.

Pistillate (1) Having a pistil. (2) Having pistiles but no stamens.

Pistillody The change of stamens into carpels.

Piston The plunger which moves within the cylinder of an engine or pump. The efficiency of compression depends very largely on the proper fitting of the piston.

Piston head The top, closed end of a piston.

Piston pin A hollow steel shaft, hardened and ground, which connects the upper end of the connecting rod to the piston. It is also called "wrist pin".

Piston pin bosses Those parts of a piston which carry the ends of the piston pins.

Piston ring A spring packing ring for a piston.

Piston rod The rod which moves the piston and is connected to the cross head or crankshaft.

Piston pump Pump used for creating head. Not very commonly used.

Piston skirt That part of a piston below the piston pin.

Piston stroke The complete length of travel of a piston in its cylinder.

Piston valve A piston operating in a cylindrical case having parts which are opened and closed by motion of the piston.

Pit (1) A cavity in the secondary wall of a plant cell, formed where secondary deposition has failed to occur, and the primary wall remains uncovered; two main types are simple pits and bordered pits. (2) The stone of a drupaceous fruit. A small hole in the surface of a metal; usually caused by corrosion or formed during electroplating operations. (3) A coal mine; the term is not commonly used by the coal industry, except in reference.

Pitcher irrigation A method of irrigation with buried pitchers (pots) where water is supplied to the roots through the pores of the pitcher.

Pitcher plant Any of various insectivorous plants of the families Sarraceniaceae and Nepenthaceae; the leaves form deep pitchers in which water collects and insects are drowned and digested.

Pitch (1) The resin from wood of conifers. (2) A heavy, dark coloured material obtained by distillation of tar derived from coal, wood, petroleum oils.

Pitch fork An implement to handle waste materials, hay and straw etc. Consists of a fork with a long handle.

Pitching, or **Beaching** A protective covering of properly packed or built-in materials on the earthen surface slopes of irrigation canals, drainage channels, and river banks, etc., to protect them from the action of water.

Pith A central zone of parenchymatous tissue that occurs in most of the vascular plants and is surrounded by vascular tissue.

Pitman A connecting rod that joins the crank of a rotary with a reciprocating part, as the knife of a cutter bar. Commonly used in machinery.

Pitot tube Device for measuring the velocity head of flowing fluid, consisting essentially of two orifices; one pointing upstream and one pointing downstream, connected to a differential manometre.

Pitting Making shallow pits of suitable capacity and so distributed so as to retain water from rainfall or snowmelt on rangeland or pastures; small cavities in a surface created by corrosion, cavitation or subatmospheric pressures.

Placenta A plant surface bearing a sporangium. A vascular organ that unites the fetus to the wall of the uterus in all mammals except marsupials and monotremes.

Placentation The management of ovules and placenta in the ovary of a plant. The formation, and fusion of the placenta to the uterine wall.

Placic Pertaining to a light black to dark reddish pan or horizon and presumably cemented by iron.

Plagiotropic complex Set of interacting meristems forming an extended plagiotropic branch system, from not necessarily plagiotropic axial components.

Plagiotropy Gravitational response which produces an oblique or horizontal axis, hence plagiotropic shoot with the complex of characters resulting from this response i.e., dorsiventral symmetry and horizontal orientation (cf. orthotropy).

Plagiotropy by apposition Character of a plagiotropic complex determined by apposition growth.

Plagiotropy by substitution Character of a plagiotropic complex determined by substitution growth.

Plain A extensive area of nearly flat or flat as Indo-Gangetic plains. It may be alluvial, coastal, desert etc.

Plains bamboo Bambusa balcooa Roxb.; Family Gramineae. Commonly used for building purposes.

Planetary gear Consists of an internal ring gear, a pinion on a short shaft through which power is applied, and three pinions carried on a plate or spider; these three pinions mesh both with the shaft pinion and the ring gear.

Planimetric map A map which presents the horizontal positions only for the features represented; distinguished from a topographic map by the omission of relief in measurable form. The natural features usually shown on a planimetric map include rivers, lakes and seas, mountains, valleys and plains and forests, prairies, marshes and deserts. The cultural features include cities, farms, transportation routes and public utility facilities, and political and private boundary lines. A planimetric map intended for special use may present only those features which are essential to the purpose to be served.

Planking, smoothing, or floating The precision touch-up in preparing land for irrigation, or the annual operation. of eliminating small irregularities caused by farming operation. Operation is carried out by a planker or a float.

Planisaic A photomap in which the planimetric detail is shown by overprints in colour. See also photomap and toposaic.

Plankton A feebly swimming and free-floating organisms unable to move against a current, mostly microscopic as plants (Phytoplankton) bacteria, algae (especially diatoms, dinoflagellates), and animals (Zooplankton) — protozoa (especially radiolarions, foraminiferans), coelenterates, centenophorans, crustaceans.

Plan, Management A written plan for the operation of a forest property, using forestry principles. It usually records data and prescribes measures designed to provide optimum use of all forest resources.

Planned economy An economy in which the problems of production, distribution and growth are resolved by a central planning authority.

Planning The plotting of a course of action. Planning may be merely 'indicative', involving the collection, processing and distribution of information of others responsible for policy making and executive action; or it may be 'normative', actually involving executive 'action or enforcement'.

Plan of work It is an outline of activities so arranged as to enable effective execution of the entire programme. It answers questions as to how, when, where and by whom the work is to be done.

Planosols A soil developed upon nearly flat upland surfaces under grass or forests in a humid or subhumid climate. The soil is with strongly eluviated surface of horizons underlain by a B horizon, more strongly illuviated, cemented or compacted.

Plant (1) An organism that take its food nutrients in liquid solution only. (2) To plant or sow crop or tree seedlings.

Plantae The plant kingdom.

Plantaginaceae The single family of the plant order Plantaginales.

Plantaginales An order of dicotyledonous herbaceous plants in the subclass Asteridae, marked by small hypogynous flowers with a persistent regular corolla and four petals.

Plant breeder Person or organisation actively engaged in the breeding and maintenance of varieties of plants.

Plant breeding The improvement of plants by selection, inbreeding, hybridization and other scientific methods.

Plant ecology Science which deals with plants in relation to their environment.

Plant geography A major division of botany, concerned with all aspects of the spatial distribution of plants. Also known as geographical botany; phytochorology; phyto-geography.

Planting ratio The recommended ratio in which the male and female parental lines are planted to make a crossing block in hybrid seed production.

Planting stock Seedlings whose kind, age, certification, source of seed and other necessary information is known.

Plagues Visible areas of cellular damage caused by a virus inoculated into susceptible tissue culture cells, analogous to colonies of bacteria on an agar plate.

Plant key An analytical guide to the identification of plants, based on the use of contrasting characters to subdivide a group under study into branches.

Plant kingdom The worldwide array of plant life constituting a major division of living organisms.

Plant pathology The branch of botany concerned with disease of plants.

Plant physiology The branch of botany concerned with the processes which occur in plants.

Plant societies Assemblages of plants which constitute structural parts of plant communities.

Plant tissue Material which consists of cells of similar origin and character.

Plantanaceae A small family of monoecious dicotyledonous plants in which flowers have several carpels which are separate, three or four stamens, and more or less orthotropous ovules, and leaves are stipulate.

Plerome Central Core of primary meristem which gives rise to all cells of the stepe from the pericycle inward.

Plasm Living matter of a cell. Protoplasm, especially general body of this as distinct from nucleus.

Plasmagene See Cytogene.

Plasma membrane A thin membrane that encloses the protoplasm of all living cells.

Plasmin A protein obtained from the plasma of blood.

Plasmodesma Protoplasmic connections between cells.

Plasmology Study of corpuscles of living matter.

Plasmodium The mass of naked multinucleate protoplasm usually having amoeboid movement which is the vegetative stage of a myxomycete.

Plasmolysis Contraction of the living protoplasm from the cell wall as the result of exosmosis.

Plastic Capable of undergoing deformation without breaking.

Plasticity Property which describes the spreading quality of the materials in plastering.

Plasticity constants See liquid limit, plastic limit and plasticity number.

Plasticity number The numerical difference between the liquid limit and the plastic limit, or synonymously, between the lower plastic limit and the upper plastic limit. Sometimes called "plasticity index".

Plasticity range The range of moisture weight percentage within which small sample of soil exhibits plastic properties.

Plastic limit (Atterberg limit) (1) The water content corresponding to an arbitrary limit between the plastic and the semi-solid states of consistence of a soil. (2) The water content at which a soil will just begin to crumble when rolled into a thread approximately 3 mm in diametre.

Plastic soil A soil capable of being molded or deformed continuously and permanently into various shapes by moderate pressure. See also consistence.

Plastic soil-cement Soil cement which is mixed and placed similar to concrete.

Plastid Protoplasmic bodies in a cell which are the centre of chemical activity.

Plastochrone Time interval between two successive similar events, as in leaf development often used in a descriptive sense for its morphological result.

Plateau A table land.

Plate budding A method of budding. Two vertical cuts and one cut across the top are made and rectangle of the bark of the scion is cut to fit in it. The flap of bark is put over the bud and tide.

Plate clutch A clutch which transmits power through two or more plates which are held in contact by the pressure of springs.

Plate count A count of the number of colonies formed on a culture medium that has been inoculated with a small amount of soil in order to estimate the member of certain organisms present in the soil sample.

Platy Consisting of soil aggregates that have developed predominantly along the horizontal axes; laminated; flaky. See also structure types, soil.

Platyedra gossypiella See pink bollworm.

Pleocyta sacchari See sugarcane diseases.

Pleat A defect in plywood due to a veneer being folded parallel to the grain.

Plectostele A specialised type of protostele in which the xylem seems to consist of separate plates with phloem in and around the xylem areas in cross section.

Plenum An air chamber from which air is distributed or collected. It is frequently a main duct of a drying system, where the velocity pressure is reduced to a desired minimum.

Pleonanthy Development of an axis which is not determinate by flowering, i.e. flowers of inflorescences are lateral; hence pleonanthic shoot, a shoot with lateral flowers.

Plerome A group of embryonic cells in the tip of stems and roots that gives rise to the stele.

Pleurocarpous Having the sporophyte in leaf axils along the side of the stem or on lateral branches; refers specifically to mosses.

Plocaederus ferrugineus See Cashew borer.

Plot A measured piece of land.

Plough See plow.

Plough draught The power needed to pull a plough bottom through a soil under specified conditions. A dynamometre is used to measure draught.

Plough size It is the perpendicular distance from wing of the share to the line joining the point of the share and heel of the landside. Light plough 100-150 mm width, Medium plough 100-200 mm width., Heavy plough 200 mm or more width.

Plough pan (Plough sole) See pressure or induced pan.

Plow An implement consisting of a share, mold board, and land side attached to a frame, used to cut, lift, turn and pulverise soil in preparation of a seedbed. Also written as plough.

Plucking Process of erosion whereby blocks of rocks are removed from bedrock.

Plum Any of various shrubs or small trees of the genus Prunus that bear smooth-skinned, globular to oval, drupaceous stone fruit. Family Rosaceae.

Plumbaginaceae The single family of the order Plumbaginales.

Plumbaginales An order of dicotyledonous plants in the subclass (Caryophyllidae); flowers are pentamerous with fused petals, trinucleate pollen, and a compound ovary containing a single basal ovule.

Plum blotch A fungus disease of plums caused by Phyllosticta congesta and characterised by minute brown or gray angular leaf spots and brown or gray blotches on the fruit.

Plum leaf spot Disease characterised by droopings of crop. It is caused by Coccomyces prunophorae Higgins.

Plum pocket A mild fungus disease of plums, caused either by Taphrina prunt or T. Communia, in which the stone of the fruit is aborted.

Plum sun scald Physiological disease characterised by appearance of injury on sides of main stem and branches on south west side.

Plumule The major young bud of the embryo within a seed or seedling from which will develop the aerial portions of the plant. It usually occurs at the tip of stem like structure called the epicotyl; that part of the embryonic plant axis above the cotyledons.

Plunger type oil pump Similar in operation to any cylinder and plunger pump. On one plunger stroke oil is drawn into the cylinder and a check valve closes after it. On the return stroke the piston forces the oil out of the cylinder into the oil lines of the engine past another check valve.

Plurennial Plant that can grow and live for many years.

Plusia orichalcea See pea — insect pests.

— **peponis** See snake gourd — insect pests.

Plutella maculipennis See Diamond black moth.

Plutonic rocks One of two kinds of igneous rocks.

Pluviilignosa A tropical rain forest.

Plywood A board formed of three or more odd layers of veneers cemented or glued usually with the grain of adjacent veneers running at angles to each other.

Pneumatic Pertaining to, or operated by, air pressure.

Pneumatic brakes Brakes operated either by air pressure or vacuum; the pressure being on a large piston to get the proper force.

Pneumatic sprayer A compressed air sprayer for spraying insecticides, pesticides etc.

Pneumatic tire A tire usually consisting of an outer casing or shoe and an inner tube which, when inflated to proper pressure, enables the tire to retain its shape under load.

Pneumatophore (1) An air bladder in marsh plants. (2) A submerged or exposed erect root that functions in the respiration of certain marsh plants, best known in woody swamp or mangrove plants.

Pod A dry dehiscent fruit or seed vessel.

Pod blight A fungus disease of legumes caused by Diaporthe species.

Pod borer Adisura atkinsoni. Caterpillar bores into tender pods to eat away seeds. Major insect pest of pulses.

Podestemaceae The single family of the order Podostemales.

Pod maize Zea mays vari tunicata, family Gramineae. An unusual type of maize each kernel of which is enclosed in a pod or husk.

Podosphaera leucotricha See Apple powdery mildew.

Podosolization The process by which soils become depleted of bases and increase in acidity, and leached surface layers are developed and the clay is removed.

Podostemales An order of dicotyledonous plants in the subclass Rosidae; plants are submerged aquatics with modified branching shoots and small perfect flowers having a superior ovary and united carpels.

Pod rot A fungus disease of cacao caused by Monilia rareri and characterised by lesions on the pods.

Podzol Before 1968, Podzol was a great group in the Podzolic order that included soils having Ae and podzolic B horizons. See also podzolic.

Podzolic An order of soil having podzolic B horizon (Bh, Bhf, or Bf) in which amorphous combinations of organic matter (dominantly fulvic acid). Al, and usually Fe are accumulated. The sola are acid and the B horizons have a high pH dependent charge. The great groups in the order are Humic Podzol, Ferro Humic Podzol and Humo-Ferric Podzol.

Podzolic B Consists of one or more of Bf, Bhf, and Bh. See also horizon, Soil.

Podzolization A process of soil formation resulting in the genesis of Podzolic soils.

Pogonochore A type of plant that produces plumed disseminules.

Point density The relative frequency, or probability mass, which may be located at different point values of a discontinuous variate.

Pointed gourd Trichosanthes dioica Roxb. Family Cucurbitaceae. A climber whose fruit is used as a vegetable. It is called *Parwal* in Hindi.

Point gauge Sharp pointed rod attached to a graduated staff of vernier scale for measuring water surface elevation.

Point integrating sediment sampler A device to take a sample of suspended sediment in a stream. The sampler is lowered to a given point below the stream surface, the pressure within and outside the sampler is equalised to prevent a sudden gushing of water in the sampler is taken out over a period of time, the sampler is closed and raised to the surface so the sample can be removed from the sampler.

Point precipitation Precipitation at a particular site in contrast to the mean precipitation over an area.

Point rainfall A term applied to the rainfall at a single station to distinguish it from average rainfall over an area.

Point throat, or Line throat Throat having no length.

Poison hemlock Conium maculatum. A branching biennial poisonous herb that contains a volatile alkaloid, coniine, in its fruits and leaves.

Poison oak Any of several bushing poison plants or shrubby poison sumaces.

Poisonous plant Any of about 400 species of vascular plants containing principles which initiate pathological conditions in men and animals.

Poison sumac Rhus vermix. A tall bush of the sumac family (Anacardiaceae) bearing pinnately compound leaves with 7-13 entire leaflets, and drooping, axillary clusters of white fruits that produce an irritating oil.

Polar bodies Deeply staining bodies found in one or both ends of certain species of bacteria.

Polarity Morphological and physiological direction in plants, largely based on a gradient of metabolic activities.

Polar nucleus One of two free nuclei in an embryo sac that unite to form the fusion nucleus.

Polder A tract of low land reclaimed from the sea, or other body of water, by dikes, etc. In the polder, the runoff is controlled by sluicing or pumping and the water table is independent of the water table in the adjacent areas.

Pole Extremity of main axis of any spherical or oval substance or diagram.

Pole blight A destructive disease of white pines characterised by shortening of the needle-bearing stems, yellowing and shortening of needles, and copious flow of resin.

Polemoniaceae A family of autotrophic dicotyledonous plants in the order Polemoniales distinguished by lack of internal phloem, corolla lobes that are convolute in the bud, three carples, and axile placentation.

Polemoniales An order of dicotyledonous plants in the subclass Asteridae, characterised by sympetalous flowers, a regular, usually five-lobed corolla, and stamens equal in number and alternate with the petals.

Polishers The machine used for polishing seeds to improve their lustre.

Poll To cut back the crown of a tree.

Pollen The male germ cells produced in the anther.

Pollen chamber A cavity developed at the apex of the nucellus in certain gymnosperms.

Pollen count The number of grains of pollen that collect on a specified area (often taken as one square centimetre) in a specified time.

Pollen grain A microspore which germinates to form the male gametophyte containing 3 nuclei, one fertilizing the ovum, second forming the endosperm and third degenerating.

Pollen parent The parent that furnishes the pollen which fertilises the ovules of the other parent in the production of seed.

Pollen sac One of the pollen containing cavities in an anther; a microsporangium in higher plants.

Pollen shedder In hybrid seed production involving male sterility, the plants of 'B' line present in 'A' line are termed as pollen shedders.

Pollen sterile A species of plants which have fertile pistils but sterile stamens. Inbreeding does not occur.

Pollen tube A tube which is formed by a pollen grain and which transports sperms to the eggs in ovules. A pollen grain and its mature tube are a male gametophyte.

Pollination The transference of pollen from the anthers to the stigmas of flowers, fertilization in flowering plants.

Pollination constant Variety of fruits in which pollination does not alter some of its characters including flavour. Pollination variant are those where alteration occur.

Polliniser (1) An insects which helps in pollination. (2) A plant in fruit planting which provides pollen for self unfruitful varieties.

Pollution Destruction or impairment of the purity of the environment.

Polyandrous With 20 or more stamens per flower.

Polyandrous families In which husbands are more than one e.g. the people of Khasa Tribe of Jaunsar-Baber, U.P. (Dehradun distt.) where several brothers marry one wife without any exclusive right of cohabitation for any one spouse.

Polyarch Characterised by a primary xylem with many points or arms radiating from the central cells.

Polyclimax A climax community under the controlling influence of many environmental factors including soils, topography, fire, and animal interactions.

Polycotyledonous Having many cotyledons, as in certain gymnosperms.

Polycross Open pollination of selected plants or lines inter se in isolation following which progenies of each may be grown comparatively to evaluate the parents.

Polyembryonous Having many embryos.

Polygalaceae A family of dicotyledonous plants in the order polygalales distinguished by having a bicarpellate pistil and monadelphous stamens.

Polygamous Both perfect and imperfect flowers present on the same plant.

Polygenous composed of various kinds of rocks.

Polygonal soil A soil characterised by irregular prisms formed by irregular shrinkage of dispersed homogeneous soil materials.

Polygynous (1) Where a man marries more than one wife. (2) Having numerous pistils.

Polymerous Composed of many parts.

Polymorphic Having many forms.

Polypetalous Having distinct petals, in reference to a flower or a corolla. Also known as choripetalous.

Polyphyletic Derived from more than one ancestral stock; showing several lines of descent.

Polyphyllous Having many or separate leaves.

Polyploid An organism with more than two sets of a basic or monoploid number of chromosomes, e.g. triploid, tetraploid, pentaploid, hexaploid, heptaploid, octoploid, or various aneuploids.

Polypodiales The true ferns; the largest order of modern ferns, distinguished by being leptosporangiate and by having small sporangia with a definite number of spores.

Polypodiophyta The ferns, a division of the plant kingdom having well-developed roots, stems, and leaves that contain xylem and phloem and show well-developed alternation of generations.

Polypodiopsida A class of the division polypodiophyta; stems of these ferns bear several large, spirally arranged compound leaves with sporangia grouped in sori on their undermargins.

Polypore Any of a large number of members of the Agaricales whose basidia are borne in pores instead of gills.

Polysaccharide A carbohydrate, such as starch or cellulose containing more than three monosaccharide units per molecule.

Polysepalous Having separate sepals known as chorisepalous.

Polysiphonous Composed of many cylindrical filaments, as a Thallus.

Polystele A stale consisting of vascular units in the parenchyma.

Polytrichales An order of ascocarpous perennial mosses; rigid, simple stems are highly developed and arise from a prostrate subterranean rhizome.

Pomace Residue of fruit and plant parts after juice extraction.

Pome A fruit of the apple or pear type, in which the true fruits are surrounded by an enlarged, fleshy floral tube and receptacle.

Pomegranate Punica granatum L. A small deciduous ornamental tree of the order Myrtales cultivated for its fruit, which is a reddish, pome- like berry containing numerous seeds embedded in crimson pulp. Called *Anar* in Hindi.

Pomology The art and science of growing and handling fruits, especially fruit trees.

Ponderosa pine Pinus ponoderosa. A hard pine tree of Western North America; attains a height of 150-225 feet (46-69 metres) and has long, dark-green leaves in bundles of two to five and twiny, yellowish bark.

Ponding The condition produced by surface water collecting in shallow pocket in an area.

Poonspar Calophyllum apetalum Willd. (calophyllum wightianum T. Anders.) Family Guttiferae. A tree. Fruit edible. Wood used for packing cases and furniture and for making plywood.

Popcorn Zeamays everta. Any kind of maize that pops on heating.

Poppet valves The disk and stem type of valves as used in an automobile motor.

Poppy Any of various ornamental herbs of the genus papaver, family papaveraceae, with large, showy flowers, opium is obtained from the fruits of the opium poppy (Papaver Somniferum).

Poppy-diseases Poppy downy mildew caused by Peronospora arborescens and poppy powdery mildew caused by Erysiphe polygoni are the main diseases.

Population parametre A parametre in a frequency distribution defining all the possible observation.

Population soil All the organisms living in the soil, including plants and animals.

Pore pressure, Pore water pressure, or **Neutral stress** The pressure in the fluid within the void of the soil.

Pore space The total space not occupied by soil particles in a bulk volume of soil.

Pore-size-distribution The volume of the various sizes of pores in a soil. They are expressed as percentages of the bulk volume (soil plus pore space).

Pore, soil The part of the bulk volume of soil not occupied by soil particles. Interstices or voids.

Porogamy Passage of the pollen tube through the micropyle of an ovule in a seed plant.

Porosity The volume percentage of the total bulk not occupied by solid particles.

Port An opening through which fuel may be admitted to the combustion chamber of the engine or one from which exhaust gases are released.

Portable That which may be easily moved or carried.

Portable hydrants Hydrants installed on the top of alfalfa valves to direct flow in a definite direction or to permit connection of small portable pipes.

Portable sprinkler method The method made up entirely of portable pipe lines from the pumping or water pressure source to the last sprinkler.

Portulacaceae A family of dicotyledonous plants in the order Caryophyllales distinguished by a syncarpous gynoecium, few cyclicpetals and stamens, two sepals, and two to many ovules.

Position control It is a method of the hydraulic control system wherein constant depth is maintained by automatic adjustment of the draft of the tractor.

Positive displacement pump A pump which moves a set quantity of fluid with each stroke or rotation; such as a gear pump.

Positive ion A cation.

Possible water expenditure Potential transpiration less allowance for root constant.

Post-climacteric stage The stage of the respiration curve of mature fruit during which the respiration rate is decreasing after the climacteric peak.

Post-construction surveys Surveys made after the completion, or partial completion, of a structure or a physical feature of a project. In general, such surveys fall into two main classes: (i) record for as-built drawing or as executed; and (ii) surveys to determine and record the behaviour of a structure following its completion.

Potassium A major nutrient of plants absorbed as potassium ion (K^+). Potassium chloride commercially known as muriate of potash and potassium nitrate are commonly used fertilizers. Browing of tips, marginal scorching of leaves are few symptoms of deficiency.

Potassium fixation The process of converting exchangeable or water soluble potassium to moderately soluble potassium; i.e. to a form not easily exchanged from the adsorption complex with a cation of a neutral salt solution.

Potassium-supplying power of soil The capacity of soil to supply potassium to growing plants from both the exchangeable and the moderately available forms.

Potato Solanum tuberosum. An erect herbaceous annual that has a round or angular aerial stem, underground lateral stems, pinnately compound leaves, and white, pink, yellow or purple flowers scourring in cymose inflorescences. Produces an edible tuber which is a shortened, thickened underground stem having nodes (eyes) and internodes. Also known as Irish potato.

Potato-diseases Potato brown rot also known as potato bacterial wilt, or potato ring disease or potato bangle disease is caused by Pseudomonas Solanacearum (Smith) Dowson; Potato black heart, a non-parasitic disease; potato black leg and soft rot caused by Erwinia carotovora, potato black scruf caused by Pellicularia filamentosa, potato charcoal rot caused by Macrophomina phaseoli, potato dry rot caused by Fusarium coeruleum, potato early blight caused by Alternaria solani, potato golden nematode caused by Synchytrium endobioticum, potato late blight caused by Phytophthora infestans, potato leaf curl and potato mosaic, virus diseases, potato powdery scab caused by Spongospora subterranea, potato scab caused by Streptomyces scabies and potato wart caused by Synchytrium endobioticum are some of the diseases of potato.

Potato — insect pests Few insects pest of potato are: potato beetles (Epilachana vigintioctopunctata Fabr., E. dodecastigma Muls., E. Ocellata Rdt., E. circularis Kobs.) whose grub and adult feeds on potato leaves; potato jassid (Empoasca kerri var. motti Pruthi) which suck saps from tender parts of the plants and potato tuber moth (Gnorimoschema operculella zell) which attack both in field and store. Caterpillar bores into tubers.

Potential The work done to move a unit mass from one place to another against a force.

Potential acidity In a soil, the amount of exchangeable hydrogen ion that can be made free or active in the soil solution by cation exchange. It is usually expressed in milliequivalents per unit mass of soil.

Potential cation exchange capacity The total number of cations that a soil can adsorb at a specified pH.

Potential evapotranspiration The amount of water utilised by plant growth plus evaporation from the soil if the soil contained sufficient moisture for plant growth at all times. See evapotranspiration.

Potential transpiration The amount of water transpired by a green crop of about the same colour as grass, which completely covers the ground, and which has an adequate supply of water.

Potometre Instrument for measuring the rate of moisture absorption by a plant.

Pottiales An order of mosses distinguished by erect stems, lanceolate to broadly ovate or aborate leaves, a strong, mostly percurrent or excurrent casta, and a cucullate calyptra.

Poultry Domesticated fowl grown for their meat and eggs.

Pound A unit of weight. 0.4536 kg (avoirdupois).

Pour point It is the temperature at which the oil just flows under prescribed conditions.

Poverty adjustment The range of supply of a limiting element through which an increase in that element brings about an increase in plant growth. There is no increase in percentage of the element in the plant.

Powdery scab A fungus disease of potato tubers caused by Spongospora subterranea and characterised by nodular discoloured lesions, which burst and expose mosses of powdery fungus spores.

Pox shallow sunken spots on rinds of citrus fruits when stored under improper temperature and humidity.

PPM Commonly used abbreviation for parts per million in a solution; one unit of solute to 1 million part of solvent.

Power brakes Brakes operated hydraulically through a booster system. When engine is running, brakes respond to light pressure on brake pedal.

Power function The inverse of the performance function is the 'power function' that relates the probability of rejecting an incorrect null hypothesis to various alternative values of the parametre. The power function can be drawn either by using the information about the 'power of the test' or by simply inverting the 'performance function'.

Power of the test In decision making, the situations of acceptance of a correct null hypothesis, and of the rejection of an incorrect null hypothesis make for good decisions. The probability that an incorrect null hypothesis is rejected is called the 'power of the test'.

Power take-off An attachment to transmit power from a tractor to an attached implement.

Prairie Level or rolling area of treeless land, naturally covered with grass, under which fertile soils usually have developed.

Prasinovolvocales An order of green algae in which there are lateral appendages in the flagellum.

Pratinealous (1) Living in meadows. (2) Living in low grass.

Precast concrete Concrete which is cast into (sections) forms or moulds and then transported or hoisted and placed in position. Used principally where many of the same units are required.

Precipitation (1) Process of clumping of proteins in solution caused by the addition of a specific precipitation. (2) Rainfall.

Precipitation-evaporation ratio For a given locality and month, an empirical expression devised for the purpose of classifying climates numerically on the basis of precipitation and evaporation. Abbreviated P.E. ratio. Also known as precipitation effectiveness.

Precipitation inversion As found in some mountain areas, a decrease of precipitation with increasing elevation of ground above see level. Also known as rainfall inversion.

Precipitation interception The stoping, interrupting, and temporary holding of precipitation in any form by a vegetative canopy or vegetation residue.

Precipitation station A station at which only precipitation observations are made.

Precipitins Antibodies that cause precipitation.

Precision A quality associated with a class of measurements and refers to the way in which repeated observations conform to themselves; and in a somewhat narrower sense refers to the dispersion of observations, or some measure of its, whether or not the mean value around which the dispersion is measured, approximates to the true value.

Precision grader This machine is used to upgrade seed quality. It makes separations on the basis of differences in width and thickness.

Preconsolidation pressure (or prestress) The greatest effective pressure to which a soil has been subjected.

Pre-construction surveys Surveys required to obtain the necessary data for authorization reports and for contract or final designs of structures or other project works, as well as for the establishment of the irrigable area of the project.

Predation The killing and eating of an individual of one species by an individual of another species.

Predatory price-cutting The selling of a commodity at unreasonably low prices in order to injure or destroy the competition of a competitor.

Prediction The process of forecasting the magnitude of statistical varieties at some future point of time.

Prediction interval The interval between the upper and lower limits attached to a predicted value to show, on a probability basis, its range of error.

Pre-emergence Before emergence; often refers to the treatment of the soil with weed control chemicals after planting and before the crop plants appear above ground.

Pre-formation Formation of the whole

of an unexpanded shoot system prior to its subsequent complete expansion (cf. neoformation).

Pre-ignition A condition of detonation brought about by either hot carbon deposits, or faulty ignition, in which gas is exploded prematurely.

Pre-irrigation Pre-sowing irrigation; generally for field preparation.

Preliminary location survey Preliminary survey made for the selection of site and preliminary design of various features of a proposed project.

Preliminary surveys The collection of survey data in the early stages of project planning. They are further classed as reconnaissance surveys and preliminary location surveys.

Preliminary watering Watering given to the land before sowing the seeds either for summer or for winter crops.

Prelogging Cutting down and removing small trees before large trees are logged.

Premium The difference between the price at which a security was issued and the price at which it now stands, if the latter is the higher.

Preservative Substance that is added or applied to a product to prevent bacterial growth and consequent spoiling.

Preservative, silage Material added to the forage crop at ensiling to quickly develop the acidity essential for preservation.

Preserve In wildlife management, a game shooting area on which game species are propagated, released, or otherwise maintained.

Pressed brick A high grade brick used for exposed surface work.

Press mud A by-product from sugar mills; commonly used as manure.

Pressure or Induced Pan A subsurface horizon or soil layer having a higher bulk density and a lower total porosity than the soil directly above or below it, as a result of pressure that has been applied by normal tillage operations or other artificial means. It is also referred to as plow pan, plow sole, or traffic pan.

Pressure conduit Any type of closed conduit which entirely confines and guides the movement of water under pressure.

Pressure head The head on any point in a conduit represented by the height of the hydraulic grade line above that point.

Pressure membrane A membrane, permeable to water and only very slightly permeable to gas when wet, through which water can escape from a soil sample in response to a pressure gradient.

Pressure reducing valves Valve installed in a sprinkler lateral to reduce pressure when it is extended to serve land considerably lower in elevation than the surrounding area; in such cases the pressure gained by the additional head would be excessive for normal sprinkler-type irrigation.

Pressure system Method of air movement for drying in which air is forced through the product by the air duct or ducts being at a pressure above atmospheric pressure. Forcing system of air movement.

Pressure system for fuel supply The system used before the invention of the vacuum tank, for supplying gas when the tank was lower than the carburetor. A small hand pump is used to provide initial pressure. The engine, when running, operates a small pump thus maintaining pressure. In this system, the gas tank must not be vented.

Pre treatment Any physical or chemical laboratory treatment of the working sample that is given solely to facilitate testing.

Previous crop The crop grown in the season immediately preceding the one in which the seed crop is grown.

Price The value of a commodity or service in terms of money.

Price level An average of individual prices in any given market.

Price market The price of a commodity during a period in which its supply is fixed.

Primary and secondary (Group) They are distinguished from each other by their type of social contact and degree of formal organisation. In the primary group, there is face-to-face association, the contacts are personal, individuals live close to one another socially and do not need the formal framework of a constitution to achieve their purpose.

Primary bud The largest and strongest bud in the compound bud on the short or cane of grape vine.

Primary drainage It is the drainage in a drainage watershed as a whole.

Primary flow irrigation A system of irrigation with border strip or furrow layout in which a constant flow is used.

Primary Infection First infection by a pathogen after going through a resting or dormant period.

Primary leaf The just true leaf of the seedling.

Primary meristem Meristem which is derived directly from embryonic tissue and which gives rise to epidermis vascular tissue, and the cortex.

Primary mineral A mineral that has not been altered chemically since deposition and crystallization from molten lava.

Primary particles Individual soil particles after a standard dispersion treatment.

Primary phloem Phloem derived from apical meristem.

Primary productivity The amount of organic matter produced by photosynthesis.

Primary root The first plant root to develop; derived from the radicle; seminal root.

Primary sample A primary sample is a small portion taken from one point in the lot.

Primary sewage sludge A semi-liquid waste resulting from sedimentation with no additional treatment.

Primary soil Soil developed in place through the disintegration and decomposition of rocks.

Primary tillage implements The implements used for primary tillage are called primary tillage implements.

Primary tissue Plant tissue formed during primary growth.

Primary xylem The first woody tissue developed by a tree or shrub.

Primales The order of mammals to which man belongs; characterised in terms of evolutionary trends by retention of a generalised limb structure and dentition, increasing digital mobility, replacement of claws by flat nails, development of stereoscopic vision, and progressive development of the cerebral cortex.

Priming (1) A term to designate the harvesting of ripe tobacco leaves from the stalk as they ripen, beginning at the bottom and progressing upward. (2) Process of filling water in the suction side of a centrifugal pump before starting.

Priming losses Absorption losses in a canal during the process of priming. After a canal has reached stable condition, these losses are called 'seepage losses'.

Priming rain The amount of rain that does not produce runoff. In semi-arid and arid regions the rain that falls weeks or months before the first runoff.

Primordia Earliest stage at which differentiation of organs in a plant can be perceived.

Primordium The rudiment or beginning of a part.

Primulaceae A family of dicotyledonous plants in the order primulales; characterised by a herbaceous habit and capsular fruit with two to many seeds.

Primulales An order of dicotyledonous plants in the subclass Dilleniidae distinguished by sympetalous flowers, stamens located opposite the carolla lobes, and a compound ovary with a single style.

Principle Ultimate cause; a fundamental truth.

Principle of independent assortment Mendel's genetic theory that differentiating characters recombine at random.

Principle of segregation Mendel's first law of genetics; it states that the alternatives in the reproductive cells remain separate.

Prismatic A soil structure type having prismlike aggregates that have vertical axes much longer than the horizontal axes. See also structure types, soil.

Private stream Any stream which diverts part or all of the drainage of another stream.

Probability The likelihood of the occurrence of any event; the ratio, sometimes expressed as a percentage of the frequency of the occurrence of a certain event to the frequency of occurrence of all events.

Probability density function An alternative term for the frequency function, when the distribution concerned is considered as one of probability.

Probability distribution A distribution giving the probability of a value *x* as a function of *x*, or more generally, the probability of joint occurrence of a set of variates $X_1, X_2 \ldots X_p$ as a function of those quantities.

Probability limits Upper and lower limits assigned to an estimated value for the purpose of indicating the range within which the true value is supposed to lie according to some statement of probabilistic character. For instance, confidence limits, fiducial limits are probability limits.

Probability paper A graph paper with the grid along one axis specially ruled so that the distribution function of a specified distribution can be plotted as a straight line against the variate as abscissa.

Probable error An indication of the precision attained in series of measurements on a given object. Probable error is 0.6745 times the standard error.

Procerus indicus See Mango leaf gall makers.

Processing of sand Improving quality and grading of sand by classifiers or dry processing method, depending upon the sand.

Procumbent Having stems that lie flat on the ground but do not have root at the nodes.

Prodenia litura See Tobacco leaf caterpillars.

Producers (1) Green plants, photosynthesisers. (2) Manufacturers of goods and services.

Production Activities resulting in a supply of useful goods and services. These activities include mineral extraction, manufacturing, agriculture, transport and communication, wholesale and retail distribution, and the work of civil servants and local government employees.

Production function The relationship between input of production services and output of products, per unit of time. Production functions are descriptive of techniques or systems of organisation of production services.

Productive soil A soil in which the chemical, physical and biological conditions are favourable for the economical production of crops suited to a particular area.

Productivity, soil (1) The capacity of a soil, in its normal environment, to produce a specified plant or sequence of plants under a specified system of management. The "specified" limitations are needed because no soil can produce all crops with equal success and a single system of management cannot produce the same effect on all soils. productivity means the capacity of soil to produce crops and is expressed in terms of yields. (2) The efficiency with which productive resources i.e. labour, capital and land are used.

Proembryo A group of cells, derived by division of the fertilised egg, which will differentiate into the suspensor and embryo.

Profile, Soil A vertical section of the soil through all its horizons and extending into the parent material.

Profile line In land levelling, the vertical section of the surface of the ground along any fixed line.

Profitable water requirement Depths of water that produce most profitable yields.

Profundal zone The region occurring below the limnetic zone and extending to the bottom in lakes deep enough to develop temperature stratification.

Progamete The enlarged initial cell before division, separating the basal cell from the gamete, as in Rhizopus.

Progenitor An individual plant or animal which is recognised as a source of a certain type or character in its offsprings.

Progeny Offspring, plants grown from the seeds produced by parent plant.

Programme It is a systematic statement of situation, objectives, problems and solutions. It is relatively permanent but requires constant revision. It forms the basis for extension plane.

Programme planning It is a process of working with rural people in an effort to recognise the problems and determine the possible solutions.

Project It is an outline of procedure and pertains only to some phase of extension work.

Project intensity The annual intensity aimed at in a project.

Prolamines Group of simple proteins insoluble in water, characteristic proteins of cereal grains chief of them being gliadin of wheat and rye, the zein of maize and the hordein of barley. Also found in vegetables.

Prolepsis Development of a lateral branch only after a period of dormancy as a lateral bud; hence proleptic branch, a branch developed by prolepsis.

Promeristem Undifferentiated cells in growing regions which will shortly become meristem.

Promycelium The basidium of smuts and rusts.

Prong budding It is similar to shield budding. Instead of shield, spur or prong is used.

Pronucleus The nucleus of a pollen grain, egg or sperm.

Propagate To multiply by sexual or asexual means.

Propagula Special reproductive bodies of a vegetative origin produced in sphacelaria.

Propeller or axial flow pump A pump which develops most of its pressure by the lifting action of vanes on the water.

Property That to which a person has legal title; protected in that title by public authority.

Prophase An early stage in mitosis in which the chromosomes become distinct and organised and in which their longitudinal splitting begins.

Propolis See Bee glue.

Proportion The proportion of two quantities is their ratio.

Proportional weir A weir in which flow rate varies directly with the head.

Proportion as mean Proportion, though ordinarily conceived as a fraction of percentage, can be reckoned as a special case of the arithmetic mean, by the device of letting a variable be 1 when the characteristic is present, 0 when it is absent, then $x = \Sigma\ x/n$, the mean, is the sample proportion.

Prop root A root that serves to support or brace the plant. Also known as brace root.

Prosenchyma Tissue of elongated cells placed with their ends interpenetrating especially fibro-vascular tissue.

Prosopis juliflora See Mesquite.

— **spicigeara** See jand.

Prostrate Flat on the ground.

Prostrate habit The tendency in many plants to grow horizontally along the ground, as the dorsiventral body of the liverwort.

Protandry The maturing and functioning of stamens before pistils in hermaphroditic flowers or in different flowers of the same plant in a monoecious species.

Proteaceae A large family of dicotyledonous plants in the order proteales, notable for often having a large cluster of small or reduced flowers.

Proteales An order of dicotyledonous plants in the subclass Rosidae marked by its strongly perigynous flowers, a fourlobed, often corolla like calyx, and reduced or absent true petals.

Protection The adoption by the State of special measures to protect an industry from competition.

Protein Complex combination of amino acids, always containing carbon, hydrogen, oxygen, and nitrogen and sometimes phosphorus and sulfur; essential part of all living matter and the feed rations of animals.

Protein, crude All nitrogenous substances contained in feedstuffs (% crude protein = % N × 6.25).

Protein equivalent A term used in calculating feeding standards for animals. It is sum of the digestible pure protein and half the digestible non-protein nitrogen.

Proteolysis Decomposition of protein through the action of organisms. Also called proteoses.

Prothallium The gametophyte of a pteridophyte in the form of a flat green thallus with thizoids.

Prothallus The gametophyte of ferns and similar plants.

Protocecaceae A monogeneric family of green algae in the suborder Ulotrichineae in which reproduction is entirely vegetative.

Protogynous With stigmas becoming receptive before the anthers.

Protogyny The reverse of protandry.

Protonema A green, filamentous structure that originates from the asexual spore of mosses and some liverworts and that gives rise by buddinging to a mature plant.

Protophloem The first-formed cells in the primary phloem of a root, stem, or other plant organ.

Protophyta A division of the plant kingdom, according to the system of classification, set up to include the bacteria, the blue-green algae, and the viruses.

Protoplasm A colloidal substance consisting of water, proteins and traces of 12 essential elements. It makes up all plant and animal cells.

Protostele A stele with a solid core of xylem, thus lacking pith, and with a surrounding mass of phloem.

Protoxylem The part of the primary xylem that differentiates from the procambium and is formed during elongation of an embryonic plant organ.

Protozoa Unicellular organisms belonging to the animal kingdom.

Provenance The region where seed was harvested.

Proximate analysis Analytical system that includes the determination of ash, crude fibre, crude protein, ether, extract, moisture (dry matter), and nitrogen-free extract.

Pruning The removal of live or dead branches from standing trees. This may be done stratigically or naturally. Natural pruning results from such causes as decay, snow ice, deficiency of light, etc.

Prunus armeniaca See Apricot.

— **avium** See Sweet cherry.

— **cerasoides** See Wild Himalayan cherry.

— **cerasus** See Sour cherry.

— **communis var. insititia** See Plum.

— **domestica var. insititia** See Plum.

— **persica** See Peach.

— **puddum** See Wild cherry

— **salicina** See Japanese plum.

— **triflora** See Japanese plum.

Prussic acid See Hydrocyanic acid.

Prygeometre An instrument that measures the effective terrestrial radiation.

Psammophytes Plants that grow best in or tolerate sand, particularly fine to medium sand.

Psara bipunctalis See Brinjal leaf caterpillars.

Psarasis A virus disease of tangerine, grape fruit, and sweet orange trees characterised by scaly bark, a gunny exudate, retarded growth, small yellow leaves, and dieback of twigs. Also known as scaly bark.

Psendopodium A slender, leafless branch of the gametophyte in certain Bryatae. Temporary projection of the protoplast of amoeboid cells in which cytoplasm streams actively during extension and withdrawal. Foot of a rotifer.

Pseudaletia albistigma See Paddy climbing cutworm.

— **unipunctata** See Paddy climbing cutworm.

Pseudaulacaspis pentagona See Peach-diseases.

Pseudococcus brimeliae See Pineapple — insect pests.

— **lilacinus** See Custard apple mealy bug.

— **virgatus** See Custard apple mealy bug.

Pseudomembrane Fibrinous exudate forming a tough membranous structure on the surface of the skin or mucous membrane.

Pseudomonas mangiferae — indicae See Mango bacterial leaf spot.

— **rubrilineans** See Sugarcane disease.

— **solanacearum** See Brinjal bacterial wilt, tobacco bacterial wilt and Potato-diseases.

Pseudoparenchyma Fungus tissue in which the hyphae have lost their filamentous identity and become isodiametric.

Pseudoperianth An involucral envelope enclosing certain fruiting structures, but not composed of sepals or petals.

Pseudopod Temporary protoplasmic process put forth by a protozoan for purposes of locomotion or obtaining food.

Pseudosperm A false seed or carpel.

Pseudosphaeriales An order of the class Ascolichenes, shared by the class Ascomycetes; the ascocarp is flash-shaped and lined with a layer of inter woven, branched pseudo-paraphyses.

Pseudowhorl A close set series produced morphogenetically in sequence but stimulating a whorl.

Psilate Lacking ornamentation; generally applied to pollen.

Psilopsida A sub division of the Tracheophyta.

Psitotophyta A division of the plant kingdom represented by three living species, the life cycle is typical of the vascular cryptogams.

Psychroenergetics Science dealing with the effect of ambient temperatures and humidity upon conversion of feed into bodily heat or energy.

Psychrometric chart The device which simplifies the measurement of air properties and eliminates many time consuming and tedious calculation which would otherwise be necessary.

Psychrometric constant The ratio of the specific heat of air to the latent heat of evaporation of water.

Psychorphile A microorganism capable of thriving temperatures as low as 0°C.

Pterochore A type of plant that produces winged disseminules.

Pteropsida A large group of vascular plants characterised by having parenchymatous leaf gaps in the stele and by having leaves which are thought to have originated in the distant past as branched stem systems.

Ptomaines Basic substances resembling alkaloids formed during the decomposition of dead organic matter.

Ptychomniaceae A family of mosses in the order Isobryales distinguished by an eight-ribbed capsule.

Ptyxis The form in which young leaves are folded in the bud.

Pubescent Hairy, in a general sense; in special use, covered with short, soft hairs.

Public domain Territory over which a commonwealth has dominion or control; used in connection with the land owned by the central government.

Public finance The provision of money by the community through taxes and rates to be spent by national and local government authorities on projects and schemes of national and local benefit.

Public sector The 'public sector' of the economy usually denotes the combination of the central Government, the local authorities and the nationalised industries and other public corporations.

Public speaking It is one mode of communication; systematic, practical discourse which aims through speech and gesture, to amuse listeners, to add to their knowledge, to increase their understanding, or to influence their attitude and their action (Bryant and Wallace).

Public works Roads, bridges, public baths, theatres, multi-storey car parks and other projects; undertaking by the government or local authority to provide services for the community.

Puccinia carthami See Safflower diseases.

— **chrysanthemai** See Pyrethrum diseases.

— **erianthi** See Sugarçane disease.

— **kuehnil** See Sugarcane disease.

— **glumarum** See wheat stripe rust.

— **graminis-avenal** See Oats stem rust.

— **graminis tritici** See Barley and wheat stem rust.

— **hordei** See Barley leaf rust.

— **parri** See Lady's finger rust.

— **purpurea** See Sorghum diseases, Johnson grass rust.

— **rubigovera tritici** See Wheat brown rust.

— **sarghi** See Maize leaf rust.

— **striformis** See Barley stripe rust.

Puddle To deflocculate or destroy the granular structure of the soil.

Puddled clay A clay or a mixture of clay and sand, kneaded or worked, when wet, to render it impervious to water.

Pull It is the total force required to pull an implement.

Puller Any mechanical or hydraulic device for removing, by pulling action, parts which are tightly fitted, e.g. wheel puller or gear puller.

Pulley A wheel used to transmit or receive power through a belt which travels over its face.

Pulley block A sheave pulley or series of such pulleys, enclosed between metal or wooden side plates which carry the shaft or pin on which the pulleys revolve.

Pulp wood Wood cut or prepared primarily for manufacture into wood pulp, for subsequent manufacture into paper, fibre, board, or other products, depending largely on the species cut and the pulping process.

Pulse Large-seeded legumes.

Pulse beetle Collosobruchus chinensis L., Bruchus anatis Fabr., B. Phaseoli Gyll., Laria affinis Frol. A pest of many stored grains.

Pulvinate Cushion-like in form.

Pulvinus A cushion-like enlargement of the base of a petiole which functions in turgor movements of leaves.

Pummelo Citrus maxima (Burm.), (Citrus grandis Osbeck; citrus decumenus L.; C. aurantium subsp., Aurantium decumana Mill., A maximum (Burm.). A citrus fruit of family Rutaceae rich in vitamins A, B and C.

Pump A machine for lifting or forcing liquids, either by means of a bucket or of a piston working in a closed cylinder.

Pump characteristics Speed head discharge relationships for a given pump.

Pump efficiency Ratio of the water horsepower produced by the pump, based on the total head, to the horsepower delivered to the pump by the power unit.

Pump house A building in which are housed pumps that supply an irrigation system, a power plant, a factory, a reservoir, a farm, a home and so on.

Pumping station A building in which two or more pumps operate to supply fluid flowing at adequate pressure to a distribution system.

Pumpkin Any of several prickly vines with large lobed leaves and yellow flowers in the genus Cucurbita of the order violales; the fruit is orange-coloured and large, with a firm rind. Called *kaddu* in Hindi.

Pump submergence Vertical distance between the inlet of the pump and the surface of the water supply.

Puna An alpine biological community in the central portion of the Andes Mountains of South America characterised by low-growing, widely spaced plants that lack much green colour most of the year.

Pupa The third or chrysalis stage in insect life; insect enclosed in a case.

Pupiparous To produce young one which has already advanced to pupal stage.

Pure culture A nutrient medium containing the growth of a single strain of an organism free from other living species or strains.

Pure line A strain of an organism that is comparatively pure genetically (homozygous) because of continued inbreeding or through other means.

Pure live seed(PLS) Percentage of the content of a seed lot that is pure and viable; determined by multiplying the percentage of pure seed by the percentage of viable seed and dividing by 100.

Pure profit Payment for taking risks. This is a necessary payment to those who invest capital other than in 'safe' investment, such as Government stock.

Pure seed The pure seed refer to the seeds (including broken seeds more than half the original size) of the species stated by the sender, or found to predominate in the purity test. It includes all botanical varieties and cultivars of that species even if immature, undersized, shrivelled, diseased or germinated, provided they can be definitely identified as of that species.

Purity The composition by weight of the sample being tested, by inference, the composition of seed lot; the identity of various species of seeds and inert particles constituting the samples.

Purple blotch A fungus disease of onions, garlic, and shallots caused by Alternaria porri and characterised by small white spots which become large purplish blothces.

Purple-top A virus disease of potato plants characterised by purplish or chloritic discolouration of the top shoots, swelling of axillary branches, and severe wilting.

Push rods In overhead valve engines, the rods between the outer end of the rocker arms and the cam followers.

Pustule A blister-like mark on a leaf due to rupture of surface tissues overlying spore masses of a parasitic fungus.

Put and take animals Used in a grazing experiment to graze excess forage beyond that needed for tester animals and to accumulate animal days.

Putrefaction The anaerobic decomposition of organic substances by micro-organisms.

Putrescible wastes Wastes which consist mainly of plant or animal residues and which undergo degradation by bacteriological action; cannery wastes, fats, fish residues, fruit wastes, vegetable wastes, waxes-animal and plant.

Pycnidium Flask-like fruiting body containing conidia.

Pycnium Spermagonium in the rusts, the O stage, resembling a pycnidium.

Pyrene A fruit stone or kernel.

Pyrenoid A starch forming body in the chloroplast of many algae.

Pyrethrin Active principle of pyrethrum flowers; used as an insecticide.

Pyrethrum A naturally occurring contact insecticide obtained from flowers of Chrysanthemum species.

Pyric Designating modifications in the soil and vegetation caused by fire.

Pyriform Pear-shaped.

Pyrology The scientific art of protecting forests from fire.

Pyrophyte A plant characterised by unusual resistance to fire (e.g., many grasses, sedges, pines, saw palmetto).

Q

Quadrant In growth, initial division of the zygote. It often results in four cells grouped together.

Quadraspidiotus perniciosus See San Jose Scale.

Quadrivalent Association of homologous chromosomes in group of four.

Quagmire Soft marshy ground.

Quail grass See cocks comb.

Quadraspidiotus perniciosus See San Jose Scale.

Quagmire Soft marshy ground.

Quail grass See cocks comb.

Qualitative analysis An analysis in which the qualitative effect is studied or the estimates of the elements constituting a product are made; the effects or constituents are not quantified.

Quality Several factor or characteristics of a product which will enhance or degrade the value of the product in the eyes of the consumers. usually designated as superior, inferior or medium quality. The characteristics include colour texture, cracks, age, maturity and blemishes etc. depending upon the product under consideration.

Qualitative characters Characters that are qualitatively different so that separation is relatively easy. Usually placed in categories.

Quality of waters As referred to in this dictionary 'quality of water' means the suitability of water for irrigation use.

Quantitative characters Characters that show a continuous range in variability, making separation into distinct classes difficult. Usually described numerically.

Quantum an elemental unit of energy with no electric charge and very little mass. In its interactions with matter, light acts as though it were composed of quanta.

Quantitative analysis In this case exact analysis is done in which the effect or proportion of each factor or element is evaluated. This analysis is more rigorous in comparison to Qualitative analysis.

Quarantine (1) Regulatory control and restriction imposed on the production and movement of plants and material so as to reduce the possibility of introduction and spread of a disease or pest from one place to another. It helps in controlling the spread of the already introduced pest. (2) Isolation of the animal, plants or seeds till they are free from the disease or pest which is feared. (3) Place of isolation.

Quart A British unit of measurement and is equal to 0.25 gallons.

Quarter turn belt The arrangement of belt to drive two shafts which are at right angles to each other.

Quartet A group of four spores formed by the two meiotic divisions.

Quartiles The three variate values which separate the total frequency of a distribution into four equal parts. The central value is called the 'median' and the other two upper and lower quartiles respectively.

Quartz (SiO_2) One of the major constituent of most common rocks. It has hardness of 7 and specific gravity of 2.65, chief constituent of sandy fractions of soils.

Quartzite See metamorphic rocks.

Quebracho Any of a number of South American trees belonging to different genera in the order Sapindales, but all being a valuable source of wood, bark and tannin.

Quebracho bark Bark of the white quebracho tree Aspiodosperma quebracho of Chile and Argentina; main components are aspidospermine, tannin, and quebrachine, used in medicine and tanning.

Queen bee A bee in a beehive which is mated and its main function being to produce eggs. She does not collect honey.

Queensland arrowroot Canna edulis Ker Gawl. Family Cannaceae. A herb native of South America, now found in South India. Rhizomes used to prepare starch.

Queensland nut Macadamia ternifolia. F. Miiell. Family Proteaceae. Nut tree producing small, oval-shaped fruit with small beak. Contains hard seed with whitish kernel of fine flavour and taste.

Quenching The dipping of heated steel into,water, oil, or other bath, to impart necessary hareness.

Quercus incana See oak.

— **lanata** See wooly oak.

Quick clay Clayey material having the tendency to change from a relatively stiff condition to a liquid mass when it is disturbed. See also liquefaction (spontaneous liquefaction).

Quick coupler A connecting or disconnecting device for two lengths of pipes in a pipeline assemblage in a sprinkler method which allows quick assembly and separation of movable pipes.

Quick freezing A method of preserving fruits, vegetable, poultry products, meat etc. The method consists in abruptly lowering the temperature of the environment of storage from about 28°F to -15°F. The time of lowering is about 30 min.

Quick-growing A plant which grows and matures at a comparatively faster rate from other species or within the same species.

Quick lime See Calcium oxide.

Quick condition See quicksand.

Quicksand Sand of low bearing capacity caused by the upward flow of water and the resultant decrease in intergranular pressure. Also called quick condition.

Quick setting cement A cement which sets more rapidly than ordinary or rapid hardening Portland cement.

Quick tests Standardised tests of evaluating chemical status of a soil in a relatively smaller time. Not much accurate but are useful in making recommendations on nutrients requirements or similar purposes.

Quiescent See dormant.

Quillwort The common name for plants of the genus Isoctes.

Quince Cydonia oblonga. A deciduous tree of the order Rosales characterised by crooked branching, leaves that are densely hairy on the underside and solitary white or pale pink flowers, fruit is an edible pear-or apple-shaped tomentose pone.

Quinquefoliate Of a leaf, having five leaflets.

Quincunx system of planting See diagonal system of planting.

Quinine Drug for Malaria prepared from Cinchona bark.

Quinoa (henopodium quinoa. An annual herb of the family Chenopodiaceae grown at high attitudes in South America for the highly nutritious seeds.

Quintal A metric system of measurement, used for quantifying weight of the object. It is equal to 100 kg.

R

Rabi and monsoon sections of a canal These terms are used in Bombay-Deccan (India). The 'rabi section' is the cross-sectional area of a canal in which assured supply is available for maturing two-season and rabi crops only. The rest of the cross-section is the 'monsoon section' in which water may be supplied when available in the monsoon and rabi seasons without any assurance.

Raceme An inflorescence on which flowers are borne on stalks of equal length on an unbranched main stalk that continues to grow during flowering. Inflorescence in which the spikelets are arranged singly along a common main axis.

Rachilla Small rachis; axis of a spikelet in grasses.

Rachis In compound leaves, the extension of the petiole bearing the leaflets; a raceme.

Racking Setting back the end of each course of brick or stone from the end of the preceding course.

Rad The unit of absorbed radiation dose; a dose which results in the absorption of 100 ergs per gram of specified material.

Radar climatology The statistics in time and space of radar weather echoes.

Radar meteorology The study of the scattering of radar waves by all types of atmospheric phenomena and the use of radar for making weather observation and forecasts.

Radar storm detection The detection of certain storms or stormy conditions by means of radar; liquid or frozen water drops within the storm reflect radar echoes.

Radar storm-detection equation The equation which relates the variables involved in the radar detection of precipitation.

Radial drainage pattern Drainage pattern resembling the spokes of a wheel in which drainage channels flow radially either outward from a peak or inward toward a basin.

Radian It is the unit of plane angle and is defined as the angle between two radii of a circle which cut off on circumference an arc equal in length to the radius.

Radiation injury Absorbed ionizing radiation at sufficient levels of exposure to injure, damage to living cells includes nuclear and cytoplasmic damage, heritable damage to genes and chromosomes and somatic cell injury, which may appear immediately or remain latent.

Radiation preservation Exposure of food products to ionizing radiation, such as electrons, and X-rays in order to destroy micro-organisms and thereby aid preservation.

Radiator hose The hose which connects radiator and engine.

Radicidation Destruction by radiation of micro-organisms in food that are significant to public health.

Radical Basal leaves and peduncles arising from root close to the ground.

Radicle The rudimentary root in the embryo of a seed plant.

Radioactive decay The process whereby nuclei of unstable isotopes break down, releasing radiation in the form of X-rays, Alpha, Beta and Gamma particles or alpha particle.

Radioactivity The property of certain elements spontaneously disintegrating into simpler elements accompanied by the emission of alpha particles, beta particles, or gamma rays.

Radio-ecology The interdisciplinary study of organisms, radionuclides, ionizing radiation, and the environment.

Radioisotope A radio-active isotope; generally used as tracer in scientific research.

Radio meteorology That branch of the science of meteorology which embraces that propagation of radio energy through the atmosphere, and the use of radio and radar equipment in meteorology; this is the most general term and includes radar meteorology. Also known as radio-electric meteorology.

Radionuclide Any radioactive nuclide, a nuclide being an atom characterised by the constitution of its nucleus, i.e. by the atomic number and the mass number.

Radiosande A balloon-borne instrument for the simultaneous measurement and transmission of meteorological data the instrument consists of transducers for the measurement of pressure, temperature, and humidity, a modulator for the conversion of the output of the transducers to a quantity which controls a property of the radio-frequency signal, a selector switch which determines the sequence in which the parametres are to be transmitted, and a transmitter which generates the radio-frequency carrier.

Radish Raphanus sativus. (1) An annual or biennial crucifer belonging to the order Capparales. (2) The edible thickened hypocotyl of the plant. Called *Mooli* in Hindi.

Radium A metal named so from its radio-active power. Resembles barium.

Radium poisoning Poisoning from radioactive radium salts in which bone necrosis, fatal anemia, and bone malignancy occur.

Radon A radioactive gas that is released from the soil into the atmosphere. It is also released during the combustion of fossil fuel, and for this reason there is more radon in the air over cities than in country air.

Raffinase An enzyme which digests certain complex sugars, as raffinose.

Rafflesales A small order of dicotyledonous plants, members are highly specialised non-green rootless parasites which grow from the roots of the host.

Railing A method of brush removal from farms employing the use of railway rail.

Rain Precipitation in the form of liquid water drops with diametres greater than 0.5 millimetre, or if widely scattered the drops may be smaller; the only other form of liquid precipitation is drizzle.

Rainers (1) Nozzles larger than 12.5 mm diametre. (2) The apparatus which distributes the water over the field.

Rainfall (1) A fall of rain; precipitation in the form of liquid water. (2) The amount of rain, usually expressed in millimetres (inches) depth of water on an area, that reaches the surface of the earth. The term is sometimes expanded to include other forms of atmospheric precipitation, such as snow, dew, but technically the term precipitation would be used in this broader scope.

Rainfall area Geographical extent of a given storm.

Rainfall distribution coefficient The distribution coefficient for any storm is obtained by dividing the maximum rainfall at any point by the mean on the basin.

Rainfall excess (Hydraulics) The volume of rainfall that will result in runoff.

Rainfall intensity The rate at which rainfall occurs expressed in depth units per unit time. It is the ratio of the total amount of rain to the length of the period in which the rain falls.

Rainfall intensity frequency or Rainfall recurrence interval The average time interval between the occurrence of rainfall of a given intensity and that of an equal or greater intensity.

Rainfall interception See precipitation interception.

Rainforest A forest of broad-leaved, mainly evergreen, trees found in continually moist climates in the tropics, sub-tropics, and some parts of the temperate zones.

Rain gauge An instrument designed to collect and measure the amount of rain that has fallen. Also known as ombrometre; pluviometre.

Rain making Popular term applied to all activities designed to increase, through any artificial means, the amount of precipitation released from a cloud.

Rainout A process in which particles in the higher troposphere q.v. may act as nuclei on which water droplets form, later to fall as rain.

Rain shadow An area of diminished precipitation on the lee side of mountains or other topographic obstacles.

Rain stage The thermodynamic process of condensation of water from moist air in an idealised saturation adiabatic or pseudoadiabatic lifting, at temperatures above the freezing point; begins at the condensation level.

Rain storm Storm associated with rainfall.

Rain tree Pithecolobium saman Benth. (Enterolobium saman Prain ex King.). A ornamental quick growing plant with plenty of foliage. Called *Vilaiti siris* in Hindi.

Rainy season In certain types of climate, an annually recurring period of one or more months during which precipitation is a maximum for that region. Also known as wet season.

Ramentum A thin brownish scale consisting of single layer of cells and occurring on the leaves and young shoots of many ferns.

Ramie Boehmeria nivea. A shrub or half-shrub of the nettle family (Urticaceae) cultivated as a source of a tough, strong, durable, lustrous natural woody fibre resembling flax, obtained from the phloem of the plant; used for high quality papers and fabrics. Also known as China grass.

Ramiflory Production of flowers on older branches.

Ramularia areola See Cotton areolate mildew.

Random error An error, that is to say, a deviation of an observed value from a true value, which behaves like a variable in the sense that any particular value occurs as though chosen at random from a probability distribution of such errors.

Random event An event with probability of occurrence determined by a probability distribution.

Random field ditches Ditches for the drainage of scattered depressions where the depth and length of cut are not excessive.

Random forecast A forecast in which one of a set of meteorological contingencies is selected on the basis of chance, it is often used as a standard of comparison in determining the degree of skill of another forecast method.

Randomization In an investigation where there is no experimental control, the principle of randomization is satisfied by the adoption of a good sampling technique making sure that no bias affects the differences between the two means, choosing as statistical individuals clusters that are statistically independent, and ensuring that the probabilities of being subjected to like treatment are equal for every relevant pair of objects in the experiment.

Randomised block An arrangement where the experimental material is divided into blocks of relatively homogeneous series, and the treatment are distributed at random within the block. This is done with a view to eliminating from the errors the order effect by assuring that each treatment is associated equally frequently with the variable, more commonly once in each block.

Randomness Essentially, the class of phenomenon to which the laws of probability apply by assuring each group of n members of a population the same chance of constituting the sample.

Random numbers A listing of members which is non repetitive and satisfies no algorithm.

Random order An order of a set of objects when the ordering process is carried out in such a way that every possible order is equally probable.

Random sample A sample which has been selected by a method of random selection.

Random sampling error Sampling error in cases where the sample has been selected by a random method.

Random sampling numbers Sets of numbers used for the drawing of random samples.

Random selection A method of selecting sample units such that each possible sample has a fixed and determinate probability of selection.

Random series A series of numbers which may be regarded as drawn at random from a fixed distribution.

Range (1) Land and native vegetation (climax or natural potential) that is predominantly grasses, grasslike plants, forbs, or shrubs suitable for grazing or browsing and present in sufficient quantity to justify grazing or browsing use. Includes lands revegetated naturally or artificially to provide a forage cover managed like native vegetation. (2) A measure of the variability of a quantity, being the difference between the largest and the smallest values of the quantity.

Range condition The state and health of the range based on what it is naturally capable of producing.

Range condition class One of the series of arbitrary categories used to classify range condition, usually expressed as either excellent, good, fair, or poor.

Range management Producing maximum sustained use of range forage without jeopardy to other resources or uses of land.

Range nozzle One of the nozzles in a two-nozzle-sprinkler head, to throw water to the outer part of the wetted circle. The other nozzle called 'spreader nozzle', is small and applies water to the area close to the sprinkler head.

Ranger An administrative officer in charge of a unit of forest land, usually a Sub-division of public forest or park. Various classifications are recognised, as forest ranger, district ranger, park ranger, and country ranger.

Rank The rank of a single observation among a set is its ordinarily a number when the set is ordered or arrayed according to some criterion, such as values of the variate borne by the individuals.

Ranney well A well that has a centre caisson with horizontal perforated pipes extending radially into an aquifer, particularly applicable to the development of thin aquifers at shallow depths.

Ranunculaceae A family of dicotyledonous herbs in the order Ranunculales distinguished by alternate leaves with net venation, two or more distinct carpels, and numerous stamens.

Ranunculales An order of dicotyldeous in the subclass Magnoliidae, characterised by its mostly separate carpels, triaperturate pollen, herbaceous or only secondarily woody habit, and frequently numerous stamens.

Rape Brassica napus. (Brassica compestris) A plant of the cabbage family of the order Capparales, the plant does not form a compact head, the leaves are bluish-green, deeply lobed, and curled and the small flowers produce black seeds, grown for forage. Also called Indian rape.

Rape oil A fatty, non-drying or semi-drying, viscous, dark brown to yellow oil with unpleasant taste and aroma, obtained from the seed of rape and turnip, soluble in other carbon-disulfide, and chloroform, solidifies at -2 to -10°C; used to make lubricants and rubber substitutes, as an illuminant, and in steel heat treatment,; Also known as colza oil, rape-seed oil.

Raphe In diatoms, a longitudinal slit extending along the median line of a value.

Raphide One of the long, needle shaped crystals, usually consisting of calcium oxalate, occurring as a metabolic by-product in certain plant cells.

Raphimatopus ablutella See sugarcane — insect pests.

Rare elements Trace elements.

Rasp A file-like tool having coarse projections for abrasion.

Raspberry Any of several species of upright shrubs of the genus Rubus, with perennial roots and prickly biennial stems, in the order Rosales; the edible black or red juicy berries are aggregate fruits, and when ripe they are easily separated from the fleshy seceptacle.

Rat Rattus sp. Family Muridae. A serious field pest.

Rate (1) The quantity or amount of a thing measured per unit of time. (2) Fixed relation of quantity, amount or degree between two things.

Rate of growth The rate of increase (growth) of objects say wood, animals etc.

Rate of interest The amount of interest due each year in respect of a sum of money borrowed, expressed as a percentage of the total sum borrowed.

Rate of runoff The volume of water running off in a unit of time.

Rating A technique often used as basis for incentive schemes. It involves three related producers (a) to determine the time required to perform a task; (b) to measure the amount of work in it; and (c) to relate the work value to some scale of cash rewards.

Rating curve A graphic or sometimes tabular representation or performa or output under a stated series of conditions; for example, a rating curve for a flume shows volume of flow per unit time at various stages of depth of flow.

Ratio of consumptive use of water to evaporation Coefficient determined experimentally, and used in determining consumptive use of water from evaporation records from free water surface, or evaporation potential determined through the use of atmometre cups.

Ratio, proportion or percentage The quotient of one magnitude divided by another magnitude of the same kind. When taken as a fraction of 100, the proportion or ratio becomes a percentage.

Rationing The official control of the right to purchase essential commodities in times of acute scarcity to ensure that everyone shall have a minimum share and yet restrict overall community consumption.

Ration A-24 hour allowance of a feed or a mixture of feedstuffs making up the animal diet.

Ratoon The shoot from the base of the stem or from the root of a plant that has been cut. Used for taking crops in second and subsequent years as in sugarcane.

Rattan Any of several long-stemmed; climbing palms, especially of the genera Calamius and Deamonothops; stem material is used to make walking sticks, wickerwork, and cardage.

Raunkiaer system A classification system for plant life-forms based on the position of buds in relation to the soil surface.

Rauwalfia A genus of mostly poisonous, tropical trees and shrubs of the dogbone family (Apocynaceae); certain species yield substances used as emetics and catharties, while R. serpentina is a source of alkaloids used as tranquilisers.

Ravine A deep narrow gorge or mountain cleft. It is larger than a gully and smaller than a canyon and usually worn down by running water.

Raw humus See mor.

Ray The corolla of a marginal flower of a composite inflorescence; also a vascular ray.

Ray flower One of the small flowers with a strap-shaped corolla radiating from the margin of the head of a capitulum.

Ray initial One of the cells of the cambium which divide to produce new phloem and xylem ray cells.

Rayleigh scattering Scattering by particles whose radii are considerably small than the wave length of the incident radiation.

Reach A portion of a waterway between two locks or gages.

Reaction, Soil The degree of acidity or alkalinity of a soil, usually expressed as a pH value. Descriptive terms commonly associated with certain ranges in pH are: extremely acid, less than 4.5; very strongly acid, 4.5-5.0; acid, 6.1-6.5; neutral, 6.6-7.3; slightly alkaline, 7.4-7.8; moderately alkaline, 7.9-8.4; strongly alkaline, 8.5-9.0; and very strongly alkaline greater than 9.0.

Reagent A substance used to effect chemical change in another substance.

Reaction wood An abnormal development of tree and therefore its woods as the result of unusual forces acting on it, such as atypical gravitational pull.

Ream To enlarge or clean out a hole.

Rear axle Consists of rear axle housing, gears, axle shafts, and drive, also such accessory parts as are needed for operation, as bearings, washers, etc.

Rebate A reduction in the price of a commodity, allowed perhaps for prompt payment or the purchases of a large quantity.

Recapitulation The repetition of ancestral stages in the development of an individual.

Recent soil A soil which has a profile without definite horizons of eluviation or illuviation.

Receptacle The stem of an inflorescence; The swollen part of a plant which forms the base of flower.

Recession The falling limb of a hydrograph after a flood event representing the withdrawal of water from storage in the valley, stream channel and the subsurface runoff. A recession curve may sometimes be referred to as a 'recession hydrography'.

Recessive A term applied to one member of an allelic pair lacking the ability to manifest itself wholly or in part when the other or dominant member is present.

Recharge (1) The processes involved in the replenishment of water to the zone of saturation. (2) The amount of water added or absorbed. Also known as groundwater increment; groundwater recharge; ground water replenishment, increment, intake.

Reciprocal crosses Crosses where the parental plants or lines are used as both male and female.

Reciprocal levelling A variant of straight differential levelling applied to long distances in which levels are taken on two points and the average of the two elevation differences is the true difference.

Reclamation Act or process of reclaiming swampy, marshy, deteriorated, desert and virgin lands and making them suitable for cultivation or habitation; also conversion of foreshores into properly drained land for any purpose, either by enclosure and drainage, or by deposition of material thereon.

Recombination The observed new combinations of characters different from those combinations exhibited by parents. Percentage of recombination equal percentage of crossing over only when the genes are relatively close together. Cytological crossing over refers to the process; recombination or genetic crossing over refers to the observed genetic result.

Reconnaissance survey A rapid survey or critical inspection of an area or site made for furnishing data (i) to find out the feasibility and extent of a development project or scheme; (ii) to permit preliminary examination of alternative sites and location; (iii) to provide the base maps for preliminary estimates of the extent and location of potential lands; or (iv) to derive a preliminary plan on which to base further studies and surveys.

Record observation A type of aviation weather observation; the most complete of all such observations and usually taken at regularly specified and equal intervals.

Recovery The quantity of converted timber of specified quality obtained from a piece, expressed as a percentage of its volume, calculated according to the conventional practices.

Rectangular drainage pattern Drainage pattern showing the influence of the angular pattern of rock joints.

Rectangular weir A weir having rectangular notch.

Recumbent Of or pertaining to a plant or plant part that tends to rest on the surface of the soil.

Recurrence interval The average time interval between actual occurrences of a hydrological event of a given or greater magnitude, e.g.in an annual flood series the average interval in which a flood of a given size recurs as an annual maximum.

Recurrent parent Used in backcrosses to refer to the parent to which the first cross and backcrossed plants are crossed.

Recurrent selection A method of determining which plants of a composite make the best hybrids in combination with the particular single cross used as a tester; used in maize breeding.

Red ant Oecophylla amaragdina Fabr. Pest of mango which destroys fresh leaves. Also attacks other fruit trees such as citrus, guava, cashewnut etc.

Red algae The common name for members of the phylum Rhodophyta.

Red brass An alloy of copper and zinc with 85:15 composition.

Red hairy caterpillar Amsacta moorei But. A. albistriga Moore. Insect pest on many cereals, cotton, pulses, groundnut and sunhemp. Its caterpillar feeds on plant leaves.

Red pumpkin beetle Aulacophora foveicollis Lucas, A. inter-media Jac. A pest of cucurbit family, cuts irregular holes in leaves on which it feeds.

Red silk cotton Bombax ceiba L. (B. malabaricum DC., Salmalia malabarica (DC) Schoff and Endl.). Large deciduous tree of family Bombacaceae. Floss used for stuffing quilts and pillows.

Red soils Soils group covering Peninsular India. Light textured, low salt content, coloured red by the presence of iron peroxide. Clay fraction rich in kaolinite. Differ greatly in depth and fertility.

Reducer Any one of the various pipe connections so constructed as to permit the joining of pipes of different sizes, such as reducing sleeve, reducing ell, reducing tee, etc.

Reduction division; heterotypic division Terms formerly applied to the one of the meiotic mitoses at which reduction and segregation were thought to occur.

Reduction gears Gears used on a shaft to reduce the speed.

Reed Any tall grass with a slender jointed stem.

Reface To true up a valve face by griding or turning.

Reference group Is a group of which an individual refers when forming an opinion, making a judgment or deciding to act. Neighbourhoods and family constitutes the most important reference groups in farm practice adoption behaviour.

Reflection coefficient The ratio of the amount of solar radiation reflected by body to the amount incident upon it.

Reforestation The natural or artificial restocking of an area with forest trees; most commonly used in reference to the latter.

Refractions All components in wheat which differ from sound grains are known as refractions.

Refractory wood Wood showing tendency to develop defects during seasoning.

Regeneration See Reproduction.

Regime (Hydraulics) Applies to streams that make at least part of their boundaries from their transported load and part of their transported load for their boundaries, carrying out the process at different places and times in the stream in a balanced or alternating manner that permits unlimited growth or removal of boundaries.

Regimen The stability of a stream and its channel. A river or canal is "in regimen" if its channel has reached a stable form as a result of its flow characteristics.

Regional outlet channel A creek, river, or other natural channel carrying the drainage from an entire large watershed or region.

Regional strain Strain developed under a given environmental condition as a result of the survival of the best adapted individuals through many seed generations; botanically an ecotype.

Registered Seed Seed that is the progeny of foundation seed, traces directly to it, and is approved and certified to comply with specified standards of purity and quality.

Regosol The only great group in the Regosolic order. The soils in the group have insufficient horizon development to meet the requirements of the other orders. Derived from deep soft mineral deposits, such as loess or glacial drift.

Regosolic An order of soil having no horizon development or development of the A and B horizons insufficient to meet the requirements of the other orders. Regosol is the only great group in this order.

Regolith The unconsolidated mantle of weathered rock and soil material overlying solid rock.

Regression A statistical method for studying and expressing the change in one variable associated with and dependent on changes in another related variable or set of variables.

Regression coefficient A coefficient that gives the rate of change in one variable (dependent variable) per unit rate of change in another (independent variable).

Regression curve Diagrammatic exposition of the regression equation.

Regulated markets Markets which protect the farmer from exploitation and ensure to him fair and competitive price.

Regulating reservoir A reservoir that regulates the flow in a water-distributing system.

Regulator (1) A structure built on canals at suitable points to control and regulate and or measure water supplies during conveyance to a location of usage or disposal and at head of canals to control, regulate and admit supplies from the parent canal or river. Also called regulation structure or control structures. (2) A chemical other than nutrients which increase or decrease plant growth.

Reinforcement Strengthening concrete, plaster, or mortar by embedding steel rods or wire mesh in it.

Reinforcing bars Steel rods that are embedded in building materials such as concrete for reinforcement.

Reiteration Development of shoots outside the normal expression of the architectural model of a tree, as a specialised environmental response.

Reiterative complex The complex of meristems produced by reiteration.

Rejuvenation See dedifferentiation.

Rejuvenation rangeland Accomplished by soil-plant-animal management practices favourable to forage plants.

Rejuvenated stream A mature stream that has reverted to the behaviour and form of a more youthful stage due to rejuvenation, usually as a result of uplift. Also known as revived stream.

Relative biological effectiveness The ratio of the dose (in rads) of gamma radiation from cobalt-60 to the dose (in rads) of any type of ionizing radiation which produces the same, specified, biological effect.

Relative branch order See branch order, relative.

Relative frequency, proportional frequency, or percentage frequency Where the frequency is expressed as a proportion of the total number of members, it is called the 'relative' or 'proportional' frequency. When expressed as a percentage of the total number, it is called 'percentage frequency'.

Relative humidity The amount of water present in the air at a given temperature in proportion to its maximum water holding capacity at that temperature.

Relay axis An axis taking over growth in a tree, either according to the model or as a result of reiteration.

Relay intercropping A form of intercropping in which two or more crops grow simultaneously during part of the life cycle of each, that is a second crop is planted before the first crop matures.

Release Refers to the release of varieties by appropriate authorities such as the Central Variety Release Committee; State Variety Release Committee.

Release cutting A cutting of larger individual trees that are over-topping young trees, for the purpose of freeing the young trees to permit them to make good growth.

Relief Elevations or inequalities of a land surface, considered collectively, land having no unevenness or differences of elevation is called level; gentle relief is called undulation, strong relief, rolling, and very strong relief, hilly. See also microrelief.

Relief drain Open or subsurface drain provided to relieve a saturated soil of its superfluous water which percolates from the soil into it. Also called 'seepage drain'.

Relief system A system which consists of a series of lateral tile lines laid out in a grid or herringbone pattern, each line being connected to a main trunk which in turn outlets into an open ditch.

Relief valve It is provided in the lubrication system to control the quantity of oil circulation and to maintain correct pressure in the lubricating system.

Relief well A well that drains a previous stratum, to relieve waterlogging at the surface.

Relogging An operation in which small trees are salvaged, often for pulpwood, after the large trees are logged.

Randzina A group of soils having brown to black surface horizons, which have developed on parent material that contains more than 40 per cent calcium carbonate equivalent. Not used in Canadian taxonomy.

Repair forecast The quantity of items estimated to be repaired or rebuilt for issue during a stated future period.

Repair kit A group of parts and tools, not all having the same basic name, used for repair or replacement of the worn or broken parts of an item, it may include instruction sheets and material, such as sand paper, tape, cement, gaskets and the like.

Repair parts list List approved by designated authorities, indicating the total quantities of repair parts, tools and equipment necessary for the maintenance of a specified number of end items for a definite period of time.

Repand Having a margin that undulates slightly, referring to a leaf.

Repeating spore The urediospore, the only spore in the life cycle of rust which is able to reinfect the same species of plant as that on which it is produced.

Repent Of a stem, creeping along the ground and rooting at the nodes.

Replication (1) Process by which genetic determinants are duplicated during cell multiplication so that identical genetic characters are passed into the next generation. (2) One or more repetitions of an experiment. Variations due to soil and other differences are lessened by replications.

Replum A thin wall separating the two valves or chambers of certain fruits.

Reproduction The process by which a forest or range is renewed, including (1) artificial; renewal by direct seeding or planting (reforestation); and (2) natural: renewal by self-sown seeds, sprouts, rhizomes, etc. (Regeneration). Also seedlings or saplings of any origin (young growth), the result of reproduction.

Repulsion The condition in linked inheritance in which an individual heterozygous for two pairs of linked factors received the dominant member of one pair and the recessive member of the other pair from one parent and the reverse condition from the other parent; e.g. AAbb x aaBB.

Rerolle system A method of underdrainage to attain a lowering of the ground water table in which the lines of tiles are water tight, and water enters them from below through pits, filled with gravel and placed at regular interval along the line.

Rescue grass Ceratochloa cathartica (Vahl) Hert (Bromus catharticus Vahl.). A herb of family Gramineae used for fodder. Also known as rescue brome.

Reset To refinish a valve seat by means of a special reamer, by lapping process, or by specially mounted abrasive stones.

Resedaceae A family of dicotyledonous herbs in the order Capparales having irregular, hypogynous flowers.

Reseeding cultivar Perpetuates itself by volunteering from shattered seed; usually made possible because of a high percentage of hard seed or seed with high dormancy.

Reseeding, Range Sowing of seed or range lands to restore or increase forage production.

Resequent stream A stream whose direction follows an original consequent stream, but is generally lower; resequent streams are generally tributary to a subsequent stream.

Reserve food Plant food stored in various parts.

Reserve price A minimum price at or below which a seller will withdraw an article from sale.

Reservoir A pond or lake built for storage of water, usually by the construction of a dam across a river.

Reservoir routing The flood routing when the flood wave is passed through a reservoir.

Reservoir sedimentation survey Survey to obtain information on the quality, quantity and distribution of the sediment accumulations in the reservoir formed by the dam and the effect of reservoir on the stream channel upstream and downstream.

Residual (1) A general term denoting a quantity remaining after some other quantity has been substracted. In particular, it denotes that if a mathematical model is fitted to the data, the values by which the observations differ from the model values are called residuals. (2) A quantity of material or energy which is left over or wasted when, in the course of human activity, inputs are converted into outputs in production and consumption.

Residual error The difference between any value of a quantity in a series of observations, corrected for known systematic errors, and the value of the quantity obtained from the combination or adjustment of that series. In practice, it is the residual errors which enter into a computation of probable error.

Residual material Unconsolidated and partly weathered minerals materials formed by the distintegration of consolidated rock in place.

Residual shrinkage The decrease in volume after the proportionaity between waterloss and volume change ceases.

Residual Soil Soil formed from, or resting on, consolidated rock of the same kind as that from which it was formed and in the same location.

Residue A sediment or deposit. That which remains after burning or pouring.

Resin A solid or semi-solid organic compound obtained from the exudate of some plants.

Resinosis An abnormal exudation of resin from conifers. Also abnormal impregnation of their tissues by resin.

Resin duct A cell, tube, or cylinder containing resin.

Resistance The ability of a plant to develop and function normally despite adverse environmental conditions or the attacks or disease or insects.

Resistance block A block made of a porous inert material used to measure the moisture content of a soil, moisture percentage curve is made for each block beforehand.

Respiration A chemical oxidative process whereby living protoplasm breaks down certain organic substances with the release of energy which is used in various anabolic activities, movements, etc. Process of absorption of oxygen and liberation of carbon dioxide.

Resting spore A spore, often thick-walled, that can remain alive in a dormant condition for some time, later germinating and capable of initiating infection.

Restionaceae A large family of monocotyledonous plants in the order Restionales characterised by unisexual flowers, wholly cauline leaves, unilocular anthers, and a more or less open inflorescence.

Restionales An order of monocotyledonous plants in the subclass Commelinidale having reduced flowers and a single pendulous, orthotropous ovule in each of the one to three locules of the ovary.

Restocking Applied to an area on which the forest is being re-established by natural means.

Restoration A conservation measure involving the correction of past abuses that have impaired the productivity of the resources base.

Restorer line An inbred line that, when crossed on a male sterile strain, causes the resulting hybrid to be male fertile (R line).

Resupinate Flat on the substratum with hymenium on outer side, inverted usually through 180° so as to appear upside down.

Resuspension The return of settled deposited sediment to a state of suspension in the water body.

Ret Soaking in water for extraction of fibre from plants.

Retailer A dealer selling goods in relatively small quantities to private consumers.

Retained water The water remaining in rock or soil after gravity groundwater has been drained out.

Retaining wall A wall designed to maintain differences in ground elevations by holding back a bank of material.

Retarding basin A basin designed and operated to provide temporary storage and thus reduce the peak flood flows of a stream.

Retention That part of the precipitation falling on a drainage area which does not escape as surface stream flow, during a given period. It is the difference between the total precipitation and total runoff during the period, and represents evaporation, transpiration, subsurface leakage, infiltration, and when short periods are concerned, temporary surface or underground storage on an area. When periods of several years are considered, it approximates consumptive use.

Retentivity profile, soil A graph showing the retaining capacity of a soil as a function of depth. The retaining capacity may be for water, for water at any given tension, for cations, or for any other substances held by soils.

Reticulate Netlike, as the chloroplasts of certain algae, or as spore walls, form of a soil structure.

Reticulate mottling A network of coarse streaks of different colours, characteristics of parent material in laterite soils. Also called vermiculate mottling.

Reticulum (1) Delicate network of cell protoplasm. (2) Cross-fibres about the base of petioles in palms.

Return period Strictly speaking, it is the period as in a time series, for a value to return to the same value. More generally, it is used in the sense of the average time interval within which a given event will be equalled or exceeded.

Retuse Having a rounded apex with a slight, central notch.

Reverse-T-budding See Shield budding.

Reversible elbow An elbow installed in a pipeline system which enables farmers to take off at right angles from either main line or sprinkler line.

Reversible plough A plough whose mould board position can be changed from left to right or vice-versa. Can be used in making irrigation channels.

Reversing gear The gear by which the reversal of an engine or machine is accomplished.

Reversion Changing of essential plant nutrient elements from soluble to less soluble forms by interaction with or reactions in the soil. Reversion is usually restricted to the conversion of mono-calcium phosphate to the less soluble di-calcium phosphate.

Reverted phosphoric acid That part of the total phosphoric acid in a fertiliser that is insoluble in water but soluble in a neutral solution of citrate of ammonia. Mostly in the form of di-calcium phosphate, $Ca_2H_2(PO_4)_2$.

Revetment Material, such as rock, concrete blocks or mattresses, placed on the bottom or banks of a river to prevent or minimise scour by water. Also a facing made on a soil or rock embankment to prevent scour by weather or water.

Revolution counter Also called "speed indicator". It is a device for counting the revolutions of a shaft. By pressing a pointer against the end of a shaft, the revolutions are registered on a dial.

Revolutions per minute An expression referring to rate of speed of machines, Abbreviation, r.p.m.

Revolving screen Trash screen or rack in the form of a cylinder or continuous belt revolved mechanically or by the power of the water passing through it.

Reworked Descriptive of material modified after its preliminary depositions, commonly by water or wind.

Rhabdolith A minute cocolith having a shield surrounded by a long stem and found at all depths in the ocean, from the surface to the bottom.

Rhacopilaceae A family of mosses in the order Isobryales generally having dimorphous leaves with smaller dorsal leaves and a capsule that is plicate when dry.

Rhamnaceae A family of dicotyledonous plants in the order Rhamnales characterised by a solitary ovule in each locule, free stamens, simple leaves and flowers that are hypogynous to perigynous or epigynous.

Rhamnales An order of dicotyledonous plants in the subclass Rosidae having a single set of stamens, opposite the petals, usually a well-developed instra-stamental disk, and two or more locules in the ovary.

Rheophile Living or thriving in running water.

Rhipidium A fan shaped inflorescence with cymose branching in which branches lie in the same plane, and are suppressed alternately on each side.

Rhizanthous Producing flowers directly from the root.

Rhizautoicous Of mosses, having the antheridial branch and the archegonial branch connected by rhizoids.

Rhizobia Small heterotrophic bacteria of the genus Rhizobium capable of forming symbiotic nodules on the roots of leguminous plants. In the nodules the bacteria fix atmospheric nitrogen that is used by the plants. The bacteria receive their energy from the plants.

Rhizocarpous Pertaining to perennial herbs, having perennating underground parts from which stems and foliage arise annually.

Rhizoctonia bataticola See Cotton root rot, and cardamom dump rot.

— **soloni** See Coffee damping off cotton rot, and cotton root rot.

Rhizoid In some fungi, mosses, liverworts, one of the hairlike appendages which penetrate the soil or other substrate, anchoring the plant and absorbing water and other substances.

Rhizome An underground horizontal stem, often thickened and tuber-shaped, and possessing buds, nodes, and scale-like leaves, capable of producing new shoots and roots at the nodes. Synonymous with rootstock.

Rhizomorph A root-like structure, characteristics of many basidiomycetes, consisting of a mass of densely packed and intertwined hyphae.

Rhizophoraceae A family of dicotyledonous plants in the order cornales distinguished by opposite, stipulate leaves, two ovules per locula, folded or convolute bud petals, and a berry fruit.

Rhizophore A leafless downward growing dichotomous selaginella shoot.

Rhizoplane The external surface of roots and of the soil particles and debris adhering to them.

Rhizoplast A fibre connected to the base of the flagella.

Rhizopus artocarpi See Bread fruit soft rot, See jack fruit inflorescence rot.

— **nigricans** See Sweet-potato diseases.

— **soloni** See Coffee damping off and See cotton rot, See cotton root rot.

Rhizosphere Interfacial layer of soil between the root and soil bulk that is under the influence of the plant root.

Rhodes grass Choris gayana Kunth. Family Gramineae. A perennial grass, used as fodder. Drought resistant, winter hardy but responds well to irrigation.

Rhodophyceae A class of algae belonging to the division or subphylum Rhodophyta.

Rhodophyta The red algae, a large diverse phylum or division of plants distinguished by having an abundance of the phycoerytherin.

Rhodoplast A reddish chromatophore occurring in red algae.

Rhoipteleaceae A monotypic family of dicotyledonous plants in the order Juglandales having pinnately compound leaves, and flowers in triplets with four sepals and six stamens, and the lateral flowers female but sterile.

Rhubarb Rheum rhaponticum. A herbaceous perennial of the order polygoniales grown for its thick, edible petioles.

Rhynchocoris humeralis See Citrus green bug.

Rhyncophorus ferrugineus See Coconut red palm weevil.

Rhythmic branching Production of branches in a regular sequence associated with rhythmic growth.

Rhythmic growth (episodic growth), Growth of an axis determined by a rhythm which results in periodic shoot extension alternating with dormancy.

Rhytidome The furrowed, wrinkled outer bark of stems and roots.

Rhyzopertha dominica See Lesser grain borer.

Ribonucleic acid (RNA) One of the two nuclei acids (see also DNA); when of viral origin, RNA is infectious for susceptible cells.

Rice Oryza sativa. An annual cereal grass plant of the order Cyperales, cultivated as a source of human food for its carbohydtrate-rich grain known as *chawal* in Hindi.

Rice bran oil Clear, combustible, liquid derived by solvent extraction of oil from fresh rice bran, used to make soaps and animal feeds, salad and cooking oils, and hydrogenated shortening.

Rice bacterial leaf spot Disease caused by Xanthomonas oryze Dowson. Characterized by leaves turning straw yellow when sever, field look like of burnt up appearance.

Rice blast Major diseases caused by Pyricularia oryzae Cav. Characterized by appearance of spindle-shaped spots with grey centres and brownish red margins on leaves. Grains remain unfixed.

Rice brown eye spot Disease of rice caused by Helminthosporium oryzae Breda de Haan; Cochliobolus miyabeanus Ito et Kurib. characterized by brown irregularly rectangular spots with a dark centre on aerial parts of the plant.

Rice caseworm Paddy pest, Nymphula depunctalis Guen. Calterpillar feeds on leaves.

Rice-diseases Rice bacterial leaf blight caused by Xanthomonas oryzae Dowson, rice blast caused by Pyricularia oryzae Car, rice brown eye spot caused by Helminthosporium oryzae or cochliobolus miyabeanus Ito et Kurib, rice false smut caused by Ustilaginoidea virens, rice foot rot or seedling blight caused by Fusarium moniliniformes Held or Gibberella fujikuroi Saw, rice irregular stem rot caused by Helminthosporium sigmoideum var. irregularae, rice leaf smut caused by Entyloma oryzae H and P., rice narrow brown leaf spot caused by Cercospora oryzae., rice stack burn caused by Trichoconis padwickii G., rice stem rot caused by Sclerotium oryzae Cat., rice udbatta disease caused by Balansia oryzae and rice pan sukh a virus disease are common in rice.

Rice false smut Disease caused by utilaginoidea virens. Few grains in the panicle gets transformed into large compact black mass.

Rice foot rot Disease of rice caused by Fusarium moniliniformes Held. Gibberella fujikuroi saw. wr. Makes seedling pale, collapse and die. Also plants rot at the base.

Rice glue A paste made from ground rice boiled in soft water; used in molded objects such as statuary.

Rice grass Leersia hexandra Sw. A herb of family Gramineae. It is used as fodder.

Rice-insect pests Rice bug (Leptocorisa Varicornis Fabr., L. accuta Thunb) whose nymphs and adults suck sap from ears; rice grasshoppers (Hieroglyphus banian Fabr. Oxya velox Fabr) whose nymphs and adults eat leaves and adults feed on ears; rice hispa (Hispa armigera oliv) whose grub burrows into leaf tissue; rice moth (Corcyra cephalonica Staint) whose larva bores into grains; rice stem borer (Tryporyza incertellus Wlk) whose caterpillar bores into plant stem and rice weevil (Sitophilus Oryzae L.) whose grub and adults feed on grass are some insect pests of rice.

Rice irregular stem rot Disease caused by Helminthosporium siymoideum var. irregularae. Characterised by occurrence of pale olive brown patches on the lower internodes.

Rice leaf smut Disease caused by Entyloma oryzae H. and P. Small, cingular black spots below the epidermis on leaves could be seen.

Rice narrow brown leaf spot Disease caused by Cercospora oryzae. Characterised by light brown lesions with dark centre on leaf and leaf sheaths.

Rice stack burn Disease caused by Trichoconis padwickii G. It is characterised by seedlings showing small, dark brown or black spots on roots, pale brown spots on leaves.

Rice stem rot Disease caused by Sclerotium oryzae cat. Affected plants show excessive tillering with base of stem becoming discoloured.

Rice volbatta disease Disease caused by Balansion oryzae. Characterised by ears looking like straight spikes as these come out of leaf sheath.

Rice wax Wax deposited when rice bran oil is allowed to stand for a day or two.

Richardson number A non-dimensional index of stability arising in the study of sheering flows of a stratified fluid.

Rick A rectangular stack of hay, straw, or small-grain bundles.

Rid meristem A type of primary meristem which contributes to the growth in length of young internodes. In stem species it is between the central mother cell zone and the maturing pith.

Ridged gourd Luffa acutangula L. Roxb (Cucumis acutangulus L.). Family Cucurbitaceae. A vegetable grown for early market. Known as *torai* in Hindi.

Ridger An implement to make ridges.

Ridges The saddle that divides the terrain into two or more gently sloping planes, making irrigation and distribution of water possible on both sides of the canal aligned more or less along it. The ridges are called 'main ridges' if their slope is generally the least of all other possible ridges in the terrain, the other being called 'secondary ridges'.

Ridging Making small embankments or borders in fields to control irrigation water.

Ridging plough It is a plough which balances itself on wheels and on which the operator rides during the operation.

Riffle Shallow rapids in an open stream, where the water surface is broken into waves by obstructions wholly or partly submerged.

Rill A narrow, very shallow, intermittent water course having steep sides. It presents no obstacles to tilling.

Rill erosion See erosion.

Rim blight A fungus disease of tea caused by members of the genus Cladosporium and characterised by yellow discolouration of the leaf margins followed by the browning.

Rim pull It is a term used to designate the tractive force between the rubber tyres of the driving wheel and the surface on which they travel.

Rind (1) The bark of a tree. (2) The thick outer covering of certain fruits.

Rind gall A localised lack of cohesion between growth incretments on woody plants caused by the growth of healing tissues over an area of dead cambium exposed by mechanical injury or fire.

Ring annual The growth layer of one year, as viewed on the cross-section of a stem, branch, or root.

Ring budding A method of budding commonly used in peach, plum, mulberry and *ber*.

Ring cell A cell in the annulus of a sporangium.

Ring gauge An gauge in the shape of a ring used for checking external diametres.

Ring gear (1) The large gear on the flywheel which is engaged by the pinion gear when the starting motor is in action. (2) Any gear in the form of a ring, having no hub or central bore.

Ringing A method of increasing fruit and formation as in grapes also called girdling.

Ring-porous wood Wood in which the pores of one part of an annual ring are distinctly different in size or number (or both) from those in the other of the ring.

Ring rot (1) A fungus disease of the sweet potato root caused by Rhizopus stolonifer and marked by rings of dry rot. (2) A bacterial disease of potatoes caused by Corynebacterium sepedonicum and characterised by brown discolourations of the annular vascular tissue.

Ring spot Disease symptoms characterised by yellowish or necrotic rings with green tissues inside the ring, as in virus and fungus diseases.

Riparian It refers to land which borders a river or a body of water. Riparian rights are commonly used while discussing distribution of water between states.

Ripe Of fruits, fully developed, having mature seed and so usable as food. Also a mature seed.

Ripper A towed machine equipped with teeth, used primarily for loosening hard soil and soft rock.

Riprap Broken stone placed on earth surfaces, and large stones against a fill of small rocks, etc., placed on the upstream slopes of embankment particularly against wave wash.

Rise It is the period during and following rainfall from the time when the hydrograph first departs from the normal depletion curve until it again becomes coincident with the normal depletion curve. A rise consists of a period of increasing flow, which may results either from increased groundwater flow or from surface runoff, or both. This culminates in the crest or peak of the rise which is followed by a 'recession period'.

Riser outlet, Riser turnout, or Riser An offtake device installed in the sprinkler lateral to feed water supply to the sprinkler head through a riser, they can either be at the coupler or at the centre of the pipe length.

Rising tide The portion of the tide cycle between low water and the following high water.

River-bank effect Growth response of trees unevenly illuminated, as on a river-bank.

River forecast A forecast of the expected stage or discharge at a specific time, or of the total volume of flow within a specific time interval, at one or more points along a stream.

River red gum Eucalyptus camaldulensis. Family Myrtaceae. Medium to large sized tree of which timber is quite durable.

River system The aggregate of stream channels draining a river basin.

River tide A tide that occurs in rivers emptying directly into the sea, showing three characteristic modifications of ocean tides, the speed at which the tide travels upstream the depends on the depth of the channel, the further upstream the longer the duration of the falling tide and shorter the duration of the rising tide, and the range of the tide decreases with distance upstream.

River-valley project Project directed to developing, controlling or utilizing water resources of a river basin.

Riverwash Barren, usually coarse-textured, alluvial soil in and along waterways, exposed at low water levels and subjected to shifting during flood periods. A miscellaneous land type.

River water Water having carbonate, sulfate, and calcium as its main dissolved constituents, distinguished from sea-water by its chloride and sodium content.

Rivet A short, metal boltlike fastening, without threads, which is clinched by hammering, Rivets are designed by the shape of the heads as flat or panhead, buttonhead, countersink, mushroom, or swollen neck.

Rivulose Marked by irregular, narrow lines.

R layer Underlying consolidated bedrock. See also horizon, soil.

Road drag Any one of the various devices dragged over a soft surfaced road for the purpose of levelling; distinguished from road scraper.

Robinia grandiflora See Agati sesbania.

Roccilaceae A family of fruticose species of Hysterales that grow profusely on trees and rocks along the coastlines of Portugal, California, and Western South America.

Roche mountonnee A rounded hummock (boss) of rock smoothed and striated by glacial action and often showing evidence of plucking on the lee side.

Rock Solid part of earth's crust underlying soil, include loose incoherent as well as solid or hard masses, comparatively homogeneous.

Rock bee Apisdorsata F. A bee species usually ferocious and sensitive to smoke. Not useful for domestication.

Rock candy Large, transparent, hydrated crystals of cane sugar.

Rock disintegration The rock disintegration by natural processes including expansion, contraction, chemical and mechanical actions of moving water, and plant roots.

Rock drumlin An elongated hill having a veneer of glacial drift over a rock core.

Rock-fill dam A dam constructed of loosely placed rock or stone.

Rock it corn grader A width and thickness grader which uses flat screens to make width and thickness separations.

Rockland An area of which usually 25 per cent to 90 per cent is occupied by rock outcrops and most of the remainder by shallow soils. A miscellaneous land type.

Rock phosphate Natural deposits of rock containing 25-35 per cent of phosphoric acid which is produced and used as fertilisers. Used in manufacture of superphosphate.

Rock weathering The natural processes which cause the physical and chemical disintegration and decomposition of rock to form soils.

Rod A British unit of linear measurement and is equal to 5.5 feet or 5.0292 metres. Also called pole or perch.

Rod assembly Consists of connecting rod, piston, piston pin, and piston rings ready for installation.

Rod ends The end portion of connecting rods which carry the bearings or process. The strap, solids head and marine are the main types.

Rodenticides Chemicals which are used to control rats and other rodents. For example zinc phosphate, barium carbonate, calcium cyanide, warfarin etc.

Rodents Animal which belong to Rodentia a large order of mammals.

Rod float Rod or staff designed to float in a practically vertical position for the purpose of observing stream velocities.

Rod row A type of field plot approximately 1 rod long. Used particularly with small grains where the seed is sown without definite spacing.

Rod weeder A type of equipment used to prepare the soil during harrowing; it is a power-driven rod, usually square in cross-section, which also operates below the surface of loose soil, killing weeds and maintaining the soil in loose mulched condition; adopted to large operations and used in dry areas in the north western United States.

Rogue (1) An off type plant, undesirable plant. (2) A seedling of inferior quality.

Roguing The act of removing undesirable individuals from a varietal mixture in the field by hand selection.

Rolling land Land having gradual, rounded undulations of surface.

Rolling resistance It is the drawbar pull of the tractor required to move tractor over a given surface.

Rood Basic unit of area in the British unit system. It is equal to 40 square rods or 0.25 acres. In metric unit it is equal to 0.101171 hectare.

Rookery A communal breeding site of shore birds or seals.

Room germinator It is a modification of the cabinet. It is constructed on the same principals, but is large enough to permits workers to enter it and place the tests along either side of a central passage.

Root The portion of the plant which goes down in the soil and allows the plant to stay in position. Supplies moisture and nutrients through root hairs. Roots may also be aerial or partially aerial.

Root antagonism The injury to roots of one species by the roots of other species of plant when grown side by side or one after another.

Root aphids Insects belonging to family aphidae which suck root sap.

Root cap A thimbl-like mass of cells (Parenchymal) which fits over the apical meristem of a root and protects it.

Root climbers The plants which climb by roots developed from stems.

Root collar Transition zone between root and stem.

Root constant The maximum soil moisture deficit that can be built up without checking transpiration.

Root crops A group of crops grown for their edible enlarged roots. Also called root vegetables.

Root crown The basal portion of the stem of a plant where the transition or change from stem to root tissue occurs.

Rooted cutting Cuttings which have developed roots used for transplanting as in grapes, *ber* etc.

Root-gall Any abnormal enlargement or swelling of, or on, the root of a plant, commonly due to a parasitic organism.

Root grafting Grafting of a scion on root or a portion of it.

Root hairs Extensions of epidermal root cells; the absorption area for water and nutrients.

Root knot Any of various plant disease caused by root-knot nematodes which produce gall-like enlargements on the roots.

Rootlet A small root.

Root mean square deviation The square root of the second moment of a set of observations taken about some arbitrary origin. The minimum value of the root mean square deviation occurs when the origin coincides with the arithmetic mean it is then called 'standard deviation'.

Root nodule An abnormal nodular growth on a plant root system caused by the establishment of symbiotic nitrogen fixing bacteria in the host tissue.

Root pressure The force set up by the root which assists water to ascend the root and stem. It is caused partly by the establishment of an osmotic gradient from the root hair to the pericycle of the roots.

Root rot Any of certain diseases characterised by decay of the roots.

Root stock A root or part of a root used as the stock for grafting. A rhizome.

Root tuber A swollen root containing reserve food material.

Rosaceae A family of dicotyledonous plants in the order Rosales typically having stipulate leaves hypogynous, slightly perigynous, or epigynous flowers, numerous stamens, and several or many separate carpels.

Rosales A morphologically diffuse order of dicotyledonous plants in the subclass Rosidae.

Rose A member of the genus Rosa in the rose family (Rosaceae); plants are erect, climbing, or trailing shrubs, generally prickly stemmed and bear alternate, oddpinnate single leaves.

Rose hip The ripened false fruit of a rose plant.

Rosemary Rosemarenus officinalis. A fragrant evergreen of the mint family from France, Spain, and Portugal, leaves have a pungent, bitter taste and are used in herb and in perfumes.

Rosidae A large subclass of the class Magnoliatae; most have a well-developed corolla with petals separate from each other, binucleate pollen, and ovules usually with two integuments

Rosette (1) A circular cluster of leaves or other organs; as the arrangement of leaves on a dandelion. (2) Disease symptoms with stem shortened to produce a bunchy growth habit.

Rosette plant A plant whose leaves radiate from a centre close to or on the ground.

Rosin A hard, brittle, natural resin obtain from the cleoresin exudate of certain trees. Rosin is a particular kind of resin. Rosin is obtained either from gum that exudes from the living pine trees or from wood by extraction. Wood rosin and gum rosin are kinds of resin.

Rosulate A condition in which leaves are packed over each other like rose petals.

Rotametre A variable area, constant-head, rate-of-flow volume metre in which the fluid flows upward through a tapered tube, lifting a shaped weight to a position where upward fluid forces just balances its weight.

Rotary harrow An implement which consists of a series of curved, spike-like teeth rotating on the main axle, young weeds are destroyed while crust is broken.

Rotating sprinkler system A sprinkler system which discharges the water through rotating sprinklers.

Rotation The growing of two or more crops in a sequence and according to a definite plan on a single piece of land. It may be a short term or long term rotation.

Rotational flow Flow of a fluid in which the curl of the fluid velocity is not zero, so that each minute particle of fluid rotates about its own axis. Also known as rotational motion.

Rotation grazing Grazing more than one pasture in regular sequence with rest periods to allow for the recovery of the forage.

Rotation pasture A pasture crop grown in sequence with other crops in a planned rotation.

Rotenone A naturally occurring contact insecticide from roots of Derris and Lonchocarpus species.

Rot, Heart A decay characteristically confined to the heartwood. It usually originates in the living tree.

Roughage Plant materials that are relatively high in crude fibre and low in digestible nutrients, such as straw and stover. See also Forage.

Rough bark (1) Any of various virus diseases of woody plants characterised by roughening and often splitting of the bark. (2) A fungus disease of apples caused by Phomopsis mali and characterised by rough cankers on the bark.

Rough broken land An area having steep slopes and many intermittent drainage channels, but usually covered with vegetation.

Roughness ccefficient (Hydraulics) A factor in velocity and discharge formulae; represents the effect of channel roughness on energy losses in flowing water. Manning's "n" is a commonly used roughness coefficient.

Roughness parametre A measure of the roughness of a surface over which a fluid is flowing.

Row crop A crop planted in rows, normally to allow cultivation between rows during the growing season.

Row planter An implement used for planting crops in a row.

Rowen Second growth of grass after a crop of hay has been harvested; another term for aftermath.

Royalty Compensation for the use of a patent, copyright, or another property, often calculated as a percentage of the sales values of the article or service.

R selection Selective conditions which favour high reproductive rates at expense of individual vegetative longevity.

R strategist A species adapted to high rates of sexual reproduction at expanse of individual longevity.

Rubber-diseases Rubber abnormal leaf fall caused by Phytophthora plamivora, rubber bird's eye spot caused by Helminthosporium heveae rubber brown rot caused by Fomes noxius, rubber charcoal rot caused by Ustulina zonata, rubber dry rot caused by Ustulina deusta, rubber pink disease caused by pellicularia salmonicolour (B. & Br) Dastur, rubber powdery mildew caused by Odium heveae Steinm, rubber spearhead drying caused by Diplodia sp. are few diseases.

Rubber termite Ghyptotermis dilatatus Hail. Termite attacks the woody part of rubber.

Rubber tree Hevea brasiliensis. A tall tree of the family Euphorbiaceae from which latex is collected and coagulated to produce rubber.

Rubiaceae The single family of the plant order Rubiales.

Rubiales An order of dicotyledonous plants marked by their inferior ovary, regular or nearly regular corolla, and opposite leaves with interpetiolar stipules or whorled leaves without stipules.

Rubidium A rare element found in soils and plants in traces. Functions are similar to potassium.

Rules (Seed Testing) The ISTA rules for testing seeds.

Rumen First compartment of the stomach of a ruminant or cub-chewing animal.

Runner A horizontally growing, sympodial stem system; adventitious roots from near the apex, and a new runner emerges from the axil of a reduced leaf.

Runner plant Plants which can be propagated through runners.

Running gear Consists of front and rear axles, wheels, springs and frame.

Runoff (1) Portion of total precipitation from a given area that appears in natural or artificial surface streams. (2) Also the total quantity of runoff during a specified period. (3) The discharge of water in surface streams above a particular point.

Runoff coefficient (1) The ratio of the maximum rate of runoff to the uniform rate of rainfall with a duration equalling or exceeding the time of concentration which produced this rate of runoff. (2) Also used incorrectly for 'percentage runoff'.

Runoff Cycle Portion of the hydrologic cycle between incident precipitation over land areas and subsequent discharge through streams channels or direct return to the atmosphere through evapo- transpiration.

Runoff pit Catchment area to which spillage from classifiers, thickeners, and slurry pumps can gravitate if it becomes necessary to dump their contents. Also known as spill pit.

Runoff plots Area of land, usually small, arranged so the portion of rainfall or other precipitation flowing off and perhaps carrying soluble materials and soil may be measured. Usually the flow runoff plots includes only surface flow.

Runs In a series, a sequence of identical signs (or symbols) followed and preceded either by a different sign (or symbol) or no sign (or symbol) at all, is called a run. In a random pattern, few like signs (or symbols) will follow one another and we have 'short runs' while in periodic or cyclical oscillations, positive and negative signs are likely to occur in bunches or clusters and we have long runs of like signs.

Runs draws or sloughs Natural waterways in many flat prairie sections, through which surface waters find their way, in the spring and at times of very heavy rains, into drainage channels.

Run through A light watering run rapidly over farms in order to wet the surface soil either to act as a frost protection measure in the frost-affected areas or to prevent deterioration of certain crops.

Rural group Two or more people who have lived together with common goals and interests sufficiently long enough to become organised and to consider themselves and being considered as a unit.

Rural sociology It is the study of the sociology of life in the rural environment which systematically studies rural communities to discover their conditions and tendencies and to formulate principles of progress.

Russet Brownish roughened areas on skins of fruit, from abnormal production of cork caused by disease, insect, or spray injury.

Rust An infections parasite of the Uredinales which produces reddish or brownish spot on infected host tissue.

Rusty blotch A fungus disease of barley caused by Helminthosporium californicum and characterised by brown blotches on the foliage.

Rusty mottle A virus disease of cherry characterised by retarded development of blossoms and leaves in the spring followed by necrotic spotting and short-holding of the foliage with considerable defoliation.

Rutabaga Brassica napobrassica. A biennial crucifer of the order Capparales probably resulting from the natural crossing of cabbage and turnip and characterised by a large, edible, yellowish fleshy root.

Rye Secala cereale L. Family Gramineae. Cereal grain.

Rye ergot A fungal disease of rye caused by Claviceps purpurea.

Rye whiskey One of the potable alcoholic beverages, obtained by distilling the alcohol-containing liquid resulting from the fermentation of a rye mash by yeast, the distilled product is aged for number of years in new-charred oak barrels.

S

Saba A plant (Musa sapicntum var. compressa) common in the Philippines; the fruit is a cooking banana. A textile made from fibres of the saba plant.

Saccate Having a sac-like or pouch-like form.

Sacchariferous Sugar bearing.

Saccharify Convert into sugar.

Saccharimetre A device to measure amount of sugar in a solution; particularly some form of polarimetre.

Saccharometre A device like a hydrometre to measure sugar in solutions. It is based on the principals of specific gravity.

Saccharose Any of the groups of sugars distinguished from the glucose group. Sucrose.

Saddle grafting Method of grafting in which the top of the stock is cut first transversely, and then two upward cuts are given on either side to form a wedge, and a cleft is made in the scion so that it fits tightly over the stock, joint then sealed, used more in herbaceous than on woody plants.

Safety factor The allowance made for overloads and unforeseen circumstances in order to ensure ample strength in a machine, material or type of construction under all conditions.

Safety release hitch An attachment to release a pulled implement when the implement strips an obstruction.

Safety valve A safety device for allowing steam or water to escape from a boiler when the pressure may become dangerous. Usually adjusted to permit not more than 5 pounds of pressure above maximum allowable working pressure of the boiler.

Safflower Carthamus tinctorius. An annual thistle like herb belonging to the composite family (Compositae); the leaves are edible, flowers yield dye, and seeds yield a cooking oil.

Safflower caterpillar Perigoea capensis G. Green caterpillar feeds on leaves.

Safflower fly Acanthiophilus helianthi Rossi. Light green fly who maggots damage buds and flowers.

Safflower leaf spot Disease caused by cercospora carthami; appearance of small irregular dark brown spots which ultimately turn into holes.

Safflower oil Non yellowing oil derived from safflower seed and similar to linseed oil, contains high proportion of linoleic acid; used as a drying oil and in food and medicine, cake used for a cattle feed.

Sufflower rust Disease caused by Puccinia carthami. Appearance of scattered, small, round, rust pustules on leaves.

Safflower yellow One of the two main colouring matters found in safflower florets.

Saffron Crocus Crocus sativus. A crocus of the iris family (Tridaceae); the source of a yellow dye used for colouring food and medicine.

Sage Salvia officinalis. A herb, shrub of mint family (Labiatae); the leaves are used as spices.

Sage brush Any of various hoary undershrubs of the genus Artemisia found on the alkaline plains of the Western United States.

Sage bush A bush growing in arid regions, olive green in colour; growing generally in strong rich soil, it does not thrive on poor, or strongly alkaline soil, nor in extremely dry desert regions.

Sagittate Shaped like an arrow.

Sagwan See teak

Sailab irrigation Irrigation through flood waters let out into the areas adjoining the rivers. The water flows over the land without any control and any surplus goes to waterways. Crops are then raised on the moisture of the soil.

Sal Shorea robusta Gaertn. Family Dipterocarpaceae An important tree grown in forests of India for its timber.

Salad oil Superior quality of olive oil.

Salad plant Any food plant grown primarily for its edible leaves, stalks or other vegetative parts. Lettuce, parsley, endive, cress, celery, chicory are some such plants.

Sales tax A tax imposed at a uniform rate on all goods at the point of retail sale, other than exempt goods.

Saline-alkali soil A soil containing enough exchangeable sodium to interfere with the growth of most crop plants, and containing appreciable quantities of soluble salts. The exchangeable sodium percentage is greater than 15, the conductivity of the saturation extract is greater than 4 dS/m at 25°C and the pH is usually 8.5 or more in the saturated soil. A saline-alkali soil has a combination of harmful quantities of salts and either a high alkalinity or high content of exchangeable sodium, or both, so distributed in the profile that the growth of most crop plants is reduced. It is also called saline-sodic soil.

Saline soil A non-alkali soil that contains enough soluble salts to interfere with the growth of most crop plants. The conductivity of the saturation extract is greater than 4 dS/m, the exchangeable sodium percentage is less than 15, and the pH is usually less than 8.5. Presently all soils whose electrical conductivity exceeds 2 dS/m are designated as saline soils.

Salinity, Soil The amount of soluble salts in a soil, expressed in terms of percentage, parts per million, or other convenient units. Presently most commonly used unit is dS/m which is equivalent to formerly used unit mmhos/cm.

Salinisation The process of accumulation of salts in soil. It can occur in two ways namely, (i) by capillary rise of underground water ultimately evaporating at the soil surface leaving the salts behind and (ii) through irrigation with saline water.

Salinometre Salometre.

Salmon gum Eucalyptus salmonopholia, Family Myrtaceae. Eucalyptus species from Australia.

Salometre A device used to measure the strength of salinity in a solution. Measures in degrees or percentages based on hydrometre principles. Also called salimetre, salinometre.

Salt (1) common salt known as sodium chloride, (NaCl). (2) Compounds of basic and acid radicals, acid with whole or part of its hydrogen replaced by a metal.

Salt-affected soil Soil that has been adversely modified for the growth of most crop plants by the presence of certain types of exchangeable ions or of soluble salts. It includes soils having an excess of salts, or an excess of exchangeable sodium, or both. See also saline-alkali soils, saline soil, and sodic soil.

Saltation Process by which a particle is picked up by turbulence in a stream and carried forward by a series of leaps and bounds.

Salt balance The salt balance of an irrigated area is commonly defined as the difference between the total dissolved solids brought to the land annually by the irrigation water and the total solids carried away annually by the drainage water.

Salt index A formula for ascertaining whether water is suitable for irrigation.

Salt sage A dwarf variety of sage bush.

Salt tree Athel tamarix.

Salt-water wedge A wedge-shaped intrusion of salty ocean water into fresh-water due to its greater density.

Salvadora persica See mustard tree.

Salvia officinalis See sage.

Salviniales A small order of heterosporous, leptosporangiate ferns (division Polypondiophyta) which float on the surface of the water.

Samanea saman See Enterolobium saman.

Sambucus nigra See European elder.

Sample A small separated part of soil or something else illustrating the quality of the mass from which it is picked up.

Sample size The number of sampling units which are to be included in the sample.

Sampling plan The procedure for acceptance on the basis of sampling information, e.g. single-sampling plan, double sampling plan, multiple-sampling plan and sequential-sampling plan.

Sampling variance The variance of a sampling distribution.

Sand (1) A soil particle between 0.05 and 2.0 millimetres in diametre. (2) Any one of five soil separates; very coarse sand, coarse sand, medium sand, fine sand, and very fine sand. (3) A soil textural class that contains 85 per cent or more sand; the percentage of silt plus 1.5 times the percentage of clay does not exceed 15.

Sandalwood Santalum album L. Any species of the genus Santalum of the sandalwood family (Santalaceae) characterised by a fragrant wood.

Sandalwood oil Pale-yellow essential oil with harsh taste and faint aromatic scent; soluble is mixed in oil; unsoluble used in glycerin; derived from the sandalwood, Santalum album.

Sand bar A ridge or narrow shoal of sand formed in a river or along a shore by the action of currents or tides.

Sand bearing method Method of testing the crushing strength of drain tile in which the tile is bedded in sand according to ASTM specifications.

Sand binder Any plant which withhelds the sand from drifting by the grip action of its roots.

Sand-carrying capacity (Lime mortar) The property expressed by the number of parts of sand which may be added to the lime without rendering the mortar too stiff to work well under the trowel.

Sand core Sand-filled depth in a bank to prevent the burrowing animals making holes.

Sand drain A vertical boring through a clay or silty soil filled with sand or gravel to facilitate drainage.

Sand drift Movement of wind blown sand along the surface of a desert or shore.

Sand dune A mound of loose wind-blown sand commonly found along low-lying seashores above high-tide level.

Sandoricum indicum Cav. Sandoricum koetjape (Burm. f.) Merr. A tree of family Meliaceae grown for its edible fruits.

Sand stone A sedimentary rock composed of sand particles, mainly quartz bound together by cementing materials such as silica and calcium carbonate, mostly sand or sandy loam soils are formed.

Sand storm A strong wind carrying sand through the air whose particles are 0.08-1 mm in diametre; in contrast to dust storms, the particles are usually confined to about 3 m above the ground.

Sand trap A device in a channel for arresting the sand, silt etc., carried by the water, and generally including means of ejecting them from the system. When the sediment load includes gravel, it is called 'sand and gravel trap'.

Sandy Containing a large amount of sand. It may be applied to any one of the soil classes that contains a large percentage of sand. See also class, soil and texture, soil.

Sandy clay A soil texture class; soil material that contains 35 per cent or more clay and 45 per cent or more sand. See also texture, soil.

Sandy clay loam A soil textural class soil material that contains 20-35 per cent clay, less than 28 per cent silt and 45 per cent or more sand.

Sandy loam A soil texture class, soil material that contains either 20 per cent or less clay, with a percentage of silt plus twice the percentage of clay that exceeds 30 per cent and 52 per cent or more sand; or less than 7 per cent clay, less than 50 per cent silt, and between 43 per cent and 52 per cent sand. See also class, soil and texture soil.

Sanguification Food conversion into blood or blood formation.

Sanguin blood red.

Sanguivorous Subsisting on blood.

Sanicle An umbeliferous plant.

Sanitary landfill A disposal area wherein solid wastes are spread in layers, compacted, and covered with soil each working day.

Sanitation The act or process of making healthy environmental conditions.

Sanhemp Crotalaria juncea L. Family Papilionaceae. Important crop grown for its fibre. Also grown for green manuring in kharif season and crop is ready in 2 to 2-1/2 months for green manuring.

Sanhemp fibre extraction It is the process by which sanhemp is extracted from the plants. The plants are kept for 2-3 days on ground after which plants are tied up and put into standing or flowing water for about a week. After taking them out, fibres are peeled off from which adhering tissues are removed. After drying for about 2 to 3 days the fibre is ready.

Sanhemp hairy caterpillar Utethesia pulchella L. Caterpillar feeds on leaves of sanhemp.

Sanhemp mosaic Disease of sanhemp; leaves of affected plants shows mosaic mottling.

Sanhemp powdery mildew Disease caused by Oidium erysiphoides and Leveillula taurica. White shiny powdery patches develop on leaf undersurface. Sanhemp rust. Disease caused by uromyces decoratus. Rust postules appear on aerial parts of plants.

Santaline dye Vegetable dye prepared from the wood of Pterocarpus santalinus L.

Sap (1) Juice circulating in plants. (2) Moisture and its soluble constituents in unseasoned wood.

Sapindus detergens See Chinese soap berry.

— **emarginatus** See soapnut.

— **mukorossi** See Chinese soap berry.

— **trifoliatus** See soapnut.

Sapling A young tree with a trunk less than 4 inches (10 centimetres) in diametre at a point approximately 4 feet (1.2 metre) (breast height) above the ground. The minimum size of sapling is usually, though not variably placed at 2 inch (5 cm) dia at breast height.

Sapotaceae A family of dicotyledonous plants in the order Ebenales characterized by a well-developed latex system.

Saprogenesis Period in the life cycle of a pathogenic organism in which it is not directly associated with a living host; either in dormant or lives as a saprophyte.

Saprobic Living on decaying organic matter; applied to plants and micro organisms.

Saprophyte A plant that gets its food from plants or animals that have died; Also an organism which feeds on lifeless organic matter.

Saproplankton A mass of decaying aquatic plants and animal material which is floating on water.

Sapwood The younger, softer, more porous outer layers of a woody stem, between the cambium and heartwood. Also known as alburnum. The sapwood is composed of living cells and serves to conduct water and minerals to the crown.

Sarcochare A plant dispersing minute, light disseminules.

Sargassum A genus of brown algae characterised by branching thalls with lateral outgrowths; they develop along tropical shores; than break away to drift in the ocean.

Sarmentose Producing slender, stems or runners.

Sarraceniaceae A small family of dicotyledonous plants in the order Sarraceniales in which leaves are modified to form pitchers, placentation is axile, and flowers are perfect with distinct filaments.

Sarraceniales An order of dicotyledonous herbs or shrubs in the subclass Dilleniidae; plants characteristically have alternate, simple leaves that are modified for catching insects, and grow in waterlogged soils.

Sarsaparilla Any of various tropical American vines of the genus Smilax (family Liliaceae) found in dense, moist jungles; a flavoring material used in medicine and soft drinks is obtained from the dried roots of atleast four species.

Sassafras Sassafras albidum. A medium sized tree of the order Magnoliales; recognised by the bright-green colour and aromatic odor of the leaves and twigs.

Saturate (1) To fill all the voids between soil particles with a liquid. (2) to form the most concentrated solution possible under a given set of physical conditions in the presence of an excess of the solute. (3) To fill to capacity; as the adsorption complex with a cation species; for example, H-saturated. The condition of saturation is termed saturated.

Saturated air Air saturated with water vapours.

Saturation deficit The amount of water vapour required to bring non-saturated air at a given temperature and pressure to the point of saturation.

Saturation extract The extract from a soil sample, that has been saturated with water.

Saturation light intensity The light intensity beyond which additional light will not produce an increase in photosynthesis.

Saturation point (1) That point at which a soil or aquifer will no longer absorb any amount of water without losing an equal amount. (2) (Wildlife) The maximum density under which a species will normally live.

Saturation vapour pressure The vapour pressure of a parcel of saturated air at a given temperature.

Sauropus androgynous Merr. Family Euphorbiaceae. Small perrenial shrub; comes up well in the tropics. Leaves eaten raw or steamed; high nutritive value. Propagated by seed.

Sausuraceae A family of dicotyledonous plants in the order Piperales distinguished by mostly alternate leaves, two to ten ovules per carpel, and carples distinct or united into a compound ovary.

Sausage dam Dam of loose rock wrapped with wire where the resulting cylindrical bundles are either laid in a horizontal or vertical position.

Savanna Any of variety of physiognomically or environmentally similar vegetation types in tropical and extratropical regions; all contain grasses and one or more species of trees of the families Leguminosae, Bombacaceae, Bignoniaceae, or Dilleniaceae. Often a transitional type between true grassland and forest.

Saving The withholding of money from expenditure on goods and services.

Sawfly An insect of Tenthredinidae, Pamphiliidae, Cimbicidae, Diprionidae or similar families whose females puncture foliage and other parts of plants to lay eggs.

Saw timber Timber stands in which trees of sawlog size make up most of the volume.

Saxifragaceae A family of dicotyledonous plants in the order Rosales which are scarcely or not at all succulent and have two to five carpels usually more or less united, and leaves not modified into pitchers.

Scab Crustlike disease lesion, or a disease in which scabs are prominent symptoms.

Scalariform Ladderlike; having bars or markings like the rungs of a ladder.

Scalding A heat treatment to fruits used to help remove peels from them.

Scaler principle or process The hierarchical principle is called the scalar principle or process. The work 'scalar' is derived from 'scale' which means ladder with several steps.

Scales The structures arising, alongwith rhizoids, from the ventral surface of certain members of the Bryophyta, usually multicellular.

Scale scar A mark left on a stem after bud scales have fallen off.

Scalper A seed processing machine designed to remove the bulk of trash from seed so as to facilitate elevating and processing operations. It is basically a vibrating or rotating screen or sieve.

Scarified seed A seed whose hard and impervious seed coat is made pervious by abrasion or chemical action to aid in germination.

Scandent Climbing by stem-roots or tendrils.

Scarcity A relative term, indicating shortage in relation to demand. It is only because of scarcity that price exists.

Scarification The process of mechanically scarring or roughing the hard seed coat to make it more permeable to water.

Scarifier huller The machine used for scarification of hard seeds.

Scarify To tear up earth by disking or dragging to prepare for seeding.

Scarious Having a thin, membranous texture.

Schistostegiales A monospecific order of mosses; the small, slender, glaucous plants are distinguished by the luminous protonema.

Schizocarp A dry fruit derived from a compound pistil in which the one seeded carples separate from one another at maturity as in umbelliferae.

Schizogoniales A small order of the chlorophyta containing algae that are either sub-microscopic filaments or macroscopic ribbons or sheets a few centimetres wide and attached by rhizoids to rocks.

Schizomycete A class of minute, often single celled organisms between algae and fungii, including bacilli, bacteria, etc.

Schizomycophyta The designation for bacteria in those taxonomic systems that consider bacteria as plants.

Schizophyta The prokaryotes, a division of the plant subkingdom Thallobionta; includes the bacteria and bluegreen algae.

Schoolmarm A term used by loggers to describe a tree forked above the first log.

Scion A section of a plant, usually a stem or bud, which is attached to the stock in grafting.

Sciophyte A plant that thrives at lowered light intensity.

Scirpophaga nivella See Sugarcane insect pests.

Scirtothrips dorsalis See Chilli thrip.

Sclereid A thick-walled, lignified plant cell typically found in sclerenchyma.

Sclerenchyma A strengthening tissue composed of thick walled, elongated cells (fibres) or shorter cells (stone cells).

Scleriasis Hardening by increase of connective tissue or of lignin.

Sclerocarp The hard seed coat or stone; generally endocarp of fruits.

Sclerocaulous Having a hard, dry stem because of exceptional development of sclerenchyma.

Sclerochore A plant that disperses disseminules without apparent morphological adaptations.

Sclerogen Hard matter deposited on the inner-surface of plant-cells such as lining walnut shell.

Sclerophyllous Characterised by thick, hard foliage due to well-developed sclerenchymatous tissue.

Sclerospora graminicola See Common millet downy mildew, See Italian millet downy mildew.

— **philippinesis** See Maize downy mildew.

— **sorghi** See Sorghum diseases.

Sclerotinia fructigena See Apple brown rot.

— **sclerotiorum** See Gram stem rot.

Sclerotium Resting mass of fungus tissue, consisting of a mass of the mycellium or plasmodium often more of less spherical, normally having no spores. Used as an asexual method of reproduction.

Sclerotium rolfsii See Brinjal blight; tomato foot rot and wilt; pepper diseases and cucurbit stem rot.

Science Soil The science dealing with soil as a natural resource. It includes; soil formation, classification, and mapping; the physical, chemical and biological properties of soils; and the management of soils for various purposes such as the production of agricultural and forest crops, the construction of roads and others.

Scorch Burning of tissue, from infection or weather conditions.

Scorpioid cyme A cyme with a curved axis and flowers arising two-ranked on alternate sides of the axis.

Scotch pine Pinus sylvestris. A hard pine of North America having two short, bluish needles in a cluster.

Scour The process in which channel bed is scoured because of excessive velocity.

Scour and fill The process of cutting and filling of channels with variations in velocity of flow.

Scraper (1) A machine which scrapes. (2) A digging, hauling and grading machine having a cutting edge, a carrying bowl, a movable front wall (apron) and a dumping or ejecting mechanism.

Screen dams Strips of wood, etc., fastened across the screen to retard flow of seed down the screen.

Screened pan An evaporation pan, the top of which is covered by wire-mesh screening (1/4 inch or six millimetre mesh); the screening reduces air circulation and insulation, and results in a pan coefficient nearer to unity than that for unscreened pan.

Screw A screw may be considered as a helix wound around a cylinder; the helix may be single as in a common screw, or double, or multiple.

Scrophulariaceae A large family of dicotyledonous plants in the order

Scrophulariales, characterised by a usually herbaceous habit, irregular flowers, axile placentation, and dry, dehiscent fruit.

Scrophulariales An order of flowering plants in the subclass Astoriidae distinguished by a usually superior ovary and, generally, either by an irregular corolla or by fewer stamens than corolla lobes, or commonly both.

Scrub, Temperature Evergreen or deciduous plants requiring considerable moisture as snow melts, mists, or seasonal downpours.

Scutellum The cotyledon of an embryo of a grass, a food absorbing structure, plate or shieldlike cover as in Microthyriales.

Seabrous A rough surface.

Sea breeze A coastal, local wind that blows from sea to land, caused by the temperature difference when the sea surface is colder than the adjacent land, it usually blows on relatively calm, sunny summer days, and alternates with the oppositely directed usually weaker, night-time land breeze.

Sea gate A gate which serves to protect a harbour or tidal basin from the sea, such as one of a pair of supplementary gates at the entrance to a tidal basin exposed to the sea.

Sealing Production of seal or impervious crust by high intensity rainfall in which spaces between larger particles is filled with colloids.

Seamless tubing Drawn tubing without joints, as distinguished from pipe or tubing with welded longitudinal joints.

Season A division of the year according to some regularly recurrent phenomena, usually astronomical or climatic.

Seasonal consumptive use Depths of water consumed by evapo-transpiration during crop growth, including water used by accompanying weed growths.

Seasonal stream A stream whose flow is not constant because it has water in its course only during certain seasons.

Seasoned timber Timber moisture content of which has been reduced to the specified minimum, under more or less controlled processes of drying.

Season, Fire The period or periods of the year during which fires are likely to occur, spread, and to do sufficient damage or otherwise warrant organised fire control.

Seasoning A process involving the reduction of moisture content in timber under more or less controlled conditions towards or to an amount suitable for the purpose for which it is to be used.

Sea weed A marine plant, especially algae.

Secernent A organism which can secrete.

Second (1) A fundamental unit of time and is defined as 1/31556925.975 of the length of the tropical year for 1900.0, the year commencing at 12.00 hours universal time on first day of January 1900.0 (2) It is the smallest unit of the plane angle and 60 seconds is equal to Π/10800 radian or one minute.

Secondary fermentation A fermentation process produced by adding sugar to wine and used to create natural carbonation for example, in champagne.

Secondary flow irrigation A system used with furrow or border strip layout. When the water front reaches the end of the run, the flow is set at a reduced rate to avoid runoff, yet maintain storage over the length of the run.

Secondary mineral A mineral resulting from the decomposition of a primary mineral or from the precipitation of the products of decompostion of a primary mineral. See also primary mineral.

Secondary periderm Any layer of the periderm except the first and outermost layer.

Secondary phloem Phloem produced by the cambium, consisting of two interpenetrating system, the vertical or axial and the horizontal or ray.

Secondary root A branch from a main or taproot.

Secondary tillage implements These are the implements used for secondary tillage operations.

Secondary wall The portion of a plant cell wall produced internal to and following deposition of the primary wall, usually consists of several anisotropic layers, and often has prominent internal rings, spirals, bars, or reticulations.

Secondary xylem Xylem produced by cambium, composed of two interpenetrating systems, the horizontal (ray) and vertical (axial).

Second bottom The first terrace above the normal floodplain of a stream.

Second crop A crop succeeding one already harvested during a growing season; either a regrowth of the harvested crop, or a newly planted crop.

Second growth Timber growth which comes up after removal of the old stand by cutting, fire, or other causes. Typical second growth conditions may come about in a forest that is untouched so far as lumbering is concerned.

Secretory structure Plant cells or organisations of plant cells which produce a variety of secretions.

Section A unit of land measurement, 640 acres or 6,400 square chains. One mile or 80 chains square; 1/36 of a township.

Secund Having lateral members arranged on the side only.

Sedentary soil Soil formed immediately over the underlying parent rock.

Sedge Grass-like plant of family Cyperaceae, growing in marshes.

Sedimentary rock A rock formed form materials deposited from suspension or precipitated from solution and usually more or less consolidated. The principal sedimentary rocks are sandstones, shales, limestones, and conglomerates.

Sedimentation Deposition of sediments by natural or mechanical means.

Sediment yield The quantity of sediments carried out of a watershed by streams.

Seed A fertilised ovule containing an embryo which forms a new plant upon germination. Also used for the seed-like fruits of cereals.

Seed analyst Appointed under section 13 of Seeds Act, 1966; the person supervising or carrying out the seed analysis work.

Seed blower This equipment is used to separate light weight material such as chaff, and empty florets in grasses, from the heavier seeds.

Seed borne Transmitted by seeds, usually diseases.

Seed coat The envelope which encloses the seed except for a tiny pore, the micropyle.

Seed disinfection Ridding the seed surface of organisms which are potentially disease producing.

Seed disinfestation Dipping the seeds in chemical soaps, and fungicide to kill pathogens.

Seed down To sow seeds for grass or forage legumes.

Seed-drill An implement for sowing seeds; seeds are sown at uniform depth and spacing; if fertiliser is also drilled than it is called seed-cum-fertiliser drill.

Seed harrow A harrow employed to refine the soil for drilling.

Seed health Health of seed refers primarily to the presence or absence of disease causing organisms such as fungi, bacteria and viruses; and animal pests, such as worms and insects; physiological conditions such as trace element deficiency may also be involved.

Seed Inspector Appointed under section 13, of the Seeds Act. Their main function is to take samples from sale points where seeds of notified varieties are being sold, got them analysed, if necessary, and initiate prosecution against offenders of the seeds Act.

Seeding (1) A plant grown from seed. (2) A tree younger and smaller than a sapling. (3) A tree grown from a seed.

Seed lot A uniformly blended quantity of seed designated by a proper number or mark.

Seed lot certificate (Orange or Green certificate) The form of international seed analysis; certificate used when the sample is drawn officially from the lot, under the authority of a member station, and the procedure followed identifies the certificate with the seed lot.

Seed parent The strain from which seed is harvested in the hybrid seed field. Also, commonly used to designate the female parent in any cross fertilisation.

Seed piece This term is applied to pieces cut from stem tissue for purpose of vegetative multiplication.

Seed protection Treating the seed with chemicals which by their presence prevent attack by seed-borne or soil-borne organisms on seed after planting.

Seed purity The percentage of well developed and whole seeds in a given seed sample.

Seeds Act As used in this book, Seeds Act refers to the Indian Seeds Act, 1966. An Act to regulate the quality of seed offered for sale.

Seed sample certificate (Blue certificate) The form of International seed analysis; certificate used when sampling from the lot is not under the responsibility of a member station (Seed Laboratory).

Seed stalk The erect stalk on a plant that produces flowers and seeds. Applied particularly to root crops and leafy vegetable crops that produce seed after the desired product has fully developed.

Seed technology The science which deals with production, harvesting, processing, testing, packaging, storing and marketing of seeds.

Seed tree A tree that produces seed; usually trees reserved in a cutting operation to supply seed.

Seepage (1) The escape of water downward through the soil. (2) The emergence of water from the soil along an extensive line of surface in contrast to a spring where the water emerges from a local spot.

Seepage head Difference between the water level in a canal or any other source of water supply and the basic subsoil pressure level or water table.

Seepage intensity coefficient A coefficient representing seepage losses from a unit area of wetted perimetre per unit of seepage head.

Seepage metre Device for measuring rate of seepage.

Seep drains A system of shallow ditches in the form of surface cuts to remove surplus water from lands composed of impervious materials on high lying sites.

Segregate polyploidy Inflorescence in which each floret within common calyx has its own perianth also.

Segregation The separation of the paternal from maternal chromosomes at meiosis and the consequent separation of differences as observed genetically in the offspring.

Seiche Oscillation of the water surface of inland bodies of water due to unequal atmospheric pressure or to wind forces on the water surface. The amplitude may amount to a foot or more and the period may reach several hours.

Sejugous With six pairs of leaflets.

Selection K. Alternate expressions of selection on traits that determine fecundity and survivorship; K-selection favouring a more efficient utilization of resources.

Selection, recurrent Breeding system involving repeated cycles of selection and recombination with the objective of increasing the frequency of favourable genes for yield or other characteristics.

Selective logging or cutting The removal of selected mature, large, or diseased trees as single, scattered trees or in small groups of trees. Young trees start in the openings thus made; the result of this type of cutting is an uneven-aged forest.

Selectivity Selectivity of data may be of two kinds: (i) If the data are not from experiment with a probability base, but from uncontrolled experience; and (ii) if in the process of obtaining sample observations from the population, observations of the one kind may be more difficult than those of another kind, and the degree of differential selectivity may be unknown.

Select seed Select seed is unique to the Canadian certification system. It is the approved progeny of breeders or select seed, produced in a manner to ensure its specific genetic identity and purity by those growers authorised by the certifying agency for the production of this class. Select seed is not a seed of commerce.

Selenium A non-metallic element present in soil in traces; related to sulphur; when absorbed by plants, toxic to animals.

Selenothrips rubrocinctus See Cashew nut leaf thrip.

Selfed Said of a pistil that is fertilised with pollen from the same plant that bears the pistil; also applied to seed resulting from such fertilization.

Self fertilization The process by which the female egg cells of a plant are fertilised by the male sperm cells of the same plant.

Self incompatible Inability of a plant to set seed as a result of fertilization by its own pollen; also referred to as self sterility. Self compatible are those which can self fertilise.

Self mulching, soil A soil in which the surface layer becomes so well aggregated that it does not crust and seal under the impact of rain, but instead serves as a surface mulch when it dries.

Self-pollination Transfer of pollen from the anther to the stigma of the same flower or of anther flower on the same plant.

Self sterile Self incompatibility. Lack of ability in a plant to set seed when fertilised by its own pollen.

Selling agency, common An agency serving a group of otherwise independent firms, the purpose of which is to avoid undercutting of prices.

Selling costs Costs incurred in creating, retaining or expanding a market for a product.

Semi-arid A term applied to regions or climates where moisture is normally greater than under arid conditions but still definitely limits the growth of most crops. Dry land farming methods or irrigation are generally required for crop production. The upper limit of average annual precipitation in the cool semi-arid regions is as low as 15 inches (37.5 cm), whereas in tropical regions it is as high as 45 or 50 inches (112.5-125 cm).

Semi-collective villages In such villages, the land belongs to the collective body of the villagers. People work on their farm as a collective body, i.e. all production sources are owned and worked collectively, annual quota fixed according to the income of the whole village and farmer purchases his supplies quota as he likes.

Semi-diurnal tides Tides requiring approximately one-half lunar day to complete their cycle, that is, two high waters and two low waters in each lunar day; predominating type of the tide throughout the world is semi-diurnal.

Semi-dormancy Decrease in plant growth rate; may be seasonal or associated with unfavourable environmental conditions.

Semi-double Pertaining to a flower that has more than the usual number of petals or disk florets while it retains same pollen bearing stamens or some perfect disk florets.

Semi-hypogeal germination In which the hypocotyl elongates during germination, e.g. durian.

Semilate Pertaining to a plant whose growing season is intermediate between midseason forms and late forms.

Semi-mounted type implements These are the implements which are attached to the tractor along a hinge axis and are controlled directly by the tractor steering unit. Weight is partly supported by the tractor.

Seminal Pertaining to seed or germ. Seminal organs are those already developed in the embryo within the seed.

Seminal root A root arising from the base of the hypocotyl.

Semiology Branch of pathology concerned with symptoms.

Semi-permanent villages The inhabitants stay at a place for a few years and afterwards, when they feel that the place has exhausted its capacity to provide food supplies and the soil has been also depleted to produce better crops due to their wrong way of cultivation, they move to another place, and settle in new villages of the same type. They keep animals like goats and cows, graze them and eat them and their products.

Semi-range A statistical term, equal to one half of the range.

Senesce To age.

Sensible temperature The temperature at which air with some standard humidity, motion, and radiation would provide the same sensation of human comfort as existing atmospheric conditions.

Sensitive Term concerns with tolerance of crops and varieties to salt, waterlogging etc. Those which can not tolerate are termed as sensitive crops.

Sensitive market A market in which market prices fluctuate widely in response to good and bad news.

Sensitivity The effect of remoulding on the shear strength of an undrained cohesive soil.

Sensor The component of an instrument that converts and input signal into a quantity that is measured by another part of the instrument.

Sentiments These are complex, emotional states directed toward some specific object of person. Love and hate are two important sentiments.

Sepal One of the division of a calyx; one of the outermost floral organs.

Separates, soil Mineral particles, less than 2.0 mm in equivalent diametre, ranging between specified size limits. The names and size limits of separates recognised by pedologists in Canada and United States are very coarse sand, 2.0 to 1.0 mm; coarse sand, 1.0 to 0.5 mm; medium sand 0.5 to 0.25 mm; fine sand 0.25 to 0.10 mm; very fine sand, 0.10 to 0.05 mm; silt, 0.05 to 0.002 mm; and clay, less than 0.002 mm. The separates recognised by the International Society of Soil Sciences are (i) coarse sand, 2.0 to 0.2 mm (ii) fine sand, 0.2 to 0.02 mm (iii) silt, 0.02 to 0.002 mm and (iv) clay, less than 0.002 mm.

Separation disk A layer of gelatinous material between two adjacent negative cells in some blue-green algae; associated with hormogonium formation.

Septate Having cross-walls, septa.

Septoria nodorum See wheat glume blotch.

— **tritici** See wheat leaf blotch.

Sequential branching See branching, sequential.

Sequential cropping A form of multiple cropping in which crops are grown in sequence on the same field, with the succeeding crop planted after the preceding crop is harvested.

Sequential experimentation When the experiment is done in a number of stages, and at each stage the data so far obtained are used to determine whether sufficient precision is attained, at the required point the experiment ceases.

Sequoia A genus of conifers having overlapping, scale-like evergreen leaves and vertical groves in the trunk; the giant sequoia (Sequoia gigantea) is the largest and oldest of all living things.

Sequum Sequence of an eluvial horizon and its related illuvial horizon in a soil.

Serial dilution Successive dilution of a specimen, for example 1:10 dilution equals 1 ml of a specimen plus 9 ml of diluent (e.g. sterile water); 1:100 dilution equals 1 ml of a 1:10 dilution plus 9 ml of diluent.

Serial correlation The correlation between members of a time series (or space series) and those members lagging behind, or leading, by a fixed distance in time (or space). The deviations of time series observations from their population means are dependent upon one another. Serial correlation refers to correlation among values of dependent variates other than the correlation accounted for by the independent variable, namely time in a time series.

Sericeous Of, pertaining to, or consisting of silk.

Series (1) Data arranged according to the values of a variable character, e.g. 'time series' arranged according to time values. (2) A major subdivision of a system.

Series dynamo A direct current dynamo in which the armature and field are connected in series internally.

Series motor A direct current motor with the armature and field connected in series. Used on elevators and other machines where variable loads occur. The speed varies with the load.

Series parallel circuit A circuit made up of two or more simple parallel circuits all joined in series.

Series soil A category in the Canadian system of soil classification. This is the basic unit of soil classification, and consists of soils that are essentially alike in all major profile characteristics except the texture of the surface.

Serotinous Plants which flower or develop late in a season.

Serrate Saw-toothed.

Service sample A sample submitted to the Central Seed Laboratory, or a State Seed Laboratory for testing, the result of which to be used as information for seeding, selling or labelling purposes.

Sesame Sesamum indicum L. (Sesamum orientale L.) Family Pedaliaceae. Important oil seed crop whose oil is edible called *Til* in Hindi. Also called sesamum.

Sesame oil A non-volatile oil extracted from sesame seeds.

Sesamum-diseases Sesamum blight caused by Cercospora sesami Zimm.; Sesamum dry root rot caused by Macrophomina phaseoli; Sesamum rosette a virus disease also known as, sesamum phyllody; sesamum wilt caused by Fusarium vasiinfectum are few diseases.

Sesamum-insect pests Sesame leaf and pod caterpillar (Antigastra catalaunalis Diys.) whose caterpillar feeds on tender foliage and bores into shoots and pods; sesamum gall fly (Aspondylia sesami Felt.) whose larva forms galls on flow bud are the major insect pests.

Sesbania Sesbania bispinosa (Jacq.) W.F. Wight. (Sesbania aculeata Pers.). Family Papilionaceae. A green manuring crop, wood used as fuel. Also called dhaincha which differs from dhaincha.

Sesbania-diseases Sesbania bacterial disease caused by Xanthomonas sesbaniae; sesbania leaf spot caused by Cercospora sesbaniae and sesbania seeding blight caused by Colletotrichum capsici are the few diseases.

Sessile Lacking of a stalk or petiole.

Seston The entire mass of suspended matter in a volume of free water; may be divided into the biosestone (the living portion) and tripton (the non-living portion).

Set (or ensemble) Group of trees within a forest with equivalent developmental potential (e.g. set of future, present and past).

Seta The stalk that supports the capsule in bryophytes; a slender bristle like structure, as the hair-like tip of a moss leaf, generally dark-coloured.

Setting time (cement) Time required by a type of cement in setting up.

Sewage farm A farm in which sewage is used for irrigation and as fertiliser.

Sewage sickness Condition developed in soils with continuous use of sewage water. Crop yield starts declining. Remedy lies in scrapping of sewage matter and proper drainage.

Sex ratio The relative proportion of males and female in a population.

Sexuality In higher plants the transposition of the individual or an axis to the flowering condition.

Shade demander Tree species requiring some degree of shade for its normal development.

Shade density The degree of shade in forests.

Shade tolerant A plant that grows well in sunshine yet shows satisfactory growth even in moderate shade.

Shadscale Atriplex confertifolia. A small shiny shrub found in the Great Basin Desert.

Shaft An axle or bar, usually cylinderical, used to support rotating pieces, or to transmit power or motion by rotation.

Shaftal See White clover and Persian clover.

Shake A lengthwise separation of wood which usually occurs between and parallel to the growth layers.

Shale (1) A finely grained sedimentary rock produced from clay and characterised by a fine stratification that easily splits into layers. (2) Silver-like lining of a carpel which before the cotton ball opens serves to separate the lock.

Shallot Allium ascalonicum. A bulbous onion like herb. Also known as scallion.

Shallow rooted crops The crops whose all or nearly all roots are near the surface of soil.

Shallow soil A soil not very deep. 15-30 cm over parent material or bed rock.

Shallow water Water of such a depth that bottom topography affects surface waves.

Shaly (1) Containing a large amount of

shale fragments. (2) A soil phase; for example, shaly phase (3) Kind of fragment; See coarse fragments.

Shama millet Echinochloa colonum (L.) Link. (Panicum colonum L.). A fodder grass of family Gramineae. Grains could also be eaten.

Share (1) It is the part of the plough bottom that penetrates into the soil and cuts the soil in the horizontal direction below the soil surface. (2) The individual portion of a company's capital owned by a share-holder.

Share point The forward end of the cutting edge which actually penetrates into the soil is known as the share point.

Shear failure, or failure by rupture Failure in which movement caused by shearing stresses in a soil mass is of sufficient magnitude to destroy or seriously endanger a structure.

Shears A type of cutting instruments.

Shear stress (1) The force per unit area acting tangentially to a given plane within a soil mass. (2) The force of retardation per unit horizontal area in the direction of the wind called shearing stress.

Sheath The basal portion of grass leaf that wholly encircles the stump. Sometimes called the "boot".

Sheave wheel A grooved wheel in a block over which a rope or chain runs.

Shedding tassel Refers to the tassels in female parent rows shedding pollen or that shed pollen in hybrid maize plots. During field inspection, a tassel whose main spike of any side branch, or both, have shed pollen, or are shedding pollen, in more than 5 cm of branch length is counted as a shedding tassel.

Sheet composting Addition of large amounts of organic residue to a soil, extra nitrogen is usually added for faster decomposition.

Sheet erosion Removal of layers of soil by wind or water action. See Erosion.

Sheetflood A broad expanse of moving storm-borne water that spreads as a thin, continuous, relatively uniform film over a large area for a short distance and duration. Also known as sheet; sheet wash.

Sheet piling A diaphragm made up of meshing or interlocking members of wood, steel, concrete, etc. driven individually to form an obstruction to percolation, to prevent movement of material, for coffer dams, stabilisation of foundations, etc.

Shelter belt A wind barrier of living trees and shrubs maintained for the purpose of protecting farm fields. As applied to individual farmsteads, termed windbreak, also called belt.

Shield budding A method of budding commonly used in citrus, jujube and apple. Also known as T. budding.

Shift The normal period of attendance at a mine or factory by one wage earner in one day.

Shifting cultivation A commonly used method among tribes. Forest is burnt and used for cultivation process is repeated at other place. Come back to the first place after few years and cultivate. Not a recommended practice.

Shim share It is a share in which shin is provided as an additional part.

Shoaling Reduction of depth of a water basin.

Shoe (tranching machinery) Device or attachment, immediately following the digging mechanism on a trenchor, which smooths the trench bottom.

Shoot The aerial portion of a plant, including stem, branches, and leaves. Also to produce roots.

Shorelands The terrain of the coastal watershed down to the upper margin of the wetlands (lower margin of coastal flood plain).

Short grass plains Area of the Great Plains or other similar areas covered with relatively low growing native grasses.

Short grass prairie The area of the Great Plains covered with comparatively low-growing native grasses.

Short shoot Where there is shoot dimorphism, an unextended, often specialised axis bearing leaves and usually flowers, not contributing to the overall architecture of the tree.

Shothole A disease symptom in which small round fragments drop out of leaves, making them look as if riddled by shot.

Shrinkage (Timber) The reduction in dimensions of timber which takes place during drying.

Shrub A low woody plant with several stems.

Shunt Usually a comparatively low resistance connected in parallel, with or across a device or part of a machine.

Sib Brother-sister relationship.

Sibling (Sib) One or two or more offsprings of the same parents but of different birth.

Sib mating Crossing of siblings, two or more individuals of the same parentage (brother-sister mating). Sibling.

Sickle The cutting mechanism of a binder, reaper, or combine.

Side bund In soil conservation, an embankment constructed on a contour.

Sidecast disposal The simplest technique for overboard disposal of dredged material whereby the spoil is discharged directly to the water.

Side draft A sidewise pull produced when the centre of the load or resistance is out of the true line of hitch or draft.

Side slope The slope that the side of a canal or embankment makes with the horizontal.

Sieve tube A phloem element consisting of a series of thin-walled cells arranged end to end, in which some sieve areas are more specialised than others.

Sieve tube cell A phloem cell with perforated end walls which occur in series, forming tubes.

Sigatoka Leaf spot disease of banana caused by the fungus Mycosphaerella musicola.

Sight rod Rod on a tile trencher used by the operator to sight at grade targets to establish grade.

Sign Any indication of disease other then reaction of the host piant-spores, mycellium, exudate, or fruiting bodies of the pathogen.

Significance of effect or significant effect An effect is said to be significant if the value of the statistic used to test it lies outside acceptable limits as due to chance alone. In statistical usage, a significant effect means — signifying a characteristic of the population from which the sample is drawn, regardless whether the characteristics is important.

Significant A term applied to differences, correlation, etc., to indicate that within a given probability they are not due to chance alone. 'Significant' ordinarily indicates a probability of not less than 95 per cent, while 'highly significant' indicates a probability of not less than 99 per cent.

Silage Green or mature fodder that is fermented to retard spoilage and produce a succulent winter feed for livestock.

Silage additive Material added to forage at the time of ensiling to enhance either its preservation or feed value.

Silage preservative Material added to silage at the time of ensiling to enhance the favourable, fermentation process.

Silica-Alumina ratio The molecules of silicon dioxide (SiO_2) per molecule of aluminium oxide (Al_2O_3) in clay minerals or soils.

Silica-Sesquioxide ratio The molecules of silicon dioxide (SiO_2); molecule of aluminium (Al_2O_3) plus ferric oxide (Fe_2O_3) in clay minerals or soils.

Silicaflagellata A class of unicellular flagellates of the plant division chrysophyta represented by a single living genus, Dictyocha.

Silicle A many seeded capsule formed from two united carpels, usually of equal length and width, and divided on the inside by a replum.

Silk (1) The stigma and style of the female maize flower and through which the pollen tube grows to reach the embryo sac. (2) The five, shiny fibre produced for cocoons of silkworms.

Sill dam A dam, generally of strong brushwood mattress weighted with heavy rip-rap, constructed on the sea side in the foreground, for the purpose of gaurding against scouring of the sea bottom by the overflowing water and consequent damage to the structures and other works.

Silique A silicle-like capsule, but usually at least four times as long as it is wide, which opens by rutures at either margin and has parietal placentation. Also called siliqua.

Silo A storage tight-walled structure for making and preserving silage whose height is more than the width or diametre. See vertical storage.

Silt (1) A soil separate consisting of particles between 0.05 and 0.002 millimetre in equivalent diametre. (2) A soil textural class; Soil material that contains 70 per cent or more silt and less than 12 per cent clay.

Silting The deposition of water-borne sediments in stream channels, lakes, reservoirs, or on floodplains, usually resulting from decrease in the velocity of the water.

Silt loam A soil texture class; soil material that contains 50 per cent or more silt and 12 per cent to 27 per cent clay, or 50 per cent to 80 per cent silt and less than 12 per cent clay. See also texture, soil and class, soil.

Silt stone A very fine, grained, consolidated, classic rock composed predominantly of particles of silt grade.

Silt clay A soil texture class; soil material that contains 40 per cent or more clay and 40 per cent or more silt. See also texture, soil and class, soil.

Silt clay loam A soil texture class, soil material that contains 27 per cent to 40 per cent clay and less than 20 per cent sand. See also texture soil and class, soil.

Silurian An early Paleozoic period following the Ordovician and preceding the Devonian.

Silver oak Grevillea robusta A. Cunn. Proteaceae family. A medium sized tree grown as avenues and as windbelt alongwith Eucalyptus globulus. It is a drought hardy tree.

Silver wattle Acacia decurrens Willd. var. dealbatta F. Muell. (Acacia dealbata Link.). Family Mimosaceae. Exotic tree ideal for afforesting unstable hill slopes.

Silvicide Contact chemicals capable of killing trees.

Silvics The life history and general characteristics of forest trees and stands, with particular reference to environmental factors.

Similitude A known and usually limited correspondence between the behaviour of a model and that of its prototype with or without geometric similarity.

Simple layering A method of layering in which shoot of the plant is bent down, given several cuts at places to be buried. A method of propagation.

Simple leaf A leaf having one blade, or a lobed leaf in which the separate parts do not reach down to the midrib.

Simulate To mimic some or all of the behaviour of one system with a different, dissimilar system, particularly with computers, models, or other equipment.

Single action disc harrow It is a harrow with two gangs placed end to end, which throw the soil in opposite direction such that right gang throws soil towards right and left gang towards left.

Single compound pistil A single seed bearing organ of a flower which is composed of two or more carpels.

Single cross A cross between two inbred lines as in corn, etc.

Single cross parent The F_1 offspring of two inbred parents which in turn is used as a parent- usually with another line or single cross parent to produce commercial hybrid seed, as in maize.

Single grain structure A soil structure in which the soil particles occur almost completely as individual or primary particles; secondary particles or aggregates are seldom present. It is usually found only in extremely coarse-textured soils.

Single super phosphate Grade of superphosphate containing 16-20 per cent phosphoric acid.

Single plate clutch A popular clutch consisting of three plates, usually two driving plates and one driven plate. So called because of the single driven plate.

Single-use goods Goods used up in a single process of consumption or production.

Sinistrorse Pertains to a plant whose stems twine or turn spirally from right to left.

Sinuate Having a wavy margin with strong indentations.

Siphon (1) A tubular element in various algae. (2) A tube. pipe or hose through which a liquid can be moved from a higher to a lower level by atmospheric pressure facing it up the shorter leg while the weight of the liquid in the longer leg causes continuous downward flow.

Siphonales A large order of green algae (Chlorophyta) which are caenocytic, non-septate, and mostly marine.

Siphonaptera An order of insects to which flea belong.

Siphonocladaceae A family of green algae in the order Siphonocladales.

Siphonocladales An order of green algae in the division chlorophyta including marine, mostly tropical forms.

Siphonogamous In plants, especially seed plants, the accomplishment of fertilization by means of a pollen tube.

Siphonostele A type of stele consisting of pith surrounded by xylem and phloem.

Siphon spillway A spillway operating on siphon principle.

Sisal hemp anthracnose Disease of sisal hemp caused by colletotrichum agaves. Whole leaf deries up after circular spots appear on leaves.

Sissoo Dalbergia sissoo Roxb. Family Papilionaceae. A timber tree, frost hardy. Wood used for making furniture and agricultural implements.

Site (1) In ecology, an area described or defined by its biotic, climatic and soil conditions in relation to its capacity to produce vegetation. (2) An area sufficiently in form biotic, climatic, and soil conditions to produce a particular vegetation.

Site index (1) A quantitative evaluation of the productivity of a soil for forest growth under the existing or specified environment. (2) The height in feet of the dominant forest vegetation taken at or calculated to an index age, usually 50 or 100 years.

Sitophilus Oryzae See rice insect pests.

Skeletal soil Soil consisting of coarse materials formed by predominant influence of physical weathering.

Skew distribution A distribution which is not symmetrical.

Skew throat, inclined throat or **oblique throat** A throat whose inlet end, or outlet end, or both are skewed to the flow.

Skid To pull logs from the stump to the skidway, landing or mill.

Skidway Two skids laid parallel at right angles to a road, usually raised above the ground at the end nearest the road. As they are brought from the stump, logs are usually piled upon a skidway for loading upon wagons or cars.

Slab The exterior portion of log or bulk which is removed by the saw in the process of converting it to a square-sawn timber.

Slaking, soil The breakdown of aggregates resulting from the addition of water to soil.

Slaty Containing a considerable quantity of slate fragments. It is used to modify soil texture class names, such as slaty clay loam. See also coarse fragments.

Slickens Fine textured materials separated in mining and in ore-mill operations; the materials may be detrimental to plant growth and so should be confined in specially constructed basins. A miscellaneous land type.

Slickenside Smoothed surface along planes of weakness resulting from the movement of one mass of soil against another in soils dominated by swelling clays.

Slick spots Small areas in a field that are slick when wet, because of a high content of alkali or exchangeable sodium.

Slide gate A crest gate which has high frictional resistance to opening because it slides on its bearings in opening and closing.

Sling Psychrometre A sling psychrometre has two thermometres mounted on a base plate. The one with the sock is the wet bulb thermometre, the other is the dry bulb. The wet bulb extends below the dry bulb. This is done purposely, so that the 'sock' can be dipped in water without wetting the dry bulb thermometre.

Slip It is the relative movement of the wheel or track in the direction of travel for a given distance under load and at no load conditions.

Slip erosion Erosion of soil caused by

hydraulic pressure exerted by water seeping into man-made slopes; mass of soil and other materials come down bodily.

Slip hose share It is the share which is provided with a separate point of share could be detached and replaced.

Slip share It is one piece share with curved cutting edge having no additional parts.

Slipping clutch A clutch which, from wear or glazing or improper pedal adjustment, is not able to transmit full engine power due to slipping between the pressure plate and the driven plate.

Slope Degree of deviation of a surface from the horizontal usually expressed in per cent of degrees.

Slope characteristics Slopes may be characterised as concave (decrease in steepness in lower portion), uniform, or convex (increase in steepness at base). Erosion is strongly affected by shape, ranked in order to increasing erodibility from concave to uniform to concave.

Slope culture A method of growing plants in which surplus nutrient fluid is allowed to run through the sand or other medium in which the plants are growing.

Sloping bench terrace A bench terrace having no fall from end to end, but a measurable slope from the back edge to the front edge of the bench.

Sloping zone The portion of the border below the level zone.

Slough, slew, sloo, or **slue** A wet or marshy place, a swamp; a marsh land crack. Also, a side channel or inlet as from a river, a bayou. In the Mississippi Valley and in California, a tidal flat or bottom land creck.

Sloughing side slopes Side slopes that are sliding into the channel.

Slope wash Soil moved downslope by the action of gravity assisted by water.

Sludge The solid portion separated from sewage; used as manure.

Sluice A passage fitted with a vertical sliding gate or valve to regulate the flow of water in a channel or lock.

Sluice gate The vertical slide gate of a sluice.

Slurry Thick suspension of chemical; used for seed treatment.

Slack water or **slack tide** Tide water during the period where there is no horizontal motion of water at the surface.

Slash Branches, bark, tops, chunks, cull logs, uprooted stumps and broken or uprooted trees left on the ground after logging, also large accumulation of debris after wind or fire.

Slavery A system in which men, who have no civil rights, work for and are kept by their masters and owners, either badly or well, receiving no wages.

Slime flux Flow of liquid from diseased fruit or forest trees due to the attack of various fungi producing a fermentation of the cortical elements down to the cambium zone.

Smear Very thin layer of material spread on a glass slide.

Smectite See montmorillonite group.

Smilax Genus of climbing shrubs.

Smokechaser A member of a fire-fighting crew.

Smoothing (1) The process of removing fluctuations in an ordered series so that the results shall be 'smooth' in the sense that first differences are regular and higher order differences small. (2) To smoothen the soil surface to remove minor irregularities, commonly done after levelling job.

Smother A method to bill or retard growth by a dense or more vigorously growing crop (smother crop).

Smudge Any of several fungus diseased of cereals and other plants characterised by dark, sooty discolourations.

Smut A fungus (infections parasite) of the Ustilaginales, characterised by sooty spore masses; the name also used for the disease caused by the smut.

Snag A standing, dead tree from which the leaves and most of the branches have fallen, or a standing section of the stem of a tree broken off at a height of 20 ft or more. If less than 20 ft high, it is properly termed a stub.

Snail Viriparus variatus, Pila vireus, Achatina fulica. A snail causing damage to paddy and vegetable crops.

Snow Precipitation composed of white or translucent ice crystals, chiefly in complex branched hexagonal form and often agglomerated into snow flakes.

Snow cover All accumulated snows on the ground including that derived from snowfall, snowslides, and drifting snow. Also known as snow mantle.

Snow gauge An instrument for measuring the amount of water equivalent in a snow-pack. Also known as snow sampler.

Snow mantle See snow cover.

Social ability Persons who understand persons, mixes with people, such as salesmen, ministers, diplomats have social ability.

Social change It is any alteration of the previous position or condition of social phenomena which are the forms of association and of social interaction.

Social cliques Are composed of a small number of persons who accepts each other as social equals and associate as close friends. Clique members communicate the farm and home information to their members.

Socialism A social system in which the means of production, distribution and exchange are wholly or substantially owned by the State.

Social relationship It is a relationship in which the members involved are aware of the presence of each other.

Social system It is defined as a population of individuals who are functionally differentiated and engaged in collective problem solving behaviour.

Sociation A recurring plant community of essentially uniform composition with at least certain dominant species in each layer of vegetation; a useful unit only in a floristically poor area.

Society It can be defined as the system of social relationships in and through which we live.

Sociology It can be defined as the study of society.

Soda lake An alkali lake rich in dissolved sodium salts, especially sodium carbonate, sodium chloride and sodium sulphate. Also known as natron lake.

Sod bound Used to describe the unproductiveness of grass sod due to lack of nitrogen.

Sod culture A soil management system in which plants are grown in permanent grass without tillage and addition of litter.

Sod grass Any grass that reproduces by stolons or rhizomes or otherwise forms a more or less continuous sod or turf.

Sodic soil (1) A soil containing sufficient sodium to interfere with the growth of most crop plants. (2) A soil having an exchangeable-sodium percentage of 15 or more.

Sodium percentage The ratio of the total sodium cations or the total sodium plus potassium cations, to the total cations held in solution, multiplied by 100.

Softwood One of the botanical group of trees that generally have needle or scale like leaves — the conifers; also, the wood produced by such trees, regardless of texture or density.

Soil (1) The unconsolidated mineral and organic material on the immediate surface of the earch that serves as a natural medium for the growth of land plants. (2) The naturally occurring unconsolidated material on the surface of the earth that has been unconsolidated material on the surface of the earth that has been influenced by parent material, climate (including the effects of moisture and temperature), macro- and micro-organisms, and relief, all acting over a period of time to produce soil that may differ from the material from which it was derived in many physical, chemical mineralogical, biological, and morphological properties. (3) For the purpose of the Canadian taxonomic system, the earth's surface (the material that is to be classified) is divided into soil and non-soil. Soil is the naturally occurring, unconsolidated, mineral or organic material at the earth's surface that is capable of supporting plant growth. It extends from the surface to 15 cm (6 inches) below the depth at which properties produced by soil forming processes can be deleted. These properties differ from those found in any underlying unconsolidated material. The soil-forming processes are defined as an interaction between climate, living organisms, and relief acting on soil and soil parent material. Unconsolidated material includes material cemented or compacted by soil-forming processes. Soil may have water covering its surface to a depth of 60 cm (24 inches) or less in the driest part of the year. Non-soil is the collection of soil material or soil-like material that does not meet the preceding definition of soil. It includes soil displaced by unnatural processes and unconsolidated material unaffected by soil-forming processes, except for the material that occurs within 15 cm (6 inches) below soil as defined. Non soil also includes unconsolidated mineral or organic material thinner than 10 cm (4 inches) overlying bedrock; organic material thinner than 40 cm (16 inches) overlying a hydric layer; and soil covered by more than 60 cm (24 inches) of water in the driest part of the year. The soil or soil is a collective term used for all soils equivalent to the term vegetation for all plants.

Soil analysis Chemical, mechanical or minerological determinations of a soil.

Soil association A group of soils with or without common characteristics, geographically associated in an individual pattern. May form a natural land type or geographical unit in soil mapping.

Soil binder A grass or plant whose roots hold the soil in position through the binding action of its roots.

Soil buffer compounds The compound which help soil to resist changes in pH clay, organic, matter, carbonates and phosphates are some such compounds.

Soil cement lining Lining consisting of a mixture of cement and soil, i.e. the soil replacing the sand and gravel used in concrete.

Soil classification The systematic arrangement of soils into classes in one or more categories or levels of classification for a specified objective. Broad groupings are made on the basis of general characteristics and subdivisions on the basis of more detailed differences in specific properties.

Soil cohesion It is the tendency of soil particles to stick together. Generally determines cracking of soils.

Soil colloids Small sized particles of inorganic or organic matter; have relatively large surface area per unit mass.

Soil conservation The preservation of soils against deterioration particularly against erosion. Maintenance of fertility and productivity of soils. Water management is included.

Soil consistence Degree and kind of cohesion and adhesion of soil material or resistance offered to force tending to deform or rupture the soil mass.

Soil creep The slow downward movement of soil mass on slopes under the force of gravity.

Soil depth The thickness of the solum or soil proper.

Soil family A taxonomic group of soils composed of a soil series.

Soil formation factors The variable usually inter-related natural agencies that are responsible for the formation of soil. The factors are parent rock, climate, organisms, relief, and time.

Soil genesis Mode of origin of soil; Generally the processes responsible for the development of solum from parent material.

Soil horizons The succession of horizontal layers of soil section up to unweathered rock. A horizon is surface soil containing organic matter; horizon B contains enriched material whereas horizon C contains partially weathered parent material.

Soiling crop One that is cut green and fed green to livestock.

Soil map A map showing the distribution of soil types or other soils mapping units in relation to the prominent physical and cultural features of the earth's surface. The following kinds of soil maps are recognised in the United States; detailed, detailed reconnissance, reconnaissance, generalised and schematic.

Soil maturity The ultimate stage of development of a soil from the raw parent material when it is able to express the full effect of environmental factors.

Soil monolith A vertical section of the soil profile mounted for display.

Soil morphology The kind and arrangement of horizons or separate natural layers in a soil profile.

Soil pipe A vertical cast-iron or plastic pipe for carrying sewage from a building into the soil drain.

Soil series A taxonomic group of soils. It is intermediate between family and type.

Soil slip Lesser in dimension than a landslide. The downward movement of soil on hill slopes when completely saturated.

Soil survey A general term for the systematic examination of soils in the field and in laboratories; their description and classification; the mapping of kinds of soil; the interpretation of soils according to their adaptability for various crops, grasses, and trees; their behaviour under use or treatment for plant production or for other purposes; and their productivity under different management systems.

Soil type (1) A subdivision of a soil series based on surface texture. At the present time in the United States a soil type is considered as a kind of phase and is not part of the soil classification system presently being used. (2) In Europe, a class roughly equivalent to a great soil group.

Soil water The hydraulic conductivity divided by the differential water capacity (using consistent units), or the flux of water per unit gradient of moisture content in the absence of other force fields.

Solanaceae A family of dicotyledonous plants in the order Polemoniales having internal phloem, mostly numerous ovules and seeds on axile placentae, and mostly cellular endosperm.

Solar constant The rate at which solar radiation is received outside the earth's atmosphere on a surface normal to the incident radiation, and at the earth's mean distance from the sun. The value of the mean solar constant is about 2.0 g cal/cm^2/min.

Solar day The interval between two successive transits of the sun's centre over the same meridian.

Solar radiation The total electromagnetic radiation emitted by the sun.

Solar spectrum The part of the electromagnetic spectrum occupied by the wavelengths of solar radiation.

Solclime The temperature and moisture conditions of the soil; the soil climate.

Solenostele An amphiphloic siphonostele. The xylem tissue has both internal and external phloem in the vascular cylinder.

Solifluction A type of creep that takes place in regions where the ground freezes to a considerable depth and as it thaws during the warm season the upper thawed position creeps downhill over the frozen material. The soil moves as a viscous liquid down slopes of as little as 2 or 3 degrees and may carry rocks of considerable size in suspension.

Solod A great group of soils in the Solonetzic order occurring most commonly in the grassland and parkland regions. The soils have a dark-coloured surface (Ah) horizon, a prominent eluvial (Ahe or Ae) horizon at least 5 cm (2 inches) thick, a prominent transitional (AB) horizon that breaks readily into blocky aggregates, and a darkly stained B (Bnf) horizon over a C horizon that is saline and usually calcareous.

Solodised soil A soil that has been subjected to the processes responsible for the development of a Solod and has at least some of the characteristics of a Solod.

Solodised Solonetz A great group of soils in the Solonetzic order, occurring most commonly in the grassland and parkland regions and consisting of soils with a variable surface (Ah, Ahe, or Ae) horizon that is underlain by a well-developed As horizon, a compact prismatic or columnar Bnt horizon, and a C horizon that is saline and usually calcareous.

Solonetz A great group of soils in the solonetzic order, occurring most commonly in the grassland and parkland regions and consisting of soils with a variable surface (Ah, Ahe, or Ae) horizon that breaks abruptly into a hard, compact prismatic or columnar B (Bnt, rarely a Bn) horizon underlain by one or more saline and usually calcareous (Bs, Cs, Csa, Csk, Cca) horizons. They lack a continuous Ae horizon 2.5 cm (one inch) or more thick.

Solonetzic An order of soils developed mainly under grass or grass-forest vegetative cover in semi-arid to sub-humid climates. The soils have a stained brownish solonetzic B(Bnt or Bn) horizon and a saline C horizon. The surface may be one or more of Ap, Ah or Ae horizons. The order includes the Solonetz, Solodised Solonetz, and Solod great groups.

Soluble-sodium percentage (SSP) The proportion of sodium ions in solution in relation to the total cation concentration, defined as follows:

$$SSP = \frac{\text{Soluble-sodium concentration (meq/litre)}}{\text{total cation concentration (meq/litre)}} \times 100$$

Solum The upper horizons of a soil in which roots are contained. It usually consists of A and B horizons.

Solution, soil The aqueous liquid phase of the soil and its solutes consisting of ions dissociated from the surface of the soil particles and of other soluble materials.

Somatic Referring to body tissues; having two sets of chromosomes, one set normally coming from the female parent and one from the male, as contrasted with germinal tissue that will give rise to germ cells.

Somatophyte A plant composed of distinct somatic cells that develop especially into mature or adult tissue.

Somatoplastic sterility The collapse of fertilised ovules during the early developmental stages due to disturbed embryo-endosperm relations.

Sombric brunisol A great group of soils in the Brunisolic order. The soils have moder Ah horizons more than 3 cm (2 inches) thick and Bm horizons in which the base saturation (NaCl) is usually 65 per cent to 100 per cent and the pH ($CaCl_2$) is usually about 5.5.

Sonne-strip photograph A continuous photo-strip in contrast to conventional series of separate and overlapping exposures.

Soredium A structure comprising algal cells wrapped in the hyphal tissue of lichens, as in certain Lecanorales; when separated from the thallus, it grows into a new thallus.

Sore shin A fungus disease of cowpea, cotton, tobacco, and other plants, beyond the seedling stage, marked by annular growth of the pathogen on the stem at the groundline.

Sorghum Any of a variety of widely cultivated grasses especially sorghum bicolour in the United States, grown for grain and herbage; growth habit and stem form are similar to Indian corn, but leaf margins are serrate and spikelets occur in pairs on hairy rachis.

Sorghum-diseases the common diseases in sorghum are Sorghum downy mildew caused by Sclerospora sorghi (Kulk) West and Uppal; sorghum grain smut caused by Sphacelotheca sorghi (Link) Clinton; sorghum head smut caused Sphacelotheca reiliana; sorghum long smut caused by Tolyposporium ehrenbergii; sorghum loose smut caused by Sphacelotheca cruenta (Kuhn.) Potter; sorghum red leaf spot caused by Colletotrichum graminicolum; sorghum rust caused by Puccinia purpurea and sorghum sooty stripe caused by Ramulispora sorghi.

Sorghum-insect pests Sorghum earhead bug (Calocoris angustatus Leith) whose adult and nymph suck sap from tender portions of plants is major insect pest of sorghum.

Sorghum poisoning Sorghum containing hydrocyanic acid; when eaten by livestock dangerous in some circumstances.

Sorus (pl. Sori) Fungus spore mass, especially of rusts and smuts; occasionally, a group of fruiting bodies.

Soursop Annona muricata L. Family Annonaceae. A tree bearing largest fruits in the Annonaceae family of trees.

Sourwood Oxydendrum arboreum. A deciduous tree of the heath family (Ericaceae); having long simple, finely toothed, long-pointed leaves that have an acid taste, and white, urn-shaped flowers. Also known as screl tree.

South-west monsoon Cause of 75 per cent of rainfall in India. Heavy rain brought about by south-westerly air current that blows over west coast and the Deccan and central parts of India. Rainfall occurs from June to September.

Sow To place seeds in the soil.

Soybean Glycine max. (L.) Marrill. (Glycine Soja (L.) Sieb. & Zucc; G. hispida-Maxim; Soja max piper). Family papilionaceae. A legume crop rich in protein raised in *kharif*; also grown for forage. Also oil extracted which is used for cooking.

Soybean-diseases Soybean bacterial disease caused by Xanthomonas phaseoli Sojense; soybean downy mildew caused by Peronospora spp.; soybean dry root not caused by Macrophomina phaseoli, soybean leaf spot caused by Cercospora sojina are few diseases.

Soybean-insect pests Few insect pests are Soybean grey weevil (Myllocerus sp.) weevil that feeds on leaves; soybean leaf miner (Stomopterynerteria M.) caterpillar that folds leaves and feeds on them and soybean stem borer beetle (Oberea brevis S.) beetle whose grub bores into stem.

Spadix A fleshy spike that is enclosed in a leaflike spathe and is the characteristic inflorescence of palms and arums.

Span of control It refers to the number of sub-ordinates an officer can effectively supervise.

Spar Kinds of crystalline mineral, easily cleavable and non-lustrous, as calcareous spar.

Sparganiaceae A family of monocotyledonous plants in the order Typhales distinguished by the in florescence of globose heads, a restigial perianth, and achenes that are sessile or nearly sessile.

Spark plug A device for conducting the high-tension current to the combustion space where the charge is fired by the spark as it jumps the gap of the plug.

Spark plug electrodes The metal points between which the electric spark jumps. Also the metal conductor which passes through the centre of the spark plug insulator.

Spate irrigation A method of irrigation practised by building earth diversion banks across a normally dry watercourse. The banks divert spate water into canals leading to embanked fields where the water is ponded until absorbed.

Spathe A large, usually coloured bract or pair of bracts enclosing an infloresescence, especially a spadix, on the same axis.

Spear A sprout or shoot or plant.

Spear grass Heteropogon contortus L.) Beauv. ex. Roem. and Schult. (Andropogon contortus L.). Family Gramineae. A grass commonly used as fodder and in soil conservation works.

Spearing Technique of connecting crossing mole drains in which a spear is forced down at the point of intersection.

Species One sort of plant or animal; sufficiently similar that is may be assumed they had a common ancestor; abbreviated as sp. = singular, and spp. = plural. A genus name followed by sp. means that the particular species is undetermined. Spp. following genus name means that several species are grouped together without being named individually.

Species population A group of similar organisms residing in a defined space at a certain time.

Specification A detailed statement of particulars.

Specific gravity The ratio of the weight of a substance to that of an equal volume of water.

Specific heat The amount of heat in calories required to raise the temperature of one gram of a substance by 1°C.

Specific humidity The ratio of the weight of the water vapour in a "parcel" of the atmosphere to the total weight of the air (including the water vapor) staged in gram of water vapor per kilogram of air.

Specific volume It is defined as the volume per unit mass of the substance.

Speck A fungus or bacterial disease of rice characterised by speckled grains.

Spectrometre An instrument which helps in determining small amounts of organic compounds and inorganic elements in plant and animal tissues;

works on principles of light.

Speed drop See percentage regulation of speed.

Speltoid Mutants occurring in cultures of common wheat, Triticum vulgare, and possessing characters of T. spelta; spelt wheat.

Sperm Spermatozoan, or sperm cell, of the male.

Spermagonium A flask shaped walled structure in which spermatia are produced.

Spermatophore A specialised hypha that bears spermatia.

Spermatium. pl. Spermatia, a sex cell (+ or -), a phycniospore. In red algae, a naked nonmotile gamete, a tiny sporelike cell in rusts which may function as a gamete.

Spermatogenous Sperm producing.

Spermatophyta Flowering and seed bearing plants.

Sphaerophoraceae A family of the Ascolichenes in the order Caliciales which are fruticose with a solid thallus.

Sphaeropleineae A suborder of green algae in the order Uttriolales distinguished by long, coenacytic cells, numerous bandlike chlorophasts, and heterogametes produced in undifferentiated vegetative cells.

Sphagnaceae The single monogeneric family of the order Sphagnales.

Sphagnales The single order of mosses in the subclass sphagnobrya containing the single family Sphagnaceae.

Sphagnobrya A sub class of the Bryopsida; plants are greyish-green with numerous, spirally arranged branches and grow in deep tufts or mats, commonly in bogs and in other wet habitats.

Sphagnum bog A bog composed principally, of mosses of the genus Sphagnum (Sphagnales) but also of other plants, especially acid-tolerant species, which tend to form peat.

Sphenopsida A group of vascular cryptogams characterised by whorled, often very small leaves and by the absence of true leaf gaps in the stele, essentially equivalent to the division Equisetophyta.

Spider gears The two, three or four gears (as the case may be) fitted free running on the spider, by means if which the differential section is obtained in the rear axle.

Spike An unbranched inflorescence or head in which the individual spikelets are attacked directly to the control axis or rachis, such as wheat.

Spike knot or splay knot A knot sawn lengthwise.

Spikelet Unit of the inflorescence in grasses, consisting of two glumes and one or more florets.

Spike-tooth harrow An implement with steel spikes extending downward from a frame and pulled by a tractor to pulverize and smooth plowed soil.

Spiles, or **furrow tubes** Portable tubes or spouts used to deliver water from temporary ditches to furrows. They may be laid over the ditch bank to convey water by siphonic action, or laid through the bank.

Spillway A hydraulic structure used to convey water from some higher level to a lower level.

Spillway outlet channel A channel connecting the stilling basin or the downstream and of the spillway either to the parent channel or other drainage.

Spinach Spinacia oleracea L. Family Chenopodiaceae. A vegetable crop grown for its green leaf called *Palak* in Hindi.

Spindle tuber A virus disease of the potato characterised by spindliness of the tops and tubers.

Spinney A small grove of trees or a thicket with undergrowth.

Spinning count A number indicating the fineness of yarn which can be spun from a given sample of wool.

Spiral (leaf) arrangement Alternate, but radially arranged, i.e., forming a helix.

Spiral separator A machine used to upgrade seed quality. It makes separation on the basis of differences in relative ability to roll.

Spirilla Bacteria of a spiral shape.

Spirochaete Bacterium with a wavy, thread-like form.

Spirodistichous (leaf) arrangement Alternate, two-ranked, but the ranks each describing a helix.

Splash erosion See erosion.

Split A crack extending from face to

another face.

Split cotter pin A pin made of wire whose section is semicircular, bent around until the flat surface meet. It is inserted in a hole near the end of a shaft or bolt to prevent the nut from backing off.

Spoil bank Rock waste, banks, and dumps, from the excavation of ditches.

Sponge gourd Luffa cylindrica L. M. Roem. (Luffa aegyptiaca Mill., Momordica cylindrica L.) Family Cucurbitaceae. A vegetable crop. Called *ghia torai* in Hindi.

Spongy mesophyll A system of loosely and irregularly arranged parenchymal cells with numerous intercellular spaces found near the lower surface in well-differentiated broad leaves. Also known as spongy parenchyma.

Spontaneous combustion The ignition or combustion of a substance by heat generated within itself or by the chemical action of its constituents. As in high moisture hay storage.

Spontaneous generation Theory that held that organisms arose spontaneously.

Sporangiole Small sporangium without a columella and with a small number of spores.

Sporangium Organ producing nonsexual spores in a more or less spherical wall.

Spore A single-to many celled reproductive body, in the fungi and lower plants, which can develop a new plant.

Sporidium, pl. Sporidia Basidiospore or rusts and smuts.

Sporocarp Any multicellular structure in or on which spores are formed.

Sporocyst A unicellular resting body from which asexual spores arise.

Sporocyte A spore mother cell of plants.

Sporodochium, pl. Sporodochia Cluster of conidiophores interwoven on a stroma or mass of hyphae.

Sporophore Spore-producing or supporting structure-fruit body; used especially in the Basidimycetes.

Sporophyte The spore forming generation in the life cycle of plants.

Sport A mutation. A sharp deviation from normal or parent type of plant.

Sporulate To produce spores.

Spot drainage Draining of small wet or springy areas in an otherwise well drained field.

Spot grading Removing humps or filling hollows, without establishing a uniform grade in any direction.

Spotted wilt A virus disease of various crops and wild plants, especially tomato, characterised by bronzing and downward curling of the leaves.

Sprayer Apparatus for applying chemicals in liquid form.

Spreading agent A material/chemical that reduces surface tension of a liquid and causes it to spread or disperse in a thin film over a surface.

Sprigging Vegetative propagation by planting stolons or rhizomes in furrows or holes in the soil.

Spring tide Maximum tide following the second and fourth quarters of the moon, that is, the new and full moon.

Springwood The less dense, larger-celled, first formed part of a growth layer.

Sprinkle A very light shower of rain.

Sprinkler head The part of the sprinkler fitted with one or more nozzles, through which water discharges out in the form of a spray.

Sprinkler intensity, or Rain intensity The amount of water which falls on a unit of land in a given unit time as artificial rain.

Sprinkler irrigation or irrigation by sprinkler A method of irrigation in which water (under adequate pressure) is sprinkled over the land through nozzle lines, perforated, pipes or sprinklers. Also called "spray irrigation" and sometimes referred to as 'overhead irrigation'.

Sprinkler lateral A pipeline taking off from a main or sub-main to feed the sprinklers. In portable or semi-portable systems, also called 'flying line'.

Sprinkler oscillator A device causing water motion which turns a sprinkler or nozzle line from side to side.

Sprinkler pattern Precipitation pattern from a single sprinkler head.

Sprinkler pressure Actual pressure available at the nozzle. The pressure that must be available at the sprinkler nozzle for proper operation and distribution, of the sprinkler is called 'correct sprinkler pressure'.

Sprinkler system A fire-protection system of pipes and outlets in a building, mine or other enclosure for delivering a fire extinguishing liquid or gas; usually automatically by the action of heat on the sprinkler head. Also known as fire sprinkling system.

Sprocket A toothlike projection on a wheel shaped so as to engage with the links of a chain.

Sprout A young shoot.

Spruce An evergreen tree belonging to the genus Picea characterised by single, four-sided needles borne on small peglike projections, pendulous cones, and resionous wood.

Spud A hand tool used in stripping bark from felled trees.

Spur .A structure in a river/canal or channel to induce deposition. It is a bank protection work and depending upon cost, it is made up of piles, rock or other materials.

Spur wheel A gear wheel whose teeth are on the outer diametre and at right angles with the sides of the wheel.

Square A cotton bud before the flower has opened.

Square metre Area equivalent to that of a square of sides equal to one metre.

Square timber Timber formed by slabbing a log on four sides.

Square yard Area equivalent to that of a square of sides equal to one yard and is equal to 0.8361 square metres.

Squash melon Citrullus vulgaris schrad var. fistulosus. A vegetable of summer season of family Cucurbitaceae. A source of vitamin A.

Stabilization soil Chemical or mechanical treatment designed to increase or maintain the stability of a mass of soil or otherwise to improve its engineering properties.

Stabilised grade Slope of a channel at which neither erosion nor deposition occurs.

Stable An atmosphere in which temperature decrease with height is less than the dry adiabatic lapse rate.

Stable humus The part of humus that is resistant to microbial attack (and is insoluble in acetyl bromide).

Stack A well-built pile of grain, forage, or straw, which usually has a square or circular base. A rick is rectangular.

Stachylidium theobromae See Banana cigar-end disease.

Stacking Refers to the arrangement of seed bags in a manner such as to prevent their falling.

Staddle A small tree; generally a forest tree.

Staff gauge Graduated scale mounted on a plank, pier, wall or other like objects from which the water surface elevation may be read.

Stage Elevation of a water surface above any chosen plane, often above an established low water plane; gage height.

Stain (1) Dye used on bacteriologic and histologic techniques. (2) A discolouration occurring on a piece of timber.

Stainless steel An alloy steel containing about 12 per cent or over of chromium with or without nickel together with other elements, and is characterised by its high resistance to corrosive media.

Stakes Wood pieces of small diametre sharpened at one end. Used for marking in levelling and other field works.

Stalk A stem-like supporting structure.

Stamen The part of the flower bearing the pollen; pollen bearing organ; each stamen is composed of a stalk (filament) and pollen sac (anther).

Staminade A stamen with no functional anther.

Staminate Possessing stamens but no pistils.

Stand An aggregation of trees or other growth occupying a specific area and sufficiently uniform in composition (species), age, arrangement, and conditions as to be distinguishable from the forest or other growth on adjoining areas.

Standard (1) The part to which the plough bottom and the handle is attached is called the standard. (2) A selected unit for comparison.

Standard deviation A measure of variability in terms of the units of measurement sample. Frequently refers to the infinite population.

Standard error (1) An estimate of the standard deviation of the means of samples drawn from some single statistical population. Often calculated from a single set of samples. (2) The positive square root of the variance of the sampling distribution of a statistic.

Standard error of estimate An expression for the standard deviation of the observed values about a regression line.

Standard normal distribution A single standard normal distribution is sufficient to cover all normal distributions whatever their mean and standard deviations, because the probability corresponding to a certain value of a normally distributed variable depends only on how far that value is from the mean, using the standard deviation as the unit for measuring its distance from the mean.

Standard soil-cement Soil cement which is mixed at optimum moisture for maximum compaction as determined by the Proctor method.

Staphylococcus Any of certain bacteria whose spherical cells are arranged in irregular clumps, resembling bunches of grapes.

Staple Fibre of cotton wool etc. especially with reference to length. As the length increases quality is superior. Short staple 11/16" to superior long staple 1" (2.54 cm) and above.

Stapling Grading of cotton on the basis of length of fibre.

Starter A series electric motor used on automobiles to turn over the gasoline engine until it starts under its own power.

Starterator A linkage which makes it possible to operate the starting motor by depressing the accelerator pedal.

Starting crank A crank formerly used for turning over the engine in order to start it. Since the invention of self-starters, the starting crank is rarely used except during certain types of repair jobs.

Starting motor The electric motor by means of which the engine is cranked by closing the electric circuit.

Startix A solenoid which automatically closes the starting motor witch as soon as the ignition switch is turned on.

Starved basin A sedimentary basin in which the rate of subsidence exceeds the rate of sedimentation.

State seed laboratory In any State, the State Seed Laboratory established or declared as such under sub-section (2) of Section 4 of Seeds Act, 1966.

Static head Pressure of a fluid due to the head of fluid above some reference point.

Static level Elevation of the water level or a pressure surface at rest.

Static suction lift The vertical distance between the centre line of pump to free level of the liquid to be pumped.

Statistical design A statistical design of a plan for the collection and analysis of data requiring forethought on what is wanted, visualization of possible ways of getting it, and selection of one way.

Statocyst A cell containing statoliths in a fluid medium. Also known as statocyte.

Statolith A sand grain or other solid inclusions which moves readily in the fluid contents of a statocyst, comes to rest on the lower surface of the cell, and is believed to function in gravity perception.

Statospore A type of spore with two silicified segments produced by certain members of the Chrysophyta.

Steady state A condition when no change is expected with respect to time.

Steady state solution Solution derived for steady state conditions.

Steam fog Fog formed when water vapor is added to air which is much colder than the vapor's source; most commonly, when very cold air drifts across relatively warm water. Also known as frost smoke.

Steckling Small sugar beet stored over winter and planted for the production of seed.

Steel An iron-carbon alloy, with carbon chemically combined as iron carbide and contains varying amounts of other elements some being added intentionally to develop specially derived properties.

Steel brakes A mechanical brake system by which the brake pedal transmits effective braking force to the brake shoes by means of cables enclosed in protective armor, to each wheel.

Steering gear A term applied generally, to all that assemblage of parts from steering wheel to axle, by means of which steering is accomplished.

Steering knuckle arms Two irregular shaped forgings, one left and one right, attached to the steering knuckles in order to make connection with the steering mechanism.

Steering wheel The hand wheel on an automobile, which through a system of gears and levers connecting it with the front wheels makes possible the steering control of the automobile.

Stefan Boltzman constant A universal constant of the sun caused by a lowered surface temperature.

Stele A collective name for the vascular and closely associated tissues in stems, roots, and other plants parts; xylem, phloem, pericycle, endodermis, and pith.

Stem The organ of vascular plants that usually develops branches and bear leaves and flowers.

Stem canker Any canker disease affecting the stem.

Stem rust Any of several fungus diseases, especially of grasses, affecting the stem and marked by black or reddish-brown lesions.

Stenohaline Unable to endure wide variation of osmotic pressure of environment, e.g. wide variation in the salinity of water. Opposite of euryhaline.

Step cut ring A piston ring with the ends cut to a rabbet or half lap joint.

Steppe Vast semiarid grass covered plain, usually lightly wooded.

Steppe climate the type of climate in which precipitation though very slight, is sufficient for growth of short, sparse grass, typical of the steppe regions of south-central Eurasia. Also known as Semi-arid climate.

Steradian The steradian is the unit of a solid angle and is defined as the solid angle which having its vertex in the centre of a sphere cuts off an area of the surface of the sphere equal to that of a square with sides of length equal to the radius of the sphere.

Sterculiaceae A family of discotyledonous trees and shrubs of the order Malvales distinguished by imbricate or contorted petals, bilocular anthers, and ten to numerous stamens arranged in two or more whorls.

Stereophotogrammetry Photogrammetry with the aid of stereoscopic equipment and methods.

Sterigma (pl. Sterigmata) A slender tip (projection) arising from a bidium and bearing a basidiospore.

Sterile Incapable of producing offspring.

Sterility Inability to produce viable offspring.

Steriloid Mutant types occurring in cultivated oats and resembling Avena sterilis.

Sterilization The process of making sterile, the killing of all forms of life.

Sthene 10^8 dynes is equal to one sthene. It is the force which, when applied to a body of mass one tonne, gives it an acceleration of one metre per second per second.

Sticking of valves Failure of the valves to open and close properly, caused by poor lubrication or by carbon deposits.

Sticky point (1) A condition of consistence at which the soil barely fails to stick to a foreign object. (2) Specifically and numerically, the weight moisture percentage of a well mixed, kneaded soil that barely fails to adhere to a polished nickel or stainless steel surface when the shearing speed is 5 cm/sec. The measurement is made rarely now.

Stigma The part of the female flower which receives the pollen from the anthers.

Stinkhorn A slender phalloid type of mushroom, with an offensive odor.

Stilling basin A depressed area in a channel or reservoir that is deep enough to reduce the velocity of the flow. Also known as stilling box.

Stilt root A prop root of a mangrove tree.

Still water A portion of a stream having a very slight gradient and no visible current.

Still-water level The level that the sea surface would assume in the absence of wind waves.

Still well A device, used in evaporation pan measurements, which provides an undisturbed water surface and support for the hook gage; the U.S.

Weather Bureau model consists of a brass cylinder, 8 inches (20.32 centimetres) high and 3.5 inches (8.89 centimetres) in diametre, mounted over a hole in a triangular galvanised iron base which is provided with levelling screws.

Stion A plant which consists of a stock and a scion growing in combination.

Stipe A stem or stalk; the supporting stalk of a mushroom.

Stipule One of the usually small, paired leaflike appendages at the base of a leaf or leafstalk in certain plants.

Strips A classificatory group in zoology.

Stock (1) The rooted plant onto which a graft is made. (2) A rhizome or root stock. (3) The main stem of a plant.

Stocker Beef animal being backgrounded prior to finishing or entering the breeding herd.

Stock, growing The sum (in number and volume) of all the trees in a forest.

Stock guard A frame for closing the ditches under a farm road. It keeps clear itself of weeds or other debris and also prevents the stock from entering the fields. It is usually used on irrigated pastures.

Stocking continuous See Grazing, continuous.

Stocking rate Number of animal units per unit of land area at a specific time.

Stockpiled Accumulated growth of forage for later use.

Stokes' law An equation relating the terminal settling velocity of a smooth, rigid sphere in a viscous fluid of known density and viscosily to the diametre of the sphere when subjected to a known force field. It is used in the particle-size analysis of soils by the pipette, hydrometre, or centrifuge methods. The equation is:

$V = (2gr^2)(d_1 - d_2)/9n$

Where

V = Velocity of fall (cm sec^{-1})

g = acceleration due to gravity (cm sec^{-2})

r = 'equivalent' radius of particle (cm),

d_1 = density of particle (g cm^{-3})

d_2 = density of medium (g cm^{-3}), and

n = viscosity of medium (dyne sec cm^{-2})

Stolon A slender branch or runner which grows as a prostrate stem, giving rise to new roots and aerial shoots or leaves in vegetative propagation.

Stoloniferous Bearing stolons or runners or shoots that take root.

Stoma Opening at the epidermis, especially on the under surface of the leaves, through which pass gases and water vapor. Stomata is plural of stoma.

Stomopteryx nertaria See groundnut leaf miner; and soybean insect pest.

Stone dust Artificial sand produced from stone.

Stone gall A clay concretion found in certain sandstones.

Stoneline A concentration of stones, boulders, gravels or concretions which occur in the soil and appears in profile as a horizontally disposed line or layer.

Stoner The stoner is essentially a modified gravity separator designed to make a two part separation by differences in specific gravity.

Stones Rock fragments greater than 25 cm (10 inches) in diametre if rounded and greater than 38 cm (15 inches) along the greater axis if flat. In engineering practice these fragments are included with boulders, which are considered to be greater than 20 cm (8 inches) in diametre.

Stone wort The common name for algae comprising the class Charaphyceal, so named because most species are lime-encrusted.

Stoniness The relative proportion of stones in or on the soil. The term is used in the classification of soils.

Stony Containing sufficient stones to interfere with or prevent tillage. To be classified as stony, more than 0.1 per cent of the surface of the soil must be covered with stones. The term is used to modify soil class, as stony clay loam or clay loam, stony phase.

Stony land Areas containing sufficient stones to make the use of machinery impractical; usually 15 per cent to 90 per cent of the surface soil is covered with stones. See also stoniness.

Stoody mold Dark fungus growing in insect honeydew.

Stool A group of stems arising from a

single plant. The stem and its attached tillers.

Stop valve A valve that can be opened or closed to regulate or stop the flow of fluid in a pipe.

Storage irrigation system Irrigation system in which water supplies are drawn from storage reservoirs. The canal carrying these may be either perennial, non-perennial or a combination of both.

Storm (1) A meteorological disturbance accompanied by such phenomena as rain, snow, hail, thunder and lightening, and wind, although wind storms may be described in accordance with the material, such as dust, sand which is carried in suspension in the air. (2) A meteorological disturbance which is either unusual or of great magnitude, rate, or intensity, ranging from tornadoes (1 km across) to extra tropical cyclones (2-3000 km across).

Storma (pl. Stromata) Mass of fungus hyphae often including host tissue containing or bearing spores.

Storm centre The area of lowest atmospheric pressure of a cyclone, this is a more general expression than eye of the storm, which refers only to the centre of a well developed tropical cyclone, in which there is a tendency of the skies to clear.

Storm distribution pattern or **storm smear** The manner in which depth of rainfall varies from station to station throughout the area for a given storm.

Storm drain A drain which conducts storm surface, or wash water or drainage after a heavy rain from a building to a storm or a combined sewer. Also known as storm sewer.

Storm runoff That portion of the total runoff from storm rainfall which reaches the point of measurement within a relatively short period of time subsequent to the occurrence of rain. It essentially excludes the base flow.

Stover Mature, cured stalks of such crops as corn or sorghum from which grain has been removed, a type of roughage.

Straight canals Canals having straight or slightly curved alignment as is typical of canals in flat plain.

Strain An organism or group of organisms differing in origin or minor aspects from other organisms of same species or variety.

Strain building The improvement of cross-pollinated plants by any one of several methods of selection. Applied especially to forage plants.

Strategy The mechanisms by which an organism's genotype is preserved (cf. r-strategist, K-strategist).

Stratification The arrangement of sediments in layers or strata marked by a change in colour, texture, dimension of particles, and composition. Stratification usually means layer of sediments that separate readily along bedding planes because of different sizes and kinds of material or some interruption in deposition that permitted changes to take place before more material was deposted.

Stratified drift (or sorted drift) Materials that are distinctly sorted according to size and weight of their component fragments, indicating a medium of transport (water or wind) more fluid than glacier ice.

Stratified estuary An estuary with a surface fresh water layer flowing out and a heavier saltwater flowing in.

Stratum, plural strata A layer characterised by certain unifying characteristics, properties, or attributes distinguishing it from adjacent layers.

Stratus Continuous horizontal sheet of cloud.

Straw Grain stalks after threshing and usually mixed with leaves and chaff. A stem of grain, such as wheat or oats.

Strawberry A low growing perennial of the genus Fragaria, order Rosales, that spreads by stolons; the juicy, usually red, edible fruit consists of a fleshy receptacle with numerous seeds in pits or nearly superficial on the receptacle.

Strawberry virus disease Virus disease characterised by leaves turning yellow red, getting crinkled and developing brownish purple spots.

Straw-walker A set of reciprocating notched bars inside a thresher or combine that push the straw to the rear.

Stream (1) A body of flowing water. The term is usually applied to water flowing in a natural surface channel, but is also applied to water flowing in an open or closed conduit, and to a jet of water issuing from an opening sometimes also for a body of underground flowing water. (2) A well-defined sea or oceanic current.

Stream flow A type of channel flow, applied to surface runoff moving in a stream.

Stream flow depletion The amount of water that flows into a valley, or into a particular land area, minus the amount that flows out of the valley or off the particular land area.

Stream routing The flood routing of a stream when the only storage is that furnished by the stream channel and valley.

Strelitziaceae A family of monocotyledonous plants in the order Zingiberales distinguished by perfect flowers with five functional stamens and without an evident hypanthium, penniveined leaves, and symmetrical gaurd cells.

Streptococcus Any of several spherical bacteria typically arranged in chains.

Streptomycin An antibiotic obtained from streptomyces griseus.

Stress A directional force acting within a material.

Strike The voluntary and collective withdrawal of labour from a place of employment. Opposite of lockout.

String of tile A single line of tile that receive no branches.

Strip cropping The practices of growing crops that require different types of tillage, such as row and sod, in alternate strips along contours of across the prevailing direction of wind.

Strip survey Estimating timber by strips running through the stand.

Strobilus A cone-like structure made up of sporophylls, or spore-bearing leaves, as in Equiselales.

Stroke The motion of a piston accomplished by the 'throw' of the crankshaft.

Stroma A compact, cushion-like mass of mycelium in which perithecia are produced.

Strontium A bivalent metallic element of the calcium group found in plants and soils only in traces. It is not essential element.

Stropanthus A genus of woody climbers of the dogbane family (Apocynaceae), natives of tropical Asia and Africa, the source of arrow poisons and of strophanthin.

Strophiole A crest like excrescence around the hilum in certain seeds.

Structure classes soil A grouping of soil structural units or peds on the basis of size.

Structure grades, soil A grouping or classification of soil structure on the basis of inter and intra aggregate adhesion, cohesion, or stability within the profile. Three grades of structure designated from 1 to 3 are: (i) Weak structure frible poorly formed, indistinct peds, barely evident in place. (ii) Moderate structure moderate well formed, distinct peds, moderately duration and evident, but not distinct, in undisturbed soil. (iii) Strong structure forte durable peds that are quite evident in undisturbed soil, adhere weakly to one another withstand displacement, and become separately when the soil is disturbed.

Structure index Any measurement of a soil physical property such as aggregation, porosity, permeability to air or water or bulk density, that denotes or indicates the structural condition of a soil.

Structure soil The combination or arrangement of primary soil particles into secondary particles, units, or peds. These peds may be, but usually are not, arranged in the profile in such a manner as to give a distinctive characteristic pattern. The peds are characterised and classified on the basis of size, shape, and degree of distinctness into classes, types, and grades. See also structure classes, soil; structure grades, soil and structure types, soil.

Structure types, soil A classification of soil structure based on the shape of the aggregates or peds and their arrangement in the profile.

Stubble The basal portions of plants that remain after cutting off the upper parts.

Stubble mulch The stubble of crops or crop residues left essentially in place on the land as a surface cover before and during the preparation of the seedbed and at least partly during the growing of a succeeding crop. Synonymous with trash cover.

Stud bolt A bolt threaded at both ends with blank space between to permit gripping with a pipe wrench.

Stuffing box A recessed chamber through which piston rods or valve stems pass, surrounded the rein with packing, and used to prevent leakage.

Stunt Any of several plant diseases marked by reduction in size of the plant.

Style In flowers, a cylindrical structure which rises from the top of an ovary and through which pollen tubes grow. In corn it is known as the 'Silk'.

Stylopodium A conical or disk-shaped enlargement at the base of the style in plants of the family Umbelliferae.

Sub-arid or **semi arid** A term applied to an area or climate, neither strictly arid nor strictly humid, in which inferior crops can be grown without irrigation. According to Thornthwaite, areas having moisture index between 20 and 40 Thornthwaite.

Sub-class A primary division of a class.

Subepidermal Beneath the epidermis.

Suber Cork tissue.

Suberin A waxy, waterproof substance deposited in the walls of certain types of cells, such as cork cells.

Suberized Containing suberin.

Sub-group soil A category in the Canadian classification system. These soils are subdivisions of the great groups and therefore, each soil is defined more specifically.

Sub-humid or **semi-humid** A term applied to an area or climate that has, on the whole, sufficient moisture to support all crops but irregularity of precipitation during the year makes it essential to provide irrigation facilities to raise better crops. According to Thornthwaite, areas having moisture index above 0.

Sub-irrigation See irrigation methods.

Sub-lateral or **Branch** A line collecting the water from the soil and discharging it into a lateral.

Sublimation The passage of water vapour directly into the solid state, by passing the liquid state.

Sub-main A subsidiary drain which discharges into a main drain and which serves as an outlet to all the laterals which discharge into it. Also a pipeline taking off from the main line and feeding sprinkler laterals.

Submerged soil A raised, horizontal formation along the bottom of an estuary mouth separating the estuarine basin from the ocean; 'the lower lip' of the estuary mouth.

Submerged weir A dam which, when in use, has the down stream water level at an elevation equal to or higher than the crest of the dam.

Submersion irrigation Controlled flood irrigation in which water applied remains ponded on the soil for a time sufficient for its infiltration to the desired depth form the surface of the land. This includes basin irrigation or check irrigation.

Sub-minors or **Quarternary canals** Canals taking off from minors and supplying water to outlets or turnouts.

Submitted sample The sample submitted to a seed testing laboratory. It comprises the composite sample reduced as necessary.

Subsequent drainage Drainage by a stream developed subsequent to the system of which it is a part; drainage follows belts of weak rocks.

Subsere A secondary community that succeeds on interrupted climax.

Subsidy A government contribution to cost incurred by public on many items.

Sub soil Soil below plow depth in weak profile soils. The B horizon of soils with distinct profiles.

Sub-soiling The breaking of compact sub-soils, without inverting them, with a special knifelike instrument (Chisel), which is pulled through the soil usually at depths of 30 to 60 cm (12 to 24 inches) and spacings of 60 to 150 cm (2 to 5 ft). Also called chieseling.

Subspecies A division of a species.

Substitution The replacement of a commodity or service by another commodity or services.

Substitution growth See growth, substitution.

Substrate (1) That which is laid or spread under; an underlying layer, such as the sub-soil. (2) The substance, base, or nutrient on which an organism grows. (3) Compounds or substances that are acted upon by enzymes or catalysts and changed to other compounds in the chemical reaction.

Substratum The material upon which seeds are placed for a germination test.

Subsurface drain, Underdrain, Covered drain or **Buried drain** A conduit below the surface of the ground.

Subsurface drainage, Underdrainage, Subsoil drainage, Underground drainage, or **Covered drainage** Drainage, either natural or artificial beneath the surface of the earth.

Subsurface flow Underground flow. Generally referred to seepage into streams, drains or river.

Subsurface geology The study of structure, thickness of rock formations beneath land.

Sub-surface tillage A method of stirring the soil with blades that leaves stubble on or just below the surface.

Subtend To live adjacent to and below another structure often enclosing it.

Subterranean Developing or living beneath the surface of the ground.

Subtropical Bordering on the tropical, the geographic region between the temperate and tropical area.

Subulate Linear, delicate, and tapering to a sharp point.

Sub-watercourses Ditches, flumes or pipelines taking off from the watercourse from which water is taken and applied to the field.

Subwatershed A watershed subdivision of unspecified size that forms a convenient natural unit.

Succession The progressive development of the vegetation towards its highest ecological expression, the climax. The replacement of one plant community by another.

Succession of crops (1) Growing a crop over a long season by either repeated sowings or a single sowing of varieties of the crop that mature at different rates. (2) Growing two or more crops in a season on the same land by planting them in succession.

Succulent Describing a plant having juicy, fleshy tissue.

Sucker An off-shoot that develops from an adventitious bud located on the roots or lower stem of a plant. Usually at the expense of the plant.

Sucker Tillers or shoots produced from a crown or rhizomes.

Suction (1) For proper penetration and efficient work by the mouldboard plough, some clearances are provided in the plough which are called suction. See horizontal and vertical suction. (2) The suction head as in a centrifugal pump. (3) Force exerted by the plant to take water from the soil.

Suction adapter An eccentric adapter which allows the use of a suction pipe larger than normal.

Suction pump A pump used to draw water in which suction is created by a piston and valves in a vertical or horizontal cylinder.

Sudan Grass Sorghum sudanense (Piper) Staff, family Gramineae. Annual grass grown as fodder utilised as hay, silage or pasture crop.

Sudan grass diseases Sudan grass ascochyta leaf spot caused by Ascochyta sorghi, sudan grass red leaf spot caused by Colletotrichum graminicolum and sudan grass sooty stripe caused by Ramulispora sorghi are few disease.

Suction stroke The intake stroke on which the fuel mixture is drawn into the cylinder.

Suffrutescent Of or pertaining to a stem intermediate between herbaceous and shrubby, becoming partly woody and perennial at the base.

Suffruiticese Low stems which are woody, grading into herbaceous at the top.

Sugarbeet Beta vulgaris. A beet characterised by a white root and cultivated for the high sugar content of the roots.

Sugarcane Saccharum officinarum. A stout, perennial grass plant characterised by two-ranked leaves, and a many-joined stalk with a terminal inflorescence in the form of a silky panicle, the source of more than 50 per cent of the world's annual sugar production.

Sugarcane-Diseases Sugarcane brown leaf spot caused by Cercospora longipes; sugarcane eye spot caused by Helminthosporium sacchari, sugarcane gummosis caused on Xanthomonas vasculorum; sugarcane pineapple disease caused by Ceratostomella paradoxa; sugarcane red rot caused by Glomerella tucumanensis Ark and Muller and Colletotrichum falcatum Went; sugarcane red stripe disease caused by Pseudomonas rubrilineans; sugarcane rind disease caused by Pleocyta sachhari, sugarcane rust caused by Puccinia kuehnii and P. erianthi; sugarcane stem canker caused by Vstilago sacchari; sugarcane whip smut caused by Ustilago scitaminea; sugarcane wilt caused by Cephalosporium sacchari, sugarcane yellowing disease (a physiological disease); sugarcane yellow leaf spot caused by Cercospora kopkei and virus diseases such as sugarcane grassy shoot and sugarcane musaic are the sugarcane diseases.

Sugarcane-insect pests Sugarcane black bug (Macropes excavatus Dist.) whose nymph and adult suck sap; sugarcane bunga (Aeginetia indica), a plant parasite on roots; sugarcane leaf hopper (Pyrilla perpusilla Wlk.) Whose nymphs and adults suck sap, sugarcane mealy wings known as sugarcane white fly (Aleurolobus barodensis Mark). fly and nymphs suck sap; sugarcane mite (oligonychus indicus Hirst), red mite feeds on underside of leaves; sugarcane pink mealy bug (Saccharicoccus sacchari Ckll). whose nymphs and adults suck juice; sugarcane root borer (Emmalocera depressella Swinh) which dries up central shoot; sugarcane shoot borer (Chilotraea in fuscatellus Shell) whose caterpillar bores into stem; sugarcane stalk borer (Chilotraea auricilia Ddgn.) whose caterpillar feeds on unopened leaves and bores into stem; sugarcane stem borer (Bissetia steniella Hmps.) whose caterpillar attack cane 60-120 cm above ground; sugarcane termites which feed on buds of planted sugarcane sets are some insect pests.

Sulfating The formation of lead sulfate on the plates of lead-acid storage batteries reducing the energy-storing ability of the battery and eventually causing failure.

Sulfofication, or **sulphofication** The oxidation of sulphur (sulphur), either free or from organic or inorganic compounds, into oxides of sulphur largely through the action of microorganisms. In the soil these oxides are quickly changed into sulphites or sulphates.

Sulphur Yellow mineral substance used as an amendment to reclaim alkali soils. Plant food element absorbed as sulphate (SO_4) ion.

Sulphur test It is conducted to measure the quantity of free and combined sulphur which is corrosive to metals. A sample of fuel is burnt in test bombs of wick lamps to measure the sulphur.

Summation curve (of particle sizes) A curve showing the accumulative percentage by weight of particles within increasing or decreasing size limits as a function of diametre; the percentage by weight of each size fraction is plotted accumulatively in the ordinate as a function of the total range of diametres represented in the sample plotted on the abscissa.

Summer annual A plant that completes its life cycle during a summer period, such as soybeans.

Summer savory Satureia hortensis (Calamintha hortensis L.) Family Labiatae. A herb used for flavouring food products.

Summerwood The less porous, usually harder portion of an annual ring that forms in the latter part of the growing season.

Sundew Any plant of the genus *Drosera* of the family Droseraceae; the genus comprises small, herbaceous, insectivorous plants that grow in all continents, especially Australia.

Sunflower Helionthus annuus. An annual plant native to the United States characterised by broad, ovate leaves growing from a single, usually long (3-20 feet or 1-6 metres) stem, and large, composite flowers with yellow petals.

Sunscald An injury to woody plants

which results in local death of the plant tissues; in summer, it is caused by excessive action of the sun's rays; in winter, by the great variation of temperature on the side of trees that is exposed to the sun in cold weather.

Sunspot A dark area in the photosphere of the sun caused by a lowered surface temperature.

Superadiabatic lapse rate An environmental lapse rate greater than the dry adiabatic lapse rate.

Supplemental heat Adding a small amount of heat for a limited temperature rise, usually less than 20°F, to dry within the maximum permissible time before spoilage.

Supply The amount of a commodity or service which will be offered for sale at a given price per unit of time.

Supply scheduled A table giving the supply of a commodity or service at various prices, while all other factors determining supply remain unchanged.

Surface-cum seepage drains Drains which, though carrying a continuous seepage discharge, are designed to deal also with storm water in the area they traverse.

Surface detention Water in temporary storage as a thin sheet over the soil surface during the occurrence of overland flow.

Surface drainage Preventing unwanted surface water from getting into surface of soil, or removing it from soil surface.

Surface drains Drains normally used for the removal of excess irrigation water from the lower portions of the field, or for the disposal of storm waters.

Surface irrigation Application of water to the soil by means of pipes or furrows along the surface.

Surface runoff See runoff.

Surface sealing The orientation and packing of dispersed soil particles in the immediate surface layer of the soil to render the surface fairly impermeable to water.

Surface slope, or **slope** (1) The inclination of the water surface expressed as change of elevation per unit of slope length. (2) The size of the angle which the surface makes with the horizontal. The tangent of the angle is ordinarily used, no appreciable error resulting except for the steepest slopes. Also called 'Water slope' in Australia.

Surface soil The uppermost part of the soil that is ordinarily moved in tillage, or its equivalent in uncultivated soils. It changes in depth from 7.5 cm to 25 cm (3 to 10 inches) and is frequently designated as the "plow layer," the Ap layer," or the "Ap horizon".

Surface storage The part of precipitation retained temporarily at the ground surface as interception or depression storage so that it does not appear as infiltration or surface runoff either during the rainfall period or shortly thereafter. Also known as initial detention, surface retention.

Surface tension The inward pull exerted on the particles of water surrounding soil particles in the soil which keep them at a stretch; increases in the increase in the curvature of the surface of the soil particles.

Surface tier The upper 40 cm (16 inches) of peat.

Surface water All bodies of water on the surface of the earth.

Surfacing Installation of impervious material over the land.

Surficial geology Geology of the characteristic of the material pertaining to, formed on, situated at, or occurring on, the earth's surface, especially consisting of unconsolidated residual, alluvial, or glacial deposits lying on bed-rock.

Surge bin (Holding Bin) The bins mounted above the processing machine.

Surinam cherry Eugenia uniflora L. A fruit tree of family Myrtaceae. Also known as Brazil cherry.

Surplus In the nationalised industries, another term for "profit".

Survey designs Survey designs are distinguished from experiments, in that the former try to measure, even if the study had not taken place, while the latter measure the effects of responses to stimuli deliberately introduced. Examples of survey designs are 'simple random sampling', 'proportional stratified sampling' 'cluster sampling' 'systematic sampling' and 'area sampling'.

Survey for as-built record drawings Surveys to collect the information of any changes, additions or deletions to that shown on sanctioned drawings or contract drawings which were used for construction of the project or any of its features.

Survey, Forest An inventory of forest land to determine area, condition, timber volume, and species for specific purposes such as timber purchase, forest management, or as a basis for forest policies and programmes.

Survey for structural behaviour Surveys performed after the completion of the structure to determine what distortion, deflection, or other movements take place in the structure as a whole, or its component parts, due to the various loads imposed by the storage of water, temperature change or uplift pressure.

Survey soil (Pedology) the systematic examination, description, classification, and mapping of soils in an area. Soil surveys are classified according to the kind and intensity of the field examination.

Suscept A living organism attacked by, or susceptible to, a given disease or pathogen; in many cases a more precise term than host but less familiar.

Suspense chart or **strip-tease chart** The full story is depicted in writing or in a picture, and is kept covered with strips of paper. As we talk we remove the strips of paper as we proceed in our talk. These are fixed with tape so that they can be removed easily.

Suspension (1) A state wherein finely divided solid particles are dispersed but not dissolved in a medium. (2) A type of movement caused by wind in wind erosion.

Sustained yield A continual annual or periodic yield of plants or plant material from an area; implies management practices that maintain the productive capacity of the land.

S-value Sum of exchangeable Ca, Mg, Na and K, expressed in meq/100 g soil.

Swale A lowying area frequently moist or marshy; an intermittent drainage way; a slough.

Swamp An area saturated with water throughout much of the year, but with the surface of the soil usually not deeply submerged. It is generally characterised by tree or shrub vegetation. See also marsh.

Sward Sward, turf, and sod may seem nearly synonymous. Sward usually refers to the grassy surface or a pasture, lawn, or playing field, Vegetation may be of pure or mixed species; turf and sod refer to the stratum of earth filled with grass roots as well as to the surface.

Swarmspore See zoospore.

Swath A strip of cut herbage lying on the stubble.

Swelling clays Refer either to clay minerals such as the smectites that exhibit interlayer swelling when wetted, or to clayey soils which, on account of the presence of swelling clay minerals, swell when wetted and shrink with cracking when dried. The latter are also known as soils which heave.

Sweet cherry Prunus avium L. Deciduous tree of family Eurasia which produces globe shaped cherry.

Sweet clover Melilotus parviflora Desf. (M. indica All.) Family Papilionacea. Legume fodder with low water requirement. Also used as green manure.

Sweet clover-diseases Sweet clover bacterial disease caused by Xanthomonas alfalfae, sweet clover downy mildew caused by Peronospora, aestivalis or M. meliloti and sweet clover powdery mildew caused by Erysiphe polygoni are few common diseases.

Sweet corn Zeamays var. saccharata. A group of maize characterised by translucent horny appearance of seed when immature and wrinkled condition when dry, contains a glossy sweetish starch.

Sweet grain Liquidambar styracifulua. A deciduous tree of the order Hamamelidales found in the south-eastern United States, and distinguished by its five-lobed, or star-shaped, leaves, and by the corky ridges developed on the twigs.

Sweet lime Citrus limettioides Tanaka.

(C. medica var. Limella Wight and Arn.) Family Rutaceae. Lime mildly sweet with practically no acidity, used as root stock for sweet oranges and mandarins. Leaves yield essential oil.

Sweet orange Citrus Sinensis (L.) Osbeck. (Citrus aurantium, Aurantium sinensis Mill; C. aurantium Mill., citrus aurantium var. aurantium). A fruit tree of family Rutaceae. Produces sweet orange oil.

Sweet orange oil A sweet, yellow, mild essential oil expressed from the peel of an orange, citrus aurantium, soluble in glacial acetic acid, somewhat in alcohol; used in flavors, perfumes, and medicine. Also known as orange oil.

Sweet potato Ipomoea batatas. A tropical vine having variously shaped leaves, purplish flowers, and a sweet fleshy edible luberous root.

Sweet potato-diseases Few diseases such as sweet potato black rot caused by Ceratostomella fimbriata; sweet potato charcoal rot caused by Macrophomina phaseoli; sweet potato leaf spot caused by Cercospora batatae and C. bataticola and sweet potato soft rot caused by Rhizopus nigricans are common.

Sweet potato-insect pests Sweet potato leaf beetles (Aspidomorpha miliaris F., Metriona circum-data H., chridia sernotata B., Oncocephala tuberculata O.) which feeds on leaves; sweet potato sphinx (Herse convolvuii L.) whose caterpillar completely defoliates vines and sweet potato weevil (Cylas formicarius Fabr.) whose grub bores into stem and tubers making them rot are common insect pests.

Swept volume See piston displacement.

Swidden agriculture Also known as "slash and burn" or shifting cultivation or shifting agriculture, a type of agriculture involving the clearing of land by cutting and burning the vegetation to open the soil for planting, it is the major agricultural technique used by the people of Central and South America, Asia, Africa and other parts of the world. After limited use the land is abandoned, and a fresh clearing made elsewhere; thus only a fraction of land area is under cultivation at any one time. Most agriculturists consider swidden agriculture to be wasteful and inefficient, while large amounts of land are required to operate it.

Swollen seeds Seeds which have imbibed water and, although healthy in appearance, have not germinated during the prescribed test period.

Sword bean Canavalia gladiata (Jacq) D.C. (C. ensiformis Baker Non DC, Dolichos gladiatus Jacq.) Family Papilionaceae. Bean cultivated for cover and green manure crop. Pod used as vegetable. Also called jack bean or *sem* in Hindi.

Sword bean-diseases Two diseases which are common are, sword bean anthracnose caused by Glomerella lindemuthianum and Colletotrichum capsici and sword bean leaf spot caused by Cercospora canavaliae.

Syconium A fleshy fruit, as a fig, enlarged pulpy seceptacle internally lined with minute flowers.

Sylepta derogata See cotton leaf roller.

Syllepsis Development of a lateral branch without a period of dormancy, i.e. contemporaneous with its parent axis; hence sylleptic branch, a branch developed by syllepsis.

Sylvigenesis The process of forest formation.

Sylvigenetic cycle One of the repeating processes in sylvigenesis.

Sylvigenetic model The analogue of a tree model in the development of a forest.

Sylvigenetic phase A sub-division of a sylvigenetic cycle, not necessarily repeated.

Symbiosis (1) An inter-relationship between two different species. (2) An inter-relationship between two different organisms in which the effects of that relationship is mutually advantageous (not parasitic). Also known as consortism.

Symbiosis relationship Mutual helpfulness. Organisms so dependent on each other that they cannot live apart.

Symbiotic bacteria Organisms that live in the nodules on the roots or inoculated legumes and take nitrogen from the soil air and supply it to the host plant on which they live.

Symbiotic nitrogen fixation Fixation of atmospheric N by rhizobia growing in nodules on roots of legumes.

Sympatric Of a species, occupying the same range as another species but maintaining identity by not interbreeding.

Symphile An organism, usually a beetle, living as a guest in the nest of a social insect, such as an ant, where it is reared and bred in exchange for its exudates.

Sympodium A branching system in trees in which the main axis is comprised of successive secondary branches, each representing the dominant fork of a dichotomy. Sympodial growth, growth from successive lateral meristems.

Symmetry concentration The quantity of cations or anions equivalent to the exchange capacity of a soil. For example, if the cation exchange capacity of a soil is 10 meq/100 gm of soil, then 1 symmetry concentration is 10 meq of any cation.

Symmetry value The percentage of the adsorbed ion released when one symmetry concentration of another ion is added.

Synanthous Flower and leaves appearing simultaneously.

Synandrour Having several united stamens.

Synangium A group of Sporangia developed as a single structure typical of the fruiting structure of certain ferns.

Synapsis The conjugation of homologous chromosomes.

Syncarpous Descriptive of a gynoecium having the carpels united in a compound fleshy fruit.

Synchytrium endobioticum See Potato - diseases.

Syncytium Mass of protoplasm with several nuclei but forming one cell.

Synecology The study of environmental relations of groups of organism, such as communities.

Synergid Either of two small cells lying near the egg at the micropylar end of the embryo sac in an ovule.

Synergism The ability of two or more organisms to bring about changes (usually chemical) that neither can accomplish alone.

Syngamy Union of gametes in fertilization. Also synagamy.

Synnema (pl. Synnemata) Groups of hyphae sometimes joined together, generally upright and producing spores; coremium.

Synthesis The building up of elements and simple compounds into complicated tissue material by the microorganisms.

Synthetic cultivar Advanced generation progeny of a number of clones or lines, or of hybrids among them, obtained by open pollination.

Synthetic unit hydrographs A unit graph developed on the basis of estimation of coefficients expressing various physical features of a catchment.

Synthetic variety A term used particularly with cross-pollinated plants to refer to a variety produced by the combination of selected lines or plants and subsequent normal pollination.

Systemic disease Term applied to disease in which single infection leads to general spread of the pathogen throughout the plant body; or to a chemical that acts through the vascular system.

Systems ecology The combined approaches of systems analysis and the ecology of whole ecosystems and subsystems.

T

Table land Extensive elevated region with level surface; Plateaux.

Table top terrace Bench terrace sloping inward or outward; with or without longitudinal slope.

Table, Volume A table showing the average contents of trees by diametre and merchantable length, in a specified unit of volume.

Tabular Arranged in, computed by means of tables.

Tabular interpolation Method of finding from a table the values of the dependent variable for intermediate values of the independent variable.

Tabulate Arrange in tabular form.

Taccavi Indian term for loan for agricultural purposes advanced by Government.

Tachometre An instrument that measures the revolutions per minute or the angular speed of a rotating shaft.

Taeniothrips caradamoni See cardamom thrips.

Tagua palm Phytelephas macrocarpa. A palm tree of tropical America; the endosperm of the seed is used as an ivory substitute.

Taiga A zone of forest vegetation encircling the Northern Hemisphere between the arctic-sub-arctic tundras in the north of the steppes, hardwood forests, and prairies in the south.

Taiga climate In general, a climate which produces taiga vegetation, that is, too cold for prolific tree growth but milder than the tundra climate and moist enough to promote appreciable vegetation. Also known as sub-arctic climate.

Tail (1) The end of a canal. (2) The term is also applied to the work constructed at the end of a canal for the distribution of water there at. (3) Tail feathers in a bird.

Tail drain, or **Waste ditch** A furrow or small ditch at the lower edge of an irrigated field for collecting surplus surface water for either redistribution or removal.

Tail gate The downstream gate of a canal.

Tailings Partly threshed material that has passed through the coarse shakers or straw walkers of a threshing machine, and has passed over the fine sieve.

Tail reach The last reach of a canal, i.e. the portion from the last control point to the tail.

Tail tank A reservoir supplied with water from a canal whenever in excess of canal requirements, having its own command and usually situated near the tail of a canal.

Tail water Water level at the head of the watercourse or farm lateral just below the outlet.

Tail grass prairie Area in the western corn belt originally covered with tall native grasses, or other similar areas.

Take The uniting of a scion with a stock after grafting or budding.

Take root To establish a healthy and normal root system.

Talc A finely ground powder from the mineral talc; used as a carrier for poison dusts as a filler.

Talinum triangulare See Ceylone spinach.

Tall coconut palm The ordinary or common variety of coconut grown in India. Long lived and starts bearing in 8-10 years after plantation.

Talud A short, steep slope formed gradually at the downslope margin of a field by deposition against a hedge, a stone wall, or other similar barrier.

Talus A sloping heap of loose rock fragments lying at the foot of a cliff or steep slope.

Tamarind Tamarindus indica L. Family Caesalpiniaceae. Plants grown as avenue; fruits edible. Unripe fruits source of tartaric acid. Polyose obtained from seeds good substitute for fruit proteins. Bark and leaves used for tanning.

Tamarindus indica See tamarind.

Tamarisk Any tree or shrub of the genus Tamarix, family Tamaricaceae. Also called Tamarix.

Tamarix aphylla See athel tamarisk.

— **articulata** See athel tamarisk.

— **dioica** Tamarix dioica Roxb. Branches used for making baskets.

— **Troupii** Tamarix troupii Hole. (Tamarix gallica anct. non L.) A bushy shrub which has galls on its branches which are used in tanning.

Tan (1) To convert raw hide into leather. (2) A coat colour in some animals.

Tan disease A physiological disorder of plants caused by excess of moisture in the soil. Swollen roots and aerial portions and peeling or breaking of outer bark are some of the characteristics in this disease.

Tangelo A tree that is hybrid between a tangerine or other mandrin and a grape fruit or shaddock; produces an edible fruit.

Tangerine Any of several, trees of the species *Citrus reticulata;* the fruit is a loose-skinned mandrin with a deep orange or scarlet rind.

Tank Natural or artificial pond for storing water.

Tankage The contents of a storage tank. Slaughter-house entrails and scraps used as fertiliser.

Tank agriculture Hydroponics. Agriculture with the help of tank water.

Tannic acid Acid obtained from some trees, soluble in water. Also called gallic acid.

Tannin Broad class of soluble polyphenols with a common property of condensing with protein to form a leather like substance that is insoluble and of impaired digestibility.

Tap To make a cut in the bark for obtaining the sap.

Tap bolt A bolt usually threaded for its entire length. It is finished only on the point and the underside of the head. Tap bolts are made with both square and hexagon heads.

Taper A gradual and uniform decrease in size, as a tapered socket, a tapered shaft, a tapered shank.

Tapetum One or more layers of nutritive cells surrounding the sporogenous tissue in a young sporangium.

Tapioca A food, high in starch, that is made from the cassava plant. Manihot utilissimma Pohl. (M. esculenta Crantz)

Tapoun A root segment.

Tappet rod A rod carrying a tappet or tappets, as one for opening or closing the valves in a steam or an internal combustion engine.

Tapping (1) Forming an internal screw thread in a hole or other part by means of a tap. (2) Collecting sap by making cuts as in turpentine, rubber etc.

Tappoon A portable dam or stop in the form of a canvas on a frame or metal panel put across ditches when it is desirable to use ditches in sections, filling new parts lower down as the irrigation progress.

Tappet A reciprocating part between the cam and the valve.

Tappet valve Valve with a disk head from which extends a spindle to serve as a pilot. Commonly used in internal combustion engines.

Tap root The single main root which grows downward and from which branch roots rise.

Tare The weight of the packing or container of an article. This must be deducted from the gross weight to find the actual or net weight of the article.

Target spot Any plant disease characterised by lesions in the form of concentric markings.

Tarragon Artemisia dracunculus L. family compositae. A prennial herb grown as a source of extract which is used in *Salads.* The extract is aromatic, pungent and flavouring.

Tartaric acid An acid abundantly occurring in plants and used in dyeing, photography and medicines.

Tartory buckwheat One of three buckwheat species grown commercially; the leaves are narrower than the other two species and arrow-shaped, and the flowers are smaller with inconspicuous greenish-white sepals. Also known as duck wheat and hullers buck-wheat.

Tassel (1) The flower cluster at the tip of a corn plant comprising pollen bearing flowers. The staminate inflorescence of maize. (2) The staminate or male inflorescence produced at the top of the plant.

Taxales A small order of gymnosperms in the class Pinatae; members are trees or shrubs with evergreen, often needle like leaves, with a well-developed fleshy covering surrounding the individual seeds, which are terminal or sub-terminal on short shoots.

Taxable income That part of income liable to tax. It is arrived at by deducting from assessable income all allowable deductions.

Tax avoidance Term used to describe not only illegal evasion of tax obligations but also the entirely legal ways by which a person may so arrange his affairs as to minimise, or even eliminate, tax liability on his property and income.

Taxes A compulsory financial contribution by a person or body of persons towards the expenditure of a public authority.

Tax, Marginal rate of The amount of tax a person pays on the last unit or increment of income received.

Taxon (Plural texa) A class at any categorical level of a taxonomic system of classification.

Taxonomist A scientist who classifies plants and animals according to their morphological relationships.

Taxonomy The science of the classification and arrangement, according to relationships, of living organisms.

Tax, regressive A tax which takes a larger percentage of income from people with smaller income.

'T' budding See shield budding.

T.D.N. Total digestible nutrient.

Tea Camellia sinensis (L.) O. Kuntze (C. thea Link. C. theifera Griff. Thea sinensis L.). Evergreen bush of family Ternstroemiaceae cultivated for its tender leaves. Leaves are processed to prepare tea leaves used for preparing common beverage also known as tea.

Tea bagworm Clania crameri. Insect pest whose caterpillars defoliate tea plantation.

Tea birds eye spot Disease caused by cercosporella theae. Plants in nursery are mainly affected with minute black or brown spots appearing on young flush.

Tea black root rot Disease caused by Rosellinia spp. leading to black strands of woolly mycelium on roots.

Tea black rot Disease caused by corticium invisum. The tender leave turns black and soft and gets rotten.

Tea blister blight Disease caused by Exobasidium vexans Massee. Small yellow spots on leaf upper surface; lower surface turns into greyish white spore bearing convex.

Tea branch canker Disease caused by fungi leading to gnarled formation on branches.

Tea brown blight Disease caused by collectotrichum camelliae Yellow green spots appear on leaves, stems etc. which turn dark brown later.

Tea brown root disease Disease caused by Fomes noxius. Leaves turn yellow; roots show white and brown mycelium covering.

Tea charcoal rot Disease caused by Ustulina zonata. Bushes die gradually.

Tea copper blight Disease caused by Guignardia camelliae. Large irregular spots appear on leaves.

Tea die back Disease caused by Nectria cinnabarina. Bushes and stem become thin, small pink cushionlike fruit body appear after shoots die.

Teak Tectona grandis L.f. A very useful timber tree of family Verbenaceae. Thrives on a variety of soils but require well drained condition. Wood used for multifarious activities.

Tea looper Biston suppressaria Guen. Destructive insect pest on tea whose caterpillar eats away leaves.

Tea mosquito Helopeltis theifera Wlk. H. antonii sign. Nymph and adult suck sap from young leaves and buds. Punctured leaves develop brown small spots which turn black giving out brown fluid.

Tea pink disease Disease caused by Pellicularia salmoni-colour. Young woody branches lose leaves and showing die back.

Tea red mite Tetranychus bioculatus wood mason; Oligonychus coffeae

Niet. Nymphs and adult such plant sap.

Tea red rust and **algal red spot** Disease caused by Cephaleuros parasiticus. Water soaked green spots and yellow hairs appear on leaf and stems; hardening of twigs are the major characters.

Tea red slugs Caterpillars of Heterusia cingala or H. magnifica defoliate tea plants.

Tea stump rot Disease of tea caused by Irpex destruens. Rotting of stumps of bushes occur.

Technical efficiency The combination of the best and most modern production techniques, efficient management and a highly skilled work-force.

Technician, Forestry One who is familiar with the principal field activities connected with the practice of forestry. Usually trained in vocational, technical, or ranger school as distinguished from professional forestry course.

Tedder A kicking implement that loosens hay in the swath or windrow to facilitate drying.

Teff love grass Eragrostis tef (zuccagni) Trotter (E. abessinica Link). A fodder grass of family Gramineae. Grains are also edible.

Tee A fitting for connecting pipes of unequal sizes, or for changing direction of pipe runs. A bullhead tee is one with outlet larger than opening on run; straight tee has all openings of same size.

Telementeorometry The study of the making meteorological observations at a distance.

Telemetre (1) The complete measuring, transmitting, and receiving apparatus for indicating or recording the value of a quantity at a distance. Also known as telemetreing system. (2) To transmit the value of a measured quantity to a remote point.

Teliospore Winter or resting form of rust spore, from which basidium is produced.

Telium Sorus producing teliospores.

Telome A morphological unit consisting of the stalks or a stalk and a sporangium or two sporangia.

Temperate fruit Fruits grown at elevations of 1220 m or above.

Temperament This is the characteristic mood of an individual which may bring elation, depression or confidence in him.

Temperate and cold sarannah A regional vegetation zone, very extensively represented in North America and in Eurasia at high altitudes; consists of scattered or clumped trees (very often conifers and mostly needle-leaved evergreen) and a shrub layer of varying coverage, masses and, even more abundantly, lichens form an almost continuous carpet.

Temperate and cold scrub Regional vegetation zone whose density and periodicity vary a good deal; requires a considerable amount of moisture in the soil, whether from mist, seasonal downpour, or snowmelt, shrubs may be evergreen or deciduous; and undergrowth of ferns and other large leaved herbs are quite frequent, especially at sub-alpine level, wind shearing and very cold winters prevent tree growth. Also known as bosque; tourre; heath.

Temperate belt A belt around the earth within which the annual mean temperature is less than 29°C (68° F) and the mean temperature of the warmest month is higher than 10°C (50°F).

Temperate climate The climate of the middle latitudes; the climate between the extremes of tropical climate and polor climate.

Temperate fruit Fruits grown at elevations of 1220 m or above.

Temperate mixed forest A forest of the North Temperate zone containing a high proportion of conifers with a few broadleafed species.

Temperate rainforest A vegetation class in temperate areas of high and evenly distributed rainfall characterised by comparatively few species with large populations of each species, evergreen are somewhat short with small leaves, and there is an abundance of large tree ferns. Also known as cloud forest; laurel forest; laurisilva; moss forest; subtropical forest.

Temperate wood land A vegetation class similar to tropical wood land in spacing, height and stratification, but it can be either deciduous or evergreen, broad-leaved or needle-leaved. Also known as parkland, woodland.

Temperature The amount of heat or cold measured in degrees. Farenheit (F) and centigrade (C) are commonly used units. At sea level water freezes at 0° (32° F) and boils at 100° C (212° F).

Temperature belt The belt which may be drawn on a thermograph trace or other temperature graph by connecting the daily maxima with one line and the daily minima with another.

Temperature constant In germination testing, a specific temperature which should not vary by more than 1°C.

Temperature control A control used to maintain the temperature of an oven, furnace, or other enclosed space within desired limits.

Temperature inversion An increase of air temperature with height, so that warmer air overlies colder, contrary to the normal lapse rate.

Tempering (1) Equalisation of moisture or temperature throughout kernel or products. (2) To bring grain to a desired moisture or temperature for processing. (3) Treating tools etc. to impart hardness or softness to metals.

Temporary pasture One grazed for only a short time.

Tenant One who rents in most of his cultivated land and cultivates less than the average peasant.

Tendril A slender, coiling organ which aids in the support of plant stems. A tendril may be a modified stem leaf, leaflet, or stipule.

Tenor The period of time stated on a bill of exchange before payment is due.

Tensile Of, or pertaining to extension or tension, as tensile action, tensile strength; capable of being stretched or drawn out.

Tensile strain A strain or pull in a longitudinal direction; the reverse of a crushing strain.

Tensile strength The strength necessary to enable a bar or structure to resist a tensile strain. The amount of directly applied pull which is a part will stand before it breaks. It is expressed as the number of pounds of force required to break a bar 1 sq. in. in area.

Tensile stress The stress to which a bar or structure is subject when in tension.

Tensiometre A device for measuring the tension of water in soil in situ; a porous, permeable ceramic cup connected through a tube to a manometre.

Tension A pulling force; the opposite of compression.

Tension spring Any spring designed to be operated under a pulling strain.

Teosinte Euchlaena mexicana schrad. Tall annual tropical grass of family Gramineae resembling maize plant. Grown as fodder. Tolerant to water-logging.

Terete Of a stem, cylindrical in section, but tapering at both ends.

Tergeminate A leaf having at its base a pair of leaflets and forking with a pair on each branch.

Terminal bud A bud that develops at the apex of a stem. Also known as apical bud.

Ternate Composed of three subdivisions, as a leaf with three leaflets.

Terraces (1) A flat, level or nearly level, narrow, area of land bordering a river bunded on at least one side by a definite steep slope rising upward from it. (2) A low embankment or ridge of earth constructed across a slope to control surface runoff and minimise soil erosion. (3) Sloping ground cut into a succession of benches and steep inclines for purpose of cultivation. Often the inclines are made quite steep and are protected by riprap, or retaining walls are substituted, thus giving greater areas for cultivation. (4) Areas bordered by low broad ridges constructed on cultivated land of such alignment; height, and spacing as to conform to the topography and to permit travel by cultivating and harvesting machinery, the object being to prevent loss of soil by erosion.

Terramycin An antibiotic obtained from Streptomyces rimosus.

Terrestrial frozen water Seasonally or perennially frozen waters of the earth, exclusive of the atmosphere.

Terrestrial photogrammetry Photogrammetry with the aid of ground photographs.

Terrestrial radiation The long wave radiation emitted from the Earth's surface.

Terric layer An unconsolidated mineral sub-stratum underlying organic soil material.

Tertiary The geological period which precedes the Quaternary.

Testa The external coating of a seed.

Tester animals Used in grazing experiments to measure animal performance or pasture quality.

Tetrachloroethane. (CH_2CCl_4). A chemical used as soil fumigant, nemacide and herbicide.

Tetrad A group of four cells produced by meiotic division of a mother cell: also, a group of four chromosomes formed at meiosis.

Tetraploid An organism whose cells contain four haploid (Monoploid) sets of chromosomes.

Tetrasporangium A sporangium in which the spores are borne in tetrads as in the Dictyotales.

Tetraspore In certain algae, one of the nonmotile spores produced in groups of four.

Textural classification See texture, soil.

Texture Term to indicate size and distribution of elements as determined on a cross section.

Texture, Soil The relative proportions of the various soil separates in a soil as described by the classes of soil texture. The names of textural soil classes may be modified by adding suitable adjectives when coarse fragments are present in substantial amounts; for example, "Gibraltar stony sandy loam", "or" Saint-Jovite silt loam, stony phase". For other modifications see coarse fragments. The sand, loamy sand, and sandy loam are further sub-divided on the basis of the proportions of the various sand separates present.

Thallobionta One of the two sub kingdoms of plants, characterised by the absence of specialised tissues or organs and multicellular sex organs.

Thallophyte One of the simpler plants, belonging to the algae, bacteria, fungi, slime molds or lichens.

Thallus Vegetative body of a thallophyte.

Thatch grass Saccharum spontaneum L. A deep rooted perennial grass of family Gramineae. It is hardy and could withstand drought, frost and disease. Love wet conditions.

Thaw To melt, in agriculture relates to melting of ice, snow, frost etc. in soils.

Theaceae A family of dicotyledonous erect trees or shrubs in the order Theales; characterised by alternate, extipulate leaves, usually five petals, and mostly numerous stamens.

Theales An order of dicotyledonous mostly woody plants in the sub-class Delleniidae with simple or occasionally compound leaves, petals usually separate, numerous stamens, and the calyx arranged in a tight spiral.

Thea sinensis See tea.

Themeda grass Themeda triandra forsk. (Anthistiria imberbis Retz.). A prennial grass of family Gramineae. Good fodder, grains also edible.

Themeda triandra See themeda grass.

Theophrastaceae A family of tropical and sub-tropical dicotyledonous woody plants in the order Primulates characterised by flowers having staminodes alternate with the corolla lobes.

Thermal analysis (or differential thermal analysis) A method of analysing a soil sample for constituents, based on the differential rate of heating of the unknown and standard samples when a uniform source of heat is applied.

Thermal belt Any one of several possible horizontal belts of a vegetation type found in mountainous terrain, resulting primarily from vertical temperature variation. Also known as thermal zone.

Thermal burn Tissue reaction to or injury resulting from application of heat.

Thermal conductivity The rate of transfer of heat to or from a point in the soil.

Thermal convection Atmospheric currents, predominantly vertical, arising from the release of gravitational visibility, commonly produced by solar heating of the ground; the cause of convective (cumulus) clouds. Also known as free convection; gravitational convection.

Thermal Convection Storm A storm

caused by local inequalities in the temperature in which the rainfall is of high intensity short duration, and limited to only a smaller area.

Thermal death point Degree of heat necessary to kill a liquid culture of a given species of bacteria in 10 minutes.

Thermal fog applicator A spraying device in which item to be sprayed is finely atomized by a blast of heated air or gas.

Thermal induction The change in growth and development of plants brought about by exposure to a given temperature, usually applied to the process resulting in flowering of biennial plants.

Thermo-couple A device that uses the voltage developed by the junction of two dissimilar metals to measure temperature differences.

Thermodusic bacteria Bacteria which can tolerate high temperatures but do not enjoy.

Thermogenic Soils Soils with properties that have been influenced primarily by high soil temperature, as the dominant soil formation factor, developed in sub-tropical and equatorial region.

Thermoperiodicity Effects of temperature differences between light and dark periods upon plants. It is the totality of response of a plant to appropriately fluctuating temperatures.

Thermophile An organism that grows best at temperatures of 50°C or higher.

Thermophilic bacteria Bacteria which enjoy or require high temperatures.

Thermopile A transducer for converting thermal energy directly into electrical energy. It is composed of pairs of thermocouples which are connected either in series or in parallel.

Thermophyte A plant that is tolerant to heat.

Thermosequence A sequence of related soils that differ one from the other, primarily as a result of temperature as a soil formation factor.

Theta pinch A device for producing a controlled nuclear fusion reaction, in which plasma in a long tarus or skimmy tube is confined by a magnetic field produced by current-carrying coils, and in shock-heated and compressed by pulses in this field to produce the high temperatures at which fusion reactions take place, the magnetic field is then sustained in order to maintain the plasma confinement.

Thiessen polygon The points of rain gauges on a map are joined by straight lines and their perpendicular bisectors are drawn. The polygon formed round each rain gauge station by these perpendiculars is known, after its originator, as Thiessen polygon.

Thickener A food adjunct, as gelatine added to food preparations to improve the viscosity or to give it a firm body.

Thinking The process of using symbols to manipulate implicitly ideas or objects not physically present, to the senses is called thinking. These symbols include words, numbers, gestures, pictures and visual images.

Thinning Cutting in an immature stand to increase its rate of growth, to foster quality growth, to improve composition, to promote sanitation, to aid in litter decomposition, to obtain greater total yield, and to recover and use material that would be lost otherwise.

Thin soil Soil lacking in depth; unproductive.

Thin stand Small number of plants in an area or poor establishment of plants.

Thionate dust A pesticide dust containing a compound of sulphur.

Thomas phosphate basic slag also known as thomas slag.

Thistle Any of the various prickly plants comprising the family compositae.

Thixotropy A property of a material that permits it to become solid on standing for a short time, but on agitation its consistence becomes soft or it changes to a highly viscous fluid. It is a reversible process.

Thorn A sharp pointed aborted branch, generally considered as of endogenous origin, as opposed to the more superficial spine.

Thorny amaranth Amranthus spinosus L. Family Amaranthaceae. An annual herb used as a vegetable and a fodder. It is drought resistant.

Thorny bamboo Bambusa arundinacea (Retz.) Willd. (Bambusa spinosa Roxb.) Family Gramineae. A tall woody bamboo growing throughout India. Culms used for making paper and young buds and grain as food.

Thornthwaite Moisture Index or **Moisture Index** In the system developed by Thornthwaite, an index based upon the adequacy of rainfall to the need of the plants.

Thread blight A fungus diseases of a number of tropical and semitropical woody plants, including cocoa and tea, caused by species of Pellicularia and Marasmius which form filamentous mycelia on the surface of twigs and leaves.

Three-point suspension A system of automobile engine mounting in which the weight is carried at only three points.

Three-way Cross A cross between a single-cross hybrid and an inbred line.

Thermatology Science of breeding animals and plants.

Thresher A machine that separates grain or seeds from straw. Also known as threshing machine.

Threshold moisture content The minimum moisture condition measured in terms of either moisture content or moisture stress, at which biological activity just becomes measureable.

Thriftiness The extent to which future, as distinct from present consumption, is valued.

Thrifty Designating a plant or animal which is thriving. A productive soil.

Thrip injury Injury due to thrip infestation. Terminal buds often destroyed; leaves appear torn, cupped or distorted. Plant is more susceptible to infection.

Thrips Very small insect of the order Thysanoptera. Harmful to plants.

Thrips tabaci See onion thrip.

Throat, or **Bottle neck** A contracted section of a channel or structure or constricted section of a closed conduit with a minimum flow section which is always less than the normal cross section.

Throat width The minimum contracted/constricted width of the throat section of a structure or open conduit/closed conduit.

Throttle Choke; the control which regulates the flow of fuel to the engine.

Throttle flow Flow which is forced to pass through a restricted area, where the velocity must increase.

Throttle valve A choking device to regulate flow of a liquid, for example, in a pipeline, to an engine or turbine, from a pump or compressor.

Thunderstorm Storm associated with lightening and thunder.

Thur soil Kallar soil.

Thyme A perennial mint plant of the genus Thymus; pungent aromatic herb is made from the leaves.

Thymelaeceae A family of dicotyledonous woody plants in the order Myrtales characterised by a superior ovary with a solitary ovule and petals, if present are scalelike.

Thyrse An inflorescence with a racemose primary axis and cymose secondary and lateral axes.

Thysanoptera Order of insects to which thrips belongs.

Tidal bench mark A bench mark set as reference to a tide staff at a tidal station and the elevation of which is determined with relation to the local tidal datum.

Tidal bore High crested roaring wave caused by rushing of flood tide up a river or estuary or by meeting of tides.

Tidal current Current caused by the tide-producing forces of the moon and the sun and is a part of the same general movement of the sea that is manifested in the vertical rise and fall of the tides.

Tidal datum A datum defined by a phase of the tide.

Tidal flats Areas of nearly flat, barren mud periodically covered by tidal waters. Normally these materials have an excess of soluble salt. A miscellaneous land type.

Tidal flood current Current due to flood.

Tidal flood interval Time between transit of moon over meridian of a place and the time of the following strength of flood.

Tidal flood strength Tidal flood current at the time of maximum velocity.

Tidal marsh Low, flat marsh land traversed by interlacing channels and tidal sloughs and usually inundated by high tides.

Tidal outlet, or **Tidal sluice** A structure to control the outflow of water.

Tidal range Difference in height between high and low tidal water level.

Tide The periodic rising and falling of the water that results from the gravitational attraction of the moon and sun acting upon the rotating earth. The term tide should not be used to designate the horizontal movement of the water which is the tidal current.

Tide day The interval between the arrival of any two consecutive high waters of the direct tide (or of the opposite tide) at any given place, its average length being 24 hours 51 minutes.

Tideflat An unvegetated inter-tidal area.

Tide gate An opening through which water may flow freely in one direction, but which closes automatically and prevents the water from flowing in the other direction.

Tide gauge A gauge for showing the height of the tide.

Tide head The head of the stretch of land up to which the influence of the tide is felt or the utmost limit of land coming within the zone of influence of the tide.

Tide land land which is inundated during flood tide.

Tide mark A high water, or rarely low water, mark left by tidal water or a flood, the point it reaches; also, a mark placed to indicate this point.

Tide rip A shearing of two adjacent currents causing a noticeable surface discontinuity.

Tide tables Tables which give daily predictions of the times and height of the tide.

Tide-way A channel in which the tide runs; also, a rush of tidal water through a channel or stream.

Tie A connection from a point of known position of a point whose position is desired. A tie is made to determine the position of a supplementary point whose position is desired for mapping or reference purpose, or to close a survey on a previously determined point. To 'tie in' is to make such a connection.

Tied shop A shop of which the proprietor is financed by a company on condition that he supplies only the products of that company.

Tie point Point of closure of a survey either on itself or on another survey.

Tier (1) A group of closely inserted branches or branch complexes (whorled or pseudowhorled) widely separated from other group of branches. (2) A depth sub-division used in the classification of organic soils.

Tie rod The transverse rod connecting the front wheels of an automobile in order to permit them to act as a unit in steering.

Tige Stem, stalk.

Tight knot Almost synonymous with live knot.

Tight soil A compact, relatively impervious and tenacious soil or sub-soil, which may or may not be plastic.

Tight times Times prescribed by management for the completion of certain tasks, which are barely adequate.

Tile drain Pipe placed at suitable depths and spacings in the soil or subsoil to provide water outlets from the soil. The pipe may be concrete, ceramic, fibre, plastic, or any other suitable materials. When it is a continuous pipe such as plastics, it is commonly known as pipe drain.

Tile hook Hook-shaped tool used to lower tile into the trench.

Tile joint Opening between two drain tiles through which water from the surrounding soil flows.

Tile storage Volume of water that can be stored in the tile main without reducing the effectiveness of the tile drains.

Till An unstratified glacial deposit of clay, silt, sand, gravel, and boulders covering part of the rock surface in those regions which were glaciated during the ice age.

Tillage The use of implements through the soil to prepare seedbeds and root beds for crops. Also the operation, practice or art of tilling land to improve it for agricultural purposes.

Tiller (1) A branch arising from the base of a monocot plant, especially in the grass family. (2) A sprout that rises up from a tree stump. (3) One who tills land.

Tillering Development of more than one stem on a single plant from near the base of the primary stem.

Tillering stage A growth stage.

Tilletia caries See wheat bunt.

— **foetida** See wheat bunt.

Tiling Installation of drain tile, usually for an entire field.

Tilth The physical condition of soil as related to its ease of tillage, fitness as a seedbed, and impedance to seedling emergence and root penetration.

Till, to To plough and prepare for seeding; to seed or cultivate the soil.

Timber A commercial wood, often in a converted form.

Timber line The highest extension of woody or forest growth on the mountains above which there is no further tree growth.

Timber marking The marking of trees in a forest stand which are to be cut or retained in a cutting operation.

Timber lining A type of lining consisting of mud sills laid at right angles to the direction of flow and planed plank flooring spiked to mud sills. The mud sills may rest upon and may be held down by bearing piles or be held down by rod to masonry anchors.

T.S.I. (Timber stand improvement) Usually applied to intermediate cutting.

Time and motion study An analysis of the methods, materials, tools and equipments used, or to be used, in the performance of a piece of work with the purpose of (a) finding the most economical way of doing the work; (b) standardising the methods, materials, tools and equipment; and (c) accurately determining the time required by a qualified and properly trained person working at a normal pace to complete the task.

Time efficiency The fraction of the total time available during which a plant is in productive operation.

Time factor The ratio of the number of days the canal is in flow to the base days.

Time lag The difference in time by which one observation lags behind (is later than) another. Contrast 'time lead'.

Time of concentration Time required for water to flow from the most remote point of a watershed, in a hydraulic sense to the outlet.

Time of one half response Time required to obtain a change in moisture content from the original moisture content half of the way to equilibrium, usually expressed in hours.

Time series A time series is a set of ordered observations on a quantitative characteristic of an individual or collective phenomena taken at different points of time. The essential quality of the series is the order of observations according to the time variable.

Timing (1) The adjustment of valves and crankshaft of an engine in their relative position to produce the greatest effective output in h.p. (2) The point at which ignition breaker is set in relation to top dead-centre position of the piston to secure the greatest effective expansive force upon head of piston.

Timing gear The gears that drive the camshaft so that the valves operate in correct time with the pistons. since the camshaft has to turn once for every two revolutions of the crankshaft, these gears must be in the ratio of 2 is to 1.

Tine A tooth, spike or prong.

Tined harrow Implement used for cultivating and interculturing, tines are rigid or adjustable.

Tip (1) The exposed part of the wool fibre in the fleece. (2) The end of a branch or twig. (3) To remove the ends of branches, stem, twigs etc.

Tip burn A disease in plants in which leaf margins and tips burn or turns brown.

Tip cutting Soft wood cutting.

Tip layering A method by which tips of the shoots are used to grow a new plant.

Tire (Tyre) Band of rubber placed round rim of wheel to strengthen it or prevent jar.

Tire tool A flat bar of iron or steel commonly spoken of as tire iron used for applying and removing tires. Any tool used for the same purpose can be considered a tire tool.

Tissue A structure in animals or plants body consisting of cells which are separated by intercellular space; Any fine woven fabric.

Tissue culture Cultivation of tissue cells away from the human or animal body.

Titanium oxide TiO_2, one of the more common mineral in the earth's crust; present in soil but its role in plant growth is undetermined.

Titer Concentration of infective microben in a medium; amount of one substance to correspond with a given amount of another substance.

Toadstool The fruiting structure of a fungus of class Basidiomycetes, order Agaricales. Popularly (but not scientifically) the name is applied to poisonous species of mushroom non-edible ones.

Tobacco Nicotiana tabacum L. Nicotiana rustica L. (1) Any plant of the genus *Nicotinia* cultivated for its leaves, which contain 1-3% of the alkaloid nicotine. (2) The dried leaves of the plant.

Tobacco bacterial wilt Disease caused by Pseudomonas solanacearum. Drooping and wilting of lower leaves occur. When stem is cut exudation of slimy bacterial matter is received.

Tobacco black shank Disease caused by phytophthora parasitica Dastur. Formation of dark brown or black lesions on basal part of stem, appearance of soft rot in stem, dying off of seedlings are some characters.

Tobacco borer beetle Lasioderma serricorne Fabr. Beetle and grubs bore into tobacco in storage, cigars, cigarettes etc.

Tobacco caterpillar Prodenia litura Fabr. Insect pest whose caterpillars feed on leaves and defoliate plants.

Tobacco damping off. A disease caused by Pythium aphanidermatum and P. debaryanum.

Tobacco frog-eye leaf spot Disease caused by cercospora nicotianae. Brown lesion on leaves appear. Dying of seedlings also occur.

Tobacco ground beetle Tenebrionidae sp. Beetle pest grows at the stems just above the ground level.

Tobacco leaf caterpillars Insect pest of tobacco. Laphygma exigua and Prodenia litura cause damage to foliage of crop.

Tobacco leaf curl Virus disease characterised by curling, dwarfing and distortion of leaves.

Tobacco mosaic Any of a complex of virus disease of tobacco and other solanaceous plants in which the leaves are mottled with light-and dark green patches, some times interspersed with yellow.

Tobacco powdery mildew Disease caused by Erysiphe cichoracearum. White powdery patches appear.

Tobacco stem borer Gnorimoschema heliopa. Caterpillar of the insect pest bores and reaches stem. Plants get stunted and start unusual branching at top.

Tobacco stem rot Disease caused by Pellicularia rolfsii. Yellowing of plant occur.

Tocopherol Any of a group of four chemically related compounds differing slightly in structure that constitute vitamin E.

Tolerance Ability of a plant to thrive under adverse circumstances.

Tolerant (1) Ability of tree to grow in shade of other trees. (2) The plants which grow in adverse circumstances.

Tolerant Capable of sustaining disease without serious injury or crop loss.

Tolerance limits Where in a fixed probability tolerance limits are the limits within which 100 (1-x) per cent of the values must fall.

Tomato Lycopersicon esculentum Mill. Solanum lycopersicum L.) Family Solanaceae. A warm season crop. Rich in vitamin A and Calcium. Also contains vitamin C. Grows well in loam soil with good clay sub-soil. In arid and semi-arid regions irrigation is essential.

Tomato Diseases Tomato bacterial leafspot caused by Xanthomonas vesicatoria; tomato bacterial disease caused by Pseudomonas solanacearum; tomato blossom-end rot, tomato damping off and foot rot caused by Pythium or Phytophthora spp.; tomato early blight caused by Alternaria solani; tomato foot rot and wilt caused by Pellicularia rolfsii; tomato foot rot caused by Phytophthora spp.; tomato fusarium wilt caused by Fusarium oxysporum; tomato late blight caused by phytophthora infestans; virus diseases tomato leaf curl and tomato mosaic; tomato leaf mould caused by Cladosporium fulvum; tomato powdery mildew caused by Macro-phomina phaseoli and tomato stem and fruit canker caused by Corynebacterium michiganense are some of the common diseases of tomato.

Tomentose Covered with densely matted hairs.

Ton A unit of weight generally equal to 2000 lbs. Long ton is equal to 2240 lbs.

Tongue grafting A method of grafting in which tongue shaped cuts are given in the scion and the stock similar to whip grafting.

Toothed bur clover Medicago denticulata Willd. A common herb of family Papilionaceae used as fodder.

Tonoplast The membrane surrounding a plant cell vacuole.

Top cross See Inbred-variety cross.

Top cross progany From out-crossed seed or selections. clones, or lines crossed with a single variety or line that serves as a common pollen parent.

Top dead centre The dead-centre position of an engine piston and its crankshaft arm when at the top or outer end of its stroke.

Topiary The art of clipping shrubs into ornamental shape.

Top necrosis The relatively rapid killing of the entire top of plant.

Top dressing Application of fertiliser anytime after establishment of a crop.

Top grafting Grafting a scion of one variety of tree on to the main branch of another.

Topographic climax A climax plant community under a uniform macroclimate over which minor topographic features such as hills, rivers, valleys, or undrained depressions exert a controlling influence.

Topoclimatology The study of local climate as affected by slope aspect, and surface characteristics.

Topographic map A map which presents the horizontal and vertical positions of the features presented; distinguished from a planimetric map by the addition of relief in measurable form. A topographic map usually shows the same features as a planimetric map but uses contours or comparable symbols to show mountains, valleys and plains, and in the case of hydrographic charts-symbols and numbers to show depths in bodies of water.

Topography The physical features of a district or region, such as those represented on a map, taken collectively; especially, the relief and contours of the land.

Topophysis The organisational status of meristem in a plant body determined by its position and stable through vegetative propagation.

Toposaic A photomap on which topographic or terrain-form lines are shown as on topographic quadrangles. See also photomap and planisaic.

Toposequence A sequence of related soil that differ, one from the other primarily because of topography as a soil formation factor.

Topsoil (1) The layer of soil moved in cultivation. (2) The A horizon. (3) The Ah horizon. (4) Presumably fertile soil material used to top dress roadbanks, gardens, and lawns.

Top width, or **surface width** Designates the width of the cross section at the water surface.

Top work A procedure employed to propagate seedless varieties of fruit and hybrids, to change the variety of fruit, and to correct pollination problems, using any of three methods, root grafting, crown grafting and top grafting.

Torpedo Channel forming head of a mole plough.

Torpedograss Panicum repens L. A grass of family Gramineae used for colonising coastal flats.

Tornado Circular whirl of great intensity and small horizontal extent, with winds or hurricane force, accompanied by heavy rain or hail, thunder and lightening; it is shortlived, usually not lasting more than an hour or two, but causes heavy damage.

Torque The turning effort of a rotor (Engine). The turning effort or twist which a shaft sustains when transmitting power.

Torque arm An arm or pair of arms used to prevent any turning of the rear-axle housing when power is applied to the rear axle.

Torque converter A specially designed transmission used to increase torque at the rear wheels, providing a quicker pick up.

Torque stand A test stand on which the engine torque is measured.

Torrent A stream of water flowing with great velocity or turbulence, as during a freshet or down a steep incline.

Torrid zone Part of earth's surface between the tropics.

Total correlation The correlation between the original data rather than between residuals after some common variation has been abstracted.

Total digestible nutrients(TDN) Sum total of all digestible organic nutrients, i.e., proteins, nitrogen free extract, fibre, and fat. Fat is multiplied by 2.25 to put its energy value on the same basis as the other nutrients. On the average for all feeds 1 g of TDN = 4.4 kcal (kilo-calorie).

Total dynamic head Head required to pump water from source to the point of discharge, equal to the static lift plus the friction head plus the velocity head in metres.

Total potential of soil water The amount of work that must be done per unit quantity of pure water in order to transport reversibly and isothermally an infinitesimal quantity of water from a pool of pure water, at a specified elevation and at atmospheric pressure to the soil water at the point under consideration. Total potential of soil water is sum total of matric potential and the osmotic potential.

Total pressure The pressure (positive or negative) in relation to the external gas pressure on the soil water, to which a pool of pure water must be subjected in order to be in equilibrium through a semi-permeable membrane with the soil water. Total pressure is therefore, equal to the sum of soil water pressure and osmotic pressure. Total pressure may also be derived from the measurement of the partial pressure of the water vapour in equilibrium with the soil water. It may be identified with the total potential defined above when gravitational and external gas pressure potentials can be neglected. See Water, soil.

Total radiation pyrometre A pyrometre which focuses heat radiation emitted by a hot object on a detector (usually a thermopile or other thermal type detector), and which responds to a broad band of radiation limited only be absorption of the focusing lens, or window and mirror.

Total runoff The runoff, from all sources, from a drainage area for a definite period of time or for the duration of a particular storm.

Tote road A term used for smaller logging roads in some parts of the country.

Toxic (Poisonous). Tending to reduce the yields of crops below the normal.

Toxic element (Irrigation water) Elements in irrigation water which when present in excess of a threshold value, cause reduction in yield. Boron, flourine, sodium, chloride and in some cases even nitrate are the few toxic elements present in irrigation water.

Toxin Poison formed by an organism.

Trabecula A plate of cells bridging or separating an intercellular space, as the separating or partitioning cell plates formed in certain sporangia.

Trace element (Obsolete) See micronutrient.

Tracheal tube A xylem vessel; a xylem element lacking end walls, typical of angiosperms.

Tracheid An elongate, spindle-shaped xylem cell, lacking protoplasm at maturity, and having secondary walls laid in various thicknesses and patterns over the primary wall.

Tracheophyta A large group of plants characterised by the presence of specialised conducting tissues (xylem and phloem) in the roots, stems, and leaves.

Traction efficiency It is defined as the ratio of the drawbar horsepower to the horsepower input to the rear axle of the tractor. However, it is generally used as a ratio of the drawbar pull to the weight of the tractor.

Traction sprayer/duster A spraying or

dusting machine which gets power from the wheels on which these are mounted.

Trade association A body of persons formed for the purpose of furthering the trade interests of its members.

Trade bill A commercial bill used in transactions actually involving goods.

Trade mark The use of a brand name or other device to relate a commodity to the firm owning, producing or distributing it.

Trading account A business account the purpose of which is to indicate gross profit on trading alone, without taking into account the overhead expenses.

Tradition The sum of all the ideas, habits and customs that belong to the people and are transmitted from generation to generation; the social 'heritage'.

Traffic pan See pressure or induced pan.

Trailed type implements Those implements whose weight is not supported by the tractor and are pulled and guided from single hitch point are known as trailed type implements.

Trailer A travelling plant.

Trailing plug Plug following the mole drain torpedo, smoothing and strengthening the walls of the mole channel.

Train To grow a plant to desired shape by pruning, cutting or by supporting its branches to grow in desired way.

Trainer A tree intermediate in size which shades lower branches of adjacent larger trees.

Training wall A wall built along the bank of a river or estuary parallel to the direction of flow to direct and confine the flow.

Trait A synonym of character with respect to function and performance, but less so with respect to form.

Transect Cross-section of an area usually used as a sample in surveying operations or studying spatial variability. Also for recording, mapping and studying the vegetation of an area.

Transfer cost What must be offered to attract a supply or a factor of production away from alternative avenues of employment. This sum of money is the necessary supply price of that factor in its new employment.

Transfer deed A legal document by which ownership of stocks or shares is transferred from a holder to a purchaser.

Transfer earnings The amount that any factor of production could earn in the best paid alternative use.

Transfer payments Payments made by public authorities which are not made in consideration of goods and services currently produced.

Transgressive segregation The appearance in the F_2 (or later) generation of individuals showing a more extreme development of a character than either parent. Assumed to be due to cumulative and complementary effects of genes contributed by the parents of the original hybrid. Adequate testing of variation in the parents is required to establish its occurrence.

Transitional soil A soil with properties intermediate between those of two different soils and genetically related to them.

Transitory or permanent(Group) It may vary from a group of momentary interest in a passing event to the relatively permanent interests as in a family.

Translocation Movement of water, mineral salts, and organic substances from one part of a plant to another. The transfer of a chromosome segment from its usual position to a new position in the same or in a different chromosome.

Translocation The change in position of segment of a chromosome to another part of the same chromosome or of a different chromosome.

Transmissibility The rate or flow of groundwater, at the prevailing temperature, through a vertical strip of aquifer one foot wide with a height equal to the saturated thickness of the aquifer and under a unit hydraulic gradient. It is hydraulic conductivity times the aquifer thickness.

Transmission The name given to the arrangement of gears, contained in a housing attached to the rear end of motor, by which variations of speed ratios and forward and reverse motion are obtained.

Transparent Permitting passage of radiation or particles.

Transpiration efficiency The dry weight of plant product per 100 weight-units of water transpired.

Transpiration factor The ratio of the amount of moisture transpired by the crop to the amount of water consumptively used by the crop.

Transpiration The passage of a gas or liquid (in the form of vapor) through the skin; a membrane, or other tissue.

Transpiration The process by which water moves up through the living plant and vapour leaves the plant and enters the atmosphere. Only a small fraction of total irrigation water need of the plant.

Transplant To replant a nursery seedling in another part of the nursery for further development.

Transplanter A special kind of equipment designed for the planting of cuttings or small plants; it transport one or more men who assist the action of the machine in placing plants in a furrow and covering them; it commonly supplies a small quantity of water to each plant.

Transplanter An implement for transplanting.

Transported soil Soil formed in places other than at the site of its existence. The soil transported by water, wind etc. and deposited at other places.

Trapa bisponasa See water chestnut.

— **natans** See water chestnut.

Trap crop A crop grown to save the nearby crop from insect, pest. Trap crop attracts the insects, pests, fungi etc.

Trash rack Screen or grate at the intake of a channel, drainage or pump structure for the purpose of stopping debris.

Treatment Any process, physical or chemical, to which a seed lot is submitted.

Tree A perennial woody plant at least 20 feet (6 metres) in height at maturity, having an erect stem or trunk and a well-developed crown or leaf canopy.

Tree fern The common name for plants belonging to the family Hylidae characterised by expanded digital adhesive disks.

Treelet A small tree.

Tree of the future Tree with potential for future expansion within the forest mosaic.

Tree of the past Over nature, decaying or damaged tree in process of elimination from the forest mosaic.

Tree of the present Dominant tree which codetermines forest architecture, but without further capability for expansion.

Tree or stream chart In it, the divisions and sub-divisions are represented by: a trunk of a tree with branches and sub-branches, or a stream (river) with its tributaries or sub-tributaries.

Tree weed A tree of a species with relatively little or no value.

Tree-wolf A tree occupying more space than its silvicultural value warrants, curtailing better neighbours. A term usually applied to broad-crowned, short-stemmed trees.

Trellis drainage pattern Drainage pattern in which tributary streams flow nearly parallel to one another and join trunk streams at right angles. Similar to the pattern of a garden trellis. Also Grapevine pattern.

Trench excavator, Trencher, Ditcher, or Trenching machine A self propelled machine, usually on crawler tracks, specialised for digging trenches. It is operated by a bucket ladder excavator or by a chain or buckets on a wheel.

Trend A long-term movement in an ordered series, say a time series, which may be regarded, together with the oscillation and random component, as generating the observed values. In practice, trend is usually represented by some smooth mathematical function, such as polynomials in time variables.

Triangular notch weir A measuring weir with a V-shaped notch for measuring small flows. Also known as V-notch weir.

Triarch Having a primary xylem with three points in cross section.

Triassic The geological period that constitutes the earliest principal division of the Mesozoic era.

Tributary A stream that feeds or flows into or joins a larger stream or a lake. Also known as contributary, feeder, side stream; tributary stream.

Tributary water way Any body of water that flows into a larger body, that is, a creek in relation to a river, a river in relation to a bay, and a bay in relation to the open sea.

Trichogyne A hair-like extension of the carpogonium in red algae; also, a hair-like extension of the ascogonium in the ascomycetes.

Trichome A plant hair or hair-like appendage.

Trifoliate Having three leaves or leaflets.

Trigonous (1) Having three corners. (2) Having a triangular cross section.

Triploid An organism whose cells contain three haploid or monoploid sets of chromosomes.

Tripogon bromoides A fast tillering and spreading perennial grass forming a close, carpet like cover on ground. An effective soil binder.

Tripping Release of the staminate and pistillate portions from the keel of the leguminous flower in the process of pollination.

Triuridaceae A family of monocotyledonous plants in the order Triuridales distinguished by unisexual flowers and several carples with one seed per carpel.

Trivalent An association of three homologous chromosomes at meiosis.

Trophallaxis Exchange of food between organisms, not only of the same species but between different species, especially among social insects.

Trophozoite Active, motile, feeding stage of a protozoan organism.

Trophic Of or pertaining to food or feeding.

Tropical rainforest A vegetation class consisting of tall, close-growing trees, their columnar trunks more or less unbranched in the lower two-thirds, and forming a spreading and frequently flat crown, occurs in areas of high temperature and high rainfall.

Tropical year A unit of time equal to the period of one revolution of the earth about the sun measured between successive vernal equinoxes; it is 365.2422 mean solar days or 365 days 5 hours 48 minutes 46 seconds.

Tropics Any portion of the earth characterised by a tropical climate. The area of earth's surface lying between Tropic of Cancer and Tropic of Capricorn.

Tropic of cancer The latitude of 23°27′ N marking the southern limit that the sun appears directly overhead at noon.

Tropic tide The tide occurring semimonthly when the effect of the moon's maximum declination is greatest.

Tropic zone The region lying between the Tropic of Cancer and the Tropics of Capricorn.

Trophogenic Originating from nutritional differences rather than resulting from genetic determinants, such as various castes of social insects.

Tropholytic Pertaining to the deep zone in a lake where dissimulation of organic matter predominates.

True value The value of a quantity which is completely free from bias and errors. Because the errors to which physical measurements are subject cannot be known exactly, it follows that the true value of the quantity cannot be known with exactness.

Truncated Having lost all or part of the upper soil horizon or horizons.

Trust A combination or amalgamation of large firms exercising monopoly control over the commodity produced. Government legislation may be introduced to protect consumers.

Tsunami A great sea wave produced by submarine earth movement or volcanic eruption.

T test A method for testing the significance of a difference.

Tube nucleus One of the nuclei in a pollen tube, influencing the growth and behaviour of the tube and not involved in sexual fusion.

Tube outlet A subsurface drainage for removing water from a ponded area where the drains or field ditches are nor practicable.

Tuber Enlarged, fleshy, underground stem, commonly borne at the end of a rhizome with many buds or eyes.

Tube well A well drilled and fitted with tube and screens to draw water from the groundwater aquifer.

Tubular axle Axle made of steel tubing.

Tubular radiator A radiator made up of many tubes through which the water of the cooling system circulates.

Tuff Volcanic ash usually more or less stratified and in various states of consolidation.

Tumble weed Any of various plants that break loose from their roots in autumn and are driven by the wind in rolling masses over the ground.

Tundra An area supporting some vegetation between the northern upper limit of trees and the lower limit of perennial snow on mountains, and on the fringes of the Antarctic continent and its neighbouring islands.

Tundra soil Any soil in the tundra region.

Tunica The outer growth zone in a stem tip, composed of one to several layers of embryonic cells.

Tunica corpus The apical histogen of a stem tip consisting of an outer growth zone and the central embryonic growth zone.

Turbine pump A pump for lifting water. Usually good for high suction heads. The impeller is built on a vertical shaft which rotates on a compact casing or bowl; each unit is called a stage. Bowl submerged in water and hence priming is not required. For high lift number of stages are used.

Turbulence An irregular eddying flow, in contrast to a smooth laminar flow.

Turf A sod. The upper portion of soil filled with the roots and stems of low growing, living plants.

Turgid Swollen with water.

Turgor Distension of a plant cell wall and membrance by the fluid contents.

Turgor movement A reversible change in the position of plant parts due to a change in turgor pressure in certain specialised cells, movement of *Mimosa* leaves when touched is an example.

Turgor pressure The hydrostatic pressure in the liquid system within the plant cell.

Turion A scaly shoot, such as asparagus, developed from an underground bud.

Turmeric Curcuma domestica Valet. (C. longa Auct. non L.). An important condiment belonging to family zingiberaceae. It is also a useful dye. Grows on variety of well drained soils.

Turmeric leaf spot Disease caused by colletotrichum capsici. Oblong yellowish spots with greyish white centres on leaves appear.

Turmeric rhizome and root rot Disease caused by Phythium raminicolumn. Characterised by drying up of leaves along margins, rotting of roots and rhizomes.

Turning The process of moving seed through the air from one bin or storage structure to another, or back to the same storage.

Turnip Brassica rapa L. (B. compestris L. var. rapa Hartm.). A biennial plant of the mustard family (Family Cruciferae) whose fleshy root is used as vegetable. It is also a good cattle feed. Rich in vitamins A, B and C as well as in calcium and iron.

Turnip Brassica rapa or **B. compestris** var. *rapa*. An annual crucifer of Asiatic origin belonging to the family Brassiaceae in the order Capparales and grown for its foliage and edible root.

Turn interval, or **Frequency of turn** The lapse of time in days from the start of one irrigation to the start of the next to a particular co-sharer on roster system. He may apply the water of his turn to the same field or to another field.

Turnover The total receipts for goods sold by a trader during a given period.

Turnover ratio In a business, the ratio of the value of annual sales to the value of fixed investments.

Turnover tax A tax imposed on the sale of a commodity every time it changes hands.

Turn strip (1) Area for turning implements at the end of field. (2) A long narrow place of land between the collector ditch and the fence line of a field.

Turpentine The oleoresin derived from various pine species which is source of turpentine oil. The oil is used in paint and varnishes as well as in insecticides.

T-value Cation exchange capacity expressed as meq/100 g. The value can vary according to the method of analysis used. See cation exchange capacity.

Twiner A climbing stem that winds about its support, as pole beans.

Two-speed rear axle In this type, two speeds are incorporated in the axle, giving twice the number of speeds obtainable with the single speed axle. Especially adapted to trucks, making possible a saving in wear and tear on the engine and in gasoline consumption.

Two trunk kniffen system A system of training grape vines wherein one trunk reaches to the lower wire, the other to the top wire and a cane is tied to the right and to the left along each wire.

Two way plough See reversible plough.

Tylosis Cell outgrowth into cavity of xylem vessel plugging it.

Type, forest A descriptive term used to group stands of similar character as regards composition and development due to certain ecological factors, by which they may be differentiated from other groups of stands.

Type, soil (1) A unit in the natural system of soil classification; a subdivision of a soil series consisting of or describing soils that are alike in all characteristics including the texture of the horizon. (2) In Europe, the term is roughly equivalent to a great soil group.

Typhales An order of marsh or aquatic monocotyledons in the sub class commelimidae with emergent or floating stems and leaves and reduced, unisexual flowers having a single ovule in an ovary composed of a single carpel.

Typhoon A severe tropical cyclone in the western pacific.

Typing (classification) Determination of the category to which an individual, object microbe etc., belongs with respect to a given standard of reference.

Tyrosinase An enzyme capable of digesting certain amino acids.

TZ The term is applied to tetrazolium test for evaluating viability of seeds.

2, 4-D. 2, 4 - dichlorophenoxyacetic acid, a synthetic plant hormone used as weedicide.

2, 4, 5 - T. 2, 4, 5 - trichlorophenoxyacetic acid also a hormonal weedicide.

U

U-blades Curved bulldozer blades designed to increase moving capacity of tractor equipment.

U-bolt A U-shapped bolt with thread's at the ends of both arms to receive nuts.

U-clamp A clamp shaped like letter 'U', used for clamping down work on planer beds etc.

Udalf A suborder of the soil order Alfisol; brown soil formed in a udic moisture regime and in a mesic or warmer temperature regime.

Udert A suborder of the soil order vertisol; formed in a humid region so that surface cracks remain open only for 2-3 months.

Udoll A suborder of the Mollisol soil order; found in humid, temperate, and warm regions where maximum rainfall comes during growing season; has thick, very dark A horizons, brown B horizons, and paler C. horizons.

Udometre Rain gauge

Udult A sub-order of the soil order Ultisol; organic-carbon content is low, argillic horizons, are reddish or yellowish, formed in a udic moisture regime.

Uliginose Designating plants growing in muddy places.

Ullage The amount that a container, such as a fuel tank, lacks of being full.

Ulmaceae A family of dicotyledonous trees in the order Urticales distinguished by alternate stipulate leaves, two styles, a pendulous ovule, and lack of a latex system.

Ulmin Alkali-soluble organic substances derived from decaying vegetable matter; occurs as a morphous brown to black gel material.

Ulotrichaceae A family of green algae, in the suborder Ulotrichineae; contains both attached and floating filamentous species with cells having parietal, plate like or band like chloroplasts.

Ulotrichales A large artificial order of the chlorophyta composed mostly of fresh-water, branched or unbranched filamentous species with mostly cylindrical, uninucleate cells having cellulose, but often muci-laginous walls.

Ulotrichineae A suborder of the Ulotrichales characterised by short cylindrical cells.

Ultimate analysis Analysis of a substance so as to determine the compounds contained in it is known as proximate analysis where as analysis of a substance or a compound to determine the elements is ultimate analysis.

Ultimate strength The tensile stress, per unit of the original surface area; at which a body will fracture, or continue to deform under a decreasing load.

Ultracentrifuge A laboratory centrifuge that develops centrifugal fields of more than 100,000 times gravity.

Ultrafiche A sheet of film, usually 10 by 15 cm in size, containing either negative or positive images, or frames, of printed material reduced more than 100 times by photographic reduction.

Ultra filtration Separation of colloidal or very fine solid materials by filtration through microporous or semi-permeable mediums.

Ultramicroscrope An instrument for investigating particles of submicroscopic dimensions. It consists of a high intensity illumination system for producing a tyndall cone in a colloidal system, coupled with a compound microscope to examine the points of light scattered from the individual particles.

Ultramotivity Power of spontaneous movement.

Ultrasonic atomizer An atomizer in which liquid is fed to, or caused to flow over, a surface which vibrates at an ultrasonic frequency; uniform drops may be produced at low feed rates.

Ultrasonic depth finder A direct reading instrument which employs frequencies above the audible range to determine the depth of water; it measures the time interval between the emission of an ultrasonic signal and the return of its echo from the bottom.

Ultrasonic microscope A special type of microscope which employs ultrasonic radiation.

Ultrasonic soldering A method used for aluminium soldering by ultrasonically vibrating the soldering iron to disrupt the oxide film on the metal.

Ultrasonic thickness gauge A thickness gauge in which the time of travel of an ultrasonic beam through a sheet of material is used as a measure of the thickness of the material.

Ultratorming A proprietary catalytic reforming process used to increase the octane ratings of petroleum naphthas.

Ultraviolet imagery The imagery produced as a result of sensing ultraviolet radiations reflected from a given target surface.

Ultraviolet lamp A lamp providing a high proportion of ultraviolet radiation, such as various forms of mercury vapour lamp.

Ultraviolet light The invisible rays of spectrum composed of light waves just shorter than violet light, used in irradiation.

Umbel An arrangement of flowers or of polyps, with stalks springing from a common centre and forming a flat, concave, or convex surface.

Umbellifer Carrot etc., the plants belonging to the family umbelliferae.

Umbraculum Appendages shaped like an umbrella.

Umbrella kniffin system A system of training grapes; a two wire trellis is used with the trunk carried to or almost to the top wire and the canes carried obliquely from the head to the lower wire. Also called umbrella system.

Umbrella system See umbrella kniffin system.

Umbriferous Affording shade.

Unarmed A term referring plants which do not have thorns.

Unavailable A common term designating non-availability of a substance in the market or from some place; commonly refers to chemical, water etc. in a soil which are not in the available form so that the plants may make their use.

Unavoidable farm losses (water) Principally those losses which are due to normal field evaporation and deep percolation.

Uncap To remove the caps from the cells from a honeycomb to obtain honey.

Uncapping knife Plain or double jacketed steel knife used for uncapping honeycomb cells. Heated with steam before use.

Uncased observation holes Uncased holes usually for measuring the watertable elevations.

Uncertainty In economics, non-insurable business risk; the unpredictability of the outcome of much of business activity.

Uncinate Hooked; crooked.

Uncinula necato See grape vine powdery mildew.

Unconditional inequality An inequality which holds true for all values of the variables involved, or which contains no variables, for example, Y + 2 Y, or 4 > 3. Also known as absolute inequality.

Unconfined groundwater Groundwater not under hydrostatic pressure. The watertable rises or falls depending upon change in the volume of water surface.

Unconsolidated sedimentary rocks See sedimentary rocks.

Uncontrolled mosaic A mosaic composed of uncorrected photographs, the details of which have been matched from point to point without ground control or other orientation, accurate measurement and direction can not be accomplished.

Uncorrelated random variables Two random variables whose correlation coefficient is zero.

Unctuous Mineral substances which are greasy or soapy to touch.

Uncured Not cured; used for cement concrete work.

Uncured products Raw products; not

cured for use.

Under cut The notch made in the trunk of a tree below the level of the major cut and on the side to which the tree is to fall; a saw cut made on the underside of branch prior to actually making the primary cut to prevent splitting or tearing.

Undercutting Removing unwanted lower branches of a tree.

Unit normal A unit vector in the direction of the principal normal to a surface of space curve.

Undate Undulating or wavy in nature.

Underbrush Small woody plants growing in a forest beneath tree species in a forest.

Underdrained (1) Ground provided with an underdrain or a system of subsurface tile drain is said to be underdrained. (2) Land is also said to be underdrained or underdesigned if tile laterals are spaced wider than desirable,

Underground Runoff (or Seepage) water flowing toward stream channels after infiltration into the ground.

Undergrowth Growth of small trees or shrubs under tall trees.

Underground water basin A valley area or plain underlain by groundwater which has established an operational regime from the prevailing hydrologic and climatic conditions, to behave as a unit and which can usefully be developed as a unit.

Underirrigation Irrigation inadequate to meet the soil and plant needs.

Undermelting The melting of floating ice from below.

Under planting To grow seedling in a garden below the old trees so as to remove the ageing trees when the new one starts fruiting.

Under population A condition when the marginal product of labour is greater than the existing average output per head for the country as a whole.

Under stock Root stock.

Understory That portion of the trees in a forest stand below the overstory.

Undifferentiated soil map unit Soil mapping unit in which two or more soil units occur, but not in a regular geographic association. For example, the steep phases of two or more soils might be shown as a unit on a map because slope is the prime characteristic. See also association, Soil and complex, soil.

Undulating Not flat. (1) Common term to designate a land surface which has depressions and hills i.e. is wavy in nature. (2) Wavy margins of leaves.

Unearned income Income from capital investments, ownership of land and savings. It includes interest and dividends, rent and income derived from deposits in the Post Office, Trustee and other savings banks, National Savings Certificates and bank deposits.

Unearned increment A phrase used by John Stuart Mill to describe the increase in the value of a piece of land which is not due to any expenditure on improvements by the owner, but due to other factors.

Unemployment A condition of involuntary idleness.

Unemployment benefit Financial assistance to those genuinely seeking work, but unable to find suitable employment.

Unemployment rate The number of persons capable of working and willing to work but unable to find suitable employment, expressed as a percentage of the total number of persons available for employment at any time.

Uneven-aged A term applied to a stand in which there are considerable differences in age of trees and in which three or more age classes are represented.

United screw thread Three series of threads; coarse, fine and extra fine: a 1/4 inch (6.35 mm) diametre thread in the UNC series has 20 threads per inch (25.4 mm), while the UNF series it has 28.

Unfree water Chemically bound water.

Unguent A substance which can be used as a lubricant.

Ungulates Mammals having hoofs.

Unhitch A term opposite to hitch; meaning to disconnect the animals or the tractor from the implement after the work has been completed or to give rest to the animals.

Unhulled seed Ripe seed still retaining pods, glumes, or other outer coverings.

Uniaxial With a single axis.

Unicameral With a single chamber; also called uniloculate.

Uniloculate See unicameral

Unicapsular Of a single capsule.

Uniflorous Bearing one flower only.

Uniform channel A channel whose cross- section does not change along its length, is straight in alignment and has a constant grade.

Uniformity (Seed lot) Refers to the uniformity of a seed lot; absence of apparently significant variation in a seed lot.

Uniform load A load distributed uniformly over a portion or over the entire length of a beam; measured in kg per cm.

Unilateral horizontal cordon A method of training vines and dwarf fruit trees. Consists in developing a horizontal arm from the trunk at about 60 cm above the ground and extending this arm on a wire from the top of the trunk.

Unimodal An adjective describing a frequency distribution having a single mode.

Unisexual One sex.

Uniform flow Open channel flow is said to be uniform if the depth of flow is the same at every section of the channel.

Union (1) Mating of two individuals. (2) A trade union. (3) Proper joining of a stock and a scion during grafting.

Uniparous (1) Producing only one egg or only one off-spring at a time. (2) Having only one axis or branch.

Unipartite Undivided.

Unipolar With only one pole.

Uniradiated Having only one arm or ray or process.

Uniserial Set in one row.

Unit (1) A quantity of a substance or phenomena chosen as a standard by which the other quantities of the substance or phenomena could be expressed such as metre, metre/sec, metre/sec^2 etc. (2) In sociology a unit consists of an individual object or a group which for calculation or observational purposes could be taken as single and complete such as family.

Unit character Term used for a character believed to be determined by the allelas at a single-gene locus. The term is now largely abandoned.

Unit circle The locus of points in the plane which are precisely one unit from the origin.

Unit hydrograph or Unitgraph Hydrograph of storm runoff at a given point on a given stream which will result from an isolated rainfall excess of unit duration occurring over the contributing drainage area and resulting in a unit of run-off.

Unit of Currency The standard monetary unit in a country.

Unit of extension A morphologically discrete growth increment.

Unit of morphogenesis A unit of growth determined by a single cycle of mitotic activity of the apical meristem.

Unit pipe The unit supporting and supplying one crown unit in the pipe model theory.

Unit-rainfall duration The duration of runoff producing rainfall or excess rainfall that results in a unit hydrograph.

Unit record device Any piece of equipment such as punch card readers, card punch, and line printers.

Unit Trust A method of investment by which money subscribed by many people is invested on their behalf by a management company; the investment and management is subject to the strict legal provisions of a trust deed.

Unit water requirement The amount of water actually used by plants in producing unit weight of dry matter.

Unity power factor Power factor of 1.0 obtained when current and voltage are in phase, as in a circuit at resonance.

Univalent A chromosome unpaired at meiosis.

Universal motor A motor that may be operated at approximately the same speed and output on either direct current or single phase alternating current. Also known as AC/DC motor.

Universal veil A thin veil like membrane covering certain young mushrooms.

Univoltine A silkworm which gives only one crop in a year.

Unona longiflora See Desmos longiflorus.

Unpaired terraces Terraces on opposite sides of a valley which are unmatched in elevation.

Unpolished rice Unhusked rice before polishing. See polishing (rice).

Unripe honey See green honey.

Unsex To deprive a male of its tester or a female of its ovary/ovaries.

Unslaked lime Burnt lime.

Unsaturated flow The movement of water in a soil not filled to capacity with water.

Unstable An atmosphere in which temperature decrease with height is greater than the dry adiabatic lapse rate.

Unstable channel An unlined channel which either silts or scours over a prolonged period.

Unsteady uniform flow In this type of flow in the open channel the surface fluctuation occurs from time to time whereas the water surface remains parallel to the channel bottom.

Unsteady varied flow In this case the depth of flow varies with time and also the water surface does not remain parallel to the channel surface. As the unsteady uniform flow can occur only rarely unsteady varied flow is usually designated as unsteady flow.

Unsteady flow See unsteady varied flow.

Unwind In computers to rearrange and code a sequence of instructions to eliminate red-tape operations.

Up-country The interior region or the region inland from the sea-coast.

Up-budding Shield budding

Upbreeding To upgrade the offspring of a inferior breed by crossing it to a superior one; also called upgrading.

Update (1) In computers, to modify an instruction so that the address numbers it contains are increased by a stated amount each time the instruction is performed. (2) To change a record by entering current information.

Updraft carburettor For a gasoline engine, a fuel air mixing device in which both the fuel jet and the airflow are upward.

Updraft furnace A furnace in which volumes of air are supplied from below the fuel bed or supply.

Upgrading See upbreeding.

Upkeep To maintain the plants, tractor animals, implements etc.; also refers to the maintenance expenditure.

Upland An elevated surface; hilly area.

Upland rice Unirrigated rice crop.

Upland cotton Gossypium hirsutum L(G. mexicanum Tod.). A cotton plant of family Malvaceae.

Upper plastic limit See liquid limit.

Upright silo A vertical silo.

Upright branch An erect branch attached to a prostrate stem of a trailing plant.

Upright training method Method of training grapevine in which two or more arms are laid off from a vertical main stem or trunk to the right and left on a horizontal wire.

Upright tree A tree holding its limbs upright.

Upset welding Pressure butt welding in which heat is generated by resistance to the passage of current across the joint.

Upstream Moving towards the head of the river or stream viz against the flow direction.

Upstream end (1) The extreme or head end of a stream. (2) The ends of a border or furrow or corrugations towards the stream (Head ends).

Up to date (1) In economics the complete records to date. (2) A potato variety.

Urate Substance formed by the combination of chemical elements with uric acid.

Urban A term which refers to living in a city or town in contrast to rural which signifies the living in a rural atmosphere (villages).

Urceolate An object with a large body and small mouth; a pitcher-shaped object.

Urea [$CO(HN_2)_2$] (1) An organic chemical and a rich source of artificial nitrogen; white crystalline hydroscopic quick acting and easily soluble in water; contains 45-46 per cent of nitrogen. Acidic with equivalent acidity 80 per cent. Applied as a basal dose, top dressing or even sprayed on the crops. Quick leaching of nitrate-N occur if water is kept standing. Multiple split recommended if leaching feared during irrigation or applied after the irrigation has been just completed. To reduce its hydroscopicty non-hydroscopic inert materials are used and it is available in pellet form with inert material coating. Used with dung coating also to slow down its action. (2) A compound of nitrogen in the urine.

Uric acid A white crystalline compound and is a chief constituent of urine of the birds and reptiles.

Urine earth A bedding of earth used in cattle yards so as to absorb the urine. Used as a manure.

Urine formaldehyde compound A compound nitrogenous material with 38.0 per cent of nitrogen with an equivalent acidity of 68 per cent.

Urban land Areas so altered or obstructed by urban-works or structures that identification of soils is not feasible. A miscellaneous land type.

Urease An enzyme that catalyzes the degradation of urea to ammonia and carbon dioxide; obtained from the seed or jack bean.

Uredium Sorus producing urediospores; summer spore of rusts; one-celled, verrucose.

Urediospore In rust fungi, a unicellular spore, commonly called spring spore or spore of the red stem stage.

Urocystis cepulae See onion smut.

— **tritici** See wheat flag smut.

Uromyces anthylidis See fenugreek rust.

— **appendiculatus** See lablab rust; hourse gram rust, cowpea and bean rust.

— **cicerisorietini** See gram rust.

— **decoratus** See sanhemp rust.

— **fabae** See lathyrus, pea, lentil and broad bean rust.

— **minor** See clover rust.

— **mucunae** See beet leaf rust.

— **pisi** See pea rust.

— **setariae italicae** See Italian millet rust.

— **Striatus** See lucenne rust.

— **trifoli** See clover rust.

Ursine Thickly covered with bristles.

Use dilution test A bioassay method for testing disinfectants for use on surfaces where a substantial reduction or bacterial contamination is not achieved by prior cleaning; the test organisms Salmonella cholerasuits and Staphylococcus aureus are deposited in stainless steel cylinders which are then exposed to the action of the test disinfectant.

Ustilaginoidea virens See rice false smut.

Ustilago avenae See oat loose smut.

— **crameri** See Italian millet rust.

— **crusgalli** See barnyard millet shoot smut.

— **kolleri** See oat covered smut.

— **hordei** See barley covered smut.

— **nuda** See barley loose smut.

— **panicifrusmentacei** See barnyard millet smut.

— **paradoxa** See barn yard millet smut.

— **scitaminea** See sugarcane diseases.

— **tritici** See wheat loose smut.

— **zeae** See maize smut.

Ustulina deusta See rubber diseases.

— **zonata** See tea charcoal rot; jackfruit charcoal rot; rubber diseases.

Usufruct A right of enjoying to use and enjoyment of another's property but not destruction and waste of its substance.

Utethesia pulchella See sanhemp hairy caterpillar.

Utilisation factor In electric power distribution the maximum demand of a system or part of a system divided by its rated capacity.

Utilisation, Forest That branch of forestry concerned with the operation of harvesting and marketing the forest crop and other resources of the forest.

Utilization ratio The ratio of the effective time on a computer to the total time.

Utilization value The usefulness of a seed determined by its purity and germinative energy, a term commonly used in forestry.

Utrasum bead tree Elaeocarpus sphaericus (Gaertn.) K. Schum. (Elaeocarpus ganitrus Roxb). Tree of the family Elaeocarpaceae whose fruits are edible. Hard nuts made into rosaries, necklaces etc.

Utricle Cell of animal or plant.

Utromyces appendiculatus See lima bean rust.

U-tube manometre A manometre consisting of a U-shaped glass tube partly filled with a liquid of known specific gravity; when the legs of the manometre are connected to separate sources of pressure, the liquid rises in one leg and drops in the other; the difference between the levels is proportional to the difference in pressures and inversely proportional to the liquids specific gravity. Also known as liquid column gage.

V

Vacillation Oscillation.

Vacuole Minute cavity in organ, containing air, fluid.

Vacuolated cytoplasm The living substance of cells excluding the nucleus; when filled with or containing bubble like structure.

Vacuum Space free of all matter; space, vessel from which air has been almost exhausted by air pump.

Vacuum brake A system particularly desirable for heavy passenger cars. The mechanical breaking system is operated by a vacuum taken from the intake manifold or carburettor at a point just above the throttle valve.

Vacuum breaker Device for taking in air at the high point on the discharge pipe to prevent siphonic action.

Vacuum cleaner A motor driven fan machine used for sucking up dirt and dust from rugs, etc.

Vacuum control The term may apply to any automobile devices such as brakes, clutch, etc., controlled by manifold vacuum.

Vacuum drying The removal of liquid from a solid material in a vacuum system; used to lower temperatures needed for evaporation to avoid heat damage to sensitive material.

Vacuum fuel supply Depends upon the use of a vacuum tank to raise the gasoline from the main tank to the higher level of the engine. When the engine is running, a partial vacuum is maintained in the vacuum tank by utilising the vacuum created at the carburettor.

Vacuum gauge An instrument calibrated in atmospheric inches to test the vacuum produced in the intake manifold of an engine or in the fuel line.

Vacuum pump A compressor for exhausting air and noncondensable gases from a space that is to be maintained at subatmospheric pressure.

Vagary Whimsical nature; common term used to refer the vagaries of nature such as monsoon.

Valance Combining or replacing power of an atom as compared with a hydrogen atom.

Valancy See valance.

Valerian Valeriana officinalis L. Family Valerianaceae. A herb found in Kashmir; dried rhizomes and roots source of an essential volatile oil for perfumery. Rhizomes also used as a tonic and a stimulant.

Valeriana officinalis See valerian.

— **Wallichii** See Indian valerian.

Validity check: Computer check of input data, based on known limits for variables in a given field.

Vallecula Any groove; also refers to a furrow.

Valley Any depression; however its most common usage is made for an area enclosed by hills and usually having a stream flowing through it.

Valley train An outwash terrace extending down a valley away from the ice front.

Valsoid Having groups of perithecia with beaks pointing inward, or even parallel with surface, as in vaisa.

Value (1) The amount, quantity, magnitude or number an algebraic symbol is supposed to denote. (2) An arithmetical number which defines the size or the quantity.

Value, colour The relative lightness of colour, which is approximately a function of the square root of the total amount of light. See also Munsel colour system, hue and chroma.

Value engineering The application of the technique of value analysis, the object of which is to obtain the same performance out of a product at a lower cost by achieving economies in the cycle of product design and manufacture.

Values It is the importance rating which people attach to things; conditions and circumstances.

Valve (One way) A valve which allows the passage of the substance in one direction only.

Vane A blade of a centrifugal pump or a wind mill.

Vangueria pubescens Kurz. A large shrub or small tree whose young leaves are eaten as vegetable; Fruits are also eaten.

— **Spinosa** See Meyna laxiflora.

Vanilla (1) Climbing orchid which yields the popular flavouring essence vanilla. Vanilla planifolia Andr. Vanilla fragrans (Salisb.) Ames and Vanilla pompona Family Orchidaceae, loves shade. propagated through shoot cuttings; seed propagation possible. (2) The extract derived from vanilla pods.

Vanilla bean The long bean-like capsule or vanilla.

Vanilla extract The flavouring extract of vanilla beans, containing in 100 cc, the soluble materials not less than 10 gm. of vanilla beans.

Vanilla fragrans See vanilla.

— **planifolia** See vanilla.

— **pompona** See vanilla.

Van't Hoff factor The ratio of the observed osmotic pressure of a solution to that predicted by Van't Hoffs law.

Van't Hoff's law The law that the osmotic pressure of dissolved substance equals the gas pressure it would exert if it were an ideal gas that occupied the same volumes as that of the solution.

Vapour lock Interruption of the flow of fuel in a gasoline engine caused by formation of vapour or gas bubbles in the fuel feeding system.

Vapour pressure The partial pressure of water vapour in the atmosphere. For a liquid or solid, the pressure of the vapour in equilibrium with the liquid or solid.

Vapour Moisture in the air; gaseous form of a liquid or solid substance.

Vapour pressure test Vapour pressure of the fuel is measured in this test and is an index of starting ability as well as vapour lock tendencies of the fuel.

Vapour proof storage Sealed storage.

Var. Most commonly used abbreviation for the term variety.

Variable costs Costs which vary directly with variations in the volume of output or time. The major variable costs involved are those of labour and material. Also known as prime costs.

Variability The scattering of the values of a frequency distribution, often measured by the standard deviation.

Variant An individual or species deviating in some character or characters from the type.

Variant, Soil A soil whose properties are believed to be sufficiently different from other known soils to justify a new series name, but comprising such a limited geographic area that creation of new series is not justified.

Variations The differences occurring among individuals of a group, such as species, variety, or strain.

Variegate Diversity in colour, mark with irregular patches of different colour on leaves or flowers. Generally a mosaic but sometimes caused by a disease or other environmental factors.

Variegated Bauhinia See mountain ebony.

Varietal trial A term referring to the experiments conducted to find out the suitability of a variety in an area or a region.

Variety A group of strains or a single strain which by its structural or functional characters can be differentiated from another group. Individual or group usually fertile with any other member of the species to which it belongs, but differing from the type in some qualities capable of perpetuation.

Variety hybrid A cross between two varieties of the same species.

Variform Having various forms.

Variolite Rock with concretionary structure causing on surface an appearance like small-pox pustules.

Varnish tree See candlenut tree.

Vascular Plant which has special fibrovascular tissue through which water and dissolved organic foods are conducted in the xylem and the phloem portions respectively.

Vascular ray Ribbon like aggregate of cells extending radially in stems through the xylem and often into the phloem.

Vascular trace The conductive tissue

which grows outward from the stem into a leaf or root.

Vasculose Chief substance of vessels of plants.

Vaseline A petroleum product unctuous in nature. Also see petrolatum.

V belt A belt, usually endless, with a trapezoidal cross section which runs in a pulley with a V-shaped groove, with the top surface of the belt approximately flush with the top of the pulley.

V crowder V-shaped implement used to crowd backfill into a trench or to gather soil for border dikes.

V ditcher A implement either tractor or bullock drawn used to make ditches. The ditches made are of V-shape originally but ultimately becomes trapezoidal. Also used for ditch cleaning or making border dykes.

Vector (1) An agent which transmits disease. (2) In mathematics, used to indicate the magnitude and direction of a quantity or phenomena such as a velocity vector.

Vector algebra A part of algebra which incorporates the study of vector quantities.

Vector equation An equation involving vectors.

Vegetable The edible portion of a usually herbaceous plant; customarily served with the main course of a meal.

Vegetable black Carbon made by the incomplete combustion or destructive distillation of vegetable matter, for example, wood.

Vegetable crop A plant crop from which vegetables are harvested.

Vegetable diastase See diastase.

Vegetable dye Any colourant that is obtained from a vegetable source; for example, indigo.

Vegetable fat A semi solid vegetable oil, used chiefly for food; for example, Sauri fat, ucuhuba tallow, Mahua fat, gamboge butter, Sierra Leone butter, and Mafura tallow.

Vegetable forcing To grow vegetables ahead of their normal time so as to be available during the period of short supply.

Vegetable gardening The art and science of raising vegetables.

Vegetable glue Mostly starch or dextrine based glues mixed with gums, resins, or antioxidants; tapioca paste is the most common; used on cheaper plywoods, postage stamps, envelopes, and labels.

Vegetable marrow See summer squash.

Vegetable matter Any material which is or recently was a plant tissue.

Vegetable oil An edible, mixed glyceride oil derived from plants including cottonseed, linseed and peanut; used in food oils, shortenings, soaps, and medicine, and as a paint drying oil.

Vegetable parchment A paper like material made from a base of cotton rags or alpha cellulose called waterleaf, and containing no sizing or filling materials; used for documents and food packaging.

Vegetable sponge See smooth sponge gourd.

Vegetable tanning Leather tanning using plant extracts, such as tannic acid.

Vegetable wax A waxy substance of vegetable origin, composed of fatty acids in combination with higher alcohols; includes Japan wax, jojoba oil, candelilla, and carnauba wax.

Vegetal Like a vegetable or related to vegetable.

Vegetation and ecosystem mapping An art and a science concerned with the drawing of maps which locate different kinds of plant cover in a geographic area.

Vegetation A group or association of plants; plants in general; plant life.

Vegetative Term used to designate stem and leaf development in contrast to flower and seed development.

Vegetative branch Refers to a branch which produces only leaves and no flower and fruits.

Vegetative cover Soil cover of plants irrespective of species.

Vegetative mantle The complete coverage of soil by vegetation, especially perennial ground cover vegetation.

Vegetative propagation See vegetative reproduction; also called asexual reproduction.

Vegetative reproduction Propagation

of plants in ways other than propagation by seeds. Reproduction by buds, grafts, cuttings and layers etc.

Vegetative state A term usually applied to a micro-organism active state in which they may multiply rapidly but their spores being dormant.

Vehicle (1) An automobile. (2) A carrier of drug etc.

Veil The annulus or ring of fungus tissue left on the stipe when the cap pulls away in growth.

Vein (1) A nerve. A vascular bundle forming the frame work of fibrous tissues in a leaf. (2) Branched vessels which convey blood to heart.

Vein-banding Symptom of virus disease in which regions along veins are darker green than the tissue between veins.

Vein clearing Clearing or chlorosis of the tissue in or adjacent to the veins in a leaf and is a symptom of virus disease.

Velamen The thick, spongy integument covering the aerial roots or epiphytic orchids.

Velocity See linear velocity, angular velocity.

Velocity of retreat Mean velocity in a conduit immediately down stream from a structure.

Velocity head tachometre A type of tachometre in which the device whose speed is to be measured drives a pump or blower, producing a fluid flow, which is converted to a pressure.

Velum A membranous partition or veil, as the membrane covering the gills in immature mushrooms.

Velutinous A product whose surface in soft and smooth. Also called velvety.

Velvet bean See Mucuna.

Velvet bean bacterial leaf spot A bacterial disease caused by Xanthomonas stizolobiicola. Its symptoms are, small, angular, translucent dot like spots on leaves. These dots later turn dark brown.

Velvet bean leaf spot Disease caused by Cercospora stizolobii in which circular or confluent and irregular spots appear chiefly on leaves. The colour of these spots is tan to brown.

Velvet flower See chaulai.

Velvety See velutinous.

Venation Disposition of veins in a leaf, or an insects wing.

Vendure Green vegetation and also the greenness of vegetation.

Veneer (1) A thin sheet of wood of uniform thickness used for facing furniture or, bonded to make plywood. (2) A facing, as of brick or marble, on the outside of a wall.

Veneer grafting A common method of grafting in mango.

Venom Poisonous fluid secreted by scorpions etc.

Vent Open, vertical pipe riser from a tile or irrigation line which allows air to enter or escape from the pipe system.

Venter The enlarged base of an archegonium within which an egg develops.

Ventilate To provide free circulation of air to purify the environment or to remove excess moisture in the room environment and to provide oxygen.

Ventilation front The locus of all points of equal traverse time in the product being conditioned.

Ventilator A fixed or variable opening to regulate flow of air through a material or space.

Vent pipe A pipe provided in the underground irrigation system to remove the air from the pipe line and to release the pressure.

Ventral The upper or front side. Opposite of dorsal.

Ventral surface The inner surface of an organ. In leaves of higher plants it becomes the upper surface, in liverworts, the lower surface.

Ventricle Any cavity (hollow part) of the body.

Ventricose Distended, inflated. Also called ventricous.

Ventricous See ventricose.

Venturia inaequalis See apple scab.

Verbal (communication) This includes listening, speaking, reading and writing.

Verdant Green; covered with grass. Common term used for forests or fields green with vegetation.

Vermian Worm-like or of worms.

Vermicelli A macaroni product made from wheat.

Vermicular Like worm in form or in movement; of worm eaten appearance; marked with close wavy lines.

Vermiculate Having thick tufts of parallel hairs; worm-eaten.

Vermiculite Mineral or minerals, which at high temperatures expands into worm-like scales and becomes a loose, self-absorbant mass; classified with mica. Used for insulation, as mulch and potting plants etc.

Vermiculite group A group of hydrous platy clay minerals similar to the smectite group, but having a higher net permanent charge and often larger particle size.

Vermiculose Worm-eaten; worm-shaped; full of worms. Also called vermiculous.

Vermiculous See vermiculose.

Vermiform Worm shaped; long and slender.

Vermigrade Moving like a worm.

Vermilion Red pigment made by grinding cinnabar or artificially; cinnabar.

Vermin Any external animal parasite or noxious animal such as rat, mice etc.

Vermivorous Feeding on worms.

Vernalisation Method of inducing or accelerating earlier flowering and fruiting in plants by pre-treatment of seeds with very low temperatures by shortening the vegetative period.

Vernation Arrangement of leaves within a leaf bud.

Vernier A short, auxillary scale which slides along the main instrument scale to permit accurate fractional reading of the least main division of the main scale.

Vernier caliper A caliper rule with an attached vernier scale.

Veronica Veronica beccabunga L. A herb of family Scrophulariaceae, the young stems of which are used as vegetable.

Verricule Thick tuft of upright hairs.

Verruca Wart, wart-like elevations.

Versant General slope of the land.

Versatile (1) An implement which could be used for several purposes. (2) Moving freely about or up and down on a support.

Versicolour Changing from one colour to another in different lights.

Vert A term used in forestry for plants that grow and bear green leaves.

Vertex Highest point; apex; top or crown of head.

Vertex angle Angle less than 90°.

Vertical Perpendicular to the plane of horizon.

Vertical anemometre An instrument which records the vertical component of the wind speed.

Vertical angles The two angles produced by a pair of intersecting lines and lying on opposite sides of the point of intersection.

Vertical centrifugal pump Centrifugal pump having a vertical shaft, often used in submerged operations.

Vertical clevis It is vertical plate with a number of holes at the end of the beam to control the depth of operation and to adjust the line of pull.

Vertical control, or **level control** A series of bench marks or other points of known elevation, established throughout an area.

Vertical drainage, or **Inverted well** Disposal of drainage water through wells into a porous layer of earth or an open rock formation. In sub-surface drainage vertical drainage is often associated with pumping of tubewells for draining groundwater aquifer.

Vertical fall The drop of an object by a distance measured perpendicular to the plane of horizon.

Vertical feed A card feed in which punch cards are placed in a hopper and enter and traverse a card track, all in a vertical position.

Vertical hollow shaft motor Vertical motor having hollow shaft which fits over and is keyed directly to the pump shaft; normally used on deep well turbine pumps.

Vertical or **upright storage** A storage structure whose height is greater than the width or diametre. Other terms used are silos, tanks, and deep bins.

Vertical photograph Photograph taken with a single lens camera with the axis vertical or as nearly vertical as is possible in an aeroplane.

Vertical T budding See T budding.

Vertical zonation Altitudinal zonation of soils. See also latitude zonation.

Verticillate or **whorled (leaf arrangements)** Several appendages inserted at precisely the same level.

Very coarse sand See separates, soil and texture, soil.

Very fine sand Fifty per cent or more very fine sand.

Very fine sandy loam Thirty per cent or more very fine sand or more than 40 per cent fine and very fine sand, at least half of which is very fine sand, and less than 15 per cent very coarse, coarse and medium sand. Also see texture, soil.

Vesicle A thin bubble like sac in which zoospores or aplanospores are formed or enclosed.

Vesicular structure Soil structure characterised by round or egg-shaped cavities or vesicles.

Vespertine Opening of flowers, flying of birds, in the evening.

Vespiary Nest of wasps.

Vespine Of wasps.

Vessel A container or structural envelope in which materials are processed, treated, or stored.

Vestige A well developed part or organ but of little or no utility.

Vestigial Small and imperfectly developed.

Vestiture hair, scales covering a surface.

Vetch The plants cultivated or found wild and are of bean family; used as fodder.

Vetchling A plant similar to vetch.

Vetiver Vetiveria zizanioides (L.) Nash. (Anatherum Zizanioides, Phalaris Zizanioides L. Andropogon muricatus Retz.) Family Gramineae. Important essential oil bearing plant of India; oil used in perfumes, medicine, soaps. A perennial grass, densely tufted, upright, growing luxuriantly on rich marshy soils.

Vetiveria zizanioides See vetiver.

Viable seed A seed capable of germinating and living more or less normal thereafter under a given set of conditions.

Viability Ability to live.

Vibrate Oscillate or pendulum like movement.

Vibratile In a position to vibrate.

Vice An instrument for holding a job such that the work on it could be accomplished. It has two jaws which move in opposite direction with a screw.

Vicia benghalesis See purple vetch.

— **faba** See broad bean.

— **Sativa** See common vetch.

Vicinity map, or **Locality map** Map showing the general area and other pertinent information of the area.

Vigna capensis See cowpea.

— **Catjung** See cowpea.

— **Sinensis** See cowpea.

— **Vexillata** See cowpea.

Vigor Excess of growth rate over death rate in a tissue, organ or plant. Strength, or force, or power.

Vigour Vigour is the sum total of all seed attributes which favour rapid and uniform stand establishment in the field.

Village Assemblage of houses in rural area, larger than hamlet, smaller than town.

Villager A person belonging to a village.

Villi Long, soft, hair covering fruit, flower etc.

Villiform Having form or appearance of velvet.

Viminal Of producing twigs or shoots.

Vinaceous Of wine or grapes; wine red.

Vine (1) The grape fruit. (2) The woody or herbaceous plant with slender stem which trails, climbs or creeps.

Vinery Vine green house; Plant species which have branches like a vine.

Vineyard A grape farm.

Vineyardist A grape cultivator.

Viniculture The cultivation of vines.

Vinifacteur The apparatus used for making Vine.

Viniferous Wine producing.

Vinificator Apparatus for collecting alcoholic vapours in wine-making.

Vinometre Apparatus for measuring alcohol in wine.

Vinous Wine like, due to or of wine.

Vintage Season of gathering grapes; also wine made from the season's production of grapes.

Violaceous Of violet; of violet family.

Virachola isocrates See anar butterfly.

Viral Having the nature of and pertaining to virus.

Virescence Abnormal greenness in petals.

Virgate Slim, straight and erect; refers to a plant with many branches; a varying measure of land.

Virgin soil A soil that has not been significantly disturbed by man's activities.

Viriparus variatus See snail.

Virmel engine A cat and mouse engine that employs vanelike piston whose motion is controlled by a gear and crank system; each set of pistons stops and restarts when a chamber reaches the spark plug.

Virogenic Pertaining to a disease caused by virus.

Virologist A person skilled in virology.

Virology A branch of science dealing with virus and diseases caused by it.

Virose Full of virus; poisonous.

Virosis A disease caused by a virus.

Virostatic Having the ability to arrest the multiplication of a virus.

Virucide An agent that inactivates or kills viruses.

Virulence (1) The disease producing power of virus or any other micro-organism. (2) The quality of being poisonous; caused by or containing virus.

Virulent Highly pathogenic; with strong capacity for causing disease.

Virulent culture A culture or growth of a micro-organism which is alive and not attenuated or modified, and which is capable of multiplication and producing a reaction or possible disease in susceptible hosts.

Virulent latent A form of latent micro-organism which differs from the common latent in its increased virulence and in the greater production of lesions.

Viruliferous Virus carrying; term applied particularly to virus-laden insects.

Virus antagonism See virus interference.

Virus competition See virus interference.

Virus disease Disease caused by virus, infections, systematic and generally spreading to whole body after a single infection. Rarely kills plants but devitalises it.

Virus indexing The determination of presence or absence of viruses in symptomless or dormant plants or seeds by inoculation of virus sensitive plants or by any other means.

Virus indicator plant A species or plant variety that is sensitive to the virus disease and is used for virus indexing in less sensitive ones.

Virus interference The action of a particular virus, strain or isolate which results in the partial or complete suppression of other viruses, strain or isolates.

Viscometres An instrument designed to measure the viscosity of a fluid.

Viscosity coefficient An empirical number used in equations of fluid mechanics to account for the effects of viscosity.

Viscosity (dynamic) Cohesive force existing between particles of a fluid which causes the fluid to offer resistance to flow.

— **(Kinematic)** Ratio of the dynamic or absolute viscosity to the density fo the liquid is known as kinematic viscosity.

Viscosity index An arbitrary scale used to show the magnitude of viscosity changes in lubricating oils with changes in temperature.

Viscous Sticky.

Visibility (Idea) The visualisation of an idea or practice. More the visibility more is its adoption.

Visible branch order See branch order, visible.

Visual Which can be seen.

Visual aid Instruments or objects which can be seen or helps in seeing the objects. Used in extension education.

Vital area A physical component, or feature, of such extreme importance to the functioning of an ecosystem that it should not be altered.

Vitality Vital power, ability to sustain life; ability to endure and perform functions.

Vitamin An organic compound which occurs in very small amount in most vegetable matter without which normal functions may not proceed.

Vitex negundo See chinese chaste tree.

Viticide Insects etc. which causes destruction of vines.

Viticolous Living on the vines.

Viticulture The art and science of growing grapes.

Vitreous Resembling glass in hardness, brittleness and transparency etc.

Vitrification Contracting and filling-up of the pore spaces of the clay by burning it at higher temperature.

Vitriol Sulphuric acid or its compounds.

— **(Blue)** Hydrous copper sulphate.

Vitriolation Burn with vitriol.

Vitriolise Convert into sulphate.

Vitta Oil tube in the fruit of some plants; strips of colour.

Vivacious Tenacious of life, surviving winter, perennial.

Vivarium A glass box used to keep or raise animals or plants. Also called vivary.

Vivary See vivarium.

Vivianite A hydrous ferrous phosphate and its oxidation forms found in many soils and especially in swamps.

Viviparous Germinating while still attached to the parent plant.

Viviparous reproduction Type of reproduction in insects where embryo develops within the body of the insect and young ones produced similar to adults; found in some flies.

Vivipary (1) The replacement of the floral structure by a vegetative outgrowth which can develop into a new plant. (2) Giving birth to a living offspring.

Vlasoy equation A modification of the Boltzmann transport equation for the study of a plasma, in which particles interact only through the mutually induced space charge field, and collisions are assumed to be negligible. Also known as collisionless Boltzmann equation.

Vocation Employment, trade or profession.

Vogel tephrosia Tephrosia vogelii Hook. f. A herb of family Papilionaceae introduced from east Africa. Produces the insecticide, rotenone; also grown for green manure and is a cover crop.

Void Ratio The ratio of the volume of void space to the volume of solid particles in a given soil mass.

Volatile Readily passing off by evaporation.

Volatility The quality of having a low boiling point or subliming temperature at ordinary pressure or, equivalently, of having a high vapor pressure at ordinary temperatures.

Volatilisation The conversion of a solid or liquid into vapours. Losses are ofen referred as volatilisation losses.

Void space Space between the solid particles in a bulk. Expressed as per cent of the total volume. Commonly used for soils.

Volcanic rocks One of the two classes of igneous rocks formed by the cooling of the lava from volcanoes. Glassy in structure, microscopically small crystals may be found dispersed in the glassy matrix.

Volcanic soil A soil derived from volcanic rocks.

Voltage Potential difference or electromotive force measured in volts.

Voltage rating The maximum sustained voltage that can safely be applied to an electric device without risking the possibility of electric breakdown. Also known as working voltage.

Voltage regulating transformer Saturated core type of transformer which holds output voltage to within a few per cent with input variations up to 20 per cent above or below normal; considerable harmonic distortion results unless extensive filters are employed.

Voltage regulation The ratio of the difference between no load out put voltage of a device to the full load output voltage, expressed as a percentage.

Voltage regulator A device of electromagnetically operated contact points and a resistance unit which, when connected in series with the field of an auto-generator, reduces the generator's output when the battery reaches a predetermined voltage.

Volt ampere hour A unit for expressing the integral of apparent power over time, equal to the product of 1 volt ampere and 1 hour, or to 3600 joules.

Voltine The annual brood-producing frequency of silkworms. Also see monovoltine and multivoltine.

Voltmetre An instrument for the measurement of potential difference between two points, in volts or in related smaller or larger units.

Voltmetre ammetre A voltmetre and an ammetre combined in single case but having separate terminals.

Volume Bulk or solid content of an object. In rectangular objects it is defined as the volume = length × bredth × height.

Volume fraction of liquid Is the

volume of the liquid phase per unit bulk volume of soil, expressed in m^3 liquid per m^3 bulk volume.

Volume weight (Obsolete) See bulk density.

Voluntary intake *Ad libitum* intake achieved when an animal is offered excess of a single feed or forage.

Volunteer Plants Unwanted plants growing from seed that remains in the field from a previous crop.

Volute A spiral casing for a centrifugal pump or a fan designed so that speed will be converted to pressure without shock.

Volute pump A centrifugal pump housed in a spiral casing.

Volution Spiral turn.

Vulcanization A chemical reaction of sulphur with rubber or plastic to cause crosslinking of the polymer chains; it increases strength and resiliency of the polymer.

Vulcanizing (1) Treating crude India rubber with sulphur at a high temperature, thus increasing its strength and elasticity. (2) The joining of two pieces of rubber by the application of cement and heat, as in patching an automobile tire.

Vulnerable Suspectible to change or injury.

W

Wabble Revolve with changing inclinations, vibrate from side to side, swerve, stagger. Also Wobble.

Waddle Walk with a rocking motion natural to fat, short legged bird.

Wag Shake to and fro, oscillate.

Wage Amount paid periodically, by the day, week or month.

Wage drift The gap between nationally-negotiated wage rates and actual earnings which tend to be much higher. This difference is due to overtime payments, bonuses and other incentives.

Wage rate The basic minimum rate of payment for a certain occupation, generally expressed as the rate per hour.

Wage-rate, standard Uniform wage rates agreed between employers and trade unions through the process of collective bargaining.

Wages A comprehensive term covering all the different forms of earnings of labour.

Wagners reagent An aqueous solution of iodine and potassium iodide used for microchemical analysis of alkaloids. Also known as Wagner's solution.

Wake System of eddies formed at the point of separation in flow past a solid body when motion in the boundary layer becomes turbulent.

Walden's rule A rule which states that the product of the viscosity and the equivalent ionic conductance at infinite dilution in electrolytic solutions is a constant, independent of the solvent. It is only approximately correct.

Walking implement It is a implement in which balancing and control during its operation is done by the operator by walking behind it.

Walking type tractor A tractor which is controlled by walking behind the tractor. A power tiller usually of 2.5 to 3 H.P. used for small areas. It can do almost all the agricultural operations for cultivation of crops. Also used for stationary jobs.

Wallopahaga Order of insects of which chewing lice belongs.

Wallow Roll, move easily about in mud, sand water.

Wall ratio Ratio of the outside radius of a gun, a tube, or jacket to the inside radius or ratio of the corresponding diametres.

Walnut Juglans regia L. Family Juglandaceae. Native of North-western Himalayas. Exist at 1200-1300 m above sea-level. Large decidous tree with even-grained beautifully mottled wood. Kernels of the nut used for dessert, confectionery and oil extraction. Timber used for furniture and butts of guns. Propagated by seeds; budding and grafting of superior varieties recommended.

Wampee (1) Clausena excavata Burma f. (Cookia punctata Retz.) Family Rutaceae Tree whose fruit are edible and grape-like. (2) Clausena lansium (Lour.) Skeels (Clausena wampi Oliver). Tree, native of China and is grown for its edible fruits.

Wants A craving or desire for mental and physical pleasures and satisfaction.

Warfarin A rodenticide with 3 alpha-phenyl-beta-acetylethyl 4-hydroxy-cumarin as active ingredient. Has slow action on rats.

Warm blooded Having fairly high and constant body temperature above that of the surrounding medium e.g.. Warm blooded animals.

Warm test A germination test wherein the seed is put in a germinating medium and placed at a temperature which is conductive to rapid growth.

Warm up Warming up a automobile (tractor) or stationary engine for a short time before placing it under load such that the engine attains a satisfactory operating temperature.

Warning stage The stage, on a fixed river gage, at which it is necessary to begin issuing warning on river forecasts if adequate precautionary measures are to be taken before flood stage is reached.

Warp (1) Any distortion of timber from the true or plane surface. (2) Yarn which runs the long way in a woven fabric.

Warped Tile that has been formed out of shape or that has settled out of shape while curing.

Warping or Commutation method A process of allowing the tide to overflow low lying lands so as to deposit thereon the silt which it carries in suspension.

Warm-season grass Grass species that make its major growth during the warmer part of the year. Preferred to 'hot weather'.

Warranted growth rate The rate of economic growth at which planned investment is equal to planned saving, a requirement for stable growth.

Warren An area, place or enclosure in which rabbits are bred and raised.

Washing soda (Na_2CO_3) Soluble in boiling water and used as an effective cleaning agent. The excessive presence of this salt alone or in combination with sodium bicarbonate is the source of soil alkalinity which is harmful to crop growth.

Wash load That part of the sediment load of a stream which is composed of particle sizes smaller than those found in the shifting portions of the stream bed.

Washout The process of scavenging the atmosphere q.v. by precipitation.

Washy Too watery, thin, insipid.

Wasp Kinds of Hymenopterous social or solitary insect with a a venomous siting. Many predators on harmful insects.

Waste (1) Any substance which is considered as useless such as remains, refuse, scraps, shreads etc. Many useful purposes could be served by these wastes such as in the manufacture of biogas, serves as fertiliser and amendments. (2) Desert, uncultivated soil.

Waste bank The ridge of excavated earth parallel to a drainage channel. Usually the side which is untrafficable.

Wasteland Land not suitable for or capable or producing materials or services of value. A miscellaneous.

Waste water Water that is wasted through spillways or otherwise discarded from an irrigation system after having been diverted to it. Also called operational waste.

Waste weir A weir used for the escape of waste water.

Water (H_2O) Clear, odourless, tasteless liquid that is essential for most animal and plant life and is an excellent solvent for many substances, melting point 0°C (32°F) boiling point 100°C (212°F) the chemical compound may be termed hydrogen oxide. (2) Irrigate, apply water.

Water application efficiency The ratio of the volume of water that is stored by the irrigator in the soil root zone and ultimately consumed (transpired or evaporated, or both) to the volume of water delivered at the farm.

Water application rate The rate in cm per hour or cm at which the irrigation water is applied to the fields.

Water blister A collection of watery fluid beneath the epidermis usually a result of excessive pressure from irritants.

Water borne Conveyed by water, e.g. a disease.

Water brake An absorption dynamometre for measuring power output of an engine shaft; the mechanical energy is converted to heat in a centrifugal pump, with a free casing where turning moment is measured.

Water-cement ratio Ratio of water to cement in a concrete mix or cement mortar.

Water chestnut (1) Trapa bispinosa Roxb. Family Trapaceae. An aquatic herb whose fruits are sweet mealy, soft and are edible. Eaten raw or boiled. Dried seeds source of straw. Grows and multiplies fast by vegetative propagation, covering entire water surface. Also called water nut. (2) Jesuits-nut. Trapa natans L. Family Trapaceae, Produces edible seeds.

Water check See check, check dam.

Water coconut Nipa fruitcans Thunb. Family Palmaceae. A prostrate gregarious palm, the leaves of which are used for thatching, hats, basket, mats, cigarette wrappers etc. Inflorescence a source of sugar. Also called nipa palm.

Water conservation The physical control, protection, management, and use of water resources in such a way as to maintain crop, grazing and forest land, vegetal cover, wildlife and wildlife habitat for maximum sustained benefits to people agriculture, industry, commerce and other segments of the national economy.

Water content The amount of water lost from the soil when it is dried to constant weight at 105°C expressed either as the weight of water per unit weight of dry soil or as the volume of soil. The relationship between water content and soil water pressure is called the soil moisture retention curve, or sometimes the soil water (moisture) characteristic. Depending upon whether the curve is determined with decreasing or increasing water content, it is desorption or absorption curve, respectively.

Water control Physical control of water by measures such as conservation practices on the land, channel improvements and installation of structures for water retardation and sediment detention.

Water couch See knot grass.

Water courses Ditches, flumes or pipelines deriving their supplies from the outlet or turnout and primarily conveying water to the sub-water courses and only secondary to the fields.

Water cress Nasturtium officiniale R. Br. [Nasturtium fontanum Aschers; Rorippa nasturtium-aquaticum (L.) Hayek]. Family Cruciferae. A small herb, growing in clear, cold and shallow water which is used as a leafy vegetable.

Water culture Hydroponics.

Water cushion Pool of water maintained to absorb the impact of water flowing from an overfall structure.

Water demand Water requirements for a particular purpose, as for irrigation, power, municipal supply, plant transpiration or storage.

Water dispersible clay The clay content in a soil which gets dispersed in water. The ratio of the total water dispersible clay to the total clay content is known as dispersion ratio.

Water-dispersible insecticides Insecticides in which apart from the diluent, wetting and suspending agents are also added such that on mixing with water a suspension suitable for spraying is formed.

Water disposal system Complete system for removing excess water from land with minimum erosion. For sloping land it may include a terrace system, terrace outlet channels, dams and grassed waterways. For level land it may include only surface or sub-surface or both of these drains.

Water distribution efficiency See distribution efficiency.

Water dropwart Oenanthe javanica (Bl.) (Oenanthe stolonifera DC.) A perennial herb of family Umbelliferae whose young shoots are used as condiment.

Water equivalent The depth of water that would result from the melting of the snowpack or of a snow sample thus, the water equivalent of a new snowfall is the same as the amount of precipitation represented by that snowfall.

Water erosion One of the processes of soil erosion. The soil erosion caused by water. It may be of rill, sheet or gully erosion; also wave erosion.

Water-flea Microcrustacean, largely used as food by baby fish.

Water flow Of the family Antidae includes swans, geese, ducks, brant; the "game bird".

Water front The front of the advancing sheet of water let into a field or unit of field in surface irrigation.

Water gate A gate is a reservoir or dam for the discharge of water. Flood gate through which flood water passes and waste gate through which waste water passes. Also a gate in a fence hung across a ditch to prevent livestock passing through during low water and to allow debris to pass through during high water.

Water gauge A device which indicates

the water level in an open channel, closed conduit or reservoir. It may be simple staff gauge marked in appropriate scale or a pressure gauge.

Water grass See Barnyard grass.

Water hammer Pressure rise in a pipeline caused by a sudden change in the rate of flow or stoppage of flow in the line.

Water head (1) Elevation of the water above a datum which imparts it potential energy and causes water to flow. (2) The source of a river or stream.

Water hyacinth Echhornia crassipes Solms. Family Pontederiaceae. A perennial, floated, rooted, aquatic weed. A menace to lakes, reservoirs and storage structures. Also found in ponds, ditches and canals and in rice fields. Multiply rapidly by seeds and vegetative parts. Fed to cattle and used as manure. Recently methods have been developed to use it for bio-gas production.

Water horsepower Total available energy from or transmitted to water. Most useful in calculating the horsepower required to lift the water in tubewells.

Water injury Any of the types of injury caused to plants due to excess of water on or in the soil.

Water intake rate See infiltration rate.

Water lemon Passiflora laurifolia L. Family Passifloraceae. A climbing vine, native to tropics and produces edible fruits. Also called yellow granadilla.

Water lettuce Pistia corniculata. Family Araceae. Stemless stoloniferous, floating aquatic weed anchoring by root when stranded in the mud or in shallow water. Propagated vegetatively and multiplies rapidly. Occurs in rice fields, ditches, ponds and slow flowing rivers.

Water level (1) Water surface, also its elevation above any datum, gauge height, stage (2) A levelling instrument in which water is used as level indicator.

Water level diagram A single chart showing longitudinal sections of the main canal and of the branch canal depending on it, all of them being drawn to a scale which greatly exaggerates the slopes.

Waterlogged soils Soil conditions where a high watertable is detrimental to plant growth, resulting from overirrigation or seepage or inadequate drainage, the replacement of most of soil air by water. Deficient oxygen and excess of CO_2 is common. Also includes soils where surface inundation of water occurs for a sufficiently long period of time that is detrimental to crops.

Water-logging A temporary or permanent condition of the soil resulting from water saturation. It may be surface waterlogging due to water stagnation at the surface of the soil or sub-surface which is due to rise in underground water.

Water management Application of practices to obtain added benefits from precipitation, water or water flow in any of a number of areas, such as irrigation, drainage, wildlife and recreation water supply, watershed management, and water storage in soil for crop production.

Water melon Citrullus vulgaris Schrad. ex. Eckl. & Zeyh. Family Cucurbitaceae. An annual trailing vine with light-yellow flowers and leaves having five to seven deep lobes; the edible, oblong or roundish fruit has a smooth, hard, green top filled with sweet, tender, juicy, pink to red tissues containing many seeds. A native to Africa. Requires a long growing season with high temperature.

Water nut Chestnut, water chestnut or Chinese water chestnut.

Water passageways Small channels in the soil through which water flows.

Water plant Aquatic plant.

Waterproof cement A cement which makes more impermeable mortars or concretes than ordinary portland cement. It contains as small percentage of integral water proofer, such as calcium, aluminium or some metal stearate non-saponifiable oil, gypsum treated with tannic acid.

Water quality The quality of water used for specific purposes e.g. for drinking, irrigation purposes etc. The irrigation water quality depends upon the silt and organic debris in suspension or salts in solution or both. Canal and river water contain silt and salts both whereas underground water usually contain dissolve salts. The nature and quality of material carried influences both soils and crops to be irrigated. EC, SAR of the irrigation water and the excess of specific ions such as boron fluoride, selenium or lithium describe the irrigation water quality.

Water regulation, drainage Subsurface drainage whose main purpose is to regulate the thermal and air regimes of the soil, according to the requirements of agricultural plants.

Water-repelling agent An admixture which is used to make the concrete water repellant, such chemicals are soaps, other fatty acids, compounds, petroleum oils and waxes.

Water requirement The quantity of water required by a crop in a given period of time for its normal growth; includes surface evaporation and other unavoidable wastes. Usually expressed in depth of water for a given time.

Water resources Supply of water in a given area or watershed usually interpreted in terms of availability of surface or underground water.

Water resources appraisal Assessment of, with the help of observed data or other known procedures of calculation, an overall picture of the extent, with respect to time and space, and dependability of water supplies, surface as well as ground, and the character of water, of a river basin, regions or country on which an evaluation of the opportunities for their control and utilization is to be based.

Water resources survey An orderly process, including collection of basic from known records or field studies and observations, methods and techniques used in their collection and interpretation and analysis, to provide basic information for water resource appraisal of a river basin, region or country. It may well include recommendations for improved network of data collecting stations to ensure, providing in so far as possible, data required for water development studies.

Water retention The capacity of a soil to retain water at a given suction head, also sometimes field capacity.

Water-retention curve See moisture-retention curve.

Water rose-apple Syzygium aqueum (Burm.f.) Alston (Eugenia aquea Burm. f., Eugenia javanica Lamk). A small tree of the family Myrtacaeae which produces edible fruits.

Watershed Total land area above a given point on a stream or waterway that contributes runoff to that point.

Watershed area All land and water within the confines of a drainage divide or a water problem area consisting in whole or in part of land needing drainage or irrigation.

Watershed lag Time from centre of mass of effective rainfall to peak of hydrograph.

Watershed management Use of a land in a watershed in accordance with predetermined objectives such as the control of erosion, stream flow, sedimentation, floods, water, forage or timber yield.

Water planning Planning the use and treatment of land and water for optimum long range benefit of both the individual farmers and all other residents of a watershed.

Watershed project Project consisting primarily of measures to improve ground cover and conditions including better cultural practices, shifts in rotations of crops and intensity of land use, strip cropping contour farming, fire protection and controlled grazing, stream protective measures, afforestation, gully-plugging, stabilisation of hill slopes and other means for the control of natural resources.

Water-sensitive Refers to a plant whose growth is severely hampered or plant is killed in high moisture conditions of the soil, or under waterlogging.

Water shoots See water sprouts; gormandiser sets.

Water slaked lime See calcium hydroxide.

Water-soaked Refers to lesions on plants which appear to have a clear liquid within the lesion.

Water, soil Soil water is understood to be the equilibrium solution in the soil; pure water refers to the chemically pure compound H_2O. Water in soil is subject to several force fields originating from the presence of the soil solid phase, the dissolved salts, the action of external gas pressure and the gravitational field. These effects may be quantitatively expressed by assigning an individual component potential to each of them. The sum of these potentials is called the total potential of soil water.

Water pressure, soil The pressure (positive or negative) in relation to the external gas pressure on the soil water to which a solution identical in composition with the soil water must be subjected in order to be in equilibrium through a porous permeable wall with the soil water. It may be identified with the total potential defined above.

Water spreading Application of water to lands for the purpose of increasing the growth of vegetation or to store it as groundwater for subsequent withdrawal; the diversion of surface runoff to allow it to infiltrate into the soil.

Water sprouts Extraordinary vigorous vegetative shoots which grow from high points on the main branch of fruit trees in an upright direction at the expense of the parent branches from which they arise. If not removed, soon close the centre of the tree and obstruct light. In India, all cultivated citrus species and pomegranate are particularly prone to producing water sprouts in large numbers. Also called water shoots.

Water stable aggregate A soil aggregate that is stable to the action of water, such as falling drops or agitation as in wet-sieving analysis.

Water standards Definitions of water quality established as a basis of control for various water-use classifications.

Water stress The stress of water caused to the plant due to deficiency of water. Usually expressed in bars. See wilting point also. One bar is equal to 1023 cm of water.

Water sucker See banana sucker.

Water suffocation Death of a plant due to excess water.

Water suspension spray Insecticides used in water for spraying. Available as water wettable or water dispersible powders containing nearly 50 per cent active material with 50 per cent inert ingredients. Wettable powder dilute to any desired concentration and water dispersible powder in certain proportions.

Watertable (groundwater surface; water surface; ground water elevation) Elevation at which the pressure in the water is zero with respect to the atmospheric pressure. See also perched water table.

Watertable control See controlled drainage.

Water tolerant Refers to a plant which continues to thrive in a soil in which considerable amount of moisture or waterlogging-condition occurs.

— **semi-tolerant** The plants which are able to withstand waterlogging somewhere in between the sensitive type and the tolerant ones are called semi-tolerant.

Water use efficiency The ratio of the water beneficially consumed to the water delivered hence 'project water use efficiency' 'farm water use efficiency', field water use efficiency.

Waterway A natural course or constructed channel for the flow of water.

Watery Containing too much of water, over moist.

Water year Any 12 month period, usually selected to begin and end during a relatively dry season, used as a basics for processing streamflow and other hydrologic data.

Water yield The total outflow of a drainage basin. The surplus of precipitation over infiltration, canopy interception and depressional storage.

Watt The unit in the metre-kilogram-second system of units, equal to 1 joule per second. Symbolised W.

Wattage rating A rating expressing the maximum power that a device can safely handle continuously.

Watt hour A unit of measure of electrical work, equal to one watt expended for one hour.

Watt-hour metre A metre that measures and registers the integral, with respect to time, of the active power of the circuit in which it is connected; the unit of measurement is usually the kilowatt hour.

Wattle (1) Species of trees which yield wattle bark such as black wattle, green wattle, silver and golden wattle. (2) Interlaced twigs, wickerwork.

Wave erosion Erosion by wave action. See erosion.

Wave height The vertical distance between a wave crest and an adjacent trough.

Wave length The least distance between elements moving in the same phase of oscillation in a wave disturbance.

Wave number The reciprocal of the wavelength of a wave, or sometimes 2 Π divided by the wavelength. Also known as reciprocal wavelength.

Wave period The time for two successive waves to pass a given point.

Wax (1) Compounds consisting of fatty acids and high alcohol; harder than fat and less greasy; deposited by insects or obtained from plants. (2) Material prepared from rosin and bees wax used in grafting fruit plants. Standard brush wax is prepared by melting 5 parts of rosin slowly and adding 1 part of bees wax to it, continuing melting and adding 1/4 part of linseed stirring in a little at a time.

Wax bean Butter bean, Kidney beans or string beans.

Wax cloth See budding cloth.

Wax gland The wax-secreting gland of the honey bee.

Wax gourd See ash gourd.

Waxy maize Zea mays ceratina. Type of maize with somewhat waxy appearance of the kernels; starch gummy, has characteristics of tapioca.

W-Ditch or **W-Drain** The closely spaced parallel single ditches having the spoil construction placed between them to permit unimpeded runoff into the W-ditch.

Weak crotch Conditions of crotch or joint in fruit trees when two equal branches arising from a common point are allowed to grow naturally; in the long run tends to split under the pressure of heavy winds; normally avoided and the frame work strengthened by pruning or shortening of the lateral branches.

Wealth Goods and other asserts in existence at any time which command a market value if offered for sale, this implies that such assets have utility and scarcity value and that it is possible to transfer their ownership from one person or body to another.

Wear and tear Depreciation in value due to use.

Weather (1) The general condition of the atmosphere at any given time or during a given period. (2) To be subjected to the influences of atmospheric conditions.

Weather bureau Meteorological office.

Weather chart See weather map.

Weather cock A revolving pointer which indicates the direction of the wind in which it is blowing. Name derived from the shape of the pointer which is of the shape of a cock.

Weather forecast A forecast of the future state of the atmosphere with specific reference to one or more associated weather elements.

Weather glass Barometre.

Weathering The physical and chemical distintegration, alteration, and decomposition of rocks and minerals at or near the earth's surface by atmospheric agents.

Weathering by heat A natural process of weathering wherein expansion and concentration of the rocks due to temperature variation in space and time causes the rocks to wear and tear. In more humid and high temperature conditions weathering is more.

Weathering by water A natural process of weathering by water. Several processes are involved. Rain drops having high kinetic energy dislocates solid particles from rocks and puddles the surface of some parent material. Causes soil erosion and transports the eroded material and the rocks material long distances and deposits them at low lying areas. Water collected in rock crevices gets frozen which causes the rocks to wear off. Thus it helps in the soil formation.

Weathering complex That portion of the soil which is of secondary origin. Includes both residual or alteration products and materials precipitated *in situ* from soluble products of chemical weathering.

Weather map A chart portraying the status of the atmospheric circulation and weather at a particular time over a wide area; it is derived from careful analysis of simultaneous weather observations made at many observing points in the area, also called weather chart.

Weather modification The changing of natural weather phenomena by technical means, so far, only on the microscale.

Weatherometre A device used to subject articles and finishes to accelerate weathering conditions; for example, a rich ultraviolet source, water spray, or salt water.

Weather vane A device similar to weather cock which indicates the direction in which the wind blows.

Wedge In logging, to drive a wedge into the saw-cut to prevent the saw from binding and to direct the fall of the tree.

Wedge grafting Methods of grafting whose name is derived from the shape of the scion which is of wedge shaped. The stock is split in the form of V-shape at its upper end, in which scion is inserted and tied up. Waxing of the union is completed. Used in grafting apple, pear and plum etc. Also called cleft grafting.

Weed Any plant in a place where it is a nuisance might be considered a weed. The term is most often applied to noncultivated plants that arise unwanted in cultivated area, lawns, pastures, or other areas used by man. Most weeds are prolific and persistent.

Weed control Any device, system or chemical methods used in the destruction, checking spread and growth of the weeds. The method may be physical such as cultivation, cultural practices, manual removal or chemical such as application of weedicides etc.

Weed-dwelling fauna Fauna found on weeds in water; rotifers, flatwarms, water mites, hydra, snails, bristleworms and insect larvae. These are all source of food for fish of omnivorous habit.

Weeder Any tool or device meant for removing weeds from the field.

Weedicides Chemical which are used to control the weeds. Synthetic growth-regulating chemicals are now available for use as weedicides. Selective weed-killers are most commonly used.

Weeding The dislodging and removal of undesirable plants (Weeds) from a field.

Weed killer Weedicides; herbicides.

Weed spray Herbicide.

Weed tree A tree which grows any where, it is not wanted.

Weedy An abundance of weeds in a field, a growth of high vigour in a plant usually associated with fruiting.

Week A time period of 7 days which has been accepted from ancient Babylon; the 7 days of the week were first given names of the seven celestial bodies; the sun, moon, and five visible planets.

Weep (1) To exude sap or moisture from a plant surface, or a cut or bruise. (2) Send forth or be covered with drops, exude sweat, drip. (3) Have drooping branches.

Weep Holes Openings left in retaining walls, aprons, linings, or foundations to allow drainage for preventing build-up of pressure.

Weeping love-grass Eragrostis curvula (Schred.) Nees. Family Gramineae. A densely tufted, tall, erect grass grown for fodder. Low palatability but good for soil conservation.

Weevil Beetle of the family Curculionidae, with mouth parts of the end of a prolonged snout. Injurious to plants and damages seed.

Weft Cross-threads woven into warp to make web.

Weight (1) The gravitational force with which the earth attracts a body. (2) By extension, the gravitational force with which a star, planet, or satellite attracts a nearby body.

Weir Dam across a stream to stop, raise or divert the flow; device for measuring the flow of water.

Weir box Box installed in a channel upstream from a weir to provide an enlargement of the waterway thereby reducing the velocity of approach to the weir.

Weir head Vertical distance from the

crest of a weir to the water surface in the forebay above the weir, not including the velocity head of approach.

Weir pond Pond upstream from a weir usually used to reduce the velocity of approach and allow for full contraction of flow for measurement purposes.

Weldablity Suitability of a metal to be welded under specified conditions.

Weld bead A deposit of filler metal on welding parts. Also known bead.

Weldbonding A process for joining metals in which adhesive, typically an epoxy paste, is applied to the parts, which are than clamped together, spot-welded, and put into an oven (250°F, or 121°C, for 1 hours) to cure the adhesive.

Weld gage A device used to check the shape and size of welds.

Welding rod Filler metal in the form of a rod or heavy wire.

Welding torch A gas-mixing and burning tool for the welding of metal.

Welding transformer A high current, low voltage power transformer used to supply current for welding.

Welfare economics A branch of economics concerned with studying the extent to which economic activity maximises human welfare, and the evaluation of public policies relating to the economy designed to achieve that end.

Welfare state Every man or woman has some fundamental human rights and the state, which protects such rights, is called a welfare state.

Well-bred A plant or animal having pedigree.

Well drain See vertical drain.

Well-drained Refers to a soil which has good un-impeded drainage. It includes surface and sub-surface drainage.

Wentworth scale A geometric grade scale for sedimentary particles ranging from clay particles (diametre less than 1/250 millimetre to boulders (diametres greater than 256 millimetres) in which the size classes are related to one another by a constant ratio of 1/2 [4, 2, 1, 1/2, etc.].

West Indian arrowroot Maranta arundinacea L. Family Marantaceae. An annual herb or shrub, native of tropical America and having edible rhizomes. The varieties blue and yellow which are based on root tuber colour.

West Indian gherkin Cucumis anguria L. Family Cucurbitaceae. A small-fruited cucumber, allied to the common cucumber. Suitable for spring and rainy season sowing. Suitable for pickles and Salad. Native to tropical America and introduced from Canada. Also called gherkin.

West Indian gooseberry Pereskia aculeata Mill. A woody vine of family Cactaceae with edible fruits. Also called lemon-vine and also see otaheite gooseberry.

West Indian lemongrass Cymbopogon citratus (DC.) Stapf (Andropogon citratus DC.). Family Gramineae. Grass whose leaves are a source of a aromatic oil used in perfumery, cosmetics and as flavouring substance.

West Indian lemongrass oil Lemongrass oil containing 75-86 per cent citral content and extracted from Cymbopogon citratus Stapf. Mixed with other aldehydes which lower its quality.

Wet-bulb depression The difference in degrees between the dry-bulb temperature and the wet-bulb temperature.

Wet-bulb temperature The temperature recorded on a thermometre that has its bulb surrounded by a moist muslin bag, thus lowering the temperature by loss of latent heat through evaporation. With the aid of a dry-bulb thermometre and a set of tables, relative humidity can be ascertained.

Wet bundh An Indian term for a perennial irrigation pond. Also used for breeding fish.

Wet climate A climate in which precipitation effectiveness is such that rain forest vegetation prevails.

Wet down To water a plant so that the soil surrounding the roots is soaked, to bring soil to field capacity by rain.

Wet feet Condition in plants when watering is in excess of drainage. Tree or plants having its roots in excessively wet soil.

Wet land (1) Swampy or marshy land. (2) Land which is under gravitational flow of water for irrigation; under river or tank irrigation mostly; ordinarily stagnated with water and rice cultivation is done. May also grow banana, sugarcane, turmeric in high level areas with drainage provision.

Wetland Farming Farming in which the land may be periodically inundated during the growing season.

Wetland puddler A very useful, economical and labour saving implement for producing a fine puddle in wet lands. Used in standing water. The puddler is used in a ploughed land. Consists of three angular-bladed cast iron hubs rigidly fixed to a hollow horizontal pipe which rotates. The diagonal set blades churn up the soil, break up clods and lumps and create a good puddle.

Wet mash Any mash to which milk, water or any other liquid is added for feeding purpose.

Wetness Index Numerical quantity, usually expressed in percentage, used to designate the relation of precipitation or annual runoff of a stream or drainage basin for a given year to the average over a considerable period of years.

Wet processing of coffee Processing of coffee beans, consists in pulping, fermenting, washing, drying and peeling of the berries. Involves considerable use of machinery and is practised for large scale processing.

Wet salt curing Process of curing hide. Hide first spread when freshly cured on sloping platform, salt sprinkled on next day; a second hide spread over the first with hair side down and similarly salted. Process repeated till pile is formed. Also done by immersing clean hide in saturated salt solution for 24 hours.

Wet season A season or time of the year when most of the precipitation occurs. Also a season in which the precipitation is greatly in excess of the normal.

Wet soil (1) A soil which is wet or inadequately drained for short or long periods of time. (2) Marshy or boggy soil.

Wettable powder any material in the form of powder or dust that will easily mix or disperse in water. Also called simply wettable.

Wettable sulphur Sulphur powder with particle size half to thee-fourth of that of ground sulphur, fungicide. Used in combination with other fungicides or insecticides.

Wetted perimetre The length of the wetted contact between a stream of flowing water, measured in a plane at right angles to the direction of flow; the length of the perimetre of a conduit below the water surface; the entire perimetre of a conduit flowing full. The wetted perimetre is used when computing the hydraulic radius.

Wetting agents Materials which directly facilitates contact between spray and sprayed surface; spreading agents. Soaps, which are alkali salts of fatty acids, synthetic material like alkali sulphates or sulphated alcohols, sulphonates and casein, saponin and gelatine also used as wetting agents.

Wetting power The ability of a detergent to lower the surface and interfacial tension of the water and the immersed objects; helping in the wetting and penetration of the material to be cleaned.

Wet weather A weather characterised by rain.

Wet wood Wood having a water soaked appearance because of a high water content, may be caused by bacteria or by physiological factors.

Wet year A year with above normal precipitation.

Wheat Any of the several plants of the genus Triticum, family Gramineae; only a few are important. Triticum sativum Lamk. Triticum aestivum L. The wheat plant is an annual cereal with erect, elastic and cylindrical stem. Grown extensively throughout world. It is a winter crop in India and growth period is controlled by temperature. High temperature at either end, shortens the growth period. Grown in variety of soils. Light loam soils or heavy soils with open structures are good. Rainfall in wheat growing regions varies from 10-100 cm. Black and medium black soils good for taking barani (unirrigated) crop. Sowing and harvesting season varies but sowing starts in India by mid October and continues till mid November. Four species of wheat mostly grown in India.

Wheat aphids Schizaphis (Toxoptera) graminum (R.). These suck the sap out of leaves when the plants are young. Also appear in cluster and suck juice of the soft seeds. Plants become yellow and the yield is materially affected.

Wheat black mould Disease in wheat caused by Cladosporium herbarum. Greenish black growth on glumes, stem and leaves appear.

Wheat black rust Disease of wheat caused by Puccinia graminis triticina (Pers.). Eriks and Henn. It appears in comparatively warmer conditions (20°-30°C) in the first week of March. Characterised by fairly big, elongated, scattered, orange-red pustules on green parts, mostly on stems and leaf-sheaths, then on leaves and ears. Also called wheat stem rust.

Wheat bran By-product in wheat milling used in mash as it is palatable and has laxative effect. Protein content superior to wheat flour. Ill-balanced regarding calcium and phosphorus.

Wheat broad bran The coarser part of the wheat grain removed in the milling process. Much fibre but poor in food nutrients.

Wheat brown rust Disease of wheat caused by Puccinia rubigrovera tritici. It is the earliest to appear when the crop is 5-6 weeks old. Characterised by rust pustules which develop mostly on leaves and leaf-sheaths. Favoured by temperature between 15°-25°C. also called wheat leaf rust.

Wheat bunt This is caused by Tilletia caries and T., foetida. The symptoms of the disease are not usually evident till heading stage. The smutted ears are darker green than normal. The grains are filled with a soft black pasty mass which later on changes to black powder. The grains produced are rotten and give fish like smell. Also called hill bunt or true bunt.

Wheat ear-cockle disease Disease of wheat caused by the nematode Anguina tritici G. Characterised by wrinkling, twisting and various other distortions of leaves in young seedlings, formation of galls on leaves, and hard, dark galls in the place of kernels in the ears. Disease spreads from Punjab to other states. Gall free seeds should be used which can be achieved by floatation method. The method is treating the seeds with 20 per cent brine solution and stirring vigorously. Affected seeds float and can be separated.

Wheat flag smut This is caused by Urocystis tritici Koern. The symptoms are usually found on leaf and leaf sheaths, sometimes on the stems and rarely on the ears. The leaves become twisted and assume a drooping habit which is followed by withering. The outward symptom is marked by formation of grey to greyish black swollen bands. Later on, the epidermis ruptures exposing a black powdery mass of spores. Seed treatment with mercury fungicide is recommended.

Wheat foot-rot This is caused by Helmintho-sporium sativum. Dark brown patches appear on the collar portion. The plant turns yellow and dries up. Symptoms appear 3-4 weeks after sowing.

Wheat germ The embryo of the wheat seed; rich in vitamin B_6.

Wheat germ oil A light-yellow oil extracted from wheat used as a source for vitamin E; as a dietary supplement, and in medicine.

Wheat glume blotch Disease of wheat caused by Septoria nodorum. Characterised by small, linear or oblong blotches light to dark brown in colour, studded with minute black dots appearing on floral bracts and nodal tissues of culms.

Wheat leaf blotch Disease of wheat caused by Septoria tritici. Leaves show mottled patches which later turn straw coloured and are studded with black dots. Spray with Bordeaux mixture (5:5:50) or other copper fungicides can effectively control the disease.

Wheat leaf blight In recent years this disease has caused problem. Warm and humid condition favour this disease. The disease is characterised by the appearance of brown oval spots on leaves. They become irregular in advanced stage. Later on, several spots coalesce and cover the entire leaf. The lowermost leaves are infected first.

Wheat leaf rust See wheat brown rust.

Wheat loose smut Disease caused by Ustilago tritici (Pers.) Roster. In infected plants, instead of normal grains, black shooty mass of spores is produced.

Wheat mamni Small wheat grain mixed with weed seeds removed during cleaning. Not to be fed to cattle over long periods.

Wheat medium bran By-product in wheat milling. Bran particles finer than those of broad bran but have much the same composition. May contain more of endosperm and flour.

Wheat pollards By-product in wheat milling. Similar to medium bran in composition and nature.

Wheat powdery mildew Disease is caused by Erysiphe graminis. It is not serious and damage is negligible. The white powder like masses, consisting of mycelium, conidiophores and condia appear on the upper surface of the leaf, leaf sheaths, stems and glumes are also affected. Dusting with sulphur (200 mesh) at the rate of 28 kg/ha, is the curative measure.

Wheat products Products obtained from wheat after conditioning. Atta (35%), maida (42%), semolina (6%) and wheat bran (17%).

Wheat pythium root rot Disease of wheat caused by Pythium graminicolum, characterised by roots getting stunted and rotten, leaves turning pale green or brown.

Wheat sclerotial disease It is caused by Pellicularia rolfsii. Characterised by white weft of mycelium formed near the base of the plant on which white to reddish-tan spherical sclerotia appear.

Wheat sorts Name given in the past to a mixture of varieties of wheat grown at a time when attention was paid only to a particular kind of grain and not to any distinguishing characters.

Wheat stemrust See wheat black rust.

Wheatstone bridge A four-arm bridge circuit, all arms of which are predominantly resistive, used to measure the electrical resistance of an unknown resistor by comparing it with a known standard resistance. Also known as resistance bridge; wheatstone network.

Wheat stripe rust It is caused by Puccinia glumarum (Schem.) Eriks and Henn. and Puccinia striiformis (yellow rust). Thrives well under colder conditions (20°C). It is characterised by small, yellow, elliptical pustules seen on leaves, which run in rows forming stripes. Sometimes pustules, appear on leaf sheaths and glumes, which later turn black. Also called yellow rust.

Wheat surface grasshopper Chrotogonus trachypterus (Bl.) This grasshopper is injurious to wheat seedlings both in its young and adult stage. It cuts off the plants close to the surface of the ground and feeds on the young leaves. Sometimes attack is serious and resowing has to be done. BHC 5% dust is used @ 27.5 kg/ha to control this.

Wheat tempering Process of preparing wheat grain for grinding, so that the bran may have a cleaner separation from the flour; consists in adding suitable amounts of water to wheat and allowing the mixture to stand for several hours, toughens the bran coat of the wheat kernel, causing it to separate completely out of the flour.

Wheat tundu This is a bacterial disease and is caused by Corynebacterium tritici (Hutch.) Berg et al. Wrinkling of the lower and twisting of the central leaves is the first symptom of the disease. This is accompanied by the exudation of a bright yellow sticky slime enveloping the entire ear. Damage is severe as every grain in the year is lost. The nematode which causes earcockle disease transmits this disease.

Wheat yellow ear-rot See wheat tundu.

Wheat yellow rust See wheat stripe rust.

Wheel Circular frame or disk arranged to revolve on an axis and used to facilitate the motion of vehicles or for various mechanical purposes. An important tool of the modern industrial civilization.

Wheel and axle An elementary mechanical principle for raising heavy weights. Power is applied to the circumference of the wheel which is mounted on an axle. The weight is attached by chain or rope to the axle which serves as a drum.

Wheel barrow A simple arrangement to move light loads on a A type frame on which a wooden box is fitted. At one end it is supported on one or two wheels and provided with handles on the other side.

Wheel barrow type duster Duster mounted on carrying frame with one or two wheels. Power derived from traction wheels and used for dusting row crops in large areas.

— **Sprayer** Sprayer with tank mounted on one or two wheels. Consists of double-acting hand pump filled inside tank. Operated by lever-like handle and provided with pressure chamber. Can spray continuously constantly without operating the handle. Used for spraying in orchards and nurseries.

Wheel base The distance from the centre of a front wheel to the centre of a rear wheel, as on an automobile or wagon.

Wheel hoe cultivator Garden cultivator.

Wheel hub The boss or centre of a wheel from which spokes radiate and is bored to receive the axle.

Wheel plough Any plough set up on wheels.

Wheel puller A device used for pulling automobile wheels from the axles.

Wheel slip It is the relative movement of the wheel in the direction of travel for a given distant under load and no load position (see 'slip').

Wheeze Breathe with audible friction.

Whet Sharpen by rubbing on the stone; stimulate.

Wheel type tractors Versatile all-purpose machine useful for general and agricultural work and provided with pneumatic wheels. See tractor.

Whip (1) An unbranched shoot of a woody plant, particularly in the first year's growth. (2) A tall slender tree in the upper crown colony. (3) To beat cream, eggwhite etc. so that it forms froth.

Whip and tongue graft See whip grafting.

Whip grafting The method of grafting; consists in giving a slanting out of about 3.54 cm at the basal end of the scion-shoot and giving a corresponding cut on the root-stock seedling at about 60-90 cm from the ground level; cut surfaces so placed that their cambial regions unite, tied and sometimes waxed. Union completes in a few months. Also called splice grafting and whip and tongue grafting.

Whip worm Round worm shaped like a whip.

Whirlwind plough Plough especially designed for constructing terraces or shallow drains, having a rotating screw at the rear of the mouldboard which throws the soil to the side by centrifugal action.

White Resembling a surface reflecting sunlight without absorbing any of the visible rays; characterised by the absence of any colour.

White ant See termite.

White alkali A obsolete term used to designate saline soils.

White arsenic (AS_2O_3) A chemical compound of arsenic and used as an insecticide.

White clover Trifolium repens L. A herb of family Papilionaceae grown for use as a fodder and is called shaftal.

White flowered gourd See bottle gourd.

White dead nettle Lamium album L. A perennial herb of family Labiatae whose tops of stems are used as vegetable.

White gourd See ash-gourd.

White grubs Phyllophaga longipennis Bl., Holotrichia longipennis Bl., Xylotrupes gideon L., Anomala polita Bl., A. rugosa, A. dimidiata Hope., Lachnosterna consanguinea Blach. Pests of potato, also on rice, sugarcane and maize. Cuts plants below soil surface and feed on them or bore into tubers or setts. Adults damage leaves and fruits of apple, pear, plum, peach, cherry.

White jute See capsularis jute.

White lotus Nymphaea nouchali Burm. (N. pubescens Willd.; N. lotus Hook. f. and Thoms.; N. lotus var. pubescens Hook. f. and Thoms.). Family Nymphaceae. A floating aquatic ornamental herb native of S.E. Asia; rhizomes and seeds edible.

White lupine Lupinus albus L. A herb

of the family Papilionaceae grown for fodder and green manure.

White mangrove Avicennia officinalis L. (A. tomentosa Jacq.; A. resinifera Forst.). Tree whose bark is employed in tanning.

White Mulberry Morus alba L. Family Moraceae. A moderate sized short lived deciduous tree, native of China. Coppices with vigour; frost-hardy. Does best under good shade. Propagation by direct sowing, planting entire transplants, branch cutting and stumps. Wood used as fuel, for tool handles, agricultural implements, boats, and furniture etc. Leaves for fodder and for raising silk-worms. Fruits sweet and edible.

White mustard Brassica hirta Moench. [Brassica alba (L.) Boiss.] Family Cruciferae. an oil seed crop grown in India mostly mixed with grain and wheat. Seeds mixed with black mustard and rarely for oil extraction. Young leaves and tender shoots used as vegetable.

White oak See oak.

White pepper Ripe fruit of plant Piper nigrum L.; deprived of its tender skin and the pulp.

White popinac Leucaena leucocephala (Lamk.) de Wit (Leucaena glauca (L.) Benth). A tree of family Mimosaceae, native of tropical America and found throughout India. Young fruits and seeds edible, leaves used as fodder. Good soil cover when planted thick, used as hedge and windbreak.

White phosphorus The element phosphorus in its allotropic form, a soft, waxy, poisonous solid melting at 44.5°C; soluble in carbondi-sulfide, insoluble in water alcohol; self igniting in air. Also known as yellow phosphorus.

White sapote Casimiroa edulis La Llave. Family Rutaceae. A shrub or small tree, with leathery leaves and small greenish flowers, a native of Mexico and central America. Grown for its sweet, persimmon-like grey or yellowish green edible fruits.

White sandal wood See sandal wood.

White silk cotton Cochlospermum religioum (L.) Alston (C. gossypium DC.). Family Cochlospermaceae. A tree native of India whose fibre obtained from flowers is used as stuffing material. Tree also yields edible gum. Also called *kumbi*.

White squill Urginea indica (Roxb). Kunth (Scilla indica Roxb.) A herb, the bulbs of which are used as an expectorant and stimulant. Also called Indian drug squill.

White sweet clover See hubum clover.

White yam Dioscorea alata L. Dioscoreaceae family. A climbing shrub, whose tubers are used for vegetable. Also called greater yam.

Wholesale market Chiefly market for foodstuffs and raw materials. Where items are sold on bulk basis.

Wholesaler A dealer who buys and stocks goods for resale to others; an intermediary between manufacturer or supplier and retailer.

Whorl Circle of flowers or leaves, arising from one point.

Whorled See verticillate.

Whole root graft Root-grafting in which a whole root is used for the stock instead a part of it.

Whole grain Cereal grain which is not cracked or chopped.

Wicker Plaited twigs as a material for baskets, mats protective covering etc.

Wicket A small sluice or opening for letting the water out of a canal lock.

Widow-maker Logger's term to describe a tree cut off at the stump but hung up in tops of adjacent trees which keep it from falling.

Wight's sago palm Arenga wightii Griff. Family palmaceae. A palm whose flower stalks yield toddy.

Wig wag An instrument for measuring solar radiation.

Wild Uncultivated; (1) Growing without cultivation. (2) Not domesticated.

Wild almond See hill coconut.

Wild ber Ziziphus sp. Family Rhamnaceae. A hardy perennial shrub or tree invading cultivated fields in the dry zones, propagated by seeds. Difficult to eradicate because of deep root system and power to regenerate.

Wild cherry Prunus cerasoides D. Don (Prunus puddum Roxb. ex. Wall.). Family Rosaceae. Wild variety of cherry, used as stock for grafting cultivated cherry varieties. Fruits edible; wood used for walking sticks and stones made into rosaries.

Wild date (1) Phoenix pusilla Gaertn. (Phoenix farinifera Roxb.) Family Palmaceae. A bushy palm; the pulp of the seed is eaten, trunk contains sago; leaves used for making mat etc. (2) Phoenix Sylvestris (L.). Roxb. (Elate sylvestris L.). Family Palmaceae. A tall palm native of India, the fruits of which are edible and are known as *khajur*.

Wild fence A wooden enclosure about 5 m square and 76.2 cm high with a precipitation gauge in its centre. The function of the fence is to minimise eddies around the gauge, and thus ensure a catch which will be representative of the actual rainfall or snowfall.

Wild fenugreek Trigonella polycerata L. Family Papilionaceae. A trailing, trifoliate, common winter weed, the leaves of which are used as vegetable.

Wild fire A bacterial disease of tobacco caused by Pseudomonas tabaci and characterised by the appearance of brown spots surrounded by yellow rings, which turn dark, rot and fall out.

Wild flooding, or **Free flooding** Surface flooding in which water is diverted to non-prepared areas and is let to flow down the natural slopes without controlling its distribution.

Wild jamun Syzygium fruiticosum (Roxb.) DC. (Eugenia fruticosa Roxb.). A tree of the family Myrtaceae grown as an avenue.

Wild land Uncultivated land. It may or may not be maintained by the owner for its productive vegetative cover or for wood, forage production, recreation, or wildlife.

Wild lettuce Lactuca serriola L. (Lactuca scariola L.). A herb of family Compositae, the seeds of which are a source of a semi-drying oil used in food. Also called prickly lettuce.

Wildlife area improvement Improving the area by planting, control of grazing, water, fire etc. by which food, cover and water is provided to the wildlife.

Wildlife area management The management of an area, particularly of its vegetation and water for the benefit of wildlife.

Wildlife conservation The preservation of rare or vanishing species of fauna.

Wildling (1) Any wild animal or plant. (2) A cultivated plant which has become wild and thrives without cultivation; an escape. (3) In forestry, a seedling produced naturally outside the nursery.

Wild mustard Brassica sinensis of family Cruciferae. Annual rabi season weed.

Wild oat Avena fatua. Family Gramineae. Annual weed in cultivated crops.

Wild orange tree Toddalia asiatica (L.) Lamk. (Toddalia aculeata Pers.). A shrub of family Rutaceae whose roots yield a yellow dye; leaves and fruits edible.

Wild passion flower Passiflora incarnata L. Family Passifloraceae. Native to North America. Fruits are edible. Also called maypop.

Wild rice Oryza rufipogon Griff. Family Gramineae. Species of wild rice indistinguishable in the early stages of growth from cultivated rice and hybridizing freely with it. Transmits its character of grain shedding to cultivated rices. Hand weeding only way to remove from the fields.

Wild safflower Carthamus oxycanthus M. Biep. Family Compositeae. An annual herbaceous thorny weed occurring in moderately dry places. Hardy, grows with winter crops, Seed-oil used for illumination and is edible.

Wild strawberry Fragaria sp. Temperate fruit growing wild. The fruits are of poor quality.

Wild tea Camellia Kissi Wall. (Camellia drupifera Dyer. non-Lour.). Family Ternstroemiaceae. A small tree, the leaves of which are used as substitute for tea.

Wilt Loss of freshness or dropping of plants due to inadequate water supply or excessive transpiration; a vascular disease interfering with utilisation of water.

Wilting coefficient (Obsolete) A calculated value of the approximate wilting point or permanent wilting percentage. Calculated as follows:

Wilting coefficient = hygroscopic coefficient/0.68, or

Wilting coefficient = moisture

equivalent/1.84.

Wilting point (Permanent wilting point) The moisture content of a soil at which plants (specifically sunflower plants) wilt and fail to recover their turgidity when placed in a dark, humid atmosphere. The wilting point is commonly estimated by measuring the 15-bar percentage of a soil.

Wilting range See wilting coefficient.

Winch A hand-crank for turning a machine, a hoisting machine.

Wind The horizontal movement of air on and above the earth's surface.

Wind break Planting of trees, shrubs, or other vegetation, usually perpendicular or nearly so to the principal wind direction, to protect such things as soil, crops, homesteads and roads against the effects of winds, such as wind erosion and the drifting of soil and snow.

Wind blast The current of air forced out by a blower or fan.

Wind blower A attachment of a thresher; consists of a large fan funnel, intake device, discharge pipe, for forcing the straw away from the thresher.

Wind borne Pertaining to disease, seeds, spores etc. distributed or propagated by wind.

Wind burn A withered, blasted appearance of foliage, action of violent wind on a plant; or the bronzed condition of some evergreen foliage caused by insufficient water supply associated with high transpiration ratio and low water content of soils.

Wind erosion Detachment, transportation and deposition of soil by the action of wind. The removal and redeposition may be in more or less uniform layers or as localised blowout and dunes.

Wind fall Falling of trees or fruits prematurly due to wind, also the area where trees have been blown down by wind.

Wind firm Able to withstand heavy wind.

Wind gap A natural land feature; a narrow grove or notch in a ridge which may have been cut by a stream.

Wind gauge See Anemometre.

Wind measurement The determination of three parametres; the size of an air sample; its speed, and the direction of motion.

Wind mill Any of various mechanisms, such as a mill, pump, or electric generator, operated by the force of wind against vanes or sails radiating about a horizontal shaft. Generally a low H.P. Up to $^3/_4$ H.P. is generated.

Wind power The force of the atmosphere in motion. The power may be used to serve many useful purposes such as to run wind mill etc.

Wind pollination Pollination which occurs because of the pollen carried by the wind instead of insects etc. The plants in which such pollination occurs are called wind-pollinated plants.

Wind pollinated See wind pollination.

Wind pressure The force of wind against an obstacle.

Wind puff Air under the skin.

Wind rose A diagram indicating the frequency and strength of winds in a definite locality for a given period of years. It is conventional to consider the wind direction as the direction from which the wind blows, e.g. a north-east wind will carry pollutants to the south-west of the source.

Window A row of hay or unthreshed grain raked up to dry.

Windrowing The method of protecting sugarcane buds on sets intended for seed from frost attack. Consists in pulling out plant with roots, laying them in pits covered by trash and earth. Kept for 2-3 months.

Wind storm Strong to violent winds; usually not accompanied by precipitation.

Windsor bean See broad bean.

Wind strip cropping The planting of tall growing crops and low growing crops in alternately arranged long but narrow strips laid across the direction of prevailing winds, regardless of the contour. A soil conservation measure.

Windward Lying in the direction in which the wind blows.

Wine A fermented alcoholic beverage made from miscellaneous fruits.

Wine grape Grape variety suitable for wine production.

Wine palm See toddy palm.

Wing (1) A leaf like dry or membranous expansion or appendage of a plant part. (2) Either of the lateral petals of a pealike flower. (3) The organ of flight in birds, insects etc. (4) The outer corner of the cutting edge of a plough. (5) Fan or blower blade, the vane of a windmill.

Wing bearing The flat surface on the wing of the plough share, which remains in contact with the ground and carries some of the downward thrust.

Wing case A horny cover or a modified forewing protecting the flying wings in certain insects. Also called wing sheath.

Wing of the share The outer end of the cutting edge of the share which supports the plough bottom is known as wing of the share.

Winged fruit A fruit having a wing.

Winged pea Lotus tetragonolabus L. A herb of family Papilionaceae and grown for fodder.

Winged yam See white yam.

Wing seed Any seed having a wing.

Wing sheath See wing case.

Wing walls Walls joining the abutments of a structure to an earth dike or the banks to provide a longer path of percolation around the end of structure and/or to improve flow conditions upstream and downstream of the controlling section.

Winkler titration A chemical method for estimating the dissolved oxygen in seawater; manganous hydroxide is added to the sample which reacts with oxygen to produce a manganese compound which in the presence of acid potassium iodide liberates an equivalent quantity of iodine that can be titrated with standard sodium thiosulfate.

Winnow To separate refuse or inferior specimens from a substance as to separate husk from the grain by wind or artificial current of air.

Winnower A machine used for winnowing. A fan is used to produce a current of air over the sieves which gets a rocking and jolting motion when fan is worked with the motor. The husk or the chaff is taken out from one outlet whereas seeds come out of the outlet provided below the sieves.

Winter The cold season of the year in temperature and arctic zone. In India a season between autumn and spring.

Winter annual A plant which germinates in the fall, lives over winter, produces its seed during the following spring, and then dies.

Winter crop See winter annual.

Winter dormancy A dormancy that occurs in winter.

Winter grafting Grafting performed in winter.

Winter green Gaultheria fragrantissima Wall. Family Ericaceae. An aromatic herb whose leaves give winter green oil and fruits are edible. Also called fragrant wintergreen.

Wintergreen oil Essential oil prepared from the plant Gaultheria fragrantissima from which salicylic acid used in the preparation of aspirin is obtained. Used in some insecticidal preparations. Also called oil of wintergreen.

Winter-hardy Pertaining to plants which are able to successfully withstand the rigours of the winter.

Winter injury Injury caused to plants due to rigours of the winter.

Winter irrigation Irrigation of lands during the non-crop or winter season in order to store water in the soil for subsequent use by plants.

Winter lettuce See endive.

Winter onion See top onion.

Winter plum See persimmon.

Winter pot herbs Herbs like spinach, Indian spinach, beet, fenugreek which thrive at low temperature.

Winter rape Brassica napus L. Family cruciferae. An annual herb native to Europe and Asia. Leaves used as vegetable.

Winter squash Cucurbita maxima. Duch. Family Cucurbitaceae. Fruit used a vegetable. Also called true squash, red pumpkin.

Winter storage The storage of nutrients by plants for winter maintenance and/or for the following growing season.

Winter vegetable A vegetable produced during winter months.

Wire grass Any grass with thin, long wiry leaves as Bermuda grass.

Wire fence (ordinary) A fence in which wire is used as a barrier such as barbed wire, wire mesh etc.

— **electrical** A fence of plane wire in which intermittent electrical current of high voltage is passed, gives a good but not too harmful shock. Scares animals.

Wire-cutter A plier-like tool used for cutting wires.

Wire grub Wire worm.

Wire stem A stem canker stage in damping off in cabbage.

Wire worm The larva of the click beetle of family. Elateridae, which eats seeds plant roots etc.

Wisp Small bundle or twist of straw.

Witches Broom Disease symptom with abnormal brushlike development of many weak shoots.

Witch weed See striga.

Wither Make or become dry and shrivelled; deprive of or lose vigour.

Withering Pertains to the condition of dehydration, dropping of a plant.

Wobble See wabble.

Wood (1) Technically, xylem; popularly, the hard fibrous xylem of trees and shrubs. (2) Growing trees occupying a large portion of ground, forest.

Wood apple Feronia limonia (L) Swingle (Feronia elephantum Correa). Family Retaceae. A tree found all over India whose fruits are edible. Gum obtained from the trunk and branches is a substitute for gum arabic.

Wood-apple slug caterpillar Parasa lepida Cr.; caterpillar feeding on leaves of wood-apple.

Wood ash The residue that remains after wood is burnt. A potash manure. Unleached wood ash contains 5-6 per cent of potash in the form of potassium carbonate and 25-30 per cent of lime. Both counteract acidity in soils.

Wood cellulose Residual which remains after chemical treatment of wood to remove extractives, lignin and hemicelluloses.

Wood charcoal Carbon fuel prepared by heating or charring wood out of contact with air, used as fuel, as additive to poultry mash or scratch feeds.

Wood fibre Much elongated cells, with attenuated ends and thickened walls, empty of living content at maturity, whose chief function is to support or protect, the parts of the plant. The walls are lignified.

Woodland Any land used primarily for growing trees and shrubs. Woodland includes, in addition what is ordinarily termed "forest" or "forest plantations", shelter belts, wind breaks, wide hedge rows containing woodland species for wildlife food or cover, stream and other banks with woodland cover etc. It also includes farmland and other lands on which woody vegetation is to be established and maintained.

Woodland drainage Drainage needed for the removal of excess surface water to permit the germination and production of forest seedlings.

Woodland pastures Farm woodlands used for grazing; wooded area where grass and other herbaceous plants grow in interspaces.

Woodlots Area set apart to serve as a source of fuel wood and other farm needs. Area that is fallow, uncultivated, unsuitable for agriculture is used for the purpose.

Wood flour Pulverised wood used in casting moulds to furnish a reducing atmosphere, help overcome expansion of sand, increase flowability of the metal, and improve casting finish.

Wood physics The area of wood science concerned with the physical and mechanical properties of wood and the factors which affect them.

Wood preservation Treatment of timbers with chemicals so as to enhance its durability.

Wood preservative A material used to coat wood to kill insects and fungi, but not usually classed as an insecticide; coal tar creosote and its derivatives are the most widely used wood preservatives.

Wood screw A threaded fastener with a pointed shank, a slotted or recessed head, and a sharp tapered thread of relatively coarse pitch for use only in wood.

Woody With a high proportion of lignified tissue. (1) Abounding in woods. (2) Pertaining to a plant which has some characteristic of wood.

Woolly aphis Eriosoma lanigerum Hausm. Family Aphidae. Serious insect pest of apples in the hills. Nymphs and adults suck sap from branches, trunk and roots; found in colonies covered with wooly substance. Badly attacked trees develop gall and produce poor quality fruits. Plants in nursery die-off. Also called wooly aphid and American blight.

Woolly aphid See woolly aphis.

Woolly oak Quercus lanata. Large evergreen tree with low branching habit and short bole. Grows on a variety of soils. Coppices well. Raised by direct sowings and transplanting. Wood very hard and makes good fuel.

Work Work is the physical entity which is measured by the line integral of the force along the path taken by the point of application of the force. In its common usage however, it is usually taken as a productive operation.

Workability The ability of quality of being snaped or worked.

Work angle In arc welding, the angle in a plane normal to the weld axis between the electrode and one member of the joint.

Work animal Any domestic animal primarily for the purpose for taking work such as bullock, horse etc.

Worker (1) A person who works with hands. (2) A non-fertile, female in a colony of social insects such as bees, termites etc.

Worker cell One of the cells in a beehive in which a worker is raised.

Working capital The net liquid resources of a business, i.e. the current assets minus the current liabilities.

Working classes An antiquated term used to describe those who work for weekly wages and are in the lower income groups.

Working farm capital The capital used in the working of a business. In farm business it includes, livestock, machinery, farm implements, feed, cash and other supplies etc.

Work-kinetic energy theorem The theorem that the change in the kinetic energy of a particle during a displacement is equal to the work done by the resulting force on the particle during this displacement.

Working sample A reduced sample taken from the submitted sample in the laboratory, on which one of the seed quality tests is made.

Workshop A place or shed where the machinery equipments are repaired.

Workstock The total number of animals used for work on a farm.

Work study A generic term for those techniques, particularly method study and work measurement, which are used in the examination of human work in all its context, and which lead systematically to the investigation of all the factors which effect the efficiency and economy of the situation being reviewed in order to effect improvement.

Worm (1) A small, soft bodied, invertebrate, usually apparently limbless, creeping animal especially, such as in segmented in rings or is parasitic in the intestine or tissues of animals; larva; grub, maggot earth worm, silkworm etc. (2) Spiral part of screw. (3) A gear.

Worm gear A gear with teeth cut on an angle to be driven by a worm used to connect nonparallel, non-intersecting shafts.

Worm parasite The parasite which is able to enter in his host only through the wound.

Worm wheel A gear wheel with curved teeth that meshes with a worm.

Worm wood Artemisia obsinthium L. Family Compositae. A herb from the flowers of which the drug santonin is obtained. A native of Europe.

Wormy Worm infested products, plants or animals.

Wound The disruption or continuity of an internal or external tissue, violently.

Wrapper leaf A type of tobacco used for the outer covering in the manufacture of cigars.

Wriggle Move body with short twistings, worm-like motions.

Wright's stain Mixture of eosin and methylene blue to demonstrate blood cells and malarial parasites.

Wrinkle Furrow like crease or depression or ridge in the skin or other flexible surfaces.

Wrought timber Timber the surface of which have been planed.

Wry Distorted, turned to one side.

X

Xantheine Soluble portion of the yellow colouring matter in flowers.

Xanthic Yellowish, as xanthic flowers.

Xanthin Insoluble part of yellow colouring matter of flowers, substance related to uric acid.

Xanthomonas alfalfae See lucerne bacterial disease; sweet clover disease.

— **compestris** See crucifera black rot.

— **compestris var. aramoriaceae** See cabbage bacterial blight

— **citri** See citrus canker.

— **cyamopsidis** See cluster bean bacterial leaf spot.

— **malvacearum** See cotton blackarm.

— **oryzae** See rice bacterial blight.

— **phaseoli indicus** See bacterial bean blight.

— **phaseoli sojense** See soyabean disease.

— **sesbaniae** See sesbania disease.

— **stizolobiicola** See velvet bean bacterial leaf spot.

— **vesicatoria** See chilli bacterial leaf spot, tomato bacterial leaf spot.

— **vasculorum** See sugarcane diseases.

— **vignicola** See cowpea bacterial blight.

Xanthone ($C_{13}H_8O_2$) An insecticide and ovicide used in the control of orchard mites.

Xanthophyll Yellow carotenoid pigment ($C_{43}H_{56}O_2$) found with chlorophyll and carotene in green plants and egg yolk.

Xanthoxalis corniulata See Indian sorrel.

X-axis (1) A reference axis in a quartz crystal. (2) A horizontal axis in a system of rectangular coordinates. (3) That line on which distance to the right or left (east or west) of the reference line are marked, especially on a map chart or graph.

Xenia The immediate effect of pollen on the endosperm, due to the phenomenon of double fertilisation in the seed plants.

Xenogamy Cross-fertilisation between flowers on different plants.

Xeric Dry, pertaining to soils which normally have a low moisture content and to plants which tolerate a very dry soil or arid conditions.

Xerography A printing method developed by the Xerox Corporation: a negative image is formed by a resinous powder on an electrically charged plate, and this image is transferred and thermally fixed on to a paper as a positive.

Xerophilous Adapted to hot and dry conditions.

Xerophytes Plants that grow in extremely dry soils or soil materials.

Xerophytic vegetation Vegetation commonly found in the desert region.

Xerothermic Capable of withstanding both drought and heat.

X-ray hardness The penetrating ability of X-rays; it is an inverse function of the wave length.

Xyleborus compactus See coffee shot-hole borer.

— **morstatti** See coffee shot-hole borer.

Xylem Lignified portion of vascular bundles in plants; tissue that make up most, if not all, of the lumbers, wood and wood products of commerce.

Xylene equivalent This term has the same significance as moisture equivalent except that Xylene is substituted for water.

Xylocarp Hard woody fruit.

Xylocarpous Of the nature of having

hard woody fruit.

Xylophagus Of insects which feed on wood.

Xylotrechus quadripes See coffee white stem borer.

Xylotrupes gideon See white grubs.

XY recorder A recorder that traces on a chart the relation of two variables, neither of which is time.

Y

Y Y-shaped tile junction used to connect two different tile lines.

Yam Tuber from plants of the genus Dioscorea, family Dioscoreaceae — cultivated for food or for ornament. See white yam, potato yam, lesser yam.

Yam bean See potato bean.

Yard (1) A facility for building and repairing ships. (2) A unit of length in common use in the United States and United Kingdom, equal to 0.9144 metre, or 3 feet. (3) Ground adjoining the farm buildings or house; also enclosed area where animals or birds are kept.

Yarn Thread produced by the clinging together of cotton fibres during spinning.

Yashul Mauritius raspberry; Rubus rosaefolius Sm. Family Rosaceae. A shrub cultivated in Kumaon and Khasia Hills. Fruits edible.

Y-axis (1) A line perpendicular to two opposite parallel faces of a quartz crystal. (2) A vertical axis in a system of rectangular co-ordinates. (3) That line on which distances above or below (north or south) the reference line are marked, especially on a map, chart, or graph.

Y-divider A structure in the form of Y to divide flow into two ditches.

Year Any of several units of time based on the revolution of the earth about the sun; the tropical year to which the calendar is adjusted is the period required for the sun's longitude to increase 360°; it is about 365.24220 mean solar days. Abbreviated as yr.

Year (Seed) A year in which a given species produces (over a considerable area) a seed crop greatly in excess of the normal. Applied usually to trees of irregular or infrequent seed production.

Yeast Fungus. Unicellular body of microscopic size of the family Saccharomycetaceae requiring moderate temperatures, moisture and food like starch and sugar for growth. Causes fermentation of sugars into alcohol with evolution of carbon dioxide. Easily destroyed in moderate heat and can not grow in medium with more than 60 per cent sugar or high concentration of vinegar or alcohol. Saccharomyces cerevisiae most common yeast used in preparation of bread and alcohol — and is called brewer's yeast or bread yeast. Some species parasite on man and animal. Also see fission yeast and budding yeast.

Yellow The hue evoked in an average observer by monochromatic radiation having a wave length in the approximate range from 577 to 597 *nanometres*; however, the same sensation can be produced in a variety of other ways.

Yellow bauhinia Bauhinia tomentosa L. A small tree of family Caesalpiniaceae, whose bark is used as cordage. Also called St. Thomas tree, *Kachnar*.

Yellow corn Maize which has yellow kernels due to carotenoid pigments.

Yellow granadilla See water melon.

Yellowing The loss of green colour of a plant's foliage due to degeneration or disorganisation of chlorophyll or chloroplasts. Also see chlorosis.

Yellow lupine Lupinus luteus L. A herb of the family Papilionaceae grown for green manure.

Yellow mealworm Tenebroides mauritanicus L. A pale white grub of a flat shiny black beetle that attacks broken grains and flour of wheat and oats and maida. Grub stage lasts 10 to 20 months.

Yellow myrobalan Terminalia chebula Retz. Family Combretaceae. Medium sized tree of deciduous forests. Fruits used for tanning and wood suitable for construction purposes.

Yellow Newton Variety of apple. Fruit large, roundish to slightly flat. Flesh creamy, firm, crisp, juicy and of excellent quality.

Yellow oxide Yellow mercuric oxide.

Yellow passion fruit Passiflora flavicarpa. Family Passifloraceae. Introduced variety of passion fruit from Sri Lanka to south India. Fruit has more pronounced and pleasant flavour than that of purple passion fruit. Spherical or ovoid in shape.

Yellow phosphorus The toxic component of some commercial rat poisons.

Yellow podzolic soils A taxonomic zonal group of soils, characterised by a thin layer of organic matter over a greyish leached layer underlain by a yellow B horizon.

Yellows Term applied to diseases in which yellowing or chlorisis is a principal symptom.

Yellow sarson Brassica compestris L. var. sarson Prain (Brassica compestris L. var. glauca Duth & Full.) Family Cruciferae. An oil seed crop containing 45 per cent oil, oil sweet flavoured commercially called colza oil and used for culinary, lighting and toilet purpose. Cake fed to cattle. Also called Indian colza, *sarson*.

Yellow sweet clover Melilotus officinalis (L.) Lamk. A herb of family Papilionaceae used as fodder. Native to Europe and Asia.

Yellow teak Adina cordifolia (Roxb.) Benth and Hook. f. Family Rubiaceae. Generally a large tree and requires more than 100 cm of rainfall. Wood used for building farm implements, furniture and battery separators. Also called saffron teak.

Yellow vetchling Lathyrus aphaca L. A climbing herb of family Papilionaceae grown in various states of India for use as fodder.

Yemma budding See chip budding.

Yield (crop) The quantity of, or, aggregate product resulting from cultivation or growth. For cultivation yield is measured in ha as yield per ha.

Yield (of concrete) The volume of concrete resulting from a mix of stated quantities of cement, water and aggregates.

Yielder A plant or animal which produces products. Usually referred as good or poor yielder.

Yield of Drainage Basin (Annual yield of Drainage Basin) Total volume or flow from a drainage basin for a long stipulated period time, e.g. 'annual yield of drainage basin' is the mean annual runoff.

Yield point The lowest stress at which strain increases without increase in stress.

Yield strength The stress at which a material exhibits a specified deviation from proportionality of stress and strain.

Yield, Sustained As applied to a policy, method, or plan of forest management, the term implies continuous production, with the aim of achieving, at the earliest practicable time, an approximate balance between net growth and harvest, either by annual or somewhat longer periods.

Yoke Variously shaped wooden beam or frame which is placed on the neck of two bullocks to make them work as one team and to attach implements or cart to pull.

Young ice Newly formed ice in the transitional stage of development from ice crust to winter ice.

Young's modulus The ratio of a simple tension stress applied to a material to the resulting strain parallel to the tension.

Young Soil See beginning to form.

Y Punch In a standard punched card, a hole in the top most row.

Zaffran See Saffron crocus.

Zaid Season between rabi and kharif March to June.

Zaid kharif Autumn sown crop.

Zaid rabi Spring sown crop.

Zamindar Big land owner who collects land rents from tenants and pays land revenue to Government. This system has been abolished in India after independence.

Zamin kand See elephant foot yarn.

Z-axis (1) The optical axis of a quarts crystal. (2) One of the three axis in a three-dimensional cartesian coordinate system: in a rectangular coordinate system it is perpendicular to the x and y axis.

Zanthoxylum oxyphyllum See timar mazenga.

— **rhetsa** Called badrang. A large tree of the family Rutaceae.

Zenithal rain In the tropics or subtropics, the rainy season which recurs annually or semi-annually at about the time that the sun is most nearly overhead.

Zea mays See maize.

— **mays var. ceratina** See waxy maize.

— **mays var. everta** See popcorn.

— **mays var. Saccharata** See sweet corn.

— **mays var. tunicata** See pod maize.

Zedoary Curcuma zedoaria Rosc. (Curcuma zerumbet Roxb.) Family Zingiberaceae. A herb cultivated throughout India for its tuberous rhizomes which are used in perfumery and cosmetics.

Zein An alcohol-soluble protein obtained from maize.

Zeolites A group of light coloured hydrous silicates containing potash and soda. It plays an important role in the base exchange in soils.

Zero The additive identity element of an algebraic system.

Zero-access storage Computer storage for which waiting time is negligible.

Zero adjuster A device for adjusting the pointer position of an instrument or metre to read zero when the measured quantity is zero.

Zero defects A programme for improving product quality to the point of perfection, so there will be no failures due to defects in construction.

Zero layer A reference level in the ocean, at which horizontal motion is at minimum.

Zero opportunity cost A situation which may arise in conditions of substantial unemployment of resources; these resources may be directed into public investment projects without any reduction in production elsewhere in the economy, that is without opportunity cost.

Zero-order reaction A reaction for which reaction rate is independent of the concentrations of the reactants; for example: a photochemical reaction in which the rate is determined by the intensity of light.

Zero pasture (grazing) See soiling crop.

Zero plane displacement An empirically determined constant introduced into the logarithmic wind profile to extend its applicability to tall vegetation.

Zero-point energy The kinetic energy retained by the molecules of a substance at a temperature of absolute zero.

Zero population growth A theory which advocates that there be no increase in population, that each person replace only oneself, and that birth control be practised in all nations.

Zerumbet ginger Zinziber zerumbet (L.) Rosc. ex Smith. Family Zingiberaceae. Cultivated throughout India. The rhizomes are used as spice and condiment.

Zeta potential The electrical potential that exists across the interface of all solids and liquids. Also known as electrokinetic potential.

Zeuzera coffeae See coffee red borer.

Zig-zag system System in which field drains as well as mains are constructed zig-zag to reduce high velocities.

Zinc Zn. Plant food element, absorbed as Zinc (Zn^{++}) ion. The ion is considered essential in enzyme systems; necessary for important reactions in plant metabolism or life activity of plants. Helps in formation of some growth hormones. Also helpful in reproduction of certain plants. See Zinc deficiency.

Zinc acetate [$Zn(C_2H_3O_2)_2 \cdot 2H_2O$] Poorly white crystals with an astringent taste; soluble in water and alcohol; decomposes at 200°C; used to preserve wood in textile dyeing, and as an analytical reagent, a feed additive, and a polymer cross-linking agent.

Zinc arsenate [$Zn(AsO_4)_2$] A toxic white powder that is insoluble in water, soluble in alkalies, used as an insecticide. Also known as Zinc orthoarsenate.

Zinc Arsenite [$Zn(AsO_2)_2$] A compound used as a component of certain insecticidal sprays.

Zinc bromide ($ZnBr_2$) Water and alcohol soluble, white crystals and melt at 294°C; used in medicine, manufacture of rayon, and photography, and as a radiation viewing screen.

Zinc carbonate ($ZnCO_3$) White crystals that are insoluble in water, soluble in alkalies and acids; used in ceramics and ointments, and as a fire proofing agent and feed additive.

Zinc chloride ($ZnCl_2$) Water-and alcohol soluble, white, fire-hazardous crystals that melt at 290°C, and are irritating to the skin; used as a catalyst and in electroplating, wood preservation, textile processing, petroleum refining medicine, and feed additives.

Zinc cyanide [$Zn(CN)_2$] A toxic, white powder that is insoluble in water and alcohol, soluble in alkalies and dilute acids; melts at 800°C; used as an analytical reagent and insecticide, and in medicine and metal plating.

Zinc deficiency Condition of plants resulting from deficiency of zinc. Important in many fruit trees and particularly citrus causing mottle leaf. Results in interveinal chlorosis often with necrosis, reduced leaf size, malformation of leaves and reduced food production. Greyish appearance of straw in mature plants. Essential for rice in alkali lands during reclamation.

Zinc formate [$Zn(CHO_2)_2 \cdot 2H_2O$] Toxic, white crystals that are soluble in water, insoluble in alcohol; used as a catalyst, water proofing agent and wood preservative.

Zinc naphthenate [$Zn(C_6H_5COO)_2$] A combustible, viscous, acetone-soluble solid; used in paints, varnishes, and resins and as a drier, wetting agent, insecticide and fungicide.

Zinc oxide (ZnO) A bitter-tasting, white to gray powder that is insoluble in water, soluble in alkalies and acids; melts at 1978°C; used as a pigment, mold-growth inhibiter, and dietary supplement, and in cosmetics, electronics and colour photography.

Zinc phosphide (Zn_3P_2) A toxic, alcohol insoluble, gray gritty powder that reacts violently with oxidizing agent; melts at over 420°C, decomposes in water; used as a rat poison and in medicine.

Zinc sulfate ($ZnSO_4 \cdot 7H_2O$) Efflorescent, water-soluble colourless crystals with an astringent taste; used to preserve skins and wood and as a paper bleach, analytical reagent, food additive, and fungicide. Also known as white vitriol, zinc vitriol.

Zinc sulphate-lime mixture Mixture for spraying on plants to overcome zinc deficiency, diseases or symptoms. A formula of 1.5:100 (zinc sulphate lime and water) is used in chronic cases.

Zingiber Cassumunar See cassumunar ginger.

— **officinale** See ginger.

— **zerumbet** See zerumbet ginger.

Ziram ($ZnC_6H_{12}N_2S_4$) A colourless solid; melting point is 246°C; used as a fungicide for vegetables and some fruit crops. Also known as zinc dimethyl-di-thiocarbanate.

Zircon Mineral found in soils; zirconium silicate.

Ziziphus jujube See jujube.

— **mauritiana** See jujube.

— **nummularia** See jharber.

— **oenoplia Mill** Wild species of ber found in India. Good root stock for ber. Fruit edible and bark used as a tan.

— **rotundifolia** See jharber.

— **rugosa Lamk**. Family Rhamnaceae. Wild species of ber, good root-stock for ber. Fruits edible.

— **sativa** See common jujube.

— **vulgaris** See common jujube.

— **xylocarpa** See kat ber.

Zoisia matrella See Manila grass.

Zonal centrifuge A centrifuge that uses a rotating chamber of large capacity in which to separate cell organelles by density-gradient centrifugation.

Zonal soil Any one of the great groups of soils having well developed soil characteristics that reflect the zonal influence of climate and living organisms, mainly vegetation, as active factors of soil genesis. The zonal great soil group was the name of a former taxon that was similar to the present Canadian concept of the great group, for example, Grey Brown luvisol.

Zone Any of the five divisions of the earth bounded by circles parallel to the equator, area enclosed between two exact or approximate concentric circles; any well-defined tract of more or less belt-like form.

Zone control The zoning of a process or building, and the independent heating or temperature controls for each zone.

Zone of aeration The upper member of the zone of rock fracture, includes belt of soil water, water of the intermediate belt and water of the capillary fringe.

Zone of saturation The lower member of the rock fracture; the zone of groundwater, the upper surface of which is the watertable. Generally beneath the depth of penetration of plant roots.

Zone of weathering That depth or thickness of the surface crust of earth in which weathering has taken place; includes soil and parent material of soil.

Zone, soil Area in which the dominant or zonal soils reflect the zonal influence of climate and vegetation and form a natural pattern with other soils that exhibit the zonal influence only weakly or not at all. The soil zone is not a taxonomic unit, but may be used as a cartographic unit.

Zoochemistry Study of the constituents of the animal body.

Zoodynamics Animal physiology.

Zoogamy Sexual reproduction.

Zooid Of incompletely animal nature; organic body or cell resembling but not being, animal or plant. Member of compound organism.

Zoology The study which deals with the natural history of animals, science of their structure, physiology, classification, habits and distribution.

Zoophyte Kinds of plant-like animals.

Zooplankton The animal component of the plankton. Belong to the protozoa or single celled animals. Form the food for baby fish.

Zoosperm Spermatozoon; Zoospore.

Zoosporangium A sporangium which produces zoospores.

Zoospore A swimming spore, swarm spore capable of independent movement. An asexual reproductive cell, which develops into a new plant by repeated divisions of protoplast.

Zootaxy Classification of animals.

Zootomy Dissection of anatomy of animals.

Z score A measure of how many standard deviations of raw score is form the mean.

Zygomorphic Bilaterally symmetrical.

Zygopophysis One of the processes on a vertebra serving as articulation with another.

Zygospore Resting spore formed from the union of two similar gametes.

Zygote The cell produced by the union of two cells (gametes) in reproduction, also the individual developing from such a cell.

Zygomorphous The terms refers to flowers divisible into smaller halves only in one plane.

Zygomycetes Sub-class of the Phycomycetes, characterised by gametes of equal size.

Zymogenous flora Organisms or plants found in soils in large numbers immediately after the addition of readily decomposable organic materials.

Zymosis Fermentation; Zymotic disease in general or any form of it. Zymotic means of fermentation.

Zymotic diseases Epidemic, endemic, contagious or sporadic diseases regarded as caused by multiplication of germs introduced from without.

Zymurgy A branch of applied chemistry dealing with fermentation.